HADRON SPECTROSCOPY AND STRUCTURE

2019
Guilin, China
HADRON

2019 Guilin, China

HADRON SPECTROSCOPY AND STRUCTURE

Proceedings of the XVIII International Conference
The 18th International Conference on Hadron
Spectroscopy and Structure (HADRON2019)
Guilin, China, 16 – 21 August 2019

Editors

Feng-Kun Guo
Institute of Theoretical Physics, Chinese Academy of Sciences, China

Wei-Hong Liang
Guangxi Normal University, China

World Scientific

NEW JERSEY • LONDON • SINGAPORE • BEIJING • SHANGHAI • HONG KONG • TAIPEI • CHENNAI • TOKYO

Published by

World Scientific Publishing Co. Pte. Ltd.
5 Toh Tuck Link, Singapore 596224
USA office: 27 Warren Street, Suite 401-402, Hackensack, NJ 07601
UK office: 57 Shelton Street, Covent Garden, London WC2H 9HE

Library of Congress Control Number: 2020941538

British Library Cataloguing-in-Publication Data
A catalogue record for this book is available from the British Library.

HADRON SPECTROSCOPY AND STRUCTURE
Proceedings of the XVIII International Conference

ISBN 978-981-121-919-1 (hardcover)
ISBN 978-981-121-931-3 (ebook for institutions)
ISBN 978-981-121-932-0 (ebook for individuals)

For any available supplementary material, please visit
https://www.worldscientific.com/worldscibooks/10.1142/11798#t=suppl

Typeset by Stallion Press
Email: enquiries@stallionpress.com

Preface

The 18th International Conference on Hadron Spectroscopy and Structure (HADRON2019) was held in Guilin, China, from Aug. 16 to Aug. 21, 2019. It was hosted by Guangxi Normal University and Institute of Theoretical Physics, Chinese Academy of Sciences. This series of conferences started in 1985 at Maryland, USA. It brings together experimentalists and theorists every other year to review the status and progress in hadron spectroscopy, structure and related topics and to exchange ideas for future explorations.

This conference had more than 330 participants from more than 20 countries. The conference consists of one plenary session, seven parallel sessions and one poster session. The parallel sessions are:

- Session 1: Meson spectroscopy
- Session 2: Baryon spectroscopy
- Session 3: Exotic hadrons and candidates
- Session 4: Hadron decays, production and interactions
- Session 5: Analysis tools
- Session 6: QCD and hadron structure
- Session 7: Hadrons in hot and nuclear environment including hypernuclei

More than 200 talks were presented, and the central issue discussed in the conference was exotic hadrons: the strongly interacting particles that have a structure different from the textbook definitions of mesons and baryons. Searches for exotic hadrons and studies of their properties have been a focus for many high-energy physics experiments as reported in this conference. Various phenomenological analyses and theoretical tools were discussed as well.

The present volume contains 138 conference contributions. Although not all the contributions were collected, the slides for all the talks can be found at the conference webpage `https://indico.ihep.ac.cn/event/9119/`. We would like to thank all our colleagues who made HADRON2019 a success.

Feng-Kun Guo Wei-Hong Liang

Contents

Plenary session

The spectroscopy of singly-heavy hadrons*

J. M. Yelton

University of Florida,
Gainesville, FL 32605, U.S.A.
E-mail: yelton@ufl.edu

This is a review of the experimental results in the field of singly-heavy hadrons. Patterns in the mass spectra are discussed, and recent results are presented.

1. Introduction

In this review, I generally consider heavy hadrons to be those with one charm or bottom quark. However the strange quark is strange because in some cases it can also be considered to be heavy, but in some cases it acts as if it is light. The B_c meson is also included. Although the accent is on the masses, it is also sometimes relevant to mention widths or lifetimes, decay modes, and production mechanisms of the heavy hadrons.

1.1. *The Experiments*

There are two basic experimental methods to produce heavy hadrons. Experiments at electron-positron colliders typically produce clean samples but have limited statistics. Belle II is starting to take data, and many results are due to Belle, BaBar, and CLEO. BESIII has been very active in charm decay studies, but its low energy range is less useful for spectroscopy of excited states. Hadron machines have much larger cross-sections and can produce higher mass particles. LHCb have huge statistics, and are the only hadron-collider experiment specifically designed for heavy hadron studies. Other LHC experiments (CMS and ATLAS) have even bigger integrated luminosities, and CDF and D0 have also made important contributions to the field.

---

* This work is supported in part by DOE Award Number: DE-SC0009824.

2. The Quark Model

Earlier this year the father of the quark model, Murray Gell-Mann, died. His quark theory gained traction at the time because it predicted the mass and properties of the Ω^-, but more impressive to me was his explanation of the mass difference of the Λ and the Σ. He considered them to each comprise a heavy (strange) quark and a (ud, uu or dd) light di-quark in its own specific spin state. What Gell-Mann did not know is that two heavier quarks would later be found, reproducing the pattern of the strange particles. With the input of effective quark masses of strange, charm and bottom, the existence, masses, and basic properties of all the heavy-quark hadrons shown in Fig. 1 were predicted using this model in which the hadron is considered to be a heavy quark plus a light quark or di-quark. All these states have been found,[1] and are seen to follow the following rules formulated using this model:

1) The extra mass associated with a being a baryon rather than a meson is independent of flavor

2) the mass difference between the spin-weighted Σ and the Λ is independent of flavor

3) the hyperfine mass-splittings (such as between the spin states of the $\Sigma^{(*)}$ and the between the spin-1 and spin-0 mesons) is inversely proportional to heavy quark mass

4) the hyperfine splittings are around 45% as large in baryons as they are in mesons.

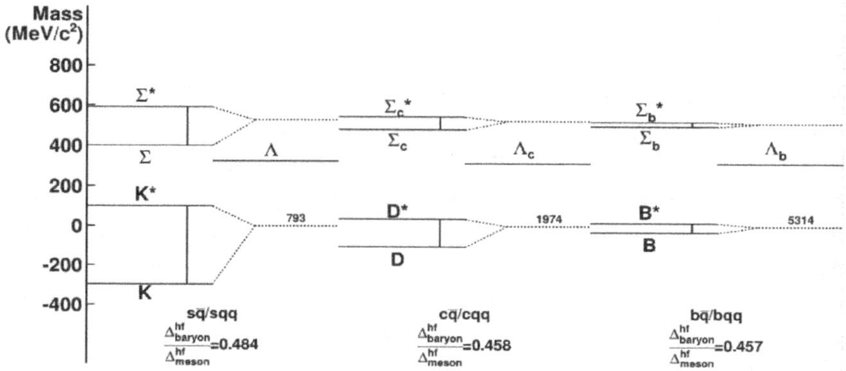

Fig. 1. The experimental mass spectrum of baryons and mesons. The zero of mass is placed at the spin-weighted average of the spin-0 and spin-1 mesons for each heavy quark system.

This remarkable regularity of the spectrum is extended to the cases where one or two of the "light" quarks is strange (Fig. 2). I will highlight some recent results concerning some individual particles in this latter sequence.

2.1. *The Ω_c*

First the Ω_c(css) baryon. LHCb have recently measured its lifetime[2] and found it to be approximately three times the previously accepted value, where the latter was based on three different measurements. However, looking at those older measurements it can be seen that E687 used a decay mode that has never been confirmed,[3] WA-89 had inconsistent mass measurements,[4] and even FOCUS had a rather poor signal-to-noise ratio.[5] It seems clear that to measure a lifetime you need to be very sure that you have a signal, as well as low background. The LHCb measurement is now generally accepted as being the definitive one.

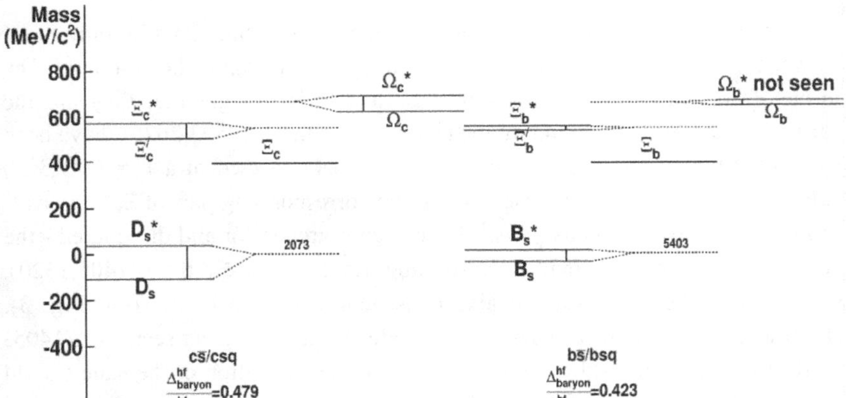

Fig. 2. The experimental mass spectrum of baryons and mesons with at least one strange (anti-) quark. The zero of mass is placed at the spin-weighted average of the spin-0 and spin-1 mesons for each heavy quark system.

2.2. *The Excited Ξ_b States*

In general the particles in Fig. 2 have all been seen, but in the excited Ξ_b system there is a small gap because although the charged Ξ_b' has been seen[6] but its neutral partner has not. This is just because the isospin mass-splittings, by an accident of nature, make it so the latter particle cannot decay strongly. Electromagnetic decays are much harder to observe. Many of the mass differences between states are close to such thresholds, some fall one way, some the other, and some coincide with a threshold. We should not interpret deep meanings into all such coincidences,

2.3. *The Ω_b States*

The mass measurement of the Ω_b shows how successful the standard quark model is. When the ground state particle was first reported[7] it was at a mass 100 MeV/c^2 higher than the simple extrapolation from the lower mass states such as the Ω_c would indicate.[†] However more recent results[8] all vindicate the model and that first measurement is no longer considered valid. The Ω_b^* is one particle still missing (again, demonstrating that an electromagnetic transition is hard to find).

3. P-wave Baryons

In a heavy baryon, a unit of orbital angular momentum can either be introduced between the heavy quark and the light di-quark, or inside the di-quark. The former (λ-excitations) are of lower mass and it is thought that they are the particles that have been found. In the charm sector the Λ_c(2593,2625)[9] have been known for two decades and have all the qualities expected of a $J^P = (1/2^-,3/2^-)$ pair. Having been found at these masses, the corresponding pair of Ξ_c baryons,[10] and more recently the Λ_b baryons,[11] have been searched for and discovered - the latter with the expected small mass splitting. Looking back at the Λ(1405,1520), we find that those masses are also consistent with this model (see Fig. 3). Despite all the theoretical work, and the odd threshold effects seen in Λ(1405) production, my view is that the default simple representation of the state should be that of a λ-excitation of the standard Λ particle.

Recently there has been much excitement concerning the LHCb result of 5 narrow states consistent with being excited Ω_c baryons.[12] Four of the five have been confirmed by Belle,[13] who have effectively much less data. As the (ss) di-quark must be in a spin-1 configuration, it is easy to see how the heavy quark, light di-quark model leads to five states and, as five have been found, it would be a remarkable coincidence if this simple explanation was found to be incorrect. It is much harder to disentangle is which of the $J^P = (1/2^-, 1/2^-, 3/2^-,$

[†]Taking the extra mass associated with a bottom quark to be 3323 MeV/c^2, adding this to the mass of the Ω_c and adjusting for the small splitting between spin states in the bottom sector, we get $M(\Omega_b) \approx$ 6050 MeV/c^2 and $M(\Omega_b^*) \approx$ 6070 MeV/c^2. The present PDG value for the ground state is 6046 MeV/c^2

3/2⁻, 5/2⁻) quintuplet of predicted states is which. One aspect which is often ignored in the models is the relative production rates. Given that one state is not seen by Belle, this could imply that it is a $1/2^-$ state, as the lower the spin, the fewer are expected to be produced – though there are complications because of decays to Ξ_c'. No doubt the J^P of each state will eventually be measured and will be a good test of the various models.

I note that there is another system with an ss di-quark, that of the standard Ξ. Here there is news that Belle have confirmed the existence of the $\Xi(1620)^0$, seen in the substructure of $\Xi_c^+ \rightarrow \Xi^-\pi^+\pi^+$ decays.[14] It is the only orbital-excitation candidate observed to decay freely to Ξπ. If the quark content is considered to be a light-quark, heavy di-quark, there should be five λ-excitations in this sequence as there are in its charmed counterpart, and only following the decay rules only one of the five can decay easily to Ξπ. However, the intrinsic widths of the states and the huge hyperfine splittings in the strange sector make it much harder to analyze than the charm sector.

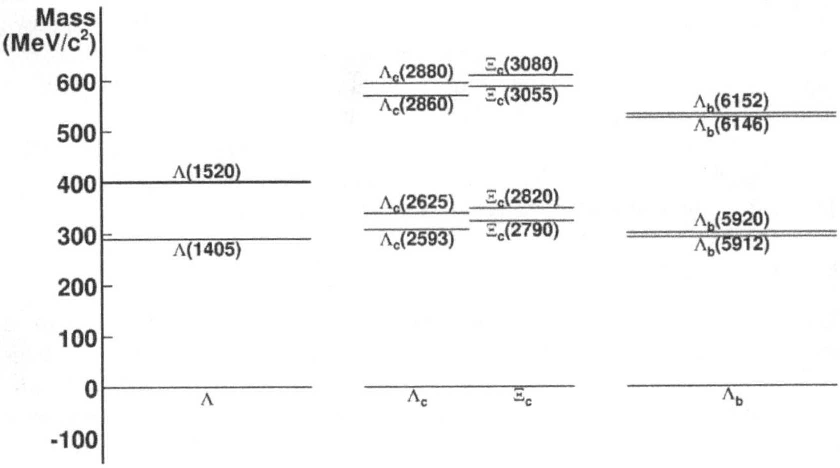

Fig. 2. The experimental mass spectrum of baryons with at least one strange quark, showing orbitally excited states. The zero of mass is placed at the mass of the weakly-decaying ground state.

4. News on Other Excited Charm and Bottom Baryons

Two states, known as the $\Lambda_c(2765)$ and the $\Lambda_c(2880)$ were first seen by CLEO 18 years ago.[15,‡] Until now there was some doubt if the former is really isospin-

‡ It is interesting to note that this paper was not cited by any theoretical work until two years after publication, showing the lack of interest in the subject two decades ago.

0. However this has now finally been confirmed by Belle.[16] LHCb have confirmed the J^P of the (2880) state to be $5/2^+$ and have shown evidence of its $3/2^+$ partner, now known as the $\Lambda_c(2860)$.[17] Rather neatly, two possibly analogous states have already been found in the Ξ_c system (the 3080 and 3055)[18] although no spin-parity analyses have been possible. Furthermore, recently another pair have been found in the excited Λ_b spectrum by LHCb.[19] This analysis shows how careful analyzers must be, because what appears at first sight to be one state has been shown to be two narrow states close together.

There are many mysteries still to be unraveled. One for which I have an educated guess is that the $\Lambda_c(2765)$ and $\Xi_c(2970)$ are the first radial excitations. This conjecture is not just because of their masses falling into the places expected by many of the theoretical models, it is because of the large production cross-sections for these particles in e^+e^- annihilations which clearly differentiates them from the orbital excitations. For the $\Lambda_c(2880)$ there is the question of why the state is so narrow and its decays to $\Sigma_c(2520)\pi$ do not dominate as would be expected by the rules of spin-parity. In the Ξ_c sector, there is the (2930) state, first reported by BaBar[20] and confirmed in both charge states by Belle[21] – all signals being found in B-decays. The spin-parity of the $\Xi_c(2930)$ is unknown and it has not been seen in continuum production. This last quality makes it appear to be very different from other excited states. Less mysteriously, there has been a new Ξ_b state[22] reported by LHCb at a mass of 6226 MeV/c^2. Could this be a radial excitation? Lastly there is a new wide Σ_b candidate[23] from LHCb at 6097 MeV/c^2. This is in the mass range we expect to see orbital excitations, but there is only one visible whereas this system should have five λ-excitations.

5. Charm and Bottom Mesons

The situation in mesons is rather more murky than it is for baryons because there are fewer narrow states. For instance, there should be 4 L=1 orbital excitations of D mesons, 2 of which are relatively narrow and have been seen for some time, and two so wide that it is very unsatisfying to represent their pole masses as well-defined values. The same problem occurs with the highest D excitation of 3214 MeV/c^2 which was previously known as the D(3000) but recent analysis has reported its mass at the higher value.[24] In the D_S sector we had a major surprise in 2003 when 2 comparatively low mass, very narrow, states were found.[25] Their masses, close to DK and D*K thresholds, make it tempting to represent them as molecular states. In the non-strange B mesons, two narrow states have been found as expected, but only one of the two wide states. One very clear experimental challenge is to find B_s equivalents of the very narrow D_s states. This serious gap in our knowledge is once again it due to the problem in finding photon transitions. It is important to know if they are at BK and B*K thresholds, or whether that is just a coincidence in the charm sector. This knowledge would clarify if these states are really best considered to be molecular states, or if the potential models simply got the mass splittings of the orbital excitations incorrect.

5.1. *The B_c Meson*

The B_c meson has been called "The Last Meson" as it was the final stable (i.e. weakly-decaying) meson possible. Its mass and lifetime have been well measured. Its spectroscopy is expected to be reminiscent of charmonium rather than classic excited mesons. There is a claim[26] by Atlas to find a (2S) state at 6842 MeV/c^2. However CMS has recently shown that this is actually two peaks close together, one of which is thought to be the $B_c(2S)$, and the other the $B_c^*(2S)$ but with a photon missing.[27] This latter interpretation has been confirmed by LHCb[28] who have, for once, less data than their competitor.

6. Conclusion

This has been a lightning review of the experimental results on heavy baryon and mesons, with the accent on their masses. A recurring theme is how well the basic charm model, dating from the time of Gell-Mann but developed using Heavy Quark Symmetry and other tools, describes a large mass of the data. The huge number of identified particles is still increasing with no end in sight. I stress that understanding these states is easier than studying lower-mass resonances, and that much can be learnt in the strange sector in particular by extrapolation of the results from charm and bottom.

References

1. M. Tanabashi *et al.,* (Particle Data Group), Phys. Rev. D98, 030001 (2018) and 2010 update.
2. P. L. Frabetti *et al.,* (E687), Phys. Lett. B358, 151 (1995).
3. M. I. Adamovich *et al.*, (WA89), Phys. Lett. B358, 151 (1995).
4. J. M.Link *et al.*, (FOCUS), Phys. Lett. B561, 41 (2003).
5. R. Aaij *et al.,* (LHCb) PRL 121, 092003 (2018).
6. R. Aaij *et al*, (LHCb) PRL 114, 062004 (2015).
7. V. M. Avazov *et al.*, (D0), PRL 101, 232002 (2008).
8. R. Aaij *et al.*, (LHCb) PRL 110, 182001 (2013), T. Aaltonen *et al.*, (CDF) Phys. Rev. D89, 072014 (2014), R. Aaij *et al.*, (LHCb) Phys. Rev. D93, 092007 (2016).

9. H. Albrecht *et al.*, (Argus) Phys. Lett. B317, 227 (1993), K. Edwards *et al.*, (CLEO) PRL 74, 3331 (1995).
10. J. Alexander *et al.*, (CLEO) PRL 83, 3390 (1999), S. Csorna *et al.*, (CLEO) PRL 86, 4243 (2001).
11. R. Aaij *et al.*, (LHCb), PRL 109, 172003 (2012).
12. R. Aaij *et al.*, (LHCb), PRL 118, 182001 (2017).
13. J. M. Yelton *et al.*, (BELLE), Phys. Rev. D97, 051102 (2018).
14. M. Sumihama *et al*, (BELLE), PRL 122, 072501 (2019).
15. M. Artuso *et al.*, (CLEO), PRL 86, 4479 (2001).
16. A. Abdesselam *et al.*, BELLE-CONF-1905, arXiv:1910.05365 (2019).
17. R. Aaij *et al.*, (LHCb), PRL 119, 101801 (2017).
18. B. Aubert *et al.,* (BaBar) Phys. Rev. D77 012002 (2008), R. Chistov *et al.,* PRL 97, 162001 (2006).
19. R. Aaij *et al.*, (LHCb), PRL 123, 152001 (2019).
20. B. Aubert *et al.*, (BaBar), Phys. Rev. D 77, 031101 (2008).
21. Y. B. Li *et al.*, (BELLE), Eur. Phys. J C78, no 11, 928 (2018), Eur. Phys. J C78, no. 3, 252 (2018).
22. R. Aiij *et al.*, (LHCb), PRL 121, 079002 (2018).
23. R. Aaij *et al.*, (LHCb), PRL 122, 012001 (2019).
24. R. Aaij *et al.*, (LHCb), Phys. Rev. D94, 072001 (2016).
25. B. Aubert *et al.*, (BaBar), PRL 90, 242001 (2003), D. Besson *et al.*, (CLEO), Phys. Rev. D68, 032002 (2003).
26. G. Aad *et al.*, (ATLAS), PRL 113, 212004 (2014).
27. A. Sirunyam *et al.*, (LHCb), PRL 122, 132001 (2019).
28. R. Aaij *et al.,* (LHCb), PRL 122, 232001 (2019).

Experimental status of the XYZ structures

Yuping Guo
Institute of Modern Physics, Fudan University,
Shanghai, 200433, P. R. China
E-mail: guoyp@fudan.edu.cn

Starting from last decade of this century, many charmonium-like states, also called as XYZ states, have been discovered at different experiments. These states are located at the charmonium energy region, but carry properties that can not be explained as conventional charmonium states. They are considered as good candidates of exotic hadrons. Here, recent experimental activities on XYZ states will be presented, including new decay and production information of $X(3872)$, new measurements of the vector states Y, and new results of the charged $Z_c(3900)$ states from e^+e^- machine and in b-flavor decays.

Keywords: Quark model; exotic hadrons; charmonium-like states.

1. Introduction

The Quark Model (QM) is successfull as it classified all the hadrons observed at that time and predicted additional hadrons which have been discovered subsequently. In QM, hadrons are mesons and baryons composed from one quark and anti-quark and three quarks, respectively. However, hadrons with other inner configurations are also allowed, such as those made of gluons only (glueball), quarks and excited gluon (hybrid), or more than three quarks (multi-quark states).[1] Searching for such kind of exotic hadrons is a long history, although no compelling evidence has been found, a group of states that do not fit into simple QM picture has been steadily accumulated from experiments in the past decades. In this paper, recent experimental progress on XYZ states from the LHCb, COMPASS, BESIII, Belle, and D0 experiments will be presented.

2. $X(3872)$

$X(3872)$ was observed in B meson decays, $B^{\pm} \to K^{\pm}\pi^{+}\pi^{-}J/\psi$, at the Belle experiment in 2003.[2] It was measured intensively from different experiments, which yield most abundant experimental information.[3] Its mass is very close to the $D^0\bar{D}^{*0}$ mass threshold, its width is smaller than 1.2 MeV, and its quantum is determined to be $J^{PC} = 1^{++}$.

2.0.1. *Production*

The $X(3872)$ is observed in B meson decays, hadron collisions, and radiative transitions from higher vector states before. Recently, it is also observed in the process of $\Lambda_b^0 \to X(3872)pK^-$ with $X(3872) \to J/\psi\pi^+\pi^-$ at LHCb experiment using 4.9 fb^{-1} proton-proton collision data.[4] A two-dimensional fit is applied to the $J/\psi\pi^+\pi^-pK^-$ and $J/\psi\pi^+\pi^-$ invariant mass distributions, as shown in Fig. 1. The fit yields (55 ± 11) $\Lambda_b^0 \to X(3872)pK^-$ signal events with a statistical significance of 7.2σ. It is found that $(58 \pm 15)\%$ of the decays proceed via the two-body $X(3872)\Lambda(1520)$ intermediate state. The process of $\Lambda_b^0 \to \psi(2S)pK^-$ is used as a reference channel, and the production ratio is measured to be:

$$\frac{\mathcal{B}[\Lambda_b^0 \to X(3872)pK^-]}{\mathcal{B}[\Lambda_b^0 \to \psi(2S)pK^-]} \times \frac{\mathcal{B}[X(3872) \to J/\psi\pi^+\pi^-]}{\mathcal{B}[\psi(2S) \to J/\psi\pi^+\pi^-]} = (5.4 \pm 1.1 \pm 0.2) \times 10^{-2}.$$

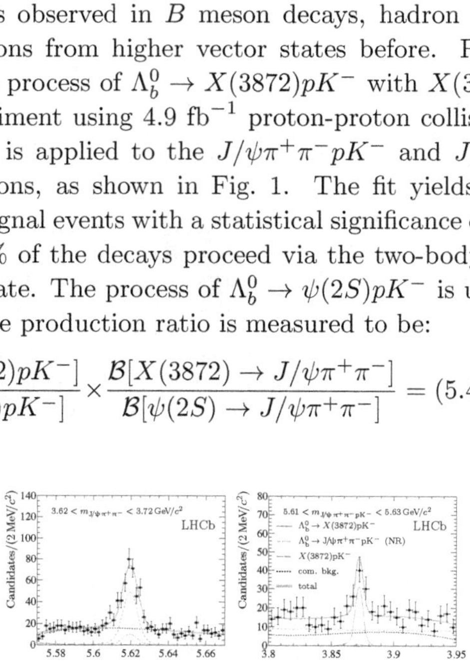

Fig. 1. Projection of the two-dimensional distribution of $J/\psi\pi^+\pi^-pK^-$ (left) and $J/\psi\pi^+\pi^-$ (right) for $\Lambda_b^0 \to X(3872)pK^-$ candidate events.

The lepton production of $X(3872)$ is studied at the COMPASS experiment in the charge-exchange (netural-exchange) reaction, $\mu^+N \to \mu^+X(3872)\pi^\pm N'$ ($\mu^+N \to \mu^+X(3872)N'$), with $X(3872) \to J/\psi\pi^+\pi^-$.[5] In the charge-exchange reaction, two peak structures are found in the invariant mass distribution of $J/\psi\pi^+\pi^-$, as shown in Fig. 2. The significance of the second peak, named as $\tilde{X}(3872)$, is 4.1σ including systematic effects. A fit to the spectrum yields (13.2 ± 5.2) events, and a mass of (3860.4 ± 10.0) MeV$/c^2$ for $\tilde{X}(3872)$. To further study its nature, the $\pi^+\pi^-$ invariant mass spectrum is examined, as shown in Fig. 2, compared with that from $X(3872) \to J/\psi\pi^+\pi^-$ process measured at the ATLAS experiment. The distribution from COMPASS study is in agreement with a three-body phase space or expectation for a state of $J^{PC} = 1^{+-}$. It may be interpreted as a neutral partner of $X(3872)$ predicted by a tetrequark model, which decays into $\sigma J/\psi$. An independent confirmation of this state from high precision measurement is needed.

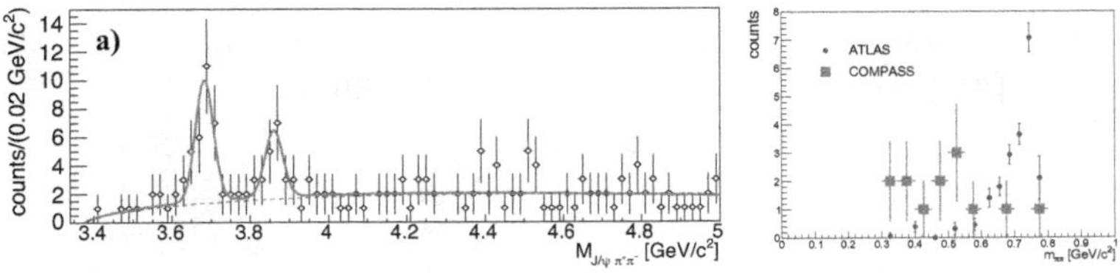

Fig. 2. (Left) The $J/\psi\pi^+\pi^-$ invariant mass distribution for the charge-exchange reaction. (Right) The $\pi^+\pi^-$ invariant mass distribution from the decay of $\tilde{X}(3872)$, compared with that from $X(3872)$.

2.0.2. *Decay Modes*

The $X(3872)$ decays into $\pi^+\pi^-J/\psi$, $\pi^+\pi^-\pi^0J/\psi$, $D^0\bar{D}^0\pi^0$, $D^0\bar{D}^{*0}$, and $\gamma\psi(1,\ 2S)$.[3] Using the radiative production of $X(3872)$ from vector states[6] with 11.6 fb^{-1} data collected at $\sqrt{s}$ between 4.0 to 4.6 GeV, the BESIII experiment studied the decay of $X(3872)$ to $\pi^0\chi_{c1}$,[7] $\omega J/\psi$,[8] $D^0\bar{D}^{*0}$, $\gamma\psi(1,\ 2S)$, and $\gamma D^+\bar{D}^-$. The first three modes are observed with significance larger than 5σ. The decay ratio of $X(3872) \to \gamma\psi(2S)$ and $X(3872) \to \gamma J/\psi$ was used to identify the nature of $X(3872)$. The combined results from BaBar and LHCb experiments is 2.6 ± 0.6,[3] while Belle only set an upper limit. The $\gamma J/\psi$ process is found at BESIII experiment with a significance of 3.5σ, while no evident signal of $X(3872)$ is found in $\gamma\psi(2S)$ process. The ratio is measured to be < 0.59 at 90% Confidence Level (C.L.). The relative branching ratios from various decay modes with respect to $X(3872) \to J/\psi\pi^+\pi^-$ are summarized in Table 1.

Table 1. Relative branching ratios from six decay modes measured at BESIII experiment with respect to $X(3872) \to J/\psi\pi^+\pi^-$.

Mode	$D^0\bar{D}^{*0}$	$\gamma J/\psi$	$\gamma\psi(2S)$	$\gamma D^+\bar{D}^-$	$\omega J/\psi$	$\pi^0\chi_{c1}$
Ratio	14.81 ± 3.8	0.79 ± 0.28	< 0.42	< 0.99	$1.7^{+0.4}_{-0.3} \pm 0.2$	$0.88^{+0.33}_{-0.27} \pm 0.10$

3. Vector States: Y

The first Y states, $Y(4260)$, is observed at BaBar experiment via $e^+e^- \to \gamma_{\rm ISR}\pi^+\pi^- J/\psi$ in the $\pi^+\pi^- J/\psi$ invariant mass distribution.[9] In the position where the $Y(4260)$ is discovered, the inclusive hadron cross section shows a dip,[10] indicates a small coupling to open-charm final states. This is abnormal for states with masses above the open-charm threshold. Similar structures are observed from the process of $e^+e^- \to \gamma_{\rm ISR}\pi^+\pi^-\psi(2S)$ in the $\pi^+\pi^-\psi(2S)$ system, which are called as $Y(4360)$ and $Y(4660)$. Before 2015, there are seven vector states observed between 4.0 and 4.7 GeV/c^2, while only five are expected from potential model calculation. In addition, those observed in $\pi^+\pi^- J/\psi$ system seems to be absent in $\pi^+\pi^-\psi(2S)$, vise versa.

From the end of 2012, the BESIII experiment started to take data above 4.0 GeV, and have accumulated about 17 fb^{-1} e^+e^- collision data at $\sqrt{s}$ from 3.8 to 4.6 GeV. The data samples include 22 energy points with integrated luminosity ($\mathcal{L}$) larger than 500 pb^{-1} each, 12 energy points with $\mathcal{L}$ around 50 pb^{-1} each, and 104 energy points with $\mathcal{L}$ about 10 pb^{-1} each. These data samples are used to study the Y states.

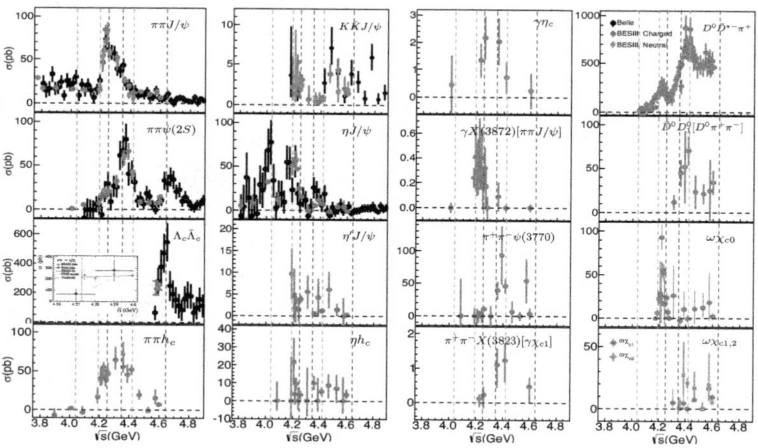

Fig. 3. The cross sections of the processes involve hidden-charm or open-charm mesons in the final states measured at BESIII experiments.

Figure 3 shows the cross sections of processes contain hidden-charm or open-charm hadrons in the final states. Comparing to the measurement from B-factories using ISR method, the precision of the BESIII measurement is much higher. The cross section line-shapes of $e^+e^- \to \pi^+\pi^- J/\psi$,[11] $e^+e^- \to \pi^+\pi^-\psi(2S)$,[12] $e^+e^- \to \pi^+\pi^- h_c$,[13] $e^+e^- \to \omega\chi_{c0}$,[14] and $e^+e^- \to \pi D^0 \bar{D}^{*}$[15] show obvious enhancements between 4.2 and 4.5 GeV. They are fitted separately with coherent sum of one or more Breit-Wigner (BW) resonances and non-resonant contribution, dependent on the processes. In $e^+e^- \to \pi^+\pi^- J/\psi$ process, two resonances are needed to describe the structure around 4.26 GeV, the previously established $Y(4260)$ from ISR measurement is splitted into two resonances. In $e^+e^- \to \pi^+\pi^-\psi(2S)$ process, a resonance is needed for structure around 4.22 GeV with significance larger than 5σ. Resonant structures around 4.22 GeV are also observed in the other three processes. The structure around 4.4 GeV is not conclusive in all the processes. An average of the masses and widths for the structure around 4.22 GeV, following the unconstrained averaging procedure from PDG,[3] leads $M = (4220.0 \pm 2.2)$ MeV/c^2 and $\Gamma = (41.5 \pm 2.9)$ MeV with $\chi^2/(N-1) = 0.95$ and 7.8, respectively. Here $N = 5$ is the number of processes used in the average. Improved description of the cross section line-shape with proper parameterization of broad resonance and consideration of coupled-channel effects will lead to a better understanding of the vector states produced in this energy region.

Using the final state of $\pi^+\pi^- D\bar{D}$, the cross sections of $e^+e^- \to \pi^+\pi^-\psi(3770)$ and $e^+e^- \to D_1(2420)\bar{D}$ are measured at BESIII.[16] Both $e^+e^- \to \pi^+\pi^-\psi(3770)$ and $e^+e^- \to D_1(2420)\bar{D}$ (with $D_1^0(2420) \to D^0\pi^+\pi^-$) are observed at $\sqrt{s} = 4.42$ GeV with significances of 5.2σ and 7.4σ, respectively. Evidences are found in $D_1(2420)^+\bar{D}^-$ mode at 4.36 and 4.42 GeV. The products of the cross section and the branching fractions for the two observed cross sections are shown in Fig. 3. Clear structure is found in the cross section distribution, but no quantitative result is given due to limited statistics. The line-shape of the cross section in $e^+e^- \to D_1(2420)\bar{D}$ process does not show a fast rise above the $D_1(2420)\bar{D}$ threshold, contradict to the prediction which explains $Y(4260)$ as a molecule of $D_1(2420)\bar{D}$.

The production of Y states from B meson decays are studied in $B^{\pm,0} \to K^{\pm,0}\pi^+\pi^- J/\psi$[17] and $B^- \to K^-\Lambda_c^+\bar{\Lambda}_c^-$[18] processes at Belle experiment using $(772{\pm}11)\times 10^6$ $B\bar{B}$ pairs. Excesses of events around 4.2 GeV are found with significance of 2.1σ (0.9σ) for charged (neutral) $B \to Y(4260)K$ decays. Upper limits on the product of branching fractions at 90% C. L. are determined, which are $\mathcal{B}[B^{+(0)} \to Y(4260)K^{+(0)}] \times \mathcal{B}[Y(4260) \to J/\psi\pi^+\pi^-] < 1.4(1.7) \times 10^{-5}$. The $\Lambda_c^+\bar{\Lambda}_c^-$ invariant mass distribution between 4.55 and 4.75 GeV is studied to search for the $Y(4660)$. No signal is observed, upper limit on the product branching fraction is determined to be $\mathcal{B}[B^- \to K^- Y(4660)] \times \mathcal{B}[Y(4660) \to \Lambda_c^+ \bar{\Lambda}_c^-] < 1.2 \times 10^{-4}$.

4. Charged Z_c States

The charged Z or Z_c states are found in the final states of a charged pion and a charmonium states, which contains at least four quarks thus could not be a conventional hadron. Eight such structures are found in two production patterns, one from B meson decay and the other from e^+e^- annihilation.

The $Z_c(3900)$ is found in e^+e^- annihilation, its production in b-flavored hadron decay or in prompt production is studied at the D0 experiment using 10.4 fb^{-1} of $p\bar{p}$ collision data.[19] The study is performed through the sequential process $Y(4260) \to Z_c(3900)^\pm \pi^\mp \to J/\psi \pi^\pm \pi^\mp$. The $Z_c(3900)$ is observed in the b-flavored hadron decay with $4.2 < M(J/\psi\pi^+\pi^-) < 4.3$ GeV, as shown in Fig. 4. The statistical significance is 5.4σ. The mass and width are measured to be $M = (3902.6^{+5.2+3.3}_{-5.0-1.4})$ MeV/c^2 and $\Gamma = (32^{+28+26}_{-21-7})$ MeV. No significant signal is observed in the prompt production mode, the upper limit on the ratio of prompt to nonprompt production at 95% C.L. is determined to be < 0.66, which is significantly lower than that of $X(3872)$.

The decay rate of Z_c to $\rho\eta_c$ is suggested to discriminate the nature of Z_c between the molecule and tetraquark scenarios. The study is performed at BESIII experiment through the process of $e^+e^- \to \pi^+\pi^-\pi^0\eta_c$.[20] $e^+e^- \to \pi^+\pi^-\pi^0\eta_c$ signal events are observed at $\sqrt{s} = 4.23$ GeV, and evidence of $Z_c(3900) \to \rho\eta_c$ signal with a statistical significance of 4.3σ is found, as shown in Fig. 4. No signal is observed for $Z_c(4020) \to \rho^\pm\eta_c$ process. The ratio, $\mathcal{B}[Z_c(3900) \to \rho\eta_c]/\mathcal{B}[Z_c(3900) \to \pi J/\psi]$, is measurement to be 2.2 ± 0.9 (< 0.56) at $\sqrt{s} = 4.23$ GeV (4.26 GeV at 90% C.L.), which is closer to tetraquark prediction. The upper limit of the ratio of $\mathcal{B}[Z_c(4020) \to \rho\eta_c]$ to $\mathcal{B}[Z_c(4020) \to \pi h_c]$ at 90% C.L. is determined to be smaller than 1.6, 0.9, and 1.4 at $\sqrt{s} = 4.23$ GeV, 4.26 GeV, and 4.36 GeV, respectively.

5. Summary and Outlook

Till now, more than 20 XYZ structures are found from experiments. A lot of new studies are achieved from experimental side. New production mechanisms of the XYZ structures are searched, $X(3872)$ is observed in Λ_b decays and $Z_c(3900)$ in b-flavored meson decays. The decay modes of $X(3872)$ are studied at BESIII using its radiative production. Its decay into $\gamma\psi(2S)$ needs further confirmation from LHCb by including the 6 fb^{-1} data

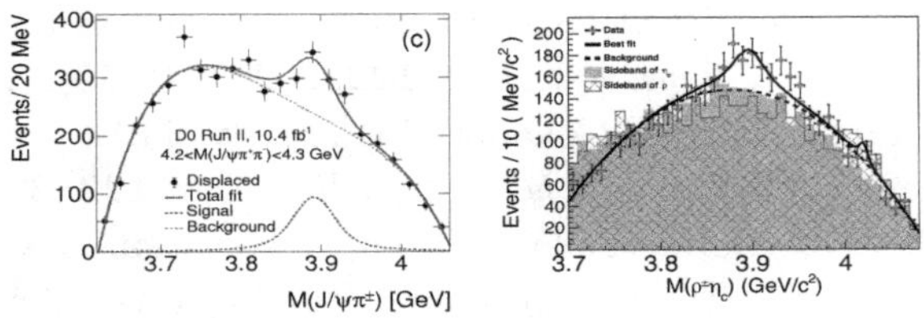

Fig. 4. (Left) The $J/\psi\pi^{\pm}$ invariant mass distributions from b-flavored decay at D0 experiment. (Right) The $\rho^{\pm}\eta_c$ invariant mass distribution from the process of $e^+e^- \to \pi^+\pi^-\pi^0\eta_c$ at the BESIII experiment at $\sqrt{s} = 4.23$ GeV.

sample collected at 13 TeV and Belle II. The vector states have been studied with large scan data samples collected at BESIII experiment. Structure around 4.22 GeV is observed in five processes. This changes previous picture that the Y states observed in $\pi^+\pi^-J/\psi$ or $\pi^+\pi^-\psi(2S)$ process do not cross. There might be structure above 4.4 GeV, measurements using the existing data ($\sqrt{s} < 4.5$ GeV) collected in 2019 and the coming data samples above 4.6 GeV will be helpful. The study of Y states from B meson decay shows no signal, improved study could be done with larger statistics from LHCb and BelleII. More information about the $Z_c(3900)$ is obtained, evidence of its decay into $\rho\eta_c$ is found. More experimental information of other Z states, and study of their properties with improved precision would lead us to a deepen understanding of their nature.

References

1. N. Brambilla *et al.*, Heavy quarkonium: progress, puzzles, and opportunities, *Eur. Phys. J. C* **71**, 1534 (2011).
2. S. K. Choi *et al.* [Belle Collaboration], Observation of a narrow charmoniumlike state in exclusive $B^{\pm} \to K^{\pm}\pi^+\pi^-J/\psi$ Decays, *Phys. Rev. Lett.* **91**, 262001 (2003).
3. M. Tanabashi *et al.* [Partice Data Group], The review of particle physics, *Phys. Rev. D* **98**, 030001 (2018).
4. R. Aaij *et al.* [LHCb Collaboration], Observation of the $\Lambda_b^0 \to \chi_{c1}(3872)pK^-$ decay, *J. High Energy Phys.* **2019:28** (2019).
5. M. Aghasyan *et al.* [COMPASS Collaboration], Search for muoproduction of $X(3872)$ at COMPASS and indication of a new state $\tilde{X}(3872)$, *Phys. Lett. B* **783**, 334 (2018).
6. M. Ablikim *et al.* [BESIII Collaboration], Observation of $e^+e^- \to \gamma X(3872)$ at BESIII, *Phys. Rev. Lett.* **112**, 092001 (2014).

7. M. Ablikim *et al.* [BESIII Collaboration], Observation of the decay $X(3872) \to \pi^0\chi_{c1}$, *Phys. Rev. Lett.* **122**, 202001 (2019).
8. M. Ablikim *et al.* [BESIII Collaboration], Study of $e^+e^- \to \gamma\omega J/\psi$ and observation of $X(3872) \to \gamma J/\psi$, *Phys. Rev. Lett.* **122**, 232002 (2019).
9. B. Aubert *et al.* [BaBar Collaboration], Observation of a broad structure in the $\pi^+\pi^- J/\psi$ mass spectrum around 4.26 GeV/c^2, *Phys. Rev. Lett.* **95**, 142001 (2005).
10. M. Ablikim *et al.* [BES Collaboration], Determination of the $\psi(3770)$, $\psi(4040)$, $\psi(4160)$, and $\psi(4415)$ resonance parameters, *Phys. Lett. B* **660**, 315 (2008).
11. M. Ablikim *et al.* [BESIII Collaboration], Precise measurement of the $e^+e^- \to \pi^+\pi^- J/\psi$ cross section at center-of-mass energies from 3.77 to 4.60 GeV, *Phys. Rev. Lett.* **118**, 092001 (2017).
12. M. Ablikim *et al.* [BESIII Collaboration], Measurement of $e^+e^- \to \pi^+\pi^-\psi(3686)$ from 4.008 to 4.600 GeV and observation of a charged structure in the $\pi^\pm\psi(3686)$ mass spectrum, *Phys. Rev. D* **96**, 032004 (2017).
13. M. Ablikim *et al.* [BESIII Collaboration], Evidence of Two Resonant Structures in $e^+e^- \to \pi^+\pi^- h_c$, *Phys. Rev. Lett.* **118**, 092002 (2017).
14. M. Ablikim *et al.* [BESIII Collaboration], Cross section measurements of $e^+e^- \to \omega\chi_{c0}$ from $\sqrt{s} = 4.178$ to 4.278 GeV, *Phys. Rev. D* **99**, 091103(R) (2019).
15. M. Ablikim *et al.* [BESIII Collaboration], Evidence of a resonant structure in the $e^+e^- \to \pi^+D^0D^{*-}$ cross section between 4.05 and 4.60 GeV, *Phys. Rev. Lett.* **122**, 102002 (2019).
16. M. Ablikim *et al.* [BESIII Collaboration], Observation of $e^+e^- \to \pi^+\pi^-\psi(3770)$ and $D_1(2420)^0\bar{D}^0$+c.c., *Phys. Rev. D* **100**, 032005 (2019).
17. R. Garg *et al.* [Belle Collaboration], Search for the $B \to Y(4360)K$, $Y(4360) \to J/\psi\pi^+\pi^-$ decays, *Phys. Rev. D* **99**, 071102(R) (2019).
18. Y. B. Li *et al.* [Belle Collaboration], Observation of $\Xi_c(2930)^0$ and updated measurement of $B^- \to K^-\Lambda_c^+\bar{\Lambda}_c^-$ at Belle, *Eur. Phys. J. C* **78**, 252 (2018).
19. V. M. Abazov *et al.* [D0 Collaboration], Properties of $Z_c^\pm(3900)$ produced in $p\bar{p}$ collisions, *Phys. Rev. D* **100**, 012005 (2019).
20. M. Ablikim *et al.* [BESIII Collaboration], Evidence for $Z_c^\pm$ decays into the $\rho^\pm\eta_c$ final state, *arXiv:1906.00831* (2019).

Heavy-light spectroscopy: A theoretical view

J. Nieves

Instituto de Física Corpuscular (centro mixto CSIC-UV), Institutos de Investigación de Paterna, Apartado 22085, 46071, Valencia, Spain
E-mail: jmnieves@ific.uv.es

We show how next-to-leading heavy meson chiral amplitudes fixed using lattice QCD for the Goldstone-boson scattering off charmed mesons not only resolve some long-standing puzzles in charm-meson spectroscopy but also are at the same time fully consistent with recent LHCb data on B decays, which provide precise experimental information for the $D\pi$ dynamics. We also extend the analysis to the analog bottom sector. Finally, we will briefly review the structure of the $\Lambda_c(2595)$ and $\Lambda_c(2625)$ resonances in the framework of an effective field theory consistent with heavy quark spin and chiral symmetries, that incorporates the interplay between $\Sigma_c^{(*)}\pi - ND^{(*)}$ baryon-meson degrees of freedom and bare P-wave $c\bar{u}d$ quark-model states.

Keywords: Heavy-quark and chiral effective theories; lattice energy levels; exotic states; final state interactions.

1. Introduction

The rich hadronic spectrum provides valuable information to disentangle the non-perturbative regime of Quantum Chromodynamics (QCD). With the recent advent of powerful experimental facilities, capable of exploring open and hidden charm (and bottom) energy ranges, several new mesons and baryons have been observed,[1] which are difficult to be accommodated within simple constituent-quark models (CQM). A clear example is the low-lying positive parity charmed-strange spectrum, which attracted the attention of the community in 2003 after the experimental observation of the $D_{s0}^*(2317)$ $[J^P = 0^+]$ and $D_{s1}(2460)$ $[J^P = 1^+]$ states by the BaBar[2] and Cleo collaborations[3] in the $D_s^+\pi^0$ and $D_s^{*+}\pi^0$ invariant mass spectra, respectively. The interest in such states arose from their low masses and widths, contrary to CQMs[4] and lattice QCD (LQCD)[5] expectations.[a] The

[a]An indirect hint for the non-perturbative nature of the DK chiral amplitudes near threshold, and the possible existence of a bound state, was already obtained in Ref. 8 from an Omnès analysis of the LQCD scalar form-factor in the $D \to \bar{K}\ell\bar{\nu}_\ell$ decay.

nature of those resonances became a matter of intense research, being interpreted as meson-meson and four-quark states (see discussion in Refs. 6, 7). The situation in the even-parity charm meson sector became more obscure when the $D_0^*(2400)$ and $D_1(2430)$ were observed by FOCUS[9] and Belle[10] collaborations, which led to two additional questions: (1) why is the mass difference between the $D_{s1}(2460)$ and the $D_{s0}^*(2317)$ differs in less that 2 MeV from that between the ground state D^* and D mesons?, and (2) why are the $D_0^*(2400)$ and $D_1(2430)$ masses almost equal to or even higher than their strange siblings, although states with an additional strange quark are typically at least 100 MeV heavier. We will show that analyses combining effective field theory (EFT) methods with LQCD allows one to resolve these puzzles, providing a scheme where all low-lying positive-parity heavy open-flavor mesons qualify as hadronic molecules.[11,12]

2. Low-lying positive-parity open charmed mesons

The appropriate EFT for the study the scattering of Goldstone bosons off heavy-light charmed mesons is heavy meson chiral perturbation theory (HChMCT), which consistently considers both chiral and heavy quark spin-flavor symmetry. Unitarized next to leading order (NLO) amplitudes are available,[13] with the low energy constants (LECs) that appear in the definition of the chiral amplitudes, determined from the lattice calculation of the $S-$wave scattering lengths in several strangeness and isospin $[(S,I)]$ sectors.[14] Such scheme received a robust support from a remarkably good agreement between a *parameter-free predicted* low-lying levels calculated using NLO HMChPT unitarized amplitudes for finite volumes[11] and the LQCD results reported in Ref. 15, as can be seen in the left top panel of Fig. 1. Furthermore within this framework, there appear two $(S = 0, I = 1/2, J^P = 0^+)$ poles in the region of the $D_0^*(2400)$ in the coupled-channel $D\pi$, $D\eta$ and $D_s\bar{K}$ scattering amplitudes (right top plot in the same figure). With the physical quark masses, the poles are located (see Ref. 16 for further details on the definition of the the second Riemann sheet of the amplitudes) at $(\sqrt{s} = E_R - i\Gamma/2) = (2105^{+6}_{-8} - i\,102^{+10}_{-12})$ MeV and $(2451^{+36}_{-26} - i\,134^{+7}_{-8})$ MeV, with the largest couplings to the $D\pi$ and $D_s\bar{K}$ channels, respectively. Since the higher pole is close to the $D_s\bar{K}$ threshold, we predicted it to show up as a threshold enhancement in the $D_s\bar{K}$ invariant mass distribution in B decay experiments. Indeed, the well-constrained amplitudes for Goldstone bosons scattering off charm mesons used in Ref. 11 are fully consistent[12] with recent high quality data on the

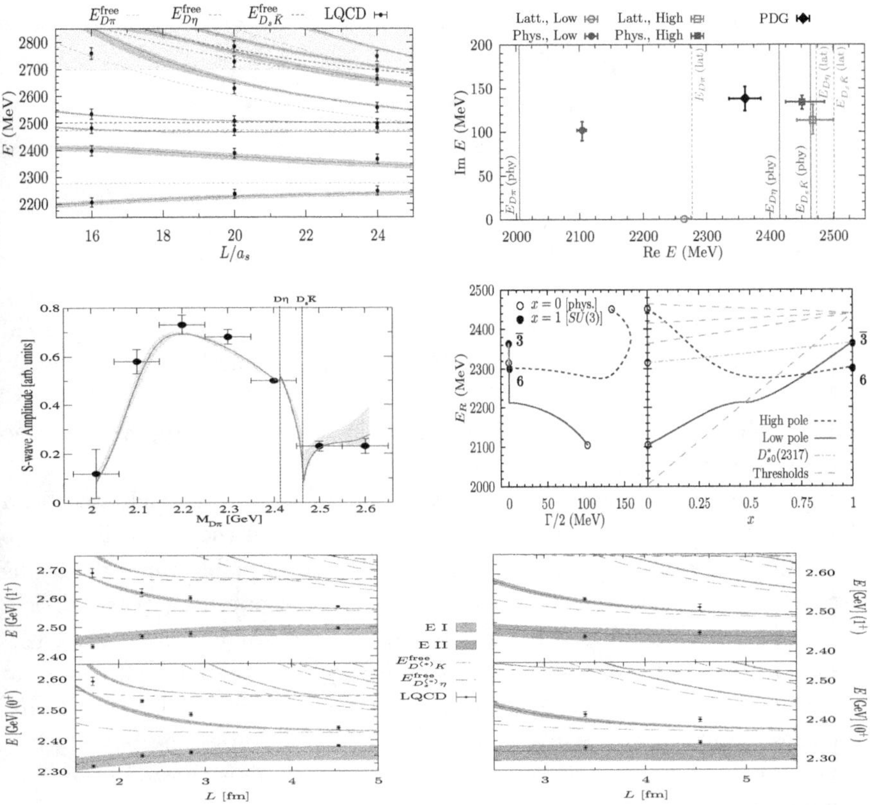

Fig. 1. Top $[(S = 0, I = 1/2)]$: Comparison (left) of Ref. 15 LQCD energy levels (black dots) in the $J^\pi = 0^+$ sector with the finite volume NLO HMChPT predictions[11,13] (red lines and bands), and complex energy plane location (right) of the two-pole-structure found in Ref. 11. In the second plot, the empty red (filled blue) symbols stand for the poles obtained when the LQCD[15] (physical) masses are used, while the black diamond represents the isospin average of the PDG values[1] for $D_0^*(2400)^0$ and $D_0^*(2400)^+$. Middle $[(S = 0, I = 1/2)]$: Comparison (left) of the NLO HMChPT S-wave amplitude with that obtained in the LHCb analysis of the $B^- \to D^+\pi^-\pi^-$ decay[17] (the red line stands for the central results[12] and the grey band quantifies the uncertainties that emerged from the fitting procedure), and evolution (right) from the physical to the flavor SU(3) symmetric cases of $(S, I) = (0, 1/2)$ and $(1, 0)$ poles $\sqrt{s} = E_R - i\Gamma/2$. In this latter plot, the blue short dashed, green solid and the orange dashed dotted lines represent the higher and lower $D_0^*(2400)$ and the $D_{s0}^*(2317)$, respectively. Both, the path followed by the poles in the complex plane and the evolution of E_R with x (departure from the SU(3) limit) are shown (the purple long dashed lines stand for the $D\pi$, $D\bar{K}$, $D\eta$, and $D_s\bar{K}$ thresholds, from bottom to top). Bottom $[(S = 1, I = 0)]$: LQCD $J^\pi = 0^+$ (bottom panels) and $J^\pi = 1^+$ (top panels) energy-levels, taken from Ref. 19, for different volumes and pion masses. LQCD data for Ensembles I and II of Ref. 19 are depicted in the left and right panels, respectively. In this plot, the solid lines stand for the energy levels calculated in 20 using the unitarized NLO HMChPT amplitudes of Refs. 13,14. In all cases, except for the S-wave amplitude in the left middle plot, the depicted 68% CL uncertainty bands are inherited from the errors on the LECs quoted in Ref. 14.

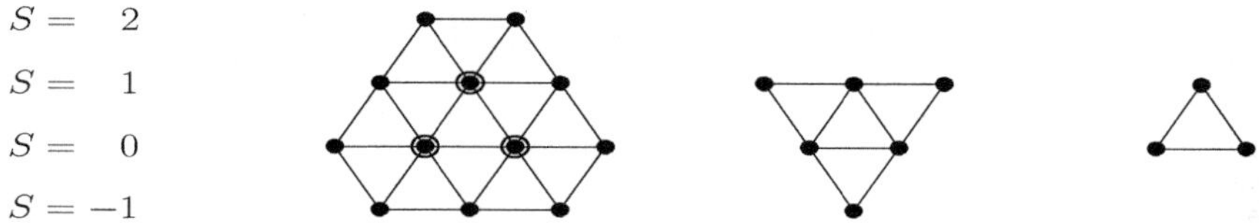

Fig. 2. Weight diagrams of the $\overline{\mathbf{15}}$, $\mathbf{6}$ and $\overline{\mathbf{3}}$ irreps.

$B^- \to D^+\pi^-\pi^-$ (see left middle plot of Fig. 1) and $B_s^0 \to \bar{D}^0 K^- \pi^+$ final states provided by the LHCb experiment in Ref. 17 and 18, respectively. This, undoubtedly, provides a strong case for the existence of two-pole pattern for the $D_0^*(2400)$, which was missed in the analysis reported in the LQCD work of Ref. 15.

The dynamical origin of this two-pole structure can be elucidated from the light-flavor SU(3) structure of the interaction. In the SU(3) limit, the heavy–light meson scattering decomposes into irreducible representations (irreps) as $\overline{\mathbf{3}} \otimes \mathbf{8} = \overline{\mathbf{15}} \oplus \mathbf{6} \oplus \overline{\mathbf{3}}$ (Fig. 2), and the potential (two particle irreducible amplitude) matrix can be diagonalized. At leading order $\mathcal{O}(p)$, the interaction in the $\mathbf{6}$ and $\overline{\mathbf{3}}$ ($\overline{\mathbf{15}}$) irreps is attractive (repulsive), being the latter the most intense. At NLO ($\mathcal{O}(p^2)$), the potentials receive corrections, but these qualitative features remain.[11] The evolution of the poles, when the physical and SU(3) symmetric cases are connected, is shown in the right middle plot of Fig. 1. The lower $D_0^*(2400)$ pole found in the physical case connects with a bound state in the SU(3) limit, whereas the higher pole connects with a virtual (the interaction in the $\overline{\mathbf{6}}$ is not attractive enough to bind) state.[b] Moreover, the analysis presented in the plot shows that the lower pole and the $D_{s0}^*(2317)$ complete the $\overline{\mathbf{3}}$ multiplet.[11] Since heavy quark spin symmetry (HQSS) relates the 0^+ and 1^+ sectors, we find in the latter a similar pattern of bound, resonant and virtual states. Hence, because the masses for the lower non-strange resonances, part of the $D_0^*(2400)$ and $D_1(2430)$ two-pole patterns, are smaller than those for the strange ones,[11,12] this scheme based on NLO HMChPT amplitudes finds a natural solution to the second of the puzzles highlighted in the introduction.

In the bottom panels of Fig. 1, we see that this framework also leads to a more than acceptable description of the $(S, I) = (0, 1/2)$ 0^+ and 1^+

[b]The sextet pole will become a bound state below threshold for large quark masses in the SU(3) limit, and can be searched for in LQCD simulations.[12]

LQCD data reported in Ref. 19 for two pion mass ensembles and different box sizes, despite having set all LECs to the values determined in Ref. 14. This provides a further piece of support for the finite-volume amplitudes obtained after unitarizing the NLO HMChPT amplitudes derived in Refs. 13, 14. Hence, the $D_0^*(2400)$ two-pole structure and the SU(3) pattern of the 0^+ and 1^+ heavy-light sectors claimed in Ref. 11 seem rather robust from the theoretical point of view. All these features have counterparts in the bottom sector as well, as expected from heavy-quark flavor symmetry.

We conclude this discussion by noting that the most attractive irrep, $\overline{\mathbf{3}}$, admits a $c\bar{q}$ $(q = u, d, s)$ configuration. Hence, one might wonder about the role played by the latter in the dynamics of the $D_{s0}^*(2317)$ and $D_{s1}(2460)$. This issue was addressed in Ref. 20, where the interplay between $D^{(*)}K$ meson-meson degrees of freedom and bare P-wave CQM $c\bar{s}$ states was taken into account. The conclusion was that these states, and also their bottom-flavor siblings,[21] have a predominantly hadronic-molecular structure. Thus, the first two of the puzzles listed above is also solved in this picture: the $D_{s1}(2460)$ and the $D_{s0}^*(2317)$ are $D^{(*)}K$ hadronic molecules, and do not correspond to the quark model $c\bar{s}$ states; HQSS predicts that the binding energies are independent of the heavy meson spin, up to 10%, as the leading spin symmetry breaking interaction is also of NLO in the chiral expansion. Thus, we understand why the $D_{s1}(2460) - D_{s0}^*(2317)$ mass difference is equal, within 2 MeV, to that between the D^* and D mesons.

3. Low-lying odd parity open charmed baryons

The experimental discovery of the five Ω_c states by the LHCb collaboration[22] triggered extensive theoretical activity also in the heavy open-flavor baryon sector, with some of the CQMs revisited in view of the new result.[23] The role of diquarks in the Ω_c spectrum was rapidly suggested in Refs. 23 and 24, while the odd-parity molecular interpretation for two or three of the states seen by LHCb was proposed in Refs. 25–27. The meson-baryon interactions used in the molecular schemes, derived in Refs. 25, 27, are consistent with both chiral and HQS symmetries, and lead to successful descriptions of the observed lowest-lying odd parity $\Lambda_c(2595)$, $\Lambda_c(2625)$ and $\Lambda_b(5912)$, $\Lambda_b(5920)$ resonances.[28–31] Thus, some of the Ω_c states observed by LHCb could be spin-flavour symmetry partners of these $\Lambda_{Q=c,b}$ baryons, and recently it has been suggested also possible connections with Ξ_c and Ξ_b odd-parity excited states[34] (see also the talk by L. Tolos).

However, the masses and decay widths of Λ_Q, at least the charmed ones, can also be accommodated within usual CQMs,[32] and thus the importance of the molecular components in their structure has not been settled yet. The structure of the $\Lambda_c(2595)$ and $\Lambda_c(2625)$ resonances was studied in Ref. 33 in the framework of an EFT consistent with HQS and chiral symmetries, and that incorporates the interplay between $\Sigma_c^{(*)}\pi - ND^{(*)}$ baryon-meson degrees of freedom and bare P-wave $c\bar{u}d$ quark-model states. It was found that these two resonances are not HQSS partners. The $J^P = 3/2^-$ $\Lambda_c(2625)$ should be viewed mostly as a dressed three quark state, whose origin is determined by a bare state, predicted to lie very close to the mass of the resonance.[32] The $J^P = 1/2^-$ $\Lambda_c(2595)$ seems to have, however, a predominant molecular structure. This is because, it is either the result of the chiral $\Sigma_c\pi$ interaction, which threshold is located much more closer than the mass of the bare three-quark state, or because the light degrees of freedom in its inner structure are coupled to the unnatural 0^- quantum-numbers. Both situations can occur depending on the renormalization procedure used.

4. Summary

The long accepted paradigm underlying open-flavor heavy spectroscopy that identifies all ground states with CQM states, is questionable. The hadron spectrum must be viewed as more than a collection of quark model states, but rather as a manifestation of a more complex dynamics that leads to an intricate pattern of various types of states that can only be understood by a joint effort from experiment, LQCD and phenomenology.

Acknowledgments

This research has been supported in part by the Spanish Ministerio de Economía y Competitividad and the European Regional Development Fund under contract FIS2017-84038-C2-1-P and by the EU STRONG-2020 project under the program H2020-INFRAIA-2018-1, grant no. 824093.

References

1. M. Tanabashi *et al.*, Review of Particle Physics, *Phys. Rev. D* **98**, 030001 (2018).
2. B. Aubert *et al.*, Observation of a narrow meson decaying to $D_s^+\pi^0$ at a mass of 2.32-GeV/c^2, *Phys. Rev. Lett.* **90**, 242001 (2003).
3. D. Besson *et al.*, Observation of a narrow resonance of mass 2.46-GeV/c^2 decaying to $D_s^+\pi^0$ and confirmation of the $D^*_{(sJ)}(2317)$ state, *Phys. Rev. D* **68**, 032002 (2003).

4. S. Godfrey and R. Kokoski, The Properties of P-wave mesons with one heavy quark, *Phys. Rev. D* **43**, 1679 (1991).
5. G. S. Bali, The $D^*_{(sJ)}(2317)$: What can the lattice say?, *Phys. Rev. D* **68**, 071501 (2003).
6. A. Cerri *et al.*, Report from WG4 : Opportunities in Flavour Physics at the HL-LHC and HE-LHC, *CERN Yellow Rep. Monogr.* **7**, 867 (2019).
7. F. K. Guo, C. Hanhart, U. G. Meißner, Q. Wang, Q. Zhao and B. S. Zou, Hadronic molecules, *Rev. Mod. Phys.* **90**, 015004 (2018).
8. J. M. Flynn and J. Nieves, Elastic S-wave $B\pi$, $D\pi$, DK and $K\pi$ scattering from lattice calculations of scalar form-factors in semileptonic decays, *Phys. Rev. D* **75**, 074024 (2007).
9. J. M. Link *et al.*, Measurement of masses and widths of excited charm mesons D^*_2 and evidence for broad states, *Phys. Lett. B* **586**, 11 (2004).
10. K. Abe *et al.*, Study of $B^- \to D^{**0}\pi^-(D^{**0} \to D^{(*)+}\pi^-)$ decays, *Phys. Rev. D* **69**, 112002 (2004).
11. M. Albaladejo, P. Fernandez-Soler, F. K. Guo and J. Nieves, Two-pole structure of the $D^*_0(2400)$, Phys. Lett. B **767**, 465 (2017).
12. M. L. Du, M. Albaladejo, P. Fernández-Soler, F. K. Guo, C. Hanhart, U. G. Meiner, J. Nieves and D. L. Yao, Towards a new paradigm for heavy-light meson spectroscopy, *Phys. Rev. D* **98**, 094018 (2018).
13. F. K. Guo, C. Hanhart, S. Krewald and U. G. Meißner, Subleading contributions to $\Gamma_{D^*_{s0}(2317)}$, *Phys. Lett. B* **666**, 251 (2008).
14. L. Liu, K. Orginos, F. K. Guo, C. Hanhart and U. G. Meissner, Interactions of charmed mesons with light pseudoscalar mesons from lattice QCD and implications on the nature of the $D^*_{s0}(2317)$, *Phys. Rev. D* **87**, 014508 (2013).
15. G. Moir, M. Peardon, S. M. Ryan, C. E. Thomas and D. J. Wilson, Coupled-Channel $D\pi$, $D\eta$ and $D_s\bar{K}$ Scattering from Lattice QCD, *JHEP* **1610**, 011 (2016).
16. J. Nieves and E. Ruiz Arriola, The $S_{11} - N(1535)$ and $N(1650)$ resonances in meson baryon unitarized coupled channel chiral perturbation theory, *Phys. Rev. D* **64**, 116008 (2001).
17. R. Aaij *et al.*, Amplitude analysis of $B^- \to D^+\pi^-\pi^-$ decays, *Phys. Rev. D* **94**, 072001 (2016).
18. R. Aaij *et al.*, Dalitz plot analysis of $B^0_s \to \bar{D}^0 K^- \pi^+$ decays, *Phys. Rev. D* **90**, 072003 (2014).
19. G. S. Bali, S. Collins, A. Cox and A. Schäfer, Masses and decay constants of the $D^*_{s0}(2317)$ and $D_{s1}(2460)$ from $N_f = 2$ lattice QCD close to the physical point, *Phys. Rev. D* **96**, 074501 (2017).

20. M. Albaladejo, P. Fernandez-Soler, J. Nieves and P. G. Ortega, Contribution of constituent quark model $c\bar{s}$ states to the dynamics of the $D_{s0}^*(2317)$ and $D_{s1}(2460)$ resonances, *Eur. Phys. J. C* **78**, 722 (2018).
21. M. Albaladejo, P. Fernandez-Soler, J. Nieves and P. G. Ortega, Lowest-lying even-parity $\bar{B}_s$ mesons: heavy-quark spin-flavor symmetry, chiral dynamics, and constituent quark-model bare masses, *Eur. Phys. J. C* **77**, 170 (2017).
22. R. Aaij *et al.*, Observation of five new narrow Ω_c^0 states decaying to $\Xi_c^+ K^-$, Phys. Rev. Lett. **118**, 182001 (2017).
23. M. Karliner and J. L. Rosner, Very narrow excited Ω_c baryons, *Phys. Rev. D* **95**, 114012 (2017).
24. A. Ali *et al.*, A new look at the Y tetraquarks and Ω_c baryons in the diquark model *Eur. Phys. J. C* **78**, 29 (2018).
25. J. Nieves, R. Pavao and L. Tolos, Ω_c excited states within a $\mathrm{SU}(6)_{\mathrm{lsf}}\times$ HQSS model, *Eur. Phys. J. C* **78**, 114 (2018).
26. G. Montaña, A. Feijoo and . Ramos, A meson-baryon molecular interpretation for some Ω_c excited states, *Eur. Phys. J. A* **54**, 64 (2018).
27. V. R. Debastiani, J. M. Dias, W. H. Liang and E. Oset, Molecular Ω_c states generated from coupled meson-baryon channels, *Phys. Rev. D* **97**, 094035 (2018).
28. C. Garcia-Recio, V. K. Magas, T. Mizutani, J. Nieves, A. Ramos, L. L. Salcedo and L. Tolos, The *S*-wave charmed baryon resonances from a coupled-channel approach with heavy quark symmetry, *Phys. Rev. D* **79**, 054004 (2009).
29. O. Romanets *et al.*, Charmed and strange baryon resonances with heavy-quark spin symmetry, *Phys. Rev. D* **85**, 114032 (2012).
30. W. H. Liang, T. Uchino, C. W. Xiao and E. Oset, Baryon states with open charm in the extended local hidden gauge approach, *Eur. Phys. J. A* **51**, 16 (2015).
31. C. Garcia-Recio, J. Nieves, O. Romanets, L. L. Salcedo and L. Tolos, Odd parity bottom-flavored baryon resonances, *Phys. Rev. D* **87**, 034032 (2013)
32. T. Yoshida, E. Hiyama, A. Hosaka, M. Oka and K. Sadato, Spectrum of heavy baryons in the quark model, *Phys. Rev. D* **92**, 114029 (2015).
33. J. Nieves and R. Pavao, Nature of the lowest-lying odd parity charmed baryon $\Lambda_c(2595)$ and $\Lambda_c(2625)$ resonances, *Phys. Rev. D* **101**, 014018 (2020).
34. J. Nieves, R. Pavao and L. Tolos, Ξ_c and Ξ_b excited states within a $\mathrm{SU}(6)_{\mathrm{lsf}}\times$HQSS model, *Eur. Phys. J. C* **80**, 22 (2020).

Coupled-channel effects in heavy hadrons

David R. Entem

Grupo de Física Nuclear and Instituto Universitario de Física Fundamental y Matemáticas (IUFFyM), Universidad de Salamanca,
E-37008 Salamanca, Spain
E-mail: entem@usal.es
www.usal.es

Heavy meson spectroscopy was very well described in terms of $Q\bar{Q}$ states since the discovery of the J/ψ in 1974 up to the discovery of the $X(3872)$ in 2003. However, the intriguing properties of this meson made evident that higher Fock components were necessary to describe heavy hadron spectroscopy for excited states. The inclusion of two hadron components induce coupled-channel effects that can have important consequences. In this contribution I will give some examples of what I see as the most relevant effects due to coupled channels in the heavy hadron spectrum.

Keywords: Coupled-channels; heavy hadrons; hadron spectrum.

1. Isospin symmetry breaking

1.1. *The X*(3872)

The $X(3872)$ was discovered by Belle[1] in the $J/\psi\pi\pi$ invariant mass distribution of the decay $B^+ \to K^+\pi^+\pi^- J/\psi$. The Belle Collaboration measured the ratio[2]

$$\frac{\mathcal{B}(X(3872) \to \pi^+\pi^-\pi^0 J/\psi)}{\mathcal{B}(X(3872) \to \pi^+\pi^- J/\psi)} = 1.0 \pm 0.4(\text{stat}) \pm 0.3(\text{syst}) \tag{1}$$

where the two-pions came from the decay of a ρ^0 meson and the three-pions came from the decay of an ω meson. This fact implies an isospin violating decay, since one channel is $I = 0$ and the other $I = 1$. The mass of the state relative to the D^0D^{*0} threshold is given by the PDG to be

$$B \equiv m_{X(3872)} - m_{D^{*0}} - m_{\bar{D}^0} = 1.1^{+0.6+0.1}_{-0.4-0.3}\ \text{MeV} \tag{2}$$

$$= (0.00 \pm 0.18)\ \text{MeV} \tag{3}$$

so is a very loosely bound state or even a resonance or virtual state. There is another important scale that measures isospin breaking, $m_{D^{*+}} + m_{D^-} -$

$m_{D^{*0}} - m_{\bar{D}^0} = 8.2 \pm 0.1$ MeV $\gg B$. Although the isospin breaking on the $D^{(*)}$ mesons is 0.13% and 0.08% respectively, the isospin breaking on the thresholds is much bigger than the binding energy and induces isospin symmetry breaking on the state.

This was already introduced by Swanson[3] doing a coupled channel calculation including D^0D^{*0}, D^+D^{*-}, $J/\psi\rho$ and $J/\psi\omega$ channels. For small binding energies the D^0D^{*0} channels is bound by one MeV or less while the D^+D^{*-} is bound by 8 MeV. The asymptotic behavior of the wave function is $\psi(r) \stackrel{r\to\infty}{\to} e^{-\gamma r}$, with $\gamma^{-1}_{D^+D^{*-}} \sim 1.5\,\text{fm}$ while, depending on the value of the binding energy, $\gamma^{-1}_{D^0D^{*0}}$ can vary from around 4 fm to infinity. So the D^0D^{*0} can extend to very large distances while the other components will die very soon. Then the probabilities on the wave function will be dominated by this component. For this reason very large isospin breaking effects on the wave function will be produced, however it will only be important for observables sensitive to the long range part.

Most of the decays are only sensitive to the short range part of the wave function and in this case isospin breaking effects will be small. There was some confusion about this point since in Quantum Field Theoretical models the couplings of the state to the DD^* channels were compatible with isospin 0.[4] In Ref. 4 the couplings obtained were $g = 2982$ and $g = 3005$ for the charged and neutral states respectively giving less than 1% isospin violation. They also checked the dependence on the isospin violation of the couplings with the binding energy finding a very small effect going up to around 1.5%. Within this model the ratio of the decay $\mathcal{R}_{\rho/\omega}$ was found to be very small, around 3.2%. However the ρ meson is very wide and including phase space effects, the measured ratio goes up to 1.4 compatible with the experiment.

This was clarified in Ref. 5 where it was shown how the couplings are not related to the total probability of the wave function in each channel but to the part of the wave function in the interacting region, which in the model used with contact interaction is the wave function at the origin.

1.2. *The pentaquarks*

Very recently the LHCb Collaboration has measured[6] three pentaquark states $P_c(4312)^+$, $P_c(4440)^+$ and $P_c(4457)^+$, updating their previous measurement in 2015.[7] From the different possible interpretations, noticing that they have been discovered close to the $\Sigma_c\bar{D}$, $\Sigma_c^*\bar{D}$ and $\Sigma_c\bar{D}^*$ thresholds, the molecular interpretation is one of the most accepted.

If this interpretation is correct, it has been proposed recently that, in analogy with the $X(3872)$, the pentaquark state $P_c(4457)^+$ could have isospin violating decays.[8] Similarities between this pentaquark state and the $X(3872)$ were already discuss in Ref. 9. This isospin violating decay is a unique feature of the $\Sigma_c \bar{D}^*$ molecular model and if measured it will be a confirmation of the nature of the pentaquarks.

The mechanism is an analog of the decay of the $X(3872)$ into $J/\psi\rho$ and again is due to a large isospin symmetry breaking on the thresholds of the charged states compared to the binding energy of one of the pentaquarks. The relevant pentaquark is the $P_c(4457)^+$ which has a mass of $M_{P_c(4457)^+} = 4457.3 \pm 0.6^{+4.1}_{-1.7}$ MeV and a width of $\Gamma_{P_c(4457)^+} = 6.4 \pm 2.0^{+5.7}_{-1.9}$. In the molecular picture many different model predicts the state as a $\Sigma_c \bar{D}^*$ with quantum number $J^P = 3/2^-$ and $I = 1/2$. The nearby thresholds are $M_{\Sigma_c^+} + M_{\bar{D}^{*0}} = 4459.8 \pm 0.4$ MeV and $M_{\Sigma_c^{++}} + M_{\bar{D}^{*-}} = 4464.23 \pm 0.15$ MeV. The binding energy of the pentaquark with respect to the lowest threshold is $2.5^{+1.8}_{-4.2}$ MeV and with the higher threshold is $6.9^{+1.8}_{-4.1}$ MeV. The charged states are combinations of $I = 1/2$ and $I = 3/2$. The state has been measured in the $J/\psi p$ channel which is $I = 1/2, I_3 = 1/2$. So the proposal is that it could decay into the $I = 3/2, I_3 = 1/2$ which corresponds to the $J/\psi\Delta^+$ channel.

The ratio

$$R_{\Delta^+/p} = \frac{|\mathcal{B}(P_c(4457)^+ \to J/\psi\Delta^+)|}{|\mathcal{B}(P_c(4457)^+ \to J/\psi p)|}$$

has been evaluated in a model where isospin breaking comes only from the different masses of the charged states. For cut-offs between 0.5 to 1 GeV the ratio varies from a few percent up to 30% with a large uncertainty coming from the pentaquark mass.

2. HQSS and HFS breaking

Heavy Quarks Spin Symmetry (HQSS) and Heavy Flavor Symmetry (HFS) are good approximate symmetries of QCD. In the heavy-light sector it means that the mass of the D and D^* mesons should be the same which is broken by $\frac{M_{D^*}-M_D}{M_{D^*}+M_D} \sim 3.6\%$. In the heavy quark sector we have that $\frac{M_{J/\psi}-M_{\eta_c}}{M_{J/\psi}+M_{\eta_c}} \sim 1.8\%$ and $\frac{M_{\chi_{c2}(1P)}-M_{\chi_{c0}(1P)}}{M_{\chi_{c2}(1P)}+M_{\chi_{c0}(1P)}} \sim 2.0\%$. HQSS and HFS has been extensively used to make predictions in different sectors.

In the case of the $X(3872)$ and its partners, considering HQSS[10–12] there is an unavoidable consequence. The interaction between $DD^*(1^{++})$ and $D^*D^*(2^{++})$ channels is the same, so if the $X(3872)$ is a $DD^*(1^{++})$

molecule it implies that there should be a bound state with very similar binding energy in the $D^*D^*(2^{++})$, which was called the $X(4012)$. However there is no candidate for it. If now we consider HFS there are again unavoidable consequences, since the interaction between charmed mesons will be the same as the interaction between bottom mesons.[13] So the molecular picture for the $X(3872)$ would imply at least the existence of three more partners. The partner in the $BB^*(1^{++})$ channels has been looked for in the $\Upsilon(1S)\pi\pi$ channel by the CMS and ATLAS Collaborations and in the $\Upsilon(1S)\pi\pi\pi$ channel by the Belle Collaboration. In all cases no indication of the existence of the state was found.

However heavy mesons above threshold can decay strongly to open charm channels, so one would expect a strong coupling between one meson and two meson states. The quark model is an appropriate framework to study such effects, since it can give with the same Hamiltonian the charmonium spectrum and the interaction between charmed (bottom) mesons. The coupling between one meson and two meson states can also be modeled and applied to the study of strong decays. In such a framework the Chiral Quark Model (CQM) was applied to the study of the $X(3872)$.[14] In this work it was found that the $X(3872)$ appears as a new state (not present in the naive quark model picture), mostly of molecular nature, while the naive quark model states were dressed by the two meson channels. In this picture both effects, the two meson dynamics and the coupling with charmonia, were important.

Although HQSS is a good symmetry, its small breaking in the masses of heavy mesons, can induce important effects due to thresholds. We have that the $M_{D^*} - M_D \sim 141$ MeV. This induces a difference on the thresholds $M_{D^*D^*} - M_{DD} \sim 282$ MeV. If we consider two meson states in S wave for the $J^{PC} = 1^{++}$ and $J^{PC} = 2^{++}$ channels only DD^* and D^*D^* channels are allowed while for the 0^{++} channel DD and D^*D^* are allowed.

With these quantum numbers the charmonium states are in a relative P wave. We have that the maximum breaking in $c\bar{c}(1P)$ states is $M_{\chi_{c2}(1P)} - M_{\chi_{c0}(1P)} \sim 141$ MeV, so one would expect $M_{\chi_{c2}(2P)} - M_{\chi_{c0}(2P)} <$ 141 MeV. In fact, in the Relativistic quark model of Ref. 15, where the masses of the $1P$ states are accurately described, this mass difference is $M_{\chi_{c2}(2P)} - M_{\chi_{c0}(2P)} \sim 80$ MeV. This difference is much smaller than the threshold breaking mentioned before, and the relative position of the relevant two meson threshold and the charmonium state in each channel can vary significantly, which has important consequences, since if the state is above(below) threshold it generates attraction(repulsion).

This effect was already pointed out in the CQM framework in Ref. 16. In Ref. 17 a systematic study of this effect was performed. Here the authors studied the existence of the $X(4012)$ in a model that preserves HQSS but the $X(3872)$ is obtained as a mixture of the DD^* and the $\chi_{c1}(2P)$. They fixed the mass of the $X(3872)$ and vary the parameter for the coupling between one meson and two meson sectors and at the same time the interaction (point-like) between DD^*. When this parameter is increased the molecular component of the $X(3872)$ decreases and the bare $\chi_{c1}(2P)$ state becomes a resonance that have a rapidly increasing width, and finally it goes into the second Riemann sheet. However for the $X(4012)$ the effective potential gets repulsion because the $\chi_{c2}(2P)$ state is below threshold. The authors fixed the value of the $\chi_{c2}(2P)$ state to the experiment varying the bare mass and studied the pole corresponding to the $X(4012)$. This state approaches the D^*D^* threshold and very soon goes into the second Riemann sheet moving quickly away from threshold along the real axis. The conclusion is that the 2^{++} HQSS partner could not be accessible to direct observation.

This study is in good agreement with the findings of the CQM.[16] However it is important to notice that other thresholds are present which can make some deviations. A careful study of the 3.9 GeV region within the CQM was carried out in Ref. 18, where again no $X(4012)$ was found.

3. Thresholds cusps

As pointed out by Wigner,[19] an enhancement of the cross section near the opening of a threshold can occur, giving what we call a threshold cusp.

In 2004 Cabibbo proposed to measured the $a_0 - a_2$ $\pi\pi$ scattering length combination[20] using this cusp-like structure in the $\pi^0\pi^0$ invariant mass distribution in the $K^\pm \to \pi^\pm\pi^0\pi^0$ decay that was first measured in Ref. 21. This threshold anomaly was already pointed out in Ref. 22. The mechanism is based on the fact that the $K^\pm \to \pi^\pm\pi^+\pi^-$ can give a contribution through the charge exchange reaction $\pi^+\pi^- \to \pi^0\pi^0$ which is proportional to this scattering length combination and shows a characteristic behavior in the vicinity of the $\pi^+\pi^-$ threshold. Due to the high accuracy of the data isospin symmetry breaking effects were included.[23] A precised determination of $(a_0-a_2)m_{\pi^+} = 0.268\pm0.010(\text{stat})\pm0.004(\text{syst})$ was given and taking into account the correlation between the two scattering lengths predicted by χPT a value of $(a_0-a_2)m_{\pi^+} = 0.264\pm0.006(\text{stat})\pm0.004(\text{syst})\pm0.013(\text{ext})$ was given.

This effect has been also used to explain strong energy dependencies near threshold of invariant mass distributions, as in the case of Z_c and Z_b states.[24] However it has been argued[25] that such big effects may not appear without the existence of a nearby pole (bound, virtual or resonance state). In particular, in the CQM the Z_c states are described due to a virtual state.[26]

4. Triangle singularities

Triangle singularities refer to singularities of a triangle diagram. There are poles, but also brunch points. The most obvious singularities are the so called normal threshold singularities. These singularities describe processes in which two intermediate states can be on-shell and the other one off-shell.

However there are also the so called anomalous threshold singularities. In this case the three intermediate particles can be on-shell and the process can be classical. However since all particles can decay they should have some width and the singularity moves from the physical region into the complex plane and the result is a finite peak in the physical region.

Triangle singularities have been extensively used to study enhancements on physical amplitudes. Recently there has been a proposal to measure the mass of the $X(3872)$ using a triangle singularity.[27] The process proposed is to produced a pair of $D^{*0}\bar{D}^{*0}$ mesons, decaying one of them into $D^{*0} \to D^0\gamma$, and the decaying D^0 into a $X(3872)$. This observable is sensitive to the long range part of the $X(3872)$ wave function.

In this case, close to the $D^{*0}\bar{D}^{*0}$ threshold, all intermediate particles are non-relativist and the loop integral can be evaluated. The triangle singularity close to threshold is located at the $X(3872)\gamma$ invariant mass

$$E_{X\gamma}^{TS} = 2m_{D^{*0}} + \frac{x^2}{2m_{D^0}} + \mathcal{O}\left(\frac{x^3}{m_{D^0}^2}\right)$$

with $x = m_{D^{*0}} - m_{D^0} - 2\sqrt{-m_{D^0}\delta} + \delta$ with δ the binding energy of the $X(3872)$. Due to the small widths of the $D^{(*)}$ mesons the singularity is located in the vicinity of the physical region and has a big impact on the $X(3872)\gamma$ line shape. The line shape depends strongly on δ and it has been shown that with enough statistics the binding energy δ could be resolved with a few keV resolution.[27]

Acknowledgments

This work has been funded by Ministerio de Economía, Industria y Competitividad under Contract No. FPA2016-77177-C2-2-P, and by the EU

STRONG-2020 project under the program H2020-INFRAIA-2018-1, grant agreement no. 824093.

References

1. S.-K. Choi *et al.*, Observation of a narrow charmoniumlike state in exclusive $B^{\pm} \to K^{\pm}\pi^{+}\pi^{-}j/\psi$ decays, *Phys. Rev. Lett.* **91**, 262001 (Dec 2003).
2. K. Abe *et al.*, Evidence for X(3872) gamma J/psi and the sub-threshold decay X(3872) omega J/psi, in *Lepton and photon interactions at high energies. Proceedings, 22nd International Symposium, LP 2005, Uppsala, Sweden, June 30-July 5, 2005*, 2005.
3. E. S. Swanson, Diagnostic decays of the x(3872), *Physics Letters B* **598**, 197 (2004).
4. D. Gamermann and E. Oset, Isospin breaking effects in the $x(3872)$ resonance, *Phys. Rev. D* **80**, 014003 (Jul 2009).
5. D. Gamermann, J. Nieves, E. Oset and E. R. Arriola, Couplings in coupled channels versus wave functions: Application to the $x(3872)$ resonance, *Phys. Rev. D* **81**, 014029 (Jan 2010).
6. R. Aaij *et al.*, Observation of a narrow pentaquark state, $P_c(4312)^+$, and of the two-peak structure of the $P_c(4450)^+$, *Phys. Rev. Lett.* **122**, 222001 (Jun 2019).
7. R. Aaij *et al.*, Observation of $j/\psi p$ resonances consistent with pentaquark states in $\Lambda_b^0 \to j/\psi K^- p$ decays, *Phys. Rev. Lett.* **115**, 072001 (Aug 2015).
8. F.-K. Guo, H.-J. Jing, U.-G. Meißner and S. Sakai, Isospin breaking decays as a diagnosis of the hadronic molecular structure of the $P_c(4457)$, *Phys. Rev. D* **99**, 091501 (May 2019).
9. T. J. Burns, Phenomenology of pc(4380)+, pc(4450)+ and related states, *The European Physical Journal A* **51**, 152 (Nov 2015).
10. J. Nieves and M. Pavón Valderrama, Heavy quark spin symmetry partners of the $x(3872)$, *Phys. Rev. D* **86**, 056004 (Sep 2012).
11. C. Hidalgo-Duque, J. Nieves and M. P. Valderrama, Light flavor and heavy quark spin symmetry in heavy meson molecules, *Phys. Rev. D* **87**, 076006 (Apr 2013).
12. Baru, V., Epelbaum, E., Filin, A. A., Hanhart, C. and Nefediev, A.V., Molecular partners of the x(3872) from heavy-quark spin symmetry: a fresh look, *EPJ Web Conf.* **137**, 06002 (2017).

13. F.-K. Guo, C. Hidalgo-Duque, J. Nieves and M. Pavón Valderrama, Consequences of heavy-quark symmetries for hadronic molecules, *Phys. Rev. D* **88**, 054007 (Sep 2013).
14. P. G. Ortega, J. Segovia, D. R. Entem and F. Fernández, Coupled channel approach to the structure of the $x(3872)$, *Phys. Rev. D* **81**, 054023 (Mar 2010).
15. D. Ebert, R. N. Faustov and V. O. Galkin, Spectroscopy and regge trajectories of heavy quarkonia and bcmesons, *The European Physical Journal C* **71**, 1825 (Dec 2011).
16. D. R. Entem, P. G. Ortega and F. Fernández, Partners of the x(3872) and heavy quark spin symmetry breaking, *AIP Conference Proceedings* **1735**, 060006 (2016).
17. E. Cincioglu, J. Nieves, A. Ozpineci and A. U. Yilmazer, Quarkonium contribution to meson molecules, *The European Physical Journal C* **76**, 576 (Oct 2016).
18. P. G. Ortega, J. Segovia, D. R. Entem and F. Fernndez, Charmonium resonances in the 3.9 gev/c2 energy region and the x(3915)/x(3930) puzzle, *Physics Letters B* **778**, 1 (2018).
19. E. P. Wigner, On the behavior of cross sections near thresholds, *Phys. Rev.* **73**, 1002 (May 1948).
20. N. Cabibbo, Determination of the a_0-a_2 pion scattering length from $K^+ \to \pi^+\pi^0\pi^0$ decay, *Phys. Rev. Lett.* **93**, 121801 (Sep 2004).
21. J. Batley *et al.*, Observation of a cusp-like structure in the p0p0 invariant mass distribution from k?pp0p0 decay and determination of the pp scattering lengths, *Physics Letters B* **633**, 173 (2006).
22. P. Budini and L. Fonda, Pion-pion interaction from threshold anomalies in K^+ decay, *Phys. Rev. Lett.* **6**, 419 (Apr 1961).
23. N. Cabibbo and G. Isidori, Pion-pion scattering and the k$\to 3\pi$ decay amplitudes, *Journal of High Energy Physics* **2005**, 021 (mar 2005).
24. E. S. Swanson, Z_b and Z_c exotic states as coupled channel cusps, *Phys. Rev. D* **91**, 034009 (Feb 2015).
25. F.-K. Guo, C. Hanhart, Q. Wang and Q. Zhao, Could the near-threshold xyz states be simply kinematic effects?, *Phys. Rev. D* **91**, 051504 (Mar 2015).
26. Ortega, Pablo G., Segovia, Jorge, Entem, David R. and Fernández, Francisco, The z_c structures in a coupled-channels model, *Eur. Phys. J. C* **79**, 78 (2019).
27. F.-K. Guo, Novel method for precisely measuring the $x(3872)$ mass, *Phys. Rev. Lett.* **122**, 202002 (May 2019).

Light meson spectroscopy at e^+e^- machines

B. J. Liu

Institute of High Energy Physics, Chinese Academy of Sciences
Beijing, 100049, China
E-mail: liubj@ihep.ac.cn

The study of light hadrons is central to the understanding of confinement–a unique property of QCD. The quark model describs mesons as bound states of quarks and antiquarks. LQCD and QCD-motivated models for hadrons, however, predict a richer spectrum of mesons that takes into account not only the quark degrees of freedom but also the gluonic degrees of freedom. A selection of recent progress in the light-quark sector with unprecedented high-statistics data sets from e^+e^- experiments are reviewed.

Keywords: Light QCD exotics; Glueball; e^+e^- experiments.

1. Introduction

Confinement is a unique property of QCD. The spectrum of light hadrons serves as an excellent probe of QCD in the confinement regime.[1–6] The quark model describs mesons as bound states of quarks and antiquarks. LQCD and QCD-motivated models for hadrons, however, predict a richer spectrum of mesons that takes into account not only the quark degrees of freedom but also the gluonic degrees of freedom. Light mesons can be produced at e^+e^- experiments through e^+e^- annihilation, through the use of the radiative return method and through two-photon fusion. Unprecedented high-statistics data sets from experiments at e^+e^- machines provide great opportunities to the quantitative understanding of confinement.

2. Search for glueballs

The spectrum of glueballs both from the quenched lattice QCD studies[8,9] and the full-QCD study[10,11] show that the lightest one having scalar quantum numbers 0^{++} and a mass between 1.5 GeV and 1.7 GeV. Also the next-higher glueball states have nonexotic quantum numbers, 2^{++} (mass 2.3–2.4 GeV) and 0^{-+} (mass 2.3–2.6 GeV), and hence will be mixed into the conventional meson spectrum and difficult to be identified experimentally. It requires systematic studies to identify a glueball by searching for out-numbering of conventional quark model states and comparing a candidates properties to the expected properties of glueballs and conventional mesons. Glueballs are expected to appear in so-called gluon-rich environments. The radiative decays of the J/ψ meson provide such a gluon-rich environment and are therefore regarded as one of the most promising hunting grounds for glueballs. Recent LQCD calculations predict that the partial width of J/ψ radiatively decaying into the pure gauge scalar glueball is 0.35(8) keV, which corresponds to a branching ratio of $3.8(9) \times 10^{-3}$;[12] the partial decay width for a tensor glueball is estimated to be 1.01(22)(10) keV which corresponds to a large branching ratio $1.1(2)(1) \times 10^{-2}$.[13] Even though the fact of a supernumerary state is suggestive for the mixing of glueball with $q\bar{q}$ state, the decay rates and production mechanisms are also needed to unravel the quark content of $f_0(1500)$ and $f_0(1710)$. In the PWA of $J/\psi \to \gamma\eta\eta$[14] and $J/\psi \to \gamma K_S K_S$[15] at BESIII, the branching fractions of the $f_0(1710)$ are one order of magnitude larger than those of the $f_0(1500)$. With the new measurements from BESIII, the known branching fraction of $J/\psi \to \gamma f_0(1710)$[23] is up to $(1.7 \times 10^{-3}$, which is already comparable to the LQCD calculation of scalar glueball $(3.8(9) \times 10^{-3})$.[12] The production property suggests $f_0(1710)$ has large gluonic component than $f_0(1500)$. The two-photon width can be used to identify glueball, even though it can also be adjusted by the glueball mixing with $q\bar{q}$ state.[16] The scalar meson $f_0(1710)$ has been seen in $\gamma\gamma \to K_S K_S$.[17,18] $f_0(1500)$ was not seen in $\gamma\gamma \to K_S K_S$ by L3,[17] or in $\gamma\gamma \to \pi^+\pi^-$, by ALEPH.[19] However, a resonance observed in $\gamma\gamma \to \pi^0\pi^0$ by Belle[20] is close to the $f_0(1500)$ mass, though it is also consistent with $f_0(1370)$ because of the large errors in the experiment and the large uncertainty in the $f_0(1370)$ mass. The assignment of scalars in two-photon fusion requires further studies with better precision. It is noticeable the nonobservation of $f_0(1710)$ and observation of $f_0(1500)$ in $B_s \to J/\psi\pi\pi$ by LHCb.[21] Because of the spectator s quark of B_s, the isosinglet scalar resonance f_0 produced in $B_s \to J/\psi f_0$ decays should have a sizable $s\bar{s}$ component.[22] The three tensors $f_2(2010)$, $f_2(2300)$ and $f_2(2340)$ observed in $\pi^- p \to \phi\phi n$[24] are also observed in $J/\psi \to \gamma\phi\phi$.[25] The large production rate of the $f_2(2340)$ in $J/\psi \to \gamma\phi\phi$ and $J/\psi \to \gamma\eta\eta$[14] indicates $f_2(2340)$ is a good candidate of tensor glueball. Significant tensor contribution around 2.4 GeV also presents in $J/\psi \to \gamma\pi^0\pi^0$,[26] and $J/\psi \to \gamma K_S K_S$.[15] However, the measured production rate of $f_2(2370)$ appears to be substantially lower than the LQCD calculated value.[13] It is

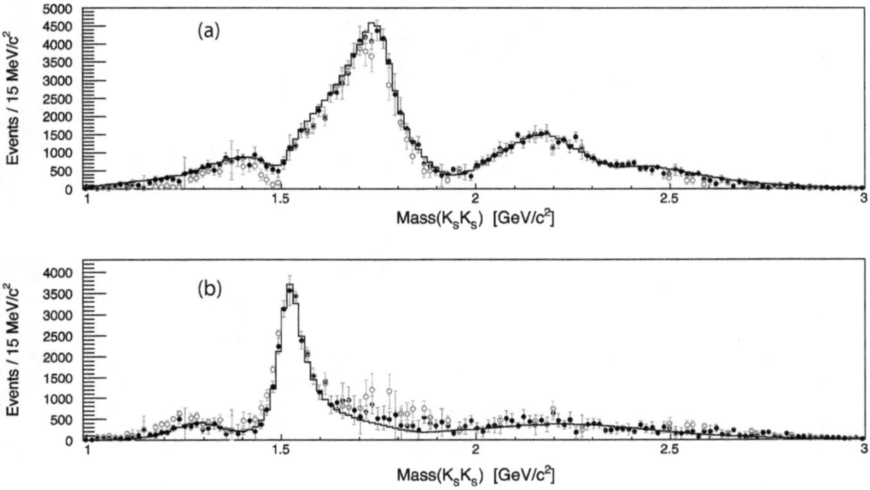

Fig. 1. (Color online) Intensities for the total (a) 0^{++} and (b) 2^{++} amplitudes in $J/\psi \to \gamma K_S K_S$ from BESIII as a function of $K_S K_S$ invariant mass for the nominal results without acceptance correction.[15] The solid black markers show one set of solutions from the mass-independent analysis, while the open red markers represent its ambiguous partner and the histogram shows the results of the mass-dependent analysis.

desirable to search for more decay modes to establish and characterize the lowest tensor glueball.

3. $a_0(980) - f_0(980)$ mixing

After the discoveries of $a_0(980)$ and $f_0(980)$ several decades ago, explanations about the nature of these two light scalar mesons have still been controversial. These two states, with similar masses but different decay modes and isospins, are difficult to accommodate in the traditional quark-antiquark model, and many alternative formulations have been proposed to explain their internal structure, including tetra-quarks,[27–31] $K\bar{K}$ molecule,[32] or quark-antiquark gluon hybrid.[33] Further insights into $a_0(980)$ and $f_0(980)$ are expected from their mixing.[34] The mixing mechanism in the system of $a_0(980) - f_0(980)$ is considered to be a sensitive probe to clarify the nature of these two mesons. In particular, the leading contribution to the isospin-violating mixing transition amplitudes for $f_0(980) \to a_0^0(980)$ and $a_0^0(980) \to f_0(980)$, is shown to be dominated by the difference of the unitarity cut which arises from the mass difference between the charged and neutral $K\bar{K}$ pairs. As a consequence, a narrow peak of about 8 MeV/c^2

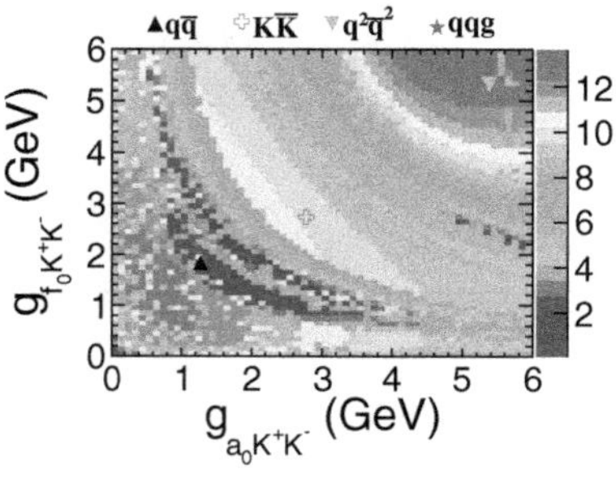

Fig. 2. The statistical significance of the signal scanned in the two-dimensional space of $g_{a_0K^+K^-}$ and $g_{f_0K^+K^-}$. The regions with higher statistical significance indicate larger probability for the emergence of the two coupling constants. The markers indicate predictions from various illustrative theoretical models.

is predicted between the charged and neutral $K\bar{K}$ thresholds. The corresponding signal is predicted in the isospin-violating processes of $J/\psi \to \phi a_0^0(980)$[35,36] and $\chi_{c1} \to \pi^0 f_0(980)$.[37] The signals of $f_0(980) \to a_0^0(980)$ and $a_0^0(980) \to f_0(980)$ mixing are first observed in $J/\psi \to \phi f_0(980) \to \phi a_0^0(980) \to \phi\eta\pi^0$ and $\chi_{c1} \to \pi^0 a_0^0(980) \to \pi^0 f_0(980) \to \pi^0\pi^+\pi^-$ at BESIII.[38,39] The statistical significance of the signal versus the values of $g_{a_0K^+K^-}$ and $g_{f_0K^+K^-}$ is shown in Fig. 2. The regions with higher statistical significance indicate larger probability for the emergence of the two coupling constants. This direct measurement of $a_0(980) - f_0(980)$ mixing is a sensitive probe to the internal structure of those ground state scalars and sheds important light on their nature. The new results from BESIII provide critical constraints to the development of theoretical models for $a_0(980)$ and $f_0(980)$. It is theorists turn to refine the calculations to understand the inner structure of the $a_0(980)$ and $f_0(980)$ mesons.

4. $X(1835)$, $X(p\bar{p})$ and structures near $N\bar{N}$ threshold in e^+e^- cross sections

An anomalously strong enhancement at the proton-antiproton ($p\bar{p}$) mass threshold, dubbed $X(p\bar{p})$, was first observed by BES in $J/\psi \to \gamma p\bar{p}$ decays[40]; this observation was confirmed by BESIII[41] and CLEO,[42] while no evidence has been seen in other channels, such as $J/\psi \to \omega p\bar{p}$[43,44] or $J/\psi \to \phi p\bar{p}$.[45] These non-observations disfavor the mass-threshold enhancement attribution to the effects of $p\bar{p}$ final state interactions. This enhancement was subsequently determined to have spin-parity $J^P = 0^-$

by BESIII.[46] The state $X(1835)$ was first observed by the BES experiment as a peak in $J/\psi \to \gamma\eta'\pi^+\pi^-$ decays.[47] This observation was later confirmed by BESIII[48] and was also observed in the $\eta K_S^0 K_S^0$ channel, where its spin-parity was determined to be $J^P = 0^-$ by a partial wave analysis.[49] No evidence of $X(1835)$ is found in $J/\psi \to \omega\eta'\pi^+\pi^-$.[50] $\eta(1475) \to \gamma\phi$ and $X(1835) \to \gamma\phi$ are observed in the decay of $J/\psi \to \gamma\gamma\phi$ at BESIII,[51] which indicates that both $\eta(1475)$ and $X(1835)$ contain a sizeable $s\bar{s}$ component. Using high-statistics J/ψ events, BESIII studied the $J/\psi \to \gamma\eta'\pi^+\pi^-$ process and observed a significant abrupt change in the slope of the $\eta'\pi^+\pi^-$ invariant mass distribution at the proton-antiproton ($p\bar{p}$) mass threshold.[52] Two models are used to characterize the $\eta'\pi^+\pi^-$ line shape around 1.85 GeV/c^2: one which explicitly incorporates the opening of a decay threshold in the mass spectrum (Flatté formula) (Fig. 3(a)), and another which is the coherent sum of two resonant amplitudes(Fig. 3 (b)). Both fits show almost equally good agreement with data, and suggest the existence of either a broad state with strong couplings to $p\bar{p}$ final states or a narrow state just below the $p\bar{p}$ mass threshold. Without

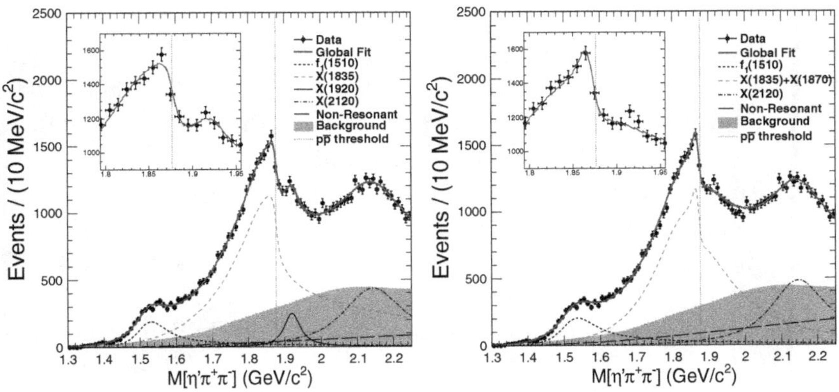

Fig. 3. An anomalous line shape of the $\eta'\pi^+\pi^-$ mass spectrum near the $p\bar{p}$ mass threshold in $J/\psi \to \gamma\eta'\pi^+\pi^-$. (a) shows the fitting results with Flatté formula and (b) shows the fitting results with the coherent sum of two Breit-Wigner amplitudes

imposing a relation to the 0^{-+} structures near $N\bar{N}$ threshold observed in J/ψ radiative decays, we should notice the dip near $p\bar{p}$ mass threshold in the six-pion photon production cross section observed by DM2,[53] Focus,[54] and BaBar,[55] which is 1^{--} . Recently, a fine structure with about 1 MeV width at the nucleon-antinucleon threshold in $e^+e^- \to 3(\pi^+\pi^-)$

and $e^+e^- \to K^+K^-\pi^+\pi^-$ production is observed by CMD-3,[56] while no such structure is seen in $e^+e^- \to 2(\pi^+\pi^-)$. However, cross sections of $p\bar{p}$ annihilation into $2(\pi^+\pi^-) > 3(\pi^+\pi^-) \gg K^+K^-\pi^+\pi^-$, which suggests a more complicated dynamics in $e^+e^- \to hadrons$ at the nucleon-antinucleon threshold.

5. Outlook

Data from e^+e^- machines with unprecedented statistical accuracy provides great opportunities to map out light mesons as complete and as precise as possible. In the next few years, many experiments (BESIII, CMD-3, SND, *etc.*) will continue to be active, while BelleII has already started data taking. BESIII collected 10 billions of J/ψ, which is unique for studying and searching for QCD exotics as a gluon rich environment with clearly defined initial and final state properties. We can expect a continuous flow of interesting results and new insights into QCD in the confinement regime from the meson spectroscopy at e^+e^- machines.

References

1. N. Brambilla *et al.*, Eur. Phys. J. C **74**, 2981 (2014).
2. C. A. Meyer and Y. Van Haarlem, Phys. Rev. C **82**, 025208 (2010).
3. V. Crede and C. A. Meyer, Prog. Part. Nucl. Phys. **63**, 74 (2009).
4. E. Klempt and A. Zaitsev, Phys. Rept. **454**, 1 (2007).
5. C. Amsler and N. A. Tornqvist, Phys. Rept. **389**, 61 (2004).
6. S. Godfrey and J. Napolitano, Rev. Mod. Phys. **71**, 1411 (1999).
7. G. S. Bali *et al.* (UKQCD Collaboration), Phys. Lett. B **309**, 378 (1993).
8. C. J. Morningstar and M. J. Peardon, Phys. Rev. D **60**, 034509 (1999).
9. Y. Chen *et al.*, Phys. Rev. D **73**, 014516 (2006).
10. W. Sun *et al.*, Chin. Phys. C **42**, 093103 (2018) .
11. E. Gregory *et al.*, JHEP **1210**, 170 (2012).
12. L. C. Gui *et al.* (CLQCD Collaboration), Phys. Rev. Lett. **110**, 021601 (2013).
13. Y. B. Yang *et al.* (CLQCD Collaboration), Phys. Rev. Lett. **111**, 091601 (2013).
14. M. Ablikim *et al.* (BESIII Collaboration), Phys. Rev. D **87**, 092009 (2013); Erratum: [Phys. Rev. D **87**, 119901 (2013)].
15. M. Ablikim *et al.* (BESIII Collaboration), Phys. Rev. D **98**, 072003 (2018).

16. H. Y. Cheng, C. K. Chua and K. F. Liu, Phys. Rev. D **92**, no. 9, 094006 (2015).
17. M. Acciarri *et al.* [L3 Collaboration], Phys. Lett. B **501**, 173 (2001).
18. S. Uehara *et al.* [Belle Collaboration], PTEP **2013**, no. 12, 123C01 (2013).
19. R. Barate *et al.* [ALEPH Collaboration], Phys. Lett. B **472**, 189 (2000).
20. S. Uehara *et al.* [Belle Collaboration], Phys. Rev. D **78**, 052004 (2008).
21. R. Aaij *et al.* [LHCb Collaboration], Phys. Lett. B **797**, 134789 (2019).
22. C. D. L, U. G. Meissner, W. Wang and Q. Zhao, Eur. Phys. J. A **49**, 58 (2013).
23. M. Tanabashi *et al.* (Particle Data Group), Phys. Rev. D **98**, 030001 (2018).
24. A. Etkin *et al.*, Phys. Rev. Lett. **41**, 784 (1978); Phys. Lett. B **165**, 217 (1985); Phys. Lett. B **201**, 568 (1988).
25. M. Ablikim *et al.* (BESIII Collaboration), Phys. Rev. D **93**, 112011 (2016).
26. M. Ablikim *et al.* (BESIII Collaboration), Phys. Rev. D **92**, 052003 (2015); Erratum: [Phys. Rev. D **93**, 039906 (2016)].
27. R. L. Jaffe, Phys. Rev. D **15**, 267 (1977).
28. M. G. Alford and R. L. Jaffe, Nucl. Phys. B **578**, 367 (2000).
29. L. Maiani, F. Piccinini, A. D. Polosa and V. Riquer, Phys. Rev. Lett. **93**, 212002 (2004).
30. L. Maiani, A. D. Polosa and V. Riquer, Phys. Lett. B **651**, 129 (2007).
31. G. 't Hooft, G. Isidori, L. Maiani, A. D. Polosa and V. Riquer, hys. Lett. B **662**, 424 (2008).
32. J. D. Weinstein and N. Isgur, Phys. Rev. D **41**, 2236 (1990).
33. S. Ishida *et al.*, in Proceedings of the 6th International Conference on Hadron Spectroscopy, Manchester, United Kingdom, 1995 (World Scientific, Singapore, 1995), p. 454.
34. N. N. Achasov, S. A. Devyanin and G. N. Shestakov, Phys. Lett. **88B**, 367 (1979).
35. J. J. Wu, Q. Zhao and B. S. Zou, Phys. Rev. D **75**, 114012 (2007).
36. C. Hanhart, B. Kubis and J. R. Pelaez, Phys. Rev. D **76**, 074028 (2007).
37. J. J. Wu and B. S. Zou, Phys. Rev. D **78**, 074017 (2008).
38. M. Ablikim *et al.* (BESIII Collaboration), Phys. Rev. D **83**, 032003 (2011).
39. M. Ablikim *et al.* (BESIII Collaboration), Phys. Rev. Lett. **121**, 022001 (2018).

40. J. Z. Bai *et al.* (BES Collaboration), Phys. Rev. Lett. **91**, 022001 (2003).
41. M. Ablikim *et al.* (BESIII Collaboration), Chin. Phys. C **34**, 421 (2010).
42. J. P. Alexander *et al.* (CLEO Collaboration), Phys. Rev. D **82**, 092002 (2010).
43. M. Ablikim *et al.* [BES Collaboration], Eur. Phys. J. C **53**, 15 (2008).
44. M. Ablikim *et al.* [BESIII Collaboration], Phys. Rev. D **87**, no. 11, 112004 (2013).
45. M. Ablikim *et al.* [BESIII Collaboration], Phys. Rev. D **93**, no. 5, 052010 (2016).
46. M. Ablikim *et al.* (BESIII Collaboration), Phys. Rev. Lett. **108**, 112003 (2012).
47. M. Ablikim *et al.* (BES Collaboration), Phys. Rev. Lett. **95**, 262001 (2005).
48. M. Ablikim *et al.* (BESIII Collaboration), Phys. Rev. Lett. **106**, 072002 (2011).
49. M. Ablikim *et al.* (BESIII Collaboration), Phys. Rev. Lett. **115**, 091803 (2015).
50. M. Ablikim *et al.* [BESIII Collaboration], Phys. Rev. D **99**, no. 7, 071101 (2019).
51. M. Ablikim *et al.* (BESIII Collaboration), Phys. Rev. D **97**, 051101 (2018).
52. M. Ablikim *et al.* (BESIII Collaboration), Phys. Rev. Lett. **117**, 042002 (2016).
53. D. Bisello, J. C. Bizot, J. Buon, A. Cordier, B. Delcourt and F. Mane, Phys. Lett. **107B**, 145 (1981).
54. P. L. Frabetti *et al.*, Phys. Lett. B **578**, 290 (2004).
55. B. Aubert *et al.* [BaBar Collaboration], Phys. Rev. D **73**, 052003 (2006).
56. R. R. Akhmetshin *et al.* [CMD-3 Collaboration], Phys. Lett. B **794**, 64 (2019).

Light-Meson Spectroscopy at Lepto- and Hadroproduction Experiments

B. Grube

Institute for Hadronic Structure and Fundamental Symmetries, Technische Universität München, Garching, Germany
E-mail: bgrube@tum.de

The excitation spectrum of light mesons, which are composed of up, down, and strange quarks, is studied since decades. However, it still holds a number of puzzles and surprises that provide new insights into the nature of the strong interaction. Recent high-quality data samples from several experiments allow us to not only study the properties of established mesons with unprecedented precision but to also search for new states. These searches aim in particular at resolving the question of the existence of so-called exotic states, such as four-quark states or states with excited gluon fields. Since light mesons have often large widths and are overlapping, the mapping of their spectrum is challenging and requires large quantities of data on different production and decay modes. The data are analyzed using a framework of interfering quantum mechanical amplitudes known as partial-wave analysis (PWA). Most excited meson states decay into multi-particle final states, for which the PWA requires extensive modeling of the dynamics of the final-state hadrons. I will give an overview on ongoing experimental studies of light mesons and discuss possible interpretations. I will also touch on novel analysis techniques and the prospects for future progress.

1. Introduction

The precision measurement of the spectrum of light mesons is the aim of several experiments. The focus of these experiments lies in particular on confirming and finding new highly excited states, on completing $SU(3)_{\text{flavor}}$ multiplets, and on searching for exotic mesons, i.e., states that cannot be composed of (just) $q\bar{q}'$. These data provide important input for theory and phenomenology and eventually help to better understand the nature of confinement. The current analyses in the field are driven not only by the high-quality data from experiments, but also by the development and application of advanced analysis techniques and of more rigorous theoretical models for partial-wave analyses (PWA).

2. Spin-Exotic Light Mesons

Spin-exotic mesons have J^{PC} quantum numbers[a] of 0^{--}, $(2n)^{+-}$, or $(2n+1)^{-+}$ with $n \in \mathbb{N}$ that are forbidden for $|q\bar{q}\rangle$ states in the non-relativistic limit. In the light-meson sector, so far three spin-exotic candidate states—all with $J^{PC} = 1^{-+}$—have been claimed by experiments:[1] $\pi_1(1400)$, $\pi_1(1600)$, and $\pi_1(2100)$. The latter one has been observed only by the BNL E852 experiment and needs confirmation.

2.1. 3π *Final State*

Some of the experimental claims are controversial, in particular the observation of the $\pi_1(1600)$ in the $\rho(770)\pi$ decay. The BNL E852 experiment claimed a $\pi_1(1600)$ signal in a PWA of 2.5×10^5 3π events diffractively produced by an 18 GeV/c π^- beam on a proton target. The PWA was performed in the range from 0.1 to 1.0 $(\text{GeV}/c)^2$ of the reduced four-momentum transfer squared t' using a PWA model containing 21 waves.[2,3] However, in a later analysis of a more than ten times larger data sample of 2.6×10^6 3π events no evidence for the $\pi_1(1600)$ was found.[4] However, the non-observation claim was based on data in the narrower range $0.1 < t' < 0.5\,(\text{GeV}/c)^2$.

Recently, the COMPASS experiment at CERN[5] published results of a PWA of a large data sample of 4.6×10^7 3π events diffractively produced by a 190 GeV/c π^- beam on a proton target in the range $0.1 < t' < 1.0\,(\text{GeV}/c)^2$.[6,7] The employed PWA model contains 88 partial waves and is an extension of the 36-wave set used in Ref. [4]. The PWA was performed in 11 narrow t' intervals. Figures 1a and 1b show the measured intensity distributions (black points) of the spin-exotic 1^{-+} $\rho(770)\pi$ P-wave for the lowest and the highest t' bin. Surprisingly, the shape of the intensity distribution changes drastically with increasing t'. At low t', COMPASS observes a broad distribution, whereas at high t', a peak emerges at 1.6 GeV/c^2. This suggests that the partial-wave amplitude has large t'-dependent contributions from non-resonant processes. The continuous curves in fig. 1 represent a resonance-model fit, where the partial-wave amplitude is modeled as the coherent sum (red curves) of a Breit-Wigner amplitude for the $\pi_1(1600)$ (blue curves) and a non-resonant amplitude (green curves). The modulation of the intensity distribution with t' is well reproduced by the model. At low t', the intensity is described mostly by the non-resonant component (green curve), whereas the peak at high t' is described nearly completely by the Breit-Wigner component of the model (blue curve). The

[a]Here, J is the meson spin and P and C are the eigenvalues of parity and charge conjugation, respectively.

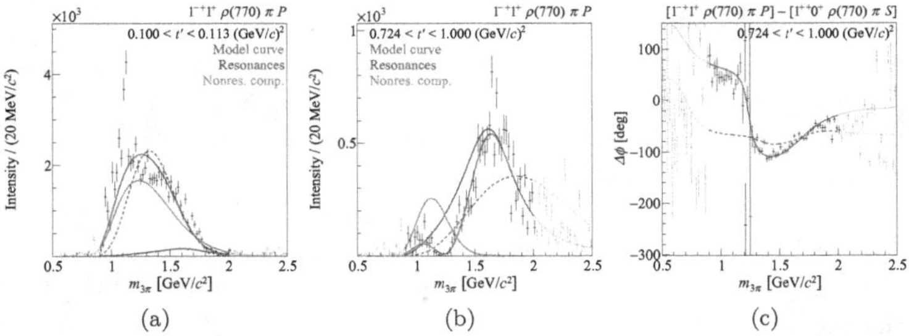

Fig. 1. (a) and (b): Intensity distributions of the 1^{-+} $\rho(770)\pi$ P-wave in the COMPASS $\pi^-\pi^-\pi^+$ proton-target data in the lowest and highest t' bin, respectively. (c) Phase of this wave relative to the 1^{++} $\rho(770)\pi$ S-wave in the highest t' bin. From Ref. [7].

measured $\pi_1(1600)$ Breit-Wigner parameters are $m_0 = 1600\,^{+110}_{-60}\,\text{MeV}/c^2$ and $\Gamma_0 = 580\,^{+100}_{-230}\,\text{MeV}/c^2$. That a resonance is indeed required to describe the data, is demonstrated by a fit, where the $\pi_1(1600)$ component is removed from the resonance model (red dashed curves in fig. 1) leaving only the non-resonant component in this wave. Using the so-called freed-isobar PWA approach, it was also verified that the observed $\pi_1(1600)$ signal is not an artifact caused by inadequate isobar parameterizations (see Ref. [8] for details).

Performing the PWA and the resonance-model fit in narrow t' bins hence resolves the long-standing puzzle of the seemingly contradictory results obtained by the BNL E852 experiment. Using an even larger wave set than the BNL analyses, the COMPASS results confirm that the prominent peak reported in the first BNL analysis in Refs. [2, 3] is an artifact caused by a too small wave set. The non-observation of the $\pi_1(1600)$ reported in the second BNL analysis in Ref. [4] is explained by the exceptionally steep t' dependence of the non-resonant component. In the range $t' \lesssim 0.5\,(\text{GeV}/c)^2$, the non-resonant component dominates and the COMPASS data show only a small $\pi_1(1600)$ signal, which becomes prominent only for $t' \gtrsim 0.5\,(\text{GeV}/c)^2$. Hence the non-observation of the $\pi_1(1600)$ in the second BNL analysis was mainly a result of the analyzed t' range.

A remaining puzzle is that although the $\pi_1(1600)$ decays into $\rho(770)\pi$, it does not seem to be observed in photoproduction.[9–11] In the future, much more precise photoproduction data from the GlueX and MesonX experiments at JLab will help to clarify the situation.

2.2. $\eta\pi$ and $\eta'\pi$ Final States

Other interesting final states to search for spin-exotic mesons are $\eta\pi$ and $\eta'\pi$. In contrast to the analysis of three-body final states, where PWA models usually employ the isobar model thereby introducing a large model dependence, the PWA of two-body final states does not require such strong model assumptions. In the $\eta^{(\prime)}\pi$ system, partial waves that correspond to non-zero, odd orbital-angular momentum between the $\eta^{(\prime)}$ and the π correspond to spin-exotic quantum numbers. The lowest such wave is the P-wave with $J^{PC} = 1^{-+}$. Previous experiments have observed the $\pi_1(1400)$ in the $\eta\pi$ P-wave and the $\pi_1(1600)$ in the $\eta'\pi$ P-wave.[1]

Recently, members of the Joint Physics Analysis Center (JPAC) have performed a coupled-channel fit of the $\eta\pi$ and $\eta'\pi$ P- and D-wave amplitudes extracted from COMPASS data[12] using a unitary model based on S-matrix principles.[13] They find two resonance poles, the $a_2(1320)$ and the $a_2(1700)$, in the D-wave amplitudes and a single pole in the P-wave amplitudes. The pole parameters of $m_0 = 1564 \pm 24\,(\text{stat.}) \pm 86\,(\text{sys.})\,\text{MeV}/c^2$ and $\Gamma_0 = 492 \pm 54\,(\text{stat.}) \pm 102\,(\text{sys.})\,\text{MeV}/c^2$ are consistent with the $\pi_1(1600)$.

This result is in so far remarkable as only a single pole is required to describe both the $\eta\pi$ and the $\eta'\pi$ P-wave amplitudes despite their rather different intensity distributions (see fig. 2). As mentioned above, this is in contrast to most previous analyses, where the broad peak at $1.4\,\text{GeV}/c^2$ in the $\eta\pi$ P-wave intensity was described by the $\pi_1(1400)$ whereas the narrower peak at $1.6\,\text{GeV}/c^2$ in the $\eta'\pi$ P-wave intensity, which is nearly identical to the peak observed in the high t' region of the $\pi^-\pi^-\pi^+$ data, was described by the $\pi_1(1600)$. As the COMPASS data are similar to those of previous experiments, the JPAC analysis raises doubts about the existence of the $\pi_1(1400)$ as a separate resonance. This would resolve the longstanding puzzle of two spin-exotic states lying unexpectedly close to each other. If interpreted as hybrid states, this would also remove the discrepancy with lattice QCD and most model calculations, which predict the lightest hybrid state to have a mass substantially higher than that of the $\pi_1(1400)$ (see e.g., Ref. [14]).

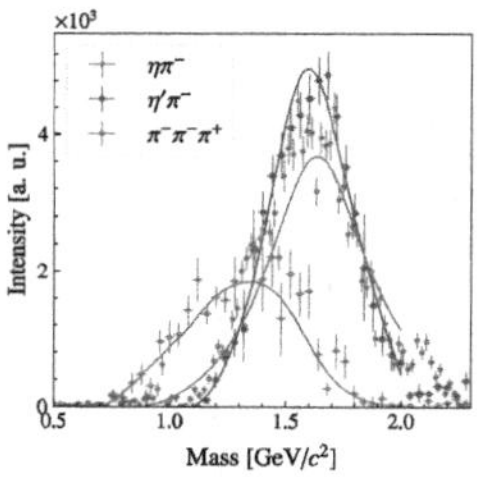

Fig. 2. Comparison of COMPASS data: Intensity distributions of the $\eta\pi$ (green points) and $\eta'\pi$ (blue points) P-wave[12] for $0.1 < t' < 1.0\,(\text{GeV}/c)^2$ and of the 1^{-+} $\rho(770)\pi$ P-wave in the $\pi^-\pi^-\pi^+$ data for $0.449 < t' < 0.724\,(\text{GeV}/c)^2$ (red points). The curves represent the fit results from Refs. [7, 13].

2.3. *GlueX Data*

The GlueX experiment at JLab[15] uses a linearly polarized photon beam to study photoproduction of light mesons on a proton target. The main goal is a precision measurement of the light-meson spectrum in the mass

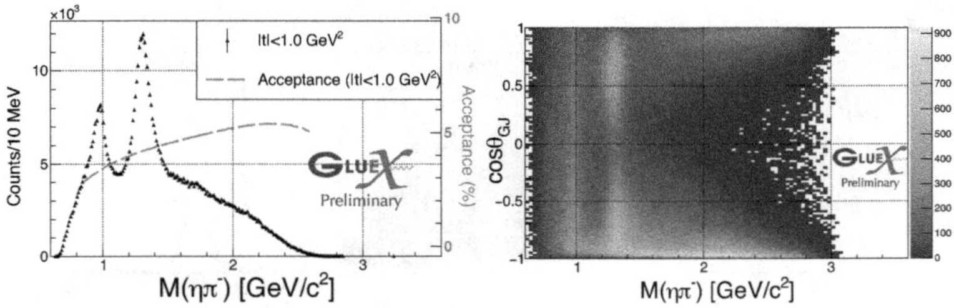

(a) (b)

Fig. 3. (a) Invariant mass distribution of the $\eta\pi^-$ system produced in the reaction $\gamma p \to \eta\pi^-\Delta^{++}$ as measured by the GlueX experiment.[17] (b) Distribution of the cosine of the polar angle of the η with respect to the beam photon in the $\eta\pi$ rest frame vs. the $\eta\pi$ mass.

range $\lesssim 3\,\mathrm{GeV}/c^2$ and in particular the search for hybrid mesons. The maximum photon polarization is reached at a photon energy of about 9 GeV. In this energy range, various exchange processes contribute and states with a wide variety of $I^G\,J^{PC}$ quantum numbers[b] are accessible. The photon polarization helps to disentangle these production processes. GlueX is hence complementary to high-energy pion-scattering experiments such as COMPASS, where only states with $I^G = 1^-$ are produced directly.

GlueX has finished its first phase of data taking and has acquired world leading data samples for many final states. First analyses showcase the potential of these data.[16,17] As an example, fig. 3 shows kinematic distributions for the reaction $\gamma p \to \eta\pi^-\Delta^{++}$ with $\eta \to \gamma\gamma$. The $\eta\pi$ invariant mass distribution in fig. 3a exhibits clear peaks of $a_0(980)$ and $a_2(1320)$. This is also consistent with the angular distribution shown in fig. 3b. The band in the $a_0(980)$ region is approximately independent of the angle indicating a spin-0 resonance, whereas the $a_2(1320)$ region shows the expected 2-bump behaviour. In total, about 10^6 $\eta\pi$ events are expected in this channel, which is about 10 times the size of the COMPASS data sample. The second data-taking campaign of GlueX that started fall 2019 is expected to enlarge the data samples by another factor of 4 to 5.

3. Kaon Spectroscopy

In order to better understand the light-meson spectrum it is important to complete the $\mathrm{SU}(3)_{\mathrm{flavor}}$ nonets. However, the kaon spectrum is not well known. Currently, the PDG lists only 25 kaon states, 12 of which need confirmation.[1]

---

[b]Here, I is the isospin and G the G-parity.

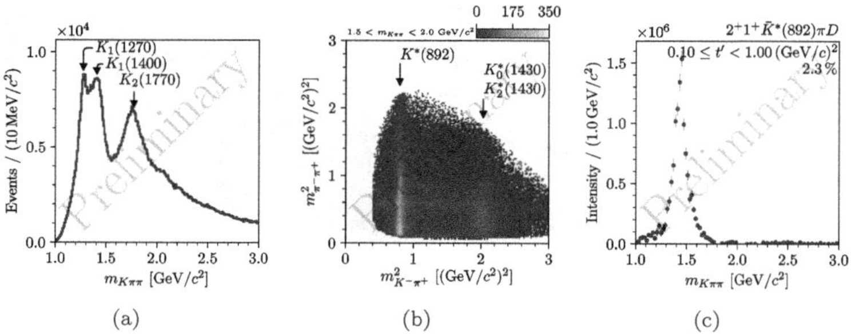

Fig. 4. (a) $K^-\pi^-\pi^+$ invariant mass distribution with potential resonance signals indicated. (b) Dalitz plot around $m_{K\pi\pi} = 1.75\,\text{GeV}/c^2$. (c) Intensity distribution of the $J^P = 2^+$ wave decaying into $K^*(892)$ and π^- in a D-wave.

The COMPASS experiment has measured kaon diffraction on a proton target using the 2.4 % K^- component in the $190\,\text{GeV}/c$ hadron beam. In particular, COMPASS has acquired the so far largest data sample of 720 000 events of the reaction $K^-p \to K^-\pi^-\pi^+p$.[18] The invariant mass distribution of the $K^-\pi^-\pi^+$ system shown in fig. 4a exhibits possible signals of known kaon resonances. The Dalitz plot in fig. 4b around $m_{K\pi\pi} = 1.75\,\text{GeV}/c^2$ shows rich structures with signals of $\rho(770)$ and $f_0(980)$ in the $\pi^-\pi^+$ subsystem and of $K^*(892)$, $K_0^*(1430)$, and/or $K_2^*(1430)$ in the $K^-\pi^+$ subsystem.

A first PWA of these data was performed using a wave set determined by employing regularization techniques.[19] To exemplify the potential of the COMPASS data, fig. 4c shows the intensity of the $J^P = 2^+$ wave decaying into $K^*(892)$ and π^- in a D-wave, which exhibits a nearly background-free signal of the $K_2^*(1430)$. More details can be found in Ref. [18].

4. Summary

Light-meson spectroscopy has entered the era of high-precision data. These data reveal new details of the light-meson spectrum and help to settle some of the controversies of the past. The currently running GlueX and VES experiments and the soon to start MesonX experiment will provide even more precise data on various final states. In the more distant future, the PANDA experiment will also contribute to this field.

However, the PWA results from these large data samples become increasingly dominated by systematic uncertainties. A large contribution to these uncertainties comes from the model assumptions employed in the PWA of multi-body final states. Reducing these systematic uncertainties will require improved PWA models that respect fundamental physical principles such as analyticity and unitarity (see e.g., Ref. [20]). For scattering experiments, in addition the production processes needs to be better understood in order to improve the treatment of non-resonant processes, which are another source of large systematic uncertainties.

References

1. M. Tanabashi *et al.*, *Phys. Rev. D* **98**, 030001 (2018) and 2019 update.
2. G. S. Adams *et al.*, *Phys. Rev. Lett.* **81**, 5760 (1998).
3. S. U. Chung *et al.*, *Phys. Rev. D* **65**, 072001 (2002).
4. A. R. Dzierba *et al.*, *Phys. Rev. D* **73**, 072001 (2006).
5. P. Abbon *et al.*, *Nucl. Instr. Meth. A* **779**, 69 (2015).
6. C. Adolph *et al.*, *Phys. Rev. D* **95**, 032004 (2017).
7. M. Aghasyan *et al.*, *Phys. Rev. D* **98**, 092003 (2018).
8. F. Krinner, *in these proceedings.*
9. M. Nozar *et al.*, *Phys. Rev. Lett.* **102**, 102002 (2009).
10. P. Eugenio and C. Bookwalter, *AIP Conf. Proc.* **1560**, 421 (2013).
11. S. Grabmüller, PhD thesis, Technische Universität München, 2012. CERN-THESIS-2012-170.
12. C. Adolph *et al.*, *Phys. Lett. B* **740**, 303 (2015).
13. A. Rodas *et al.*, *Phys. Rev. Lett.* **122**, 042002 (2019).
14. C. A. Meyer and E. S. Swanson, *Prog. Part. Nucl. Phys.* **82**, 21 (2015).
15. H. Al Ghoul *et al.*, *AIP Conf. Proc.* **1735**, p. 020001 (2016).
16. D. Mack, *in these proceedings.*
17. C. Gleason, *in these proceedings.*
18. S. Wallner, *in these proceedings.*
19. F. Kaspar, *in these proceedings.*
20. M. Mikhasenko *et al.*, *Eur. Phys. J. C* **78**, 229 (2018).

Excited light baryons from quark-gluon-level calculations

J. Segovia

Departamento de Sistemas Físicos, Químicos y Naturales
Universidad Pablo de Olavide, E-41013 Sevilla, Spain
E-mail: jsegovia@upo.es

C. Chen

Institut für Theoretische Physik, Justus-Liebig-Universität Gießen
Heinrich-Buff-Ring 16, 35392 Giessen, Germany

Z.-F. Cui, Y. Lu, and C.D. Roberts

School of Physics and Institute for Nonperturbative Physics
Nanjing University, Nanjing, Jiangsu 210093, China

The task of mapping and explaining the spectrum of baryons and the structure of these states in terms of quarks and gluons is a longstanding challenge in hadron physics, which is likely to persist for another decade or more. We review the progress made in this topic using a functional method based on Dyson-Schwinger equations. This framework provides a non-perturbative, Poincaré-covariant continuum formulation of Quantum Chromodynamics which is able to extract novel insight on baryon properties since the physics at the hadron level is directly related with the underlying quark-gluon substructure, via convolution of Green functions.

Keywords: Quantum chromodynamics; baryons; Dyson-Schwinger equations; covariant bound-state equations.

1. Introduction

At high energies, many experiments combined with perturbative calculations have established that Quantum Chromodynamics (QCD) is a valid candidate for a theory of the strong interactions.[1] However, the infrared dynamics of QCD represents a challenge of notorious complexity: the most characteristic phenomena such as dynamical generation of quark and gluon masses, formation of bound states and confinement are purely non-perturbative.

The most plausible way to study QCD's non-perturbative regime is through detailed investigations of hadron's mass, structure and reaction properties. In fact, hadron physics is a key part of the international effort in basic science. The experimental programs performed at B-factories (BaBar, Belle and CLEO), at τ-charm facilities (CLEO-c, BESIII), at proton–(anti-)proton colliders (CDF, D0, LHCb, ATLAS, CMS) and at fixed-target experiments (COMPASS, Hall-B at JLab) have provided a large amount of data that has been crucial for sustained progress in the field as well as the breadth and depth necessary for a vibrant research environment. Moreover, a new generation of facilities are either currently running such as JLab12, or scheduled for the next one-to-two decades as FAIR and EIC.

Many theoretical techniques have been developed in the last fifty years in order to tackle the problem of describing hadrons as bound-states of quarks and gluons. Among them, the QCD's Dyson-Schwinger Equations (QCD-DSEs) formalism provides an appealing tool for several reasons. QCD-DSEs are the quantum equations of motion derived from the QCD Lagrangian and thus represent a continuum approach which operates within a fully relativistic quantum field theory, provides access to both perturbative and non-perturbative regimes, and is able to cover the full quark-mass range between chiral limit and the heavy-quark domain. Note, however, that for QCD-connected analysis one relies upon a truncation of the infinite system of non-linear integral equations to a subset that captures the physical content and is solved explicitly, combined with the use of *a*nsätze for those Green functions that enter the equations but are not solved for. These *a*nsätze are constrained by symmetry properties, multiplicative renormalizability, perturbative limits, etc. The interested reader on QCD-DSEs applied to hadron physics can consult some existing reviews.[2,3]

2. Dyson-Schwinger equations

The Green functions of a quantum field theory contain all its physical properties and one can state that the theory is solved once every Green function is determined. During the past years, a cross-fertilization between QCD-DSEs and lattice-QCD has provided further insight into the non-perturbative structure of the lowest n-point Green functions: quark, gluon and ghost propagators as well as quark-gluon, three-gluon and four-gluon vertices.

This section is focused on highlighting the results recently delivered for the quark and gluon propagators, in Landau gauge. Note, however, that studies of the infrared point-wise behavior for the ghost propagator,[7] the quark-gluon vertex,[8] and the three- and four-gluon[9,10] vertices have recently released too. Another remarkable achievement of this kind of studies has been the combination of the ghost and gluon propagators in order to deliver

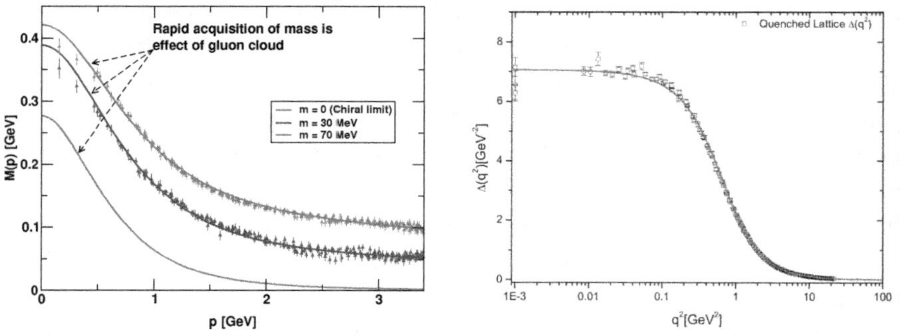

Fig. 1. Figures adapted from Ref. 4 and 5, respectively. *L*eft-panel: Data points are lattice-QCD results for $M(p^2)$ from Ref. 6; solid curves are solutions of the quark DSE (gap equation) for $M(p^2)$, obtained using the current-quark mass that best fit lattice-QCD results; and the (red) lowest solid curve represents the gap equation's solution in the chiral limit. *R*ight-panel: The quenched lattice gluon propagator $\Delta(q^2)$ and an adapted solution

a process-independent running-coupling for QCD which saturates at zero momentum.[11,12]

The dressed-quark propagator in Landau gauge can be written in the following form:

$$S_f(p) = -i\gamma \cdot p\, \sigma_V^f(p^2) + \sigma_S^f(p^2) = 1/[i\gamma \cdot p\, A_f(p^2) + B_f(p^2)]\,. \quad (1)$$

It is known that for light-quarks the wave function renormalisations, $Z_f(p^2) = 1/A_f(p^2)$, and dressed-quark masses, $M_f(p^2) = B_f(p^2)/A_f(p^2)$, receive strong momentum-dependent corrections at infrared momenta[4]: $Z_f(p^2)$ is suppressed and $M_f(p^2)$ enhanced (see left-panel of Fig. 1). These features are an expression of dynamical chiral symmetry breaking (DCSB) and, plausibly, of confinement via the violation of reflection positivity.

The gluon propagator in Landau gauge assumes the totally transverse form

$$i\Delta_{\mu\nu}(q) = -iP_{\mu\nu}(q)\Delta(q^2); \qquad P_{\mu\nu}(q) = g_{\mu\nu} - q_\mu q_\nu/q^2\,, \quad (2)$$

where the scalar form factor $\Delta(q^2)$ is related to the all-order gluon self-energy. Lattice-QCD simulations[13–16] have recently confirmed[17] that this form factor saturates in the deep infrared (see right-panel of Fig. 1) indicating gluon mass generation. A demonstration of how this occurs at the level of the gluon DSE is given in Ref. 7.

The transition from a massless to a massive gluon propagator can be implemented as $\Delta^{-1}(q^2) = q^2 J(q^2) + m^2(q^2)$, where $J(q^2)$ is the gluon's dressing function and $m^2(q^2)$ is the momentum dependent gluon mass, dy-

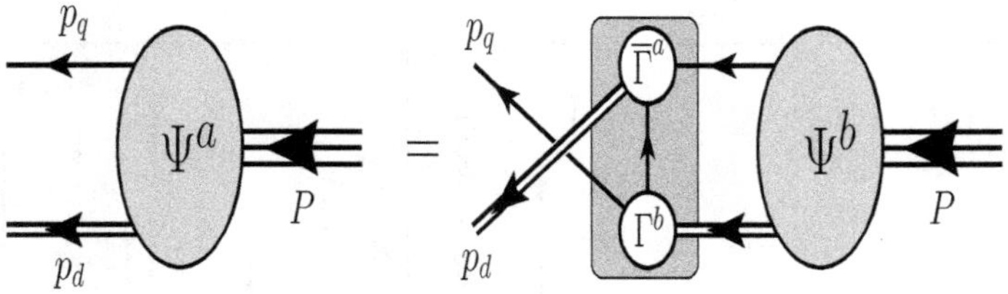

Fig. 2. Poincaré-covariant Faddeev equation: a homogeneous linear integral equation for the matrix-valued function Ψ, being the Faddeev amplitude for a baryon of total momentum $P = p_q + p_d$, which expresses the relative momentum correlation between the dressed-quarks and -diquarks within the baryon. The (purple) highlighted rectangle demarcates the kernel of the Faddeev equation: *single line*, dressed-quark propagator; *double line*, diquark propagator; and Γ, diquark correlation amplitude.

namically generated by the Schwinger mechanism.[18,19] This can be summarize as follows: the vacuum polarization of a gauge boson that is massless at the level of the original Lagrangian may develop a massless pole, whose residue can be identified with $m^2(0)$. The origin of the aforementioned poles is due to purely non-perturbative dynamics. They act as composite Nambu-Goldstone bosons which are colored, massless and have a longitudinal coupling. These features maintain gauge invariance and makes them disappear from any on-shell S-matrix element.

3. Covariant bound-state equations

Mesons appear as free-poles in the quark-antiquark scattering matrix and baryons in the complete three-quark one. Their properties are extracted upon solving homogeneous bound-state equations (Bethe-Salpeter for mesons and Faddeev for baryons) which are valid at pole positions and need information of the lowest n-point Green functions of QCD. In combination they provide a powerful tool to calculate experimentally accessible hadron observables, *e.g.* meson and baryon masses, decay constants, scattering processes, and electromagnetic properties such as form factors.

The problem of solving the Faddeev equation can be transformed into that of solving the linear, homogeneous matrix equation depicted in Fig. 2. This is because two decades of studying baryons as three-quark bound-states[20–23] have demonstrated the appearance of soft (non-pointlike) fully-interacting diquark correlations within baryons, whose characteristics are greatly influenced by DCSB.[24] Note that a baryon described by Fig. 2 can be interpreted as a Borromean bound-state where the binding energy is given by two main contributions: One part is expressed in the formation

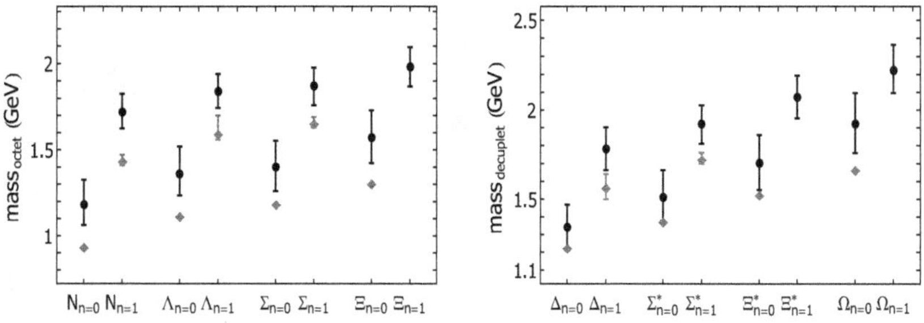

Fig. 3. (Black) solid points are the computed masses for the ground- and first-excited state of the octet (left-panel) and decuplet (right-panel) baryons.[25] The vertical riser indicates the response of our predictions to a coherent $\pm 5\%$ change in the mass-scales that appear in the baryon bound-state equation. The horizontal axis lists a particle name with a subscript that indicates whether it is ground-state ($n = 0$) or first positive-parity excitation ($n = 1$). (Green) diamonds are empirical Breit-Wigner masses take from Ref. 26, the estimated uncertainty in the location of a resonance's Breit-Wigner mass is indicated by an error bar.

of tight diquark correlations, the second one is generated by the quark exchange depicted in the highlighted rectangle of the Fig. 2. This exchange ensures that no quark holds a special place because each one participates in all diquarks to the fullest extent allowed by its quantum numbers. The continual rearrangement of the quarks guarantees that the wave function complies with the baryon's fermionic nature.

Figure 3 shows the computed masses of octet and decuplet baryons and their first positive-parity excitations.[25] It is apparent that the theoretical values are uniformly larger than the corresponding empirical ones. This is because our results should be viewed as those of a given baryon's dressed-quark core, whereas the empirical values include all contributions, including meson-baryon final-state interactions (MB FSIs), which typically generate a measurable reduction.[27] This was explained and illustrated in a study of the nucleon, its parity-partner and their radial excitations[28]; and has also been demonstrated using a symmetry-preserving treatment of a vector $\times$ vector contact interaction.[22,23] Identifying the difference between our predictions and experiment as the result of MB FSIs, then one finds that such effects are fairly homogeneous across the spectrum. Namely, they act to reduce the mass of ground-state octet and decuplet baryons and their first positive-parity excitations by roughly 0.25 GeV.

It is important to highlight here that a possible way to evade the effects of MB FSIs is studying electromagnetic transition form factors of nucleon

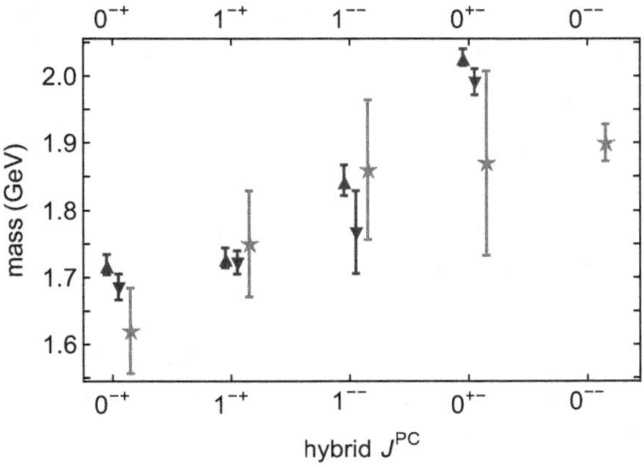

Fig. 4. Figure taken from Ref. 38. Light hybrid-meson spectrum computed with QCD-DSEs (stars, green), and the re-scaled lattice-QCD results.

resonances because their contribution to this kind of observables is soft and disappears quickly, allowing to explore the dressed-quark core of a baryon when electromagnetic proves of high-momenta are used.[29–33]

Herein, I have sketched the exploitation of QCD-DSEs to baryon physics. There are also numerous applications to topical problems in the meson sector. For example, programmes are approved at JLab12, proposed at the CERN-SPS, and possible at the electron-ion collider which would reveal much about parton distributions within QCD's Nambu-Goldstone modes.[34] A strong motivation for such measurements is the fact that the leading x-moment of the pion's (kaon's) GPD provides access to the distributions of mass and momentum within the pion (kaon). Since these distributions can be calculated,[35,36] such data can significantly influence future theoretical perspectives.

Finally, QCD-DSEs can also be applied to exotic hadrons; namely, tetra- and penta-quark bound-states[37] but also systems with valence glue, *e.g.* hybrid mesons,[38] $Q\bar{Q}G$ (see Fig. 4); hybrid baryons, $QQQG$; and even glueballs,[39] GG. Here, G is a nebulously defined "constituent gluon" degree of freedom, whose nature will only be known once such systems are detected.

4. Perspective

The Dyson-Schwinger equations provide a fully relativistic continuum approach for the calculation of hadron properties. Among its advantages, one can highlight that it provides access to infrared and ultraviolet momenta, and covers the full quark mass range from the chiral limit to arbitrarily large current masses. At the present stage, especially in the baryon sector, we are only beginning to explore its potential and possibilities.

Acknowledgments

Work supported by: Ministerio de Economia Industria y Competitividad (MINECO), under grant no. FPA2017-86380-P; Helmholtz International Center for FAIR within the LOEWE program of the State of Hesse and by the DFG grant FI 970/11-1; National Natural Science Foundation of China, grant nos. 11535005, 11805097; Jiangsu Province Natural Science Foundation grant no. BK20180323; and Jiangsu Province Hundred Talents Plan for Professionals.

References

1. W. J. Marciano and H. Pagels, Quantum Chromodynamics: A Review, *Phys. Rept.* **36**, 137 (1978).
2. C. D. Roberts and A. G. Williams, Dyson-Schwinger equations and their application to hadronic physics, *Prog. Part. Nucl. Phys.* **33**, 477 (1994).
3. R. Alkofer and L. von Smekal, The Infrared behavior of QCD Green's functions: Confinement dynamical symmetry breaking, and hadrons as relativistic bound states, *Phys. Rept.* **353**, 281 (2001).
4. M. S. Bhagwat, M. A. Pichowsky, C. D. Roberts and P. C. Tandy, Analysis of a quenched lattice QCD dressed quark propagator, *Phys. Rev.* **C68**, 015203 (2003).
5. J. Papavassiliou, Unraveling the organization of the QCD tapestry, *J. Phys. Conf. Ser.* **631**, 012006 (2015).
6. P. O. Bowman, U. M. Heller, D. B. Leinweber and A. G. Williams, Modeling the quark propagator, *Nucl. Phys. Proc. Suppl.* **119**, 323 (2003).
7. A. C. Aguilar, D. Binosi and J. Papavassiliou, Gluon and ghost propagators in the Landau gauge: Deriving lattice results from Schwinger-Dyson equations, *Phys. Rev.* **D78**, 025010 (2008).
8. A. C. Aguilar, J. C. Cardona, M. N. Ferreira and J. Papavassiliou, Non-Abelian Ball-Chiu vertex for arbitrary Euclidean momenta, *Phys. Rev.* **D96**, 014029 (2017).

9. G. Eichmann, R. Williams, R. Alkofer and M. Vujinovic, Three-gluon vertex in Landau gauge, *Phys. Rev.* **D89**, 105014 (2014).
10. G. Eichmann, C. S. Fischer and W. Heupel, Four-point functions and the permutation group S4, *Phys. Rev.* **D92**, 056006 (2015).
11. D. Binosi, C. Mezrag, J. Papavassiliou, C. D. Roberts and J. Rodriguez-Quintero, Process-independent strong running coupling, *Phys. Rev.* **D96**, 054026 (2017).
12. S. Zafeiropoulos, P. Boucaud, F. De Soto, J. Rodrguez-Quintero and J. Segovia, Strong Running Coupling from the Gauge Sector of Domain Wall Lattice QCD with Physical Quark Masses, *Phys. Rev. Lett.* **122**, 162002 (2019).
13. I. L. Bogolubsky, E. M. Ilgenfritz, M. Muller-Preussker and A. Sternbeck, Lattice gluodynamics computation of Landau gauge Green's functions in the deep infrared, *Phys. Lett.* **B676**, 69 (2009).
14. P. O. Bowman, U. M. Heller, D. B. Leinweber, M. B. Parappilly, A. Sternbeck, L. von Smekal, A. G. Williams and J.-b. Zhang, Scaling behavior and positivity violation of the gluon propagator in full QCD, *Phys. Rev.* **D76**, 094505 (2007).
15. A. Cucchieri and T. Mendes, Constraints on the IR behavior of the gluon propagator in Yang-Mills theories, *Phys. Rev. Lett.* **100**, 241601 (2008).
16. A. Cucchieri and T. Mendes, Landau-gauge propagators in Yang-Mills theories at beta = 0: Massive solution versus conformal scaling, *Phys. Rev.* **D81**, 016005 (2010).
17. J. M. Cornwall, Dynamical Mass Generation in Continuum QCD, *Phys. Rev.* **D26**, 1453 (1982).
18. J. S. Schwinger, Gauge Invariance and Mass, *Phys. Rev.* **125**, 397 (1962).
19. J. S. Schwinger, Gauge Invariance and Mass. 2., *Phys. Rev.* **128**, 2425 (1962).
20. M. Oettel, G. Hellstern, R. Alkofer and H. Reinhardt, Octet and decuplet baryons in a covariant and confining diquark - quark model, *Phys. Rev.* **C58**, 2459 (1998).
21. G. Eichmann, H. Sanchis-Alepuz, R. Williams, R. Alkofer and C. S. Fischer, Baryons as relativistic three-quark bound states, *Prog. Part. Nucl. Phys.* **91**, 1 (2016).
22. Y. Lu, C. Chen, C. D. Roberts, J. Segovia, S.-S. Xu and H.-S. Zong, Parity partners in the baryon resonance spectrum, *Phys. Rev.* **C96**, 015208 (2017).

23. P.-L. Yin, C. Chen, G. Krein, C. D. Roberts, J. Segovia and S.-S. Xu, Masses of ground-state mesons and baryons, including those with heavy quarks, *Phys. Rev.* **D100**, 034008 (2019).
24. J. Segovia, C. D. Roberts and S. M. Schmidt, Understanding the nucleon as a Borromean bound-state, *Phys. Lett.* **B750**, 100 (2015).
25. C. Chen, G. I. Krein, C. D. Roberts, S. M. Schmidt and J. Segovia, Spectrum and structure of octet and decuplet baryons and their positive-parity excitations, *Phys. Rev.* **D100**, 054009 (2019).
26. M. Tanabashi *et al.*, Review of Particle Physics, *Phys. Rev.* **D98**, 030001 (2018).
27. N. Suzuki, B. Julia-Diaz, H. Kamano, T. S. H. Lee, A. Matsuyama and T. Sato, Disentangling the Dynamical Origin of P-11 Nucleon Resonances, *Phys. Rev. Lett.* **104**, 042302 (2010).
28. C. Chen, B. El-Bennich, C. D. Roberts, S. M. Schmidt, J. Segovia and S. Wan, Structure of the nucleons low-lying excitations, *Phys. Rev.* **D97**, 034016 (2018).
29. J. Segovia, I. C. Cloet, C. D. Roberts and S. M. Schmidt, Nucleon and Δ elastic and transition form factors, *Few Body Syst.* **55**, 1185 (2014).
30. J. Segovia, B. El-Bennich, E. Rojas, I. C. Cloet, C. D. Roberts, S.-S. Xu and H.-S. Zong, Completing the picture of the Roper resonance, *Phys. Rev. Lett.* **115**, 171801 (2015).
31. J. Segovia and C. D. Roberts, Dissecting nucleon transition electromagnetic form factors, *Phys. Rev.* **C94**, 042201 (2016).
32. C. Chen, Y. Lu, D. Binosi, C. D. Roberts, J. Rodrguez-Quintero and J. Segovia, Nucleon-to-Roper electromagnetic transition form factors at large Q^2, *Phys. Rev.* **D99**, 034013 (2019).
33. Y. Lu, C. Chen, Z.-F. Cui, C. D. Roberts, S. M. Schmidt, J. Segovia and H. S. Zong, Transition form factors: $\gamma^* + p \to \Delta(1232)$, $\Delta(1600)$, *Phys. Rev.* **D100**, 034001 (2019).
34. A. C. Aguilar *et al.*, Pion and Kaon Structure at the Electron-Ion Collider (2019).
35. T. Frederico, E. Pace, B. Pasquini and G. Salme, Pion Generalized Parton Distributions with covariant and Light-front constituent quark models, *Phys. Rev.* **D80**, 054021 (2009).
36. C. Mezrag, L. Chang, H. Moutarde, C. D. Roberts, J. Rodrguez-Quintero, F. Sabati and S. M. Schmidt, Sketching the pion's valence-quark generalised parton distribution, *Phys. Lett.* **B741**, 190 (2015).
37. C. S. Fischer and G. Eichmann, Overview of multiquark states, *PoS* **Hadron2017**, 007 (2018).

38. S.-S. Xu, Z.-F. Cui, L. Chang, J. Papavassiliou, C. D. Roberts and H.-S. Zong, New perspective on hybrid mesons, *Eur. Phys. J.* **A55**, 113 (2019).
39. E. V. Souza, M. N. Ferreira, A. C. Aguilar, J. Papavassiliou, C. D. Roberts and S.-S. Xu, Pseudoscalar glueball mass: a window on three-gluon interactions (2019).

Dynamically generated hadron resonances

Tetsuo Hyodo

Department of Physics, Tokyo Metropolitan University,
Hachioji 192-0397, Japan
E-mail: hyodo@tmu.ac.jp

Recent developments in hadron spectroscopy triggers the discussion on various exotic configurations beyond the simple three-quark state for baryons and quark-anti-quark pair for mesons. In particular, the states observed near a two-hadron threshold are often regarded as candidates of hadronic molecules, which are made of constituent hadrons and dynamically generated by the hadronic interactions. However, identification of the hadronic molecules involves several subtle difficulties which are often overlooked or underestimated. Here we summarize these subtleties and present a promising approach to distinguish the dynamically generated states from others using observable quantities.

Keywords: Hadron resonances; dynamically generated states; compositeness.

1. Introduction

In the history of hadron spectroscopy, more than 300 species of hadrons have been reported to exist.[1] While all these structures should eventually stem from QCD, we still do not fully understand how they are constructed. For instance, the flavor quantum numbers of the observed hadrons exhibit an empirical regularity; the isospin and strangeness of baryons (I_B, S_B) are restricted to

$$I_B = 0,\ \frac{1}{2},\ 1,\ \frac{3}{2},\quad S_B = 0,\ -1,\ -2,\ -3, \tag{1}$$

and the isospin and strangeness of mesons (I_M, S_M) have values of

$$I_M = 0,\ \frac{1}{2},\ 1,\quad S_M = 0,\ \pm\frac{1}{2},\ \pm 1. \tag{2}$$

These are exactly the quantum numbers which can be achieved by three valence quarks (I_B, S_B), and one quark and one antiquark (I_M, S_M). Corresponding flavor SU(3) representations (**1**, **3**, $\bar{\mathbf{3}}$, **6**, **8**, **10** for baryons and **1**, **3**, $\bar{\mathbf{3}}$, **8** for mesons) are classified by "exoticness" quantum numbers.[2–6]

There have been no established hadrons having the flavor quantum numbers other than those given above. This is the "exotic hadron problem", for which any reasonable explanation from QCD has not been made.

Recent developments in spectroscopy of hadrons indicate a bunch of candidates of unconventional states, which are considered to have exotic configurations such as multiquark states, gluon hybrids, and hadronic molecules. Here, we do not deal with the configurations characterized by quark-gluon degrees of freedom, but rather focus on hadronic molecules[7] which are composite objects of multiple hadrons, because of the following reasons. First of all, the number of constituent quarks and gluons is not a conserved quantum number in QCD (gluons and quark-antiquark pairs can be created or annihilated). Of course, the quark number (the number of quarks minus number of antiquarks) is conserved, but it merely gives the distinction between mesons, baryons, dibaryons, and so on. In other words, there is no well-defined "multiquark states" in QCD (except for the characterization by exoticness using flavor quantum numbers, or J^{PC} exotics using spin, parity, and charge conjugation symmetries). On the other hand, while the number of hadrons are not conserved (mesons can be created or annihilated), there is a way to distinguish the two-hadron component from others because hadrons are the asymptotic states in QCD.[8,9] In addition, near-threshold s-wave bound states have a spatially large wave function irrespective to microscopic details, thanks to the low-energy universality.[10,11] These facts imply that the structure of hadrons should be characterized by the hadronic degrees of freedom. However, there are still subtleties in identifying the hadronic molecules in the spectrum of hadrons.

1.1. *Hadron resonances*

We first note that most hadrons are unstable states, which eventually decay into two or more hadrons through the strong interaction. Among a few hundred of hadrons listed in Ref. 1, stable hadrons occupy only small fraction, the ground states in each sector. All the other hadrons, in particular the candidates of exotic hadrons, are located above two-hadron threshold. This is partly because of the light mass of pions, the Nambu-Goldstone bosons of chiral symmetry, in comparison with the internal excitation energy of hadrons. The unstable hadrons should be treated as resonances in hadron scatterings. Hence, the structure of hadrons should be discussed with keeping their unstable nature in mind.

In quantum mechanics, an unstable resonance state shares common features with a stable bound state. For instance, both are represented by an eigenstate of the Hamiltonian with the outgoing boundary condition,[12–14] which corresponds to the pole of the S matrix in scattering theory.[15] This observation encourages us to discuss the structure of a resonance by regarding it as a "state". At the same time, however, there are also some important differences. The resonance wave function is not square integrable, and its magnitude diverges at large distance. The diverging wave function is physically understood as a consequence of the accumulation of the decaying component at large distance. Formally, the non-square-integrable wave function is required to accommodate the complex eigenvalue of the Hamiltonian, but it also leads to complex expectation values for any other operators. Namely, the properties of a resonance state (norm of each component, mean squared radius, etc.) are obtained as complex numbers, whose interpretation is not straightforward.

1.2. *Dynamically generated states*

Usually, a state is regarded as a dynamically generated one, if it appears in the physical spectrum, in addition to the states prepared in the original model space. To formulate this statement in quantum mechanics, we regard the physical spectrum as the eigenstates of the full Hamiltonian H, and the original model space as the eigenstates of the free Hamiltonian H_0 which is defined as $H = H_0 + V$ with the interaction V. The dynamically generated state is identified as the one which exists only in the eigenstates of H. This classification is well defined if certain decomposition of the Hamiltonian H_0+V is prepared in advance. In hadron physics, however, the Hamiltonian at the level of hadrons are not given a priori. More specifically, while the full Hamiltonian H is determined by experiments, its decomposition into the free part H_0 and the interaction V is not unique. In fact, it is explicitly demonstrated by Weinberg[16] that one can add new degrees of freedom in the model space (eigenstates of H_0) without changing the observables (eigenstates of H), by modifying the interaction V appropriately. This is essentially a unitary transformation of the potential which leaves the on-shell quantities invariant. In this way, there is an ambiguity in identifying the dynamically generated states with respect to the model space.

1.3. *Compositeness*

A solution to the problem in the previous section was given also by Weinberg.[17] It is shown that the ambiguity of the model space can be well controlled when we deal with an weakly bound s-wave state. By establishing the weak-binding relation, the deuteron is shown to be a composite system of a proton and a neutron, as we describe below. In that work, the structure of the bound state is characterized by the wave function renormalization constant of the bare (elementary) state. Its counterpart, which represents the two-body composite component, is now called compositeness. It is recently shown that the compositeness can be evaluated from the residue of the pole of the scattering amplitude,[18,19] and the compositeness of several hadron resonances have been calculated (see Ref. 9 and references therein).

Although the weak-binding relation is a nice tool to investigate the internal structure, with its original derivation, applicability was limited to stable bound states. It is, therefore, necessary to generalize the weak-binding relation to unstable resonances, for the application to hadron physics. This direction of study has been initiated in Ref. 20. Eventually, direct generalization of the weak-binding relation for unstable resonances is developed with the effective field theory.[21,22]

In the following, we briefly revisit the original weak-binding relation for stable bound states from the viewpoint of effective field theory. Next, we present the generalized weak-binding relation for unstable resonances based on Refs. 21,22. Finally, we show an application of the generalized relation to $\Lambda(1405)$ resonance.

2. Which hadron is dynamically generated?

2.1. *Weak-binding relation*

We first consider an s-wave stable bound state in a single-channel scattering. For concreteness, let us take an example of the deuteron in the NN scattering. Weinberg's weak-binding relation is given by[17]

$$a_0 = R\left\{\frac{2X}{1+X} + \mathcal{O}\left(\tfrac{R_{\text{typ}}}{R}\right)\right\}, \quad R = \frac{1}{\sqrt{2\mu B}}, \tag{3}$$

where a_0 is the scattering length which governs the low-energy NN scattering, and R is the "radius", the length scale associated with the binding energy B with the reduced mass μ. R_{typ} is the typical length scale of the two-body interaction, which is estimated by the mass of the lightest possible exchanged particle. For the deuteron, it is given by the pion mass as $R_{\text{typ}} \sim 1/m_\pi$. The compositeness X is defined as the weight of the two-body component of the wave function, namely,

$$|\,d\,\rangle = \sqrt{X}|\,NN\,\rangle + \sqrt{1-X}|\,\text{others}\,\rangle. \tag{4}$$

Using Eq. (3), we can determine the compositeness X from a_0 and B, if the correction term of order $\mathcal{O}(R_{\rm typ}/R)$ is sufficiently small. In other words, if the binding energy is small and the condition $R \gg R_{\rm typ}$ is met, then the internal structure of the bound state (X) is reflected in the observable quantities (a_0, B). By using the experimental values of a_0 and B, it is concluded that the deuteron is dominated by the NN composite system.[17]

Originally, the quantity $Z = 1 - X$ is introduced as the wave function renormalization constant of a bare state, and interpreted as the weight of an elementary component. From the modern viewpoint, it is the compositeness X that is determined by the weak-binding relation, and $1-X$ represents the weight of all the components other than the two-body composite one. The weak-binding relation (3) can be derived in the effective field theory with the two-body continuum states coupled with a single discrete state through the contact interactions.[21,22] This formulation provides a clear interpretation of the weak-binding relation with respect to the renormalization. Another virtue of the effective field theory is that it enables us to generalize the weak-binding relation for unstable quasi-bound states.

2.2. *Generalization to hadron resonances*

The generalization of the weak-binding relation for unstable states is derived by adding a two-body decay channel in the effective field theory. We consider the same setup as before, and call the original two-body continuum channel 1. The threshold energy of the decay channel is set at $E = -\nu$ with $\nu > 0$ measured from the threshold of channel 1. This changes the stable bound state to an unstable quasi-bound state with finite decay width. Representing the wave function of the quasi-bound state as $|\,h\,\rangle$, the compositeness X is given by

$$|\,h\,\rangle = \sqrt{X}|\,\text{channel 1}\,\rangle + \sqrt{1-X}|\,\text{others}\,\rangle. \tag{5}$$

Note that X represents the composite component of channel 1, and anything else are represented by $|\,\text{others}\,\rangle$. For instance, the two-body component of the decay channel is included in $|\,\text{others}\,\rangle$. In addition, because of the unstable nature of $|\,h\,\rangle$, the compositeness defined above becomes in general complex. Following the similar procedure, we can derive the weak-binding relation for the quasi-bound state $|\,h\,\rangle$ as

$$a_0 = R\left\{\frac{2X}{1+X} + \mathcal{O}\left(\left|\frac{R_{\rm typ}}{R}\right|\right) + \mathcal{O}\left(\left|\frac{\ell}{R}\right|^3\right)\right\} \tag{6}$$

with

$$R = \frac{1}{\sqrt{-2\mu E_h}}, \quad \ell = \frac{1}{\sqrt{2\mu\nu}} \tag{7}$$

Several comments are in order. First, a_0 is the scattering length of channel 1, which is in general complex. Because the eigenenergy E_h is complex, the "radius" R is also a complex number. ℓ is the length scale associated with the threshold energy deference ν. The first two terms in Eq. (6) are essentially the same with Eq. (3), and the last term can be neglected when the decay channel is located far away from the threshold of channel 1. For a fixed $R_{\rm typ}$ and ℓ, the correction terms becomes small if the magnitude of the eigenenergy $|E_h|$ is decreased. In this case, the compositeness X of the unstable state can be determined once the eigenenergy E_h and the scattering length a_0 are given.

2.3. *Application to* $\Lambda(1405)$

Finally, we show an application of the generalized weak-binding relation in hadron physics. Here we consider $\Lambda(1405)$ resonance, for which two eigenstate poles are associated.[1,23] Here we consider the high-mass pole, whose energy is located close to the $\bar{K}N$ threshold. We collect the eigenenergy E_h measured from the $\bar{K}N$ threshold and the $\bar{K}N$ scattering length a_0 in the analyses adopted in PDG[1] in Table 1. Applying the weak-binding relation (6), we determine the $\bar{K}N$ compositeness $X_{\bar{K}N}$. For the interpretation, we evaluate the real-valued compositeness $\tilde{X} = (1-|1-X|-|X|)/2$ with the estimated uncertainties from the correction terms (see Refs. 21,22 for details). In all cases, the evaluated $\bar{K}N$ compositeness is close to unity. This indicates that the high-mass pole of the $\Lambda(1405)$ is dominated by the $\bar{K}N$ component.

Table 1. Compositeness of the high-mass pole of $\Lambda(1405)$.

Set	E_h [MeV]	a_0 [fm]	$X_{\bar{K}N}$	$\tilde{X}_{\bar{K}N}$
Set 1 (Refs. 24, 25)	$-10 - i26$	$1.39 - i0.85$	$1.2 + i0.1$	$1.0^{+0.0}_{-0.4}$
Set 2 (Ref. 26)	$-13 - i20$	$1.30 - i0.85$	$0.9 - i0.2$	$0.9^{+0.1}_{-0.4}$
Set 3 (Ref. 27)	$2 - i10$	$1.21 - i1.47$	$0.6 + i0.0$	$0.6^{+0.3}_{-0.1}$
Set 4 (Ref. 27)	$-3 - i12$	$1.52 - i1.85$	$1.0 + i0.5$	$0.8^{+0.2}_{-0.2}$

3. Summary

We have discussed that the investigation of the structure of hadrons contains several subtleties, such as the treatment of unstable resonances and the ambiguity of the model space. The weak-binding relation and its generalization to unstable states can determine the structure of near-threshold s-wave states, avoiding the latter problem. At present, there is still no consensus on the characterization of the structure of unstable states. It will therefore be an important subject in future hadron spectroscopy to establish the prescription to characterize the structure of unstable states.

Acknowledgments

The author is grateful to Feng-Kun Guo for kind invitation to Hadron 2019 conference. This work is supported in part by the Grants-in-Aid for Scientific Research from JSPS (Nos. 19H05150 and 16K17694).

References

1. M. Tanabashi *et al.*, Review of Particle Physics, *Phys. Rev. D* **98**, 030001 (2018).
2. V. Kopeliovich, On exotic systems of baryons in chiral soliton models, *Phys. Lett. B* **259**, 234 (1991).
3. D. Diakonov and V. Petrov, Exotic baryon multiplets at large number of colours, *Phys. Rev. D* **69**, 056002 (2004).
4. E. Jenkins and A. V. Manohar, Baryon exotics in the quark model, the Skyrme model and QCD, *Phys. Rev. Lett.* **93**, 022001 (2004).
5. T. Hyodo, D. Jido and A. Hosaka, Exotic Hadrons in s-Wave Chiral Dynamics, *Phys. Rev. Lett.* **97**, 192002 (2006).
6. T. Hyodo, D. Jido and A. Hosaka, Study of exotic hadrons in S-wave scatterings induced by chiral interaction in the flavor symmetric limit, *Phys. Rev. D* **75**, 034002 (2007).
7. F.-K. Guo, C. Hanhart, U.-G. Meissner, Q. Wang, Q. Zhao and B.-S. Zou, Hadronic molecules, *Rev. Mod. Phys.* **90**, 015004 (2018).
8. C. Hanhart, How and when can one identify hadronic molecules in the baryon spectrum, *Eur. Phys. J. A* **35**, 271 (2008).
9. T. Hyodo, Structure and compositeness of hadron resonances, *Int. J. Mod. Phys. A* **28**, 1330045 (2013).
10. E. Braaten and H.-W. Hammer, Universality in few-body systems with large scattering length, *Phys. Rept.* **428**, 259 (2006).

11. P. Naidon and S. Endo, Efimov Physics: a review, *Rept. Prog. Phys.* **80**, 056001 (2017).
12. G. Gamow, Zur Quantentheorie des Atomkernes, *Z. Phys.* **51**, 204 (1928).
13. A. J. F. Siegert, On the Derivation of the Dispersion Formula for Nuclear Reactions, *Phys. Rev.* **56**, 750 (1939).
14. N. Moiseyev, *Non-Hermitian Quantum Mechanics* (Cambridge University Press, Cambridge, 2011).
15. J. R. Taylor, *Scattering Theory: The Quantum Theory on Nonrelativistic Collisions* (Wiley, New York, 1972).
16. S. Weinberg, Elementary particle theory of composite particles, *Phys. Rev.* **130**, 776 (1963).
17. S. Weinberg, Evidence That the Deuteron Is Not an Elementary Particle, *Phys. Rev.* **137**, B672 (1965).
18. T. Hyodo, D. Jido and A. Hosaka, Compositeness of dynamically generated states in a chiral unitary approach, *Phys. Rev. C* **85**, 015201 (2012).
19. F. Aceti and E. Oset, Wave functions of composite hadron states and relationship to couplings of scattering amplitudes for general partial waves, *Phys. Rev. D* **86**, 014012 (2012).
20. V. Baru, J. Haidenbauer, C. Hanhart, Y. Kalashnikova and A. E. Kudryavtsev, Evidence that the a0(980) and f0(980) are not elementary particles, *Phys. Lett. B* **586**, 53 (2004).
21. Y. Kamiya and T. Hyodo, Structure of near-threshold quasi-bound states, *Phys. Rev. C* **93**, 035203 (2016).
22. Y. Kamiya and T. Hyodo, Generalized weak-binding relations of compositeness in effective field theory, *PTEP* **2017**, 023 (2017).
23. T. Hyodo and D. Jido, The nature of the $\Lambda(1405)$ resonance in chiral dynamics, *Prog. Part. Nucl. Phys.* **67**, 55 (2012).
24. Y. Ikeda, T. Hyodo and W. Weise, Improved constraints on chiral SU(3) dynamics from kaonic hydrogen, *Phys. Lett. B* **706**, 63 (2011).
25. Y. Ikeda, T. Hyodo and W. Weise, Chiral SU(3) theory of antikaon–nucleon interactions with improved threshold constraints, *Nucl. Phys. A* **881**, 98 (2012).
26. Z.-H. Guo and J. A. Oller, Meson-baryon reactions with strangeness -1 within a chiral framework, *Phys. Rev. C* **87**, 035202 (2013).
27. M. Mai and U.-G. Meißner, Constraints on the chiral unitary $\bar{K}N$ amplitude from $\pi\Sigma K^+$ photoproduction data, *Eur. Phys. J. A* **51**, 30 (2015).

Review of light baryon spectroscopy with CLAS at JLAB

S. Diehl[1,2,*], K. Joo[2,†] for the CLAS Collaboration

[1] *2nd Physics Institute, Justus Liebig University Giessen, 35410 Giessen, Germany*

[2] *Department of Physics, University of Connecticut, Storrs, CT 06269, USA*

[*] *E-mail: stefan.diehl@exp2.physik.uni-giessen.de*

[†] *E-mail: kyungseon.joo@uconn.edu*

Baryon spectroscopy and structure studies are the key to a deeper understanding of the processes which happened during the first few microseconds of the universe. The understanding of the fundamental degrees of freedom inside a proton or a neutron and how they change with varying quark masses can help us to obtain a deeper understanding on the origin of confinement and its connection to chiral symmetry breaking. While photoproduction experiments are have established as a very effective tool to search and identify new resonances, a hard scale, provided by electroproduction is needed to probe the internal structure of the N* states and to reveal their true nature. The paper will give an overview on the spectroscopy of light baryon states, with a special focus on recent progress in hadron structure studies with CLAS at JLAB. The CLAS spectrometer offers a unique dataset to determine the electrocouplings of N* and Δ resonances using both single and double pion electroproduction. As an outlook, an overview of the planed baryon resonance studies with CLAS12 at the recently upgraded 12 GeV CEBAF electron accelerator will be presented.

Keywords: Baryon spectroscopy; electromagnetic interactions; form factors; nucleon structure; excited nucleon states; CLAS.

1. Introduction

The study of the baryon spectrum can help us to obtain a more complete understanding of the processes which happened within the first microseconds of the universe. Up to 10^{-8} s after the big bang the universe consisted of a quark gluon plasma. A short time later after ~ 10^{-6} s confinement emerges, the chiral symmetry is broken, light quarks acquire mass and form the first baryon resonances. Only ~ 10^{-4} s later all these resonances already decayed and only protons and neutrons remained, which finally formed the first nuclei after ~ 100s. The interaction of electromagnetic probes in the energy region of a few GeV with baryons can give us access to the processes which took place during this early era of the universe only 10^{-6} s after the big bang and help us to understand the underlying mechanisms, like the origin of confinement and its connection to chiral symmetry breaking. The study of excited baryons can especially help us to investigation of the fundamental degrees of freedom inside a proton or a neutron and how they change with varying quark masses.

Experimentally, the N* program can be divided in two parts. On the one side the spectroscopy, which is mainly driven by real photon scattering and on the other side the study of the structure of excited baryons, which is mainly driven by electron scattering. While the spectroscopy search for unobserved or so-called missing resonances to provide a more complete and more precise picture of the N* spectrum, the structure measurements study resonance from factors and their Q^2 dependence to investigate the underlying symmetries of the hadron system and to provide information on the confining/effective forces of the 3-quark system.

2. Experimental facilities

Around the world there are several experiments, which contribute to the exploration of the excited baryon spectrum. Besides others, the most important are the CB-ELSA and BGO-OD experiment in Bonn (Germany), the CLAS and the new CLAS12 spectrometer as well as the GlueX experiment at JLAB (USA), the crystal ball experiment in Mainz (Germany) and the BESIII experiment in Beijing (China).

This review will focus on the results obtained with CLAS at JLAB during the recent years, since many detailed studies of resonance electroexcitations in exclusive meson electroproduction off nucleons became possible only after dedicated experiments were carried out with the CLAS detector [1] in Hall B at Jefferson Lab (see Fig. 1).

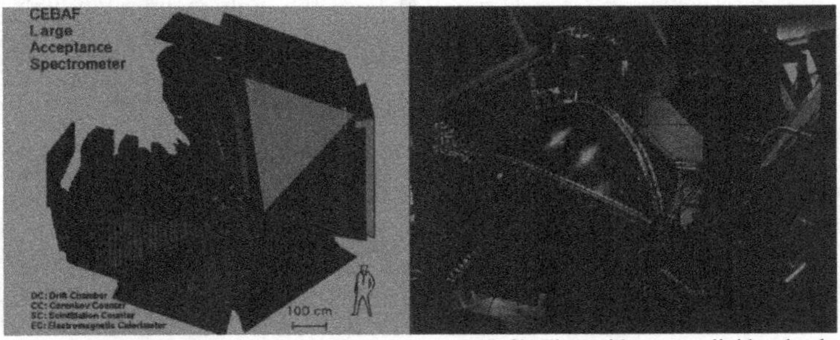

Fig. 1. A schematic drawing of the CLAS spectrometer (left). The torid magnet divides the detector in six sectors, Each equipped with series of drift chambers, a cherkenkov counter, a scintillation counter an an electromagnetic calorimeter. The picture on the right shows the CLAS detector in an open maintenance position exposing the large drift chambers covering nearly the full polar angle range. The time-of-flight scintillator bars and the forward electromagnetic calorimeters are also visible.

3. Search for new excited baryon states

To analyze high precision data of the hadron spectrum, accurate analysis procedures have to be established. A close interplay between the underlying QCD with predictions based on lattice QCD, Dyson-Schwinger equations and relativistic quantum mechanics and the parameters extracted from experimental data based on an amplitude analysis in combination with dispersion relations from reaction theory has to take place.

Over the recent years, the combination of different experimental observables besides the cross section, like the beam asymmetry, the recoil polarisation, the transverse target asymmetry as well as other polarization observables has led to more and more precise fits of the data. Based on the most recent data, several new states could be discovered and previously poorly known states could be fully established (see Tab. 1).

Table 1. Newly discovered and fully established states in PDG 2018 compared to 2012 [2]. **** - existence is certain *** - existence is likely ** - evidence of existence is fair * - evidence of existence is poor

State $N((mass)J^P$	PDG pre 2012	PDG 2018 evidence	Mass (Pole)	State $N((mass)J^P$	PDG pre 2012	PDG 2018 evidence	Mass (Pole)
$N(1710)1/2^+$	***	****	1700	$N(1875)3/2^-$		***	1900
$N(1880)1/2^+$		***	1860	$N(2120)3/2^-$		***	2100
$N(2100)1/2^+$	*	***	2100	$N(2060)5/2^-$		***	2070
$N(1895)1/2^-$		****	1910	$\Delta(2200)7/2^-$	*	***	2150
$N(1900)3/2^+$	**	****	1920				

If we ignore the absolute mass scale, the newly discovered states nicely agree with the quantum numbers J^P predicted by lattice QCD in Ref. [3]. However, the absolute mass values are several hundred MeV too high, since a too high pion mass has to be used for the calculations to overcome mathematical constraints in the calculation.

4. Structure of excited baryon states

While the photoproduction of N* states is very effective for the search and identification of new resonances, a hard scale is needed to probe the internal structure and relevant degrees of freedom versus distance scale to finally reveal the true nature of the discovered states. Such a hard scale can be provided by electroproduction experiments, which can study the structure of the nucleon resonances in the domain where dressed quarks are the major active degree of freedom and explore the dynamics of the non-perturbative strong interaction behind the resonance formation [4, 5]. Nucleon resonance electroexcitation amplitudes of the $\gamma_V pN^*$ electrocouplings have been turned out to be an excellent source of information on many facets of non-perturbative strong interactions in the generation of the excited proton states from quarks and gluons.

The major part of the available world data on all meson electroproduction channels off the nucleon in the resonance region for Q^2 up to 5.0 GeV^2 has been produced by CLAS and can be found in the CLAS physics database [6]. Within the framework of phenomenological reaction models information on electrocouplings of most excited nucleon states in the mass range up to 1.8 GeV could be extracted from this data set [7]. The data base provides detailed information on the Q^2-evolution of the $\gamma_V pN^*$ electrocouplings of the excited nucleon states in the resonance region. The $\gamma_V pN^*$ electrocouplings of these resonances were determined from independent studies of $N\pi$ [8, 9], $N\eta$ [10] and $\pi^+\pi^- p$ [11-13] electroproduction off protons. Further results on $g_V pN^*$ electrocouplings from the exclusive meson electroproduction with CLAS are stored in the web databases listed in Ref. [14, 15].

The extracted electrocoupling values for the resonances from the dominant $N\pi$ and $\pi^+\pi^- p$ exclusive channels are fully consistent, even the channels show completely different non-resonant mechanisms, which can be seen as a strong evidence for a reliable nearly model independent extraction of these fundamental quantities and demonstrates the capability of the developed reaction models. Several $\gamma_V pN^*$ electrocouplings for different nucleon resonances determined from the CLAS can be found in the recent PDG edition [16]. Analyses of the CLAS results on $\gamma_V pN^*$ electrocouplings within the framework of continuum QCD Dyson-Schwinger equation approach and quark models showed, that the N* structure is a complex interplay between the inner core of three dressed quarks and the external meson-baryon cloud. In combination with theoretical models, CLAS results for nucleon resonance electrocouplings can help to obtain a better understanding of active degrees of freedom in the N* structure at different distances and the strong QCD dynamics underlying the generation of excited nucleon states. However, to obtain a more complete understanding, the full spectrum of excited nucleon states within a larger range of photon virtualities as to be studied. Exploring all prominent resonances will provide a deeper understanding of the diversity of qq-correlations in the structure of the excited nucleons of different quantum numbers as well as into dynamical chiral symmetry breaking through analyses of the results on the electroexcitation amplitudes of the pairs of resonances which are the chiral parity partners.

4.1 *The current status of experimental measurements with CLAS*

Based on the already available data of $N\pi$ and $\pi^+\pi^- p$ electroproduction off protons from CLAS, the JLab-Moscow (JM) model has been developed to extract resonance electrocouplings as well as $\pi\Delta$ and ρp hadronic decay widths. Examples for extracted structure functions of $p\pi^0$ are shown in Fig. 2 in comparison with predictions based on resonance electrocouplings and hadronic decay parameters from previous CLAS based analyses [8, 12, 13] and resonant contributions calculated from the JM model.

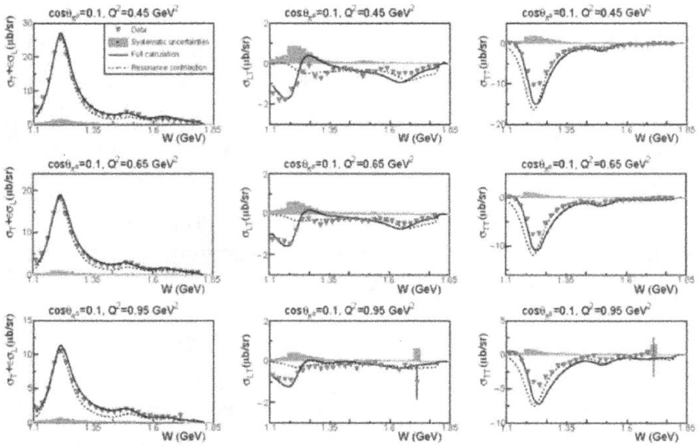

Fig. 2. W dependencies of the exclusive structure functions $\sigma_T + \varepsilon\sigma_L$, σ_{LT}, and σ_{TT} for a fixed cosθ oφ 0.1 and different bins of Q^2. For comparison, the exclusive structure function within the framework of the JM model and with the resonance parameters determined from the CLAS exclusive meson electroproduction data [8, 12, 13] are shown (solid lines), while the blue dashed lines show the resonant contributions. Modified figure from Ref. [19].

All relevant reaction mechanisms in the final-state resonance region are included into the model, including the $\pi^-\Delta^{++}$, $\pi^+\Delta^0$, $\rho^0 p$, $\pi^+N(1520)\ 3/2^-$ and $\pi^+N(1685)\ 5/2^+$ meson-baryon channels, as well as the direct production of the $\pi^+\pi^-p$ final state without considering the formation of intermediate unstable hadrons. The contributions from the well-established N* states in the mass range up to 2.0 GeV were included into the amplitudes of the $\pi\Delta$ and ρp meson-baryon channels using a unitarized version of the Breit-Wigner ansatz. The comparison with experimental data shows, that the JM model provides a good description of the CLAS measurements at $W < 2.0$ GeV and $0.2\ \text{GeV}^2 < Q^2 < 5.0\ \text{GeV}^2$. Considering only statistical uncertainties a χ^2/d.o.f. < 3.0 can be achieved.

Fig. 3 shows experimental results for the fully integrated $\pi^+\pi^-p$ electroproduction cross sections in comparison to the predicted values and the contributions from the meson-baryon mechanisms of the JM model. The determined resonant/non-resonant contributions are located within well-defined ranges. Employing the unitarized Breit-Wigner ansatz [12], which fully accounts for the unitarity restrictions on the resonant amplitudes, it becomes possible to determine the resonance photo-electrocouplings along with the $\pi\Delta$ and ρN decay widths, based on the isolation of the resonant contributions. The obtained results can be used as a valuable input for global multi-channel analyses in the resonance excitation region within advanced coupled channel approaches for their extensions towards the extraction of $\gamma_v pN^*$ electrocouplings from exclusive meson electroproduction off nucleon data.

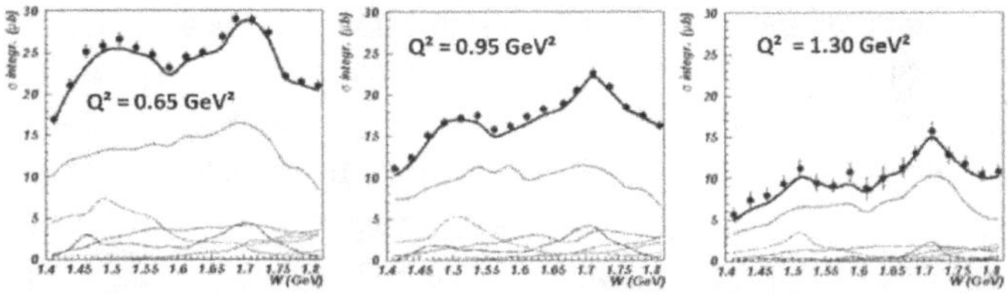

Fig. 3. Fully integrated $\pi^+\pi^-p$ electroproduction cross section in comparison to calculations achieved within the framework of the JM model. In addition, the cross sections for the various contributing mechanisms are shown: full cross section (black solid), $\pi^-\Delta^{++}$ (red thin solid), ρ^0p (green thin solid), $\pi^+\Delta^0$ (blue thin dashed), $\pi^+N(1520)\ 3/2^-$ (black dotted), direct 2π mechanisms (magenta thin dot-dashed), and $\pi^+N(1685)\ 5/2^+$ (red thin dashed). Figure from Ref. [19].

4.2 *Resonance parameters from the exclusive πN and $\pi^+\pi^-p$ channels*

The results for the γ_vpN^* electrocouplings of the $N(1440)1/2^+$ and $N(1520)3/2^-$ resonances, determined in independent analyses of the dominant meson electroproduction channels πN and $\pi^+\pi^-p$ are shown in Fig. 4.

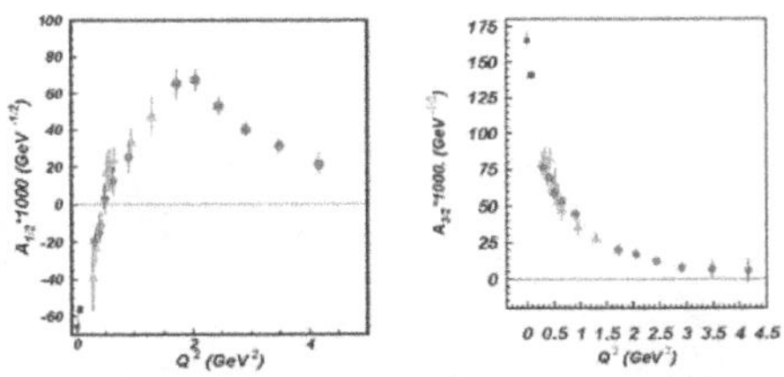

Fig. 4. $A_{1/2}$ γ_vpN^* electrocouplings of the $N(1440)1/2^+$ (left), $A_{3/2}$ γ_vpN^* electrocouplings of the $N(1520)3/2^-$ (right) from analyses of the CLAS electroproduction data off protons in the πN (red circles) and $\pi^+\pi^-p$ channels [12, 13] (green triangles). The photocouplings were taken from the RPP [17] (blue filled triangles) and the CLAS data analysis [18] of πN photoproduction (blue filled squares). Modified figure from Ref. [19].

The overlap of the results from the different channels / analyses demonstrates the quality and reliability of the extraction method for these quantities.

For low lying excited nucleon states in the mass range up to 1.6 GeV for which the dominant decay is into a πN final states, the extraction of resonance electrocouplings is mainly driven by the data on single-pion exclusive electroproduction. The CLAS data for the $\pi^+\pi^-p$ channel plays a critical role in the extraction of the γ_vpN^* electrocouplings of excited nucleon states in the mass range above 1.60 GeV, which decay preferentially to πN final states, for example $\Delta(1620)1/2^-$, $\Delta(1700)1/2^-$, $N(1720)3/2^+$, and the not fully established $N'(1720)3/2^+$ state. Based on the presently available data, the electrocouplings of these states can only be determined from $\pi^+\pi^-p$ exclusive electroproduction off protons, while the πN channels do not show enough sensitivity.

The information on the excited nucleon state photocouplings could be extended recently based on CLAS photoprduction measurements to the photocouplings of most nucleon resonances with masses above 1.6 GeV from the exclusive πN and $\pi^+\pi^- p$ photoproduction off protons. Many of the investigated resonances decay preferentially to the $\pi\pi N$ final states. The cross sections of $\pi^+\pi^- p$ photoproduction off protons dominates the $\pi\pi N$ photoproduction off proton channels. A total of 400 million $\pi^+\pi^- p$ events were collected exceeding the statistics of previous experiments by a factor of ~ 50. Within the framework of the JM meson-baryon reaction model a good description of the new data could be achieved ($1.15 < \chi^2/\mathrm{d.o.f.} < 1.3$). The resonance photocouplings extracted from this work are listed in Tab. 1. A comparison with the resonance photocoupling ranges and the results of the multichannel analysis included in the PDG2018 is given in the table.

Table 2. Resonance photocouplings determined from analysis of the $\pi^+\pi^- p$ photoproduction data from this work in comparison with the previous results from the PDG average and from multichannel analysis.

resonances	$A_{1/2} \times 10^3$ from $\pi^+\pi^- p$ $\mathrm{GeV}^{-1/2}$	$A_{1/2} \times 10^3$ PDG ranges $\mathrm{GeV}^{-1/2}$	$A_{1/2} \times 10^3$ multi-channel Analysis $\mathrm{GeV}^{-1/2}$	$A_{3/2} \times 10^3$ from $\pi^+\pi^- p$ $\mathrm{GeV}^{-1/2}$	$A_{3/2} \times 10^3$ PDG ranges $\mathrm{GeV}^{-1/2}$	$A_{3/2} \times 10^3$ multi-channel analysis $\mathrm{GeV}^{-1/2}$
Δ(1620) 1/2⁻	29.0 ± 6.2	30 – 60	55 ± 7			
N(1650) 1/2⁻	60.5 ± 7.7	35 – 55	32 ± 6			
N(1680) 5/2⁺	-27.8 ± 3.6	-18 – -5	-15 ± 2	128 ± 11	130 – 140	136 ± 5
N(1720) 3/2⁺	80.9 ± 11.5	80 – 120	115 ± 45	-34.0 ± 7.6	-48 – 135	135 ± 40
Δ(1700) 3/2⁻	87.2 ± 18.9	100 – 160	165 ± 20	87.2 ± 16.4	90 – 170	170 ± 25
Δ(1905) 5/2⁺	19.0 ± 7.6	17 – 27	25 ± 5	-43.2 ± 17.3	-55 – -35	-50 ± 5
Δ(1950) 7/2⁺	-69.8 ± 14.1	-75 – -65	-67 ± 5	-118.1 ± 19.3	-100 – 80	-94 ± 4

A good agreement in the magnitude and sign of the photocouplings with the photocoupling ranges in the PDG listings can be observed. However, for several resonances, the photocouplings determined from the multichannel analysis are different. Implementing the new $\pi^+\pi^- p$ photoproduction data from CLAS into the global multichannel analyses will improve our knowledge on the photocouplings and hadronic decay parameters of the resonances in the mass range up to a W of 1.6 GeV significantly. The analysis shows, that for a complete description of the $\pi^+\pi^- p$ photo- and electroproduction measured with CLAS based on Q^2-independent resonance hadronic decay parameters, the new baryon states $N(1720)3/2^+$ is needed.

5. Outlook

The upgraded CLAS12 spectrometer started data taking in the beginning of 2018. Based on the already obtained high quality and high statistics data, the exploration of the structure of excited nucleon states in exclusive πN, KY, and $\pi^+\pi^- p$

electroproduction off protons with photon virtualities Q^2 up to 12.0 GeV^2 has already started. The new data will significantly extend the accessible Q^2 range and improve the statistics in the already accessible range. The distance scale accessible at high Q^2 correspond to the still unexplored regime for N* electroexcitations where the resonance structure is dominated by the quark core with almost negligible meson-baryon cloud contributions providing access to the properties of dressed quarks inside N* states of different quantum numbers. The extraction of the dressed quark mass function from the data on the $\gamma_V pN^*$ electrocouplings of the resonances with different structure, such as radial excitations, spin-isospin flip, and orbital excitations, will the study of the fundamental assumptions of strong QCD with experimental data. The new data will provide access to the dressed quark mass function in the range of quark momenta up to 1.5 GeV for the first time. In this regime the transition from quark-gluon confinement to pQCD takes full effect, which will allow us to address several open problems of the Standard Model, like the nature of the hadron mass and quark-gluon confinement.

Acknowledgments

The authors are grateful to V. I. Mokeev, I. G. Aznauryan, C. D. Roberts, E. Santopinto, and J. Segovia for helpful discussions. This work was supported partly by the U.S. Department of Energy (DE-FG-04ER41309).

References

[1] B.A. Mecking et al., Nucl. Inst. and Meth. A 503, 513 (2003).
[2] Particle data group, http://pdg.lbl.gov/2019/reviews/rpp2018-rev-n-delta-resonances.pdf
[3] R. Edwards et al., Phys. Rev. D 84, 074508 (2011).
[4] I. G. Aznauruan and V. D. Burkert, Progr. Part. Nucl. Phys. 67, 1, (2012).
[5] V. D. Burkert et al., Rev. Mod. Phys. (in press), arXiv: 1710.02549.
[6] JLAB Experiment CLAS Database, http://clasweb.jlab.org/physicsdb/
[7] V. I. Mokeev, Few Body Syst., 59, 46 (2018).
[8] I. G. Aznauryan et al., Phys.Rev. C 80, 055203 (2009).
[9] I. G. Aznauryan, Phys.Rev. C 67, 015209 (2003).
[10] H. Denizli et al. (CLAS Collaboration), Phys. Rev. C76, 015204 (2007).
[11] V. I. Mokeev et al. (CLAS Collab.), Phys. Rev. C 80, 045212 (2009).
[12] V. I. Mokeev et al., Phys. Rev. C 86, 035203 (2012).
[13] V. I. Mokeev et al. (CLAS Collaboration), Phys. Rev. C 93, 025206 (2016).
[14] V. I. Mokeev et al., https://userweb.jlab.org/~mokeev/resonance_electrocouplings/
[15] V. I. Mokeev, Few Body Syst. 59, 46 (2018).
[16] M. Tanabashi et al. (PDG), Phys. Rev. D 98, 010001 (2018).
[17] C. Patrignani et al. (PDG), Chin. Phys. C 40, no. 10, 100001 (2016).
[18] M. Dugger et al. (CLAS Collaboration), Phys. Rev. C 79, 065206 (2009).
[19] K. Joo, JPS Conf. Proc. 26, 011006 (2019).

Hadron spectroscopy from lattice QCD

Jozef J. Dudek

Jefferson Lab, 12000 Jefferson Avenue, Newport News, VA 23606, USA
College of William and Mary, Williamsburg, VA 23187, USA
E-mail: dudek@jlab.org

I will present a brief overview of where lattice QCD currently sits with regard to the challenge of understanding the excited hadron spectrum in terms of the quarks and gluons of QCD. No attempt is made to be encyclopaedic, rather a sample of topics of contemporary interest are supplied as illustration. An emphasis is placed on the need to use rigorous methods where possible with the intention of eventually making definitive statements about the spectrum of QCD.

Lattice QCD is a first-principles numerical technique to study QCD making only the controlled approximation of placing the fields on a discrete space-time grid of finite size. Modern lattice QCD calculations now routinely consider light quark masses such that the physical pion mass is reproduced, and in some cases even include the small effects of electromagnetism and the breaking of isospin symmetry through $m_u \neq m_d$. Coupled with systematic control over the lattice spacing and lattice volume, these calculations allow for direct comparison with experimental observables, but so far only for the simplest quantities such as stable hadron masses and current matrix elements.[1]

Progress of more immediate relevance to excited spectroscopy lies somewhat outside this area of complete systematic control, rather concerning the ability to compute previously inaccessible quantities.[2] Examples include the use of large diverse bases of hadronic operators,[3–6] including those resembling systems of multiple hadrons,[7] to allow the computation of large matrices of two-point correlation functions. When diagonalized, these matrices give variational best-estimates of the energies of many excited states. Inclusion of $q\bar{q}$ annihilation effects, in the form of 'disconnected' diagrams, has allowed access to the spectrum of isoscalar mesons,[8,9] as well as making possible studies of hadron-hadron scattering, and hence the decay properties of hadron resonances.

Lattice QCD should ultimately provide a faithful representation of QCD as we observe it in hadron physics. In that sense we expect excited hadrons to appear as short-lived resonances, and the simplest way to investigate resonances is through scattering amplitudes. It has long been known[10,11] that one can access scattering amplitudes indirectly using the discrete spectrum of eigenstates of QCD within the finite volume defined by the periodic spatial boundaries of the lattice, but it is only recently with implementation of the technical progress mentioned above that it has become practical to perform the relevant computations.[12]

In this proceedings I will provide a flavor of the calculations being currently pursued, making as a point of focus the need to build up gradually to some of the more complex cases being considered experimentally, by first studying simpler systems where the level of rigor that can be maintained within the calculation is high.

1. Quasi flavor-exotic hadrons

There is much contemporary interest in the spectrum of quasi flavor-exotic hadrons, spurred by the observation of many 'XYZ' enhancements in the charmonium region, whose appearance was not predicted in otherwise broadly successful QCD-inspired models. An illustrative example[13] is the $Z_c(3900)$ observed as a peak-like enhancement in the $J/\psi\,\pi$ invariant mass in the process $e^+e^- \to \pi\pi J/\psi$. The final state in which the enhancement appears has led many to suggest that it could be explained as a 'tetraquark' containing a hidden charm pair, e.g. $Z_c^+ \sim c\bar{c}u\bar{d}$. Of course in this picture, the state is actually a resonance, decaying to $J/\psi\,\pi$ and possibly other final states, and if this is the case, one would want to calculate the relevant coupled-channel scattering matrix in QCD, and see it features the behavior expected for a resonance. As mentioned in the previous section, this can be done within lattice QCD by first calculating the spectrum of eigenstates in a finite-volume, and this can be achieved if an appropriate basis of operators is constructed and a matrix of correlation functions computed and diagonalized.

The first attempt to compute such a spectrum was presented in Ref. 14, where a basis featuring a number of meson-meson-like constructions as well as tetraquark-like constructions was used. The resulting spectrum[a] is presented in Figure 1. What is notable is how close the extracted energy levels

[a]An earlier iteration of Ref. 14 presented a calculation that used a somewhat smaller basis of operators which led to an erroneous claim that a Z_c candidate was present.

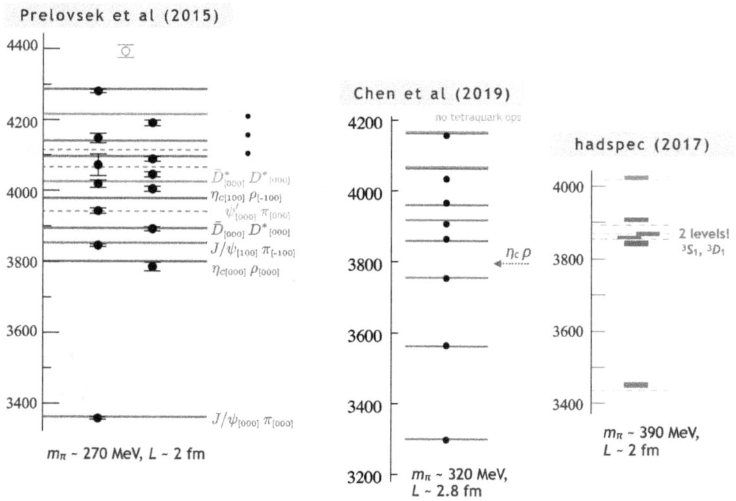

Fig. 1. Z_c channel finite-volume energy levels in three lattice QCD calculations, Refs. 14–16. Horizontal lines indicate the non-interacting meson-meson spectrum in the volume relevant to each calculation.

are to the expected spectrum for a system in which the pairs of mesons have no residual interactions (and in which there would not be a resonance). A more recent calculation, at a slightly different pion mass, in a somewhat larger volume[15] shows very similar behavior. In the period between these calculations, the *hadspec* collaboration computed the spectrum in this channel,[16] and as can be seen in Figure 1, the qualitative behavior is the same, but details differ. Notably there are two energy levels observed near the expected energy of $J/\psi\,\pi$ with minimum allowed non-zero back-to-back momentum, where in the other two calculations only one such level appears. This reflects the use in Ref. 16 of a more complete basis of operators capable of accounting for the fact that in pseudoscalar-vector scattering, the $J^P = 1^+$ partial wave can actually be constructed in two possible ways, as ${}^{2S+1}\ell_J = {}^3S_1$ and ${}^3D_1$. From the energy levels shown in Figure 1 we can state that there is no obvious signal for a narrow Z_c resonance, but beyond this it is very hard to draw any definitive conclusions. It should be clear that this is a rather complicated sector in which to begin studying resonances in lattice QCD, given that several scattering channels are kinematically accessible, and the non-zero spin of the J/ψ (and ρ and D^*) complicates the partial-wave separation. In short, **this is a tough place to start, so take a step back and compute where you have more control ...**

2. Light elastic and coupled-channel resonances

The simplest possible case is where a resonance appears in *elastic* scattering, that is, where only one final state is kinematically accessible. An example, now widely studied in lattice QCD, is the ρ in P-wave $\pi\pi$ scattering, where at heavier than physical pion masses, the ρ lies below $\pi\pi\pi\pi$ threshold.[b]

The finite-volume formalism in this case, to a very good approximation simplifies, down to a procedure which converts finite-volume energy values, $E_n(L)$, into values of the P-wave elastic scattering phase-shift at that energy, $\delta_1(E_n)$. By determining multiple energy levels in several volumes and/or several frames moving with respect to the lattice, the energy dependence of the scattering phase-shift can be mapped out, and it is found to feature the characteristic rapid rise associated with a narrow resonance.[c]

Progress studying elastic resonances like the ρ has motivated the extension into the *coupled-channel* case, an extension which is vital as this is where virtually all interesting hadron resonances lie. This case is significantly more complicated, not in obtaining the finite volume spectrum within lattice QCD, which proceeds in a manner quite similar to the elastic case just with a larger set of appropriate operators, but rather in the conversion of the spectrum into scattering information. In the coupled-channel case scattering is described by a t-matrix, meaning that at each energy there are multiple unknown quantities (the complex-valued elements of the matrix, constrained only by unitarity and time-reversal synmetry). Because of this, the relation between spectrum and amplitude is no longer a one-to-one mapping. The *hadspec* collaboration has pioneered an approach[20] which considers parameterizations of $\mathbf{t}(E)$, where the free parameters are adjusted to obtain the best description of the computed lattice QCD finite-volume spectra. By considering a range of possible parameterizations capable of describing the spectra, an estimate of the parameterization sensitivity can be obtained, which in cases featuring narrow resonances is found to be small. A notable feature of this approach is that with explicit forms for the scattering amplitudes, the analytic continuation into the complex energy plane can be carried out, and a rigorous determination of the resonance content made, through the presence of pole singularities.

[b]Experimentally it is known that the inelasticity into $\pi\pi\pi\pi$ is extremely small.

[c]See the relatively recent review, Ref. 12, for a comprehensive list of references. The same technique has also been applied in the $\pi\pi$ S-wave where the phase-shift has been found,[17–19] but the resonant interpretation is somewhat less clear.

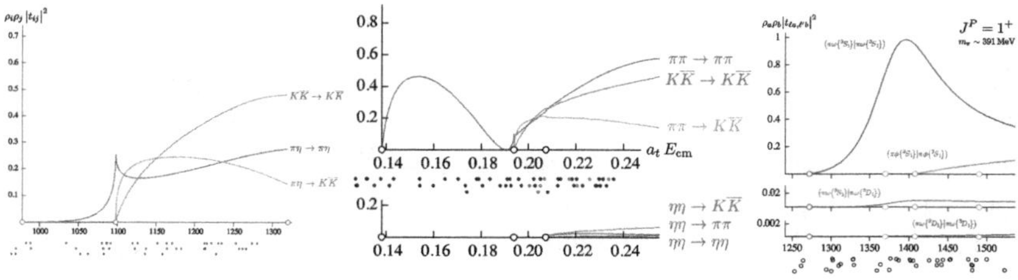

Fig. 2. Scattering amplitudes with $m_\pi \sim 400$ MeV, for $\pi\eta, K\overline{K}$ (left, Ref. 25), $\pi\pi, K\overline{K}, \eta\eta$ (center, Ref. 26) and $\pi\omega, \pi\phi$ (right, Ref. 27).

Using lattices describing approximately 400 MeV pions with three volumes and several moving frames, the *hadspec* collaboration[d] have studied a number of coupled-channel scattering systems.[21–26] In Ref. 25, S-wave $\pi\eta, K\overline{K}$ scattering was computed in which an analogue of the $a_0(980)$ resonance was found, appearing as a highly asymmetric peak in $\pi\eta$ at the $K\overline{K}$ threshold. In isospin-0 $\pi\pi, K\overline{K}, \eta\eta$ scattering in S-wave,[26] analogues of the σ and $f_0(980)$ were observed, with the σ appearing as a stable bound-state lying below $\pi\pi$ threshold, and the $f_0(980)$ being visible through a dip in the $\pi\pi \to \pi\pi$ amplitude just below $K\overline{K}$ threshold. The resulting scattering amplitudes can be seen in Figure 2. The low-lying scalar states including the σ, $f_0(980)$ and $a_0(980)$ are longstanding challenges for simple models seeking to describe QCD, and while these first lattice QCD calculations do not yet resolve questions surrounding the structure of these states, the ability of a first principles method in QCD to observe them as they appear experimentally, as unstable resonances, represents significant progress.

These first calculations of coupled-channel scattering considered only pseudoscalar-pseudoscalar channels, and recalling that it is necessary to study pseudoscalar-vector scattering[28] in order to understand our motivating case, the $Z_c(3900)$, as well as many others, *hadspec* proceeded to compute a relatively simple case in which a resonant behavior is expected, namely $\pi\omega$ scattering, where the ω is stable with 400 MeV pions.[27] A rather complete basis of operators was used, allowing access to the coupled $^3S_1$ and $^3D_1$ partial-waves, and indeed as expected a clear signal for an axial vector resonance, the b_1, was observed, as can be seen in Figure 2.

With the relevant techniques and formalisms now demonstrated in simpler systems, it is to be expected that in the near future coupled-channel

[d] www.hadspec.org

systems like those housing the Z_c will be studied with comparable levels of rigor.

3. Exotic hadrons in the light quark sector

Mesons lying outside the simple $q\bar{q}$ categorization have long been suggested as being natural features one would expect within a strongly coupled theory of quarks and gluons. As well as higher quark Fock states like the tetraquarks mentioned above, other suggestions include *hybrid mesons* in which a $q\bar{q}$ pair is partnered with an excitation of the gluonic field,[9] or the *glueballs* in which gluons alone form a color-singlet state. The experimental search for a clear glueball candidate in the isoscalar meson spectrum continues, and a process of longstanding interest is the radiative decay of charmonium states, e.g. $J/\psi \to \gamma X$, as the intermediate state formed from $c\bar{c}$ annihilation is expected to be 'gluon rich' and hence may preferentially couple to glueballs.

Theoretically, most of what we know about glueballs comes from pure Yang-Mills theory, in which quarks are excluded completely, and the theory is just the gluon sector of QCD. Lattice studies of this theory show a clear picture of glueball bound-states, with a scalar being the lightest.[29] Several recent papers[30–32] have presented lattice estimates of the matrix elements for $J/\psi \to \gamma G$ for some of the lowest lying glueball states, G. These calculations compute three-point correlation functions making use of the fact that in pure Yang-Mills, the glueballs are the only bound-states in the theory, and the heavy charm quarks can be considered as purely valence entities. These calculations provide an interesting baseline for how strongly glueballs might appear in radiative decays, but ultimately we have to ask how glueballs manifest in true QCD, which features light quarks as well as gluons. Clearly any role that glueballs play will have to be as part of the spectrum of isospin-0 resonances, and indeed if the mass scale suggested by the Yang-Mills calculations is a guide, they will appear above several 'conventional' resonances. A lattice QCD calculation will then have to do a scattering determination, along the lines described in the previous section, but going somewhat higher in energy (where more final states will be accessible) to find a candidate resonance. Once that resonance is determined, the three-point correlation functions can be computed to investigate its coupling in the J/ψ radiative decay process. How one goes about rigorously identifying the glueball content of any given resonance in QCD is not obvious. It should be clear that **this is a tough place to start, so take a step back and compute where you have more control ...**

Transition form-factors of resonances have been studied in lattice QCD, using a finite-volume formalism described in Refs. 33–35. In particular the relatively simple process $\pi\gamma^\star \to \pi\pi$ in which the ρ resonance is found to appear has been successfully computed,[36–38] as has the process $\gamma^\star \to \pi\pi$, also known as the timelike pion form-factor.[39,40] The next stage of progress required to address processes of the complexity indicated above will be to consider coupled-channel transition amplitudes.

4. Truly flavor exotic hadrons

One very interesting sector, where current lattice QCD calculations seem to be suggesting a robust prediction, is that of the $J^P = 1^+$ double-bottom channel, where a stable bound-state lying below the BB^* threshold is observed.[41,43,44] These calculations, using slightly different techniques, all observe a deeply bound state, and while the binding energy is not yet pinned down with precision, and the set of operators used is typically not as complete as one would like, the presence of a bound-state appears to be a solid conclusion (see Fig. 3). Of course, observing such a heavy flavor-exotic state experimentally is very challenging, so it is natural to ask if there is a double-charm analogue, which would be much lighter and potentially accessible in several current and near-future experiments. In Ref. 42, the dependence of the binding energy under variation of the heavy quark mass

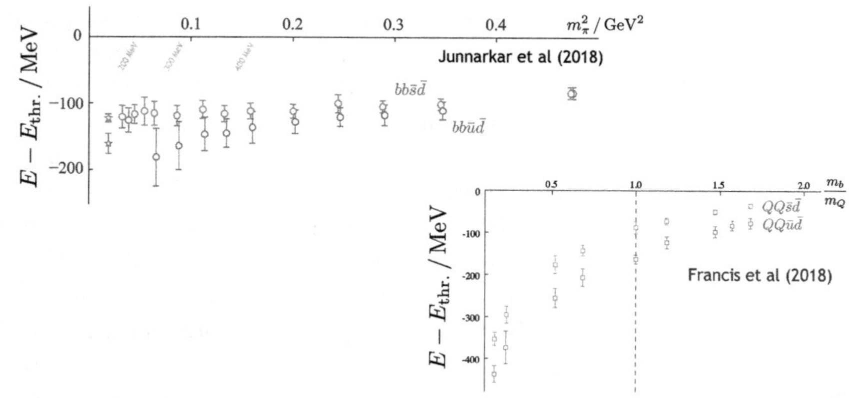

Fig. 3. (Top left) Binding energy (below BB^*, or B_sB^* threshold) of a bound-state in the double-bottom sector as computed in lattice QCD as a function of light quark mass expressed via m_π^2 – results from Ref. 41. (Bottom right) Variation in binding energy as the heavy quark mass is varied – results from Ref. 42.

was studied, with the result that the binding energy decreases as the heavy quark mass decreases toward the charm mass. As the state approaches the DD^* threshold it becomes vitally important to include meson-meson operators, which was not done in the exploratory calculation presented in Ref. 42. Modulo this caveat, the general trend observed poses a simple question: what happens to the bound-state pole in the meson-meson scattering amplitude (BB^* becoming DD^* as the heavy quark mass changes), as the 'external parameter' of the heavy quark mass is varied toward the charm mass? There would seem to be two possibilities, either it merges with a virtual bound-state pole to form a pair of resonance poles lying symmetrically off the real energy axis, which could manifest as a narrow peak above threshold, or it might double back on real energy axis on the unphysical sheet as a virtual bound-state pole, which depending upon how close it lies to threshold could lead to no significant enhancement in the DD^* mass spectrum.

The relevant sector was computed (with 400 MeV pions) by *hadspec* using a basis of operators featuring both tetraquark-like and meson-meson-like constructions.[16] The obtained finite-volume spectrum, reproduced in Figure 4 shows some attraction above threshold, but no obvious sign of a narrow resonance. An important observation about the importance of using a complete basis of operators can be made here: Figure 4 shows that if the meson-meson-like constructions are not included, a completely different, and incorrect, spectrum is obtained, one that might lead an incautious observer to conclude the presence of a tetraquark state above threhsold.

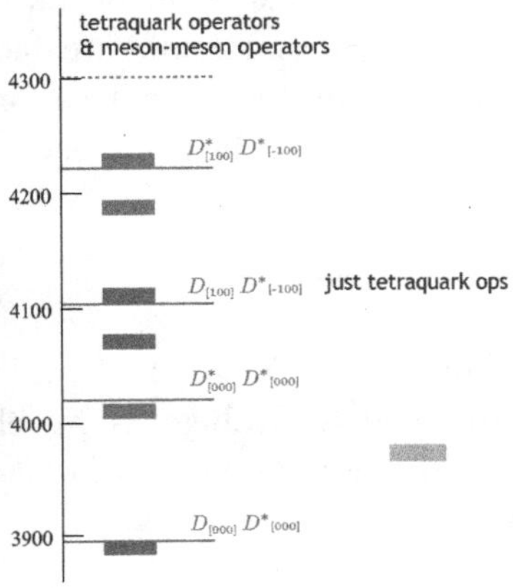

Fig. 4. Double-charm finite-volume energy spectrum in a $\sim (2\,\mathrm{fm})^3$ box with $m_\pi \sim 400\,\mathrm{MeV}$ reported in Ref. 16 determined using a basis of meson-meson and tetraquark operators. Horizontal lines show the expected non-interacting meson-meson spectrum. Also shown (in gray) the result when a correlation matrix featuring only tetraquark operators is diagonalized.

5. Summary

A sample of recent results obtained using lattice QCD of relevance to hadron spectroscopy have been presented. Several examples of exotic sectors, investigated because of contemporary experiment interest, such as the $Z_c(3900)$ and the glueballs, are shown to be rather complicated in QCD, and perhaps not the optimum places to develop theoretical techniques – a pathway toward studying them using simpler processes has been outlined, with significant successes presented. All these processes consider observing excited hadron states as *resonances* in meson-meson scattering, which can be accessed using the finite-volume spectrum computable in lattice QCD. A current limitation lies in the lack of a tested finite-volume formalism for *three hadron or higher final states.* As well as restricting how high in energy one can go when determining scattering amplitudes, this is also a practical restriction on how *low* one can go toward the physical pion mass, as one inevitably introduces low three-hadron thresholds featuring pions. Progress implementing three-body finite-volume formalisms is underway.[e] Lattice QCD calculations in the exotic flavor double-bottom sector show the novel result of at least one likely QCD-stable state – whether this state has an observable double-charm analogue remains an interesting question to be studied using the meson-meson scattering formalism within lattice QCD.

Acknowledgments

The author acknowledges support from the U.S. Department of Energy contract DE-SC0018416 and contract DE-AC05-06OR23177, under which Jefferson Science Associates, LLC, manages and operates Jefferson Lab. Jefferson Lab preprint number: `JLAB-THY-19-3103`. Special thanks are due to Prof. Dai-Hui Wei and the rest of the local organizing committee for their very kind practical assistance and hospitality during the conference.

References

1. S. Aoki *et al.*, FLAG Review 2019 (2019).
2. M. Peardon, J. Bulava, J. Foley, C. Morningstar, J. Dudek, R. G. Edwards, B. Joo, H.-W. Lin, D. G. Richards and K. J. Juge, A Novel quark-field creation operator construction for hadronic physics in lattice QCD, *Phys. Rev.* **D80**, 054506 (2009).

---

[e]See for example Ref. 45 and references therein.

3. J. J. Dudek, R. G. Edwards, M. J. Peardon, D. G. Richards and C. E. Thomas, Highly excited and exotic meson spectrum from dynamical lattice QCD, *Phys. Rev. Lett.* **103**, 262001 (2009).
4. J. J. Dudek, R. G. Edwards, M. J. Peardon, D. G. Richards and C. E. Thomas, Toward the excited meson spectrum of dynamical QCD, *Phys. Rev.* **D82**, 034508 (2010).
5. C. E. Thomas, R. G. Edwards and J. J. Dudek, Helicity operators for mesons in flight on the lattice, *Phys. Rev.* **D85**, 014507 (2012).
6. L. Liu, G. Moir, M. Peardon, S. M. Ryan, C. E. Thomas, P. Vilaseca, J. J. Dudek, R. G. Edwards, B. Joo and D. G. Richards, Excited and exotic charmonium spectroscopy from lattice QCD, *JHEP* **07**, 126 (2012).
7. J. J. Dudek, R. G. Edwards and C. E. Thomas, S and D-wave phase shifts in isospin-2 $\pi\pi$ scattering from lattice QCD, *Phys. Rev.* **D86**, 034031 (2012).
8. J. J. Dudek, R. G. Edwards, B. Joo, M. J. Peardon, D. G. Richards and C. E. Thomas, Isoscalar meson spectroscopy from lattice QCD, *Phys. Rev.* **D83**, 111502 (2011).
9. J. J. Dudek, R. G. Edwards, P. Guo and C. E. Thomas, Toward the excited isoscalar meson spectrum from lattice QCD, *Phys. Rev.* **D88**, 094505 (2013).
10. M. Luscher, Volume Dependence of the Energy Spectrum in Massive Quantum Field Theories. 1. Stable Particle States, *Commun. Math. Phys.* **104**, 177 (1986).
11. M. Luscher, Volume Dependence of the Energy Spectrum in Massive Quantum Field Theories. 2. Scattering States, *Commun. Math. Phys.* **105**, 153 (1986).
12. R. A. Briceno, J. J. Dudek and R. D. Young, Scattering processes and resonances from lattice QCD, *Rev. Mod. Phys.* **90**, 025001 (2018).
13. M. Ablikim *et al.*, Observation of a Charged Charmoniumlike Structure in $e^+e^- \to \pi^+\pi^- J/\psi$ at $\sqrt{s}$ =4.26 GeV, *Phys. Rev. Lett.* **110**, 252001 (2013).
14. S. Prelovsek, C. B. Lang, L. Leskovec and D. Mohler, Study of the Z_c^+ channel using lattice QCD, *Phys. Rev.* **D91**, 014504 (2015).
15. T. Chen, Y. Chen, M. Gong, C. Liu, L. Liu, Y.-B. Liu, Z. Liu, J.-P. Ma, M. Werner and J.-B. Zhang, A coupled-channel lattice study on the resonance-like structure $Z_c(3900)$, *Chin. Phys.* **C43**, 103103 (2019).
16. G. K. C. Cheung, C. E. Thomas, J. J. Dudek and R. G. Edwards, Tetraquark operators in lattice QCD and exotic flavour states in the charm sector, *JHEP* **11**, 033 (2017).

17. R. A. Briceno, J. J. Dudek, R. G. Edwards and D. J. Wilson, Isoscalar $\pi\pi$ scattering and the σ meson resonance from QCD, *Phys. Rev. Lett.* **118**, 022002 (2017).
18. M. Mai, C. Culver, A. Alexandru, M. Doring and F. X. Lee, A cross-channel study of pion scattering from lattice QCD (2019).
19. D. Guo, A. Alexandru, R. Molina, M. Mai and M. Doring, Extraction of isoscalar $\pi\pi$ phase-shifts from lattice QCD, *Phys. Rev.* **D98**, 014507 (2018).
20. P. Guo, J. Dudek, R. Edwards and A. P. Szczepaniak, Coupled-channel scattering on a torus, *Phys. Rev.* **D88**, 014501 (2013).
21. J. J. Dudek, R. G. Edwards, C. E. Thomas and D. J. Wilson, Resonances in coupled $\pi K - \eta K$ scattering from quantum chromodynamics, *Phys. Rev. Lett.* **113**, 182001 (2014).
22. D. J. Wilson, J. J. Dudek, R. G. Edwards and C. E. Thomas, Resonances in coupled $\pi K, \eta K$ scattering from lattice QCD, *Phys. Rev.* **D91**, 054008 (2015).
23. D. J. Wilson, R. A. Briceno, J. J. Dudek, R. G. Edwards and C. E. Thomas, Coupled $\pi\pi, K\bar{K}$ scattering in P-wave and the ρ resonance from lattice QCD, *Phys. Rev.* **D92**, 094502 (2015).
24. G. Moir, M. Peardon, S. M. Ryan, C. E. Thomas and D. J. Wilson, Coupled-Channel $D\pi$, $D\eta$ and $D_s\bar{K}$ Scattering from Lattice QCD, *JHEP* **10**, 011 (2016).
25. J. J. Dudek, R. G. Edwards and D. J. Wilson, An a_0 resonance in strongly coupled $\pi\eta$, $K\overline{K}$ scattering from lattice QCD, *Phys. Rev.* **D93**, 094506 (2016).
26. R. A. Briceno, J. J. Dudek, R. G. Edwards and D. J. Wilson, Isoscalar $\pi\pi, K\overline{K}, \eta\eta$ scattering and the σ, f_0, f_2 mesons from QCD (2017).
27. A. J. Woss, C. E. Thomas, J. J. Dudek, R. G. Edwards and D. J. Wilson, b_1 resonance in coupled $\pi\omega$, $\pi\phi$ scattering from lattice QCD, *Phys. Rev.* **D100**, 054506 (2019).
28. A. Woss, C. E. Thomas, J. J. Dudek, R. G. Edwards and D. J. Wilson, Dynamically-coupled partial-waves in $\rho\pi$ isospin-2 scattering from lattice QCD (2018).
29. C. J. Morningstar and M. J. Peardon, The Glueball spectrum from an anisotropic lattice study, *Phys. Rev.* **D60**, 034509 (1999).
30. L.-C. Gui, Y. Chen, G. Li, C. Liu, Y.-B. Liu, J.-P. Ma, Y.-B. Yang and J.-B. Zhang, Scalar Glueball in Radiative J/ψ Decay on the Lattice, *Phys. Rev. Lett.* **110**, 021601 (2013).

31. Y.-B. Yang, L.-C. Gui, Y. Chen, C. Liu, Y.-B. Liu, J.-P. Ma and J.-B. Zhang, Lattice Study of Radiative J/? Decay to a Tensor Glueball, *Phys. Rev. Lett.* **111**, 091601 (2013).
32. L.-C. Gui, J.-M. Dong, Y. Chen and Y.-B. Yang, Study of the pseudoscalar glueball in J/ψ radiative decays, *Phys. Rev.* **D100**, 054511 (2019).
33. L. Lellouch and M. Luscher, Weak transition matrix elements from finite volume correlation functions, *Commun. Math. Phys.* **219**, 31 (2001).
34. R. A. Briceno, M. T. Hansen and A. Walker-Loud, Multichannel $1 \to 2$ transition amplitudes in a finite volume, *Phys. Rev.* **D91**, 034501 (2015).
35. R. A. Briceno and M. T. Hansen, Multichannel $0 \to 2$ and $1 \to 2$ transition amplitudes for arbitrary spin particles in a finite volume, *Phys. Rev.* **D92**, 074509 (2015).
36. R. A. Briceno, J. J. Dudek, R. G. Edwards, C. J. Shultz, C. E. Thomas and D. J. Wilson, The resonant $\pi^+\gamma \to \pi^+\pi^0$ amplitude from Quantum Chromodynamics, *Phys. Rev. Lett.* **115**, 242001 (2015).
37. R. A. Briceno, J. J. Dudek, R. G. Edwards, C. J. Shultz, C. E. Thomas and D. J. Wilson, The $\pi\pi \to \pi\gamma^\star$ amplitude and the resonant $\rho \to \pi\gamma^\star$ transition from lattice QCD, *Phys. Rev.* **D93**, 114508 (2016).
38. C. Alexandrou, L. Leskovec, S. Meinel, J. Negele, S. Paul, M. Petschlies, A. Pochinsky, G. Rendon and S. Syritsyn, $\pi\gamma \to \pi\pi$ transition and the ρ radiative decay width from lattice QCD, *Phys. Rev.* **D98**, 074502 (2018).
39. X. Feng, S. Aoki, S. Hashimoto and T. Kaneko, Timelike pion form factor in lattice QCD, *Phys. Rev.* **D91**, 054504 (2015).
40. C. Andersen, J. Bulava, B. Horz and C. Morningstar, The $I = 1$ pion-pion scattering amplitude and timelike pion form factor from $N_{\mathrm{f}} = 2+1$ lattice QCD, *Nucl. Phys.* **B939**, 145 (2019).
41. P. Junnarkar, N. Mathur and M. Padmanath, Study of doubly heavy tetraquarks in Lattice QCD, *Phys. Rev.* **D99**, 034507 (2019).
42. A. Francis, R. J. Hudspith, R. Lewis and K. Maltman, Evidence for charm-bottom tetraquarks and the mass dependence of heavy-light tetraquark states from lattice QCD, *Phys. Rev.* **D99**, 054505 (2019).
43. A. Francis, R. J. Hudspith, R. Lewis and K. Maltman, Lattice Prediction for Deeply Bound Doubly Heavy Tetraquarks, *Phys. Rev. Lett.* **118**, 142001 (2017).

44. L. Leskovec, S. Meinel, M. Pflaumer and M. Wagner, Lattice QCD investigation of a doubly-bottom $\bar{b}\bar{b}ud$ tetraquark with quantum numbers $I(J^P) = 0(1^+)$, *Phys. Rev.* **D100**, 014503 (2019).
45. F. Romero-Lopez, S. R. Sharpe, T. D. Blanton, R. A. Briceno and M. T. Hansen, Numerical exploration of three relativistic particles in a finite volume including two-particle resonances and bound states, *JHEP* **10**, 007 (2019).

Lattice results on dibaryons and baryon–baryon interactions

Sinya Aoki

Center for Gravitational Physics, Yukawa Institute for Theoretical Physics, Kyoto University, Kitashirakawa Oiwakecho, Sakyo-ku, Kyoto 606-8502, Japan
E-mail: saoki@yukawa.kyoto-u.ac.jp

We present the recent study on dibaryons at the almost physical pion mass in lattice QCD by the HAL QCD potential method.

Keywords: Dibaryon; HAL QCD potential method; almost physical pion mass.

1. Introduction

A diabaryon is a bound state (or resonance) with a baryon number $B = 2$. The deuteron, made of a proton and a neutron, is only a stable dibaryon observed in Nature so far. Thus it is interesting to ask whether other dibaryons exist in Nature or not.

A dibaryon can be classified in the flavor SU(3) representation as

$$\mathbf{8} \otimes \mathbf{8} = \mathbf{27} \oplus \mathbf{8_s} \oplus \mathbf{1} \oplus \overline{\mathbf{10}} \oplus \mathbf{10} \oplus \mathbf{8_a} \tag{1}$$

for the octet–octet baryons, where a deuteron belongs to the $\overline{\mathbf{10}}$ representation, while a H–dibaryon was predicted in the $\mathbf{1}$ representation,[1] and was recently studied in lattice QCD.[2–6] Classifications including decuplet ($\mathbf{10}$) baryons are

$$\mathbf{10} \otimes \mathbf{8} = \mathbf{25} \oplus \mathbf{8} \oplus \mathbf{10} \oplus \mathbf{27}, \tag{2}$$

where $N\Omega$ and $N\Delta$ dibaryons were predicted in $\mathbf{8}$ and $\mathbf{27}$ representations, respectively,[7,8] and

$$\mathbf{10} \otimes \mathbf{10} = \mathbf{28} \oplus \mathbf{27} \oplus \mathbf{35} \oplus \overline{\mathbf{10}}, \tag{3}$$

where $\Omega\Omega$ was predicted in the $\mathbf{28}$ representation,[9] while $\Delta\Delta$ dibaryon were predicted in the $\overline{\mathbf{10}}$ representation,[10,11] whose candidate $d^*(2380)$ has indeed been observed.[12] Note however that only Ω is a stable decuplet baryon against strong decays.

2. HAL QCD potential method

A fundamental quantity in the HAL QCD method[13–15] is an equal-time Nambu–Bethe–Salpeter (NBS) wave function in the center of mass system, which is given for a two nucleon system as

$$\varphi_{\mathbf{k}}(\mathbf{r}) = \langle 0|N(\mathbf{r}/2,0)N(-\mathbf{r}/2,0)|NN, W_{\mathbf{k}}\rangle, \tag{4}$$

where $|NN, W_{\mathbf{k}}\rangle$ is a two-nucleon eigenstate in QCD having the relative momentum $\mathbf{k}$ and the center of mass energy $W_{\mathbf{k}} = 2\sqrt{\mathbf{k}^2 + m_N^2}$ with the nucleon mass m_N, and $N(x)$ with $x = (\mathbf{x}, t)$ is the nucleon operator. In our study, we usually take the local operator in terms of quark fields as $N(x) = \epsilon^{abc}(u_a^T(x)C\gamma_5 d_b(x))q_c(x)$, where u_a (d_a) is a up (down) quark field with color a while $q = u$ (d) corresponds to a proton (neutron), and $C = \gamma_2\gamma_4$ is a charge conjugation matrix acting on spinor indices. A choice of the nucleon operator is a part of the definition (or scheme) for the potential. In the HAL QCD method, we restrict the total energy below the lowest inelastic threshold as $W_{\mathbf{k}} < W_{\text{th}} \equiv 2m_N + m_\pi$ with a pion mass m_π, so that only the elastic NN scatterings can occur.

It can be shown[16,17] that an asymptotic behavior of the NBS wave function at large $r \equiv |\mathbf{r}|$ is given by

$$\varphi_{\mathbf{k}}(\mathbf{r}) \simeq \sum_{l,m} C_l \frac{\sin(kr - l\pi/2 + \delta_l(k))}{kr} Y_{lm}(\Omega_{\mathbf{r}}), \quad k \equiv |\mathbf{k}|, \tag{5}$$

where Y_{lm} is a spherical harmonic function for the solid angle of $\mathbf{r}$ ($\Omega_{\mathbf{r}}$). For simplicity, we ignore the spin of nucleons here, but you can find a complete formula in Refs. 14, 18. It is important to note that $\delta_l(k)$ is the phase of the S-matrix in QCD for the partial wave with the angular momentum l, which is encoded in the asymptotic behavior of the NBS wave function, similar to the scattering phase shift of the scattering wave in quantum mechanics.

Using this property, we define the energy–independent *potential* with derivatives from NBS wave functions as

$$[E_{\mathbf{k}} - H_0]\,\varphi_{\mathbf{k}}(\mathbf{r}) = V(\mathbf{r}, \nabla)\varphi_{\mathbf{k}}(\mathbf{r}), \quad E_{\mathbf{k}} = \frac{\mathbf{k}^2}{m_N}, \quad H_0 = \frac{-\nabla^2}{m_N}, \tag{6}$$

for $W_{\mathbf{k}} < W_{\text{th}}$. The potential at the next-to-leading order (NLO) takes a form as

$$V(\mathbf{r}, \nabla) = V_0(r) + V_\sigma(r)(\sigma_1 \cdot \sigma_2) + V_{\text{T}}(r)S_{12} + V_{\text{LS}}(r)\mathbf{L}\cdot\mathbf{S} + O\left(\nabla^2\right), \tag{7}$$

where σ_i is a spin operator acting on the i-th nucleon, $S_{12} = 3(\sigma_1\cdot\hat{\mathbf{r}})(\sigma_2\cdot\hat{\mathbf{r}}) - (\sigma_1\cdot\sigma_2)$ with $\hat{\mathbf{r}} = \mathbf{r}/r$ is the tensor operator, $\mathbf{L} = \mathbf{r}\times\nabla$ and $\mathbf{S} = (\sigma_1 + \sigma_2)/2$.

The Schrödinger equation with the potential $V(\mathbf{r}, \nabla)$ give a correct QCD *phase shift* $\delta_l(k)$, since NBS wave functions are solutions to the equation by construction. Note that non-relativistic approximation is not employed here since the Klein–Gordon operator reduces to the Helmholtz operator for a given center of mass energy as $-\Box - m^2 = (W_k/2) + \nabla^2 - m^2 = \mathbf{k}^2 + \nabla^2$.

We determine local functions such as $V_X(r)$ $(X = 0, \sigma, \mathrm{T}, \mathrm{LS})$ order by order. For example, the leading order (LO) potential $V_0^{\mathrm{LO}}(r) \equiv V_0(r) + V_\sigma(r)(\sigma_1 \cdot \sigma_2) + V_\mathrm{T}(r) S_{12}$ can be approximately determined from one NBS wave function $\varphi_\mathbf{k}$ as

$$V_0^{\mathrm{LO}}(r; \varphi_\mathbf{k}) = \frac{[E_\mathbf{k} - H_0]\varphi_\mathbf{k}(\mathbf{x})}{\varphi_\mathbf{k}(\mathbf{x})}, \tag{8}$$

where an argument $\varphi_\mathbf{k}$ of the potential represents an input for its determination. If $V_0^{\mathrm{LO}}(r; \varphi_\mathbf{k}) \simeq V_0^{\mathrm{LO}}(r; \varphi_\mathbf{q})$ for $|\mathbf{k}| < |\mathbf{q}|$, it turns out that the LO approximation is good at $\mathbf{p}$ in $|\mathbf{k}| \le |\mathbf{p}| \le |\mathbf{q}|$. If $V_0^{\mathrm{LO}}(r; \varphi_\mathbf{k}) \neq V_0^{\mathrm{LO}}(r; \varphi_\mathbf{q})$, on the other hand, the NLO term can be determined from two equations given by

$$[E_\mathbf{k} - H_0]\, \varphi_\mathbf{p}(\mathbf{x}) = \left[V_0^{\mathrm{NLO}}(r) + V_{\mathrm{LS}}^{\mathrm{NLO}}(r) \mathbf{L} \cdot \mathbf{S}\right] \varphi_\mathbf{p}(\mathbf{x}), \quad \mathbf{p} = \mathbf{k}, \mathbf{q}, \tag{9}$$

where a superscript NLO represent the order of the approximation to determine these terms. We can continue this procedure to increase accuracy of the determination. Once the potential is approximately obtained, physical observables such as scattering phase shift can be extracted.

In lattice QCD, a NBS wave function is extracted from a 4-pt correlation function as

$$\begin{aligned} F(\mathbf{r}, t) &\equiv \langle 0 | N(\mathbf{r}/2, t) N(-\mathbf{r}/2, t) \bar{\mathcal{J}}_{NN}(0) | 0 \rangle = \sum_n A_n \varphi_{\mathbf{k}_n}(\mathbf{r}) e^{-W_{\mathbf{k}_n} t} + \cdots \\ &\simeq A_0 \varphi_{\mathbf{k}_0}(\mathbf{r}) e^{-W_{\mathbf{k}_0} t}, \quad t \to \infty, \end{aligned} \tag{10}$$

where $\bar{\mathcal{J}}_{NN}(t)$ is an operator which creates two-nucleon elastic states at t with an overlap factor $A_n = \langle NN, W_{\mathbf{k}_n} | \bar{\mathcal{J}}_{NN}(0) | 0 \rangle$, an ellipsis represents contributions form inelastic states, and $W_{\mathbf{k}_0}$ is an energy of the NN ground state. In practice it is very difficult to take a large t due to a bad signal-to-noise ration for two baryons, but a use of smaller t may introduce large systematic errors due to contaminations from elastic excited states to the grand state, which is a very serious problem for the conventional method.[19–24]

In Ref. 25, an improved method to extract potentials has been proposed. We define the normalized 4-pt function as

$$R(\mathbf{r}, t) \equiv \frac{F(\mathbf{r}, t)}{G_N(t)^2} = \sum_n \bar{A}_n \varphi_{\mathbf{k}_n}(\mathbf{r}) e^{-\Delta W_{\mathbf{k}_n} t} + \cdots, \quad \Delta W_{\mathbf{k}_n} = W_{\mathbf{k}_n} - 2m_N,$$

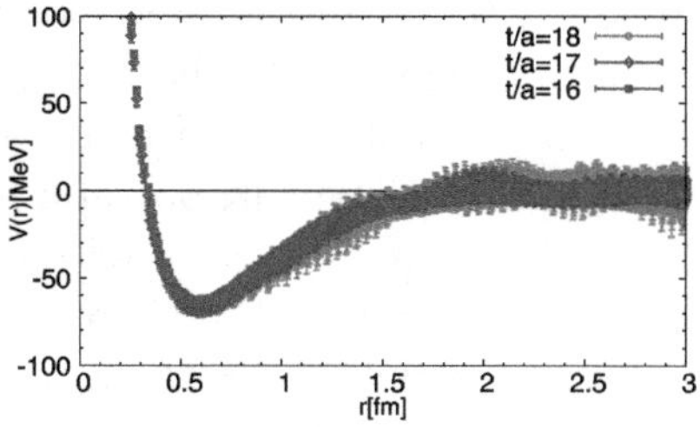

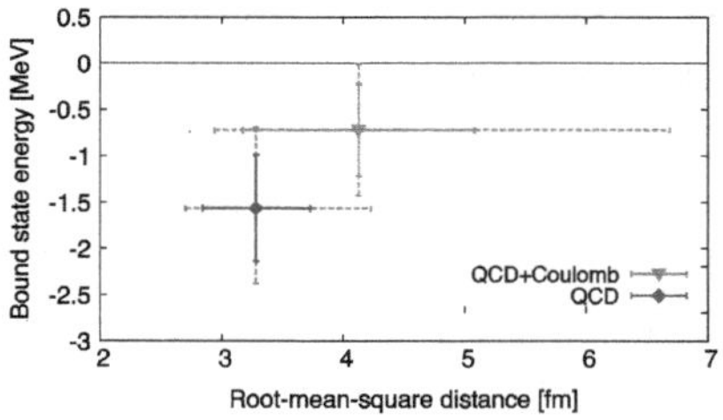

Fig. 1. (Left) The $\Omega^-\Omega^-$ potential $V(r)$ in the ${}^1S_0$ channel at $t/a = 16, 17, 18$, in $2+1$ flavor QCD at almost physical pion mass. (Right) The binding energy of the $\Omega^-\Omega^-$ system and the root-mean-square distance between two Ω^-'s without and without the Coulomb repulsion (blue solid diamond and red solid triangle, respectively). Taken from Ref. 26

where $G_N(t)$ is a nucleon 2-pt function at rest, which behaves as $Ze^{-m_N t}$ as long as inelastic contributions to the 2-pt function can be neglected, and $\bar{A}_n = A_n/Z^2$. Since all NBS wave functions, $\phi_{\mathbf{k}_n}$, below inelastic threshold satisfy the same Shrödinger equation (6), we obtain

$$\left\{ -H_0 - \frac{\partial}{\partial t} + \frac{1}{4m_N^2}\frac{\partial^2}{\partial t^2} \right\} R(\mathbf{r}, t) = V(\mathrm{r}, \nabla) R(\mathbf{r}, t) \simeq V_0^{\mathrm{LO}}(\mathrm{r}) R(\mathbf{r}, t), \quad (11)$$

where we use a relation $\Delta W_{\mathbf{k}} = \mathbf{k}^2/m_N - (\Delta W_{\mathbf{k}})^2/(4m_N^2)$, and we need to take a moderately large t satisfying $W_{\mathrm{th}} t \gg 1$ to ignore inelastic contributions. Note that $V(\mathbf{r}, \nabla)$ extracted from the above equation should be t independent. Therefore, the t dependence for the LO potential $V_0^{\mathrm{LO}}(\mathrm{r})$, for example, indicates either an existence of inelastic contributions or contributions from higher order terms in the derivative expansion.

3. Dibaryons at the almost physical pion mass

As an application of the HAL QCD potential method, we presents some results on dibaryons.

In our studies on dibayons, we employ (2+1)-flavor gauge configurations generated on a $L^3 \times T = 96^3 \times 96$ lattice with the RG-improved Iwasaki gauge action and non-perturbatively $\mathcal{O}(a)$-improved Wilson quark action, at $a \simeq 0.085$ fm (thus $La \simeq 8.1$ fm) with $(m_\pi, m_K, m_N) \simeq (146, 525, 955)$ MeV. which correspond to the almost physical point.

We first consider the $\Omega^-\Omega^-$ system in the ${}^1S_0$ channel, which belongs to the **28** representation.[26]

In Fig. 1 (Left), we show $\Omega^-\Omega^-$ potentials at $t/a = 16, 17, 18$, which has qualitative features similar to the central potentials for NN. We notice,

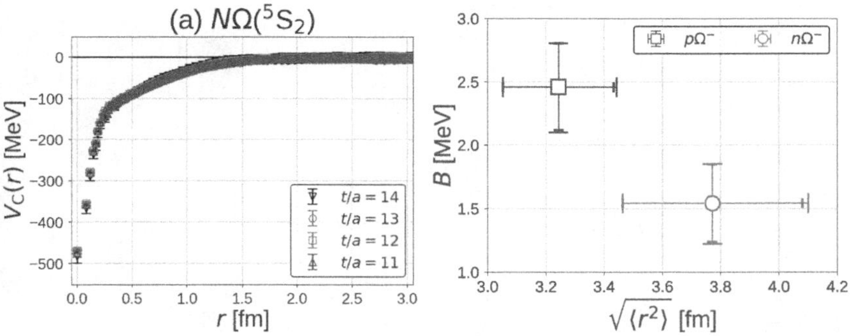

Fig. 2. (Left) The $N\Omega$ potential $V_C(r)$ in the $^5S_2$ channel at $t/a = 11, 12, 13, 14$, with the same lattice setup for $\Omega\Omega$. (Right) The binding energy and the root-mean-square distance for the $n\Omega^-$ (red open circle) and $p\Omega^-$ (blue open square). Taken from Ref. 28.

however, that its repulsion is weaker and attraction is shorter-ranged than the NN case. With this potential, we obtain one shallow bound state, whose binding energy is shown in Fig. 1 (Right) as a function of the root-mean-square distance, with and without Coulomb repulsion between $\Omega^-\Omega^-$ as α/r, denoted by red circle and blue square, respectively. Such a bound state may be searched experimentally by two-particle correlations in future relativistic heavy-ion collisions.[27]

We next consider the $N\Omega^-$ system with $S = -3$ in the $^5S_2$ channel, which belongs to the **8** representation.[28] At the almost physical pion mass, $N\Omega(^5S_2)$ may couple to D-wave octet-octet channels below the $N\Omega$ threshold such as $\Lambda\Xi$ and $\Sigma\Xi$. We thus assume that such couplings are small.

In Fig. 2 (Left), we plot the $N\Omega^-$ potential at $t/a = 11$–14, showing attraction at all distances without repulsive core. Thus there is a chance to form a bound state, and indeed one bound state is found to exist in this channel. Fig. 2 (Right) shows the binding energy as a function of the the root-mean-square distance for $n\Omega^-$ with no Coulomb interaction (red) and $p\Omega^-$ with Coulomb attraction (blue). These binding energies are found to be much smaller than $B = 18.9(5.0)(^{+12.1}_{-1.8})$ MeV at heavier pion mass $m_\pi = 875$ MeV.[29] Such a $N\Omega$ state can be searched through two-particle correlations in relativistic nucleus-nucleus collisions,[27] if indeed exists. Actually, some indications in experiments were recently reported,[30] and more will be expected to come.

From potentials we obtain for $\Omega\Omega(^1S_0)$ and $N\Omega(^5S_2)$, we calculate the scattering length a_0 and the effective range r_{eff} for these systems. Fig. 3 shows the ratio r_{eff}/a_0 as a function of r_{eff} for $\Omega\Omega(^1S_0)$ (blue diamond)

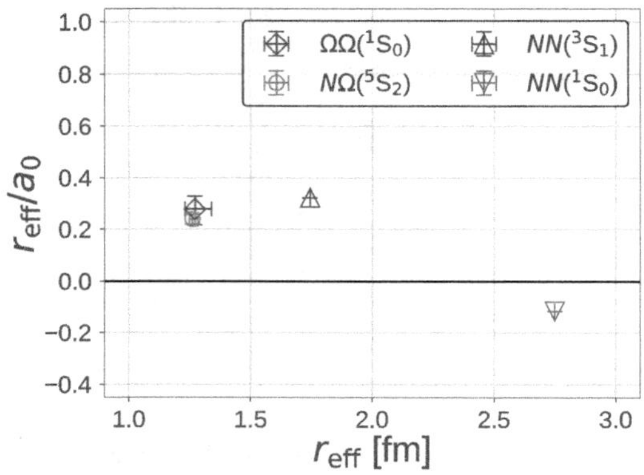

Fig. 3. The ratio of the effective range and the scattering length $r_{\rm eff}/a_0$ as a function of $r_{\rm eff}$ for $\Omega\Omega(^1S_0)$ (blue open diamond) and $N\Omega(^5S_2)$ (red open circle) obtained in lattice QCD, as well as for $NN(^3S_1)$ (purple open up-triangle) and $NN(^1S_0)$ (green open down-triangle) in experiments. Taken from Ref. 28.

and $N\Omega(^5\mathrm{S}_2)$ (red circle) obtained in lattice QCD near the physical pion mass, together with the experimental values for $NN(^3\mathrm{S}_1)$ (deuteron, purple up-triangle) and $NN(^1\mathrm{S}_0)$ (di-neutron, green down-triangle). For all cases, $|r_{\rm eff}/a_0|$ is small, indicating that these systems are located close to the unitary limit. It will be interesting to understand why dibaryons or dibaryon candidates appear in the unitary region near the physical pion mass.

Acknowledgments

This work is supported in part by the Grant-in-Aid of the Japanese Ministry of Education, Sciences and Technology, Sports and Culture (MEXT) for Scientific Research (Nos. JP16H03978, JP18H05236), by a priority issue (Elucidation of the fundamental laws and evolution of the universe) to be tackled by using Post “K” Computer, and by Joint Institute for Computational Fundamental Science (JICFuS).

References

1. R. L. Jaffe, Perhaps a Stable Dihyperon, *Phys. Rev. Lett.* **38**, 195 (1977), [Erratum: *Phys. Rev. Lett.* **38**, 617 (1977)].
2. T. Inoue, N. Ishii, S. Aoki, T. Doi, T. Hatsuda, Y. Ikeda, K. Murano, H. Nemura and K. Sasaki, Bound H-dibaryon in Flavor SU(3) Limit of Lattice QCD, *Phys. Rev. Lett.* **106**, 162002 (2011).

3. T. Inoue, S. Aoki, T. Doi, T. Hatsuda, Y. Ikeda, N. Ishii, K. Murano, H. Nemura and K. Sasaki, Two-Baryon Potentials and H-Dibaryon from 3-flavor Lattice QCD Simulations, *Nucl. Phys.* **A881**, 28 (2012).
4. T. Inoue, N. Ishii, S. Aoki, T. Doi, T. Hatsuda, Y. Ikeda, K. Murano, H. Nemura and K. Sasaki, Baryon-Baryon Interactions in the Flavor SU(3) Limit from Full QCD Simulations on the Lattice, *Prog. Theor. Phys.* **124**, 591 (2010).
5. S. R. Beane *et al.*, Evidence for a Bound H-dibaryon from Lattice QCD, *Phys. Rev. Lett.* **106**, 162001 (2011).
6. A. Francis, J. R. Green, P. M. Junnarkar, C. Miao, T. D. Rae and H. Wittig, Lattice QCD study of the H dibaryon using hexaquark and two-baryon interpolators, *Phys. Rev.* **D99**, 074505 (2019).
7. J. T. Goldman, K. Maltman, G. J. Stephenson, Jr., K. E. Schmidt and F. Wang, STRANGENESS -3 DIBARYONS, *Phys. Rev. Lett.* **59**, 627 (1987).
8. M. Oka, Flavor Octet Dibaryons in the Quark Model, *Phys. Rev.* **D38**, 298 (1988).
9. Z. Y. Zhang, Y. W. Yu, P. N. Shen, L. R. Dai, A. Faessler and U. Straub, Hyperon nucleon interactions in a chiral SU(3) quark model, *Nucl. Phys.* **A625**, 59 (1997).
10. F. Dyson and N. H. Xuong, Y=2 States in Su(6) Theory, *Phys. Rev. Lett.* **13**, 815 (1964).
11. T. Kamae and T. Fujita, Possible Existence of a Deeply Bound Delta-Delta System, *Phys. Rev. Lett.* **38**, 471 (1977).
12. P. Adlarson *et al.*, ABC Effect in Basic Double-Pionic Fusion — Observation of a new resonance?, *Phys. Rev. Lett.* **106**, 242302 (2011).
13. N. Ishii, S. Aoki and T. Hatsuda, The Nuclear Force from Lattice QCD, *Phys. Rev. Lett.* **99**, 022001 (2007).
14. S. Aoki, T. Hatsuda and N. Ishii, Theoretical Foundation of the Nuclear Force in QCD and its applications to Central and Tensor Forces in Quenched Lattice QCD Simulations, *Prog. Theor. Phys.* **123**, 89 (2010).
15. S. Aoki, T. Doi, T. Hatsuda, Y. Ikeda, T. Inoue, N. Ishii, K. Murano, H. Nemura and K. Sasaki, Lattice QCD approach to Nuclear Physics, *PTEP* **2012**, 01A105 (2012).
16. C. J. D. Lin, G. Martinelli, C. T. Sachrajda and M. Testa, $K \to \pi\pi$ decays in a finite volume, *Nucl. Phys.* **B619**, 467 (2001).
17. S. Aoki *et al.*, I=2 pion scattering length from two-pion wave functions, *Phys. Rev.* **D71**, 094504 (2005).

18. N. Ishizuka, Derivation of Luscher's finite size formula for N pi and NN system, *PoS* **LAT2009**, 119 (2009).
19. T. Iritani *et al.*, Mirage in Temporal Correlation functions for Baryon-Baryon Interactions in Lattice QCD, *JHEP* **10**, 101 (2016).
20. S. Aoki, T. Doi and T. Iritani, Lüscher's finite volume test for two-baryon systems with attractive interactions, *PoS* **LATTICE2016**, 109 (2017).
21. T. Iritani, S. Aoki, T. Doi, T. Hatsuda, Y. Ikeda, T. Inoue, N. Ishii, H. Nemura and K. Sasaki, Are two nucleons bound in lattice QCD for heavy quark masses? Consistency check with Lüschers finite volume formula, *Phys. Rev.* **D96**, 034521 (2017).
22. S. Aoki, T. Doi and T. Iritani, Sanity check for NN bound states in lattice QCD with Lüscher's finite volume formula – Exposing Symptoms of Fake Plateaux –, in *35th International Symposium on Lattice Field Theory (Lattice 2017) Granada, Spain, June 18-24, 2017*, 2017.
23. T. Iritani, S. Aoki, T. Doi, S. Gongyo, T. Hatsuda, Y. Ikeda, T. Inoue, N. Ishii, H. Nemura and K. Sasaki, Systematics of the HAL QCD Potential at Low Energies in Lattice QCD, *Phys. Rev.* **D99**, 014514 (2019).
24. T. Iritani, S. Aoki, T. Doi, T. Hatsuda, Y. Ikeda, T. Inoue, N. Ishii, H. Nemura and K. Sasaki, Consistency between Lüscher's finite volume method and HAL QCD method for two-baryon systems in lattice QCD, *JHEP* **03**, 007 (2019).
25. N. Ishii, S. Aoki, T. Doi, T. Hatsuda, Y. Ikeda, T. Inoue, K. Murano, H. Nemura and K. Sasaki, Hadron-hadron interactions from imaginary-time Nambu-Bethe-Salpeter wave function on the lattice, *Phys. Lett.* **B712**, 437 (2012).
26. S. Gongyo *et al.*, Most Strange Dibaryon from Lattice QCD, *Phys. Rev. Lett.* **120**, 212001 (2018).
27. K. Morita, S. Gongyo, T. Hatsuda, T. Hyodo, Y. Kamiya and A. Ohnishi, Probing $\Omega\Omega$ and $p\Omega$ dibaryons with femtoscopic correlations in relativistic heavy-ion collisions (2019).
28. T. Iritani *et al.*, $N\Omega$ dibaryon from lattice QCD near the physical point, *Phys. Lett.* **B792**, 284 (2019).
29. F. Etminan, H. Nemura, S. Aoki, T. Doi, T. Hatsuda, Y. Ikeda, T. Inoue, N. Ishii, K. Murano and K. Sasaki, Spin-2 $N\Omega$ dibaryon from Lattice QCD, *Nucl. Phys.* **A928**, 89 (2014).
30. J. Adam *et al.*, The Proton-Ω correlation function in Au+Au collisions at $\sqrt{s_{NN}}$=200 GeV, *Phys. Lett.* **B790**, 490 (2019).

Session 1

Meson Spectroscopy

Quark model explanation of $\Upsilon\,(10860)$

P. González and R. Bruschini

Departamento de Física Teórica, Universidad de Valencia (UV) e IFIC (UV - CSIC), Spain

We argue that the assignment of the experimental resonance $\Upsilon\,(10860)$ to a mixing of the conventional $\Upsilon\,(5s)$ quark model state with the lowest hybrid state may allow for a good description of its mass and decay properties.

1. Introduction

The low lying bottomonium $(b\bar{b})$ spectrum can be well described from a conventional nonrelativistic quark model Hamiltonian containing a spin independent Cornell like quark-antiquark potential, the difference with experimental masses being explained by spin dependent potential corrections that can be implemented via first order perturbation theory, see for instance Ref. 1 and references therein. The experimental resonance $\Upsilon\,(10860)$[2] is naturally assigned to the $\Upsilon\,(5s)$ $b\bar{b}$ state. However, the observed decay properties put into question such an assignment. Specifically, if we identify $\Upsilon\,(10860)$ with $\Upsilon\,(5s)$, then i) the Heavy Quark Spin Symmetry (HQSS) violating decays $\Upsilon\,(10860) \rightarrow \pi^+\pi^- h_b\,(np)$ with $n = 1,2$ have a similar production rate to the HQSS conserving decays $\Upsilon\,(10860) \rightarrow \pi^+\pi^-\Upsilon\,(n_f s)$ with $n_f = 1,2,3$ and ii) the measured ratios for decays to open bottom meson-meson states are in clear disagreement with quark model expectations. Moreover, the dipion decay widths $\Upsilon\,(10860) \rightarrow \pi^+\pi^-\Upsilon\,(n_f s)$ are somewhat anomalous in the sense that they are about two orders of magnitude larger than those for $\Upsilon\,(n's) \rightarrow \pi^+\pi^-\Upsilon\,(1s)$ $(n' = 2,3,4)$. This has led to different proposals for the nature of the $\Upsilon\,(10860)$. Among them, the authors of Ref. 3 suggested, on qualitative arguments, that a tetraquark structure could give rise to a non resonant amplitude for $\Upsilon\,(10860) \rightarrow \pi^+\pi^-\Upsilon\,(n_f s)$ two orders of magnitude bigger than for $\Upsilon\,(n's) \rightarrow \pi^+\pi^-\Upsilon\,(1s)$ $(n' = 2,3,4)$.

On the other hand, in Ref. 4, the author followed heavy quark spin symmetry arguments to propose a mixture of a $b\bar{b}$ (1^{--}) configuration with a p-wave $B_s^*\overline{B}_s^*$ state as the possible nature of $\Upsilon\,(10860)$. However, no

Table 1. Calculated $0^{-}\left(1^{--}\right)$ bottomonium masses, M_{nl}, from $V_C(r)$ with $\sigma = 873$ MeV/fm, $\zeta = 100$ MeV.fm and $m_b = 4793$ MeV. Experimental masses, M_{PDG}, from Ref. 2.

nl States	M_{nl} (MeV)	M_{PDG} (MeV)
$1s$	9463	9460.30 ± 0.26
$2s$	10023	10023.26 ± 0.31
$1d$	10169	10163.7 ± 1.4
$3s$	10358	10355.2 ± 0.5
$2d$	10455	
$4s$	10628	10579.4 ± 1.2
$3d$	10703	
$5s$	10865	$10889.9^{+3.2}_{-2.6}$
$4d$	10926	

quantitative dynamical calculation supporting this proposal was presented.

In this talk we propose as an alternative that the mixture of the $\Upsilon(5s)$ quark model state with the lowest hybrid state may give quantitative dynamical account of the observed properties of $\Upsilon(10860)$.[5] For this purpose we revise in Sec. 2 the 1^{--} bottomonium $(b\bar{b})$ spectrum and the decay properties of $\Upsilon(5s)$. Then, the possible consequences of a mixing of $\Upsilon(5s)$ with the lowest hybrid state that lies pretty close in mass are examined.

2. Bottomonium spectrum and decay properties

For the description of bottomonium $(b\bar{b})$ we use a non relativistic quark model defined by the Cornell like potential

$$V_C(r) = \sigma r - \frac{\zeta}{r} \tag{1}$$

where r is the b–$\bar{b}$ distance and the parameters σ and ζ stand for the string tension and the chromoelectric coulomb strength respectively. From it, a nice description of the spectral masses can be obtained as shown in Table 1.

The natural mass assignment of $\Upsilon(10860)$ with a mass of $10889.9^{+3.2}_{-2.6}$ MeV to the $\Upsilon(5s)$ state is perfectly compatible with data from leptonic widths, as can be checked from Table 2. Notice that this also precludes a significant mixing of the $\Upsilon(5s)$ with the $\Upsilon(4d)$ state.

As for the HQSS conserving dipion processes $\Upsilon(n_i s) \to \pi^+\pi^-\Upsilon(n_f s)$ they can be evaluated as $E1$–$E1$ double electric dipole transitions in the QCD Multipole Expansion (QCDME) framework.[6] The calculation of the amplitude involves an unknown constant and the sum over intermediate

Table 2. Calculated leptonic width ratios from $\Upsilon(5s)$ from Ref. 5. Experimental ratios from Ref. 2.

n	$\frac{\Gamma_{\Upsilon(5s)\to e^+e^-}}{\Gamma_{\Upsilon(ns)\to e^+e^-}}$	$\left(\frac{\Gamma_{\Upsilon(10860)\to e^+e^-}}{\Gamma_{\Upsilon(ns)\to e^+e^-}}\right)_{Exp}$
1	0.19	0.23 ± 0.05
2	0.51	0.52 ± 11
3	0.71	0.70 ± 0.16

Table 3. Calculated dipion widths, Γ, from Ref. 5 as compared to data, Γ_{PDG}, from Ref. 2. Data for $\Upsilon(2s) \to \pi^+\pi^-\Upsilon(1s)$ is used to fit the unknown QCDME constant.

Process	Γ (keV)	Γ_{PDG} (keV)
$\Upsilon(2s) \to \pi^+\pi^-\Upsilon(1s)$	5.7	5.7 ± 0.6
$\Upsilon(3s) \to \pi^+\pi^-\Upsilon(1s)$	0.94	0.89 ± 0.10
$\Upsilon(3s) \to \pi^+\pi^-\Upsilon(2s)$	0.58	0.57 ± 0.09
$\Upsilon(4s) \to \pi^+\pi^-\Upsilon(1s)$	6.9	1.7 ± 0.3
$\Upsilon(4s) \to \pi^+\pi^-\Upsilon(2s)$	4.0	1.7 ± 0.4
$\Upsilon(5s) \to \pi^+\pi^-\Upsilon(1s)$	660	270 ± 70
$\Upsilon(5s) \to \pi^+\pi^-\Upsilon(2s)$	120	400 ± 120
$\Upsilon(5s) \to \pi^+\pi^-\Upsilon(3s)$	20	240 ± 130

color singlet $(b\bar{b} + gluon)$ hybrid states. These can be obtained from hybrid potentials derived in the Born-Oppenheimer approximation from quenched lattice results.[7] Actually, the lowest hybrid potential $V_{hybrid}(r)$, from which the 1^{--} hybrid spectrum comes out, is completely equivalent at intermediate and long distances to the deepest vibrational string potential,[8] $V_{vib}(r)$, so that we can use

$$V_{hybrid}(r) \simeq V_{vib}(r) = \sigma r \sqrt{1 + \frac{2\pi\hbar}{\sigma r^2}} \tag{2}$$

to get the spectrum. Thus, the lowest hybrid state, $\mathcal{M}_{1p} = 10888$ MeV is pretty close to the calculated mass of the $5s$ state, $M_{5s} = 10865$ MeV (the masses for the higher hybrid states are $\mathcal{M}_{2p} = 11082$ MeV, $\mathcal{M}_{3p} = 11267$ MeV...) This gives rise to an enhancement of the amplitudes for $\Upsilon(5s)$ as compared to $\Upsilon(n_i s)$ $(n_i < 5)$, as shown in Table 3.

We see that the calculated widths from $\Upsilon(3s)$ are in perfect agreement with data; from $\Upsilon(4s)$ the calculated widths are bigger (by at most a factor 4) as should be expected if the experimental resonance has some $3d$ mixing (for the suppression of dipion decays from d states see Ref. 9).

Regarding $\Upsilon(5s)$ the calculated (nonresonant) dipion widths to $\Upsilon(1s)$ and $\Upsilon(2s)$ have the correct order of magnitude differing from data by at most a factor 3 whereas in the case to $\Upsilon(3s)$ the calculated width is one order of magnitude lower than data. These disagreements may be qualitatively understood as due to the lost of accuracy of the QCDME when increasing n_i and, in the dipion decay into $\Upsilon(3s)$, for the presence of a dominant resonant contribution from $f_0(500)$ (notice that $M(f_0(500)) \simeq M(5s) - M(3s)$).

The order of magnitude of the nonresonant amplitude for the HQSS violating dipion processes $\Upsilon(10860) \to \pi^+\pi^- h_b(np)$ can be also evaluated as a $E1$–$M1$ transition in the QCDME framework. As the lack of data prevents the fixing of the QCDME constants in this case, one can approximate hadronic transition rates by gluon emission rates in order to have some predictive power.[5] Thus, one gets $\Gamma(\Upsilon(5s) \to \pi^+\pi^- h_b(1p)) \approx (1.1 \times 10^2)\, \Gamma(\Upsilon(3s) \to \pi^+\pi^- h_b(1p))$ which is more than two orders of magnitude lower than data $\Gamma(\Upsilon(1860) \to \pi^+\pi^- h_b(1p))_{Exp} \approx (7.3 \times 10^4)\, \Gamma(\Upsilon(3s) \to \pi^+\pi^- h_b(1p))_{Exp}$ as could be expected from approximate HQSS.

With respect to the decays to open bottom two meson states the widths can be calculated from our Cornell potential model wave functions by applying the well known $^3P_0$ decay model (for the final meson states we use Gaussian wave functions as usual). Thus, for instance, we get $\Gamma(\Upsilon(5s) \to B\overline{B}^*) = 2.3\,\Gamma(\Upsilon(5s) \to B\overline{B})$ to be compared to $\Gamma(\Upsilon(1860) \to B\overline{B}^*) = (2.5 \pm 0.5)\,\Gamma(\Upsilon(1860) \to B\overline{B})$ and $\Gamma(\Upsilon(5s) \to B^*\overline{B}^*) = 0.4\,\Gamma(\Upsilon(5s) \to B\overline{B})$ to be compared to $\Gamma(\Upsilon(1860) \to B^*\overline{B}^*) = (6.9 \pm 1.4)\,\Gamma(\Upsilon(1860) \to B\overline{B})$. We see that whereas the $\Upsilon(5s)$ decay to $B\overline{B}^*$ (normalized to the decay to $B\overline{B}$) agrees with data there is a strong disagreement for the $\Upsilon(5s)$ decay to $B^*\overline{B}^*$. This disagreement puts into question the interpretation of $\Upsilon(1860)$ as the $\Upsilon(5s)$ state.

By realizing that the lowest hybrid state, $H_b(1p)$, is close in mass to the $\Upsilon(5s)$ state, some mixing between them may occur (notice that the mixing implies an interaction that breaks HQSS). This mixing alternative has been explored in Ref. 5. In order to preserve the excellent description of the ratios of leptonic (e^+e^-) widths the probability of the hybrid component in $\Upsilon(1860)$ should be at most of a few percent. This would imply that the HQSS conserving widths $\Gamma(H_b(1p) \to \pi^+\pi^- h_b(1p, 2p))$ should be bigger than 1 MeV.[5] On the other hand, as the dominant $H_b(1p)$ decay could be to open bottom two-meson channels, this mixing might provide a solution to the measured width of $\Upsilon(1860)$ to $B^*\overline{B}^*$. For this purpose a thorough revision of the existing decay models for these processes,[10,11] in order to consistently incorporate the Born-Oppenheimer hybrid description coming out from lattice results, is needed before extracting any definite conclusion.[12]

Acknowledgments

This work has been supported by MINECO of Spain and EU Feder Grant No. FPA2016-77177-C2-1-P, by SEV-2014-0398, and by EU Horizon 2020 Grant No. 824093 (STRONG-2020). R. B. acknowledges a FPI fellowship from the MICINN of Spain under Grant No. BES-2017-079860.

References

1. N. Brambilla *et al.*, Heavy quarkonium physics, CERN Yellow Report No. CERN-2005-005, 487 (2005).
2. M. Tanabashi *et al.*, Review of Particle Physics, *Phys. Rev.* **D98**, 030001 (2018).
3. A. Ali, C. Hambrock and M. J. Aslam, A Tetraquark interpretation of the BELLE data on the anomalous Upsilon(1S) pi+pi- and Upsilon(2S) pi+pi- production near the Upsilon(5S) resonance, *Phys. Rev. Lett.* **104**, 162001 (2010), [Erratum: Phys. Rev. Lett.107,049903(2011)].
4. M. B. Voloshin, Heavy quark spin symmetry breaking in near-threshold $J^{pc} = 1^{--}$ quarkoniumlike resonances, *Phys. Rev. D* **85**, 034024 (Feb 2012).
5. R. Bruschini and P. González, A plausible explanation of Υ (10860), *Phys. Lett.* **B791**, 409 (2019).
6. Y.-P. Kuang, T. Barnes, C. Yuan and H.-X. Chen, Charmonium transitions, *Int. J. Mod. Phys.* **A24S1**, 327 (2009).
7. E. Braaten, C. Langmack and D. H. Smith, Born-Oppenheimer Approximation for the XYZ Mesons, *Phys. Rev.* **D90**, 014044 (2014).
8. R. Giles and S. H. H. Tye, The Application of the Quark-Confining String to the psi Spectroscopy, *Phys. Rev.* **D16**, 1079 (1977).
9. P. Moxhay, Hadronic Transitions of d Wave Quarkonium, *Phys. Rev.* **D37**, 2557 (1988).
10. A. Le Yaouanc, L. Oliver, O. Pène, J. C. Raynal and S. Ono, $q\bar{q}g$ Hybrid Mesons in $\psi \to \gamma +$ Hadrons, *Z. Phys.* **C28**, 309 (1985).
11. F. Iddir, S. Safir and O. Pène, Do 1- c anti-c g hybrid meson exist, do they mix with charmonium?, *Phys. Lett.* **B433**, 125 (1998).
12. R. Bruschini and P. González, work in preparation.

Spectroscopy of a_1 mesons from lattice QCD with the truncated overlap fermions

Masayuki Wakayama[1,2,3,*], Yuko Murakami[4], Atsushi Nakamura[5,6,3], Motoo Sekiguchi[7] and Hiroaki Wada[7]

[1] *Center for Extreme Nuclear Matters (CENuM), Korea University, Seoul 02841, Republic of Korea*

[2] *Department of Physics, Pukyong National University (PKNU), Busan 48513, Republic of Korea*

[3] *Research Center for Nuclear Physics (RCNP), Osaka University, Ibaraki, Osaka 567-0047, Japan*

[4] *Research and Development Laboratory, Seikow Chemical Engineering & Machinery, LTD,Akashi 674-0093, Japan*

[5] *School of Biomedicine, Far Eastern Federal University, 690950 Vladivostok, Russia*

[6] *Theoretical Research Division, Nishina Center, RIKEN, Wako 351-0198, Japan*

[7] *School of Science and Engineering, Kokushikan University, Tokyo 154-8515, Japan*

* *E-mail: wakayama@rcnp.osaka-u.ac.jp*

We study the ground state and next radial excitation of the a_1 mesons from a quenched lattice QCD simulation with the truncated overlap fermions formalism based on domain wall fermions. Our results are consistent with the experimental values for $a_1(1260)$ and $a_1(1640)$.

Keywords: Hadron spectroscopy; lattice QCD; chiral fermion action.

1. Introduction

In hadron physics, determining the properties of excited light mesons will provide important information on the chiral dynamics of quantum chromodynamics (QCD). We study the structures of a_1 mesons determined using the truncated overlap fermion (TOF) formalism by Boriçi[1] based on the domain wall fermion formalism.[2,3] The TOF formalism is classified into lattice chiral fermions,[2–5] and exhibits good chiral symmetry. Our objective is to reveal the relationship between the nature of the a_1 meson associated with the chiral partner of the ρ meson and dynamical chiral symmetry breaking, as for the π and the chiral partner of the σ meson.[6] In the conventional constituent quark model, the σ meson and the a_1 meson are assigned excited states.

Lattice simulations of a_1 mesons have been previously conducted. Wingate *et al.* were the first to measure the mass of the a_1 meson using two-flavor lattice QCD.[7] Their result agrees with the experimental value for $a_1(1260)$. A decade later, a study of the a_1 meson was performed with the Lüscher–Weisz gauge action and the chirally improved Dirac operator in the quenched approximation.[8] Although the ground state of the a_1 meson was improved by using various interpolators including derivative quark sources in the simulations, the obtained mass of the ground state is close to $a_1(1420)$, instead of $a_1(1260)$. Moreover, the mass of the first excited state of a_1 was observed to be above 2 GeV.

Recently, Prelovsek *et al.* presented results for the mass of the a_1 meson and its coupling constant.[9] They performed a simulation for a full QCD lattice with clover-improved Wilson quarks. This work was continued in Ref. 10, in which they extracted the resonance mass of the ground state of the a_1 meson $m_{a_1} = 1.435(53)(^{+0}_{-109})$ GeV and the coupling $g_{a_1\pi\rho} = 1.71(39)$ GeV by simulating the corresponding scattering channel $\pi\rho$. Their obtained value of the a_1 meson mass is higher than the experimental result of $a_1(1260)$.[11,12] In our previous work, we investigated the mass of the ground state of the a_1 meson by a quenched lattice QCD using TOF.[13] Our result is consistent with the experimental value of $a_1(1260)$.

2. Truncated overlap fermions

The TOF[1] are defined by

$$D_{TOF} = \epsilon^\dagger P^\dagger D_{PV}^{-1} D_{DWF} P\epsilon \,, \tag{1}$$

$$\epsilon_{x_5} = \delta_{1,x_5} \,, \quad P_{x_5 y_5} = P_L \delta_{x_5,y_5} + P_R \delta_{x_5+1,y_5} + P_R \delta_{x_5,N_5}\delta_{y_5,1} \,, \tag{2}$$

where the five dimensional projection operator P is constructed from the four-dimensional projection operators $P_{R/L} = (1 \pm \gamma_5)/2$. The indexes x_5 and y_5 represent the fifth-dimensional lattice sites, which are defined in $x_5, y_5 \in [1, N_5]$. The domain wall fermion operator D_{DWF}[2,3] is defined by

$$\begin{aligned} D_{DWF\,x_5 y_5}(x,y) &= D_{WF}(x,y)\delta_{x_5,y_5} - (P_L\delta_{x_5+1,y_5} + P_R\delta_{x_5-1,y_5})\delta_{x,y} \\ &\quad + \delta_{x,y}\delta_{x_5,y_5} + m_f(P_R\delta_{x_5,1}\delta_{y_5,N_5} + P_L\delta_{x_5,N_5}\delta_{y_5,1})\delta_{x,y} \,, \end{aligned} \tag{3}$$

where D_{WF} is the Wilson fermion operator,

$$D_{WF}(x,y) = (4 - M_5)\delta_{x,y} - \frac{1}{2}\sum_{\mu=\pm 1}^{\pm 4}(1-\gamma_\mu)U_\mu(x)\delta_{x+\hat{\mu},y} \,. \tag{4}$$

The parameters of TOF are m_f and M_5, which correspond to the bare quark mass and the height of the domain wall, respectively. The Pauli–

Villars matrix D_{PV} is given by $D_{PV} = D_{DWF}(m_f = 1)$. In the $N_5 \to \infty$ limit, the lattice chiral symmetry is exactly reproduced in TOF.

3. Simulation setup and Lattice QCD results

In this work, we simulate the spectroscopy of a_1 mesons on $8^3 \times 24$ quenched lattice with the plaquette gauge action with $\beta = 5.7$. Gauge configurations are generated with the pseudo-heat-bath method. After 20000 thermalization iterations, we start to save gauge configurations every 1000 sweeps. The propagators of the π, ρ and a_1 mesons are calculated with TOF. The fermion parameters are set to $N_5 = 32$, $M_5 = 1.65$ and $m_f a = 0.04$–0.08.

We calculate the meson propagators and their effective masses (see Fig. 1). We estimate the statistical errors using the jackknife method. By performing a single- or double-pole fit to the effective masses, we obtain the meson masses as listed in Table 1.

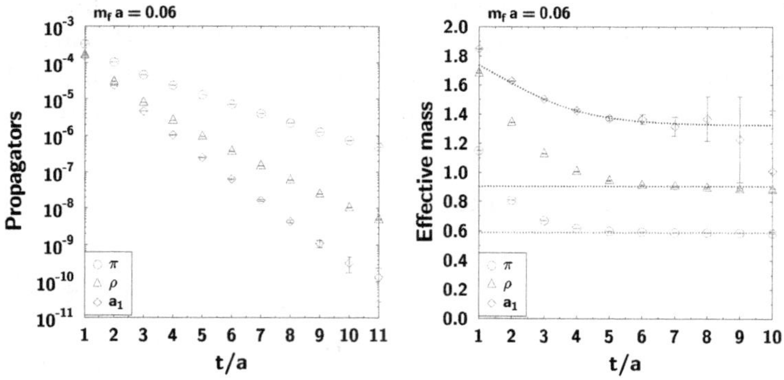

Fig. 1. Time dependences of the propagators (left) and the effective masses (right) for $m_f a = 0.06$. In the right figure, the dotted lines and dotted curve represent the single-pole and double-pole fitting results, respectively.

Table 1. Masses of the π and ρ mesons, mass ratio m_π/m_ρ and number of configurations for each quark mass.

$m_f a$	$m_\pi a$	$m_\rho a$	m_π/m_ρ	Confs.
0.08	0.6668(7)	0.9496(18)	0.702(2)	3000
0.07	0.6283(7)	0.9249(21)	0.679(2)	3000
0.06	0.5895(8)	0.9042(24)	0.652(3)	3000
0.05	0.5478(8)	0.8816(27)	0.621(3)	3600
0.04	0.5028(6)	0.8614(24)	0.584(2)	7864

Fig. 2 shows the quark mass dependence of the meson masses. We linearly extrapolate the meson masses to the chiral limit, $(m_\pi a)^2 = 0$. By tuning the ρ meson mass in the chiral limit to $m_\rho = 775$ MeV, we obtain a lattice spacing of $a = 0.1893(15)$ fm. Note that calculations for TOF have a residual mass such as $m_f a = -m_{\rm res} a$ in the chiral limit due to the finite N_5 effect. In our calculation with $N_5 = 32$, we obtain $m_{\rm res} a = 1.29(4) \times 10^{-2}$, that is $m_{\rm res} = 13.4(5)$ MeV, which is negligible. We estimate the masses of the ground state and first excited states of the a_1 mesons to be $m_{a_1} = 1158(42)$ MeV and $m_{a_1}^{\rm ext} = 1667(202)$ MeV. Our results are consistent with the experimental values of $a_1(1260)$ and $a_1(1640)$, $m_{a_1(1260)}$=1230(40) MeV and $m_{a_1(1640)}$=1654(19) MeV.[12]

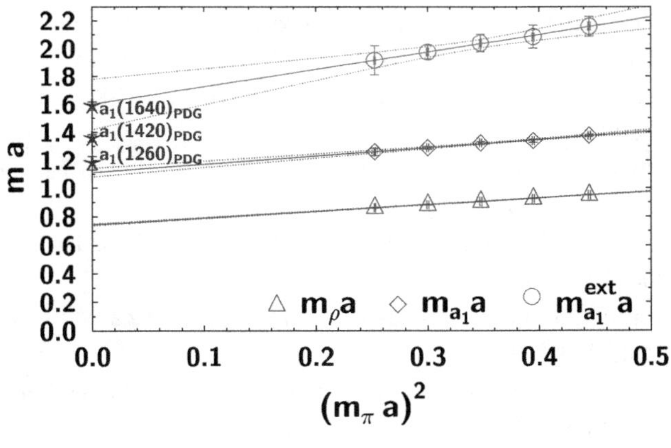

Fig. 2. Squared pion mass dependence of the meson masses.

4. Conclusion

We have investigated the masses of the ground and first excited states of a_1 mesons from a quenched lattice QCD with the TOF action. We have obtained the masses of the a_1 mesons to be 1158(42) MeV and 1667(202) MeV, which are in good agreement with the experimental values of $a_1(1260)$ and $a_1(1640)$. Since our simulation was performed in the quenched approximation and the $\bar{q}q$ source and sink, in which virtual intermediate states such as $\bar{q}\bar{q}qq$ states are highly suppressed, our results suggest that $a_1(1260)$ and $a_1(1640)$ are the simple $\bar{q}q$ states whereas $a_1(1420)$ may have a more complicated structure than the $\bar{q}q$ state.

Acknowledgments

This work was supported by RSF grant 15-12-20008 and the National Research Foundation of Korea (NRF) grant funded by the Korean government (MSIT) (No. 2018R1A5A1025563). The simulations were performed on the supercomputer system SX-ACE at RCNP and the Cybermedia Center, Osaka University, and were conducted using the Fujitsu PRIMEHPC FX10 System (Oakleaf-FX, Oakbridge-FX) at the Information Technology Center, The University of Tokyo. This work was supported by "Joint Usage/Research Center for Interdisciplinary Large-scale Information Infrastructures" in Japan (Project ID: jh180053-NAJ and jh190048-NAH).

References

1. A. Boriçi, Nucl. Phys. Proc. Suppl. **83**, 771 (2000) [hep-lat/9909057].
2. D. B. Kaplan, Phys. Lett. B **288**, 342 (1992) [hep-lat/9206013].
3. V. Furman and Y. Shamir, Nucl. Phys. B **439**, 54 (1995) [hep-lat/9405004].
4. R. Narayanan and H. Neuberger, Phys. Rev. Lett. **71**, no. 20, 3251 (1993) [hep-lat/9308011].
5. R. C. Brower, H. Neff and K. Orginos, Nucl. Phys. Proc. Suppl. **140**, 686 (2005) [hep-lat/0409118].
6. Y. Nambu and G. Jona-Lasinio, Phys. Rev. **122**, 345 (1961); Phys. Rev. **124**, 246 (1961).
7. M. Wingate, T. A. DeGrand, S. Collins and U. M. Heller, Phys. Rev. Lett. **74**, 4596 (1995) [hep-ph/9502274].
8. C. Gattringer, L. Y. Glozman, C. B. Lang, D. Mohler and S. Prelovsek, Phys. Rev. D **78**, 034501 (2008) [arXiv:0802.2020 [hep-lat]].
9. S. Prelovsek, C. B. Lang, D. Mohler and M. Vidmar, PoS LATTICE **2011**, 137 (2011) [arXiv:1111.0409 [hep-lat]].
10. C. B. Lang, L. Leskovec, D. Mohler and S. Prelovsek, JHEP **1404**, 162 (2014) [arXiv:1401.2088 [hep-lat]].
11. M. Alekseev *et al.* [COMPASS Collaboration], Phys. Rev. Lett. **104**, 241803 (2010) [arXiv:0910.5842 [hep-ex]].
12. M. Tanabashi *et al.* [Particle Data Group], Phys. Rev. D **98**, 030001 (2018).
13. M. Wakayama, Y. Murakami, S. Muroya, A. Nakamura, C. Nonaka, M. Sekiguchi and H. Wada, JPS Conf. Proc. **26**, 031007 (2019).

Constructing J/ψ family with updated data of charmonium-like Y states*

Takayuki Matsuki

Tokyo Kasei University, 1-18-1 Kaga, Itabashi, Tokyo 173-8602, Japan
Theoretical Research Division, Nishina Center, RIKEN, Wako, Saitama 351-0198, Japan
E-mail: matsuki@tokyo-kasei.ac.jp

Dian-Yong Chen

School of Physics, Southeast University, Nanjing 210094, China
E-mail: chendy@seu.edu.cn

Jun-Zhang Wang† and Xiang Liu‡

School of Physical Science and Technology, Lanzhou University, Lanzhou 730000, China
Research Center for Hadron and CSR Physics, Lanzhou University and Institute of Modern Physics of CAS, Lanzhou 730000, China
†*Email: wangjzh2012@lzu.edu.cn*
‡*E-mail: xiangliu@lzu.edu.cn*

Based on the updated data of charmonium-like state $Y(4220)$ reported in the hidden-charm channels of the e^+e^- annihilation, we propose a $4S$-$3D$ mixing scheme to categorize $Y(4220)$ into the J/ψ family, and also predict its charmonium partner $\psi(4380)$, whose evidence is found by analyzing the $e^+e^- \to \psi(3686)\pi^+\pi^-$ data from BESIII. In this work, we also introduce a $5S$-$4D$ mixing scheme, and obtain the information of the resonance parameters and partial open-charm decay widths of $\psi(4415)$, which do not contradict the present experimental data. Additionally, we predict a charmonium partner $\psi(4500)$ of $\psi(4415)$, which can be accessible at future experiments, especially, BESIII and BelleII.

1. Charmonium spectrum

In Fig. 1, we collect the corresponding information of the observed charmonia with the year for their first discoveries. It is obvious that the year 1978 is an important time point since most of charmonia are listed in the latest PDG were announced.[2]

*This talk is based on the paper Ref. 1 recently published.

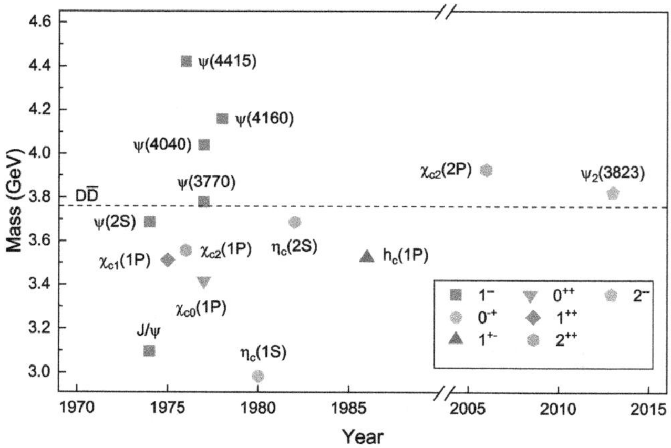

Fig. 1. The observed charmonia with the corresponding first observed year 2. Here, the $D\bar{D}$ threshold is also given.

Our scenario of this topic is as follows: 1) First, we calculate the charmonium spectrum. 2) Then, we approximately identify $Y(4220)$ in the calculated table. 3) Next, we use the idea of S-D mixing[3] so that we can adjust a mixing angle to obtain physical mass close to 4220 MeV, 4) and we also predict its partner as $\psi(4380)$. 5) The similar procedure is applied to the state $\psi(4415)$ and predict its partner as $\psi(4500)$.

In Fig. 2, we list the resonance parameters of 4.2 GeV structures in $e^+e^- \to h_c\pi^+\pi^-$, $e^+e^- \to J/\psi\pi^+\pi^-$, $e^+e^- \to \psi(3686)\pi^+\pi^-$, and $e^+e^- \to \chi_{c0}\omega$. We collectively refer this structure to as $Y(4220)$. Using the modified Godfrey-Isgur potential model, we calculate the charmonium spectrum listed in Table 1, some of which are omitted for brevity. We identify $Y(4220)$ in this Table as $\psi(4^3S_1)$ with mass 4274 MeV.

2. *S*-*D* mixing

We introduce the following equation,

$$\begin{pmatrix} |\psi'_{4S-3D}\rangle \\ |\psi''_{4S-3D}\rangle \end{pmatrix} = \begin{pmatrix} \cos\theta & \sin\theta \\ -\sin\theta & \cos\theta \end{pmatrix} \begin{pmatrix} |4^3S_1\rangle \\ |3^3D_1\rangle \end{pmatrix} \tag{1}$$

to describe the $4S$-$3D$ mixing. Here, θ denotes the mixing angle. The mass eigenvalues of ψ'_{4S-3D} and ψ''_{4S-3D}, i.e., physical masses, are determined

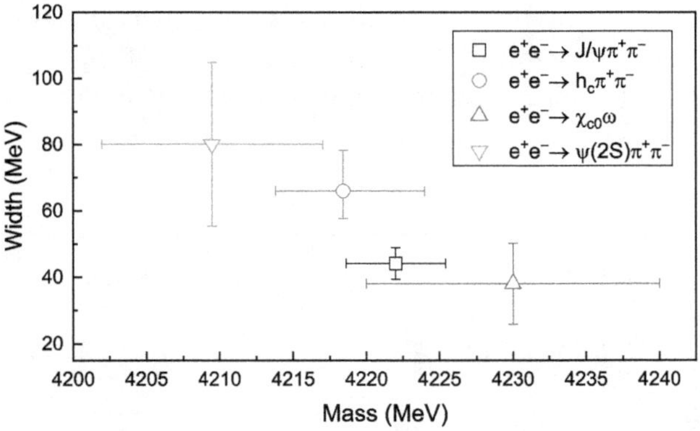

Fig. 2. The resonance parameters of $Y(4220)$ in hidden-charm processes $e^+e^- \to J/\psi\pi^+\pi^-$, $e^+e^- \to h_c\pi^+\pi^-$, $e^+e^- \to \chi_{c0}\omega$, and $e^+e^- \to \psi(3686)\pi^+\pi^-$ measured at BESIII.

Table 1. The charmonium mass spectrum calculated by the unquenched potential model. $Y(4220)$ is identified as $\psi(4^3S_1)$ in our calculation. All results are in units of MeV. Here, we also list the experimental data for comparison with the theoretical results.

State	Mass	Expt.[2]	State	Mass	Expt.[2]
$\eta_c(1^1S_0)$	2981	2983.9±0.5	$\psi(1^3D_1)$	3830	3778.1±1.2
$\psi(1^3S_1)$	3096	3096.9±0.006	$\psi_2(1^3D_2)$	3848	3822.2±1.2
$\eta_c(2^1S_0)$	3642	3637.6±1.2	$\psi_3(1^3D_3)$	3859	⋯
$\psi(2^3S_1)$	3683	3686.097±0.01	$\eta_{c2}(2^1D_2)$	4137	⋯
$\eta_c(3^1S_0)$	4013	⋯	$\psi(2^3D_1)$	4125	4159±20
$\psi(3^3S_1)$	4035	4039±1	$\psi_2(2^3D_2)$	4137	⋯
$\eta_c(4^1S_0)$	4260	⋯	$\psi_3(2^3D_3)$	4144	⋯
$\psi(4^3S_1)$	4274	4230±8	$\eta_{c2}(3^1D_2)$	4343	⋯
$\eta_c(5^1S_0)$	4433	⋯	$\psi(3^3D_1)$	4334	⋯
$\psi(5^3S_1)$	4443	⋯	$\psi_2(3^3D_2)$	4343	⋯
$h_c(1^1P_1)$	3538	3525.38±0.11	$\psi_3(3^3D_3)$	4348	⋯
$\chi_{c0}(1^3P_0)$	3464	3414.71±0.3	$\eta_{c2}(4^1D_2)$	4490	⋯
$\chi_{c1}(1^3P_1)$	3530	3510.67±0.05	$\psi(4^3D_1)$	4484	⋯
$\chi_{c2}(1^3P_2)$	3571	3556.17±0.07	$\psi_2(4^3D_2)$	4490	⋯
$h_c(2^1P_1)$	3933	⋯	$\psi_3(4^3D_3)$	4494	⋯
$\chi_{c0}(2^3P_0)$	3896	3918.4±1.9	$h_{c3}(1^1F_3)$	4074	⋯
$\chi_{c1}(2^3P_1)$	3929	-	$\chi_{c2}(1^3F_2)$	4070	⋯
$\chi_{c2}(2^3P_2)$	3952	3927.2±2.6	$\chi_{c3}(1^3F_3)$	4075	⋯

by two basis masses m_{4S}, m_{3D} in Table 1 and the mixing angle θ as,

$$m^2_{\psi'_{4S-3D}} = \frac{1}{2}\left(m^2_{4S} + m^2_{3D} - \sqrt{(m^2_{3D} - m^2_{4S})^2 \sec^2 2\theta}\right), \tag{2}$$

$$m^2_{\psi''_{4S-3D}} = \frac{1}{2}\left(m^2_{4S} + m^2_{3D} + \sqrt{(m^2_{3D} - m^2_{4S})^2 \sec^2 2\theta}\right). \tag{3}$$

Fitting parameters in these equations to experimental data, we obtain ranges of each parameter. The central mass value of ψ''_{4S-3D} is given by 4380 MeV and the range of mixing angle is given by $\theta = \pm(30^\circ \sim 36^\circ)$, which can be explained by Fig. 3.

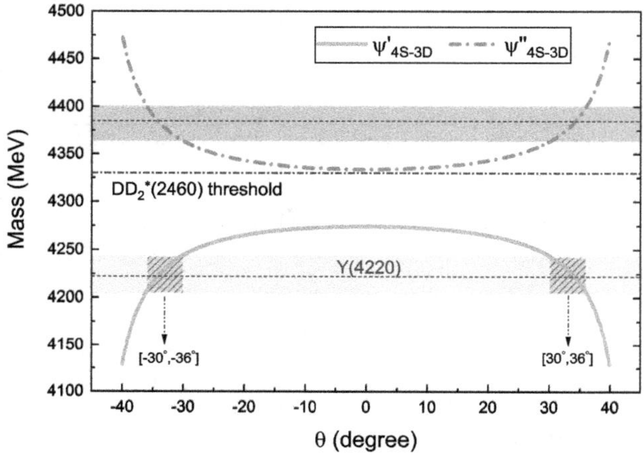

Fig. 3. The masses of ψ'_{4S-3D} and ψ''_{4S-3D} with dependence on the $4S$-$3D$ mixing angle θ. The yellow and cyan bands denote the measured mass range of $Y(4220)$ and the predicted mass range of $\psi(4380)$, respectively, where the dashed horizontal lines in bands correspond to each average value. The two shaded regions represent the mixing angle interval, in which the theoretical results of ψ'_{4S-3D} meet the measurements of $Y(4220)$.

A very strong support for existence of $\psi(4380)$ is given by fitting the experimental data of $e^+e^- \to \psi(2S)\pi^+\pi^-$ in Fig. 4. The $3R$ fit is obtained by including $\psi(4160)$, $Y(4220)$, and $\psi(4415)$, while the best fit of the $4R$ is obtained by including $\psi(4380)$ (mixing partner of $Y(4220)$) together with these states.

3. ψ(4415) and its partner

Similar to the discussion on $Y(4220)$, we identify $\psi(4415)$ as $\psi(5S)$ in Table 1.

The mass spectrum in Table 1 shows that the mass of $\psi(4415)$ is close to that of $\psi(5S)$. Thus, we propose a $5S$-$4D$ mixing scheme to study $\psi(4415)$, which is the same idea as for $Y(4220)$. To depict this mixing scheme, we

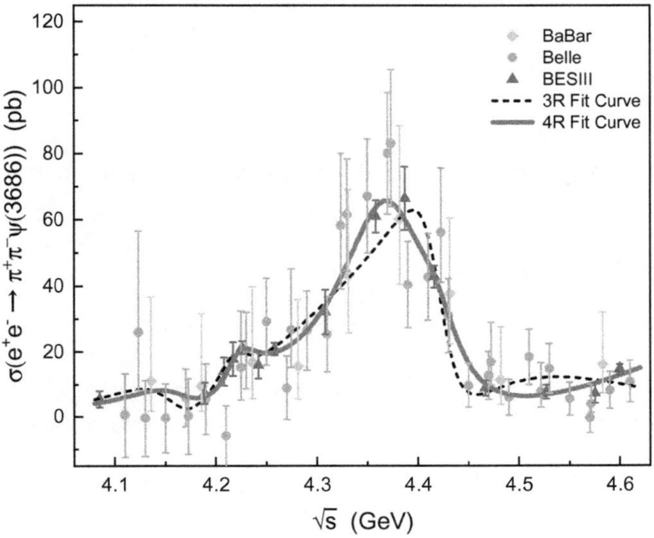

Fig. 4. The fit to the cross section for the $e^+e^- \to \psi(2S)\pi^+\pi^-$ reaction in the Fano-like interference picture under $3R$ and $4R$ fit schemes. Here, the data of BaBar, Belle, and the latest BESIII are included.

have an expression

$$\begin{pmatrix} |\psi'_{5S-4D}\rangle \\ |\psi''_{5S-4D}\rangle \end{pmatrix} = \begin{pmatrix} \cos\phi & \sin\phi \\ -\sin\phi & \cos\phi \end{pmatrix} \begin{pmatrix} |5^3S_1\rangle \\ |4^3D_1\rangle \end{pmatrix}, \tag{4}$$

where ϕ is a mixing angle. The masses of $\psi(5S)$ and $\psi(4D)$ are taken from our calculations listed in Table 1. The dependence of the masses of ψ'_{5S-4D} and ψ''_{5S-4D} on ϕ determines the central value of the mixing angle and mass of ψ''_{5S-4D} to be $\phi = \pm 30^\circ$ and 4503 MeV, respectively. Hence, we name ψ''_{5S-4D} as $\psi(4500)$.

References

1. J. Z. Wang, D. Y. Chen, X. Liu and T. Matsuki, Phys. Rev. D **99**, no. 11, 114003 (2019).
2. M. Tanabashi *et al.* [Particle Data Group], Phys. Rev. D **98**, no. 3, 030001 (2018).
3. J. L. Rosner, Phys. Rev. D **64**, 094002 (2001).

Investigation of $\phi(2170)$ in a hybrid meson scenario

Wei Chen*
School of Physics, Sun Yat-Sen University
Guangzhou 510275, China
E-mail: chenwei29@mail.sysu.edu.cn

J. Ho, R. Berg and T. G. Steele
Department of Physics and Engineering Physics, University of Saskatchewan
Saskatoon, SK, S7N 5E2, Canada

D. Harnett
Department of Physics, University of the Fraser Valley
Abbotsford, BC, V2S 7M8, Canada

We study the $\phi(2170)$ state as a vector $\bar{s}gs$ strangeonium hybrid meson with $J^{PC} = 1^{--}$ by using the QCD Gaussian sum-rules method. In a two-resonance model, the relative resonance strength of the $\phi(2170)$ state was found to be less than 5% that of a heavier 2.9 GeV state, which disfavors the hybrid interpretation of the $\phi(2170)$.

Keywords: Hybrid meson; QCD sum rules.

1. Introduction

The $\phi(2170)$ resonance was first observed in 2006 by the BaBar Collaboration in the $\phi(1020)f_0(980)$ invariant mass spectrum via the ISR process,[1] and confirmed later by BaBar,[2–4] BES,[5] and Belle[6] and BESIII.[7,8] Its mass and decay width are collected as $M = (2188 \pm 10)\,\text{MeV}$ and $\Gamma = (83 \pm 12)\,\text{MeV}$ in PDG.[9]

To date, many theoretical configurations have been studied to explain the nature of $\phi(2170)$, such as a conventional strange quarkonium $3\,^3S_1$ or $2\,^3D_1$ $s\bar{s}$ state, a $qs\bar{q}\bar{s}$ or $ss\bar{s}\bar{s}$ tetraquark state, a $\Lambda\bar{\Lambda}$ molecular state, or a $\bar{s}gs$ strangeonium hybrid meson. However, none of these interpretations has been concluded since the absence of experimental information, especially the decay modes and rates of $\phi(2170)$. In this study, we shall investigate

*Speaker

the $\phi(2170)$ as a strangeonium hybrid meson in the method of QCD sum rules. In a previous QCD Laplace sum rules (LSRs) study,[10] the authors found a much heavier vector strangeonium hybrid mass of (2.9 ± 0.3) GeV than the mass of $\phi(2170)$. Here we shall extend the previous calculation by including the 5d and 6d condensate contributions to the correlation function in a method of QCD Gaussian sum rules (GSRs).[11] We probe the relative coupling to the hybrid current of the $\phi(2170)$ resonance via a double-narrow resonance model for the hadronic spectral function.

2. Formalism of QCD Gaussian sum rules

We start from the two-point correlation function

$$\Pi(q^2) = \frac{i}{D-1}\left(\frac{q_\mu q_\nu}{q^2} - g_{\mu\nu}\right)\int d^D x\, e^{iq\cdot x}\langle\Omega|\tau j_\mu(x) j_\nu(0)|\Omega\rangle, \tag{1}$$

where the strangeonium hybrid meson current j_μ is given as

$$j_\mu = \frac{g_s}{2}\bar{s}\gamma^\rho\gamma_5\lambda^a\widetilde{G}^a_{\mu\rho}s,\ \widetilde{G}^a_{\mu\rho} = \frac{1}{2}\epsilon_{\mu\rho\omega\zeta}G^a_{\omega\zeta}. \tag{2}$$

We calculate the correlator up to dimension-six condensate via the operator product expansion(OPE) method

$$\Pi^{\rm QCD}(q^2) = \left[\frac{\alpha_s}{\pi}\left(-\frac{q^6}{240\pi^2} + \frac{5m_s^2q^4}{48\pi^2} - \frac{4q^2}{9}\langle m_s\bar{s}s\rangle\right) + \frac{q^2}{36\pi}\langle\alpha G^2\rangle \right.$$
$$\left. + \frac{19\alpha_s m_s}{72\pi}\langle g\bar{s}\sigma Gs\rangle\right]\log\left(\frac{-q^2}{\mu^2}\right). \tag{3}$$

At hadron side, $\Pi(q^2)$ can be written as a dispersion relation

$$\Pi(Q^2) = \frac{Q^8}{\pi}\int_{t_0}^{\infty}\frac{{\rm Im}\Pi(t)}{t^4(t+Q^2)}\,{\rm d}t + \cdots \tag{4}$$

where t_0 is a hadron production threshold and $\cdots$ represents subtraction constants. The hadronic spectral function can be described using a resonance-plus-continuum decomposition,

$$\frac{1}{\pi}{\rm Im}\Pi(t) = \rho^{\rm had}(t) + \theta(t-s_0)\frac{1}{\pi}{\rm Im}\Pi^{\rm QCD}(t), \tag{5}$$

where $\rho^{\rm had}(t)$ represents the resonance contribution to ${\rm Im}\Pi(t)$ and $\theta(t-s_0)$ is a Heaviside step function shifted to the continuum threshold s_0.

Different from the Laplace sum rules, here we define the Gaussian sum rules to enhance the lowest-lying resonance contributions

$$G(\hat{s},\,\tau)=\sqrt{\frac{\tau}{\pi}}\lim_{\substack{N,\Delta^2\to\infty\\ \tau=\Delta^2/(4N)}}\frac{\left(-\,\Delta^2\right)^N}{\Gamma(N)}\left(\frac{d}{d\Delta^2}\right)^N\left\{\frac{\Pi(-\hat{s}-i\Delta)-\Pi(-\hat{s}+i\Delta)}{i\Delta}\right\}. \tag{6}$$

Substituting Eq. (5) into Eq. (4) and subtracting the continuum contribution, we finally obtain the GSRs

$$G^{\rm QCD}(\hat{s},\,\tau,\,s_0)\equiv\frac{1}{\sqrt{4\pi\tau}}\int_0^{s_0}e^{-\frac{(\hat{s}-t)^2}{4\tau}}\frac{1}{\pi}{\rm Im}\Pi^{\rm QCD}(t)\,{\rm d}t \tag{7}$$

$$=\frac{1}{\sqrt{4\pi\tau}}\int_0^{s_0}e^{-\frac{(\hat{s}-t)^2}{4\tau}}\left(\frac{\alpha_s}{\pi}\left(-\frac{t^3}{240\pi^2}+\frac{5m_s^2t^2}{48\pi^2}-\frac{4t}{9}\langle m_s\bar{s}s\rangle\right)\right.$$

$$\left.+\frac{t}{36\pi}\langle\alpha G^2\rangle+\frac{19\alpha_s m_s}{72\pi}\langle g\bar{s}\sigma Gs\rangle\right){\rm d}t. \tag{8}$$

Note that the kernel of the subtracted GSRs in Eq. (7) is a Gaussian of width $\sqrt{2\tau}$ centred at $\hat{s}$, which is particularly well-suited to study the ground state and excited resonances with similar sensitivity. We then define the normalized GSRs (NGSRs) to isolate the information in the GSRs that is independent of the FESR constraint

$$N^{\rm QCD}(\hat{s},\,\tau,\,s_0)\equiv\frac{G^{\rm QCD}(\hat{s},\,\tau,\,s_0)}{\int_{-\infty}^{\infty}G^{\rm QCD}(\hat{s},\,\tau,\,s_0)\,{\rm d}\hat{s}}$$

$$=\frac{\frac{1}{\sqrt{4\pi\tau}}\int_{t_0}^{\infty}e^{-\frac{(\hat{s}-t)^2}{4\tau}}\rho^{\rm had}(t)\,{\rm d}t}{\int_{t_0}^{\infty}\rho^{\rm had}(t)\,{\rm d}t}. \tag{9}$$

3. Numerical analysis

Assuming the $\phi(2170)$ to be a strangeonium hybrid meson ($m_1 = m_{\phi(2170)}$)), we use a double-narrow resonance model for the hadronic spectral function

$$\rho^{\rm had}(t)=f_1^2\delta\left(t-m_1^2\right)+f_2^2\delta\left(t-m_2^2\right)\,, \tag{10}$$

which provides the hadronic contribution to the NGSRs in Eq. (9),

$$N^{\rm had}\left(\hat{s},\,\tau\right)=\frac{1}{\sqrt{4\pi\tau}}\left(re^{-\frac{(\hat{s}-m_1^2)^2}{4\tau}}+(1-r)e^{-\frac{(\hat{s}-m_2^2)^2}{4\tau}}\right), \tag{11}$$

where the normalized couplings are defined as

$$r = \frac{f_1^2}{f_1^2 + f_2^2}, 1 - r = \frac{f_2^2}{f_1^2 + f_2^2}, 0 \le r \le 1. \tag{12}$$

The value of the normalized coupling r provides the possibility of the $\phi(2170)$ to be a strangeonium hybrid meson.

To determine the unknown quantities $\{m_2, r, s_0\}$, we seek for the best fit of the $\hat{s}$ dependence of the QCD prediction and hadronic model for the NGSRs in the region $-10\,\text{GeV}^2 < \hat{s} < 30\,\text{GeV}^2$, as shown in Fig. 1. This fitted region encloses the regions where the NGSRs are numerically significant. The predictions of the quantities are given as

$$s_0^{\text{opt}} = 9.7 \pm 1.0\,\text{GeV}^2 \tag{13}$$

$$m_2 = m_{\text{fit}} = 2.90 \pm 0.16\,\text{GeV} \tag{14}$$

$$r \le 0.033\,. \tag{15}$$

The small relative resonance coupling $r \le 3.3\%$ seems to preclude the possibility of the $\phi(2170)$ as a strangeonium hybrid meson. We can also consider the s_0 dependence of r in Fig. 2. Within the chosen region of s_0, a more conservative bound is obtained as $r \le 5\%$.

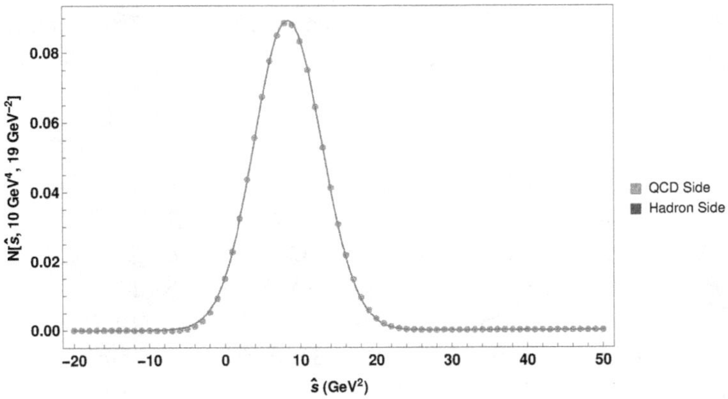

Fig. 1. Fit of the $\hat{s}$ dependence of the QCD prediction and hadronic model for the NGSRs.

4. Summary

In summary, we investigate the possibility of the $\phi(2170)$ as a strangeonium hybrid meson by using the QCD Gaussian sum rules method. In

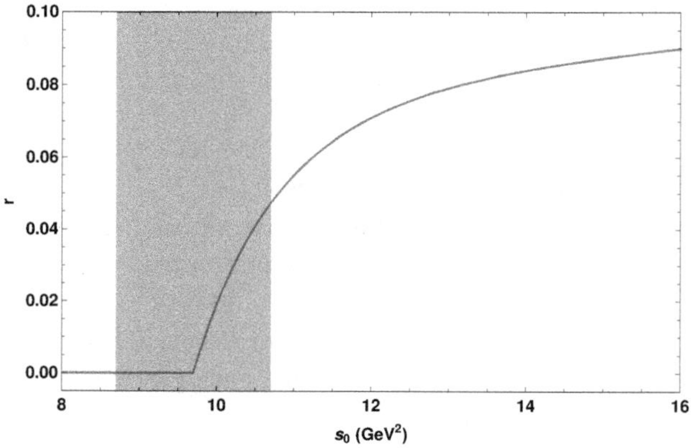

Fig. 2. The s_0 dependence of the relative resonance coupling r. The vertical band highlights the optimized value of continuum onset s_0^{opt}.

our calculation, we include the 5d and 6d condensate contributions to the correlation function. We adopt a double-narrow resonance model for the hadronic spectral function to probe the relative couplings to the hybrid current of the $\phi(2170)$ resonance and an additional unknown resonance. For excellent fitting between the QCD prediction and hadronic model for NGSRs, we find a small relative resonance coupling $r \leq 5\%$. Such a result does not support the strangeonium hybrid interpretation for the $\phi(2170)$.

References

1. B. Aubert et al. (BaBar), Phys. Rev. **D74**, 091103 (2006).
2. B. Aubert et al. (BaBar), Phys. Rev. **D76**, 012008 (2007).
3. B. Aubert et al. (BaBar), Phys. Rev. **D77**, 092002 (2008).
4. J. P. Lees et al. (BaBar), Phys. Rev. **D86**, 012008 (2012).
5. M. Ablikim et al. (BES), Phys. Rev. Lett. **100**, 102003 (2008).
6. C. P. Shen et al. (Belle), Phys. Rev. **D80**, 031101 (2009).
7. M. Ablikim et al. (BESIII), Phys. Rev. **D91**, 052017 (2015).
8. M. Ablikim et al. (BESIII) (2017), arXiv:1709.04323.
9. M. Tanabashi et al. (Particle Data Group), Phys. Rev. **D98**, 030001 (2018).
10. J. Govaerts, L. J. Reinders, and J. Weyers, Nucl. Phys. **B262**, 575 (1985).
11. J. Ho, R. Berg, T. G. Steele, W. Chen, and D. Harnett, Phys. Rev. **D100**, 034012 (2019).

Production of X resonances in B_c^- decays

N. Ikeno

Department of Agricultural, Life and Environmental Sciences, Tottori University, Tottori 680-8551, Japan,
Departamento de Física Teórica and IFIC, Centro Mixto Universidad de Valencia-CSIC Institutos de Investigación de Paterna, Aptdo.22085, 46071 Valencia, Spain
E-mail: ikeno@tottori-u.ac.jp

M. Bayar

Department of Physics, Kocaeli University, Izmit 41380, Turkey
E-mail: melahat.bayar@kocaeli.edu.tr

E. Oset

Departamento de Física Teórica and IFIC, Centro Mixto Universidad de Valencia-CSIC Institutos de Investigación de Paterna, Aptdo.22085, 46071 Valencia, Spain
E-mail: oset@ific.uv.es

We report the results for the study of the semileptonic decay of B_c^- meson into $\bar{\nu}\ell^-$ and the $X(3930)$, $X(3940)$, $X(4160)$ resonances. In the reaction, these resonaces appear as dynamically generated from the vector-vector interaction in the charm sector, and couple strongly to $D^*\bar{D}^*$ and $D_s^*\bar{D}_s^*$. We see the $B_c^- \rightarrow \bar{\nu}_l\ell^- D^*\bar{D}^*$ and $B_c^- \rightarrow \bar{\nu}_l\ell^- D_s^*\bar{D}_s^*$ reactions close to threshold and relate the $D^*\bar{D}^*$ and $D_s^*\bar{D}_s^*$ mass distribution to the rate of production of the X resonances.

Keywords: X resonances; semileptonic decays.

1. Introduction

The X, Y, Z states, that challenge the constituent quark model picture of mesons have been one of the most spectacular findings in hadron spectroscopy recently.[1] Tetraquark pictures have been proposed as well as molecular pictures stemming from the interaction of more elementary mesons.[2]

One of these pictures deals with the interaction of vector mesons with charm, leading to hidden charm quasibound meson states.[2] Some quasibound states were found which could be associated to known resonances.

One state was around 3943 MeV with $I^G[J^{PC}] = 0^+[0^{++}]$, which was associated to the $X(3940)$. Another state was around 3922 MeV with $0^+[2^{++}]$, which was associated to the $X(3930)$. And a third one was at 4169 MeV with $0^+[2^{++}]$, which was associated to the $X(4160)$. In Ref. 2, the $X(3940)$ was found to couple mostly to $D^*\bar{D}^*$, the $X(3930)$ also had its strongest coupling to $D^*\bar{D}^*$, and the $X(4160)$ had its strongest coupling to $D_s^*\bar{D}_s^*$.

Here, in order to see the nature of X resonances, we investigate the semileptonic decay of $B_c^- \to \bar{\nu}_l \ell^- X_i$, with X_i any of the three resonances, $X(3930)$, $X(3940)$, $X(4160)$ in Ref. 3. The use of semileptonic weak decays aiming at determining the structure of resonances has been already studied in different cases.[4] In our approach, these X resonances are generated from the interaction of vector meson ($D^*\bar{D}^*$, $D_s^*\bar{D}^*{}_s$). We calculate the $B_c \to \nu_l \ell^+ X_i$ decay rates and compare them with the $B_c \to \nu_l \ell^+ D^*\bar{D}^*$, $\nu_l \ell^+ D_s^*\bar{D}_s^*$ decays.

2. Formalism

We use the formalism of the semileptonic B decays in Ref. 4. To produce X states in the semileptonic B_c^- decays, we consider three steps. The $B_c^- \to \bar{\nu} e^- X$ process proceeds at the quark level through a first step shown in Fig. 1. The process involves the bc weak transition. Next, to obtain two mesons, the $c\bar{c}$ quarks of Fig. 1 must hadronize into two mesons components. One introduces an extra $\bar{q}q$ pair with vacuum quantum number, $\bar{u}u + \bar{d}d + \bar{s}s + \bar{c}c$, and then the two quarks after the weak process participate in the formation of the two mesons. With two quarks after the weak vertex, as in Fig. 2, the new $\bar{q}q$ pair can be placed in between these $c\bar{c}$ quarks. Then, we obtain the vector mesons of $D^*\bar{D}^*$ and $D_s^*\bar{D}_s^*$. Finally, these mesons are allowed to undergo final state interaction from where three resonances appear. The final state interaction of $D^*\bar{D}^*$ is depicted in Fig. 3, and we use the interaction of Ref. 2 where, using an extension of the local hidden gauge approach, the interaction of $D^*\bar{D}^*$ generates several resonances, and some X states were dynamically generated.

Here, we study the coalescence process which produces the resonances after rescattering. This process has a three-body final state with a lepton, its neutrino and the resonance R. The resonance R stands for the some X resonances. The hadronization factor V_{had} can be obtained as

$$V_{\text{had}} = C(\sqrt{2}\, G_{D^*\bar{D}^*}\, g_{R,D^*\bar{D}^*} + G_{D_s^*\bar{D}_s^*}\, g_{R,D_s^*\bar{D}_s^*}), \tag{1}$$

for the resonance R in $J = 0$ which requires $L = 0$. For the constant C, we use the same value of the semileptonic B decays as established in Ref.

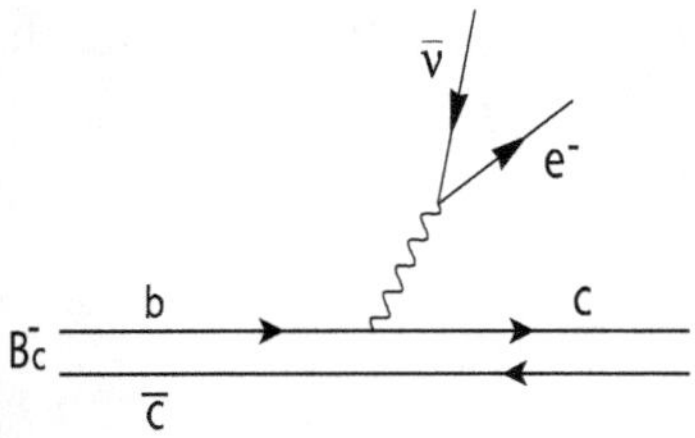

Fig. 1. Diagrammatic representation of the quark level for $B_c \to \nu_e e^- (c\bar{c})$.

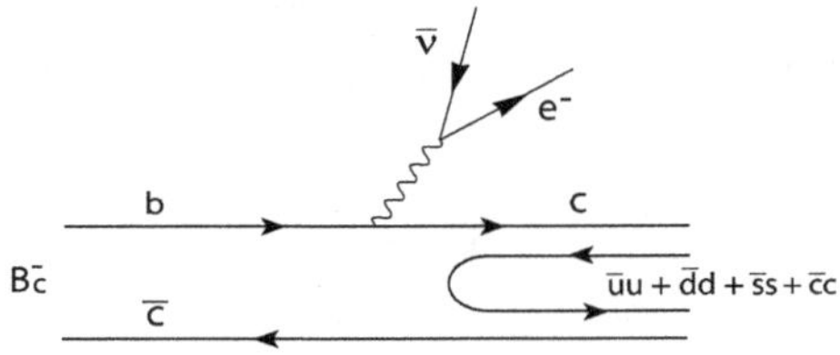

Fig. 2. Dominant mechanism for the hadronization into two mesons of the $c\bar{c}$ state after the weak process.

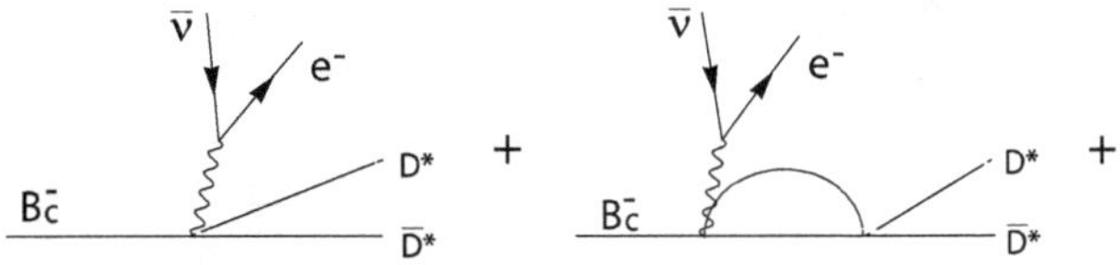

Fig. 3. Diagrams involved in the final state interaction of the primary $D^*\bar{D}^*$ mesons, (a) tree level, (b) rescattering.

4. $G_{D^*\bar{D}^*}$ and $G_{D_s^*\bar{D}_s^*}$ are the two meson loop functions, and $g_{R,D^*\bar{D}^*}$ and $g_{R,D_s^*\bar{D}_s^*}$ are the couplings of the resonance to these channels. We use the values reported in Ref. 2.

Finally we obtain a formula of the decay widths Γ_{coal} for the coalescence of the X resonances as

$$\Gamma_{\text{coal}} = \frac{|G_{\text{F}} V_{bc} V_{\text{had}}|^2}{8\pi^3 m_{B_c}^3 m_R} \int dM_{\text{inv}}^{(\nu e)} P_R^{\text{cm}} \tilde{p}_\nu |M_{\text{inv}}^{(\nu e)}|^2 \left(\tilde{E}_{B_c} \tilde{E}_R - \frac{\tilde{p}_{B_c}^2}{3} \right), \quad (2)$$

where P_R^{cm} is the momentum of the resonance R in the B_c rest frame, and $\tilde{p}_\nu$ is the momentum of the neutrino in the νe rest frame. The energies $\tilde{E}_{B_c}$ and $\tilde{E}_R$ are calculated in the νe rest frame, and $\tilde{p}_{B_c}$ is the momentum of the B_c in the νe rest frame. The integral range of $M_{\text{inv}}^{(\nu e)}$ is $[m_e + m_\nu, m_{B_c} - m_R]$. We take the Fermi coupling constant $G_{\text{F}} = 1.166 \times 10^{-5}$ GeV^{-2} and the Cabibbo-Kobayashi-Maskawa matrix element $V_{bc} = 0.0411$. Detailed calculations are shown in Ref. 3.

3. Numerical results

First, we show the results of the coalescence process of $X(3934)$ resonance. We consider $X(3934)$ as the resonance R, namely the $B_c^- \to X(3934)\bar{\nu}l^-$ process. In Fig. 4, we plot the integrand of Eq. (2). We evaluate the decay widths $\Gamma_{\rm coal}$ in Eq. (2) for the resonance $L = 0$ and $J = 0$ state using the mass $m_R = 3943$ MeV, and find

$$\Gamma_{\rm coal} = 2.6 \times 10^{-14}\ \text{MeV}. \tag{3}$$

The mean life of the B_c is 0.507×10^{-12} s, and then the branching ratio is evaluated as

$$\frac{\Gamma_{\rm coal}}{\Gamma_{\rm tot}(B_c)} = 2.0 \times 10^{-5}. \tag{4}$$

This value seems to be small. However, it is within measuring range, since boundaries of 10^{-6} for certain branching ratios have been established.

For the other resonance of 2^{++}, we can not use the formula of the decay widths in Eq. (2) because we need an $L = 2$ state and hence the matrix element would be different. Thus, we calculated the ratio of the two $L = 2$. We consider two options. In the first one the ratio is evaluated as,

$$\frac{\Gamma_{\rm coal}(2)}{\Gamma_{\rm coal}(2')} = \frac{|V_{\rm had}(2)|^2/m_R(2)}{|V_{\rm had}(2')|^2/m_R(2')} = 1.6 \tag{5}$$

where (2) indicates the $X(3930)$ resonance state of the mass $m_R = 3922$ MeV, and $(2)'$ the $X(4160)$ resonance state of the $m_R = 4169$ MeV. This is only a rough approximation that does not take into account the different masses and phase space, nor the $L = 2$ character of the interaction.

For the precise evaluations, we consider that the integral in Eq. (2) would have a different form to account for $L = 2$,

$$\int dM_{\rm inv}^{(\nu e)} (P_R^{\rm cm})^5 \tilde{p}_\nu |M_{\rm inv}^{(\nu e)}|^2 \left(\tilde{E}_{B_c} \tilde{E}_R - \frac{\tilde{p}_{B_c}^2}{3} \right). \tag{6}$$

By replacing the integral in Eq. (2) by Eq. (6), the other ratio is evaluated as

$$\frac{\Gamma_{\rm coal}(2)}{\Gamma_{\rm coal}(2')} = 3.6. \tag{7}$$

From this ratio, the $X(3930)$ has bigger production rate. The probability to produce the $X(3930)$ is about 3 times larger than for the $X(4160)$. We also show the integrand of Eq. (6) in Fig. 5 for the two tensor resonances.

4. Summary

We have studied the semileptonic decay of B_c in the reaction $B_c^- \to \bar{\nu}_e e^- X$, with X any of the resonances $X(3930)$ (2^{++}), $X(3940)$ (0^{++}) and $X(4160)$ (2^{++}). In our approach, we have treated these X resonances as dynamically generated from the vector-vector interaction ($D^* \bar{D}^*$,

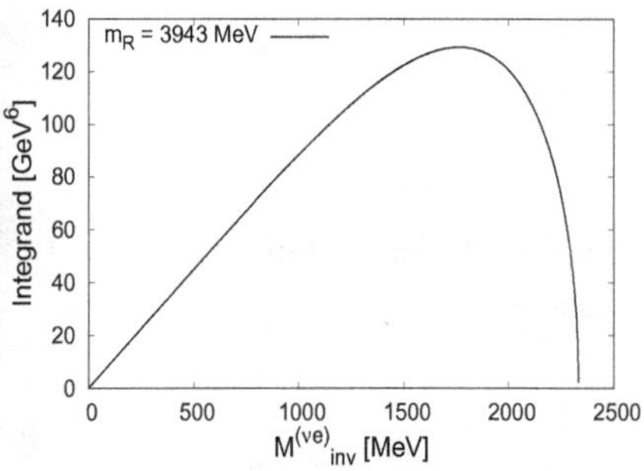

Fig. 4. The integrand of the integral that appears in Eq. (2) as a function $M_{\rm inv}^{(\nu e)}$ for the resonance mass $X(3940)$ of $M_R = 3943$ MeV.

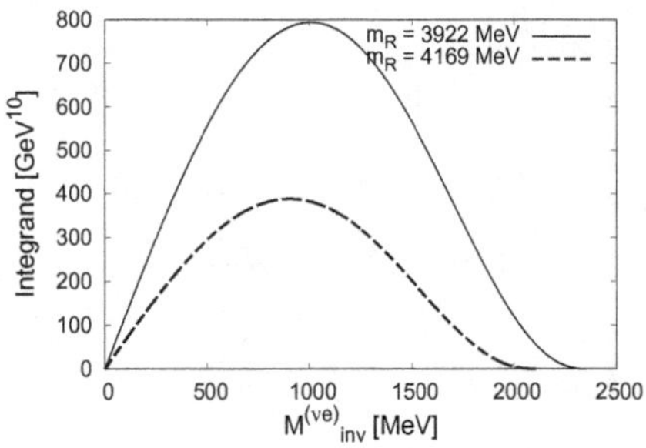

Fig. 5. The integrands of the integral that appears in Eq. (6) as a function $M_{\rm inv}^{(\nu e)}$ for the resonance mass $X(3930)$ of $M_R = 3922$ MeV and the resonance mass $X(4160)$ of $M_R = 4169$ MeV.

$D_s^*\bar{D}_s^*$) in the charm sector. We have made an estimation of the rate of $B_c^- \to \bar{\nu}_e e^- X(3940)$. For the production of the two tensor states we can not obtain the absolute rate of production, but we can obtain the ratio for the $X(3930)$ and $X(4160)$ states. In Ref. 3, we have also looked at the production of $D^*\bar{D}^*$ and $D_s^*\bar{D}_s^*$ close to threshold and we can make predictions of the ratio of this differential mass distribution to the rate of resonance production, which are tied to the nature of these resonances as dynamically generated from the vector-vector interaction in the charm sector. As more decay modes of B_c become available, it would be interesting to look into these modes which will provide good information on the nature of these X resonances.

Acknowledgment

N. I. acknowledges the supports from JSPS Overseas Research Fellowships and JSPS KAKENHI Grant Number JP19K14709.

References

1. H. X. Chen, W. Chen, X. Liu and S. L. Zhu, Phys. Rept. **639**, 1 (2016).
2. R. Molina and E. Oset, Phys. Rev. D **80**, 114013 (2009).
3. N. Ikeno, M. Bayar, E. Oset, Eur. Phys. J. C **78**, 429 (2018).
4. F. S. Navarra, M. Nielsen, E. Oset and T. Sekihara, Phys. Rev. D **92**, 014031 (2015).

The role of $X(4140)$ and $X(4160)$ in the reactions of $B^+ \to J/\psi\phi K^+$

En Wang

School of Physics and Microelectronics, Zhengzhou University, Zhengzhou, Henan 450001, China
E-mail: wangen@zzu.edu.cn

Ju-Jun Xie

Institute of Modern Physics, Chinese Academy of Sciences, Lanzhou 730000, China, and
School of Physics and Microelectronics, Zhengzhou University, Zhengzhou, Henan 450001, China

Li-Sheng Geng

School of Physics, Beihang University, Beijing 100191, China, and
School of Physics and Microelectronics, Zhengzhou University, Zhengzhou, Henan 450001, China

Eulogio Oset

Departamento de Física Teórica and IFIC, Centro Mixto Universidad de Valencia-CSIC Institutos de Investigación de Paterna, Aptdo. 22085, 46071 Valencia, Spain

We have studied the $J/\psi\phi$ mass distribution of the process $B^+ \to J/\psi\phi K^+$ from the threshold to about 4250 MeV, by considering the contribution of the $X(4140)$ with a narrow width, together with the $X(4160)$ state. Our results show that the cusp structure at the $D_s^*\bar{D}_s^*$ threshold is tied to the molecular nature of the $X(4160)$ state.

Keywords: $X(4140)$; $X(4160)$; molecular state.

1. Introduction

Recently, the state $X(4140)$ was confirmed in the measurement of the $B^+ \to J/\psi\phi K^+$ reaction at LHCb.[1,2] However, one surprise is that the $X(4140)$ deduced from the analysis, with quantum numbers $J^{PC} = 1^{++}$, has a width $\Gamma \approx 83 \pm 21^{+21}_{-14}$ MeV, substantially larger than that claimed in the former experiments.[3]

In this work, we show that the low invariant $J/\psi\phi$ mass region, requires the contributions of a narrow $X(4140)$, and a wide $X(4160)$ resonance which couples to $J/\psi\phi$ but is mostly made by a $D_s^*\bar{D}_s^*$ molecule as discussed in Ref. 4. As a consequence of analyticity and driven by the molecular nature of $X(4160)$, the $J/\psi\phi$ mass spectrum develops the strong cusp around the $D_s^*\bar{D}_s^*$ threshold.[1,2]

2. The formalism

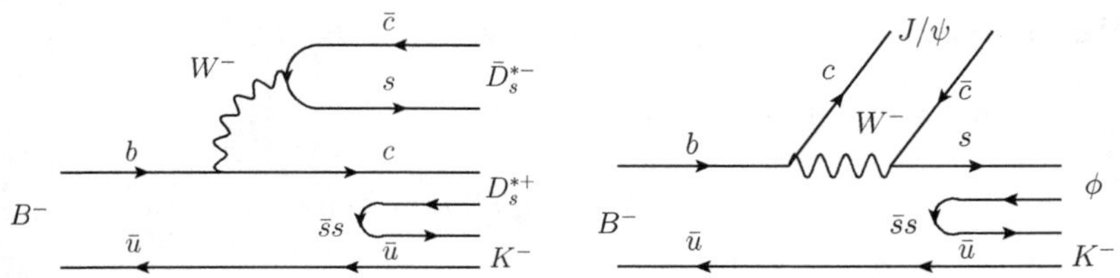

Fig. 1. Microscopic quark level production of $K^-D_s^*\bar{D}_s^*$ (left) and $B^- \to K^-J/\psi\phi$ (right) in B^- decay.

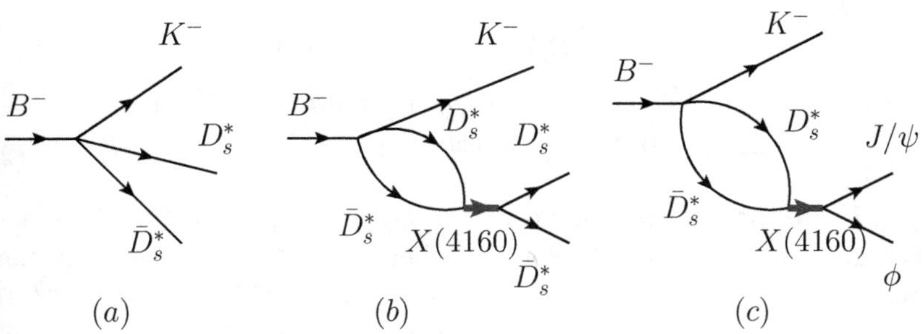

Fig. 2. (a) The tree diagram, and (b) the $D_s^*\bar{D}_s^*$ final state interaction for $B^- \to K^-D_s^*\bar{D}_s^*$ in the presence of the $X(4160)$ resonance. (c) The mechanism for $B^- \to K^-J/\psi\phi$ driven by the $X(4160)$ resonance.

As the first step, the dominant process of $B^- \to K^-D_s^*\bar{D}_s^*$ (we take the complex conjugate reaction to deal with b quark rather than $\bar{b}$ quark) at the quark level proceeds as shown in the left panel of Fig. 1, involving external emission. Obviously in the neighborhood of the resonance the tree level term of Fig. 2(a) is small compared to the resonant term of Fig. 2(b), but we keep it in the calculations. For the production of $J/\psi\phi$ with this mechanism, the tree level of Fig. 2(a) does not contribute and then we are led to the diagram of Fig. 2(c)

The $B^- \to K^- J/\psi\phi$ reaction can also proceed through the mechanism as shown in the right panel of Fig. 1 involving internal conversion. Yet, the internal conversion is penalized by color factors with respect to the external emission of the left diagram of Fig. 1, and hence this term, or rescattering of this term like that in Fig. 2(c), but with $J/\psi\phi$ intermediate state instead of $D_s^*\bar{D}_s^*$, which would involve the extra factor $g_{J/\psi\phi}/g_{D_s^*\bar{D}_s^*}$ versus the amplitude of 2(c), can be safely neglected.

We can then write the amplitude for the $B^- \to K^- D_s^*\bar{D}_s^*$,

$$\mathcal{M}_{B\to KD_s^*\bar{D}_s^*} = A\Big[1 + G_{D_s^*\bar{D}_s^*}(M_{\rm inv}(D_s^*\bar{D}_s^*)) \\ \times\, t_{D_s^*\bar{D}_s^*\to D_s^*\bar{D}_s^*}(M_{\rm inv}(D_s^*\bar{D}_s^*))\Big]\left(\vec{\epsilon}\cdot\vec{k}\,\vec{\epsilon}\,'\cdot\vec{k} - \frac{1}{3}\vec{k}^2\vec{\epsilon}\cdot\vec{\epsilon}\,'\right), \tag{1}$$

and the amplitude for the $B^- \to K^- J/\psi\phi$,

$$\begin{aligned}\mathcal{M}_{B\to KJ/\psi\phi} &= A\times G_{D_s^*\bar{D}_s^*}(M_{\rm inv}(J/\psi\phi)) \\ &\quad \times t_{D_s^*\bar{D}_s^*\to J/\psi\phi}(M_{\rm inv}(J/\psi\phi))\left(\vec{\epsilon}\cdot\vec{k}\,\vec{\epsilon}\,'\cdot\vec{k} - \frac{1}{3}\vec{k}^2\vec{\epsilon}\cdot\vec{\epsilon}\,'\right) \\ &\quad + \frac{B\,M^3_{X(4140)}(\vec{\epsilon}_{J/\psi}\times\vec{\epsilon}_\phi)\cdot\vec{k}}{M^2_{\rm inv}(J/\psi\phi) - M^2_{X(4140)} + iM_{X(4140)}\Gamma_{X(4140)}} \\ &= \mathcal{M}^{X(4160)} + \mathcal{M}^{X(4140)}\end{aligned} \tag{2}$$

where $\vec{\epsilon}$, $\vec{\epsilon}\,'$ are the polarization vectors of D_s^* and $\bar{D}_s^*$ (or J/ψ and ϕ), and we evaluate it in the frame of reference where the $D_s^*\bar{D}_s^*$ (or $J/\psi\phi$) system is at rest. For the $J/\psi\phi$ production, we have included the term $\mathcal{M}^{X(4140)}$ to account for the production of $J/\psi\phi$ via the 1^{++} $X(4140)$ resonance. Here we take $M_{X(4140)} = 4132$ MeV and $\Gamma_{X(4140)} = 19$ MeV. The A and B are free parameters to be fitted to the data. The details of the loop functions G and the transition amplitude t in Eqs. (1) and (2) are given in Ref. 5. Finally, the mass distribution can be written as,

$$\frac{d\Gamma}{dM_{\rm inv}} = \frac{1}{(2\pi)^3}\frac{1}{4M^2_{B^-}}|\vec{k}'|\,\tilde{p}|\mathcal{M}|^2, \tag{3}$$

where $\vec{k}'$ is the K^- momentum in the B^- rest frame, and $\tilde{p}$ the D_s^* (or J/ψ) momentum in the $D_s^*\bar{D}_s^*$ (or $J/\psi\phi$) rest frame.

3. Results

A suitable fit to the data is obtained as shown in the left panel of Fig. 3, Our results (red solid curve, labeled as 'Full') are in good agreement with the experimental data at low $J/\psi\phi$ invariant masses, and we also present the uncertainties of our results in the figure. As we can see in the figure, we obtain a contribution from the $X(4140)$ (blue dotted curve) that is dominant at low invariant masses, and is responsible for the peak observed in the experiment around 4135 MeV. The $X(4160)$ (green dashed curve) is responsible for most of the strength and produces a broader peak around 4170 MeV. And finally a cusp appears at the $D_s^*\bar{D}_s^*$ threshold as it shows up in the experiment. This cusp comes from the factor $G_{D_s^*\bar{D}_s^*}(M_{\rm inv})$ and reflects the analytical structure of this function with a discontinuity of the derivative at threshold. A similar feature is also found in Refs. 6, 7. One must stress that this factor appears here as a consequence of analyticity, with a large strength due to the $D_s^*\bar{D}_s^*$ molecular structure of the $X(4160)$. In an analysis like the one of Refs. 1, 2, where a sum of amplitudes for resonance excitation and some background are fitted to the data, this factor is not considered, and as a consequence the cusp around $D_s^*\bar{D}_s^*$ in the data is missed in the fit.

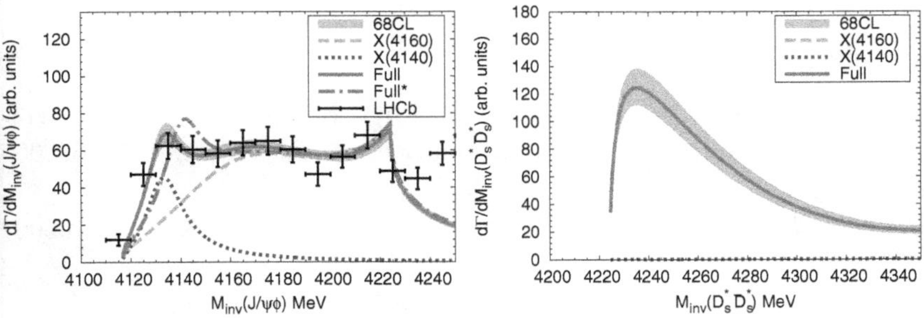

Fig. 3. Left: the $J/\psi\phi$ mass distribution of the $B^- \to K^- J/\psi\phi$ decay. Right: The $D_s^*\bar{D}_s^*$ mass distribution of the $B^- \to K^- D_s^*\bar{D}_s^*$ decay. The bands come considering the errors in the fitted parameters, and represents the 68% confidence-level.

To finish the work, and as a test of the explanation given here, we present in the right panel of Fig. 3 the $D_s^*\bar{D}_s^*$ mass distribution above threshold obtained with the same parameters as in Fig. 3. As we can see in this figure, there is a peak close to threshold, which should not be misidentified with a new resonance, but it is the reflection of the $X(4160)$ which in our fit has the mass at 4169 MeV.

4. Summary

The present work and the prediction, tied to the interpretation given for the $B^- \to K^- J/\psi\phi$ spectrum, should act as an incentive to measure this reaction and learn about properties of the $X(4140)$ and $X(4160)$. In addition, the $J/\psi\phi$ mass distribution of the $e^+e^- \to \gamma J/\psi\phi$ reaction reported by BESIII[8] is also compatible with the existence of the $X(4140)$ state, appearing as a peak, and a strong cusp structure at the $D_s^*\bar{D}_s^*$ threshold, resulting from the molecular nature of the $X(4160)$ state.[7]

Acknowledgments

This work is partly supported by the National Natural Science Foundation of China under Grant Nos. 11375024, 11522539, 11735003, 11505158, 11475015, and 11647601. It is also supported by the Youth Innovation Promotion Association CAS (No. 2016367), and Academic Improvement Project of Zhengzhou University. This work is also partly supported by the Spanish Ministerio de Economia y Competitividad and European FEDER funds under the contract number FIS2014-57026-REDT, FIS2014-51948-C2-1-P, and FIS2014-51948-C2-2-P, and the Generalitat Valenciana in the program Prometeo II-2014/068.

References

1. R. Aaij *et al.* [LHCb Collaboration], Phys. Rev. D **95**, 012002 (2017).
2. R. Aaij *et al.* [LHCb Collaboration], Phys. Rev. Lett. **118**, 022003 (2017).
3. C. Patrignani *et al.* [Particle Data Group], Chin. Phys. C **40**,100001 (2016).
4. R. Molina and E. Oset, Phys. Rev. D **80**, 114013 (2009).
5. E. Wang, J. J. Xie, L. S. Geng and E. Oset, Phys. Rev. D **97**, no. 1, 014017 (2018).
6. L. R. Dai, G. Y. Wang, X. Chen, E. Wang, E. Oset and D. M. Li, Eur. Phys. J. A **55**, no. 3, 36 (2019).
7. E. Wang, J. J. Xie, L. S. Geng and E. Oset, Chin. Phys. C **43**, no. 11, 113101 (2019).
8. M. Ablikim *et al.* [BESIII Collaboration], Phys. Rev. D **97**, no. 3, 032008 (2018).

Regge trajectories in light and heavy mesons: The pattern of appearances and possible dynamical explanations

S. S. Afonin

Saint Petersburg State University, 7/9 Universitetskaya nab., St. Petersburg, 199034, Russia
E-mail: s.afonin@spbu.ru

I briefly review the Regge approach to the hadron spectrum and advocate a dynamical emergence of principal quantum number in the known spectrum of light non-strange mesons. Further it is shown how the linear radial trajectories with universal slope can be extended to heavy quarkonia and a qualitative string interpretation is given. After that I discuss a recently proposed non-string model leading to a natural appearance of linear Regge trajectories and explaining many mass relations.

Keywords: Regge trajectories; meson spectrum.

1. Regge trajectories and degeneracy in light meson spectrum

The Regge phenomenology emerged from the dual amplitudes[1] and gave rise to various hadron string models. This approach continues to play an important role in the study of hadron spectroscopy. The dual amplitudes and some related string approaches predict the following behavior of meson masses,

$$M_n^2 = a(J + bn + c), \qquad J, n = 0, 1, 2, \ldots, \tag{1}$$

where J is the spin, n enumerates the radially excited states, a represents a universal slope and b, c are some constants. The relation (1) reproduces the classical Regge behavior $M^2 \sim J$ observed in the light baryons and mesons. Also this relation predicts the equidistant daughter Regge trajectories. The number n enumerates the states on these "radial" Regge trajectories. The available experimental data on light non-strange mesons seem to confirm the linear Regge behavior (1).[2–14]

A popular assumption about the structure of mesons is the hypothesis that the mesons represent a gluon string with a quark/antiquark at the ends. A careful analysis of light non-strange meson spectrum showed, however, that instead of spin J in Eq. (1) one should use the angular momentum L of the quark-antiquark pair,[8–10]

$$M^2 \sim a(L + bn). \tag{2}$$

In addition, the data suggest $b = 1$ that is a remarkable property leading to numerous Hydrogen-like degeneracies.[15–18]

2. Universal description

Various attempts to generalize the relation (1) to the sector of heavy quarks resulted in the emergence of strong non-linearities with respect to J and n. Let us look, however, at the experimental data. A relatively rich set of data on radial recurrences in this sector exists only for the unflavored vector heavy quarkonia. Using the relevant data from the Particle Data,[19] see Refs. 20 and 21, one can see again the emergence of linear behavior as a function of consecutive number n. Aside from the ground state, the masses approximately lie on the linear radial trajectories

$$M_n^2 = a(n + b), \tag{3}$$

where we re-denoted the vector intercept $b = 1 + c$. In general, the slope and intercept in (3) depend strongly on the quark flavor.[20,21]

The universal linearity suggests that some universal gluodynamics lies behind the observed behavior. We can propose a simple string scheme explaining this universality. The central idea is that at the conditions when a quark-antiquark pair form a resonance in the QCD vacuum, the pair may be viewed as a virtally static system. The binding is provided by the exchange of some massless particle (the pion or gluon in concrete realizations of the scheme). And one must quantize the motion of this particle (not the radial motion of quarks as in the standard hadron string approaches!). The total energy (mass) of the system is

$$M = m_1 + m_2 + p + \sigma r. \tag{4}$$

Here m_1 and m_2 are the masses of quark and antiquark separated by the distance r, p is the momentum of exchanged particle, and σ represents the standard string tension. Let us apply the semiclassical quantization to the momentum p

$$\int_0^l p\, dr = \pi(n + b), \qquad n = 0, 1, 2, \dots. \tag{5}$$

Here l is the maximal quark separation and the constant b depends on the boundary conditions. Substituting p from (4) to (5) and making use of the definition $\sigma = \frac{M}{l}$ we obtain the linear radial trajectory

$$(M_n - m_1 - m_2)^2 = 2\pi\sigma(n + b). \tag{6}$$

In our unflavored case $m_1 = m_2 \equiv m$ and the relation (6) can be simplified to

$$(M_n - 2m)^2 = a(n + b), \tag{7}$$

which is our generalization of the linear spectrum (3) to non-zero quark masses. In this relation, the universal gluodynamics (the slope a) and dependence on quantum numbers (the dimensionless intercept b) are clearly separated from the contribution of quark masses. For this reason, the parameters a and b in the relation (7) should be flavor-independent. The analysis of Refs. 20 and 21 indeed confirmed this expectation. A successful check of relation (6) in the case heavy-light systems was performed in Ref. 22.

3. A possible non-string mechanism for linear Regge trajectories

Usually the observation of linear trajectories is interpreted as an evidence for string picture of mesons. In spite of many attempts, however, a satisfactory quantized hadron string has not been constructed. Among typical flaws of this approach one can mention the absence of spontaneous chiral symmetry breaking, rapidly growing (with mass) size of meson excitations, and totally unclear role of higher Fock components in the hadron wavefunction.

A novel realization of quark model concept has been recently proposed in Ref. 23. This realization leads to a natural (and alternative to hadron strings) explanation of Regge recurrences and is potentially free of typical shortcomings inherent to string, potential and some other approaches. Stated more strictly, the given approach represents a phenomenological mass counting scheme in which the relativistic invariance, renormalization invariance, chiral symmetry breaking, higher Fock components, and linear Regge behavior are qualitatively built-in. The approach allows to classify the light mesons without use of any angular momentum associated with hadron constituents. The states π_1 (which are exotic for the standard quark model) emerge in a natural way while the other exotic quantum numbers remain forbidden. The constructed mass counting scheme permits to obtain hadron masses from very simple relations with a typical accuracy comparable to numerical calculations in complicated dynamical models.

The underlying motivation is as follows. The hadron masses must be renorminvariant. On the other hand, the light non-strange mesons can be viewed, in one way or another, as some quantum excitations of pion. Basing on the Gell-Mann–Oakes–Renner relation for the pion mass and the known formula for the stress-energy tensor in QCD it is motivated that the masses of these mesons can be represented as

$$m_h^2 = \Lambda(E_h + 2m_q) = \Lambda E_h + m_\pi^2. \tag{8}$$

The energy parameter E_h seems to be related to the renorminvariant gluon condensate, $E_h \sim \alpha_s \langle G_{\mu\nu}^2 \rangle / \langle \bar{q}q \rangle$.[23] A phenomenological mass counting scheme based on the relation (8) and a demonstration how it describes the meson spectroscopy is scrutinized in Ref. 23.

It is also suggested that the parameter E_h can be interpreted as an effective energy of some constituents different from the current quarks. The problem is to propose a model for these constituents which should represent some excitations inside the pion. Three basic excitations are postulated. The first one appears when one of quarks absorbs a gluon of certain energy $E_h = E_\rho$. The spin of excited quark changes its direction to the opposite one, converting the original spin-singlet $q\bar{q}$-pair (π-meson) to the spin-triplet one (ρ-meson). The second kind of excitation emerges due to formation of spin-singlet $q\bar{q}$-pair with effective mass $E_h = E_0$ (the lower index stays for total spin of S-wave $q\bar{q}$-pair). The third basic excitation is the formation of spin-triplet $q\bar{q}$-pair with effective mass $E_h = E_1$. The formation of these constituent pairs are likely a QCD analogue of formation of para- and ortho-positronia from photons. By assumption, any excitation inside pion leading to an observable resonance can be represented as a combination of these basic excitations so that the total effective energy E_h in (8) is just a sum (appropriate number of times) of E_ρ, E_0, and E_1. Simultaneously this will dictate the quantum numbers of constructed resonance. In a sense, the introduction of these constituent pairs reflects the excitations of higher Fock components in the pion wavefunction. These pairs in excited hadrons, in some sense, bear a superficial resemblance to neutrons and protons in atomic nuclei. The given approach represents thus a new realization of the quark model concept, a realization in which highly excited states appear due to multiquark components in hadron wavefunctions.

The inclusion of strange quarks into this approach is straightforward. It would be interesting to extend the underlying ideas to light baryons and to hadrons with heavier quarks.

Acknowledgment

The present work was supported by the Saint-Petersburg State University travel grant (Id: 40114404).

References

1. P.D.B. Collins, *Phys. Rept. C* **1**, 103 (1971).
2. A. V. Anisovich, V. V. Anisovich and A. V. Sarantsev, *Phys. Rev. D* **62**, 051502(R) (2000).
3. D. V. Bugg, *Phys. Rept.* **397**, 257 (2004).
4. E. Klempt and A. Zaitsev, *Phys. Rep.* **454**, 1 (2007).
5. D. M. Li, B. Ma, Y. X. Li, Q. K. Yao and H. Yu, Eur. Phys. J. C **37**, 323 (2004).
6. S. S. Afonin, *Phys. Lett. B* **639**, 258 (2006).
7. S. S. Afonin, *Eur. Phys. J. A* **29**, 327 (2006).
8. S. S. Afonin, *Phys. Rev. C* **76**, 015202 (2007).
9. S. S. Afonin, *Mod. Phys. Lett. A* **22**, 1359 (2007).
10. M. Shifman and A. Vainshtein, *Phys. Rev. D* **77**, 034002 (2008).
11. P. Masjuan, E. Ruiz Arriola and W. Broniowski, *Phys. Rev. D* **85**, 094006 (2012).
12. J. Sonnenschein and D. Weissman, *JHEP* **1408**, 013 (2014).
13. C. Q. Pang, B. Wang, X. Liu and T. Matsuki, *Phys. Rev. D* **92**, 014012 (2015).
14. D. Jia, C. Q. Pang and A. Hosaka, *Int. J. Mod. Phys. A* **32**, 1750153 (2017).
15. S. S. Afonin, *Int. J. Mod. Phys. A* **22**, 4537 (2007).
16. E. H. Mezoir and P. Gonzalez, *Phys. Rev. Lett.* **101**, 232001 (2008).
17. S. S. Afonin, *Int. J. Mod. Phys. A* **23**, 4205 (2008).
18. S. S. Afonin, *Mod. Phys. Lett. A* **23**, 3159 (2008).
19. M. Tanabashi *et al.* (Particle Data Group), *Phys. Rev. D* **98**, 030001 (2018).
20. S. S. Afonin and I. V. Pusenkov, *Phys. Rev. D* **90**, 094020 (2014).
21. S. S. Afonin and I. V. Pusenkov, *Mod. Phys. Lett. A* **29**, 1450193 (2014).
22. K. Chen, Y. Dong, X. Liu, Q. F. Lu and T. Matsuki, *Eur. Phys. J. C* **78**, 20 (2018).
23. S. S. Afonin, *Adv. High Energy Phys.* **2019**, 1701939 (2019).

Strange-Meson Spectroscopy at COMPASS

S. Wallner for the COMPASS Collaboration

Physics Department, Technical University of Munich,
Garching, 85748, Germany
E-mail: stefan.wallner@tum.de

COMPASS is a multi-purpose fixed-target experiment at CERN aimed at studying the structure and spectrum of hadrons. It has collected the so far world's largest data set on diffractive production of the $K^-\pi^-\pi^+$ decay, which in principle gives access to all kaon states. We performed an elaborate partial-wave analysis, using model-selection techniques to select the wave set based on a large systematically constructed pool of allowed partial waves. The partial-wave decomposition reveals signals in the mass region of well-known states, such as $K_1(1270)$ and $K_1(1400)$. In addition, we observe potential signals from excited states, such as $K_1(1650)$.

Keywords: COMPASS; strange mesons; spectroscopy; proceedings; HADRON 2019; XVIII International Conference on Hadron Spectroscopy and Structure.

1. Meson Spectroscopy at COMPASS

COMPASS is a fixed-target multi-purpose experiment located at CERN. So far, COMPASS has studied mainly isovector resonances of the a_J and π_J families with high precision, using the dominant π^- component of the 190 GeV/c negative hadron beam.[1,2] In this analysis, we study the strange-meson spectrum up to masses of 3 GeV/c^2 using the approximately 2.4 % K^- fraction of the beam in diffractive scattering off a liquid-hydrogen target. Our flagship channel is the production of $K^-\pi^-\pi^+$, which in principle gives access to all kaon states, i.e. K_J and K_J^* mesons.[a] COMPASS acquired the so far world's largest data set of about 720 000 exclusive events for this channel.

2. Kinematic Distributions

Figure 1a shows the invariant mass spectrum of the $K^-\pi^-\pi^+$ final state. It exhibits structures in the $m_{K\pi\pi}$ region of well-known resonances, e.g.

[a]Except for $J^P = 0^+$ states.

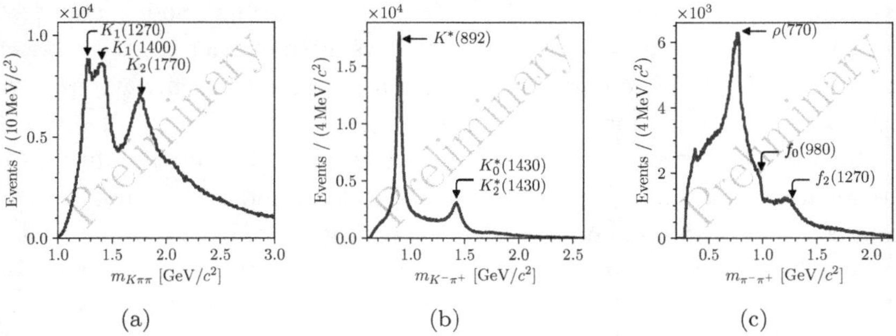

Fig. 1. Invariant mass distribution of the $K^-\pi^-\pi^+$ system (a) and of the $K^-\pi^+$ (b) and $\pi^-\pi^+$ (c) sub-systems. The distributions are not corrected for acceptance effects, which are significant in certain regions of the phase space.

$K_1(1270)$ or $K_1(1400)$, which sit on top of a broad distribution. The invariant mass distributions of the $K^-\pi^+$ and $\pi^-\pi^+$ sub-systems are dominated by well-known two-body resonances as indicated by the labels in Figs. 1b and 1c.[b] This dominance of two-body resonances justifies the application of the isobar model in the partial-wave decomposition, which is discussed in the following section.

3. Analysis Method: Partial-Wave Decomposition

We employ the method of partial-wave analysis to decompose the data into contributions from various partial waves (see Ref. 1 for details). To this end, we construct a model for the intensity distribution $\mathcal{I}(\tau)$ of the $K^-\pi^-\pi^+$ final state in terms of the five phase-space variables that are represented by τ. Using the isobar approach, $\mathcal{I}(\tau)$ is modeled as a coherent sum of partial-wave amplitudes

$$\mathcal{I}(\tau; m_{K\pi\pi}, t') = \left| \sum_{a}^{\text{waves}} \mathcal{T}_a(m_{K\pi\pi}, t')\, \psi_a(\tau; m_{K\pi\pi}) \right|^2 . \tag{1}$$

These partial waves are represented by $a = J^P\, M^\varepsilon\, \zeta\, b\, L$ and are defined by the quantum numbers of the $K^-\pi^-\pi^+$ system ($J^P M^\varepsilon$),[c] the intermediate

[b]The small spike at about $0.4\,\text{GeV}/c^2$ in the $m_{\pi^-\pi^+}$ system originates from $\phi \to K^-K^+$ decays from a small contamination of the $K^-\pi^-\pi^+$ data sample by $K^-K^-K^+$, where the kaons are misidentified as pions.

[c]Here, J is the spin of the $K^-\pi^-\pi^+$ state and P its parity. The spin projection of J along the beam axis is given by M^ε.

two-body resonance ζ,[d] and the orbital angular momentum L between the bachelor particle b and the isobar. Within the isobar model, the decay amplitudes ψ_a can be calculated. This allows us to extract the partial-wave amplitudes $\mathcal{T}_a$, which determine strength and phase of each wave from the data by an unbinned maximum-likelihood fit.

In order to extract the dependence of the partial-wave amplitudes on the invariant mass $m_{K\pi\pi}$ of the $K^-\pi^-\pi^+$ system and on the squared four-momentum transfer t' between the beam particle and the target proton, the maximum-likelihood fit is performed independently in 75 narrow bins of $m_{K\pi\pi}$ and 4 bins of t' in the analyzed kinematic range of $1.0 < m_{K\pi\pi} < 3.0\,\mathrm{GeV}/c^2$ and $0.1 < t' < 1.0\,(\mathrm{GeV}/c)^2$.

In order to construct the wave set, i.e. the partial waves that enter the sum in Eq. (1), we apply model-selection techniques. We systematically construct a large set of possible partial waves, called wave pool. We restrict ourselves to $J \leq 7$, $L \leq 7$, positive naturality of the exchange particle, and twelve isobars: two $K\pi$ S-wave amplitudes, $K^*(892)$, $K^*(1680)$, $K_2^*(1430)$, $K_3^*(1780)$, a broad $\pi\pi$ S-wave amplitude, $f_0(980)$, $f_0(1500)$, $\rho(770)$, $f_2(1270)$, and $\rho_3(1690)$.[e] This results in a wave pool of 596 allowed waves. In order to select the wave set, we fit the wave pool to the data applying regularization techniques to suppress insignificant waves (see Ref. 3 for details). This results in an individual wave set for each $(m_{K\pi\pi}, t')$ cell, which we fit again to the data without regularization terms.

4. Selected Results of the Partial-Wave Decomposition

4.1. $J^P = 1^+$ *Waves*

The $1^+0^+K^*(892)\pi S$ wave is the most dominant one in our data. It accounts for about 30 % of the total intensity. Fig. 2a shows the t'-summed intensity spectrum, i.e. the intensity spectrum incoherently summed over the four t' bins, of the $1^+0^+K^*(892)\pi S$ wave. Two peaks are visible in the mass regions of the $K_1(1270)$ and the $K_1(1400)$ in agreement with previous observations.[4] Systematic studies of our analysis revealed, that the low-mass region, indicated by the gray area in Fig. 2a, of this wave and some other waves not discussed here is affected by systematic effects. These effects arise mainly from the limited kinematic range of the COMPASS final-

[d]We consider isobar resonances in the $K^-\pi^+$ or $\pi^-\pi^+$ sub-system.

[e]We use relativistic Breit-Wigner amplitudes for all isobar line shapes, except for the S-wave isobars, where we employ K-matrix parameterizations.

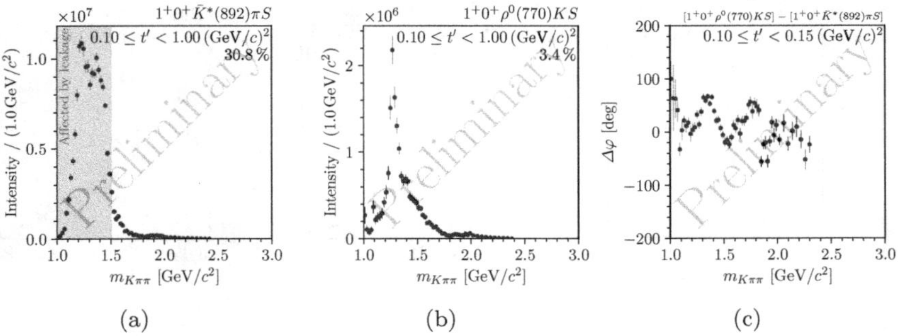

Fig. 2. t'-summed intensity distributions of the $J^P M^\varepsilon = 1^+0^+$ partial waves in the $K^*(892)\pi S$ (a) and the $\rho(770)KS$ (b) decay. (c) shows the relative phase between these two partial waves in the lowest t' bin.

state particle identification and are under further study. Therefore, the low-mass region of the $1^+0^+K^*(892)\pi S$ wave should be considered only on a qualitative level.

The intensity spectrum of the $1^+0^+\rho(770)KS$ wave, shown in Fig. 2b, exhibits only one dominant peak in the mass region of the lower-lying $K_1(1270)$. We observe a bump in its high-mass shoulder at about $1.6\,\mathrm{GeV}/c^2$. The PDG[5] lists the $K_1(1650)$ resonance in this mass region, which needs further clarification.

Figure 2c shows the relative phase between the $1^+0^+\rho(770)KS$ and the $1^+0^+K^*(892)\pi S$ wave. It exhibits rich structures. First, the phase rises in the $K_1(1270)$ mass region, which can be explained by the dominance of the $K_1(1270)$ in the $\rho(770)KS$ decay.[f] Then, the relative phase drops in the mass region of the $K_1(1400)$. This can be explained by the strong contribution of the $K_1(1400)$ to the $K^*(892)\pi S$ decay, which enters with a minus sign in the relative phase. Finally, the relative phase rises again at about $1.7\,\mathrm{GeV}/c^2$. This could be a potential indication of the $K_1(1650)$ resonance in the $\rho(770)KS$ decay.

4.2. $J^P = 2^-$ *Waves*

The 2^- waves show a large variety of different structures in various decay modes. The most dominant signal is in the $2^-0^+K_2^*(1430)\pi S$ wave shown

[f]The phase motion produced by the $K_1(1270)$ resonance in the $K^*(892)\pi S$ decay is potentially smeared out by the low-mass tail of the $K_1(1400)$.

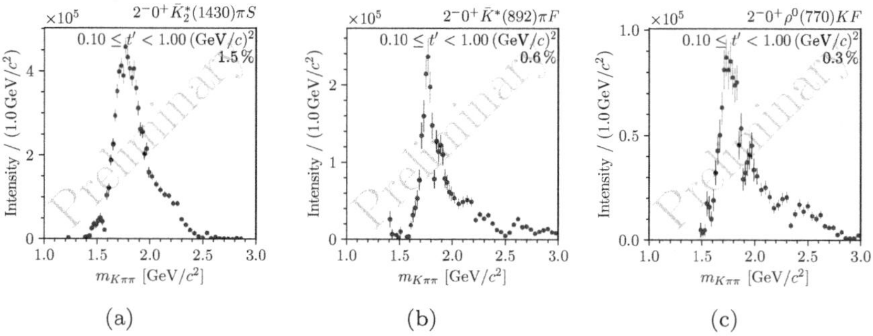

Fig. 3. t'-summed intensity distributions of the $J^P M^\varepsilon = 2^-0^+$ partial waves in the $K_2^*(1430)\pi S$ (a), $K^*(892)\pi F$ (b), and the $\rho(770)FS$ (c) decay.

in Fig. 3a, which accounts for about 1.5 % of the total intensity. It exhibits a broad peak at about $1.8\,\mathrm{GeV}/c^2$ with a bump in its high-mass shoulder. The PDG[5] lists the established resonances $K_2(1770)$ and $K_2(1820)$ in the mass region of the peak and the $K_2(2250)$, which needs further confirmation, in the mass region of the high-mass bump.

The $2^-0^+K^*(892)\pi F$ wave shows a narrow peak at about $1.8\,\mathrm{GeV}/c^2$ with a bump in its high-mass shoulder. Also the $2^-0^+\rho(770)KF$ wave exhibits a peak at about $1.8\,\mathrm{GeV}/c^2$. Its high-mass shoulder contains more complicated structures, which potentially also contain the $K_2(2250)$ resonance. To clarify the nature of the observed signals and to extract their parameters, i.e. masses and widths, a resonance-model fit similar to Ref. 2 will be the next major step in this analysis.

References

1. C. Adolph *et al.*, Resonance production and $\pi\pi$ S-wave in $\pi^- + p \to \pi^-\pi^-\pi^+ + p_{\mathrm{recoil}}$ at 190 GeV/c, *Physical Review D* **95**, 032004 (2017).
2. M. Aghasyan *et al.*, Light isovector resonances in $\pi^- p \to \pi^-\pi^-\pi^+ p$ at 190 GeV/c, *Physical Review D* **98**, 092003 (2018).
3. F. Kaspar *et al.*, Wave-selection techniques for partial-wave analysis in light-meson spectroscopy, within these proceedings.
4. C. Daum *et al.*, Diffractive production of strange mesons at 63 GeV, *Nuclear Physics B* **187**, 1 (1981).
5. M. Tanabashi *et al.*, Review of Particle Physics, *Physical Review D* **98**, 030001 (2018).

Heavy $K^*(4307)$ meson with hidden charm

X.-L. Ren[1], B. B. Malabarba[2], L.-S. Geng[3], K. P. Khemchandani[4] and A. Martínez Torres[2]

[1]*Ruhr-Universität Bochum, Fakultät für Physik und Astronomie, Institut für Theoretische Physik II, D-44780 Bochum, Germany*
[2]*Instituto de Física, Universidade de São Paulo, C.P. 66318, 05389-970 São Paulo, São Paulo, Brazil*
[3]*School of Physics & Beijing Key Laboratory of Advanced Nuclear Materials and Physics, Beihang University, Beijing 100191, China*
[4]*Universidade Federal de São Paulo, C.P. 01302-907, São Paulo, Brazil*

We report on a robust prediction of heavy K^* meson by solving the Faddeev equations with fixed-center approximation for the three-body $KD\bar{D}^*$ system. As the excited Kaon state, K^* is an exotic hidden charm meson with $M - i\Gamma/2 = 4307 \pm 2 - i9 \pm 2$ MeV and $I(J^P) = 1/2(1^-)$. We further performed the evaluation of the decay width of $K^*(4307)$ to the open two-body channels. We expect that the above findings inspire an experimental investigation of this exotic K^* meson and to study the so far unexplored heavy strange meson sector.

Keywords: Exotic hadrons; few body systems; heavy quark symmetry.

1. Introduction

Recently, the exotic hadrons with the open/hidden heavy quark components have attracted great attention in the experimental and theoretical studies.[1,2] However, in the strange sector, there is no experimental data available on heavy K or K^* meson states, leaving the heavy strange physics experimentally unexplored. In this Hadron conference, the COMPASS collaboration presented their preliminary results of the strange-meson spectrum.[3]

From the theoretical side, we have explored the possibility of forming a heavy K^* meson in the three-body $KD\bar{D}^*$ system.[4] All the two-body interactions, KD, $K\bar{D}^*$, and $D\bar{D}^*$, are stringently constrained by a large number of experimental ($D^*_{s0}(2317)$, $D^*_{s1}(2460)$, $X(3872)$, and $Z_c(3900)$) as well as the lattice QCD data, which lead to a unique system of $KD\bar{D}^*$. Using the so-called fixed-center approximation (FCA)[5] to the Faddeev equations, we found a heavy K^* meson around 4307 MeV with hidden charm. In order to provide more information, we also studied the decay width of $K^*(4307)$ from two-body decay processes.[6]

2. $K^*(4307)$ state in the $KD\bar{D}^*$ system

For a three-body system, if the mass of the third particle P_3 is much smaller than a stable cluster composed of the two other particles P_1 and P_2, one can consider the use of the FCA to the Faddeev equations. Our $KD\bar{D}^*$ system exactly satisfies such criteria. We take the light kaon as the third particle to scatter off the cluster of $D\bar{D}^*$, which can form the as $X(3872)$ or $Z_c(3900)$ states.[7,8]

Here, we briefly present the basic equations, in particular for the coupled-channel case, in the FCA framework. The details can be found in Ref. 4. The total scattering amplitude is decomposed into two Faddeev partitions,

$$\begin{aligned} T &= T_{31} + T_{32}, \\ T_{31} &= t_{31} + t_{31}G_0t_{32} + t_{31}G_0t_{32}G_0t_{31} + \cdots = t_{31} + t_{31}G_0T_{32}, \\ T_{32} &= t_{32} + t_{32}G_0t_{31} + t_{32}G_0t_{31}G_0t_{32} + \cdots = t_{32} + t_{32}G_0T_{31}, \end{aligned} \tag{1}$$

where the two-body amplitudes, t_{31} and t_{32}, are the functions of the isospin 0 and 1 s-wave interactions of the KD and $K\bar{D}^*$ subsystems.[9,10]

In the $K[D\bar{D}^*]$ system, we consider $KX(3872)$ and $KZ_c(3900)$ as coupled channels with total isospin $I = 1/2$. As shown in Ref. 4, this makes that the t_{31} and t_{32} amplitudes can be written as 2×2 matrices in the coupled channel space,

$$t_{31} = \begin{pmatrix} (t_{31})_{11} & (t_{31})_{12} \\ (t_{31})_{21} & (t_{31})_{22} \end{pmatrix}, \quad t_{32} = \begin{pmatrix} (t_{32})_{11} & (t_{32})_{12} \\ (t_{32})_{21} & (t_{32})_{22} \end{pmatrix}, \tag{2}$$

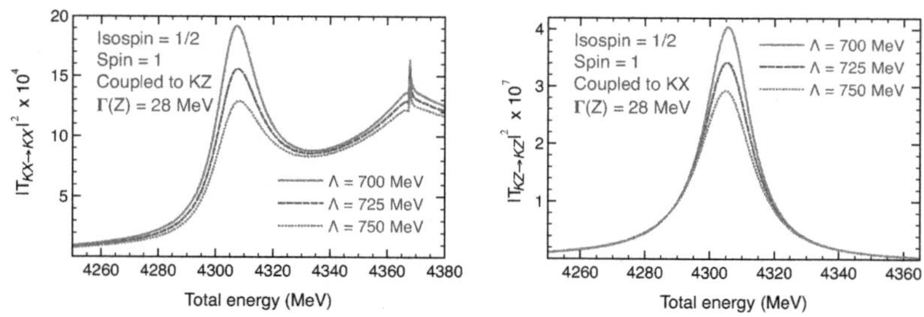

Fig. 1. Modulus squared of the KX and KZ scattering amplitudes in $I = 1/2$ with different cutoffs. A cusp related to the three-body $KD\bar{D}^*$ threshold is observed in the $KX \to KX$ amplitude.

where the element (11) denotes $KX \to KX$, the element (12) $KX \to KZ_c$, and so on. The corresponding loop function G_0 is,

$$G_0 = \begin{pmatrix} (G_0)_{11} & 0 \\ 0 & (G_0)_{22} \end{pmatrix}. \tag{3}$$

Finally, the total T-matrix appearing in Eq. (1) becomes as,

$$T = T_{31} + T_{32} = \begin{pmatrix} T_{11} & T_{12} \\ T_{21} & T_{22} \end{pmatrix}. \tag{4}$$

In Fig. 1, the modulus squared of the KX and KZ_c scattering amplitudes in $I = 1/2$ are given with the momentum cutoff varying from 700 MeV to 750 MeV. We found that the mass and width of the heavy K^* meson in the KX configuration is $M - i\Gamma/2 = (4308 \pm 1) - i(8 \pm 1)$ MeV, and of the KZ_c configuration is $M - i\Gamma/2 = (4306 \pm 1) - i(9 \pm 1)$ MeV. After averaging, the mass of the K^* meson is 4307 ± 2 MeV with a width of 18 ± 4 MeV. Note that our result is consistent with the one of Ref. 11, where the Born-Oppenheimer approximation is applied to the $KD\bar{D}^*$ system by solving the Schrödinger equation.

3. $K^*(4307)$ decay

In the following, we investigated the properties of the K^* state and calculated the decay widths of the possible two-body channels. As shown in Fig. 1, the magnitude of the squared amplitude in the KZ_c configuration

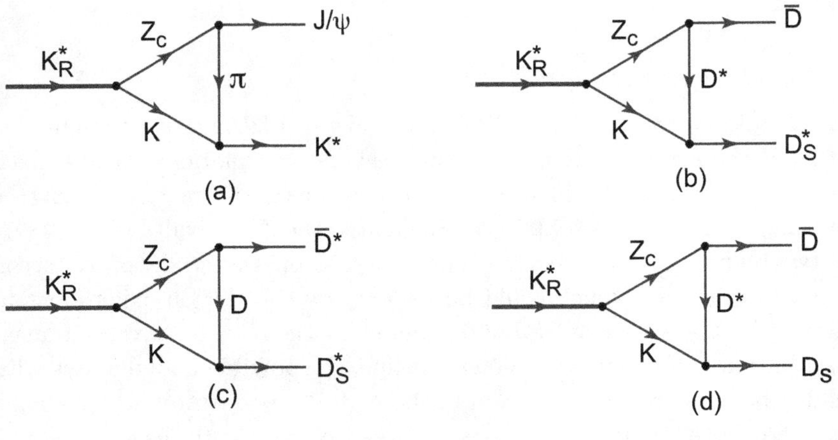

Fig. 2. Main two-body decay channels for the predicted $K^*(4307)$ meson.

is around 200 times larger than that found in the KX configuration. Thus, the decay width of $K^*(4307)$ is mainly from the subprocess $K^* \to KZ_c$. In Fig. 2, the open two-body decay channels, $J/\psi K^*(892)$, $\bar{D}D_s$, $\bar{D}D_s^*$, and $\bar{D}^* D_s^*$, via the triangle loops are presented.

For the details about the determination of the t_i $(i = a,\ b,\ c,\ d)$ amplitudes and the calculation of the triangular loops using the momentum cutoff regularization, we refer the reader to Ref. 6.

Once we have the t_i, the decay width of the $K^*(4307)$ meson to the two-body channels can be obtained

$$\Gamma_i = \int \frac{d\Omega}{4\pi^2} \frac{1}{8M_{K_R^*}^2} \frac{p_{\text{c.m.}}}{3} \sum |t_i|^2 = \frac{p_{\text{c.m.}}}{24\pi M_{K_R^*}^2} \sum |t_i|^2, \tag{5}$$

where $d\Omega$ represents the solid angle, $p_{\text{c.m.}}$ is the center of mass momentum of the particles in the final state, the factor 3 has its origin in the average over the $K^*(4307)$ meson polarizations and the symbol $\sum$ indicates summation over the polarization of the initial and final states.

With the momentum cutoff Λ changing from 700 MeV to 800 MeV, as used in Ref. 4, the decay widths from different channels are

$$\begin{aligned} \Gamma_a &= 6.97 \pm 0.27 \text{ MeV}, \quad \Gamma_b = 0.54 \pm 0.08 \text{ MeV}, \\ \Gamma_c &= 0.54 \pm 0.07 \text{ MeV}, \quad \Gamma_d = 1.14 \pm 0.17 \text{ MeV}. \end{aligned} \tag{6}$$

Besides, we also considered the width of $K^*(892)$ in the evaluation of the decay width of $K^*(4307) \to J/\psi K^*$. Since the mass of $K^*(4307)$ is far from the $J/\psi K^*(892)$ threshold, the decay width Γ_a does not change.

4. Conclusion

In this talk, we presented a prediction of heavy $K^*(4307)$ meson with hidden charm in the $KD\bar{D}^*$ system by solving the Faddeev equations with the fixed center approximation. In order to provide more information about the internal structure on $K^*(4307)$, we calculated the decay width of $K^*(4307)$ to two-body channels. We are now working on the study of B meson decay to $J/\psi\pi\pi K$, which could be handled by e.g. LHCb experiment to investigate the signal of $K^*(4307)$ meson in the $J/\psi\pi K$ invariant mass distribution.[12] Furthermore, along this line, the possible bound states with hidden bottom are also predicted in the $\bar{K}^{(*)}B^{(*)}\bar{B}^{(*)}$ systems.[13] We hope that the above studies could arouse interest to investigate the exotic states in the strange sector.

Acknowledgments

X.-L.R. acknowledge the financial support from the Hadron 2019 conference. This work was partly supported by DFG and NSFC through funds provided to the Sino-German CRC 110 "Symmetries and the Emergence of Structure in QCD" (DFG Grant No. TRR110), the NSFC under Grants No. 11735003, No. 11522539, 11375024, and No. 11775099, Fundação de Amparo à pesquisa do estado de São Paulo (FAPESP) under Grants No. 2019/16924-3 and No. 2019/17149-3, and CNPq (Grant Nos. 310759/2016-1 and 311524/2016-8).

References

1. H.-X. Chen, W. Chen, X. Liu, Y.-R. Liu and S.-L. Zhu, Rept. Prog. Phys. **80**, 076201 (2017).
2. H.-X. Chen, W. Chen, X. Liu and S.-L. Zhu, Phys. Rept. **639**, 1 (2016).
3. COMPASS Collaboration [S. Wallner, et al.,], "Strange-meson spectroscopy at COMPASS" in the Hadron (2018) proceedings.
4. X.-L. Ren, B. B. Malabarba, L.-S. Geng, K. P. Khemchandani and A. Martínez Torres, Phys. Lett. B **785**, 112 (2018).
5. S. S. Kamalov, E. Oset and A. Ramos, Nucl. Phys. A **690**, 494 (2001).
6. X.-L. Ren, B. B. Malabarba, K. P. Khemchandani and A. Martínez Torres, JHEP **1905**, 103 (2019).
7. D. Gamermann, E. Oset, D. Strottman and M. J. Vicente Vacas, Phys. Rev. D **76**, 074016 (2007).
8. F. Aceti, M. Bayar, E. Oset, A. Martínez Torres, K. P. Khemchandani, J. M. Dias, F. S. Navarra and M. Nielsen, Phys. Rev. D **90**, 016003 (2014).
9. F.-K. Guo, P. N.-Shen, H.-C. Chiang, R.-G. Ping and B.-S. Zou, Phys. Lett. B **641**, 278 (2006).
10. F.-K. Guo, P.-N. Shen and H.-C. Chiang, Phys. Lett. B **647**, 133 (2007).
11. L. Ma, Q. Wang and U.-G. Meißner, Chin. Phys. C **43**,014102 (2019).
12. X.-L. Ren, K. P. Khemchandani and A. Martínez Torres, in preparation.
13. X.-L. Ren and Z.-F. Sun, Phys. Rev. D **99**, 094041 (2019).

A possible prescription for incorporating the Nambu-Goldstone pions within the quark model

K. Yamada*, T. Komada and T. Maeda

Department of Science and Manufacturing Technology,
Junior College Funabashi Campus, Nihon University,
Funabashi, Chiba 274-8501, Japan
**E-mail: yamada.kenji@nihon-u.ac.jp*

M. Oda

Research Institute of Science and Technology,
College of Science and Technology, Nihon University,
Tokyo 101-8308, Japan

The Nambu-Goldstone boson nature of the pions in QCD is not taken into consideration in the conventional quark model at all. We present a possible prescription for incorporating its nature into the pions within the quark model by extending the Pauli spinors of quarks to the Dirac spinors. As an ideal case we take the chiral form $\gamma^5/2$ for the pion spin wave function and apply it to the decay $\rho(770) \to \pi\gamma$, then it is shown that its decay width and the $\rho(770)$ meson magnetic moment are both described consistently.

1. Introduction

The quark model has been applied with considerable success to mass spectra, strong and electromagnetic decays of hadrons. In these approaches only the degrees of freedom of the valence quark are retained, thus the pions are simply treated as interacting quark-antiquark pairs, which are identical to the $\rho(770)$ mesons, except for the internal spin structure of the quarks. Meanwhile, from the point of view of QCD the pions are the pseudo-Nambu-Goldstone (NG) bosons associated with the spontaneous breaking of chiral symmetry. This NG boson nature of the pions is not taken into consideration in the quark model at all.

In this work we present a possible prescription for incorporating the NG boson nature into the pions within the quark model by extending the Pauli spinors of quarks to the Dirac spinors. We will take the irreducible representation (1/2, 1/2) of the chiral symmetry group $SU(2)_L \times SU(2)_R$ for the pion spin wave function (WF) as an ideal case. In order to examine the validity of this chiral spin WF, it will be applied to the $\pi\gamma$ decay of the $\rho(770)$ meson in the covariant oscillator quark model (COQM).[1,2]

2. The covariant oscillator quark model

Basic framework of the COQM.[3] In the COQM quark-antiquark meson systems are described by the bilocal field $\Psi(x_1, x_2)_\alpha{}^\beta = \Psi(X, x)_\alpha{}^\beta$, where x_1^μ (x_2^μ) is the space-time coordinate, α (β) the Dirac spinor index of the constituent quark (antiquark), and the center-of-mass and relative coordinates are defined, respectively, by $X^\mu = (m_1 x_1^\mu + m_2 x_2^\mu)/(m_1 + m_2)$ and $x^\mu = x_1^\mu - x_2^\mu$ with the constituent quark (antiquark) mass m_1 (m_2).The bilocal meson field is required to satisfy the Klein-Gordon-type equation

$$\left(-\frac{\partial^2}{\partial X_\mu \partial X^\mu} - \mathcal{M}^2(x)\right) \Psi(X, x)_\alpha{}^\beta = 0, \tag{1}$$

where $\mathcal{M}^2(x)$ is the squared-mass operator, which is given by the four-dimensional harmonic oscillator in the pure confining force limit. A solution of Eq. (1) can be written as

$$\Psi(X, x)_\alpha^{\pm\beta} = N e^{\mp i P_\mu X^\mu} \Phi(v, x)_\alpha^{\pm\beta}, \quad v^\mu = P^\mu / M, \tag{2}$$

where N is the normalization constant for the plane wave, P^μ and M are the center-of-mass momentum and mass of the whole meson system, respectively, and $\Phi(v, x)_\alpha^{\pm\beta}$ is the covariant internal WF which is given by the direct product of the eigenfunctions of $\mathcal{M}^2(x)$ and the covariant spinor WFs, defined by the direct tensor product of respective Dirac spinors, with the meson four velocity v^μ, for the constituent quark and antiquark.

Covariant spin wave functions of $q\bar{q}$ meson systems. The covariant spin WFs of $q\bar{q}$ meson systems are expressed in terms of the individual Dirac spinor functions as $W_\alpha^{+\beta}(v) = u_\alpha(v)\bar{v}^\beta(v)$ and $W_\alpha^{-\beta}(v) = v_\alpha(v)\bar{u}^\beta(v)$, where $u(v)$ ($v(v)$) is the free Dirac spinor with positive (negative) energy. The $W_\alpha^{\pm\beta}(v)$ can be covariantly decomposed on the basis of the Dirac matrices $(\mathbf{1}, \gamma^\mu, \sigma^{\mu\nu}, \gamma^\mu\gamma^5, \gamma^5)$. The definite J^{PC} states with 0^{-+} and 1^{--} in the nonrelativistic quark model correspond respectively to the covariant spin WFs, $\gamma^5(1 - \not{v})/2\sqrt{2}$ and $\gamma^\mu(1 - \not{v})\varepsilon_\mu(v)/2\sqrt{2}$, where $\varepsilon_\mu(v)$ is the polarization vector for vector meson fields. In order to incorporate the NG boson nature of the pions, in addition to $\gamma^5(1 - \not{v})/2\sqrt{2}$, we introduce another spin WF, $-\gamma^5(1 + \not{v})/2\sqrt{2}$, which is orthogonal to it. These two orthogonal states can be represented in terms of another basis ($\gamma^5/2$,

$-\gamma^5 \not{p}$), in which basis vectors correspond to the irreducible representations $(1/2, 1/2)$ and $(0,1)+(1,0)$, respectively, of the chiral symmetry group $SU(2)_L \times SU(2)_R$. Hereafter we ideally consider $\gamma^5/2$ for the spin WF of the pions as a reflection of the NG boson nature.

Electromagnetic currents for $q\bar{q}$ meson systems. In the previous work[4] we have found that the dimension of bilocal meson fields is bosonic for light-quark systems, while fermionic for heavy-light and heavy-heavy systems. Since the present work is concerned with light-quark systems, we take the Klein-Gordon-like Lagrangian for bilocal meson fields. The Klein-Gordon-type equation (1) is derived from the action, in terms of the quark and antiquark coordinates, with the Klein-Gordon-like Lagrangian

$$\mathcal{L}_{\text{free}}^{(\text{KG})} = \overline{\Psi}(x_1, x_2) 2(m_1 + m_2) \left(\sum_{i=1}^{2} \frac{-1}{2m_i} \frac{\overleftarrow{\partial}}{\partial x_{i\mu}} \frac{\overrightarrow{\partial}}{\partial x_i^{\mu}} - U(x_1, x_2) \right) \Psi(x_1, x_2), \tag{3}$$

where $U(x_1, x_2)$ is the four-dimensional harmonic potential. It should be noted that the bilocal field $\Psi(x_1, x_2)$ has the dimension of bosons $[\mathrm{M}^1]$, except for the dimension of internal WFs $[\mathrm{M}^2]$. The interaction of quark-antiquark meson systems with an electromagnetic field can be obtained[5,6] by the minimal substitutions $\partial/\partial x_i^{\mu} \to \partial/\partial x_i^{\mu} + ieQ_i A_{\mu}(x_i)$ in the free Lagrangian (3), in which the heuristic prescription by Feynman, Kislinger and Ravndal[7] with some extension is adopted as the following replacements[6]

$$\frac{\overleftarrow{\partial}}{\partial x_{i\mu}} \frac{\overrightarrow{\partial}}{\partial x_i^{\mu}} \to (1 - g_M^{(i)}) \frac{\overleftarrow{\partial}}{\partial x_{i\mu}} \frac{\overrightarrow{\partial}}{\partial x_i^{\mu}} + g_M^{(i)} \frac{\overleftarrow{\partial}}{\partial x_i^{\mu}} \gamma^{\mu} \gamma^{\nu} \frac{\overrightarrow{\partial}}{\partial x_i^{\nu}}, \tag{4}$$

where Q_i $(i = 1, 2)$ are the quark and antiquark charges in units of e and $g_M^{(i)}$ are the parameters related to the anomalous magnetic moments of constituent quarks. Then the action for the electromagnetic interaction of meson systems is obtained,[2] up to the first order of e, as

$$S_{\text{int}}^{(\text{KG})} = \int d^4 x_1 \int d^4 x_2 \sum_{i=1}^{2} j_i^{(\text{KG})\mu}(x_1, x_2) A_{\mu}(x_i) \tag{5}$$

with the conserved currents

$$j_i^{(\text{KG})\mu}(x_1, x_2) = 2(m_1 + m_2)$$
$$\times \left\langle \overline{\Psi}(x_1, x_2) \frac{-ieQ_i}{2m_i} \left(\frac{\overleftrightarrow{\partial}}{\partial x_{i\mu}} - i g_M^{(i)} \sigma^{\mu\nu} \left(\frac{\overrightarrow{\partial}}{\partial x_i^{\nu}} + \frac{\overleftarrow{\partial}}{\partial x_i^{\nu}} \right) \right) \Psi(x_1, x_2) \right\rangle, \tag{6}$$

where $\langle \cdots \rangle$ means taking trace concerning the Dirac indices. The conserved currents (6) have no absolute mass scales of constituent quarks. This would seem to be natural for light-quark systems from the viewpoint of QCD in the chiral limit.

3. Magnetic moment and $\pi\gamma$ decay of the $\rho(770)$ meson

Magnetic moment of the $\rho(770)$ meson. The effective electromagnetic interaction (5) describes systematically all the electromagnetic interactions of $q\bar{q}$ meson systems. For the magnetic moment of the $\rho(770)$ meson the expression $\mu_\rho = 2g_M(e/2M_\rho)$ is obtained,[1] where M_ρ is the $\rho(770)$ meson mass and g_M the parameter related to the anomalous magnetic moment of u and d quarks, assuming the isospin symmetry, $g_M^{(u)} = g_M^{(d)}$. The $g_M = 1$ corresponds to the normal moment of quarks. Since the $\rho(770)$ meson is unstable, its lifetime being very short, there are not ever direct measurements for its magnetic moment. So, we use the lattice QCD results for this to estimate reasonable values of the parameter g_M. The values of μ_ρ in the lattice QCD calculations are tabulated in Table 1. Using the weighted and simple averages of the lattice QCD results, we obtain $g_M = 1.19$ and 1.10, respectively.

Table 1. The $\rho(770)$-meson magnetic moment in the lattice QCD calculations (in units of $e/2M_\rho$).

μ_ρ	Weighted average	Simple average	Reference
2.25 ± 0.34			8
2.39 ± 0.01			9
2.21 ± 0.08	2.38 ± 0.01	2.20	10
2.11 ± 0.10			11
2.02	—		12

The $\pi\gamma$ decay width of the $\rho(770)$ meson. For the radiative decays of light-quark mesons the decay width is obtained, following the usual procedure with the Klein-Gordon-like interaction (5). Here we restrict ourselves to a discussion on the radiative decays of S-wave isovector mesons with nonstrange quarks. Then the present radiative decay model has the only parameter of g_M. For the spin WFs of the pions we consider two ideal cases, the nonrelativistic and chiral WFs, $\gamma^5(1-\not{v})/2\sqrt{2}$ and $\gamma^5/2$, respectively. For the $\pi^\pm\gamma$ decay widths of the $\rho(770)^\pm$ meson, we tabulate the numerical results in Table 2. These results show that the chiral spin WF is favored over the nonrelativistic one for the reasonable values of g_M, that is, the $\rho(770)$-meson magnetic moment.

Table 2. Numerical results of the $\pi^{\pm}\gamma$ decay widths for the $\rho(770)^{\pm}$ mesons in keV.

g_M	Nonrelativistic spin WF	Chiral spin WF	Experiment[13]
1.19	134	67	68 ± 7
1.10	115	57	

4. Summary and future work

The calculated results for the $\pi\gamma$ decay and magnetic moment of the $\rho(770)$ meson show that they are both described consistently by the chiral spin WF of the pions. It is necessary to examine the further validity of the chiral spin WF by applying to other $\pi\gamma$ decay processes, such as those of excited light-quark $q\bar{q}$ mesons. It is also intriguing to investigate a mixture of the nonrelativistic and chiral spin WFs for excited-state mesons.

References

1. S. Ishida *et al.*, *Prog. Theor. Phys.* **71**, 806 (1984).
2. S. Ishida, K. Yamada and M. Oda, *Phys. Rev. D* **40**, 1497 (1989).
3. S. Ishida and K. Yamada, *Phys. Rev. D* **35**, 265 (1987).
4. K. Yamada and T. Maeda, PoS (Hadron2017) 080.
5. R. G. Lipes, *Phys. Rev. D* **5**, 2849 (1972).
6. S. Ishida and J. Otokozawa, *Prog. Theor. Phys.* **53**, 217 (1975).
7. R. P. Feynman, M. Kislinger and F. Ravndal, *Phys. Rev. D* **3**, 2706 (1971).
8. W. Andersen and W. Wilcox, *Ann. Phys. (N.Y.)* **255**, 34 (1997).
9. F. X. Lee, S. Moerschbacher and W. Wilcox, *Phys. Rev. D* **78**, 094502 (2008).
10. B. J. Owen *et al.*, *Phys. Rev. D* **91**, 074503 (2015).
11. E. V. Luschevskaya, O. E. Solovjeva and O. V. Teryaev, *J. High Energy Phys.* 09 (2017) 142.
12. J. N. Hedditch *et al.*, *Phys. Rev. D* **75**, 094504 (2007).
13. M. Tanabashi *et al.* (Particle Data Group), *Phys. Rev. D* **98**, 030001 (2018).

Relativistic effects in radiative charmonium transitions: A covariant quark model approach

Tomohito Maeda*, Toshihiko Komada and Kenji Yamada

Department of Science and Manufacturing Technology, Junior College Funabashi Campus, Nihon University,
Funabashi 274-8501, Japan
**E-mail: maeda.tomohito@nihon-u.ac.jp*

Masuho Oda

Research Institute of Science and Technology, College of Science and Technology, Nihon University,
Tokyo 101-8308, Japan

We investigate the radiative transitions of charmonium system in the framework of the relativistic covariant quark model, where the center of mass motion of the hadron is treated in a manifestly covariant way. We discuss the obtained results in comparison with that in the nonrelativistic model.

Keywords: Charmonium; radiative transition; relativistic quark model.

1. Introduction

The non-relativistic quark model (NRQM) has been widely applied to study of hadron spectroscopy and structure, and has achieved significant success. It is well known that the study of radiative transitions provides a useful tool to confirm the validity of the wave functions (WFs) that reproduce the mass levels of the hadrons, since the fundamental electromagnetic interaction are well understood. Calculated results of the NRQM are typically reliable for heavy quarks, where non-relativistic treatment is permissible, though not for light quarks. Thus far the charmonium system have been regarded as a typical example of a non-relativistic system, so that many authors have applied the NRQM to analyze the radiative transitions for $c\bar{c}$ mesons and have found that the models generally reproduce the overall trend of experiments. However, recent high statistical experiments[1–3] on the charmonium radiative decays might provide an opportunity to elucidate a contribution not taken into account in the simple NRQM. In this work, we study the radiative transitions of the $c\bar{c}$ system in the relativistic covariant quark model where the center of mass motion of the hadron is treated in a manifestly covariant way. This feature makes it possible to take into account the boosting effect of the spin wave function and the deformation of the form factor (including the contribution of higher-order multipoles) due to the Lorentz contraction. Although our collaboration have tackled this problem about 20 years ago,[4] the long wavelength approximation was assumed and the relativistic effect of the form factor was not evaluated properly at that time. Thus in our revisit to this problem, we focus on the relativistic effect in our approach and carefully study how it affects the calculated widths.

2. The boosted LS coupling scheme

In this study, we adopt the covariant quark model formulated in the boosted-LS (bLS) coupling scheme.[4,5] In this scheme, the internal WF of $c\bar{c}$ mesons are described as

$$\Phi(v, x^\mu)_\alpha^\beta = f(v, x)^{(nL)\nu\rho\cdots} \left(W(v)_\alpha^\beta\right)_{\nu\rho\cdots}, \tag{1}$$

where the v^μ and x^μ represent the center of mass (CM) velocity and the relative coordinates of $c\bar{c}$ system, respectively. The α and β denote the Dirac spinor indices corresponding to respective constituents. The WF consists of the spacetime part $f(v, x)$ and the spin part $W(v)$,[a] being separately extended to the covariant form from the NRQM. Due to limitations of the space, the detailed description of our formulation have to be omitted here (see, Ref. 4 for details). We emphasize that the WFs depend on the CM velocity of mesons, hence the matrix elements concerning the spacetime part and the spin part, respectively, incorporate the effect of Lorentz boost for the hadronic CM motion. If we set the v, denoting the velocity of the final state at the rest frame of the initial state, in the WF to zero, the expression of our model basically reduces to the usual simple NRQM (except for the treatment of relative-time degrees of freedom). In this study, within the framework of our model, we calculate the decay amplitudes in both cases, the non-relativistic ($v = 0$) and relativistic ($v \neq 0$) and compere them to elucidate how the relativistic contribution appears in each process of charmonium radiative transitions. Since our purpose of this study is to examine

---

[a]Concretely, we use the definite metric 4-dimensional simple harmonic oscillator functions[6,7] for the $f(v, x)$ and the irreducible product of Dirac spinors with the velocity v^μ for the $W(v)$, respectively.

how the relativistic correction affects, we take a typical set of parameters without tuning them to reproduce the experimental values. In practice, we use the following set of parameters:

$$\begin{aligned}&\text{magnetic moment parameter} : g_M^{(c)} = 1,\\&\text{quark mass} : m_c = M_{J/\psi}/2 = 1.55 \text{ GeV},\\&\text{harmonic oscillator scale parameters} :\\&\beta = 0.71, 0.66, 0.50, 0.49, 0.47 \text{ GeV for } 1^1S_0, 1^3S_1, 1^1P_1, 1^3P_{2,1,0}, 2^3S_1.\end{aligned} \tag{2}$$

3. Results and discussion

In the following, we refer to the case in the relativistic treatment as the "R-case", and that in the non-relativistic treatment as the "NR-case".

Table 1. Calculated widths for the $n^3S_1 \to 1^1S_0 + \gamma$ ($n = 1, 2$) (keV).

Initial	Final	NR → R	Exp.[8]	BGS2005[9]	DLGZ2017[10]
J/ψ	$\eta_c(1S)$	$2.35 \to 2.41$	1.58 ± 0.37	2.9/2.9/2.4 NR/NR(j_0)/GI	2.39/2.44 LP/SP
$\psi(3686)^\dagger$	$\eta_c(1S)$	$1.56 \to 3.11$	1.00 ± 0.15	4.6/9.7/9.6 NR/NR(j_0)/GI	8.08/7.80 LP/SP

Note: † Here we have supposed $\psi(3686) \equiv \psi(2S)$.

As shown in Table 1, for the most familiar M1 transition $J/\psi \to \eta_c(1S)$, we found that the width increases slightly in the R-case compared to that in the NR-case due to the boosting of the spin WF, but it remains almost unchanged. This is a natural result because a transition for the small mass difference causes almost no recoil. Similarly for M1 transition $\psi(2S) \to \eta_c(1S)$, it found that the result in the R-case is larger than that in the NR-case.[b] In this case, not only overlapping of the spin WF but also the spacetime WF contributes to increase the width in the R-case. This trend is the same as Ref. 9. Note that calculated results for both process seem to deviate from the PDG values, though similar results are obtained with other quark models. See Refs. 11–15 about recent developments on this issue.

[b]Here we show only in the case that $\beta_{\text{initial}} = \beta_{\text{final}} = 0.54$ GeV since the numerical result of the spacetime OI is not stable.

Table 2. Calculated widths for the $1^1P_1 \to 1^1S_0 + \gamma$, $1^3P_J \to 1^3S_1 + \gamma$, $2^3S_1 \to 1^3P_J + \gamma$ $(J = 0, 1, 2)$ (keV).

Initial	Final	NR → R	Exp.[8]	BGS2005[9]	DLGZ2017[10]
$h_c(1P)$	$\eta_c(1S)$	377 → 330	357 ± 208	498/352 NR/GI	361/373 LP/SP
$\chi_{c2}(1P)$	$J/\psi(1S)$	292 → 252	374 ± 20	424/313 NR/GI	284/292 LP/SP
$\chi_{c1}(1P)$	$J/\psi(1S)$	242 → 254	288 ± 16	314/239 NR/GI	306/319 LP/SP
$\chi_{c0}(1P)$	$J/\psi(1S)$	113 → 132	151 ± 10	152/114 NR/GI	172/179 LP/SP
$\psi(3686)^\dagger$	$\chi_{c2}(1P)$	24.3 → 24.9	28.0 ± 0.96	38/24 NR/GI	38/46 LP/SP
$\psi(3686)^\dagger$	$\chi_{c1}(1P)$	34.2 → 31.8	28.7 ± 1.1	54/29 NR/GI	42/45 LP/SP
$\psi(3686)^\dagger$	$\chi_{c0}(1P)$	40.7 → 30.3	28.8 ± 0.98	63/26 NR/GI	22/22 LP/SP

Note: † $\psi(3686) \equiv \psi(2S)$.

Calculated results for $S \leftrightarrow P$ transitions are shown in Table 2. The spin non-flip transition $1^1P_1 \to 1^1S_0$ is a typical E1 transition in which the spin current does not contribute. The result shows Lorentz contraction of the electric dipole matrix element reduces the width about 10% in the R-case. A similar effect occurs in the transition $^3P_J \leftrightarrow {}^3S_1$, but in these case, the spin current term also contributes constructively or destructively depending on the process, resulting widths increase or decrease. Our covariant form factor in the spin current term includes the so-called "Extra E1" and M2 contributions.[10] These corrections to the E1 main term should be different from the NRQM, but there seems to be no apparent difference in these processes. These results indicate that, in order to confirm the difference between our relativistic model and the NRQM, it is necessary to examine the transition between the levels with large mass differences.

4. Summary and remarks

Motivated by the recent precise measurements for the charmonium, we have revisited their radiative transitions in the covariant quark model based on bLS scheme. As a result, in most cases, the over/under-estimated widths of the NR-case can be reduced to the desired widths by relativistic effects. On the other hand, there are various reports[16–18] that experimental data can be reproduced well as a result of incorporating the effects of couple channels (unquenched quark model). It is interesting that both contributions (relativistic correction and the coupled channel effect) play an effective role to reproduce the experiment. In order to clarify the relativistic/couple channels effects in charmonium transitions, further studies on the decays with large mass differences would be required.

References

1. M. Ablikim *et al.* (BESIII Collaboration), Phys. Rev. Lett. **115**, 011803 (2015).
2. M. Ablikim *et al.* (BESIII Collaboration), Phys. Lett. B **753** (2016)103.
3. M. Ablikim *et al.* (BESIII Collaboration), Phys. Rev. D **91**, 092009 (2015); **95**, 072004 (2017); **96**, 032001 (2017).
4. S. Ishida, A. Morikawa and M.Oda, Prog. Theor. Phys. **99**, 257 (1998).
5. S. Ishida, K. Yamada and M.Oda, Phys. Rev. D**40**, 1497 (1989).
6. T. Takabayasi, Prog. Theor. Phys. Suppl. **67**, 1 (1979).
7. S. Ishida and M. Oda, Prog. Theor. Phys. **89**, 1033 (1993).
8. M. Tanabashi *et al.* (Particle Data Group), Phys. Rev. D **98**, 030001 (2018) and 2019 update.
9. T. Barnes, S. Godfrey and E. S. Swanson, Phys. Rev. D **72**, 054026 (2005).
10. Wei-Jun Deng, Hui Liu, Long-Cheng Gui and Xian-Hui Zhong, Phys. Rev. D **95**, 034026 (2017).
11. V. V. Anashin *et al.* (KEDR Collaboration), Phys. Lett. B **738** (2014) 391.
12. J.J. Dudek, R. G. Edwards and D. G. Richards, Phys. Rev. D **73**, 074507 (2006).
13. J. J. Dudek, R. G. Edwards and C. E. Thomas, Phys. Rev. D **79**, 094504 (2009).
14. D. Becirevic and F. Sanfilippo, JHEP01(2013)028.
15. D. Becirevic, M. Kruse and F. Sanfilippo, JHEP05(2015)014.
16. E. J. Eichten, K. Lane and C. Quigg, Phys. Rev. D **69**, 094019 (2004).
17. G. Li and Q. Zhao, Phys. Lett. B **670**, 55 (2008); Phys. Rev. D **84**, 074005 (2011).
18. C. W. Zhao, G. Li, X. H. Liu, and F. L. Shao, Eur. Phys. J. C **73**, 2482 (2013).

Overview of light hadron physics at KLOE/KLOE-2

Xiao Lin Kang

LNF-INFN, Via E. Fermi 40, Frascati (RM), 00044, Italy
E-mail: xiaolin.kang@lnf.infn.it

(on behalf of KLOE-2 Collaboration)

The KLOE-2 experiment has completed its data-taking campaign at the e^+e^- DAΦNE collider in Frascati, achieving the integrated luminosity goal of more than 5 fb^{-1} at the ϕ-meson peak. This new data sample, together with the KLOE data, corresponds to 2.4×10^{10} ϕ and 3.1×10^8 η meson decays, which supports plenty of precision measurements. An overview of the recent results and status of the light hadron physics are presented.

Keywords: KLOE/KLOE-2; e^+e^- collision; light hadron physics.

1. π^0 production from $\gamma\gamma$ scattering

$\gamma\gamma$ couplings to mesons and the partial widths provide information about their structure and can be measured in the $e^+e^- \to e^+e^-\gamma^*\gamma^* \to e^+e^-X$ processes, where X refers to the final state with $J^{PC} = 0^{\pm+}$, $2^{\pm+}$ and can be π^0, $\pi\pi$, $\eta\pi$, or η.[1] A precise knowledge of the $\gamma\gamma$ decay width provides a measurement of the transition form factor, which contributes to the hadronic light-by-light term of the anomalous magnetic moment of the muon.[2]

Two High Energy Tagger (HET) stations,[3] installed in KLOE-2[4] to tag electrons and positrons from the reaction $e^+e^- \to e^+e^-\gamma^*\gamma \to e^+e^-\pi^0$, give the opportunity to investigate $\gamma^*\gamma \to \pi^0$ physics at the ϕ resonance, aiming to achieve an accuracy of $\mathcal{O}(1\%)$ matching the current theory precision and the first measurement of the transition form factor $F_{\pi^0\gamma^*\gamma}(q^2)$ in the space-like region at low momentum transfer $q^2_{\gamma^*} \leq 0.1$ GeV^2.[2] By analysing a sample of 500 pb^{-1}, a clear statistical evidence of correlated coincidence events between the tagger station hits and KLOE calorimeter clusters has been observed for the first time, after the subtraction of continuously registered accidentals out of coincidence along the runs. Multivariate analysis is used to seperate signal from the radiative Bhabha background.

The analysis is being extended to a sample of 1.5 fb^{-1} and refined studies of the tagged sample to improve the rejection of backgrounds are ongoing.

2. Search for P and CP violating decay $\eta \to \pi^+\pi^-$

In the Standard Model (SM), the decay $\eta \to \pi^+\pi^-$ can proceed only through CP-violating weak interaction with a branching fraction of order 10^{-27} according to Ref. 5, 6. Higher branching fractions can be achieved by introducing the CP-violating QCD θ-term (10^{-17}) or allowing a CP-violation in the extended Higgs sector (10^{-15}).[5,6] By taking into account the higher order chiral Lagrangian, the couplings of $\eta(\eta')\pi\pi$ can be connected with the neutron electric dipole moment (nEDM). A recent work by using the present upper bound on the nEDM gives an estimation of the upper limit to be 5.3×10^{-17}.[7] Therefore, any detection of a larger branching fraction would indicate a new source of CP violation in the strong interaction.

With the first 350 pb^{-1} data collected at ϕ resonance peak, KLOE collaboration set the best upper limit on the branching fraction, $\mathcal{B}(\eta \to \pi^+\pi^-) \leq 1.3 \times 10^{-5}$ at 90% confidence level (CL).[8] A new preliminary limit has been determined by the KLOE-2 collaboration with an integrated luminosity of 1.6 fb^{-1}. After the acceptance cut and event selection criteria, the $\pi^+\pi^-$ invariant mass distribution, $M(\pi^+\pi^-)$, is shown in Figure 1-left, which is mainly $e^+e^- \to \pi^+\pi^-\gamma(\gamma)$ events. No peak is observed in the η mass region. The preliminary upper limit on the branching fraction is determined to be 5.8×10^{-6} at 90% CL by using a Bayesian approach. With all the KLOE and KLOE-2 data the upper limit is expected to reach the level of 2.7×10^{-6}.

3. The $\eta \to \pi^0\gamma\gamma$ decay

The doubly radiative decay $\eta \to \pi^0\gamma\gamma$ decay is an important test of Chiral Perturbation Theroy (ChPT) because of its sensitivity to the p^6 term on both the branching fraction and the $M(\gamma\gamma)$ spectrum.[9] A preliminary KLOE measurement,[10] based on 450 pb^{-1} data, is $\mathcal{B}(\eta \to \pi^0\gamma\gamma) = (8.4 \pm 2.7_{stat} \pm 1.4_{syst}) \times 10^{-5}$, which is significantly lower than the PDG value,[11] $(25.6 \pm 2.2) \times 10^{-5}$, and two standard deviations lower than the predictions from Chiral-loop, LσM and VMD.[9,12]

The analysis for this decay with four times larger data sample is in progress at KLOE-2. Kinematic fits by enforcing the energy-momentum conservation and constraints on the time-of-flight of neutral clusters under

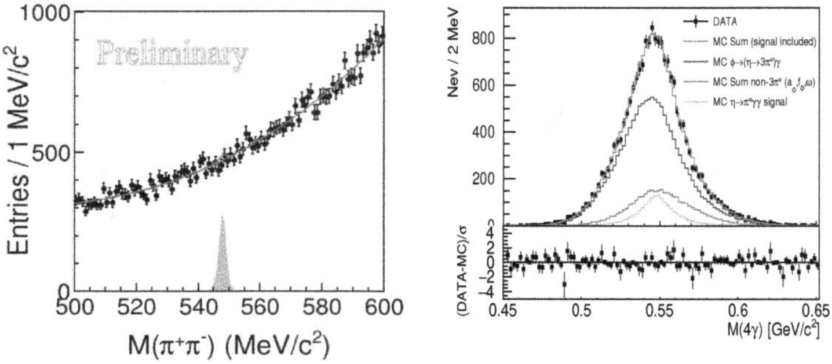

Fig. 1. Left: $\pi^+\pi^-$ invariant mass distribution for the observed $\phi \to \pi^+\pi^-\gamma$ candidates, the solid curve is a third-order polynomial function with superimposed a MC signal shape (shaded histogram in arbitrary units). Right: $\pi^0\gamma\gamma$ invariant mass spectrum of $\phi \to \eta(\pi^0\gamma\gamma)\gamma$ candidates for data (black dots), the sum of MC background contributions (red histogram), $\eta \to 3\pi^0$ (blue), the sum of the backgrounds from $\phi \to a_0(\eta\pi^0)\gamma$, $f_0(\pi^0\pi^0)\gamma$, and $e^+e^- \to \omega(\pi^0\gamma)\pi^0$ (violet), and the expected signal (green). The lower panel shows the ratio between data and total MC simulation.

different signal or background hypotheses and multivariate analysis with cluster shape information are performed to suppress backgrounds. The preliminary invariant mass distribuion of $\pi^0\gamma\gamma$ is shown in Figure 1-right. Further background subtraction and signal extraction is in progress and the preliminary result on the branching fraction is forthcoming. This decay also provides a probability of finding a leptophobic mediator B-boson between the dark sector and the SM particles via the decay chain $\eta \to B(\pi^0\gamma)\gamma$.[13]

4. Search for B-boson in $\phi \to \eta\pi^0\gamma$ process

The model presented in Ref. 13 proposes a gauge boson, dubbed B, that couples to the baryon number and arises from a $U(1)_B$ gauge symmetry in the GeV-scale, especially in the range 100 MeV< m_B <1 GeV.

In KLOE the searches for B-boson with mass below 600 MeV are performed, where the dominant decay is $B \to \pi^0\gamma$. Except the decay $\phi \to \eta(B\gamma)\gamma$ with $B \to \pi^0\gamma$, mentioned in section 3, the decay $\phi \to \eta B(\pi^0\gamma)$ is also used to search for a B-boson. In Figure 2-left, preliminary invariant mass spectrum of $\pi^0\gamma$ from the selected $\eta\pi^0\gamma$ candidates is shown. In case of formation of the B-boson, the signature is an enhancement in the corresponding $\pi^0\gamma$ invariant mass. In case no signal is observed an upper limit in the coupling of such a dark matter mediator to the SM as a function of its mass is going to be set.

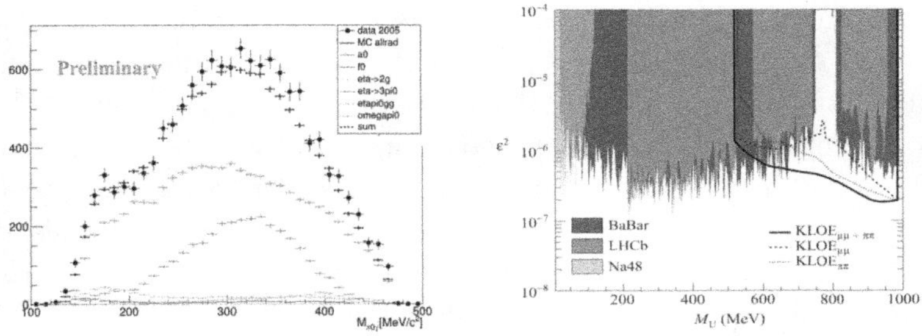

Fig. 2. Left: Invariant mass spectrum of $\pi^0\gamma$ candidates for the channel $\phi \to \eta B \to \eta\pi^0\gamma$ for data (black dots), the sum of MC simulated backgrounds (red), and different background contributions. Right: 90% CL exclusion plot for ε^2 as a function of the U-boson mass for the $e^+e^- \to U\gamma$ process with $U \to \mu^+\mu^-$ limit (dashed line), $U \to \pi^+\pi^-$ limit (dotted line), and combination of the two channels (solid line) with the competitive limits by BaBar, NA48/2, and LHCb experiments.

5. Combined limit on U-boson with U$\to$ $\pi^+\pi^-$ and $\mu^+\mu^-$

A new light vector gauge boson U is introduced as a intermediary between the Dark Matter and the SM particles, which could interact with the photon via a kinetic-mixing term $\varepsilon^2 = \alpha'/\alpha$, where $\alpha = \alpha_{em}$ and α' is the effective dark coupling constant. If the kinetic mixing appears at the one-loop level, ε can be estimated to be in the range $10^{-2} - 10^{-6}$, allowing visible effects at high luminosity e^+e^- colliders.

Using the initial state radiation (ISR) technique, KLOE experiment has extensively contributed to the U-boson searches. A previous search for the U-boson in the decay into $\mu^+\mu^-$[14] has been extended by using the full KLOE statistics at $L_{int} = 1.93$ fb^{-1},[15] updating the analysis with a new estimate of the background, analogous to the one used for the $U \to \pi^+\pi^-$ search.[16] This new search confirms the non existence of U-boson signal in the di-muon invariant mass spectrum. To increase the sensitivity in the region of the $\rho - \omega$ interference both results on the 90% upper limit for $\mu\mu$ and $\pi\pi$ have been combined, giving the up-to-date most stringent upper limit for the mixing parameter ε^2 in the U-boson mass region 519-987 MeV. The limit is shown in Figure 2-right, together with other most competitive limits in the region.

6. Combined hadronic cross section of $e^+e^- \to \pi^+\pi^-\gamma(\gamma)$

The $\sim$ 4 standard deviations discrepancy between the SM prediction,[17] $a_\mu^{\rm SM}$, and the experimental measurements of the anomalous magnetic moment of the muon,[18] $a_\mu^{\rm exp}$, provides one of the most powerful probe of physics beyond SM. A large part of the uncertainty on the theoretical estimates originates from the leading order hadronic vacuum contribution, $a_\mu^{\rm had,LO}$, which has to be evaluated with a dispersion integral using the bare hadronic cross sections $\sigma^0(e^+e^- \to \text{hadrons})$ as input. The $\pi^+\pi^-$ final state process, $a_\mu^{\pi^+\pi^-}$, contributes approximately 70% of the total estimate and is also the largest contributor to its uncertainty. The KLOE collaboration performed three precise measurements of the cross section $\sigma^0(e^+e^- \to \pi^+\pi^-\gamma(\gamma))$ below 1 GeV^2 using the method of radiative return.[19–21] All three measurements are undressed of all vacuum polarisation effects, including final state radiation and overlapping in the energy range 0.6-0.95 GeV. Using the proper statistical and systematic covariance matrices, the three measurements have been combined to provide both the two-pion cross section and the pion from factor.[22] Figure 3-left shows the KLOE combined $\pi^+\pi^-$ cross section measurement compared with results obtained by CMD-2 and SND experiments by energy scan, Babar, BESIII by radiative retrun. Figure 3-right reports the estimated $a_\mu^{\pi^+\pi^-}$ from KLOE combined in the overlapping region and all the above experiments. The combined contribution to muon magnetic moment anomaly in the full energy range ($0.10 < s < 0.95$ GeV^2), is also obtained, $a_\mu^{\pi^+\pi^-} = (489.8 \pm 5.1) \times 10^{-10}$, which is consistent with the CMD-2, SND and BESIII measurements within 1.5σ. The observed difference with the BaBar evaluation is below 3σ.

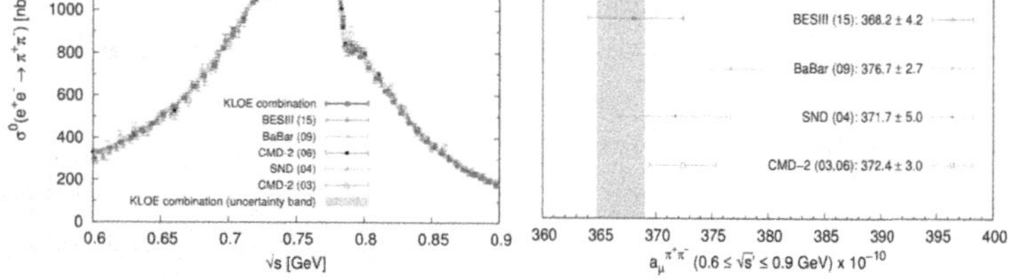

Fig. 3. Left: the $\pi^+\pi^-$ cross section from the KLOE combination compared with CMD-2, SND, BaBar and BESIII data. Right: estimated $a_\mu^{\pi^+\pi^-}$ from the KLOE combination, compared with CMD-2, SND, BaBar and BESIII in the 0.60.9 GeV range. The yellow band represents the KLOE combination uncertainty. All the uncertainties shown are the statistical and systematic uncertainties summed in quadrature.

References

1. G. Amelino-Camelia *et al.*, *Eur. Phys. J. C* **68**, 619 (2010).
2. D. Babusci *et al.*, *Eur. Phys. J. C* **72**, 1917 (2012).
3. D. Babusci *et al.*, *Acta Phys. Pol. B* **46**, 81 (2015).
4. A. Gallo *et al.*, *Conf. Proc. C* **060626**, 604 (2006); C. Milardi *et al.*, *JINST* 7, T03002 (2012); Bossi F., De Lucia E. *et al.* and KLOE Collaboration, *Nuovo Cimento.* **30**, 10 (2008).
5. C. Jarlskog, E. Shabalin, *Phys. Scr. T* **99**, 23 (2002).
6. E. Shabalin, *Phys. Scr. T* **99**, 104 (2002).
7. T. Gutsche *et al.*, *Phys. Rev. D* **99**, 031703 (2019) (R).
8. F. Ambrosino *et al.* (KLOE Collaboration), *Phys. Lett. B* **606**, 276 (2005).
9. L. Ametller, J. Bijnens, A. Bramon and F. Cornet, *Phys. Lett. B* **276**, 185 (1992).
10. B. Di Micco *et al.* (KLOE Collaboration), *Acta Phys. Slov.* **56**, 403 (2006).
11. M. Tanabashi *et al.* (Particle Data Group), *Phys. Rev. D* **98**, 030001 (2018).
12. R. Escribano, S. Gonzàlez-Solis, R. Jora and E. Royo, *arXiv:1812.08454 [hep-ph].*
13. S. Tulin, *Phys. Rev. D* **89**, 114008 (2014).
14. D. Babusci *et al.* (KLOE-2 Collaboration), *Phys. Lett. B* **736**, 459 (2014).
15. A. Anastasi *et al.* (KLOE-2 Collaboration), *Phys. Lett. B* **784**, 336 (2018).
16. A. Anastasi *et al.* (KLOE-2 Collaboration), *Phys. Lett. B* **757**, 356 (2016).
17. F. Jegerlehner, *EPJ Web of Conferences* **166**, 00022 (2018).
18. G. W. Bennett *et al.*, (Muon g-2 Collaboration), *Phys. Rev. Lett.* **89**, 101804 (2002); *Phys. Rev. Lett.* **92**, 161802 (2004); *Phys. Rev. D* **73**, 072003 (2006).
19. F. Ambrosino *et al.* (KLOE Collaboration), *Phys. Lett. B* **670**, 285 (2009).
20. F. Ambrosino *et al.* (KLOE Collaboration), *Phys. Lett. B* **700**, 102 (2011).
21. D. Babusci *et al.* (KLOE Collaboration), *Phys. Lett. B* **720**, 336 (2013).
22. A. Anastasi *et al.*, (KLOE-2 Collaboration), *J. High Energy Phys.* **03**, 173 (2018).

Studies of $\eta^{(\prime)}\pi$ final states using GlueX data

Colin Gleason, for the GlueX Collaboration

Department of Physics, Indiana University Bloomington,
Bloomington, IN 47405 United States of America
E-mail: colgleas@iu.edu

The primary goal of the GlueX experiment at Jefferson Lab is to search for and map the spectrum of light hybrid mesons. Many experiments have studied and reported evidence of exotic mesons decaying into $\eta\pi$ and $\eta'\pi$ final states. With a large acceptance to both charged and neutral particles, GlueX has access to both the neutral, $\gamma p \to \eta^{(\prime)}\pi^0 p$, and charged, $\gamma p \to \eta^{(\prime)}\pi^-\Delta^{++}$, exchanges. These proceedings will give an overview of the current studies being performed at GlueX in $\eta^{(\prime)}\pi^-$ final states. It will discuss early physics goals and outline the strategy for an amplitude analysis as GlueX begins its quest to illuminate the light hybrid meson spectrum.

1. Introduction

GlueX is a photoproduction experiment located at the Thomas Jefferson National Accelerator Facility in Newport News, VA. The goal of GlueX is to understand how hadrons are generated from quantum chromodynamics (QCD). To do this, GlueX will attempt to map the broad spectrum of hybrid mesons, as predicted by lattice QCD.[1] Some hybrids have quantum numbers (J^{PC}) that can not be constructed from a $q\bar{q}$ system. These are known as exotic hybrid mesons. The search for hybrids at GlueX begins with the lowest lying exotic hybrid, the $\pi_1(1600)$. Eventually, we will work towards measuring the properties of hybrids across their full spectrum which will lead to an understanding of how hadrons arise from QCD.

Final states with $\eta^{(\prime)}\pi$ are an ideal place to search for hybrids due to the fact that exotic J^{PC} arise if an $\eta^{(\prime)}\pi$ system is observed in an odd partial wave. Many experiments have searched for exotic J^{PC} in $\eta^{(\prime)}\pi$ final–states.[2–13] In general, these experiments report a P–wave ($J = 1$) enhancement corresponding to a $\pi_1(1400)$ in $\eta\pi$ final–states and/or a $\pi_1(1600)$ in $\eta'\pi$ final–states. The Joint Physics Analysis Center (JPAC) recently did a coupled–channel fit to the COMPASS data and showed that the $\pi_1(1400)$

and $\pi_1(1600)$ originate from the same pole, thus resolving the π_1 discrepancy seen in previous experiments.[13,14]

2. The GlueX Experiment

A schematic of the GlueX spectrometer can be seen in Figure 1. An electron beam with energies up to 12 GeV is delivered to Hall D and interacts with a thin radiator to produce the photon beam that continues down the beam line towards a LH_2 target. The target, central drift chamber, forward drift chamber, and barrel calorimeter are located inside of the solenoid magnet, which operates with a field of 2T. Further downstream and outside of the solenoid magnet are a time–of–flight detector, forward calorimeter, and a newly installed detection of internally reflected light (DIRC) detector. GlueX is unique in that it has a real linearly polarized photon beam and nearly full acceptance of both charged and neutral particles. GlueX began taking data in 2016 and completed Phase–I of its data taking in December of 2018 with $\approx$ 110 pb^{-} data collected. Phase–II begins in December 2019 and will use the newly installed DIRC.

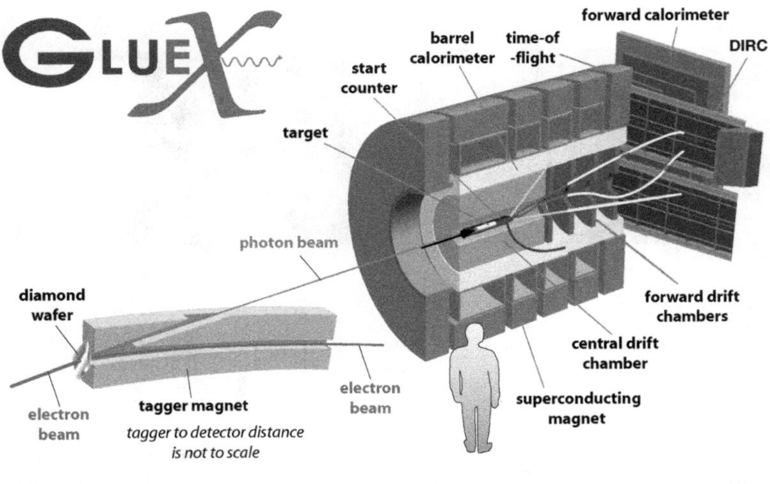

Fig. 1. The GlueX spectrometer.

3. $\eta^{(\prime)}\pi$ Systems With GlueX

GlueX has access to multiple channels and decay modes of $\eta^{(\prime)}\pi$ final–states, such as the neutral exchange $\eta^{(\prime)}\pi^0 p$ and charge exchange $\eta^{(\prime)}\pi^-\Delta^{++}$ reactions. For $\eta\pi$ systems, we are studying η decays into $\gamma\gamma$ or $\pi^+\pi^-\pi^0$. The different decay modes of the η should contain the same physics yet have different acceptances and backgrounds. Additionally, the charge exchange and neutral reactions are complimentary and should provide information on how hybrid mesons are produced.

3.1. $\gamma p \to \eta\pi^-\Delta^{++}$

Figure 2 shows $\cos\theta_{GJ}$ as a function of $\eta\pi^-$ mass for $\eta \to \pi^+\pi{-}\pi^0$ (left) and $\eta \to \gamma\gamma$ (right), not corrected for acceptance. In both distributions, clear enhancements are seen in the region of the $a_0^-(980)$ and $a_2^-(1320)$, which have the expected S and D wave structures. One can also see how the acceptances and backgrounds vary between η decay modes, particularly at large $\eta\pi$ masses. There is less acceptance as $\cos\theta \to -1$ for the 3π decay when compared to the 2γ decay. This leads to a larger amount of background at $\cos\theta \approx -1$ in $\eta \to \gamma\gamma$, which comes from double Regge exchange (e.g. Deck effect).

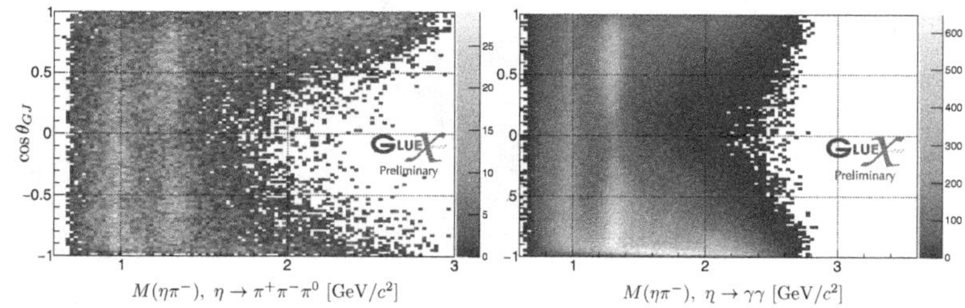

Fig. 2. $\cos\theta_{GJ}$ as a function of $M(\eta\pi^-)$ where $\eta \to \pi^+\pi^-\pi^0$ (left) and $\eta \to \gamma\gamma$ (right).

There are $\approx$ 10x more statistics in the $\eta \to \gamma\gamma$ than the $\eta \to \pi^+\pi^-\pi^0$ decay mode. The statistics in the $\eta \to \pi^+\pi^-\pi^0$ channel are comparable to the COMPASS result.[13] A short term physics goal in these channels is to study a_0 and a_2 production as a function of momentum transfer t. We hope to use this information to test our understanding of the acceptance and backgrounds. Additionally, we plan on studying double Regge exchange at large $\eta\pi$ masses. The goal here would be to constrain this background in the resonance (a_0, a_2, π_1) region.

3.2. $\gamma p \to \eta'\pi^-\Delta^{++}$

Figure 3 shows the $\eta'\pi^-$ mass (left) and $\cos\theta_{GJ}$ as a function of $\eta'\pi^-$ mass (right). In the $\eta'\pi^-$ mass plot, the red curve is an estimate of the acceptance, the black points are data in the region of the η', and the blue points are from the η' sidebands. The $\cos\theta_{GJ}$ distribution is neither sideband subtracted nor corrected for accpetance.

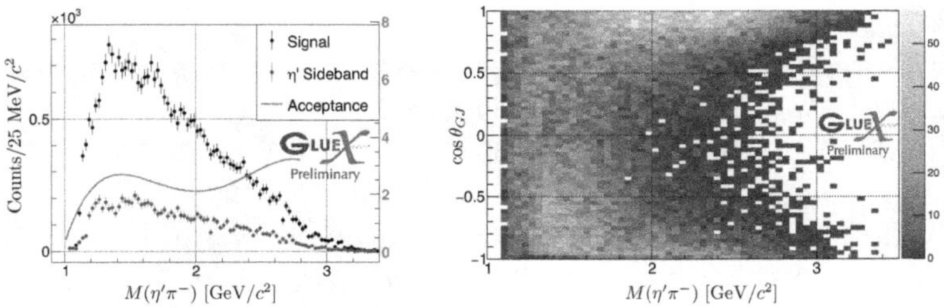

Fig. 3. Left: Invariant mass of the $\eta'\pi^-$ system for the signal region (black), sideband regions (blue), and acceptance (red). Right: $\cos\theta_{GJ}$ as a function of $\eta\pi$ mass.

Based of the previous searches for the $\pi_1(1600)$, we would expect an enhancement in the $\eta'\pi^-$ mass around 1600 MeV, which can be seen in our $\eta'\pi$ mass distribution. If there is $\pi_1 \to \eta'\pi^-$, we would expect to see a P–wave interfering with an S, D, or G–wave in the angular distribution. At around 1300 MeV, we see a D wave structure corresponding to the $a_2(1320)$. As one moves towards higher $\eta'\pi$ masses, an asymmetry between backward and forward begins to stand out. There is a broad enhancement at $\approx$1400-1700 MeV at backward angles that is not apparent at forward angles. Above $\approx$ 1800 MeV and at $\cos\theta_{GJ} \approx 1$, we have backgrounds that will need to be subtracted. Even though we see some interesting features in the angular distributions, we can not say anything concrete in regards to hybrids until an amplitude analysis is performed.

4. Summary

GlueX is beginning its search for hybrid mesons by studying $\eta^{(\prime)}\pi$ final–states. The eventual goal of this work is to confirm the π_1 pole position extracted by JPAC with different backgrounds and production mechanism than COMPASS. This work presented mass and angular distributions for the reactions $\gamma p \to \eta\pi^-\Delta^{++}$ and $\gamma p \to \eta'\pi^-\Delta^{++}$. We are currently developing the tools and procedures necessary to perform an amplitude analysis on these channels.

References

1. J. J. Dudek *et al.*, Toward the excited isoscalar meson spectrum from lattice QCD, *Phys. Rev.* **D88**, 094505 (2013).
2. D. Alde *et al.*, Evidence for a 1^{-+} Exotic Meson, *Phys. Lett.* **B205**, 397 (1988).
3. H. Aoyagi *et al.*, Study of the $\eta\pi^-$ system in the $\pi^- p$ reaction at 6.3–GeV/c, *Phys. Lett.* **B314**, 246 (1993).
4. G. M. Beladidze *et al.*, Study of $\pi^- N \to \eta\pi^- N$ and $\pi^- N \to \eta'\pi - N$ reactions at 37–GeV/c, *Phys. Lett.* **B313**, 276 (1993).
5. D. R. Thompson *et al.*, Evidence for exotic meson production in the reaction $\pi^- p \to \eta\pi^- p$ at 18–GeV/c, *Phys. Rev. Lett.* **79**, 1630 (1997).
6. A. Abele *et al.*, Evidence for a $\pi\eta$ P wave in anti-p p annihilations at rest into $\pi^0\pi^0\eta$, *Phys. Lett.* **B446**, 349 (1999).
7. G. S. Adams *et al.*, Confirmation of a π_1^0 Exotic Meson in the $\eta\pi^0$ System, *Phys. Lett.* **B657**, 27 (2007).
8. E. I. Ivanov *et al.*, Observation of exotic meson production in the reaction $\pi^- p \to \eta'\pi^- p$ at 18–GeV / c, *Phys. Rev. Lett.* **86**, 3977 (2001).
9. V. Dorofeev, New results from VES, *Frascati Phys. Ser.* **15**, 3 (1999).
10. Yu. P. Gouz *et al.*, Study of the wave with $J^{PC} = 1^{-+}$ in the partial wave analysis of $\eta'\pi^-, \eta\pi^-, f_1\pi^-$ and $\rho^0\pi^-$ systems produced in $\pi^- N$ interactions at $p_{\pi^-} = 37$ GeV/c, *AIP Conf. Proc.* **272**, 572 (2008).
11. D. V. Amelin *et al.*, Investigation of hybrid states in the VES experiment at the Institute for High Energy Physics (Protvino), *Phys. Atom. Nucl.* **68**, 359 (2005).
12. G. S. Adams *et al.*, Amplitude analyses of the decays $\chi_{c1} \to \eta\pi^+\pi^-$ and $\chi_{c1} \to \eta^{'}\pi^+\pi^-$, *Phys. Rev. D* **84**, 112009 (Dec 2011).
13. Odd and even partial waves of $\eta\pi^-$ and $\eta'\pi^-$ in $\pi^- p \to \eta^{(')}\pi^- p$ at 191 GeV/c, *Physics Letters B* **740**, 303 (2015).
14. A. Rodas *et al.*, Determination of the pole position of the lightest hybrid meson candidate, *Phys. Rev. Lett.* **122**, 042002 (2019).

Tetraquark mixing framework to explain two light-meson nonets

Hungchong Kim

Research Institute of Basic Science, Korea Aerospace University, Goyang 412-791, Korea
E-mail: hungchong@kau.ac.kr

and Center for Extreme Nuclear Matters, Korea University, Seoul 02841, Korea

In this talk, we summarize our recent works on the tetraquark mixing framework for the two light-meson nonets in the $J^{PC} = 0^{++}$ channel, the light nonet [$a_0(980)$, $K_0^*(700)$, $f_0(500)$, $f_0(980)$] and the heavy nonet [$a_0(1450)$, $K_0^*(1430)$, $f_0(1370)$, $f_0(1500)$]. We briefly explain this mixing framework and present various phenomenological signatures to support this picture.

Keywords: Tetraquark mixing; light-meson nonet; quark model.

1. Tetraquark mixing framework

Recently, the tetraquark mixing framework[1–5] has been proposed as a plausible picture to explain the two light-meson nonets in PDG,[7] the light nonet composed of $a_0(980)$, $K_0^*(700)$,[a] $f_0(500)$, $f_0(980)$ constituting the lowest-lying resonances in the $J^{PC} = 0^{++}$ channel, and the heavy nonet composed of $a_0(1450)$, $K_0^*(1430)$, $f_0(1370)$, $f_0(1500)$ lying next to the lowest-lying resonances.

In constructing this framework, we reexamined the diquark-antiquark model[8] and advocated two possible tetraquark types. The first tetraquark type, which is commonly adopted in tetraquark studies, is constructed from the spin-0 diquark with the color and flavor structure of $(\bar{\mathbf{3}}_c, \bar{\mathbf{3}}_f)$. We denote this tetraquark with its spin configuration $|000\rangle$ where the first zero is the tetraquark spin, the second the diquark spin and the third the antidiquark spin. The second tetraquark type, which was proposed as another possibility, is constructed from the spin-1 diquark with $(\mathbf{6}_c, \bar{\mathbf{3}}_f)$. This second tetraquark is denoted by $|011\rangle$. The two tetraquark types differ

---

[a] $K_0^*(700)$ was named as $K_0^*(800)$ in the previous version of PDG.[6] Also its average mass is now listed as 824 MeV changed from its old value of 682 MeV.

by spin and color configurations but both have the same flavor structure, a nonet that can be broken down to an octet and a singlet. Common characteristics are the quantum number, $J^{PC} = 0^{++}$, possible isospins, $I = 0, 1/2, 1$, and most importantly the mass ordering among the octet members, $M_{I=1} > M_{I=1/2} > M_{I=0}$. The last one is unique in a sense that this ordering cannot be generated from a flavor nonet constructed from a two-quark picture, $q\bar{q}$.

Physical candidates corresponding to the two tetraquark types must be sought from the resonances with $J^{PC} = 0^{++}$. Indeed, the two nonets in PDG above, the light and heavy nonets, are the possible candidates because both satisfy the tetraquark characteristics. Specifically, each nonet is composed of the isospin members of $I = 0, 1/2, 1$ and the tetraquark mass ordering is satisfied quite well for the light nonet and, for the heavy nonet, it is still satisfied though marginally.

Then, we have the two tetraquark types $|000\rangle$, $|011\rangle$ in the one hand and the two nonets in PDG with the tetraquark characteristics in the other hand. Thus, it is quite tempting to match the two tetraquark types with the two nonets in PDG even though there is a huge mass gap between the two nonets, around 500 MeV or more. To establish a matching, a crucial observation to make is that the two types, $|000\rangle, |011\rangle$, in each isospin channel mix through the color-spin interaction, $V_{CS} = v_0 \sum_{i<j} \frac{\lambda_i \cdot \lambda_j J_i \cdot J_j}{m_i m_j}$, and the expectation value, $\langle V_{CS} \rangle$, namely the hyperfine mass, forms a 2×2 matrix. The upshot is that *the physical resonances, namely the two nonets in PDG, can be identified by the eigenstates that diagonalize this matrix.* In other words, the two nonets in each isospin channel can be written as a linear combination of $|000\rangle, |011\rangle$ and we express them collectively as

$$|\text{Heavy nonet}\rangle = -\alpha|000\rangle + \beta|011\rangle \,, \tag{1}$$

$$|\text{Light nonet}\rangle = \beta|000\rangle + \alpha|011\rangle \,, \tag{2}$$

where the mixing parameters α, β are fixed by the diagonalization. This is the tetraquark mixing framework for the two nonets in PDG.

2. Signatures to support the tetraquark mixing framework

There are various signatures to support the tetraquark mixing framework. To explain them, we present in Table 1 the hyperfine mass, $\langle V_{CS} \rangle$, calculated in the physical basis and the mixing parameters, α, β. For the isoscalar resonances $f_0(500), f_0(980)$, we include the flavor mixing between $\mathbf{8}_f$ and $\mathbf{1}_f$ in three different ways,[3] SSC [SU(3) symmetric case], IMC [ideal

Table 1. Hyperfine masses and the mixing parameters for the two nonets.

LN	M_{exp}	$\langle V_{CS}\rangle$	HN	M_{exp}	$\langle V_{CS}\rangle$	α	β
$a_0(980)$	980	−488.5	$a_0(1450)$	1474	−16.8	0.8167	0.5770
$K_0^*(700)$	824	−592.7	$K_0^*(1430)$	1425	−26.9	0.8130	0.5822
$f_0(500)$	475	−667.5	$f_0(1370)$	1350	−29.2	0.8136	0.5814
$f_0(980)$	990	−535.1	$f_0(1500)$	1506	−20.1	0.8157	0.5784

Note: The middle value is taken for M_{exp} known with some range.

mixing case], RCF [realistic case with fitting]. Here in the table, we present the RCF results only.

First, the mixing parameters α, β support our original identification of each nonet in PDG as a flavor nonet. α, β are determined in each isospin channel separately and in principle they may depend on isospin. But as shown in Table 1, their values are almost independent of isospin. This means, the right-hand side of Eq. (1) or Eq. (2) also forms a flavor nonet consistently with the corresponding nonet in the left-hand side.

Secondly, there is a strong mixing between $|000\rangle, |011\rangle$ which can explain the huge mass gap, around 500 MeV or more, between the two nonets in PDG. To put it more clearly, the hyperfine masses, $\langle V_{CS}\rangle$, calculated in the isovector channel corresponding to $a_0(980), a_0(1450)$, are diagonalized to yield the eigenvalues,

$$\begin{array}{c|cc} & |000\rangle & |011\rangle \\ \hline |000\rangle & -173.9 & -222.3 \\ |011\rangle & -222.3 & -331.5 \end{array} \quad \rightarrow \quad \begin{array}{c|cc} & |a_0(1450)\rangle & |a_0(980)\rangle \\ \hline |a_0(1450)\rangle & -16.8 & 0.0 \\ |a_0(980)\rangle & 0.0 & -488.5 \end{array} .$$

The strong mixing, $\langle 000|V_{CS}|011\rangle = -222.3$ MeV, causes a huge separation in the eigenvalues, $-16.8 - (-488.5) = 471.7$ MeV which is comparable to the experimental mass gap, $M[a_0(1450)] - M[a_0(980)] = 494$ MeV. Indeed, in this mixing formalism, one can establish the mass splitting formula,[1,3] $\Delta M_{exp} \approx \Delta\langle V_{CS}\rangle$, between the two nonets in each isospin channel. The experimental mass splitting is found to be consistent qualitatively with this mass splitting formula.

Thirdly, the hyperfine mass for the light nonet is negatively huge around −500 MeV. As one can see from the 2×2 matrix above, this is a consequence of the mixing which pushes down substantially the hyperfine mass through the diagonalization. This certainly helps us to understand the light nonet mass lying below 1 GeV which, without mixing, seems too low to be the tetraquark mass. On the other hand, $\langle V_{CS}\rangle$ for the heavy nonet is very small around −20 MeV so the heavy nonet mass is not far from $\sim 4m_q$.

Fourthly, Table 1 shows $\alpha > \beta$ indicating that the resonances in the light nonet, through Eq. (2), have more probability to stay in $|011\rangle$ rather than in $|000\rangle$. This is quite surprising because the light nonet member is often believed to have the $|000\rangle$ configuration only. The result, $\alpha > \beta$, in fact, originates from the fact that the second tetraquark $|011\rangle$ is more compact than $|000\rangle$ when the binding is calculated from the color-spin interactions not only for the diquark (antidiquark) but also for the other pairs composed of $q\bar{q}$. For the isovector case, in particular, we have $\langle 011|V_{CS}|011\rangle = -331.5$ MeV which is more negative than $\langle 000|V_{CS}|000\rangle = -173.9$ MeV. Our finding here is also supported by a QCD sum rule calculation performed for $a_0(980)$[5] using an interpolating field involving the two tetraquark types.

Fifthly, our hyperfine masses, $\langle V_{CS}\rangle$, can give a partial explanation for the marginal mass ordering seen in the heavy nonet. In Table 1, one can see that $\langle V_{CS}\rangle$ has the same ordering as the mass $M_{I=1} > M_{I=1/2} > M_{I=0}$ both for the light and heavy nonets. This means, $\langle V_{CS}\rangle$ is also responsible for the mass ordering in addition to the quark masses. But the hyperfine splitting is much narrower in the heavy nonet than in the light nonet. For example, we have the hyperfine splitting in the heavy nonet, $\langle V_{CS}\rangle[a_0(1450)] - \langle V_{CS}\rangle[K_0^*(1430)] \approx 10$ MeV which is much smaller than the corresponding splitting in the light nonet, $\langle V_{CS}\rangle[a_0(980)] - \langle V_{CS}\rangle[K_0^*(700)] \approx 104$ MeV. Thus, the mass splitting is narrower in the heavy nonet by the reduction of the hyperfine mass splitting.

Another signature can be found from fall-apart decay of the tetraquarks. In this decay, two $q\bar{q}$ in a tetraquark simply fall apart into two mesons. This decay is allowed because the tetraquarks, $|000\rangle$ and $|011\rangle$, have a component with two color-singlet, $\mathbf{1}_c \otimes \mathbf{1}_c$, when the wave functions are rearranged into two $q\bar{q}$ from the diquark-antidiquark ($qq\bar{q}\bar{q}$) configuration. There are two modes depending on the two mesons in the final state, the PP mode for two pseudoscalars and the VV mode for two vectors. The coupling strengths are calculated by recombining the color and spin configurations of $|000\rangle$, $|011\rangle$, into two $q\bar{q}$. From this recombination, we found that the PP mode is enhanced in the light nonet but suppressed in the heavy nonet. The relative enhancement factor in the couplings is about 4. We tested this prediction from the ratios[b] of partial widths from $a_0(980), a_0(1450)$ in comparison with the two experimental analyses, one by Bugg[9] and the other based on

[b]The ratios eliminate the unknown dependence from the overall constant.

PDG data. They agree very well,[2]

Ratio	Ours	Bugg	PDG
$\frac{\Gamma[a_0(980)\to\pi\eta]}{\Gamma[a_0(1450)\to\pi\eta]}$	2.51–2.54	2.53	2.93–3.9
$\frac{\Gamma[a_0(980)\to K\bar{K}]}{\Gamma[a_0(1450)\to K\bar{K}]}$	0.52–0.89	0.62	0.61–0.81

For the VV mode, we found the opposite trend. It is enhanced in the heavy nonet but suppressed in the light nonet. The relative enhancement factor in the heavy nonet is about 15.[4] This signature, however, is difficult to confirm since most decay modes are prohibited by the kinematical constraint. A few exceptions are $f_0(1370) \to \rho\rho$, $f_0(1500) \to \rho\rho$. These barely satisfy the constraint through the high tail of their decay widths and, as a result, these partial widths are presumably suppressed substantially. But some data in PDG show nonnegligible partial widths and the enhancement reported here can give one possible explanation for them.

Acknowledgment

This work was supported by the National Research Foundation of Korea (NRF) grant funded by the Korea government (MSIT) (No. NRF-2018R1A2B6002432, No. NRF-2018R1A5A1025563).

References

1. H. Kim, M. K. Cheoun, and K. S. Kim, *Eur. Phys. J. C* **77**, 173 (2017); Erratum: *Eur. Phys. J. C* **77**, 545(E) (2017).
2. K. S. Kim and H. Kim, *Eur. Phys. J. C* **77**, 435 (2017).
3. H. Kim, K. S. Kim, M. K. Cheoun and M. Oka, *Phys. Rev. D* **97**, no. 9, 094005 (2018).
4. H. Kim, K. S. Kim, M. K. Cheoun, D. Jido and M. Oka, *Phys. Rev. D* **99**, no. 1, 014005 (2019).
5. H. J. Lee, K. S. Kim and H. Kim, *Phys. Rev. D* **100**, no. 3, 034021 (2019).
6. C. Patrignani *et al.* [Particle Data Group], *Chin. Phys. C* **40**, no. 10, 100001 (2016).
7. M. Tanabashi *et al.* [Particle Data Group], *Phys. Rev. D* **98**, no. 3, 030001 (2018).
8. R. L. Jaffe, *Phys. Rev. D* **15**, 267 (1977); R. L. Jaffe, *Phys. Rev. D* **15**, 281 (1977); R. L. Jaffe, *Phys. Rept.* **409**, 1 (2005).
9. D. V. Bugg, *Phys. Rev. D* **78**, 074023 (2008).

Vector and baryon spectra via holography in an AdS deformed background

Miguel Angel Martin Contreras* and Alfredo Vega
Instituto de Física y Astronomía, Universidad de Valparaíso, Valparaíso, ZIP 2340000, Chile
**E-mail: miguelangel.martin@uv.cl*

Eduardo Folco Capossoli
Departamento de Física, Colégio Pedro II, 20.921-903 - Rio de Janeiro-RJ - Brazil

Danning Li
Department of Physics and Siyuan Laboratory, Jinan University, Guangzhou 510632, China

Henrique Boschi-Filho
Instituto de Física, Universidade Federal do Rio de Janeiro, 21.941-972 - Rio de Janeiro-RJ - Brazil

In this work we discuss how to construct the Regge trajectories for light mesons and baryons in the context of a bottom-up holographic model, consisting on a 5-dimensional AdS background, deformed with a quadratic function in the holographic coordinate. We fit the scalar meson f_0, vector meson ρ, baryons with spin $1/2^+$ and $3/2^*$ trajectories. These results are in fine agreement with those exposed in PDG. We also found in this model a universal Regge slope near to 1.1 GeV^2.

Keywords: AdS/QCD; holographic QCD; hadronic spectra.

1. Introduction

One of the most important AdS/CFT achievements is the description of QCD systems at strong coupling regime.[1] In this manuscript, we will focus on the holographic description of hadronic spectra in the context of bottom-up models. In fact, due to the confinement process, hadronic states organize in linear Regge trajectories. Holographically, confinement is realized by deformations of the background, as in the case of the hard wall and soft wall models, where space is cut by a Dirichlet condition on the fields[3] or the presence of a static dilaton field.[2,4] The effect of both methodologies is the same: they induce a confinement potential, whose eigenvalues are the hadronic masses. The hadronic identity of each spectrum is defined by the bulk mass associated with the bulk fields, dual to the hadronic states.[5] In the case of mesons with spin S we have

$$M_{5,S}^2 R^2 = (\Delta - S)(\Delta + S - 4), \tag{1}$$

with Δ the scaling dimension of the operator that creates mesons, that is $\Delta = 3$, and R is the AdS radius. For Baryons, with $S = 1/2$, we have

$$M_{5,B} = \Delta + 2, \tag{2}$$

where $\Delta = 3/2$ is the dimension of the hadronic operator that create these baryons.

The high 1/2 spin equation for fermions has the same structure as the Sturm–Liouville one we had obtained for the nucleon case. Thus, we can use the same equation but changing the bulk mass $M_{5,B}$ since the conformal dimension of the operator that creates these hadrons has a different dimension. We will discuss the 3/2 and 5/2 cases in the next sections following this idea.

2. Geometric Background

In our case, the geometric background is deformed with a quadratic perturbation to the AdS metric, as it was introduced in Ref. 6. Therefore, each hadronic state has it own background, that can be written in a generic form

$$dS^2 = e^{2\,A_{\text{hadron}}(z)}\left[dz^2 + \eta_{\mu\nu}\,dx^\mu\,dx^\nu\right], \tag{3}$$

where the warp factor is defined as $A_{\text{hadron}}(z) = \frac{1}{2}\,k_{\text{hadron}}\,z^2 + \log(R/z)$, with R defined as the AdS radius.

3. Hadronic spectra

Hadronic states will be given by fields living in this sort of backgrounds. For each hadronic specie we can construct an action of the form

$$I = \frac{1}{\mathcal{K}} \int d^5x\,\sqrt{-g_{\text{hadron}}}\,\mathcal{L}_{\text{Hadron}} \tag{4}$$

with $\mathcal{L}_{\text{Hadron}}$ given by:

$$\mathcal{L}_V = -\frac{1}{4\,g_V^2}\, g^{m\,r}\, g^{n\,p}\, F_{mn}\, F_{rp} - \frac{1}{2}\, M_{5,V}^2\, g^{m\,n}\, A_m\, A_n, \tag{5}$$

$$\mathcal{L}_S = -\frac{1}{2\,g_S^2}\, g^{m\,n}\, \partial_m\, S\, \partial_n\, S - M_{5,S}^2\, S^2, \tag{6}$$

$$\mathcal{L}_B = \bar{\Psi}\,[\Gamma^m\, \partial_m - M_{5,B}]\;\Psi. \tag{7}$$

For mesons, scalar and vector, hadronic states come from the generic master equation of motion (Sturm-Liouville) obtained from the corresponding bulk fields:

$$\partial_z\left[e^{-B(z)}\,\partial_z\,\psi_n(z,q)\right] + M_n^2\, e^{-B(z)}\,\psi_n(z,q) + M_5^2\, e^{B(z)-2A(z)}\psi_n(z,q) = 0, \tag{8}$$

where $B(z) = \beta\, A(z)$ and $\psi(z,q)$ can be either the vector field $A_m(z,q)$ or the scalar one $S(z,q)$. Scalar mesons are defined when we fix $\beta = -3$ and $M_5^2\, R^2 = -3$. Vector mesons when $\beta = -1$ and $M_5^2\, R^2 = 0$. Regge trajectories are defined from the eigenstates of the equation (8). Each trajectory is labeled by a value of k_{hadron}.

For baryons, we define bulk chiral spinors obeying the following Sturn-Liouville equation of motion:

$$\psi_-'' + 4\,A'\,\psi_-' + \big[4\,A'^2 + 2\,A'' - m_5\,A'\,e^{A} - m_5^2\,e^{2\,A}\big]\,\psi_- + (-q^2)\,\psi_- = 0, \tag{9}$$

where the bulk mass for these baryonic state is fixed as $M_5 = 5/2$. Notice that all of the bulk masses given here for mesons and baryons are allowed by the BF bound.

4. Numerical results

The numerical results obtained by solving the boundary valued problem associated to the different hadronic species is depicted in the figure 1. The corresponding Regge trajectories obtained from this model are summarized in table 1, where we compare the holographic results with the experimental data avaliable.[7]

Notice that these states were constructed with $L = 0$ as a first approximation. So we do not fit father and daughter trajectories.[8] It is worthy to

Table 1. Summary of the fitted Regge trajectories for light scalar and vector mesons, and for baryons with spin 1/2, $3/2^+$ and $5/2^+$ compared against PDG.[7] The last column represents the RMS error fitting each hadronic family.

Hadron	**Experimental Trajectory** M^2_{Exp} (GeV^2)	**Theoretical Trajectory** M^2_{Th} (GeV^2)	**RMS** %
f_0 meson	$(1.314 \pm 0.017)\, n_r - (0.285 \pm 0.332)$	$(1.288 \pm 0.009)\, n_r - (0.117 \pm 0.024)$	3.8
ρ meson	$(0.720 \pm 0.076)\, n_r - (0.223 \pm 0.302)$	$(0.754 \pm 8 \times 10^{-7}) n_r$	7.8
N(1/2)	$(0.863 \pm 0.029)\, n_r + (0.114 \pm 0.111)$	$(0.860 \pm 0.042)\, n_r - (0.081 \pm 0.164)$	4.1
N($3/2^+$)	$(0.678 \pm 0.117)\, n_r + (1.517 \pm 0.364)$	$(1.021 \pm 0.017)\, n_r + (0.501 \pm 0.047)$	9.0
N($5/2^+$)	$(0.785 \pm 0.135)\, n_r + (1.934 \pm 0.291)$	$(0.931 \pm 0.031)\, n_r + (1.429 \pm 0.068)$	2.75

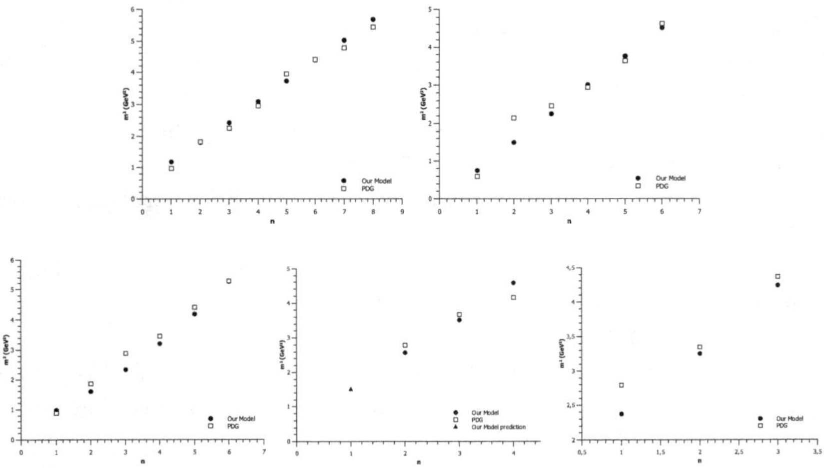

Fig. 1. This figure shows numerical results for this model. The upper left panel displays the f_0 scalar mesons trajectory with $k_s = -0.613^2$ GeV^2. The upper right panel is the ρ meson family with $k_v = -0.332^2$ GeV^2. In the lower left panel we have baryons with spin 1/2 with $k_N = 0.205^2$ GeV^2. The middle lower panel has the trajectory for $S = 3/2^+$ baryons with $k_{3/2} = 0.190^2$ GeV^2. The right lower panel has the $S = 5/2^+$ baryons with $k_{5/2} = 0.205^2$ GeV^2. All of the hadronic data read from PDG.[7]

say that the total RMS error of fitting eight scalar and six vector mesons; six N(1/2), four N($3/2^+$) and three $N(5/2^+)$ baryons with five parameters (the set of k_{hadron}) is around 12%.

5. Conclusions

Confinement can be realized via background deformations, as it was mentioned in Ref. 9. This information is hidden in the holographic potential constructed from the Sturm-Liouville equations (8) and (9). In this sense, confinement is achevided geometrically, as it happens in the context of top/down models, as the D3/D7.[10]

Acknowledgements

The authors acknowledge the financial support by FONDECYT (Chile) under Grants No. 1180753 (A. V.) and No. 3180592 (M. A. M. C.).

References

1. J. Erdmenger, N. Evans, I. Kirsch and E. Threlfall, "Mesons in Gauge/Gravity Duals - A Review," Eur. Phys. J. A **35**, 81 (2008).
2. A. Karch, E. Katz, D. T. Son and M. A. Stephanov, "Linear confinement and AdS/QCD," Phys. Rev. D **74**, 015005 (2006).
3. H. Boschi-Filho and N. R. F. Braga, "Gauge/string duality and scalar glueball mass ratios," JHEP **0305**, 009 (2003).
4. A. Vega and I. Schmidt, "Hadrons in AdS/QCD correspondence," Phys. Rev. D **79**, 055003 (2009).
5. J. M. Maldacena, "The Large N limit of superconformal field theories and supergravity," Int. J. Theor. Phys. **38**, 1113 (1999) [Adv. Theor. Math. Phys. **2**, 231 (1998)].
6. E. Folco Capossoli, M. A. M. Contreras, D. Li, A. Vega and H. Boschi-Filho, "Hadronic Spectra from Deformed AdS Backgrounds," arXiv:1903.06269 [hep-ph].
7. M. Tanabashi *et. al.* (Particle Data Group). Phys. Rev. D **98**, 030001 (2018) doi:10.1103/PhysRevD.98.030001.
8. A. V. Anisovich, V. V. Anisovich and A. V. Sarantsev, "Systematics of q anti-q states in the (n, M^2) and (J, M^2) planes," Phys. Rev. D **62**, 051502 (2000).
9. H. Forkel, M. Beyer and T. Frederico, "Linear square-mass trajectories of radially and orbitally excited hadrons in holographic QCD," JHEP **0707**, 077 (2007).
10. M. Kruczenski, D. Mateos, R. C. Myers and D. J. Winters, "Meson spectroscopy in AdS / CFT with flavor," JHEP **0307**, 049 (2003).

Session 2

Baryon Spectroscopy

Observations of new hyperons at Belle

C. P. Shen

Key Laboratory of Nuclear Physics and Ion-beam Application (MOE) and Institute of Modern Physics, Fudan University,
Shanghai, 200443, China
E-mail: shencp@fudan.edu.cn
www.fudan.edu.cn

S. X. Li

School of Physics, Beihang University,
Beijing, 100191, China

Although Belle stopped data taking more than ten years ago and Belle II has already started data taking, Belle is still producing many excited results on hyperons. Recently Belle confirmed the existence of $\Xi(1620)^0$ via $\Xi_c^+ \to \Xi^-\pi^+\pi^+$ decays and observed a new excited Ω state decaying into $\Xi^0 K^-$ and $\Xi^- K_S^0$ in $\Upsilon(1S, 2S, 3S)$ decays.

Keywords: Hyperons; Belle; $\Xi(1620)^0$; $\Omega(2012)^-$.

1. Introduction

Although the ground states of the flavor SU(3) octet and decuplet baryons have been successfully described by the constituent quark model,[1–3] some of the observed excited states do not agree well with the theoretical prediction or some excited states are still missing in experiments. It is thus important to study such unusual states and search for missing states, both to probe the limitation of the quark models and to spot unrevealed aspects of the quantum-chromodynamics description of the structure of hadron resonances. Studying of the Ξ and Ω resonances with strangeness $S = -2$ and $S = -3$ may provide important information on the latter aspect.

2. Observation of $\Xi(1620)^0$

Up to now, only a small number of Ξ states have been measured.[1] Neither the first radial excitation with the spin parity of $J^P = \frac{1}{2}^+$ nor a first orbital excitation with $J^P = \frac{1}{2}^-$ has been identified. The first excited Ξ candidate is $\Xi(1620)$ or $\Xi(1690)$. There is a lingering theoretical controversy about the interpretation of the $\Xi(1620)$ and $\Xi(1690)$ states, extending from their assignment in the quark model to their existence. Recently Belle studied the $\Xi(1620)$ and $\Xi(1690)$ in $\Xi\pi$ mass spectrum via the decay $\Xi_c^+ \to \Xi^-\pi_L^+\pi_H^+$ based on a data sample corresponding to an integrated luminosity of 980 fb^{-1}, where π_L^+/π_H^+ denotes the pion with the lower/higer momentum.[4] The charge conjugate mode is included throughout this paper.

The $M(\Xi^-\pi_L^+)$ distribution of the final selected Ξ_c^+ decay candidates is shown in Fig. 1, where peaks corresponding to $\Xi(1530)^0$, $\Xi(1620)^0$, and $\Xi(1690)^0$ are observed. A binned maximum-likelihood fit to the efficiency-corrected $M(\Xi^-\pi_L^+)$ distribution is applied for the data samples in the Ξ_c^+ signal region and the sideband region. The fitting function for the mass spectrum in the signal region includes resonances due to the $\Xi(1530)^0$, $\Xi(1620)^0$, and $\Xi(1690)^0$, a nonresonant contribution described by a phase space, and the combinatorial background described by a threshold function. The fitting function for the mass spectrum in the sideband region includes the $\Xi(1530)^0$ signal and the combinatorial background. The $\Xi(1620)^0$ signal is modeled with the S-wave relativistic Breit-Wigner (BW) function. The $\Xi(1530)^0$ and $\Xi(1690)^0$ signals are modeled with P- and S-wave relativistic BW functions convolved with a fixed Gaussian resolution determined from the MC simulation. The interference between the $\Xi(1620)^0$ and the S-wave nonresonant process is taken into account. The solid line in Fig. 1 shows the fitting result for the events in Ξ_c^+ signal region. The signal significances of the $\Xi(1620)^0$ and $\Xi(1690)^0$ are larger than 10σ and 4.5σ, respectively. The measured mass and width of $\Xi(1620)^0$ are $(1610.4 \pm 6.0^{+6.1}_{-4.2})$ MeV/c^2 and $(60.0 \pm 4.8^{+2.8}_{-7.1})$ MeV, respectively. For the first time, the $\Xi(1620)^0$ particle is observed in its decay to $\Xi^-\pi^+$ via $\Xi_c \to \Xi^-\pi^+\pi^-$ decays. The measured masses of $\Xi(1620)^0$ and $\Xi(1690)^0$ are not consistent with the predictions from the constituent quark models; therefore, it is difficult to explain the structure of the $\Xi(1620)^0$ and $\Xi(1690)^0$ as normal excited Ξ states, which implies that these states are candidates of a new class of exotic hadrons.

3. Observation of an Excited $\Omega(2012)^-$ Baryon

The Ω^- baryon comprises three strange quarks (sss). The size of the Ω^- family is much smaller than other baryons'. Only three excited Ω states have been assigned, including the $\Omega(2250)$, $\Omega(2380)$, and $\Omega(2470)$.[1] It is remarkable that these excited Ω states have the masses much higher than that of the ground state than expected. Recently Belle searched for excited

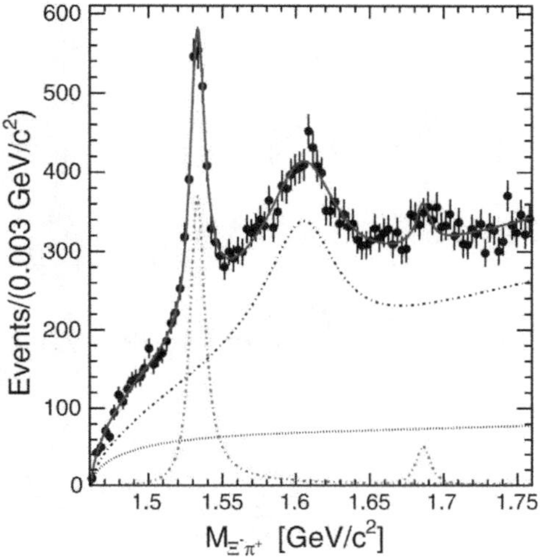

Fig. 1. The $\Xi^-\pi^+_L$ invariant mass spectrum in the Ξ_c^+ signal region from the selected $\Xi_c^+ \to \Xi^-\pi^+\pi^+$ candidates, together with the fit result (solid blue curve) including the $\Xi(1530)^0$ signal (dashed red curve), $\Xi(16900)^0$ signal (dot-dashed pink curve), $\Xi(1620)^0$ signal and nonresonant contribution (dot-dashed black curve), and the combinatorial backgrounds (dotted black curve).[4]

Ω^- decays into Ξ^0K^- and $\Xi^-\bar{K}^0$ with subsequent decays into $\Xi^- \to \Lambda\pi^-$, $\Xi^0 \to \Lambda\pi^0$, $\bar{K^0} \to \pi^+\pi^-$, $\Lambda \to p\pi^-$ and $\pi^0 \to \gamma\gamma$ using data samples collected at $\Upsilon(1S)$, $\Upsilon(2S)$, and $\Upsilon(3S)$ resonances, with integrated luminosities of 5.7 fb^{-1}, 24.9 fb^{-1}, and 2.9 fb^{-1}, respectively.[5] Figure 2 shows the Ξ^0K^- and $\Xi^-K^0_S$ invariant mass distributions after event selections. Excesses are present in both distributions at around 2.01 GeV/c^2. A simultaneous fit is applied to the two distributions, where the signal functions are Voigtian functions and the background functions second-order Chebyshev polynomials. The measured mass of the resonance is $(2012.4\pm0.7\pm0.6)$ MeV/c^2 and its width, Γ, $(6.4^{+2.5}_{-2.0}\pm1.6)$ MeV with a signal significance of 8.3σ. Taking into account the detection efficiency and signal yields of the two modes, the branching fraction ratio $\mathcal{R} = \frac{\mathcal{B}(\Omega^{*-}\to\Xi^0K^-)}{\mathcal{B}(\Omega^{*-}\to\Xi^-\bar{K}^0)}$ is obtained to be 1.2 ± 0.3, which is consistent with the expectation due to isospin symmetry. This new resonance, denoted $\Omega(2012)^-$, has a mass 340 MeV/c^2 higher than the ground state, and thus helps fill the large gap in the Ω spectrum between the ground state and the already observed excited states.

After the observation of the $\Omega(2012)$, several theoretical interpretations were offered including a $1P$ orbital excitation of the ground-state Ω baryon,

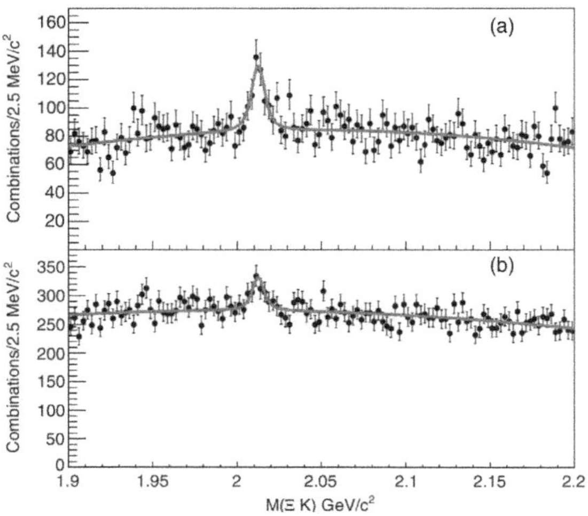

Fig. 2. The (a) $\Xi^0 K^-$ and (b) $\Xi^- K^0_S$ invariant mass distributions in the combined $\Upsilon(1S, 2S, 3S)$ data samples. The curves show a simultaneous fit to the two distributions with a common mass and width, where an excited Ω^- state is observed clearly.[5]

and a $K\Xi(1530)$ hadronic molecule with a large decay width for $\Omega(2012) \to K\pi\Xi$.[6–8] The authors in Ref. 9 also discussed the three-body decay of the $\Omega(2012)$. They regarded $\Omega(2012)$ as a member of the compact decuplet states only if the sum of branching fractions of the $\Omega(2012) \to K\pi\Xi$ and $\Omega(2012) \to \Omega\pi\pi$ is not too large ($< 70\%$).

Using the same $\Upsilon(1S, 2S, 3S)$ data samples, Belle also reported on a search for $\Omega(2012) \to K\Xi(1530) \to K\pi\Xi$ including $\Omega(2012)^- \to \Xi^-\pi^+K^-$, $\Xi^-\pi^0 K^0_S$, $\Xi^0\pi^- K^0_S$, and $\Xi^0\pi^0 K^-$ decay modes.[10] After all event selection requirements, especially $1.49\ \mathrm{GeV}/c^2 < M(\Xi\pi) < 1.53\ \mathrm{GeV}/c^2$ to selected $\Xi(1530)$ signal candidates, no obvious $\Omega(2012)^-$ signal is observed in any of searched modes. An unbinned extended maximum-likelihood simultaneous fit to all three-body decay modes is performed with the ratios of the expected observed $\Omega(2012)$ signals between each decay channel fixed according to the detection efficiency and the ratios of branching fractions of the three-body decay modes of $\Omega(2012)$. The fit result is shown in Fig. 3, corresponding to a total fit yield of 22.4 ± 14.0. The statistical significance of the $\Omega(2012)$ signal is 1.6σ. The 90% credibility level upper limit on the ratio $\mathcal{R}^{\Xi\pi K}_{\Xi K} = \frac{\mathcal{B}(\Omega(2012)\to\Xi(1530)(\to\Xi\pi)K)}{\mathcal{B}(\Omega(2012)\to\Xi K)} < 11.9\%$ is obtained, where $\mathcal{B}(\Omega(2012) \to \Xi(1530)(\to \Xi\pi)K) = \mathcal{B}(\Omega(2012)^- \to \Xi(1530)^0(\to$

$\Xi^-\pi^+)K^-) + \mathcal{B}(\Omega(2012)^- \to \Xi(1530)^-(\to \Xi^-\pi^0)\bar{K}^0) + \mathcal{B}(\Omega(2012)^- \to \Xi(1530)^-(\to \Xi^0\pi^-)\bar{K}^0) + \mathcal{B}(\Omega(2012)^- \to \Xi(1530)^0(\to \Xi^0\pi^0)K^-)$ and $\mathcal{B}(\Omega(2012) \to \Xi K) = \mathcal{B}(\Omega(2012)^- \to \Xi^-\bar{K}^0) + \mathcal{B}(\Omega(2012)^- \to \Xi^0 K^-)$. The above result is in tension with the predictions from the molecular interpretation of the $\Omega(2012)$.

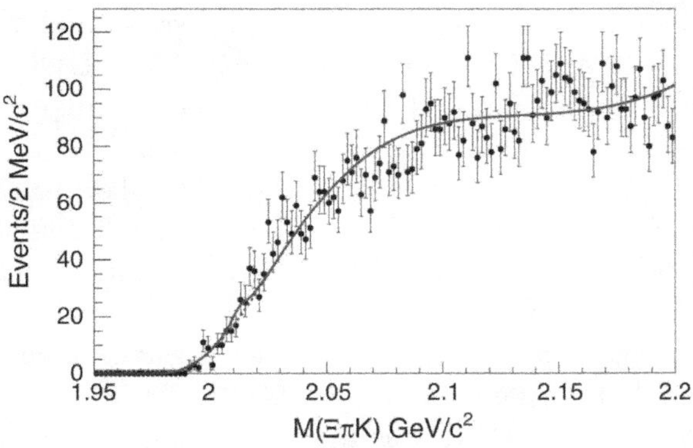

Fig. 3. The simultaneous fit result to all three-body $\Omega(2012)$ decay modes $\Omega(2012)^- \to \Xi(1530)^0(\to \Xi^-\pi^+)K^-$, $\Xi(1530)^-(\to \Xi^-\pi^0)\bar{K}^0$, $\Xi(1530)^-(\to \Xi^0\pi^-)\bar{K}^0$, $\Xi(1530)^0(\to \Xi^0\pi^0)K^-$ from the combined $\Upsilon(1S, 2S, 3S)$ data samples. The solid curve is the best fit, and the dashed line represents the backgrounds.[10]

4. Summary

It is expected that data samples accumulated at Belle will continue to be used for the study of the hyperons. Belle II has started its pioneering run with the full detector in April 2019 and is expecting its target integrated luminosity of 50 ab^{-1} in a few years.[11] In the near future, the study of hyperons will benefit a lot from the large data samples of the Belle II experiment in many different ways including charmed baryons' decays, continuum processes, B decays, $\Upsilon(nS)$ $(n = 1 - 6)$ decays, and so on.

Acknowledgments

This work is supported in part by National Natural Science Foundation of China (NSFC) under contract Nos. 11575017, 11761141009, and 11975076; the Ministry of Science and Technology of China under Contract No.

2015CB856701; and the CAS Center for Excellence in Particle Physics (CCEPP).

References

1. M. Tanabashi *et al.*, Review of Particle Physics, *Phys. Rev.* **D98**, 030001 (2018).
2. K.-T. Chao, N. Isgur and G. Karl, Strangeness -2 and -3 Baryons in a Quark Model With Chromodynamics, *Phys. Rev.* **D23**, 155 (1981).
3. S. Capstick and N. Isgur, Baryons in a Relativized Quark Model with Chromodynamics, *Phys. Rev.* **D34**, 2809 (1986).
4. M. Sumihama *et al.*, Observation of $\Xi(1620)^0$ and evidence for $\Xi(1690)^0$ in $\Xi_c^+ \to \Xi^-\pi^+\pi^+$ decays, *Phys. Rev. Lett.* **122**, 072501 (2019).
5. J. Yelton *et al.*, Observation of an Excited Ω^- Baryon, *Phys. Rev. Lett.* **121**, 052003 (2018).
6. Y.-H. Lin and B.-S. Zou, Hadronic molecular assignment for the newly observed Ω^* state, *Phys. Rev.* **D98**, 056013 (2018).
7. M. P. Valderrama, $\Omega(2012)$ as a hadronic molecule, *Phys. Rev.* **D98**, 054009 (2018).
8. R. Pavao and E. Oset, Coupled channels dynamics in the generation of the $\Omega(2012)$ resonance, *Eur. Phys. J.* **C78**, 857 (2018).
9. M. V. Polyakov, H.-D. Son, B.-D. Sun and A. Tandogan, $\Omega(2012)$ through the looking glass of flavour SU(3), *Phys. Lett.* **B792**, 315 (2019).
10. S. Jia *et al.*, Search for $\Omega(2012) \to K\Xi(1530) \to K\pi\Xi$ at Belle, *Phys. Rev.* **D100**, 032006 (2019).
11. N. Brambilla, S. Eidelman, C. Hanhart, A. Nefediev, C.-P. Shen, C. E. Thomas, A. Vairo and C.-Z. Yuan, The XYZ states: experimental and theoretical status and perspectives, *arXiv:* **1907.07583** (2019).

Experimental determination of the isospin of $\Lambda_c/\Sigma_c(2765)^+$

K. Tanida for the Belle Collaboration

Advanced Science Research Center, Japan Atomic Energy Agency
Tokai, Ibaraki 319-1106, Japan
E-mail: tanida@post.j-parc.jp

We report an experimental determination of the isospin of $\Lambda_c/\Sigma_c(2765)^+$ using 980 fb^{-1} data in the e^+e^- annihilation around $\sqrt{s} = 10.6$ GeV collected by the Belle detector located at the KEKB collider. The isospin partners are searched for in the $\Sigma_c(2455)^{++/0}\pi^0$ channels, and no evidence was obtained. Thus the isospin is determined to be zero, and the particle is established to be a Λ_c.

Keywords: Charmed baryon; isospin.

1. Introduction

$\Lambda_c/\Sigma_c(2765)^+$ is the lightest charmed baryon for which there is no assumed quark-model identification. It was first observed by CLEO[1] in the $\Lambda_c^+\pi^+\pi^-$ final state and later reported by Belle[2] in the $\Sigma_c^{++/0}\pi^{\mp}$ final state. The experimental status is very poor: the spin, parity and isospin are not determined yet and CLEO did not determine the uncertainty on its width.

In a quark model calculation by Capstick and Isgur,[3] there are six states predicted in this mass region, one state each for $I(J^P) = 0(1/2^+)$, $0(1/2^-)$, and two states each for $I(J^P) = 1(1/2^-)$, $(3/2^-)$. On the other hand, in the most recent quark model calculation by Yoshida *et al.*,[4] the $I = 0$ states are found in the higher mass region, while the $I = 1$ states are still in the range of 2765 ± 50 MeV$/c^2$. In addition, there are many theoretical calculations in various models (for example[5–8]), and almost any combination of $I = 0$ or 1, $J = 1/2$ or $3/2$, and $P = +$ or $-$ seems to be possible. Therefore, the nature of $\Lambda_c/\Sigma_c(2765)^+$ is quite controversial and experimental determination of its quantum numbers is awaited.

In this paper, we present a determination of the isospin of $\Lambda_c/\Sigma_c(2765)^+$ by searching for the possible isospin partners of the state in the decay modes to $\Sigma_c(2455)^{++/0}\pi^0 \to \Lambda_c^+\pi^{\pm}\pi^0$. We use a dataset of 980 fb^{-1} of integrated luminosity around $\sqrt{s} = 10.6$ GeV collected by the Belle detector[9] at the KEKB asymmetric-energy e^+e^- collider.[10] We use a Monte Carlo (MC) simulation to model the detector response and its acceptance to obtain the reconstruction efficiency, the mass resolution for signals and feed-down signals.

2. Reconstruction of $\Lambda_c/\Sigma_c(2765)^+$ in $\Lambda_c^+\pi^+\pi^-$

We reconstruct $\Lambda_c/\Sigma_c(2765)^+$ candidates in the $\Lambda_c^+\pi^+\pi^-$ final state (the inclusion of the charge-conjugate modes is implied throughout this paper). For the primary selection of charged particles, we use the same track quality and particle identification as the $\Lambda_c(2880)^+$ study by Belle.[2] The resulting $\Lambda_c^+\pi^+\pi^-$ invariant mass $[M(\Lambda_c^+\pi^+\pi^-)]$ spectrum is shown in Fig. 1(a), which presents clear peaks of $\Lambda_c/\Sigma_c(2765)^+$ and $\Lambda_c(2880)^+$.

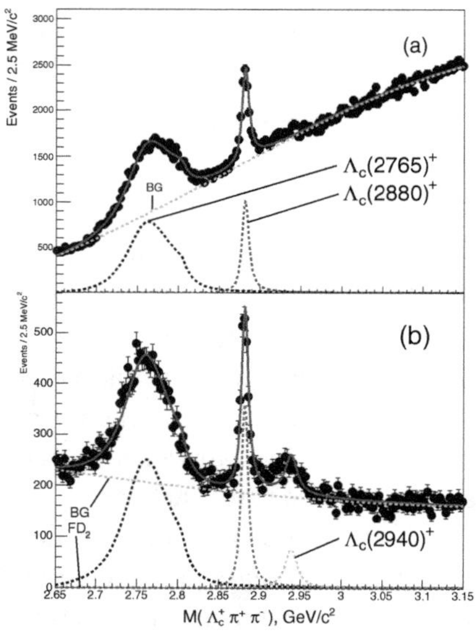

Fig. 1. The $M(\Lambda_c^+\pi^+\pi^-)$ distributions with (a) no cut on $M(\Lambda_c^+\pi)$ and (b) $\Sigma_c(2455)$ selection. The fit results (solid curves) and its components (dashed curves) are also presented. Dashed black, blue, and gray lines are for the $\Lambda_c(2765)^+$, $\Lambda_c(2880)^+$, and $\Lambda_c(2940)^+$ signals, respectively, and the dashed pink and green lines are for $\Sigma_c(2520)$ feed-down (FD_2), and the background (BG), respectively.

The yield for $\Lambda_c/\Sigma_c(2765)^+ \to \Sigma_c(2455)^{++/0}\pi^{\mp}(\to \Lambda_c^+\pi^+\pi^-)$ is determined by a fit to Fig. 1(b) using Breit-Wigner functions for $\Lambda_c/\Sigma_c(2765)^+$, $\Lambda_c(2880)^+$, and $\Lambda_c(2940)^+$ and a second-order polynomial as background.

In addition, there is a small contribution from $\Sigma_c(2520)$ feed-down which is due to an accidental coincidence of $\Sigma_c(2520)$ and a random pion that produces a rather narrow peak in the $M(\Lambda_c^+\pi^+\pi^-)$ spectrum around 2.68 GeV/c^2. The details are to be reported elsewhere.[11]

3. Search for isospin partners in $\Lambda_c^+\pi^\pm\pi^0$ final state

In order to determine the isospin, we search for the isospin partners with $I_3 = \pm 1$ in the $\Sigma_c(2765)^{++/0} \to \Sigma_c^{++/0}\pi^0 \to \Lambda_c^+\pi^\pm\pi^0$ final states. The $\Sigma_c(2765)^{++/0} \to \Sigma_c^+\pi^\pm \to \Lambda_c^+\pi^0\pi^\pm$ decay modes are not used because the momentum of π^0 in these modes is low and thus the background level is very high. The ratio for the product of cross section and branching fractions between $I_3 = 0$ and $I_3 = \pm 1$ states (R_I)

$$R_I = \frac{\sigma_{+1}\mathcal{B}_{+1} + \sigma_{-1}\mathcal{B}_{-1}}{2\sigma_0\mathcal{B}_0},$$

where σ_i is the production cross section for the $I_3 = i$ states, $\mathcal{B}_{\pm 1}$ is the branching fraction for the $I_3 = \pm 1$ states decaying into $\Sigma_c(2455)^{++/0}\pi^0$ decay modes, and $\mathcal{B}_0$ is the sum of the branching fractions of the $I_3 = 0$ state decaying into $\Sigma_c(2455)^{++}\pi^-$ and $\Sigma_c(2455)^0\pi^+$, is measured. If we assume the cross sections are the same for $I_3 = +1$ and $I_3 = -1$ and all the branching fractions are the same, this ratio simplifies to $R_I = \sigma_{\pm 1}/\sigma_0$. If resonances are observed with masses and widths close to those of the $I_3 = 0$ state, it is strong evidence that the isospin is one. If resonances are not observed and obtained upper limit for the R_I is much smaller than one, it is strong evidence that the isospin is zero.

Except for the π^0, the same event selection method is used as that for the $\Lambda_c^+\pi^+\pi^-$ analysis. A π^0 is reconstructed from two photons, energies of which are required to be greater than 50 MeV. The π^0 candidates are selected by requiring 120 MeV/$c^2 < M(2\gamma) <$ 150 MeV/c^2. In order to suppress the combinatorial background further, the π^0 helicity angle, defined as the angle between the momentum of the $\Lambda_c(2765)$ in the center-of-mass frame and momentum of the π^0 in the $\Sigma_c(2765)^{++/0}$ rest frame is required to be greater than -0.35.

3.1. *Result*

Fig. 2 shows $M(\Lambda_c^+\pi^\pm\pi^0)$ distributions for the signal regions with fit lines ovarlaid. A simultaneous unbinned extended maximum likelihood fit is performed with the constraint that the signal yields for $I_3 = \pm 1$ states are the same. The signal PDF is made based on the signal MC simulation assuming that the mass and width are the same as those of the $I_3 = 0$ state. The sum of the signal yields for $I_3 = \pm 1$ is obtained to be 230 ± 250. Note that, if we assume that the cross-section and branching fractions are the same, the expected signal yields would be 8764, estimated from the yields of $\Lambda_c(2765)^+ \to \Sigma_c(2455)^{++/0}\pi^\mp \to \Lambda_c^+\pi^+\pi^-$ taking the difference of acceptance and efficiency into account.

The $\chi^2/d.o.f.$ of the fit is calculated after filling the data in bins as 211/232 (215/232) for $M(\Lambda_c^+\pi^+\pi^0)$ $[M(\Lambda_c^+\pi^-\pi^0)]$ without including the signal component. The 95% upper limit on the R_I is obtained based on Bayesian approach, after incorporating systematic uncertainties. The obtained 95% C.L upper limit for R_I is 0.09, which is much smaller than unity, demonstrating that isospin of $\Lambda_c(2765)^+$ is indeed zero. This is consistent with the fact that the state was not seen in the $\Lambda_c^+\pi$ spectra by Belle,[12] that would be an expected decay mode of an excited Σ_c baryon.

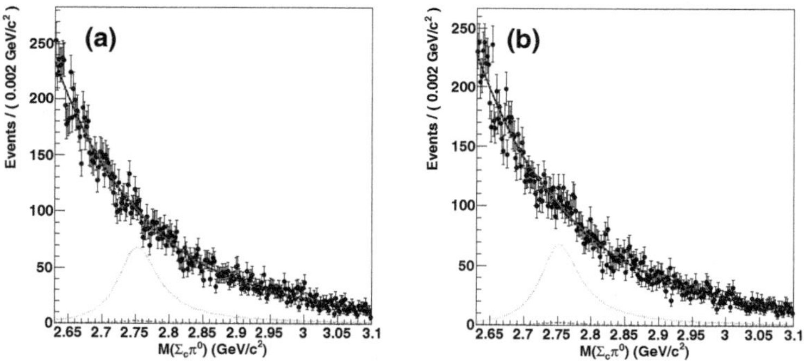

Fig. 2. (a) $M(\Lambda_c^+\pi^+\pi^0)$ and (b) $M(\Lambda_c^+\pi^-\pi^0)$ distributions with $\Sigma_c(2455)^{++/0}$ selection for data with the fit results (blue solid lines). The red dashed lines are the signal contributions from the fit, while the green dotted lines show the expected signals in case the production cross section and branching fractions are the same as $\Sigma_c(2765)^+$.

4. Summary and discussion

In summary, we determined the isospin of $\Lambda_c/\Sigma_c(2765)^+$ by searching for possible isospin partners in the decays to $\Sigma_c(2455)^{++/0}\pi^0 \to \Lambda_c^+\pi^\pm\pi^0$. The isospin is determined to be $I = 0$, therefore the name should indeed be $\Lambda_c(2765)^+$. The result is rather inconsistent with the most recent quark model calculation,[4] but not with the older ones.[3,5] Also, some theoretical predictions are ruled out by the result. However, in order to discuss the nature of this state in more detail, the determination of its other quantum numbers (i.e., spin and parity) is highly desirable and we are planning a publication on this subject, using this data set, in the near future.

References

1. M. Artuso *et al.* [CLEO Collaboration], Observation of new states decaying into $\Lambda_c^+\pi^-\pi^+$, *Phys. Rev. Lett.* **86**, 4479 (2001).
2. K. Abe *et al.* [Belle Collaboration], Experimental constraints on the possible J^P quantum numbers of the $\Lambda_c(2880)^+$, *Phys. Rev. Lett.* **98**, 262001 (2007).
3. S. Capstick, N. Isgur, Baryons in a relativized quark model with chromodynamics, *Phys. Rev. D* **34**, 2809 (1986).
4. T. Yoshida, E. Hiyama, A. Hosaka, M. Oka, and K. Sadato, Spectrum of heavy baryons in the quark model, *Phys. Rev. D* **92**, 114029 (2015).
5. W. Roberts and Muslema Pervin, Heavy baryons in a quark model, *Int. J. Mod. Phys. A* **23**, 2817 (2008).
6. Yongseok Oh and Byung-Yoon Park, Excited states of heavy baryons in the Skyrme model, *Phys. Rev. D* **53**, 1605 (1996).
7. Xian-Hui Zhong and Qiang Zhao, Charmed baryon strong decays in a chiral quark model, *Phys. Rev. D* **77**, 074008 (2008).
8. D. Ebert, R.N. Faustov, and V.O. Galkin, Masses of excited heavy baryons in the relativistic quark model, *Phys. Lett. B* **659**, 612-620 (2008).
9. A. Abashian *et al.* (Belle Collarboration), The Belle detector, *Nucl. Instrum. Methods Phys. Res., Sect. A* **479**, 117 (2002); also see detector section in J. Brodzicka *et al.*, Physics achievements from the Belle experiment, *Prog. Theor. Exp. Phys.* **2012**, 04D001 (2012).
10. S. Kurokawa and E. Kikutani, KEK-B: The KEK B-factory, *Nucl. Instrum. Methods Phys. Res., Sect. A* **499**, 1 (2003), and other papers included in this Volume; T. Abe *et al.*, Achievements of KEKB, *Prog. Theor. Exp. Phys.* **2013**, 03A001 (2013) and references therein.
11. Belle collaboration, in preparation.
12. R. Mizuk *et al.* [Belle Collaboration], Observation of an isotriplet of excited charmed baryons decaying to $\Lambda_c^+\pi$, *Phys. Rev. Lett.* **94**, 122002 (2005).

Hyperon resonances and meson-baryon interactions in isospin 1

K. P. Khemchandani

Departamento de física, Universidade Federal de São Paulo, Campus Diadema, 09913-030, São Paulo, Brazil..
E-mail: kanchan.khemchandani@unifesp.br

A. Martínez Torres

Instituto de Física, Universidade de São Paulo, C.P 66318, 05314-970 São Paulo, SP, Brazil.
E-mail: amartine@if.usp.br

J. A. Oller

Departamento de Fìsica. Universidad de Murcia. E-30071 Murcia. Spain.
E-mail: oller@um.es

In this work we extend our formalism to study meson-baryon interactions by including s- and u-channel diagrams for pseudoscalar-baryon systems. We study the coupled systems with strangeness -1 and focus on studying the isospin-1 resonance(s), especially in the energy region around 1400 MeV. By constraining the model parameters to fit the cross section data available on several processes involving relevant channels, we find resonances in the isoscalar as well as the isovector sector in the energy region around 1400 MeV.

Keywords: Light mass hyperons; $\lambda(1405)$; isovector resonance; coupled channel scattering.

1. Introduction

There exists a large consensus on the nature of the first excited state of the isoscalar strange baryon, $\Lambda(1405)$, as being a moleculelike state rather than a standard three valence quark baryon with one of the quarks in a higher angular momentum state (see Refs. 1–4 for some of the most-cited articles on the topic). In the isovector case, on the other hand, the lightest Σ with the same spin-parity as $\Lambda(1405)$, $1/2^-$, listed by the particle data group,[5] is $\Sigma(1620)$, which possesses a poor status of information on its properties. Various studies,[6–13] though, seem to indicate that the existence of a $1/2^-$ Σ, with a much lower mass ($\sim$ 1400 MeV), is needed to describe the experimental data on the processes like: $K^-p \to \Lambda\pi^+\pi^-$, $\gamma N \to K^+\pi\Lambda$, $\Lambda p \to \Lambda p\pi^0$,

$\Lambda_c^+ \to \eta\pi^+\Lambda$, $\gamma + p \to K^+ + \Sigma^{\pm,0} + \pi^{\mp,0}$. In some of these former works the existence of a Σ state with mass around 1380 MeV is discussed and in some others a scenario with two close lying poles is discussed. In such a situation, we find it useful to study the hyperon resonances with isospin 1 in the energy region around 1400 MeV. Indeed, in our previous work[14] based on a coupled channel formalism, including vector and pseudoscalar mesons, we found two Σ resonances: $1427 - i145$ MeV, $1438 - i198$ MeV, which are much wider than those claimed in Refs. 6–13. In Ref. 14, though, we considered a more elaborate formalism to obtain the lowest order vector meson-baryon (VB) interaction by calculating the s-, t-, u- channel diagrams and a contact interaction, while relying on the Weinberg-Tomozawa term alone for the pseudoscalar-baryon (PB) interaction. Inspired by the works in Refs. 11, 12, we now include s- and u-channel diagrams for the pseudoscalar interaction and solve the Bethe-Salpeter equation. Further, we constrain the parameters by fitting the amplitudes to the cross section data on the processes: $K^-p \to K^-p$, $\bar{K}^0 n$, $\eta\Lambda$, $\pi^0\Lambda$ $\pi^0\Sigma^0$, $\pi^\pm\Sigma^\mp$, on the energy level shift and width of the $1s$ state of the kaonic hydrogen from Ref. 15 and certain cross section ratios near the threshold (more details can be found in the article[16]).

2. Formalism and results

The amplitudes (for the contact interaction, s- and u-channel diagrams), for the pseudoscalar meson-baryon interactions, are obtained from the lowest order chiral Lagrangian

$$\begin{aligned}\mathcal{L}_{PB} = &\langle \bar{B} i\gamma^\mu \partial_\mu B + \bar{B} i\gamma^\mu [\Gamma_\mu, B]\rangle - M_B \langle \bar{B} B\rangle \\ &+ \frac{1}{2} D' \langle \bar{B}\gamma^\mu \gamma_5 \{u_\mu, B\}\rangle + \frac{1}{2} F' \langle \bar{B}\gamma^\mu\gamma_5 [u_\mu, B]\rangle,\end{aligned} \tag{1}$$

where $u_\mu = i u^\dagger \partial_\mu U u^\dagger$, and

$$\Gamma_\mu = \frac{1}{2}\left(u^\dagger \partial_\mu u + u \partial_\mu u^\dagger\right), U = u^2 = exp\left(i\frac{P}{f_P}\right), \tag{2}$$

with P and B representing the matrices of the pseudoscalar mesons and the octet baryons. The s-, t-, u-amplitudes for the vector meson-baryon interactions are obtained from the Lagrangian[17]

$$\mathcal{L}_{\mathrm{VB}} = -g\Big\{ \langle \bar{B}\gamma_\mu [V_8^\mu, B]\rangle + \langle \bar{B}\gamma_\mu B\rangle\langle V_8^\mu\rangle + \frac{1}{4M}\big(F\langle \bar{B}\sigma_{\mu\nu}[V_8^{\mu\nu}, B]\rangle + D\langle \bar{B}\sigma_{\mu\nu}\{V_8^{\mu\nu}, B\}\rangle\big) + \langle \bar{B}\gamma_\mu B\rangle\langle V_0^\mu\rangle + \frac{C_0}{4M}\langle \bar{B}\sigma_{\mu\nu}V_0^{\mu\nu}B\rangle \Big\}, \quad (3)$$

where V refers to the matrix of vector mesons, with the numbers 0/8 in the subscript indicating the singlet/octet part of the wave function (arising from the consideration of $\omega-\phi$ mixing). The contribution of the $V_{\mu\nu}$ tensor gives rise to yet another diagram, a contact interaction.

The transition amplitudes between the two types of meson-baryons systems are obtained from the Lagrangian[18]

$$\mathcal{L}_{\mathrm{PBVB}} = \frac{-ig_{PBVB}}{2f_v}\left(F'\langle \bar{B}\gamma_\mu\gamma_5[[P, V^\mu], B]\rangle + D'\langle \bar{B}\gamma_\mu\gamma_5\{[P, V^\mu], B\}\rangle\right). \quad (4)$$

With all the amplitudes determined from the above Lagrangians, we solve the Bethe-Salpeter equation in the on-shell factorization form for the coupled channels: $\bar{K}N$, $K\Xi$, $\pi\Sigma$, $\eta\Lambda$, $\pi\Lambda$, $\eta\Sigma$, $\bar{K}^*N$, $K^*\Xi$, $\rho\Sigma$, $\omega\Lambda$, $\phi\Lambda$, $\rho\Lambda$, $\omega\Sigma$, $\phi\Sigma$. The detailed amplitudes for the different processes are given in Ref. 16. To solve the scattering equations, we need to regularize the divergent loop functions, which is done by using the dimensional regularizations method. We, thus, have a parameter at hand, which is the subtraction constant needed for the calculation of the loop function. We treat the subtraction constant for each channel as a free parameter, which is fitted to reproduce the experimental data mentioned in the previous section. In addition to the subtraction constants, we treat the coupling between the PB and VB channels, and the decay constants for the two types of mesons as a parameter to be fixed by the data. When fitting the data, we calculate the χ^2 per degree of freedom as

$$\chi^2_{\mathrm{d.o.f}} = \frac{\sum\limits_{k=1}^{N} n_k}{N(\sum\limits_{k=1}^{N} n_k - n_p)} \sum_{k=1}^{N} \frac{\chi_k^2}{n_k}, \quad (5)$$

where N is the number of different sets of data, n_k represents the number of data points in the kth data set, n_p is the number of free parameters, and the χ^2 for the kth data set is obtained as

$$\chi_k^2 = \sum_{i=1}^{n_k} \frac{(y_{k;i}^{\mathrm{th}} - y_{k;i}^{\mathrm{exp}})^2}{\sigma_{k;i}^2}, \quad (6)$$

with $y_{k;i}^{\text{exp}}$ ($y_{k;i}^{\text{th}}$) representing the ith experimental (theoretical) data of the kth set and $\sigma_{k;i}^2$ the related standard deviation.

We find two sets of solutions with $\chi^2_{\text{d.o.f}} \sim 1$ (labelled as Fit I and II in Ref. 16). We show in Fig. 1 the cross sections, for the $K^-p \to K^-p$ process, obtained from Fit I of Ref. 16. Figure 1 also shows the relevant

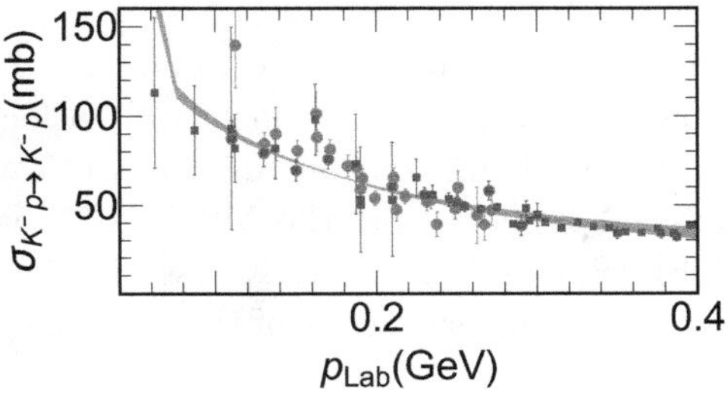

Fig. 1. Cross sections for the process $K^-p \to K^-p$. The (red) circles represent the data used in the fitting procedure. The shaded region represents the results found with Fit I (of Ref. 16).

experimental data for a comparison. The results on the cross sections for other processes obtained within the two types of fits can be found in Ref. 16. With the two types of fits at hand, we look for the resonances present in the amplitudes by studying them in the complex-energy plane. In the following table we give the isoscalar and isovector poles present in the s-

Central values of the pole positions, in MeV, of the $(J^P) = (1/2^-)$ states found in each isospin, in the two types of fits found in Ref. 16.

	$\Lambda(1405)$		Σ's around 1400 MeV	
Fit I	$1373 - i\,114$	$1426 - i\,16$	$1396 - i\,5$	$1367 - i\,57$
Fit II	$1385 - i\,124$	$1426 - i\,15$	—	$1399 - i\,36$

wave at around 1400 MeV (the uncertainties on the pole positions can be found in Ref. 16). The results obtained in isospin 0 are in good agreement with former works on the topic, including those obtained from the data on photo-/electro-production of $\Lambda(1405)$. As can be seen, we find poles around the $\bar{K}N$ threshold in the I=1 case also. Two poles appear in Fit I while one pole is found in Fit II. An observation to be mentioned in relation with these

Σ states is that their masses lie in the energy region where the left hand cut is crossed for some coupled channels. However, we find, interestingly, that the second pole in Fit I as well as the pole in Fit II continue to appear in the complex plane even if the contribution from the u-channel diagrams is switched off.

Acknowledgments

K.P.K. and A.M.T. gratefully acknowledge the financial support received from Conselho Nacional de Desenvolvimento Científico e Tecnológico (CNPq) (under Grant No. 310759/2016-1 and No. 311524/2016-8). J.A.O. thanks partial financial support from the MINECO (Spain) and FEDER (EU) Grant No. FPA2016-77313-P. K.P.K also thankfully acknowledges the support received from the organizers of the conference.

References

1. D. Jido *et al.*, Nucl. Phys. A **725**, 181 (2003).
2. J. A. Oller, Eur. Phys. J. A **28**, 63 (2006).
3. B. J. Menadue, W. Kamleh, D. B. Leinweber and M. S. Mahbub, Phys. Rev. Lett. **108**, 112001 (2012).
4. M. Mai and U. G. Meissner, Nucl. Phys. A **900**, 51 (2013).
5. M. Tanabashi *et al.*, Phys. Rev. D **98**, no. 3, 030001 (2018).
6. J. J. Wu, S. Dulat and B. S. Zou, Phys. Rev. D **80**, 017503 (2009).
7. J. J. Wu, S. Dulat and B. S. Zou, Phys. Rev. C **81**, 045210 (2010).
8. P. Gao, J. J. Wu and B. S. Zou, Phys. Rev. C **81**, 055203 (2010).
9. J. J. Xie *et al.*, Phys. Rev. C **90**, no. 5, 055204 (2014).
10. J. J. Xie and L. S. Geng, Phys. Rev. D **95**, no. 7, 074024 (2017).
11. J. A. Oller and U. G. Meißner, Phys. Lett. B **500**, 263 (2001).
12. Zhi-Hui Guo and J. A. Oller, Phys. Rev. C. **87**, 035202 (2013).
13. K. Moriya *et al.*, Phys. Rev. C **87**, no. 3, 035206 (2013).
14. K. P. Khemchandani, A. Martinez Torres, H. Nagahiro and A. Hosaka, Phys. Rev. D **85**, 114020 (2012).
15. M. Bazzi *et al.*, Phys. Lett. B **704**, 113 (2011).
16. K. P. Khemchandani, A. Martnez Torres and J. A. Oller, Phys. Rev. C **100**, no. 1, 015208 (2019).
17. K. P. Khemchandani, H. Kaneko, H. Nagahiro and A. Hosaka, Phys. Rev. D **83** (2011) 114041.
18. K. P. Khemchandani, A. Martinez Torres, H. Kaneko, H. Nagahiro and A. Hosaka, Phys. Rev. D **84** (2011) 094018.

Heavy Baryons in a pion mean-field approach

Hyun-Chul Kim

Department of Physics, Inha University,
Incheon, 22212, Korea
E-mail: hchkim@inha.ac.kr

In the present talk, we briefly review a series of recent works on the structure of singly heavy baryons based on a pion mean-field approach. We discuss the results of the mass spectra and radiative decays of the singly heavy baryons. We also give some prospects for the future direction within this scheme.

Keywords: Pion mean-field approach; singly heavy baryons.

1. Introduction

When the number of colors (N_c) is taken to be infinity, the quantum fluctuation of mesons is suppressed by $1/N_c$ and a baryon emerges as a bound state of the N_c valence quarks in a pion mean field.[1,2] Recently, this pion mean-field approach has been applied to the description of singly heavy baryons[3–6] (See also a recent review[7]). In this talk, we briefly report the results of these works and give prospects for future directions to understanding the structure of singly heavy baryons within this approach.

2. Singly heavy baryons as a bound state of $N_c - 1$ valence quarks

Since a heavy quark inside a sinlg heavy baryon is much heavier than the light quarks in it, we can take the limit of the infinitely heavy mass of the heavy quark, i.e. $m_Q \to \infty$. Then, the heavy quark can regarded as a mere static color source and will not be involved in the dynamics of the heavy baryon. In this limit, the spin of the heavy quark is conserved because of its infinitely heavy mass. It leads also to the conservation of the total spin of light quarks: $\boldsymbol{J}_L \equiv \boldsymbol{J} - \boldsymbol{J}_Q$, where $\boldsymbol{J}_L$, $\boldsymbol{J}_Q$, and $\boldsymbol{J}$ stand respectively for the spin of the light-quark pair, that of the heavy quark, and the total spin of the heavy baryon. This is called the heavy-quark spin symmetry that allows $\boldsymbol{J}_L$ to be a good quantum number. Moreover, the physics is kept intact under the placement of heavy quark flavors, which is known as the heavy-quark flavor symmetry (see a book[8] and references therein). In this picture, the dynamics of the singly heavy baryons is governed by the light degrees of freedom.

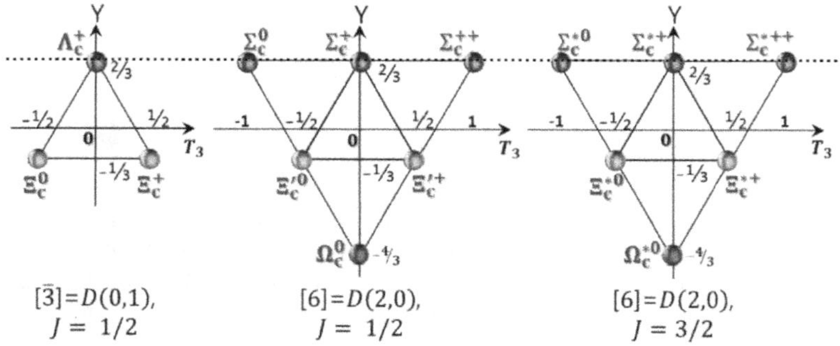

Fig. 1. The flavor SU(3) representations of the lowest-lying singly heavy baryons.

With this heavy-quark flavor-spin symmetry considered, we can represent the lowest-lying singly heavy baryons as the baryon antitriplet $(\overline{\mathbf{3}})$ with spin 1/2 and the baryon sextet with spin 1/2 and 3/2. Note that the baryon sextet are degenerate in the limit of $m_Q \to \infty$, so that we need to consider the hyperfine splitting to lift the degeneracy, which arises from the $1/m_Q$ corrections.

As we mentioned above, the heavy quark is required only for making the singly heavy baryon a color singlet, we can strip off the heavy quark from a singly heavy baryon in the course of the investigation. It will appear later in the collective wavefunctions of the heavy baryon. The presence of the $N_c - 1$ valence quarks produce the pion mean fields. However, the pion mean fields also affect the $N_c - 1$ valence quarks in a self-consistent manner. Moreover, the $N_c - 1$ valence quarks contrain the right hyper charge to be $Y' = (N_c - 1)/3$, which allows the lowest-lying representations: the baryon anti-triplet $(\overline{\mathbf{3}})$, the baryon sextet $(\mathbf{6})$, and so on, as shown in Fig. 1. One great advantage of this pion mean-field approach is that one can use all the dynamical parameters determined in the light baryon sector[9] without changing them for the heavy baryons. We can introduce only a new parameter for the hyperfine splitting between the baryon sextet with spin 1/2 and 3/2, which can be determined by defference between the center values of the sextet masses. For the detailed formalism, we refer to Ref. 7.

3. Results and discussion

We present the masses of the charmed and bottom baryons respectively in Tables 1 and 2.

Table 1. The numerical results of the charmed baryon masses in comparison with the experimental data.[10]

$\mathcal{R}_J^Q$	B_c	Mass	Experiment
$\overline{\mathbf{3}}_{1/2}^c$	Λ_c	2272.5 ± 2.3	2286.5 ± 0.1
	Ξ_c	2476.3 ± 1.2	2469.4 ± 0.3
$\mathbf{6}_{1/2}^c$	Σ_c	2445.3 ± 2.5	2453.5 ± 0.1
	Ξ_c'	2580.5 ± 1.6	2576.8 ± 2.1
	Ω_c	2715.7 ± 4.5	2695.2 ± 1.7
$\mathbf{6}_{3/2}^c$	Σ_c^*	2513.4 ± 2.3	2518.1 ± 0.8
	Ξ_c^*	2648.6 ± 1.3	2645.9 ± 0.4
	Ω_c^*	2783.8 ± 4.5	2765.9 ± 2.0

Table 2. The results of the masses of the bottom baryons in comparison with the experimental data.[10]

$\mathcal{R}_J^Q$	B_b	Mass	Experiment
$\overline{\mathbf{3}}_{1/2}^b$	Λ_b	5599.3 ± 2.4	5619.5 ± 0.2
	Ξ_b	5803.1 ± 1.2	5793.1 ± 0.7
$\mathbf{6}_{1/2}^b$	Σ_b	5804.3 ± 2.4	5813.4 ± 1.3
	Ξ_b'	5939.5 ± 1.5	5935.0 ± 0.05
	Ω_b	6074.7 ± 4.5	6048.0 ± 1.9
$\mathbf{6}_{3/2}^b$	Σ_b^*	5824.6 ± 2.3	5833.6 ± 1.3
	Ξ_b^*	5959.8 ± 1.2	5955.3 ± 0.1
	Ω_b^*	6095.0 ± 4.4	–

The results are in good agreement with the experimental data without any adjustment of the parameters except for the anomalous chromomagnetic moment entered in the hyperfine interaction. Note that there is no experimental data on the Ω_b^* mass yet. So, the present value $M_{\Omega_b^*} = (6095.0 \pm 4.4)$ MeV is the prediction for the Ω_b^* mass.

In Table 3, we list the results for the radiative decays of the singly charmed baryons in comparison with those of the lattice calculations.[11,12] The results in the first and second columns are those without and with the effects of flavor SU(3) symmetry breaking. Note that $\Xi_c^{'0}$ and Ξ_c^{*0} are forbidden to decay in the SU(3) symmetric case. It implies that the

Table 3. Numerical results of the radiative decay rates for the charmed baryons in units of keV.

$B_c \to B'_c\,\gamma$	$\Gamma_\gamma^{(0)}$ [keV]	$\Gamma_\gamma^{(\mathrm{total})}$ [keV]	Lattice results[11,12]
$\Sigma_c^+ \to \Lambda_c^+\,\gamma$	8.32 ± 0.73	12.82 ± 0.95	–
$\Xi_c^{\prime +} \to \Xi_c^+\,\gamma$	2.18 ± 0.20	2.02 ± 0.20	5.468
$\Xi_c^{\prime 0} \to \Xi_c^0\,\gamma$	0	0.06 ± 0.01	0.002
$\Sigma_c^{*+} \to \Lambda_c^+\,\gamma$	23.04 ± 2.12	35.49 ± 2.81	–
$\Xi_c^{*+} \to \Xi_c^+\,\gamma$	9.34 ± 0.82	8.66 ± 0.81	–
$\Xi_c^{*0} \to \Xi_c^0\,\gamma$	0	0.25 ± 0.06	–
$\Sigma_c^{*++} \to \Sigma_c^{++}\,\gamma$	0.31 ± 0.03	0.36 ± 0.03	–
$\Sigma_c^{*+} \to \Sigma_c^+\,\gamma$	0.02 ± 0.003	0.02 ± 0.003	–
$\Sigma_c^{*0} \to \Sigma_c^0\,\gamma$	0.08 ± 0.01	0.12 ± 0.01	–
$\Xi_c^{*+} \to \Xi_c^{\prime +}\,\gamma$	0.02 ± 0.002	0.03 ± 0.003	–
$\Xi_c^{*0} \to \Xi_c^{\prime 0}\,\gamma$	0.08 ± 0.01	0.09 ± 0.01	–
$\Omega_c^{*0} \to \Omega_c^0\,\gamma$	0.09 ± 0.01	0.06 ± 0.01	0.074

magnitude of their decay rates should be very small. Though there are not many data from lattice QCD, the present results are in agreement with them.

4. Summary and prospects

As shown above, the pion mean-field approach describes very well the masses and radiative decays of the singly heavy baryons with all the parameters taken from the light baryon sector. The pion mean-field approach has a clear virtue for the description of the light and heavy baryons in a unified way. Recently, we also have shown that the electromagnetic form factors of the singly heavy baryons are also in good agreement with the lattice data.[13,14]

So far, we have ignored the $1/m_Q$ corrections except for the hyperfine interaction. While this will be technically involved, it is worthwhile to consider them to investigate the properties of heavy baryons quantitatively. It is also of great importance to investigate excited heavy baryons within the framework of the meson mean-field approach. This will be considered as future works.

Acknowledgments

I am very grateful to J.-Y. Kim, M. V. Polyakov, M. Praszałowicz, and Gh.-S. Yang for fruitful collaborations and discussions. The present work was supported by Basic Science Research Program through the National Research Foundation (NRF) of Korea funded by the Ministry of Education,

Science and Technology (MIST) (No. 2018R1A5A1025563 and No. NRF-2018R1A2B2001752).

References

1. E. Witten, *Nucl. Phys. B* **160**, 57 (1979).
2. E. Witten, *Nucl. Phys. B* **223**, 422 (1983) and *Nucl. Phys. B* **223**, 433 (1983).
3. Gh.-S. Yang, H.-Ch. Kim, M. V. Polyakov and M. Praszałowicz, *Phys. Rev. D* **94**, 071502 (2016).
4. H.-Ch. Kim, M. V. Polyakov and M. Praszaowicz, Phys. Rev. D **96**, no. 1, 014009 (2017) Addendum: [Phys. Rev. D **96**, no. 3, 039902 (2017)].
5. H.-Ch. Kim, M. V. Polyakov, M. Praszalowicz and G. S. Yang, Phys. Rev. D **96**, no. 9, 094021 (2017) Erratum: [Phys. Rev. D **97**, no. 3, 039901 (2018)].
6. G. S. Yang and H.-Ch. Kim, to appear in *Phys. Lett. B* arXiv:1909.03156 [hep-ph].
7. H.-Ch. Kim, *J. Korean Phys. Soc.* **73**, 165 (2018).
8. A. V. Manohar and M. B. Wise, *Camb. Monogr. Part. Phys. Nucl. Phys. Cosmol.* **10**, 1 (2000).
9. G. S. Yang and H.-Ch. Kim, Prog. Theor. Phys. **128**, 397 (2012).
10. M. Tanabashi et al. (Particle Data Group), *Phys. Rev. D* **98**, 030001 (2018) and 2019 update.
11. H. Bahtiyar, K. U. Can, G. Erkol and M. Oka, *Phys. Lett. B* **747**, 281 (2015).
12. H. Bahtiyar, K. U. Can, G. Erkol, M. Oka and T. T. Takahashi, *Phys. Lett. B* **772**, 121 (2017).
13. J. Y. Kim and H.-Ch. Kim, *Phys. Rev. D* **97**, no. 11, 114009 (2018).
14. J. Y. Kim and H.-Ch. Kim, arXiv:1912.01437 [hep-ph].

Charmed excited baryons with heavy-quark spin symmetry

L. Tolos

Institut für Theoretische Physik, University of Frankfurt,
Max-von-Laue-Str. 1, 60438 Frankfurt am Main, Germany
Frankfurt Institute for Advanced Studies, University of Frankfurt,
Ruth-Moufang-Str. 1, 60438 Frankfurt am Main, Germany
Institute of Space Sciences (ICE, CSIC),
Campus UAB, Carrer de Can Magrans, 08193, Barcelona, Spain
Institut d'Estudis Espacials de Catalunya (IEEC),
08034 Barcelona, Spain
E-mail: tolos@th.physik.uni-frankfurt.de

J. Nieves and R. Pavao

Instituto de Física Corpuscular (centro mixto CSIC-UV),
Institutos de Investigación de Paterna, Aptdo. 22085, 46071, Valencia,Spain

Recently five Ω_c excited states have been reported by the LHCb Collaboration, four of them corroborated by Belle. The Belle Collaboration has also discovered in 2017 one excited Ξ_c with mass of 2930 MeV. In view of these recent detections, we analyze the possible molecular description of these states, using a unitarized baryon-meson model that incorporates both chiral and heavy-quark spin symmetries in a consistent manner. We pay a special attention to the renormalization procedure, so as to determine the robustness of our predictions. Within our model, we generate dynamically several Ω_c and Ξ_c states. We find that, at least, three of our Ω_c states can be identified with the experimental ones and have spin-parity $J = 1/2^-$ or $J = 3/2^-$. Moreover, we find a plausible molecular description of not only the $\Xi_c(2930)$ state, but also $\Xi_c(2790)$, $\Xi_c(2815)$ and $\Xi_c(2970)$, reported in the PDG. Interestingly, we determine that $\Xi_c(2930)$ and $\Xi_c(2970)$ are heavy-quark spin partners with $J = 1/2^-$ and $J = 3/2^-$, respectively, while determing that $\Xi_c(2930)$, $\Omega_c(3090)$ or the $\Omega_c(3119)$, and $\Sigma_c(2800)$ would belong to the same SU(3) multiplet.

Keywords: unitarized baryon-meson model; charmed baryonic excited states; heavy-quark spin symmetry.

1. Introduction

The study of charmed baryonic excited states has been a matter of high interest over the past years in view of several recent discoveries. The LHCb Collaboration has reported the existence of five Ω_c excited states in the

$\Xi_c^+K^-$ spectrum in pp collisions,[1] four of them confirmed by the Belle Collaboration.[2] Also, the $\Xi_c(2930)$ state has been determined by Belle[3] in its decay to $\Lambda_c^+K^-$ from $B^- \to K^-\Lambda_c^+\bar{\Lambda}_c^-$. Other excited Ξ_c states are reported around 3 GeV, such as $\Xi_c(2790)$, $\Xi_c(2815)$ and $\Xi_c(2970)$.[4]

The understanding of the nature of these newly discovered states have triggered an important theoretical effort within molecular baryon-meson models,[5–9] and, in particular, within a model that fully respects heavy-quark spin symmetry (HQSS), which is a proper QCD symmetry in the limit of large quark masses.[10–16] The starting point is a $\mathrm{SU(6)}_{\mathrm{lsf}} \times$ HQSS extension of the Weinberg-Tomozawa (WT) interaction, with "lsf" indicating the light quark-spin-flavor symmetry. In Ref. 12 five Ω_c states were generated dynamically, three $J = 1/2$ and two $J = 3/2$ bound states, stemming from the most attractive $\mathrm{SU(6)}_{\mathrm{lsf}}\times$ HQSS representations, but with masses below the LHCb predictions. Also, within the same model, nine Ξ_c were obtained, below or close in mass to the experimental $\Xi_c(2790)$ and $\Xi_c(2815)$ states, while being far below in mass from $\Xi_c(2930)$ and $\Xi_c(2970)$.

All these predictions of masses, however, depend on the adopted renormalization scheme (RS). In Refs. 17, 18 we have revisited the RS used in Ref. 12. We have payed a special attention to the flavor-symmetry content of the $\mathrm{SU(6)}_{\mathrm{lsf}} \times$ HQSS model in order to determine the possible HQSS partners and siblings among the experimental states while predicting new ones. In the present work we summarize the results of Refs. 17, 18 on the Ω_c and Ξ_c sectors.

2. Formalism

We solve the Bethe-Salpeter equation in the on-shell approximation for the scattering amplitude (T^J) for a total angular momentum J as

$$T^J(s) = \frac{1}{1 - V^J(s)G^J(s)} V^J(s), \tag{1}$$

with V^J being the S-wave baryon-meson potential resulting from the $\mathrm{SU(6)}_{\mathrm{lsf}} \times$ HQSS WT interaction between pseudoscalar or vector mesons and the low-lying $1/2^+$ or $3/2^+$ baryons. The diagonal $G^J(s)$ matrix contains the baryon-meson loop functions for each channel i, that is,

$$G_i(s) = \overline{G}_i(s) + G_i(s_{i+}). \tag{2}$$

These are logarithmically ultraviolet (UV) divergent, and need to be renormalized. The finite part, $\overline{G}_i(s)$ is obtained from Ref. 19, whereas the divergent contribution, $G_i(s_{i+})$, can be renormalized, either by one subtraction at certain scale ($\sqrt{s} = \mu$)

$$G_i^{\mu}(s) = \overline{G}_i(s) - \overline{G}_i(\mu^2), \tag{3}$$

or using a sharp-cutoff regulator Λ in momentum space, that is,

$$G_i^{\Lambda}(s) = \overline{G}_i(s) + G_i^{\Lambda}(s_{i+}). \tag{4}$$

The excited Ω_c and Ξ_c states are obtained as poles of the scattering amplitudes in each J sector (see Refs. 12, 17, 18 for details).

3. Results

3.1. Ω_c *excited states*

In Ref. 12 five excited Ω_c states with $J = 1/2^-$ and $J = 3/2^-$ were found, with masses below 3 GeV, thus, being difficult to identify them with the LHCb results. These states were obtained renormalizing with one-substraction at a certain scale. In order to study the dependence of our results on the RS and the possible identification with the experimental observations, we employ a common UV cutoff for all baryon-meson loops within a reasonable range. In this way we avoid any uncontrolled reduction of a loop and an arbitrary change of the subtraction constants.

First we need to follow the original Ω_c states in the complex energy plane as we modify our prescription from one-subtraction to a common UV cutoff renormalization. Hence, we change each loop function by

$$G_i(s) = \overline{G}_i(s) - (1-x)\overline{G}_i(\mu^2) + xG_i^{\Lambda}(s_{i+}), \tag{5}$$

with x moving adiabatically from 0 to 1. In this manner and by varying the value of UV cutoff, we find that (probably) at least three of the experimental states can be identified with three of our Ω_c.[17] As an example, for $\Lambda = 1090$ MeV, we identify the experimental $\Omega_c(3000)$ and $\Omega_c(3090)/\Omega_c(3119)$ with two of our $1/2^-$ states, and $\Omega_c(3050)$ with one of our $3/2^-$. Also, $\Omega_c(3000)$ and $\Omega_c(3050)$ would be members of the same $\mathrm{SU(6)_{lsf}} \times$ HQSS multiplet.[17]

Other models have also determined the molecular nature of some of the experimental Ω_c states. In particular, in Ref. 5 the $\Omega_c(3050)$ and $\Omega_c(3090)$ were identified as $1/2^-$ states. In Ref. 6 the $\Omega_c(3050)$ and $\Omega_c(3090)$ were also obtained as $1/2^-$ states, due to the use of the same interaction for $J = 1/2$ as Ref. 5, whereas the experimental $\Omega_c(3119)$ was obtained with $3/2^-$. The difference between the predictions of our model and these schemes comes from a different RS as well as different interactions for the channels involving D, D^* and light vector mesons with baryons, that are not fixed by chiral or HQSS symmetries. With regards to a broad structure around 3188 MeV determined by LHCb,[1] our results differ from those of Ref. 7, as we cannot make any identification, since this state would come from a less attractive $SU(6)_{lsf}\times$HQSS representation. Also the prediction of Ref. 8 of a loosely bound molecule of 3140 MeV is not confirmed in our model, as in Ref. 8 the $\Xi^{(*)}D^{(*)}$ channels were not considered.

3.2. Ξ_c *excited states*

In the Ξ_c sector we proceed in a similar way as done for the Ω_c states by revising the results of Ref. 12. There, nine states were obtained with masses below or close to the experimental $\Xi_c(2790)/\Xi_c(2815)$ states, while being far below in mass with respect to $\Xi_c(2930)/\Xi_c(2970)$. Now we identify our Ξ_c states in the cutoff scheme and assess the dependence of our predictions on the value of the cutoff.

In this manner, we arrive to several conclusions, extensively discussed in Ref. 18. We find that the experimental $\Xi_c(2790)$, $\Xi_c(2815)$, $\Xi_c(2930)$ and $\Xi_c(2970)$ states can be described as molecules. In particular, we determine that the $\Xi_c(2790)$ state has a large molecular $\Lambda_c\bar{K}$ component, with a dominant $j^{\pi}_{ldof} = 0^-$ configuration for the light degrees of freedom (ldof). We find that $\Xi_c(2790)$ could be the sibling of the narrow $\Lambda_c(2595)$ in the double pole description of the $\Lambda_c(2595)$. We determine that the $3/2^-$ $\Lambda_c(2625)$ and $\Xi_c(2815)$ states cannot be SU(3) siblings, in particular given the fact that $\Lambda_c(2625)$ is probably a constituent three quark state.[20,21] Moreover, we predict the existence of other Ξ_c–states, not experimentally detected yet, being one of them the sibling of the wide $\Lambda_c(2595)$. Interestingly, the recently discovered $\Xi_c(2930)$ and $\Xi_c(2970)$ are HQSS partners. The $\Xi_c(2930)$ would belong to a SU(3) sextet, where there is also a Ω_c state. This state would correspond to either the experimental $\Omega_c(3090)$ or the $\Omega_c(3119)$, given their decay channels. Assuming the equal spacing rule, we determine a $J = 1/2^-$ Σ_c state around 2800 MeV that will complete the sextet and fits clearly with the experimental $\Sigma_c(2800)$.[4]

The description of the excited Ξ_c states determined by the local hidden gauge model of Ref. 9 is again different to ours due to the distinct RS and the interactions involving D and D^* and light vector mesons with baryons.

4. Conclusions

In view of the recent observation of excited Ω_c and Ξ_c states by the LHCb and Belle Collaborations, we have explored their possible interpretation as molecular states. We have used a coupled-channel unitarized model based on a $SU(6)_{lsf} \times$HQSS-extended WT baryon-meson interaction, revising the predictions of Ref. 12. The present work is a summary of Refs. 17, 18.

With regards to the five Ω_c excited states, we conclude that (probably) at least three of the states observed by LHCb have spin-parity $J = 1/2^-$ and $J = 3/2^-$. Also, we find that $\Xi_c(2790)$, $\Xi_c(2815)$, $\Xi_c(2930)$ and $\Xi_c(2970)$ states can be described as molecules. Interestingly, $\Xi_c(2930)$ and $\Xi_c(2970)$ would be HQSS partners, and $\Xi_c(2930)$ would be part of a SU(3) sextet containing either the $\Omega_c(3090)$ or the $\Omega_c(3119)$, and that would be completed by the experimental $\Sigma_c(2800)$.

Acknowledgments

R. P. Pavao acknowledges support from the Generalitat Valenciana in the program GRISOLIAP/2016/071. This research has been supported by the Deutsche Forschungsgemeinschaft under Project Nr. 383452331 (Heisenberg Programme) and Project Nr. 411563442; Spanish Ministerio de Economía y Competitividad and the European Regional Development Fund, under contracts FIS2017-84038-C2-1-P, FPA2013-43425-P, FPA2016-81114-P and SEV-2014-0398; the THOR COST Action CA15213; and by the EU STRONG-2020 project under the program H2020-INFRAIA-2018-1, grant agreement no. 824093.

References

1. R. Aaij *et al.*, Observation of five new narrow Ω_c^0 states decaying to $\Xi_c^+ K^-$, *Phys. Rev. Lett.* **118**, 182001 (2017).
2. J. Yelton *et al.*, Observation of Excited Ω_c Charmed Baryons in e^+e^- Collisions, *Phys. Rev.* **D97**, 051102 (2018).
3. Y. Li *et al.*, Observation of $\Xi_c(2930)^0$ and updated measurement of $B^- \to K^- \Lambda_c^+ \bar{\Lambda}_c^-$ at Belle, *Eur. Phys. J.* **C78**, 252 (2018).
4. M. Tanabashi *et al.*, Review of Particle Physics, *Phys. Rev.* **D98**, 030001 (2018).
5. G. Montaña, A. Feijoo and A. Ramos, A meson-baryon molecular interpretation for some Ω_c excited states, *Eur. Phys. J.* **A54**, 64 (2018).
6. V. R. Debastiani, J. M. Dias, W. H. Liang and E. Oset, Molecular Ω_c states generated from coupled meson-baryon channels, *Phys. Rev.* **D97**, 094035 (2018).

7. C. Wang, L.-L. Liu, X.-W. Kang, X.-H. Guo and R.-W. Wang, Possible open-charmed pentaquark molecule $\Omega_c(3188)$ - the $D\Xi$ bound state - in the Bethe-Salpeter formalism, *Eur. Phys. J.* **C78**, 407 (2018).
8. R. Chen, A. Hosaka and X. Liu, Searching for possible Ω_c-like molecular states from meson-baryon interaction, *Phys. Rev.* **D97**, 036016 (2018).
9. Q. X. Yu, R. Pavao, V. R. Debastiani and E. Oset, Description of the Ξ_c and Ξ_b states as molecular states, *Eur. Phys. J.* **C79**, 167 (2019).
10. C. Garcia-Recio *et al.*, The s-wave charmed baryon resonances from a coupled-channel approach with heavy quark symmetry, *Phys. Rev.* **D79**, 054004 (2009).
11. D. Gamermann, C. Garcia-Recio, J. Nieves, L. L. Salcedo and L. Tolos, Exotic dynamically generated baryons with negative charm quantum number, *Phys. Rev.* **D81**, 094016 (2010).
12. O. Romanets, L. Tolos, C. Garcia-Recio, J. Nieves, L. Salcedo and R. Timmermans, Charmed and strange baryon resonances with heavy-quark spin symmetry, *Phys. Rev.* **D85**, 114032 (2012).
13. C. Garcia-Recio, J. Nieves, O. Romanets, L. Salcedo and L. Tolos, Odd parity bottom-flavored baryon resonances, *Phys. Rev.* **D87**, 034032 (2013).
14. C. Garcia-Recio, J. Nieves, O. Romanets, L. L. Salcedo and L. Tolos, Hidden charm N and Delta resonances with heavy-quark symmetry, *Phys. Rev.* **D87**, 074034 (2013).
15. L. Tolos, Charming mesons with baryons and nuclei, *Int. J. Mod. Phys.* **E22**, 1330027 (2013).
16. C. Garcia-Recio, C. Hidalgo-Duque, J. Nieves, L. L. Salcedo and L. Tolos, Compositeness of the strange, charm, and beauty odd parity Λ states, *Phys. Rev.* **D92**, 034011 (2015).
17. J. Nieves, R. Pavao and L. Tolos, Ω_c excited states within a $\mathrm{SU}(6)_{\mathrm{lsf}}\times$ HQSS model, *Eur. Phys. J.* **C78**, 114 (2018).
18. J. Nieves, R. Pavao and L. Tolos, Ξ_c and Ξ_b excited states within a $\mathrm{SU}(6)_{\mathrm{lsf}}\times$HQSS model, *1911.06089 [hep-ph]* .
19. J. Nieves and E. Ruiz Arriola, The S(11) - N(1535) and - N(1650) resonances in meson baryon unitarized coupled channel chiral perturbation theory, *Phys. Rev.* **D64**, 116008 (2001).
20. T. Yoshida, E. Hiyama, A. Hosaka, M. Oka and K. Sadato, Spectrum of heavy baryons in the quark model, *Phys. Rev.* **D92**, 114029 (2015).
21. J. Nieves and R. Pavao, On the nature of the lowest-lying odd parity charmed baryon $\Lambda_c(2595)$ and $\Lambda_c(2625)$ resonances, *1907.05747 [hep-ph]* .

Molecule pictures for the LHCb P_c states

Yong-Hui Lin

CAS Key Laboratory of Theoretical Physics, Institute of Theoretical Physics, Chinese Academy of Sciences, Beijing 100190, China

School of Physical Sciences, University of Chinese Academy of Sciences, Beijing 100049, China
E-mail: linyonghui@mail.itp.ac.cn

The first two pentaquark candidates $P_c(4380)$ and $P_c(4450)$ were observed by LHCb Collaboration in 2015. Recently, they presented their improved measurements in the P_c energy range, that a new narrow signal denoted as $P_c(4312)$ is reported and the previous observed structure $P_c(4450)$ is resolved into two narrow states, $P_c(4440)$ and $P_c(4457)$ while there is no obvious signal for the broad state $P_c(4380)$. We analyze the strong decays of these P_c states systematically in the hadronic molecule pictures. With the effective Lagrangian method, the partial widths of these P_c decaying into all allowed channels are estimated. Our results show that these P_c states are good candidates of the S-wave $\bar{D}\Sigma_c$, $\bar{D}\Sigma_c^*$ and $\bar{D}^*\Sigma_c$ molecules. Also other possible flavor and spin partners of P_c states are investigated in the present work.

Keywords: Pentaquark states; strong decay; hadronic molecule.

1. Introduction

The spectroscopy of exotic hadrons is one of the most important topics in the frontiers of hadron physics. Especially, the first observation of two hidden charm pentaquark-like structures, $P_c^+(4380)$ and $P_c^+(4450)$, by the LHCb Collaboration in 2015[1] triggered widespread further theoretical investigations on their nature, see Refs. 2, 3 for the reviews. Very recently, the more precise line shape of the $J/\psi p$ invariant mass distribution from the process $\Lambda_b^0 \to J/\psi p K^-$ was presented by the latest LHCb measurement.[4] The experimental data suggested that the previous observed structure $P_c(4450)$ is resolved into two narrow states, $P_c(4440)$ and $P_c(4457)$ while the broad state $P_c(4380)$ have not been confirmed yet. In addition, a new structure $P_c(4312)$ is discovered with 7.3σ significance. The existence of such pentaquark states around this energy range has been predicted systematically in some early theoretical works[5] before 2015. There is a very interesting fact about these P_c states that the reported masses of not only the previous $P_c^+(4380)$ and $P_c(4450)$, but the newly observed $P_c(4312)$, $P_c^+(4440)$ and $P_c(4457)$ locate just below the thresholds of $\bar{D}\Sigma_c$, $\bar{D}\Sigma_c^*$ and $\bar{D}^*\Sigma_c$ with the binding energy ranging from 5 to 20 MeV. This closeness to the thresholds makes the interpretation of hadronic molecule consisting of the corresponding meson-baryon system naturally for these pentaquark-like states. And the previous theoretical works[5] also predicted that the $\bar{D}\Sigma_c$ and $\bar{D}^*\Sigma_c$ account for a large proportion of component in lower $P_c(4312)$ and higher two P_c states, respectively.

We investigated the strong decays of these P_c states in the hadronic molecule pictures, for $P_c^+(4380)$ and $P_c(4450)$ in Ref. 6 and for $P_c(4312)$, $P_c^+(4440)$ and $P_c(4457)$ in Ref. 7. In our previous work, we assumed various possible quantum numbers for some P_c states which are allowed in the molecule scenarios, such as $3/2^-$ and $5/2^+$ for the $P_c(4450)$, quantum number set $(1/2^-, 3/2^-)$ and $(3/2^-, 1/2^-)$ for $P_c^+(4440)$ and $P_c(4457)$ since the experiment knowledge on the quantum numbers of observed P_c states are very scarce. And we expect that the difference on the decay behaviors between various quantum numbers can help us to distinguish them in the future experimental investigation. And we also extended our formalism on the P_c states to their possible flavor partners in our previous work.[8]

2. Theoretical formalism

With the effective Lagrangian approach, the partial decay widths of P_c molecules to all possible channels can be estimated consistently. Only the tree-level three-body decay and one-loop two-body decay are considered in our calculation as shown in Fig. 1. All the two-body decay channels are collected in Table 1.

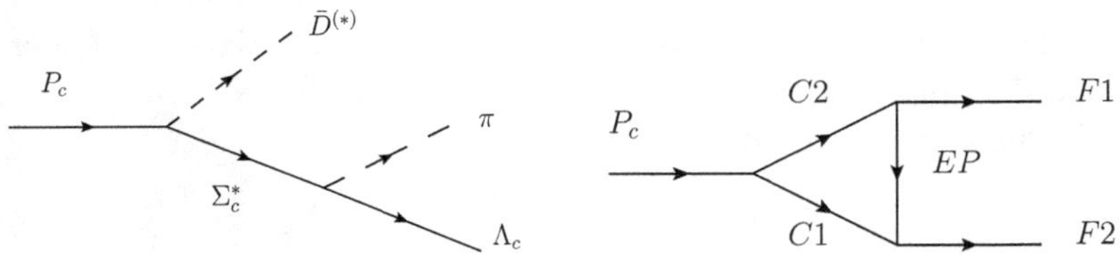

Fig. 1. The left panel is the three-body decays of the $\bar{D}^{(*)}\Sigma_c^*$ molecules. And the right panel denotes the triangle diagram for the two-body decays of the P_c states in the $\bar{D}^{(*)}\Sigma_c^{(*)}$ molecule scenarios.

Table 1. All possible decay channels for the P_c states in the $\bar{D}^{(*)}\Sigma_c^{(*)}$ molecule scenario.

Initial state	Final states	Exchanged particles
$P_c(4312)(\bar{D}\Sigma_c)$ $P_c(4376)(\bar{D}\Sigma_c^*)$	$\bar{D}^*\Lambda_c$	π, ρ
	$\bar{D}\Lambda_c$, $\bar{D}\Sigma_c$[a]	ρ
	$J/\psi N$, ωp, ρN	D^*, D
	$\eta_c N$	D^*
	πN	D^*, Λ_c, Σ_c
	$\chi_{c0} N$	D
$P_c(4440)\&P_c(4457)$ $(\bar{D}^*\Sigma_c)$ $P_c(4500)\&P_c(4511)$ $\&P_c(4523)(\bar{D}^*\Sigma_c^*)$	$\bar{D}^*\Lambda_c$, $\bar{D}\Lambda_c$, $\bar{D}\Sigma_c^*$, $\bar{D}\Sigma_c$, $\bar{D}\Sigma_c^*$[b]	π, ρ
	$J/\psi N$, ωp, ρN, ηN	D^*, D
	πN	D^*, D, Λ_c, Σ_c
	$\chi_{c0} N$	D^*

Note: [a] $\bar{D}\Sigma_c$ channel is only considered for the $\bar{D}\Sigma_c^*$ molecule.
[b] $\bar{D}\Sigma_c^*$ channel is only considered for the $\bar{D}^*\Sigma_c^*$ molecule.

For the first vertex that P_c states couple to the hadronic baryon-meson pairs, the Lorentz covariant L-S scheme proposed in Ref. 9 is used. The effective couplings $g_{\bar{D}^{(*)}\Sigma_c^{(*)}P_c}$ are estimated with the compositeness criterion. For the additional Lagrangians required to construct the one meson exchanged potential, we adopt the conventional formula used in a variety of phenomenological approaches. The specific formalism of the Lagrangians and all effective couplings we used can refer to our previous work.[6–8] One should note that most of these values can only be regarded as rough estimations, which should suffice for an order-of-magnitude estimate of the decay rates under consideration.

The following two kinds of Gaussian regulator which render all the amplitudes UV finite are employed

$$f_1(p_E^2/\Lambda_0^2) = \exp(-p_E^2/\Lambda_0^2), \qquad f_2(\boldsymbol{p}^2/\Lambda_0^2) = \exp(-\boldsymbol{p}^2/\Lambda_0^2), \tag{1}$$

where p_E is the geometric average Euclidean Jacobi momentum of the $\bar{D}^{(*)}\Sigma_c^{(*)}$ molecules. $\boldsymbol{p}$ is the spatial part of the momentums of $\bar{D}^{(*)}$ and $\Sigma_c^{(*)}$ in the rest frame of P_c states. In addition, a multipolar form factor is introduced to suppress the off-shell contributions of the exchanged mesons in our triangle diagrams. It is chosen as

$$f_3(q^2) = \frac{\Lambda_1^4}{(m^2 - q^2)^2 + \Lambda_1^4}, \tag{2}$$

where m and q is the mass and momentum of the exchanged particle. In the present work, we vary the value of Λ_0 and Λ_1 from 0.6 GeV to 1.4 GeV for a rough estimate of the two-body partial widths.

3. Numerical Results and Discussions

The partial decay widths of various P_c states in $\bar{D}^{(*)}\Sigma_c^{(*)}$ hadronic molecule pictures with $\Lambda_0 = 1.0$ GeV and $\Lambda_1 = 0.6$ GeV are displayed in Table 2, where the form factor set (f_1, f_3) is used. The masses of four spin partners of P_c are taken from the predictions in the Ref. 10. And the numerical results for the possible flavor partners of P_c are collected in Table 3. The binding energy of the S-wave $B\Sigma_b$ bound state is predicted to be around 90 MeV in Ref. 11. We take this value for all the S-wave $B\Sigma_b^*$ and $B^*\Sigma_b$ molecules approximately. The parameter dependence of our numerical decay widths can refer to the Refs. 6–8.

Table 2. Partial decay widths of the oberved P_c states and four spin partners within the HQSS framework. The form factor set (f_1, f_3) is used and $\Lambda_0 = 1.0$ GeV, $\Lambda_1 = 0.6$ GeV. The short bars denote that the corresponding channel is closed or its contribution is negligible.

Mode	Widths (MeV) with (f_1, f_3)								
	$D\Sigma_c$	$D^*\Sigma_c$				$D\Sigma_c^*$	$D^*\Sigma_c^*$		
	$P_c(4312)$	$P_c(4440)$		$P_c(4457)$		$P_c(4376)$	$P_c(4500)$	$P_c(4511)$	$P_c(4523)$
	$\frac{1}{2}^-$	$\frac{1}{2}^-$	$\frac{3}{2}^-$	$\frac{1}{2}^-$	$\frac{3}{2}^-$	$\frac{3}{2}^-$	$\frac{1}{2}^-$	$\frac{3}{2}^-$	$\frac{5}{2}^-$
$\bar{D}^*\Lambda_c$	3.8	13.9	6.2	12.5	6.1	12.4	7.1	17.0	4.5
$J/\psi p$	0.001	0.03	0.02	0.02	0.01	0.01	0.006	0.02	0.006
$\bar{D}\Lambda_c$	0.06	5.6	1.7	3.8	1.5	~ 0	10.0	0.3	1.5
πN	0.004	0.002	~ 0	0.001	~ 0	~ 0	0.003	~ 0	~ 0
$\chi_{c0}p$	-	~ 0	~ 0	~ 0	~ 0	0.003	0.01	0.002	~ 0
$\eta_c p$	0.01	~ 0	~ 0	~ 0	~ 0	0.001	0.01	~ 0	~ 0
ρN	~ 0	~ 0	~ 0	~ 0	~ 0	~ 0	0.001	0.01	~ 0
ωp	~ 0	0.001	~ 0	~ 0	~ 0	0.002	0.004	0.005	~ 0
$\bar{D}\Sigma_c$	-	3.4	0.5	2.6	1.0	~ 0	10.6	0.2	1.3
$\bar{D}\Sigma_c^*$	-	0.8	5.4	1.9	6.2	-	1.0	33.8	6.2
$\bar{D}^*\Sigma_c$	-	-	-	-	-	-	10.6	0.07	1.2
$\bar{D}\Lambda_c\pi$	-	-	-	-	-	-	-	-	-
$\bar{D}^*\Lambda_c\pi$	-	-	-	-	-	-	4.0	7.7	7.8
Total	3.9	23.7	13.9	20.7	14.7	17.5	43.3	59.1	22.5

It is found that the measured widths of $P_c(4312)$, $P_c(4440)$ and $P_c(4457)$ can be reproduced well respectively in the $1/2^-$-$\bar{D}\Sigma_c$, $1/2^-$-$\bar{D}^*\Sigma_c$ and $3/2^-$-$\bar{D}^*\Sigma_c$ molecule pictures. And the $3/2^-$-$\bar{D}^*\Sigma_c$ and $1/2^-$-$\bar{D}^*\Sigma_c$ molecule assignments for the $P_c(4440)$ and $P_c(4457)$ can not be ruled out yet. The novel difference on the decay patterns between the spin parity $1/2^-$ and $3/2^-$ $\bar{D}^*\Sigma_c$ molecule for both $P_c(4440)$ and $P_c(4457)$, such as $\Gamma_{\bar{D}\Sigma_c}/\Gamma_{\bar{D}\Sigma_c^*}$ and $\Gamma_{J/\psi p}/\Gamma_{\eta_c p}$, can be used to distinguish the quantum numbers in the future experiments. For the strange sector, the hadronic molecular scenario with $N(1875)$ and $N(2080)$ as S-wave $K\Sigma^*$ and $K^*\Sigma$ molecules, respectively, can also provide a good explanation to their measured decay behaviors. Albeit with large uncertainty, the findings here discussed can be considered

Table 3. Partial decay widths of the flavor partners of P_c states in the hidden strange and bottom sectors. The form factor set (f_2, f_3) is used and $\Lambda_0 = 1.0$ GeV, $\Lambda_1 = 1.2$ GeV for the hidden-strange molecule while $\Lambda_0 = 1.0$ GeV, $\Lambda_1 = 2.0$ GeV for the hidden-bottom molecule. The notation is same with Table. 2.

Mode	Widths (MeV)			Mode	Widths (MeV)		
	$K\Sigma^*$		$K^*\Sigma$		$B\Sigma_b^*$		$B^*\Sigma_b$
	$J^P = 3/2^-$		$J^P = 1/2^-$		$J^P = 3/2^-$		$J^P = 1/2^-$
	$N(1875)$	$N(2080)$	$N(2080)$		$N_{b\bar{b}}$ (with binding energy 90 MeV)		
$N\sigma(500)$	1.5	0.05	0.3	$B^*\Lambda_b$	271.1	19.9	167.3
πN	0.2	0.2	27.0	Υp	0.3	0.04	0.1
ρN	1.4	3.3	7.3	ρN	10.4	0.02	0.1
ωp	4.0	10.0	21.7	ωp	39.5	0.07	0.4
$K\Sigma$	0.02	1.3	10.9	$B\Lambda_b$	-	7.3	136.1
$K\Lambda$	0.4	3.3	23.0	$\eta_b p$	0.02	0.0001	0.0009
ηp	0.4	0.3	2.1	$\chi_{b0} p$	1.4	0.0008	0.2
$\pi\Delta$	117.3	73.8	55.5	πN	0.05	0.005	0.003
$K^*\Lambda$	-	2.1	9.5				
ϕp	-	17.0	32.2				
$K\Sigma^*$	-	6.5	1.6				
$K\Lambda(1520)$	-	0.04	0.5				
$K\Lambda(1405)$	-	2.4	3.6				
$K\pi\Lambda$	10.1	-	-				
$K\pi\Sigma$	-	20.8	23.1				
Total	135.2	141.1	218.1	Total	322.7	27.4	304.2

as the direct consequences of the hadronic molecular assignments and can be tested by the further experimental investigation in future. It will help us to uncover the natures of these pentaquark-like states.

We thank Bing-Song Zou for the contribution to this work. This project is supported by NSFC under Grant No. 11621131001 (CRC110 cofunded by DFG and NSFC) and Grant No. 11747601, and the Chinese Academy of Sciences (CAS) under Grant No. XDPB09.

References

1. R. Aaij *et al.*, Phys. Rev. Lett. **115**, 072001 (2015).
2. F. K. Guo *et al.*, Rev. Mod. Phys. **90**, no. 1, 015004 (2018).
3. H. X. Chen *et al.*, Phys. Rept. **639**, 1 (2016).
4. R. Aaij *et al.*, Phys. Rev. Lett. **122**, no. 22, 222001 (2019).
5. J. J. Wu *et al.*, Phys. Rev. Lett. **105**, 232001 (2010).
6. Y. H. Lin *et al.*, Phys. Rev. D **95**, no. 11, 114017 (2017).
7. Y. H. Lin *et al.*, Phys. Rev. D **100**, no. 5, 056005 (2019).
8. Y. H. Lin *et al.*, Nucl. Phys. A **980**, 21 (2018).
9. B. S. Zou *et al.*, Phys. Rev. C **67**, 015204 (2003).
10. M. Z. Liu *et al.*, Phys. Rev. Lett. **122**, no. 24, 242001 (2019).
11. C. W. Shen *et al.*, Chin. Phys. C **42**, no. 2, 023106 (2018).

Baryon spectroscopy at LHCb

A. Xu*

School of Physics State Key Laboratory of Nuclear Physics and Technology, Peking University, Beijing, 100871, China
E-mail: ao.xu@pku.edu.cn

The LHCb experiment has been continually producing interesting results of heavy baryon spectroscopy, owing to the large data sample of heavy flavor hadrons provided by the LHC and the excellent performance of the delicately designed detector. Selected latest results of conventional heavy baryons from the LHCb experiment are presented.

Keywords: Baryon; spectroscopy; heavy flavour.

1. Introduction

Study of baryon spectroscopy is essential to understand nonperturbative quantum chromodynamics. The LHCb experiment has been performing important measurements on heavy flavour baryon spectroscopy since its commissioning. The LHCb detector is a single-arm forward spectrometer designed for studies of particles containing b or c quarks.[1,2] A total pp collision data set corresponding to an integrated luminosity of $9\,\mathrm{fb}^{-1}$ is collected at centre-of-mass energies $\sqrt{s} = 7$, 8 and 13 TeV during the LHC Run 1 and Run 2 data-taking period. In the following sections, recent results of heavy baryon spectroscopy from the LHCb experiment are reported.

2. Excited beauty baryons

In the constituent quark model,[3,4] beauty baryons form multiplets according to their symmetry of flavour, spin and spatial wave functions. While a number of ground states have been experimentally observed, only a few of heavier states, where constituent quarks are radially or orbitally excited, are observed in the beauty sector. The search for and study of such excited

---

*On behalf of the LHCb collaboration.

beauty baryons are helpful to understand the internal mechanism governing the dynamics of constituent quarks.

Structures are observed in the $\Lambda_b^0\pi^\pm$ mass spectra (charge conjugation is implied throughout) using pp collision data collected at $\sqrt{s} = 7$ and 8 TeV, corresponding to an integrated luminosity of $3\,\text{fb}^{-1}$.[5] Candidate Λ_b^0 decays are reconstructed with the $\Lambda_b^0 \to \Lambda_c^+\pi^-$, $\Lambda_c^+ \to pK^-\pi^+$ decay chain. To reconstruct $\Sigma_b^\pm \to \Lambda_b^0\pi^\pm$ decays, the aforementioned Λ_b^0 candidates are combined with a prompt pion of $p_\text{T} > 1000\,\text{MeV}/c$ and the mass difference $Q \equiv m(\Lambda_b^0\pi^\pm) - m(\Lambda_b^0) - m(\pi^\pm)$ is required to be in the range $[0, 600]\,\text{MeV}/c^2$. Two new state, denoted as $\Sigma_b(6097)^\pm$, are observed in both $\Lambda_b^0\pi^\pm$ spectra, as shown in Fig. 1. Extended unbinned maximum-likelihood fits to the Q distributions are performed to evaluate the signal yields and parameters of $\Sigma_b(6097)^\pm$ states. All signal components are described by relativistic Breit-Wigner functions including Blatt-Weisskopf form factors. The resulting parameters are shown in Table 1, where the first uncertainty is statistical and the second systematic. The new $\Sigma_b(6097)^+$ and $\Sigma_b(6097)^-$ states have local statistical significances of 12.7σ and 12.6σ, respectively.

The newly observed $\Sigma_b(6097)^\pm$ states are compatible with $\Sigma_b(1P)$ excitations. Other interpretations, such as superpositions of more than one state and molecular states, may also be possible.

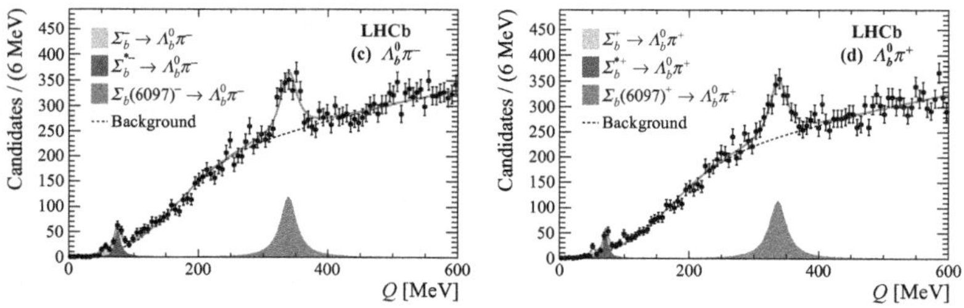

Fig. 1. Mass distributions for selected $\Lambda_b^0\pi^\pm$ candidates. The left (right) column shows $\Lambda_b^0\pi^-$ ($\Lambda_b^0\pi^+$) combinations.

Table 1. Masses and widths of $\Sigma_b(6097)^\pm$ states.

State	Mass $[\,\text{MeV}/c^2\,]$	Width $[\,\text{MeV}/c^2\,]$
$\Sigma_b(6097)^-$	$6098.0 \pm 1.7 \pm 0.5$	$28.9 \pm 4.2 \pm 0.9$
$\Sigma_b(6097)^+$	$6095.8 \pm 1.7 \pm 0.4$	$31.0 \pm 5.5 \pm 0.7$

Excited Λ_b^0 states are searched for in the $\Lambda_b^0\pi^+\pi^-$ mass spectrum with the total LHCb pp collision data set.[6] The Λ_b^0 candidates are reconstructed with the $\Lambda_b^0 \to \Lambda_c^+\pi^-$, $\Lambda_c^+ \to pK^-\pi^+$ decay chain. The selected Λ_b^0 candidates are combined with two prompt pions to search for excited states, requiring the transverse momentum of pion pairs $p_{\rm T} > 500\,{\rm MeV}/c$ to suppress combinatorial background. A new structure is observed in the $\Lambda_b^0\pi^+\pi^-$ mass spectrum near $6.15\,{\rm GeV}/c^2$. Given that the mass of the new structure is above $\Sigma_b^{*\mp}\pi^\pm$ thresholds, the $\Lambda_b^0\pi^+\pi^-$ mass spectrum is investigated in three non-overlapping $\Lambda_b^0\pi^\pm$ mass regions: candidates with a $\Lambda_b^0\pi^\pm$ mass within the width of the known $\Sigma_b^\pm$ mass, the known $\Sigma_b^{*\pm}$ mass, and the remaining non-resonant (NR) region. Figure 2 shows the $\Lambda_b^0\pi^+\pi^-$ mass spectra in these three regions, which have different shapes and indicate the presence of two narrow peaks. A simultaneous extended unbinned maximum-likelihood fit to $\Lambda_b^0\pi^+\pi^-$ mass distributions in these three regions is performed to probe the two-resonance hypothesis, where the mass and width parameters are shared across regions. The signal components are modelled by the incoherent sum of two relativistic Breit-Wigner functions convolved with the detector resolution. The statistical significance of the two-signal hypothesis is favoured with respect to the single-signal hypothesis by more than 7σ. Taking into account systematic uncertainties, the resulting masses and widths of the two narrow states, denoted as $\Lambda_b(6146)^0$ and $\Lambda_b(6152)^0$, are

$$\begin{aligned}
m_{\Lambda_b(6146)^0} &= 6146.17 \pm 0.33 \pm 0.22 \pm 0.16\,{\rm MeV}/c^2, \\
m_{\Lambda_b(6152)^0} &= 6152.51 \pm 0.26 \pm 0.22 \pm 0.16\,{\rm MeV}/c^2, \\
\Gamma_{\Lambda_b(6146)^0} &= \quad 2.9 \ \pm 1.3 \ \pm 0.3 \ {\rm MeV}/c^2, \\
\Gamma_{\Lambda_b(6152)^0} &= \quad 2.1 \ \pm 0.8 \ \pm 0.3 \ {\rm MeV}/c^2,
\end{aligned}$$

where the uncertainties are statistical, systematic and (for mass parameters) are due to the imprecise knowledge of the Λ_b^0 mass.[7]

The two narrow states can be interpreted as doublet of $\Lambda_b(1D)^0$ states with quantum numbers $J^P = \frac{3}{2}^+$ and $J^P = \frac{5}{2}^+$, given the measured masses and widths and the observed decay patterns. However, the interpretation of these states as excited Σ_b^0 states cannot be excluded.

3. Doubly charmed baryons

The constituent quark model[3,4] predicts the existence of doubly charmed baryons, which are composed of two c quarks and a light quark. These states form one isospin doublet Ξ_{cc} ($\Xi_{cc}^+ = ccd$ and $\Xi_{cc}^{++} = ccu$)

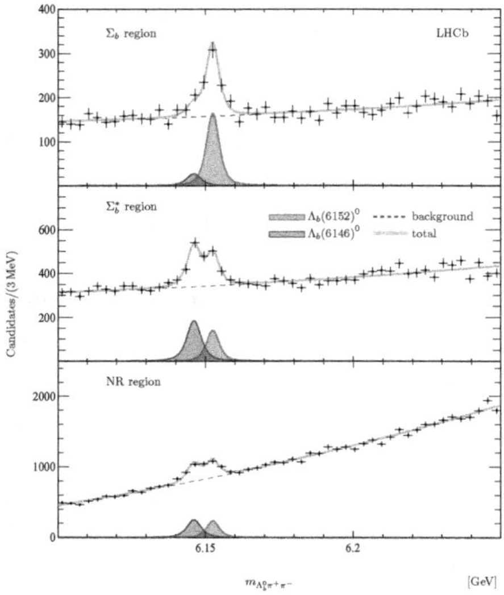

Fig. 2. Mass distributions of selected $\Lambda_b^0\pi^+\pi^-$ candidates for the three regions in $\Lambda_b^0\pi^\pm$ mass: (top) $\Sigma_b^\pm$, (middle) $\Sigma_b^{*\pm}$ and (bottom) non-resonant (NR) region.

and one isospin singlet Ω_{cc} ($\Omega_{cc}^+ = ccs$). LHCb reported the observation of the Ξ_{cc}^{++} baryon in the $\Lambda_c^+K^-\pi^+\pi^+$ final state using data collected at $\sqrt{s} = 13\,\text{TeV}$, corresponding to an integrated luminosity of $1.7\,\text{fb}^{-1}$.[8] The mass of the Ξ_{cc}^{++} baryon is measured to be $3621.40 \pm 0.72\,(\text{stat}) \pm 0.27\,(\text{syst}) \pm 0.14\,(\Lambda_c^+)\,\text{MeV}/c^2$. The Ξ_{cc}^{++} lifetime is measured with the same data set to be $0.256\,^{+0.024}_{-0.022}\,(\text{stat}) \pm 0.014\,(\text{syst})$ ps, which establishes its weakly-decaying nature.[9] This data set is also used to search for other Ξ_{cc}^{++} decay modes to improve the understanding of the decay of doubly charm baryons.

Firstly, the $\Xi_c^+\pi^+$ mass spectrum is studied to search for $\Xi_{cc}^{++} \to \Xi_c^+\pi^+$ decays.[10] The Ξ_c^+ candidates are reconstructed in the $pK^-\pi^+$ final state. A multivariate classifier based on the multilayer perceptron (MLP) algorithm, exploiting the kinematic and topological features of the signal decay, is used to suppress combinatorial background. The $\Xi_c^+\pi^+$ mass spectrum is shown in the left plot of Fig. 3, where a peaking structure is observed near the mass of the Ξ_{cc}^{++} baryon. An unbinned extended maximum-likelihood fit is performed to the $m(\Xi_c^+\pi^+)$ distribution to measure the property of this structure. The local statistical significance is found to be 5.9σ. After considering the systematic uncertainties, the resulting mass is $3620.6 \pm 1.5 \pm$

$0.4 \pm 0.3\,\text{MeV}/c^2$, where the uncertainties are statistical, systematic and are due to the imprecise knowledge of the Ξ_c^+ mass.[7] The measured mass is in a good agreement with the previous LHCb measurement in the $\Lambda_c^+ K^- \pi^+ \pi^+$ final state.[8] The branching fraction of the $\Xi_{cc}^{++} \to \Xi_c^+ (\to pK^-\pi^+)\pi^+$ decay chain is measured with respect to the $\Xi_{cc}^{++} \to \Lambda_c^+ (\to pK^-\pi^+) K^-\pi^+\pi^+$ decays, whose mass spectrum is shown in the right plot of Fig. 3. The measured branching fraction ratio is $0.035 \pm 0.009 \pm 0.003$, where the first uncertainty is statistical and the second systematic.

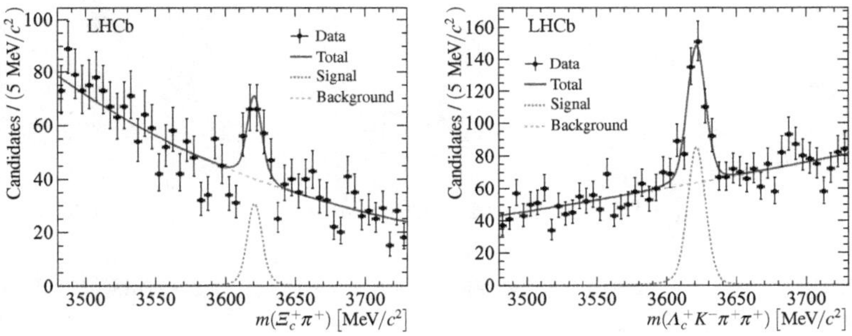

Fig. 3. Mass distributions for the selected Ξ_{cc}^{++} candidates in the (left) $\Xi_c^+\pi^+$ and (right) $\Lambda_c^+ K^-\pi^+\pi^+$ final state.

The doubly charmed baryon Ξ_{cc}^{++} is also searched for in the $D^+ pK^-\pi^+$ mass spectrum.[11] The D^+ candidates are reconstructed in the $K^-\pi^+\pi^+$ final state. A similar MLP classifier to the $\Xi_{cc}^{++} \to \Xi_c^+\pi^+$ search is used to suppress combinatorial background. A clean D^+ sample is selected, as shown in the $K^-\pi^+\pi^+$ mass spectrum in the left plot of Fig. 4. The $D^+ pK^-\pi^+$ mass spectrum is shown in the right plot of Fig. 4, where no significant signal is observed. An upper limit on the branching fraction of the $\Xi_{cc}^{++} \to D^+ pK^-\pi^+$ decay is set with respect to the $\Xi_{cc}^{++} \to \Lambda_c^+ K^-\pi^+\pi^+$ decay using the CLs method. The upper limit is found to be $1.7(2.1) \times 10^{-2}$ at 90% (95%) confidence level.

4. Summary and prospects

Recent progress of heavy baryon spectroscopy in the LHCb experiment has been reported, including the observation of several resonances of beauty baryons and the searches for doubly charmed baryons in more decay modes. The total LHCb data set of $9\,\text{fb}^{-1}$ is being actively exploited and new results are on the way.

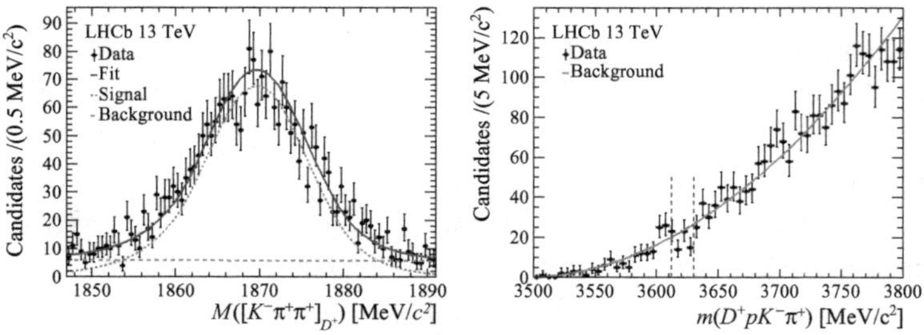

Fig. 4. Distributions of (left) $K^-\pi^+\pi^+$ and (right) $D^+pK^-\pi^+$ mass spectra for selected Ξ_{cc}^{++} candidates.

References

1. A. A. Alves Jr. *et al.*, The LHCb detector at the LHC, *JINST* **3**, S08005 (2008).
2. R. Aaij *et al.*, LHCb detector performance, *Int. J. Mod. Phys.* **A30**, 1530022 (2015).
3. M. Gell-Mann, A schematic model of baryons and mesons, *Phys. Lett.* **8**, 214 (1964).
4. G. Zweig, *An SU_3 model for strong interaction symmetry and its breaking; Version 1*, Tech. Rep. CERN-TH-401, CERN (Geneva, 1964).
5. R. Aaij *et al.*, Observation of two resonances in the $\Lambda_b^0\pi^\pm$ systems and precise measurement of $\Sigma_b^\pm$ and $\Sigma_b^{*\pm}$ properties, *Phys. Rev. Lett.* **122**, 012001 (2019).
6. R. Aaij *et al.*, Observation of New Resonances in the $\Lambda_b^0\pi^+\pi^-$ System, *Phys. Rev. Lett.* **123**, 152001 (2019).
7. M. Tanabashi *et al.*, Review of Particle Physics, *Phys. Rev.* **D98**, 030001 (2018).
8. R. Aaij *et al.*, Observation of the doubly charmed baryon Ξ_{cc}^{++}, *Phys. Rev. Lett.* **119**, 112001 (2017).
9. R. Aaij *et al.*, Measurement of the lifetime of the doubly charmed baryon Ξ_{cc}^{++}, *Phys. Rev. Lett.* **121**, 052002 (2018).
10. R. Aaij *et al.*, First observation of the doubly charmed baryon decay $\Xi_{cc}^{++} \to \Xi_c^+\pi^+$, *Phys. Rev. Lett.* **121**, 162002 (2018).
11. R. Aaij *et al.*, A search for $\Xi_{cc}^{++} \to D^+pK^-\pi^+$ decays, *JHEP* **10**, 124 (2019).

QCD sum rule study of P-wave bottom baryons

Er-Liang Cui

College of Science, Northwest A&F University, Yangling 712100, China
E-mail: erliang.cui@nwafu.edu.cn

Hui-Min Yang

School of Physics, Beihang University, Beijing 100191, China
E-mail: huiminy@buaa.edu.cn

Hua-Xing Chen

School of Physics, Beihang University, Beijing 100191, China
School of Physics, Southeast University, Nanjing 210094, China
E-mail: hxchen@buaa.edu.cn

Atsushi Hosaka

Research Center for Nuclear Physics (RCNP), Osaka University, Ibaraki 567-0047, Japan
Advanced Science Research Center, Japan Atomic Energy Agency (JAEA), Tokai 319-1195, Japan
E-mail: hosaka@rcnp.osaka-u.ac.jp

We use the method of QCD sum rules to study the mass spectrum of the $\Xi_b(6227)$ and $\Sigma_b(6097)$ recently observed by LHCb. We also use the method of light-cone sum rules to study their decay properties. Our results suggest that they can be well interpreted as P-wave bottom baryons with $J^P = 3/2^-$.

Keywords: Excited bottom baryons; QCD sum rules; light-cone sum rules.

1. Introduction

Recently, the LHCb Collaboration observed two new excited bottom baryons: the $\Xi_b(6227)^-$ in the $\Lambda_b^0 K^-$ and $\Xi_b^0 \pi^-$ invariant mass spectra,[1] and the $\Sigma_b(6097)^\pm$ in the $\Lambda_b^0 \pi^\pm$ invariant mass spectrum.[2] LHCb also measured the following relative branching ratio[1]:

$$\frac{\mathcal{B}(\Xi_b(6227)^- \to \Lambda_b^0 K^-)}{\mathcal{B}(\Xi_b(6227)^- \to \Xi_b^0 \pi^-)} \simeq 1\,. \tag{1}$$

The $\Sigma_b(6097)$ and $\Xi_b(6227)$ are good candidates of P-wave bottom baryons,[3] while there also exists the molecular interpretation for the $\Xi_b(6227)$.[4] In this paper we shall use the method of QCD sum rules to study their mass spectrum.[5] We shall also use the method of light-cone sum rules to study their decay properties.[6] Our calculations will be done within the framework of heavy quark effective theory.[7]

2. Mass spectrum of *P*-wave bottom baryons

First we briefly explain our notations. A P-wave bottom baryon consists of one bottom quark and two light quarks:

- The color representation of the two light quarks is $\bar{\mathbf{3}}_C$.
- The $SU(3)$ flavor representation of the two light quarks can be either $\bar{\mathbf{3}}_F$ or $\mathbf{6}_F$.
- The spin of the two light quarks can be either $s_l = 0$ or $s_l = 1$.
- The orbital excitation can be either between the two light quarks (ρ-type with $l_\rho = 1$ and $l_\lambda = 0$) or between the bottom quark and the two-light-quark system (λ-type with $l_\rho = 0$ and $l_\lambda = 1$).
- The total symmetry of the two light quarks is antisymmetric due to the Pauli principle.

Accordingly, we can categorize all the P-wave bottom baryons into eight multiplets, denoted as $[F(\text{flavor}), j_l, s_l, \rho/\lambda]$, where j_l is the total angular momentum of the light components, $j_l = l_\lambda \otimes l_\rho \otimes s_l$. Every multiplet contains one or two bottom baryons with the total angular momentum $j = j_l \otimes s_Q = |j_l \pm 1/2|$, as shown in Fig. 1.

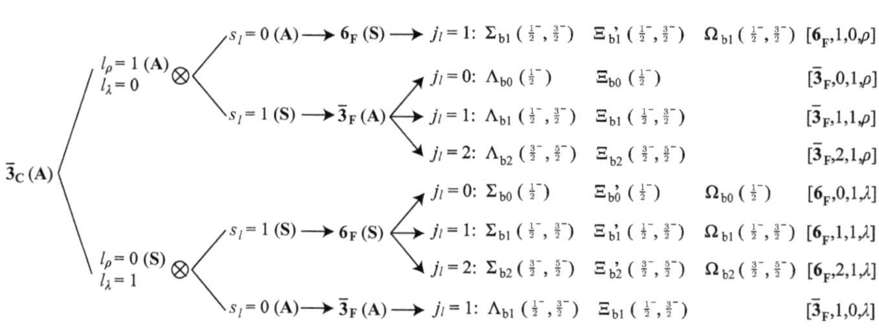

Fig. 1. Categorization of P-wave bottom baryons.

The mass spectrum of P-wave heavy baryons have been systematically investigated in Ref. 5 using the method of QCD sum rules within heavy quark effective theory, where we consider the sum rule result at the $\mathcal{O}(1/m_b)$ order, so that the mass of the bottom baryon can be written as:

$$m_{j,P,F,j_l,s_l,\rho/\lambda} = m_b + \overline{\Lambda}_{F,j_l,s_l,\rho/\lambda} + \delta m_{j,P,F,j_l,s_l,\rho/\lambda}\,. \tag{2}$$

We refer interested readers to Refs. 5, 8 for detailed explanations of the above notations and formulae.

Table 1. Mass spectrum of P-wave bottom baryons belonging to the baryon multiplet $[\mathbf{6}_F, 2, 1, \lambda]$.

Multiplet	Baryons (j^P)	Mass (GeV)	Difference (MeV)	f (GeV^4)
$[\mathbf{6}_F, 2, 1, \lambda]$	$\Sigma_b(3/2^-)$	6.10 ± 0.12	13 ± 5	0.102 ± 0.022
	$\Sigma_b(5/2^-)$	6.11 ± 0.12		0.045 ± 0.010
	$\Xi'_b(3/2^-)$	6.27 ± 0.12	12 ± 5	0.099 ± 0.021
	$\Xi'_b(5/2^-)$	6.29 ± 0.11		0.044 ± 0.009
	$\Omega_b(3/2^-)$	6.46 ± 0.12	11 ± 5	0.194 ± 0.038
	$\Omega_b(5/2^-)$	6.47 ± 0.12		0.087 ± 0.017

In the present study we use the method of QCD sum rules to reinvestigate the baryon multiplet $[\mathbf{6}_F, 2, 1, \lambda]$. The results are summarized in Table 1, suggesting that the $\Sigma_b(6097)$ and $\Xi_b(6227)$ can be interpreted as P-wave bottom baryons with $J^P = 3/2^-$ belonging to the baryon multiplet $[\mathbf{6}_F, 2, 1, \lambda]$.

3. Decay properties of P-wave bottom baryons

In this section we study decays of the $\Sigma_b(6097)$ and $\Xi_b(6227)$ into ground-state bottom baryons together with a pseudoscalar meson (π or K). We refer interested readers to Refs. 6, 9 for detailed discussions.

The baryon doublet $[\mathbf{6}_F, 2, 1, \lambda]$ contains six bottom baryons: $\Sigma_b(\frac{3}{2}^-/\frac{5}{2}^-)$, $\Xi'_b(\frac{3}{2}^-/\frac{5}{2}^-)$ and $\Omega_b(\frac{3}{2}^-/\frac{5}{2}^-)$. There are altogether four non-vanishing S-wave decay channels, whose widths are evaluated to be

$$(s)\ \Gamma_{\Sigma_b[\frac{3}{2}^-]\to\Sigma_b^*\pi} = 0.013\ ^{+0.019}_{-0.010}\ \mathrm{MeV}\,,$$

$$(t)\ \Gamma_{\Xi'_b[\frac{3}{2}^-]\to\Xi_b^*\pi} = 0.004\ ^{+0.006}_{-0.003}\ \mathrm{MeV}\,,$$

$$(u)\ \Gamma_{\Xi_b'[\frac{3}{2}^-]\to\Sigma_b^* K\to\Lambda_b\pi K} = 2\ {}^{+14}_{-2} \times 10^{-7}\ \text{MeV}\,,$$
$$(v)\ \Gamma_{\Omega_b[\frac{3}{2}^-]\to\Xi_b^* K} = 0.001\ {}^{+0.008}_{-0.001}\ \text{MeV}\,. \qquad (3)$$

We also find four non-vanishing D-wave decay channels, whose widths are evaluated to be

$$(w)\ \Gamma_{\Sigma_b[\frac{3}{2}^-]\to\Lambda_b\pi} = 46\ {}^{+58}_{-28}\ \text{MeV}\,,$$
$$(x)\ \Gamma_{\Xi_b'[\frac{3}{2}^-]\to\Xi_b\pi} = 16\ {}^{+19}_{-10}\ \text{MeV}\,, \qquad (4)$$
$$(y)\ \Gamma_{\Xi_b'[\frac{3}{2}^-]\to\Lambda_b K} = 6.5\ {}^{+7.9}_{-3.8}\ \text{MeV}\,,$$
$$(z)\ \Gamma_{\Omega_b[\frac{3}{2}^-]\to\Xi_b K} = 58\ {}^{+65}_{-33}\ \text{MeV}\,.$$

4. Summary and discussions

In this paper we have investigated the $\Xi_b(6227)^-$ and $\Sigma_b(6097)^\pm$ recently observed by LHCb.[1,2] Using the methods of QCD sum rules and light-cone sum rules, masses and decay widths of the $\Sigma_b(3/2^-)$ and $\Xi_b'(3/2^-)$ belonging to $[\mathbf{6}_F, 2, 1, \lambda]$ are extracted to be

$$M_{\Sigma_b(3/2^-)} = 6.10 \pm 0.12\ \text{GeV}\,,$$
$$\Gamma_{\Sigma_b(3/2^-)} = 46\ {}^{+58}_{-28}\ \text{MeV}\,,$$
$$M_{\Xi_b'(3/2^-)} = 6.27 \pm 0.12\ \text{GeV}\,,$$
$$\Gamma_{\Xi_b'(3/2^-)} = 23\ {}^{+27}_{-14}\ \text{MeV}\,,$$

which are consistent with the experimental parameters of the $\Xi_b(6227)^-$ and $\Sigma_b(6097)^\pm$.[1,2] The relative branching ratio

$$\frac{\mathcal{B}(\Sigma_b(3/2^-)^- \to \Lambda_b^0 K^-)}{\mathcal{B}(\Sigma_b(3/2^-)^- \to \Xi_b^0 \pi^-)} = 0.6\ {}^{+1.1}_{-0.5}\,,$$

is also consistent with the one measured by LHCb[1] (see Eq. (1)). Accordingly, we conclude that the $\Sigma_b(6097)$ and $\Xi_b(6227)$ can be well interpreted as P-wave bottom baryons with the spin-parity $J^P = 3/2^-$, both of which belong to the bottom baryon doublet $[\mathbf{6}_F, 2, 1, \lambda]$.

Acknowledgments

This project is supported by the National Natural Science Foundation of China under Grant No. 11722540, the Fundamental Research Funds for the Central Universities, Grants-in-Aid for Scientific Research (No. JP17K05441 (C)), Grants-in-Aid for Scientific Research on Innovative Areas (No. 18H05407), and the Foundation for Young Talents in College of Anhui Province (Grant No. gxyq2018103).

References

1. R. Aaij *et al.* [LHCb Collaboration], Phys. Rev. Lett. **121**, 072002 (2018).
2. R. Aaij *et al.* [LHCb Collaboration], Phys. Rev. Lett. **122**, 012001 (2019).
3. B. Chen, K. W. Wei, X. Liu and A. Zhang, Phys. Rev. D **98**, 031502 (2018); B. Chen and X. Liu, Phys. Rev. D **98**, 074032 (2018); P. Yang, J. J. Guo and A. Zhang, Phys. Rev. D **99**, 034018 (2019); K. L. Wang, Q. F. Lü and X. H. Zhong, Phys. Rev. D **99**, 014011 (2019); T. M. Aliev, K. Azizi, Y. Sarac and H. Sundu, Phys. Rev. D **99**, 094003 (2019); T. M. Aliev, K. Azizi, Y. Sarac and H. Sundu, Phys. Rev. D **98**, 094014 (2018); C. K. Chua, Phys. Rev. D **99**, 014023 (2019); M. Karliner and J. L. Rosner, Phys. Rev. D **98**, 074026 (2018).
4. Y. Huang, C. J. Xiao, L. S. Geng and J. He, Phys. Rev. D **99**, 014008 (2019); Q. X. Yu, R. Pavao, V. R. Debastiani and E. Oset, Eur. Phys. J. C **79**, 167 (2019).
5. H. X. Chen, W. Chen, Q. Mao, A. Hosaka, X. Liu and S. L. Zhu, Phys. Rev. D **91**, 054034 (2015); Q. Mao, H. X. Chen, W. Chen, A. Hosaka, X. Liu and S. L. Zhu, Phys. Rev. D **92**, 114007 (2015).
6. H. X. Chen, Q. Mao, W. Chen, A. Hosaka, X. Liu and S. L. Zhu, Phys. Rev. D **95**, 094008 (2017); E. L. Cui, H. M. Yang, H. X. Chen and A. Hosaka, Phys. Rev. D **99**, 094021 (2019).
7. B. Grinstein, Nucl. Phys. B **339**, 253 (1990); E. Eichten and B. R. Hill, Phys. Lett. B **234**, 511 (1990); A. F. Falk, H. Georgi, B. Grinstein and M. B. Wise, Nucl. Phys. B **343**, 1 (1990).
8. X. Liu, H. X. Chen, Y. R. Liu, A. Hosaka and S. L. Zhu, Phys. Rev. D **77**, 014031 (2008); H. X. Chen, Q. Mao, A. Hosaka, X. Liu and S. L. Zhu, Phys. Rev. D **94**, 114016 (2016); Q. Mao, H. X. Chen, A. Hosaka, X. Liu and S. L. Zhu, Phys. Rev. D **96**, 074021 (2017); H. X. Chen, W. Chen, X. Liu, Y. R. Liu and S. L. Zhu, Rept. Prog. Phys. **80**, 076201 (2017).
9. P. Ball, JHEP **9901**, 010 (1999); P. Ball, V. M. Braun and A. Lenz, JHEP **0605**, 004 (2006).

On the study of dibaryon resonance $d^*(2380)$

Yubing Dong[1,2,3,*], Fei Huang[2,**], Pengnian Shen[1,2,3,†], and Zongye Zhang[1,2,3,‡]

[1] *Institute of High Energy Physics, Chinese Academy of Sciences (CAS), Beijing 100049, China*

[2] *School of Physical Sciences, University of Chinese Academy of Sciences, Beijing 101408, China*

[3] *Theoretical Physics Center for Science Facilities (TPCSF), CAS, China*

[*] *E-mail: dongyb@ihep.ac.cn*

[**] *E-mail: feihuang@ucas.ac.cn*

[†] *E-mail: shenpn@ihep.ac.cn*

[‡] *E-mail: zhangzy@ihep.ac.cn*

In this talk, we summarize the present calculations for the newly observed dibaryon resonance $d^*(2380)$ with the help of a chiral constituent quark model. Due to its binding and its narrow width, we interpret it as a deeply bound 6-quark state. The important role of quark-exchange is addressed. Moreover, the calculated decays as well its electromagnetic form factors are reported. Comparing to the available experimental data, we conclude that the $d^*(2380)$ resonance can be reasonably assigned as a compact hexaquark dominant state.

Keywords: Dibaryon resonance $d^*(2380)$; hexaquark; chiral quark model.

1. Introduction

The studies of dibaryons, like d^*, d' and H particles, have been carried out for more than half century.[1] In 2009 a clear experimental evidence for d^* was released by the CELSIUS/WASA and WASA@COSY collaborations, when they studied the ABC effect.[2–5] They found that the observed peak in the cross section could not be simply understood by the contributions either from the intermediate $P_{11}(1440)$ Roper resonance or from the t-channel $\Delta\Delta$, and consequentially, they introduced an intermediate new resonance $d^*(2380)$ with the quantum numbers of $I(J^P) = 0(3^+)$, the mass of $2370 \sim 2380$ MeV, and the width of $70 \sim 80$ MeV. Very recently, this dibaryon resonance was also observed and confirmed by the Researcher Center of Electron Photon Sciences (ELPH) experiment from the $\gamma + d$ scattering process.[6,7]

It should be stressed that the mass of this dibaryon is about 80 MeV below the $\Delta\Delta$ threshold and about 70 MeV above the $\Delta\pi N$, it contains light flavor, and moreover, its decay width is narrow and is only about 1/3 of the two-Δ's width. Those phenomena are quite different from the well-known XYZ particles which usually locate very near to the threshold and contain heavy flavor.[8–10] They are also different from the recently predicted weakly bound dibaryons, like $N\Omega$ and $\Omega\Omega$ et al., by Lattice calculations.[11]

There are mainly two scenarios for the $d^*(2380)$ structure. One postulates that it has a compact structure, and may be an exotic hexaquark dominated state.[12–21] Another regards it as a $D_{12}\pi$ molecular-like hadronic state, which originates from a three-body $\Delta N\pi$ resonance assumption.[22,23] Although some of the experimental measurements, like its mass and double pionic decays, can be well-reproduced by using either scheme, the described structures of $d^*(2380)$ are quite different. In this talk, the results for $d^*(2380)$ in the hexaquark scenario, with the help of a chiral constituent quark model, will be focused on.

2. Description of $d^*(2380)$ in quark degree of freedom

We know that the quark-quark interaction has been the key issue for a long time in quark degrees of freedom and using the Resonant Group Method (RGM) approach for a dynamical calculation. Among various quark models, our chiral constituent quark model is a successful one since it may well-reproduce the low-lying baryon spectra, nucleon-nucleon scattering phase-shifts, deuteron static properties (like its binding and form factors) as well as the nucleon-hyperon scattering.[12] Recently, this phenomenological model was also applied for the dibaryon systems, like $d^*(2380)$. It was found that $d^*(2380)$ resonance might be a $\Delta\Delta + C_8C_8$ system where the quantum numbers of spin (S), isospin (I) and color (C) for Δ and C_8 clusters being $[S, I, C]_\Delta = [3/2, 3/2, (00)]$ and $[S, I, C]_{C_8} = [3/2, 1/2, (11)]$, respectively. The obtained channel wave function in the RGM approach is briefly written as

$$\left|d^*(2380)_{3^+,M_z}\right\rangle = \sum_{ch} \sum_{pw} \left[|ch >_{S_{ch},M_S} \chi_{ch}^{pw,m_l}(\vec{r}) \right]_{3^+,M_z}, \tag{1}$$

with $ch = \Delta\Delta$ or C_8C_8 standing for the constituents of the configuration, M_z representing the magnetic quantum number of total spin, $pw = 0$ and 2 depicting the $S-$ and $D-$ partial waves between the two clusters, respectively, and m_l being its magnetic quantum number.

Noticed that our obtained wave function of $d^*(2380)$ shows that the probability of the hidden-color C_8C_8 channel is about 69%, which is dominant comparing to the $\Delta\Delta$ channel (around 31%). This feature is due to the strong quark exchange effect between the two clusters and to the special quantum number of 3^+. Moreover, the inclusion of the hidden-color channel sizeably suppresses the binding energy about 20 MeV, namely, the calculated mass of $d^*(2380)$ ($IJ^p = 03^+$) is about 2380 − 2414 MeV. Our channel wave function illustrates that $d^*(2380)$ is a compact system, and the C_8C_8 wave function is very similar to a typical single Gauss-type function with a width parameter of 0.45 fm. In addition, the $D-$wave components of the two channels are both negligible small.

3. Decays and charge form factor of $d^*(2380)$

The numerical calculations for the $d^*(2380)$ double-pionic decays have been carried out by employing the numerical $d^*(2380)$ wave function, which stimulates the relative momentum distributions of the two clusters (C_8C_8 or $\Delta\Delta$) inside $d^*(2380)$, based on a non-relativistic framework. We deemed that the component of C_8C_8 cluster does not contribute to the strong decays in the leading order approximation, since the process of emitting a pion is color blind. In our calculations, the $N - N$ final state interaction and the effect of isospin symmetry breaking, due to the mass difference of charge and neutral pions, are taken into account. Our model results for the partial widths of the double-pionic decay mode are listed in Table 1. An excellent agreement between our model calculation and experimental data can been seen. Here, we emphasize that the hidden-color component plays important role to reduce the partial widths as well as the total width. This is because that it remarkably suppresses the probability of $\Delta\Delta$ component. If only the single-channel $\Delta\Delta$ component is taken into account, the results of the partial and total decay widths are all overestimated.

Table 1, Calculated double-pion decay widths of $d^*(2380)$ (MeV).

Decay Modes	$d^* \to d\pi^+\pi^-$	$d^* \to d\pi^0\pi^0$	$d^* \to pn\pi^+\pi^-$	$d^* \to pn\pi^0\pi^0$
$\Gamma_{\Delta\Delta+C_8C_8}$	16.8	9.2	20.6	9.6
$\Gamma^{Expt.}$	16.7	10.2	21.8	8.7
Decay Modes	$d^* \to pp\pi^0\pi^-$	$d^* \to nn\pi^0\pi^+$	$d^* \to pn$	Total
$\Gamma_{\Delta\Delta+C_8C_8}$	3.5	3.5	8.7	71.9
$\Gamma^{Expt.}$	4.4	4.4	8.7	74.9

The single-pionic decay mode of $d^*(2380)$ in our scenario was also calculated[18]), where the contribution of C_8C_8 is ignored as well and due to the same reason, the next-leading effect with a pion being emitted from one Δ cluster and absorbed by another Δ is considered. Our obtained branching ratio for $d^*(2380) \to pn\pi$ is about 1%, which is compatible to the upper limit ($< 9\%$) by the experimental measurement.[24] However, this branching ratio obtained in the pure $\Delta N\pi$ scenario is about 18%[25] which obviously beyond the data region.

Clearly, the strong decays of $d^*(2380)$ in our scenario is contributed only by the $\Delta\Delta$ component, and the dominant component in our wave function, C_8C_8, does not play a directly role in the leading order approximation. In addition, although the physics pictures of our compact and the $D_{12}\pi$ scenario are quite different, they both can well-reproduce the double-pionic decays of $d^*(2380)$, but the pure $D_{12}\pi$ model cannot explain the single-pionic decay of $d^*(2380)$ except this model is modified to a mixed one. In order to manifest the C_8C_8 effect, we employed the electromagnetic (EM) interaction to probe $d^*(2380)$. Since it is a spin-3 particle, it has all together 7 form factors.[19,20] We calculated its electric form factor (see Fig. 1), where both the C_8C_8 and $\Delta\Delta$ contribute to the charge radius of $d^*(2380)$. We also find that the two scenarios, one (A) with a 6-quark compact structure and another (B) with a $D_{12}\pi$ scenario, give remarkably different Q^2-dependence of the electric distributions of $d^*(2380)$ and consequentially give sizeably different charge radius of 0.7 fm and 2 fm, respectively. Those quantities may be an additional criterion for discriminating d^*'s interpretations.

4. Summary

The current studies of the dibaryon resonance $d^*(2380)$ in quark degrees freedom are briefly reported. Our dynamical calculations interpret the resonance as a compact 6-quark system with a dominant hidden-color component. This unusual structure is due to the strong quark-exchange effect and the short-ranged attractive interaction in this channel. We expect that the $d^*(2380)$ wave function is mainly localized in a compact space. In the $d^*(2380)$ strong decays, the dominant component of C_8C_8 does not directly contribute, at least, in leading order, since the quark-quark-pion or hadron-hadron-pion interaction is color blind. Thus, only the suppressed $\Delta\Delta$ component contributes to the $d^*(2380)$ decays and makes the total width narrow.

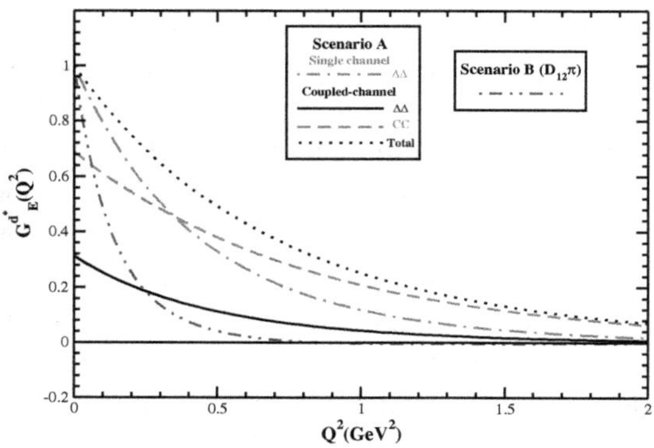

Fig. 1. The obtained charge distributions of the $d^*(2380)$ in our chiral constituent quark model. The solid(black), red(dashed), and black(dotted) curves stand for $\Delta\Delta$, C_8C_8, and the total contributions in the coupled channel approximation, respectively. The pink(dotted-dashed) curve represents the single channel case of scenario A. The blue(double-dotted-dashed) curve stands for the results of $D_{12}\pi$ scenario with binding of $\epsilon = 0.25$ MeV.

Moreover, other observables are also discussed. We find that although both two pictures of $d^*(2380)$ can well-reproduce the its mass and widths of the double-pionic strong decays simultaneously, the predicted width of the single-pionic decay, the Q^2-dependence of the EM form factors, and the charge radius are remarkably different. Of course, the real world must be much more complicated than our imagination. For example, it is commonly believed that the nucleon is a three-quark core (in quark degrees of freedom) surrounded by a meson cloudy (hadronic degrees of freedom). Similarly, one can also incorporate our 6-quark compact picture with the $D_{12}\pi$ scenario, and thus an admixture picture should fairly describe the $d^*(2380)$ properties.[25]

Acknowledgment

This work is supported by the National Natural Sciences Foundations of China under the grant Nos. 11975245, 11521505, 11565007, and 11635009, the Sino-German CRC 110 by NSFC under the grant No.11621131001, the Key Research Program of CAS, Grant No. Y7292610K1, and the IHEP Innovation Fund No. Y4545190Y2.

References

1. H. Clement, Prog. Part. Nucl. Phys. **93**, 195 (2017).
2. M. Bashkanov *et al.*, Phys. Rev. Lett. **102**, 052301 (2009).
3. P. Adlarson *et al.* [WASA-at-COSY Collaboration], Phys. Rev. Lett. **106**, 242302 (2011).
4. P. Adlarson *et al.* [WASA-at-COSY Collaboration], Phys. Lett. B**721**, 229 (2013).
5. P. Adlarson *et al.* [WASA-at-COSY Collaboration], Phys. Rev. Lett. **112**, 202301 (2014).
6. T. Ishikawa *et al.*, Phys. Lett. B **772**, 398 (2017).
7. Takatsugu Ishikawa, talk given in "QNP2018", Nov. 13-17, 2018, Tsukuba, Japan.
8. H. X. Chen, W. Chen, X. Liu, and S. L. Zhu, Phys. Rept. **639**, 1 (2016).
9. F. K. Guo *et al.*, Rev. Mod. Phys. **90**, 015004 (2018).
10. Y. Dong, A. Faessler, and V. E. Lyubovitskij, Prog. Part. Nucl. Phys. **94**, 282 (2017).
11. S. Gongyo *et al.*, Phys. Rev. Lett. **120**, no. 21, 212001 (2018).
12. X. Q. Yuan, Z. Y. Zhang, Y. W. Yu and P. N. Shen, Phys. Rev. C **60**, 045203 (1999).
13. M. Bashkanov, S. J. Brodsky, and H. Clement, Phys. Lett. B **727**, 438 (2013).
14. F. Huang, Z. Y. Zhang, P. N. Shen, and W. L. Wang, Chin. Phys. C **39**, 071001 (2015).
15. F. Huang, P. N. Shen, Y. B. Dong, and Z. Y. Zhang, Sci. China Phys. Mech. Astron. **59**, 622002 (2016).
16. Y. Dong, P. Shen, F. Huang, and Z. Zhang, Phys. Rev. C **91**, 064002 (2015).
17. Y. Dong, F. Huang, P. Shen, and Z. Zhang, Phys. Rev. C **94**, 014003 (2016).
18. Y. Dong, F. Huang, P. Shen, and Z. Zhang, Phys. Lett. B **769**, 223 (2017).
19. Y. Dong, F. Huang, P. Shen, and Z. Zhang, Phys. Rev. D **96**, 094001 (2017).
20. Y. Dong, P. Shen and Z. Zhang, Phys. Rev. D **97**, 114002 (2018).
21. Y. Dong, P. Shen, F. Huang and Z. Zhang, Int. J. Mod. Phys. A **33**, 1830031 (2018).
22. A. Gal and H. Garcilazo, Phys. Rev. Lett. **111**, 172301 (2013).
23. A. Gal and H. Garcilazo, Nucl. Phys. A **928**, 73 (2014).
24. The WASA-at-COSY Collaboration, Phys. Lett. B**774**, 599-607 (2017).
25. A. Gal, Phys. Lett. B **769**, 436 (2017).

The pseudoscalar meson and baryon octet interaction in the unitary coupled-channel approximation

Bao-Xi Sun*, Zheng-Ran Zhang, Hai-Lin Wu, Si-Yu Zhao and Fang-Yong Dong

College of Applied Sciences, Beijing University of Technology,
Beijing 100124, China
**E-mail: sunbx@bjut.edu.cn*

The pseudoscalar meson-baryon octet interaction is studied within a nonlinear realized Lagrangian, and then the Bethe-Salpeter equation is solved in the unitary coupled-channel approximation. In sector of strangeness $S = -1$ and isospin $I = 0$, only one pole is generated dynamically in the 1400MeV region, which might correspond to the $\Lambda(1405)$ particle. When the case of strangeness zero is studied, the $s-$ and $u-$ potentials are taken into account in the kernel, and a resonance state is produced in the 1500MeV region, which might be a counterpart of the N(1535) particle.

Keywords: Chiral Lagrangian; meson-baryon interaction; hadronic resonance; Bethe-Salpeter equation.

1. Introduction

The chiral perturbation theory has become a powerful tool to study the pseudoscalar meson-baryon octet interaction.[1,2] However, this theory is not suitable to the energy region where the hadronic resonance appears. Therefore, a non-perturbative resummation technique has to be taken into account. By solving the Bethe-Salpeter equation in the unitary coupled-channel approximation, the hadronic resonance state can be generated dynamically.[3–5] In the calculation, the unitarity of the scattering amplitude is conserved, which implies all terms in the expansion of the scattering amplitude are included and no truncation is performed when the Bethe-Salpeter equation is solved.

2. The pole structure of $\Lambda(1405)$

The pseudoscalar meson-baryon octet interaction with strangeness $S = -1$ and isospin $I = 0$ was studied in the unitary coupled-channel approximation by solving the Bethe-Salpeter equation, and two poles of the scattering amplitude were found on the complex energy plane between the $\pi\Sigma$ and $\bar{K}N$ thresholds, which were assumed to correspond to the $\Lambda(1405)$ particle. Since 2016, a review article on the double-pole structure of the $\Lambda(1405)$ particle had appeared in the Particle Data Group(PDG) manual, where many theoretical research works on this problem were cited.[6] However, In Ref. 7, the loop function of the intermediate pseudoscalar meson-baryon octet is derived from the dimensional regularization scheme strictly, and the relativistic kinetic effect is taken into account when the Bethe-Salpeter equation is solved. Finally, the revised loop function in the dimensional regularization scheme takes the form of

$$G_l = \frac{1}{32\pi^2\sqrt{s}}\left[(a_l+1)(m_l^2 - M_l^2) + \left(m_l^2 ln\frac{m_l^2}{\mu^2} - M_l^2 ln\frac{M_l^2}{\mu^2}\right)\right] + \left(\frac{s+M_l^2-m_l^2}{4M_l\sqrt{s}} + \frac{1}{2}\right)G_l', \tag{1}$$

where the G_l' is the original form of the loop function,[8] while $\sqrt{s}$ and P are the total energy and the momentum of the system in the center of mass frame, M_l and m_l are the masses of the intermediate baryon and pseudoscalar meson, respectively. In Eq. (1), the subtraction constant a_l and the regularization scale μ can be regarded as one parameter in the calculation.

By fitting the energy shift and width of the 1s state of kaonic hydrogen measured precisely at the SIDDHARTA experiment at $DA\Phi NE$[9] and the branch ratios at the K^-p threshold,[10,11] the subtraction constant can be determined, which are listed in Table 1, while the regularization scale $\mu =$ 630MeV is fixed.

Table 1. The subtraction constants used in the calculation with the regularization scalar $\mu = 630$MeV.

	$\bar{K}N$	$\pi\Lambda$	$\pi\Sigma$	$\eta\Lambda$	$\eta\Sigma$	$K\Xi$
a_l	-1.1	-1.6	-1.9	-2.7	-2.6	-2.8

With the loop function in Eq. (1) and the parameters in Table 1, only one resonance state is generated dynamically between the $\pi\Sigma$ and $\bar{K}N$ thresholds, which might be a counterpart of the $\Lambda(1405)$ particle, as depicted in Fig. 1. It implies that too many approximations had been made in the original formula of the pseudoscalar meson-baryon octet loop function

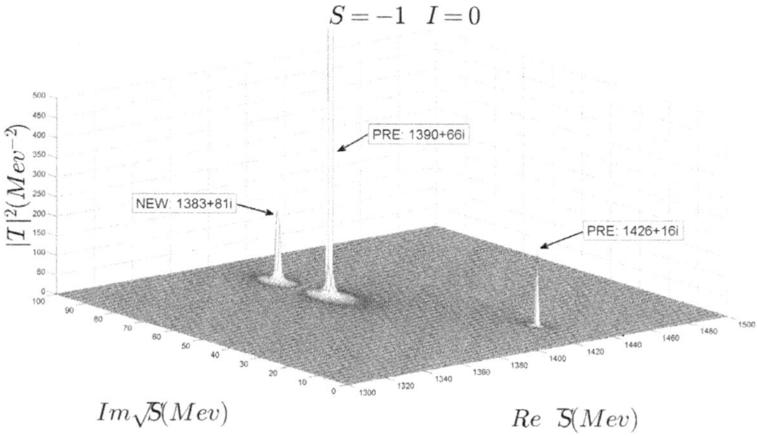

Fig. 1. Comparison of poles in the strangeness $S = -1$ and isospin $I = 0$ sector. The label NEW represents the case calculated from the loop function in Eq. (1), while the label PRE stands for the case of the original loop function G_l' with parameters in Ref. 12.

when the Bethe-Salpeter equation is solved, and the double-pole structure of the $\Lambda(1405)$ particle is model-dependent.

If one hopes the double pole structure of the $\Lambda(1405)$ particle is correct, the pseudoscalar meson-baryon octet interaction must be studied in the unitary coupled-channel approximation by solving the Bethe-Salpeter equation with the loop function where the relativistic kinetic effect is taken into account, as used in Ref. 7. Otherwise, the method without on-shell approximation can also be used, i.e., this problem should be recalculated by solving the integral formula of the Bethe-Salpeter equation in the unitary coupled-channel approximation. If the double-pole structure of the $\Lambda(1405)$ particle can be reproduced by fitting the experimental data, our conclusion made in Ref. 7 can be certified to be wrong. In this sense, any attempt to certify the double-pole structure of the $\Lambda(1405)$ particle with the original formula of the loop function by solving the Bethe-Salpeter equation is unreasonable.

3. $N(1535)$

With the loop function in Eq. (1), we continue to study the pseudoscalar meson-baryon octet interaction in the strangeness $S = 0$ sector.[13] In addition to the Weinberg-Tomozawa term, the $s-$ channel and $u-$ channel potentials are also taken into account in the calculation.

The $s-$ channel and $u-$ channel potentials are weaker than the Weinberg-Tomozawa term in the pseudoscalar meson-baryon octet interaction. However, if they are included as a kernel in addition to the Weinberg-Tomozawa term when the Bethe-Salpeter equation is solved in the unitary coupled-channel approximation, the subtraction constants must be readjusted in order to produce the resonance state at the reasonable position on the complex energy plane.

A resonance state is generated dynamically in the 1500MeV region, which lies between the ηN and $K\Lambda$ thresholds and can be regarded as a counterpart of the $N(1535)$ particle. Since the πN threshold is far lower than the energy region where the N(1535) particle might be generated dynamically, it is found that the pole position is not sensitive to the subtraction constant $a_{\pi N}$. Moreover, the mass of the $N(1535)$ particle is close to the $K\Lambda$ threshold, and the subtraction constant $a_{K\Lambda}$ has an important influence on the generation of the $N(1535)$ particle. Therefore, the $a_{K\Lambda}$ value is stable in the calculation.

4. Summary

The interaction of the pseudoscalar meson and the baryon octet is studied within a nonlinear realized Lagrangian. When the Bethe-Salpeter equation is solved in the unitary coupled-channel approximation, the relativistic kinetic correction is included in the loop function of the intermediate meson and baryon, which results in only one pole corresponding to the $\Lambda(1405)$ particle generated dynamically on the complex energy plane in the strangeness $S = -1$ and isospin $I = 0$ sector. When this method is extended to study the case of strangeness zero, the $s-$ and $u-$ channel potentials are taken into account in addition to the Weinberg-Tomozawa term. By adjusting the subtraction constants, a resonance state is generated dynamically in the 1500MeV region, which might be a counterpart of the N(1535) particle in the PDG data.

Acknowledgment

Bao-Xi Sun would like to thank Han-Qing Zheng for useful discussions.

References

1. S. Weinberg, *The quantum theory of fields: Vol.2, Modern Applications* (Cambridge University Press, London, 1995).

2. S. Scherer, Introduction to chiral perturbation theory, *Adv. Nucl. Phys.* **27**, 277 (2003).
3. N. Kaiser, P. B. Siegel and W. Weise, Chiral dynamics and the low-energy kaon - nucleon interaction, *Nucl. Phys. A* **594**, 325 (1995).
4. E. Oset and A. Ramos, Nonperturbative chiral approach to s wave anti-K N interactions, *Nucl. Phys. A* **635**, 99 (1998).
5. B. Krippa, Chiral dynamics of the low-energy kaon-baryon interactions, *Phys. Rev. C* **58**, 1333 (1998).
6. M. Tanabashi et al.(Particle Data Group), Review of particle physics, *Phys. Rev. D* **98**, 030001 (2018).
7. F. Y. Dong, B. X. Sun and J. L. Pang, The $\Lambda(1405)$ state in a chiral unitary approach with off-shell corrections to dimensional regularized loop functions, *Chin. Phys. C* **41**, 074108 (2017).
8. J. A. Oller, U.-G. Meissner, Chiral dynamics in the presence of bound states: Kaon nucleon interactions revisited, *Phys. Lett. B* **500**, 263 (2001).
9. M. Bazzi et al.(SIDDHARTA Collaboration), A new measurement of kaonic hydrogen X-rays, *Phys. Lett. B* **704**, 113 (2011).
10. D. N. Tovee et al., Some properties of the charged Sigma hyperons, *Nucl. Phys. B* **33**, 493 (1971).
11. R. J. Nowak et al., Charged Sigma hyperon production by K-meson interactions at rest, *Nucl. Phys. B* **139**, 61 (1978).
12. E. Oset, A. Ramos and C. Bennhold, Low lying $S = -1$ excited baryons and chiral symmetry, *Phys. Lett. B* **527**, 99 (2002); Erratum, *Phys. Lett. B* **530**, 260 (2002).
13. B. X. Sun S. Y. Zhao and X. Y. Wang, Pseudoscalar meson and baryon octet interaction with strangeness zero in the unitary coupled-channel approximation, *Chin. Phys. C* **43**, 064111 (2019).

Strangeness photoproduction at the BGO-OD experiment

Georg Scheluchin[1], Stefan Alef[1], Patrick Bauer[1], Reinhard Beck[2], Alessandro Braghieri[13], Philip Cole[3], Rachele Di Salvo[4], Daniel Elsner[1], Alessia Fantini[4,5], Oliver Freyermuth[1], Francesco Ghio[6,7], Anatoly Gridnev[8], Daniel Hammann[1], Jürgen Hannappel[1], Thomas Jude[1], Katrin Kohl[1], Nikolay Kozlenko[8], Alexander Lapik[9], Paolo Levi Sandri[10], Valery Lisin[9], Giuseppe Mandaglio[11,12], Roberto Messi[4,5], Dario Moricciani[4], Vladimir Nedorezov[9], Dmitry Novinsky[8], Paolo Pedroni[13], Andrei Polonski[9], Björn-Eric Reitz[1], Mariia Romaniuk[4], Hartmut Schmieden[1], Victorin Sumachev[8], Viacheslav Tarakanov[8],and Christian Tillmanns[1]

[1]*Rheinische Friedrich-Willhelms-Universität Bonn, Physikalisches Institut, Nußallee 12, 53115 Bonn, Germany*
[2]*Helmholtz-Institut fuer Strahlen- und Kernphysik, Universitaet Bonn, Nussallee 1-16, D-53115 Bonn Germany*
[3]*Lamar University, Department of Physics, Beaumont, Texas, 77710, USA*
[4]*INFN Roma "Tor Vergata", Rome, Italy*
[5]*Universit di Roma "Tor Vergata", Via della Ricerca Scientifica 1, 00133 Rome, Italy*
[6]*INFN sezione di Roma La Sapienza, P.le Aldo Moro 2, 00185 Rome, Italy*
[7]*Istituto Superiore di Sanita, Viale Regina Elena 299, 00161 Rome, Italy*
[8]*Petersburg Nuclear Physics Institute, Gatchina, Leningrad District, 188300, Russia*
[9]*Russian Academy of Sciences Institute for Nuclear Research, prospekt 60-letiya Oktyabrya 7a, Moscow 117312, Russia*
[10]*INFN - Laboratori Nazionali di Frascati, Via E. Fermi 40, 00044 Frascati, Italy*
[11]*INFN sezione Catania, 95129 Catania, Italy*
[12]*Universita degli Studi di Messina, Via Consolato del Mare 41, 98121 Messina, Italy*
[13]*INFN sezione di Pavia, Via Agostino Bassi, 6 - 27100 Pavia, Italy*

The BGO-OD experiment at the University of Bonn's ELSA accelerator facility in Germany is ideally suited to investigate photoproduction at extreme forward angles. It combines a highly segmented BGO electromagnetic calorimeter at central angles and an open dipole magnetic spectrometer in the forward direction. This allows the detection of forward going kaons, and complex final states of mixed charge from hyperon decays.

Current projects at the BGO-OD experiment include strangeness production of $\gamma p \to K^+\Lambda/\Sigma^0$ at forward angles, $K^0\Sigma^0$ with a deuteron target and $K^+\Lambda(1405)$ line shape and cross section measurements.

Keywords: $\Lambda(1405)$; BGO-OD; strangeness photoproduction.

1. Introduction

Hadron spectroscopy is used to investigate the degrees of freedom of the constituents of the nucleon. Since the conception of the quark model, there have been descriptions of baryons and mesons with more than three and two valence quarks respectively. Such hadrons could manifest as penta- and tetraquarks, or as meson-meson and meson-baryon molecular like states. Candidates for such exotic matter were found in recent years in the charm sector,[1,2] and there is evidence that similar configurations may exist in the light, strange sector.[3] To study such effects in photoproduction experiments, access to a low momentum exchange region, where the meson is produced at forward angles is crucial.

The BGO-OD experiment at the University of Bonn's ELSA accelerator facility in Germany is ideally suited for this endeavor. It combines a highly segmented BGO electromagnetic calorimeter at central angles and an Open Dipole magnetic spectrometer in the forward direction. This allows the detection of forward going kaons, and complex final states of mixed charge from hyperon decays (see section 2). Preliminary results in current projects are shown in section 3 for $\Lambda(1405)$ production, in section 4 for $\gamma n \to K^0\Sigma^0$ and in section 5 cross section measurement of $\gamma p \to K^+\Lambda/\Sigma^0$ in the extreme forward regime. The presented results are only a handful of currently pursued projects.

2. The BGO-OD experiment

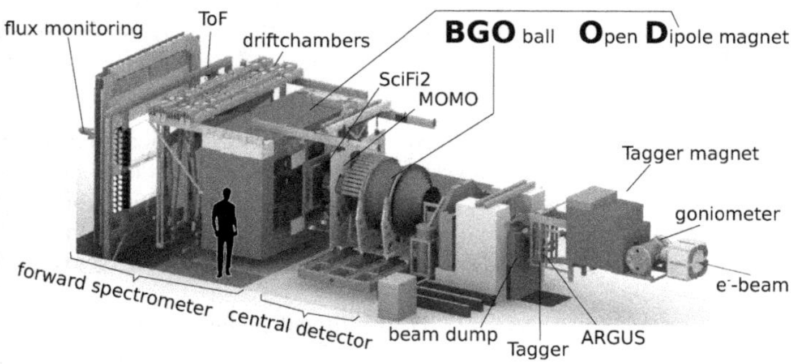

Fig. 1. Overview of the BGO-OD experiment.

The BGO-OD experiment, shown in figure 1, is located at the **E**lectron **S**tretcher **A**ccelerator (ELSA) facility in Bonn. ELSA is an electron accelerator with a quasi continuous beam up to 3.2 GeV.[4] The electron beam hits a radiator inside the goniometer tank, which creates a real photon beam via bremsstrahlung. The energy of the photons is determined by a measurement of the bremsstrahlung electron with the Tagger detector. The photon beam interacts with the target cell inside the BGO ball, which is filled with liquid hydrogen or deuterium. The final state particles of the reaction are detected with the BGO calorimeter using bismuth germinate oxide crystals between polar angles 25° to 155°. Charged particles traveling in θ <12° are detected in the forward detector. The track trajectory before the open-dipole magnet is measured with the MOMO and SciFi2 scintillating fibre detectors, and the drift chambers measure the trajectory after the magnet. The measured track curvature is used to determine the momentum, while the ToF walls at the end of the experiment measure the velocity via time of flight.

3. Λ(1405) photoproduction

Since years it is speculated that the $\Lambda(1405)$ has a molecular like structure of $N\bar{K}$.[5,6] Lattice QCD calculations give more support for the molecule like structure compared to a genuine three quark state.[7,8] BGO-OD is ideally suited to detect the $\Lambda(1405) \to \pi^0\Sigma^0$ decay. The preliminary results on the line shape of the $\pi^0\Sigma^0$ decay are compared to other experimental results in figure 2a, while figure 2b shows an energy bin of the measured differential cross section. The line shape and cross section agrees to the other experimental data within statistics, while the forward spectrometer allows the extension of the CLAS cross section to more forward angles.

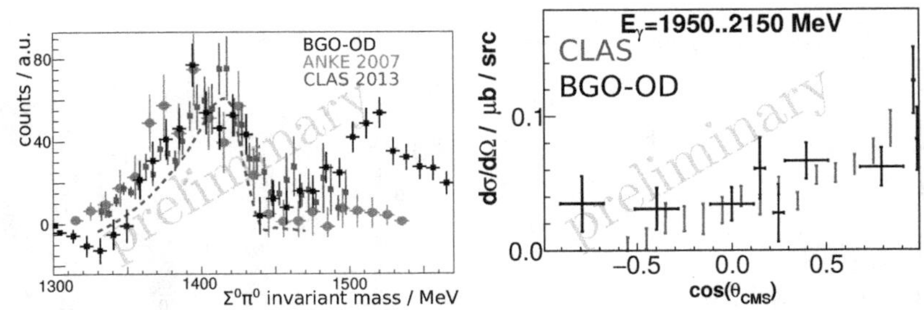

(a) Line shape results for $E_\gamma = 1500..2300$ MeV. The dashed blue line marks the predictions by J.C.Nacher.[6]

(b) differential cross section

Fig. 2. Preliminary results on $\gamma p \to K^+\Lambda(1405) \to K^+\pi^0\Sigma^0$. (a) shows the line shape, while (b) the differential cross section. Black points show the results of the full topology reconstruction and magenta for the K^+ in the forward detector. The blue data points were take from the CLAS experiment[9,10] and the red data from the ANKE experiment.[11]

4. $K^0\Sigma^0$ with deuterium target

Previous measurements of the $\gamma p \to K^0\Sigma^+$ cross section showed a cusp over the K^*Y thresholds.[12] Similar models which predicted the pentaquark at LHCb, predict a rise in the $K^0\Sigma^0$ cross section at the exact same energy.[3] In these models a K^* is created below threshold, feeding the $K^0\Sigma^+$ cross section. At BGO-OD deuteron data was analyzed on the $\gamma n(p) \to K^0\Sigma^0(p)$ reaction by K. Kohl. The preliminary results are shown in figure 3. They show consistency with the predicted structure of the interaction.

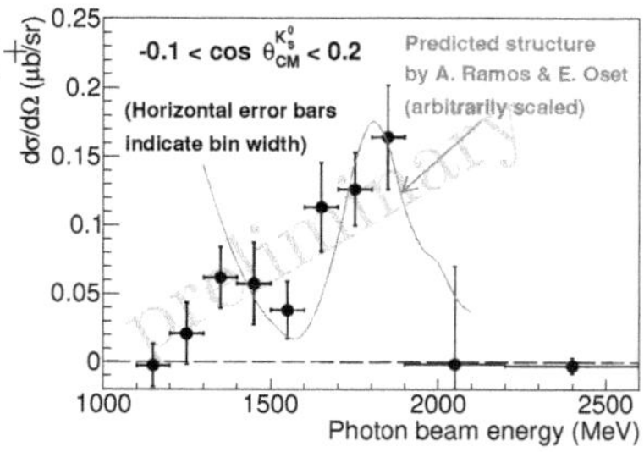

Fig. 3. Preliminary differential cross section on $\gamma n(p) \to K^0\Sigma^0(p)$ using a deuterium target. The red line shows the predictions for the total cross section.[3]

5. K^+ at extreme forward angles

The $\gamma p \to K^+\Lambda$ cross section at forward angles is virtually unconstrained by data, which correspond to low transfer momentum, t.[13] Preliminary results on the $K^+\Lambda$ and $K^+\Sigma^0$ cross section in this kinematic regime were analyzed by T.C. Jude at BGO-OD and are shown in figure 4. BGO-OD experiment shows a high statistics compared to other experimental results in the forward region. The high angular resolution of the forward spectrometer allows to map out crucial regions at low t. The $K^+\Sigma^0$ cross section shows a apparent cusp close to multiple thresholds, which is still under investigation.

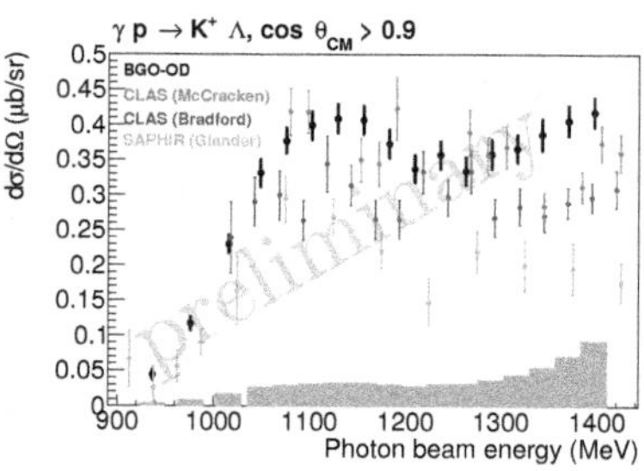

(a) $\gamma p \to K^+\Lambda$. Red data shows the CLAS results of McCracken.[14]

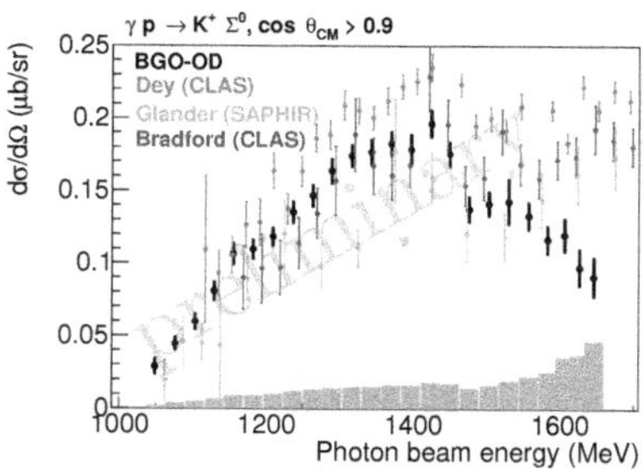

(b) $\gamma p \to K^+\Sigma^0$. Red data shows the CLAS results of Dey.[15]

Fig. 4. Differential cross section for γp to $K^+\Lambda$ (a) and $K^+\Sigma^0$ (b) for forward going K^+. Blue data shows the CLAS results of Bradford,[16] while green shows the SAPHIR data.[17]

6. Summary and Outlook

Current projects at BGO-OD investigate strangeness photoproduction. The here presented $\gamma p \to K^+\Lambda$, $\gamma p \to K^+\Sigma^0$, $\gamma p \to K^+\Lambda(1405)$ and $\gamma n(p) \to K^0\Sigma^0$ analyses are only a subset of the projects to come. The future projects will include K^* and hypernucleus production. Also non-strange analyses like the $\gamma p \to \eta' p$ are currently pursued.

Acknowledgments

This work is supported by the Deutsche Forschungsgemeinschaft project numbers 388979758 and 405882627, the RSF grant number 19-42-04132, the Third Scientific Committee of the INFN and the European Unions Horizon 2020 research and innovation program under grant agreement No. 824093.

References

1. S.-K. Choi et al. (Belle collaboration), Physical Review Letters **91**, doi:10.1103/PhysRevLett.91.262001 (2003).
2. R. Aaij et al. (LHCb collaboration), Physical Review Letters **115**, doi:10.1103/PhysRevLett.115.072001 (2015).
3. A. Ramos and E. Oset, Phys. Lett **B 727**, 287 (2013).
4. W. Hillert, Eur. Phys. J. A **28**, 139 (2006).
5. R. J. Hemingway, Nucl. Phys. B **253**, 742 (1985).
6. J. C. Nacher et al., Physical Review Letters B **455**, 55 (1999).
7. J. M. M. Hall et al., Physical Review Letters **114**, doi:10.1103/PhysRev Lett.114.132002 (2015).
8. R. Molina and M. Döring, Physical Review Letters D **94**, 056010 (2016).
9. K. Moriya et al. (CLAS Collaboration), Physical Review Letters C **87**, 035206 (2013).
10. K. Moriya et al. (CLAS Collaboration), Physical Review Letters C **88**, 045201 (2013).
11. I. Zychor et al., Physical Review Letters B **660**, 167-171 (2008).
12. R. Ewald et al. (CBELSA/TAPS Collaboration), Phys. Lett. **B 713**, 180 (2012).
13. D. Skoupil and P. Bydzovsky, Phys. Rev. **C97**, 025202 (2018).
14. M. E. McCracken et al. (CLAS), Phys. Rev. **C81**, 025201 (2010).
15. B. Dey et al. (CLAS), Phys. Rev. **C82**, 025202 (2010).
16. R. Bradford et al. (CLAS), Phys. Rev. **C73**, 035202 (2006).
17. K. H. Glander et al. (SAPHIR), Eur. Phys. J. **A19**, 251 (2004).

Implication of chiral symmetry on neutral weak pion production off a nucleon

De-Liang Yao

School of Physics and Electronics, Hunan University,
410082 Changsha, China
E-mail: yaodeliang@hnu.edu.cn
www.hnu.edu.cn

Neutral current weak pion production induced by neutrinos and antineutrinos off the nucleon is analyzed within the framework of manifestly relativistic baryon chiral perturbation theory with explicit Δ resonances up to $\mathcal{O}(p^3)$. It is found that, in the low energy region where chiral perturbation theory works, the total cross sections for the different physical channels exhibit a sizable non-resonant contribution. However, such contribution is not present in GENIE and NuWro event generators, which have been broadly used in the analyses of neutrino oscillation and cross section experiments.

Keywords: Weak pion production; chiral perturbation theory; Δ resonances.

1. Introduction

Neutrino- and antineutrino-induced single pion production dominates the contribution to the inclusive (anti-)neutrino-nucleus cross section in the relevant energy region of several experiments. Particularly, neutral current (NC) π^0 production could be misidentified as the electron-like signal in $\nu_\mu \to \nu_e$ ($\bar{\nu}_\mu \to \bar{\nu}_e$) researches. In order to overcome conflict between data sets and models,[1] uncertainties at the $20 - 30\%$ level are currently considered for single pion production when carrying out analyses of oscillation experiments. Hence, weak single pion production off the nucleon, as a fundamental ingredient of the corresponding process on nuclei, should be studied more precisely.

Although weak single pion production off the nucleon has been extensively investigated at intermediate energies with various phenomenological models, see e.g. Refs. 2–4, systematical application of baryon chiral perturbation theory (BChPT) to the study of this process is rare. The first work was done by using the heavy-baryon formalism but was only limited to low-energy theorems for the axial-vector part.[5]

Recently, charged-current (CC) weak single pion production off the nucleon has been analyzed in Ref. 6 within the framework of covariant baryon chiral perturbation theory (ChPT).[7–9] The calculation has been systematically performed up to the leading one-loop order, i.e., chiral order $\mathcal{O}(p^3)$. The so-called extended-on-mass-shell (EOMS) scheme[10] was employed to solve the power counting breaking problem.[11] The EOMS scheme has been extensively utilized in various investigations of baryonic phenomenology, see e.g. Refs. 12–14, even beyond the low-energy regime.[15] Furthermore, the contribution of Δ-resonance exchange has been explicitly implemented following the δ-counting rule proposed in Ref. 16.

In this proceeding, I will review the EOMS-BChPT study of neutral-current induced single pion production.[17]

2. Amplitudes for neutral-current single pion production

Under the one-boson exchange approximation and neglecting the momentum of the exchanged boson, the Lorentz-invariant amplitude for neutrino-induced single-pion production off the nucleon, $\nu(k_1) + N(p_1) \to \nu(k_2) + N'(p_2) + \pi^b(q)$ can be written as

$$\mathcal{M} = \frac{G_F}{\sqrt{2}} \left[\bar{u}(k_2)\gamma_\alpha(1-\gamma_5)u(k_1)\right] \langle \pi^b N' | J^\alpha(0) | N \rangle \tag{1}$$

with $G_F = 1.166 \times 10^{-5}$ GeV^{-2} being the Fermi constant. The leptonic part in square brackets is well-known in the sense that it can be straightforwardly calculated in the standard model (SM), while the term after is the hadronic matrix element that is needed to be determined in BChPT. The isospin structure of the hadronic part reads

$$\langle \pi^b N' | J_\alpha(0) | N \rangle = \chi_f^\dagger \left[H_\alpha^+ \delta^{b3} + i H_\alpha^- \epsilon^{b3c} \tau^c + H_\alpha^0 \tau^b \right] \chi_i \ ,$$

where τ^b stands for a Pauli matrix and $\chi_{i,f}$ are isospinors of the initial and final nucleon states, respectively. $H_\alpha^{\pm,0}$ are isospin amplitudes. Physical amplitudes for all the possible reactions can be deduced from the three isospin amplitudes.

The hadronic amplitudes can be derived order by order in BChPT. The same framework used in Ref. 6 for the CC single pion production can be extended to the NC case here. The relevant chiral Lagrangian is given in Ref. 18. The neutral-current weak interaction enters into the chiral Lagrangian via the following way

$$r_\mu = e \tan(\theta_W) Z_\mu^0 \frac{\tau^3}{2} \ , \quad v_\mu^{(3)} = \frac{3\tan(\theta_W)}{2} Z_\mu^0 \ ,$$

$$l_\mu = -\frac{g}{\cos(\theta_W)} Z^0_\mu \frac{\tau^3}{2} + e\tan(\theta_W) Z^0_\mu \frac{\tau^3}{2} \tag{2}$$

with θ_W the Weinberg angle. Compared to the Lagrangian used for CC productions, one has three additional pieces proportional to c_7, d_7 and d_9.

The calculation of the hadronic amplitudes can be readily performed following the procedure demonstrated in Ref. 6 and using the same topologies of Feynman diagrams depicted therein. Furthermore, the δ-counting rule[16] is applied to deal with the explicit inclusion of explicit Δ resonances at low energies near threshold. The chiral amplitude at one-loop level is renormalized in the EOMS scheme so as to remove ultraviolet divergence and restore correct power counting. More details can be found in Refs. 6, 17.

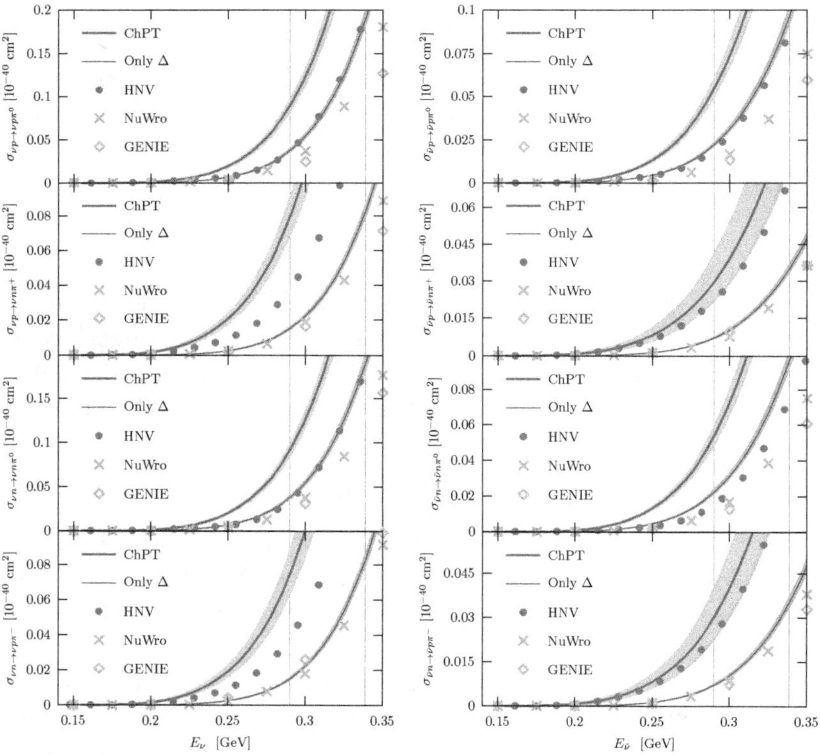

Fig. 1. Total cross sections for neutral-current weak pion production off the nucleon as a function of the Laboratory neutrino energy.

3. Numerical results on total cross section

The total cross section in the center-of-mass frame of the initial states has the form

$$\sigma(s) = \frac{1}{2(4\pi)^4\sqrt{s}\,|\mathbf{k}_1|}\int_{\omega_\nu^-}^{\omega_\nu^+}\mathrm{d}\omega_\nu\int_{\omega_\pi^-}^{\omega_\pi^+}\mathrm{d}\omega_\pi\int_{-1}^{+1}\mathrm{d}x_1\int_0^{2\pi}\mathrm{d}\phi_{12}\,|\mathcal{M}|^2 . \quad (3)$$

In above, the unpolarized $\mathcal{M}$-matrix squared reads

$$|\mathcal{M}|^2 = \frac{G_F^2}{2}L^{\alpha\beta}H_{\alpha\beta}\,, \quad (4)$$

where $L^{\alpha\beta}$ and $H_{\alpha\beta}$ are leptonic and hadronic tensors respectively. We refer readers to Ref. 17 for specific definitions and the values of the involved parameters. In Fig. 1, total cross sections for all the physical channels are shown as a function of E_ν, which is the energy of the incoming (anti)neutrino in the laboratory frame. It can be seen from Fig. 1 that both the NuWro and GENIE results agree with the Δ contribution. The sizable non-resonant contribution is not accounted by the event generators.

4. Summary

The neutral-current weak pion production off the nucleon has been studied in covariant ChPT with explicit $\Delta(1232)$ up to $\mathcal{O}(p^3)$ by applying the δ-counting rule and using the EOMS scheme. Our predictions for the Δ-exchange mechanism agree well with the corresponding output from the NuWro and GENIE neutrino event generators. Non-resonant contributions are sizable but not implemented in those event generators.

Acknowledgments

I would like to thank L. Alvarez-Ruso, A. Hiller-Blin and M. J. Vicente-Vacas for sharing their insights into this project. This work has been supported by National Science Foundation of China under Grant No. 11905258.

References

1. T. Katori and M. Martini, Neutrinonucleus cross sections for oscillation experiments, *J. Phys.* **G45**, 013001 (2018).
2. S. L. Adler, Photoproduction, electroproduction and weak single pion production in the (3,3) resonance region, *Annals Phys.* **50**, 189 (1968).
3. D. Rein and L. M. Sehgal, Neutrino Excitation of Baryon Resonances and Single Pion Production, *Annals Phys.* **133**, 79 (1981).

4. E. Hernandez, J. Nieves and M. Valverde, Weak Pion Production off the Nucleon, *Phys. Rev.* **D76**, 033005 (2007).
5. V. Bernard, N. Kaiser and U. G. Meißner, Low-energy theorems for weak pion production, *Phys. Lett.* **B331**, 137 (1994).
6. D.-L. Yao, L. Alvarez-Ruso, A. N. Hiller Blin and M. J. Vicente Vacas, Weak pion production off the nucleon in covariant chiral perturbation theory, *Phys. Rev.* **D98**, 076004 (2018).
7. S. Weinberg, Phenomenological Lagrangians, *Physica* **A96**, 327 (1979).
8. J. Gasser and H. Leutwyler, Chiral Perturbation Theory to One Loop, *Annals Phys.* **158**, 142 (1984).
9. J. Gasser and H. Leutwyler, Chiral Perturbation Theory: Expansions in the Mass of the Strange Quark, *Nucl. Phys.* **B250**, 465 (1985).
10. T. Fuchs, J. Gegelia, G. Japaridze and S. Scherer, Renormalization of relativistic baryon chiral perturbation theory and power counting, *Phys. Rev.* **D68**, 056005 (2003).
11. J. Gasser, M. E. Sainio and A. Svarc, Nucleons with Chiral Loops, *Nucl. Phys.* **B307**, 779 (1988).
12. Y.-H. Chen, D.-L. Yao and H. Q. Zheng, Analyses of pion-nucleon elastic scattering amplitudes up to $O(p^4)$ in extended-on-mass-shell subtraction scheme, *Phys. Rev.* **D87**, 054019 (2013).
13. D.-L. Yao, D. Siemens, V. Bernard, E. Epelbaum, A. M. Gasparyan, J. Gegelia, H. Krebs and U.-G. Meißner, Pion-nucleon scattering in covariant baryon chiral perturbation theory with explicit Delta resonances, *JHEP* **05**, 038 (2016).
14. D.-L. Yao, L. Alvarez-Ruso and M. J. Vicente-Vacas, Extraction of nucleon axial charge and radius from lattice QCD results using baryon chiral perturbation theory, *Phys. Rev.* **D96**, 116022 (2017).
15. E. Epelbaum, J. Gegelia, U.-G. Meißner and D.-L. Yao, Baryon chiral perturbation theory extended beyond the low-energy region, *Eur. Phys. J.* **C75**, 499 (2015).
16. V. Pascalutsa and D. R. Phillips, Effective theory of the delta(1232) in Compton scattering off the nucleon, *Phys. Rev.* **C67**, 055202 (2003).
17. D.-L. Yao, L. Alvarez-Ruso and M. J. Vicente Vacas, Neutral-current weak pion production off the nucleon in covariant chiral perturbation theory, *Phys. Lett.* **B794**, 109 (2019).
18. N. Fettes, U.-G. Meißner, M. Mojzis and S. Steininger, The Chiral effective pion nucleon Lagrangian of order p**4, *Annals Phys.* **283**, 273 (2000), [Erratum: Annals Phys. **288**, 249 (2001)].

Session 3

Exotic Hadrons and Candidates

Recent BESIII results of X(3872)

Kai Zhu (for the BESIII Collaboration)

Institute of High Energy Physics
Beijing 100049, People's Republic of China
E-mail: zhuk@ihep.ac.cn

$X(3872)$ has been observed for more than 15 years, but its nature is still unclear. It is considered as an exotic charmonium-like candidate. BESIII has collected about 12 fb^{-1} data at center-of-mass energies from 4.15 to 4.60 GeV. Using these data samples, the decay of $X(3872)$ has been studied extensively, including final states of $\pi^0\chi_{cJ}$, $\omega J/\psi$, $\gamma\psi(1S, 2S)$, $D^0\bar{D}^{*0} + c.c.$, and γD^+D^-. New results from these studies will be presented.

Keywords: X(3872); BESIII.

1. Introduction

$X(3872)$ was first observed in process $B \to K(\pi^+\pi^- J/\psi)$ by Belle[1] in 2003. Its mass is close the the $D^{*0}D^0$ threshold with $M_{X(3872)} = 3871.68 \pm 0.17$ MeV, the difference to the threshold is only 0.01 ± 0.20 MeV. Its width is very narrow, till now only the upper limit < 1.2 MeV has been presented. CDF[2] and LHCb[3] have determined its quantum number to be $J^{PC} = 1^{++}$, and no iso-spin partner of this state has been found yet. It can be produced in various processes such as pp collision, $p\bar{p}$ annihilation, B decays, γ^*N collision, $\pi^\pm Li$ collision, and radiative process $e^+e^- \to \gamma X(3872)$ at BESIII. About half of it decays are into open charm channels, while the final states involving charmonia take about $O(\%)$ contributions. Since the first day of its discovery, much interest are on it since it is a good exotic multi-quark candidate. However, 15 years later its nature is still mysterious, even many different proposals are suggested such as D^0D^{0*} molecule, hybrid, charmonium $\chi_{c1}(2P)$, Tetra-quark, cusp effect, and so on even mixture of these states. To solve this puzzle, more and better experimental measurements are required. Recently, BESIII using about 12 fb^{-1} e^+e^- collision data taken above 4.15 GeV to study the X(3872). Some interesting results are obtained and will be presented in this talk.

BESIII is a general purpose detector with a 93% solid angle coverage operating on the electron-positron collider BEPCII, that runs at center-of-mass energies from 2.0 to 4.0 GeV with reached luminosity of $1 \times 10^{33}\ cm^{-2}s^{-1}$. BESIII is mainly composed by a multi-layer drift chamber, a electro-magnetic calorimeter, a time-of-flight system, and a muon counter, that provide good momentum measurement of charged tracks, particle identification, and energy measurement of neutral clusters. The detailed parameters and performance of BESIII can be found in paper.[4] Recently, BESIII has accumulated large data samples dedicated mainly for the studies of XYZ states.

2. Results at BESIII

2.1. $X(3872) \to \pi^0\chi_{c1}$

In conventional $c\bar{c}$ hypothesis, the decay width of $X(3872) \to \pi^0\chi_{c1}$ is small, that is estimated to be $\Gamma\left(X(3872) \to \pi^0\chi_{c1}(1P)\right) \sim 0.06$ keV; however, in the tetra-quark/molecular state hypothesis, the decay width could be sizeable.[5,6] Using 9.0 fb^{-1} e^+e^- collision data with center-of-mass energies between 4.15 and 4.30 GeV, BESIII searched $e^+e^- \to \gamma_1 X(3872)$, $X(3872) \to \pi^0\chi_{cJ}$, $\chi_{cJ} \to \gamma_2 J/\psi$, $J/\psi \to e^+e^-/\mu^+\mu^-$.[7] From Fig. 1 we can see after select $X(3872)$ signal region, clear $\chi_{c1}(1P)$ signal is observed; and vise verse. It is the first observation of $X(3872) \to \pi^0\chi_{c1}$, and the significance is 5.2σ. The ratio between the branching fractions of $B(X \to \pi^0\chi_{cJ})$ and $B(X \to \pi^+\pi^- J/\psi)$ is $0.88^{+0.33}_{-0.27} \pm 0.10$. Using $3.2\% < B(X(3872) \to \pi^+\pi^- J/\psi) < 6.4\%$, it is found that $3\% < B(X(3872) \to \pi^0\chi_{c1}(1P)) < 6\%$. Using $\Gamma_{X(3872)} \sim 1.2$ MeV, conventional $c\bar{c}$ hypothesis predicts $B(X(3872) \to \pi^0\chi_{c1}) \sim 0.5\%$. So our results disfavor the $c\bar{c}$ interpretation of $X(3872)$. From Fig. 2 we also can see the hint of relation between $Y(4260)$ and $X(3872)$.

2.2. $X(3872) \to \omega J/\psi$

Belle and BaBar reported 4σ evidence for the process $X(3872) \to \omega J/\psi$,[8,9] and give the ratio between the branching fractions of iso-spin conserved process $B(X(3872) \to \pi^+\pi^-\pi^0 J/\psi)$ and the iso-spin violated process $B(X(3872) \to \pi^+\pi^- J/\psi)$ is $1.0 \pm 0.4 \pm 0.3$. BESIII is expected to find about 70 $X(3872) \to \omega J/\psi$ events with the data accumulated around 4.2 GeV, and has searched this channel via $e^+e^- \to \gamma_1 X(3872)$, $X(3872) \to \omega J/\psi$, $\omega \to \pi^+\pi^-\pi^0$, $\pi^0 \to \gamma\gamma$, $J/\psi \to e^+e^-/\mu^+\mu^-$.[10] From Fig. 3 we can see at least one additional BW-formed resonance except $X(3872)$ is needed to

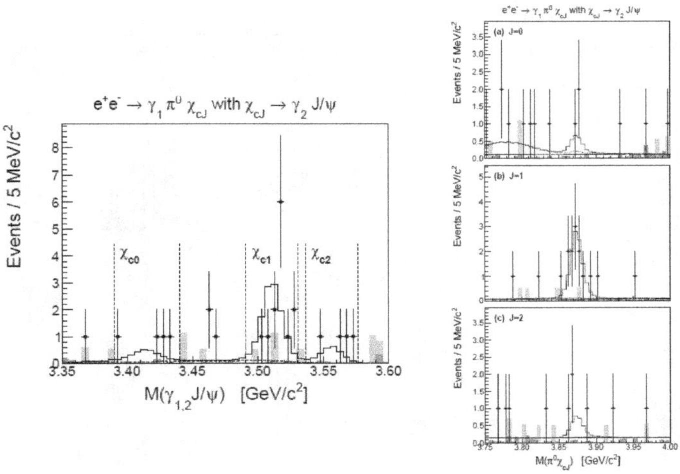

Fig. 1. The left plot is the invariant mass of $\gamma_{1,2}J/\psi$. The right plots are the invariant masses of $\pi^0\chi_{cJ}$. The dark histograms are peaking J/ψ backgrounds from MC simulation, the light histograms are the sideband of J/ψ, the dots with error bars are data.

fit the invariant mass of $\omega J/\psi$. And it is hard to distinguish between the two hypotheses of additional one and two resonances, since there is only 2.5 σ difference between them. The fitted mass and width of the $X(3872)$ are 3873.3 ± 1.1 MeV and < 1.2 MeV, respectively. Figure 4 shows the simultaneous fit to the cross sections of $e^+e^- \to \gamma X(3872) \to \gamma\omega J/\psi$ and $e^+e^- \to \gamma X(3872) \to \gamma\pi^+\pi^- J/\psi$. The fitted mass and width of the resonance are $M = 4200.6^{+7.9}_{-13.3} \pm 3.0$ MeV and $\Gamma = 115^{+38}_{-26} \pm 12$ MeV, respectively. And the ratio between $B(X(3872) \to \omega J/\psi)$ and $B(X(3872) \to \pi^+\pi^- J/\psi)$ is $1.6^{+0.4}_{-0.3} \pm 0.2$, that agrees with the previous measurements. And it will be the input for the hadronic molecule interpretation for the $X(3872)$ resonance. From Fig. 4 we can see hint of the relation between $Y(4260)$ and $X(3872)$, again.

2.3. $X(3872) \to \gamma J/\psi$, $\gamma\psi(2S)$, $D^0\bar{D}^{*0} + c.c.$, γD^+D^-

Combined the results from BaBar,[11] Belle,[12] and LHCb,[13] the ratio between $B(X(3872) \to \gamma\psi(2S))$ and $B(X(3872) \to \gamma J/\psi)$ is 2.31 ± 0.57, also the ratio between $B(X(3872) \to \gamma J/\psi)$ and $B(X(3872) \to \pi^+\pi^- J/\psi)$ is 0.24 ± 0.05. BESIII using the data sets from 4.18 GeV to 4.28 GeV searched the following processes: $X(3872) \to D^0\bar{D}^{*0} + c.c.$, $D^{*0} \to \gamma D^0/\pi^0 D^0$;

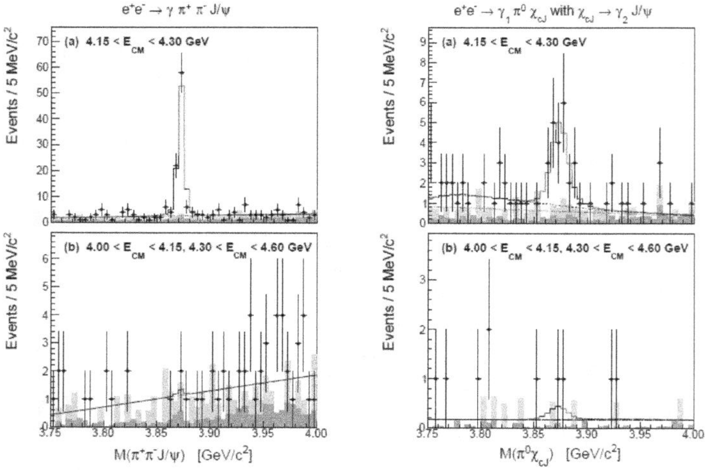

Fig. 2. The left plots are the invariant masses of $\pi^+\pi^- J/\psi$. The right plots are the invariant masses of $\pi^0\chi_{cJ}$. The dark histograms are peaking J/ψ backgrounds from MC simulation, the light histograms are the sideband of J/ψ, the dots with error bars are data.

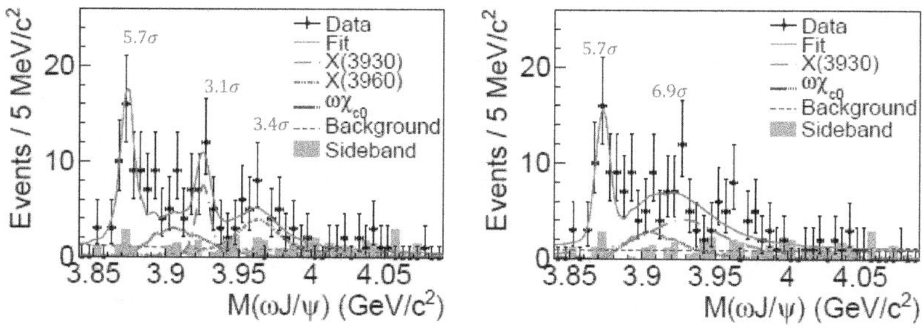

Fig. 3. Invariant masses of $\omega J/\psi$. The left plot shows the fit with two additional resonances except $X(3872)$. The right plot shows the fit with only one additional resonance.

$X(3872) \to \gamma J/\psi$, $J/\psi \to \mu\mu$ *or* ee; $X(3872) \to \gamma\psi(2S)$, $\psi(2S) \to \pi^+\pi^- J/\psi$ *or* $\mu\mu$; $X(3872) \to D^+D^-$, $D^\pm \to K\pi\pi$ *or* $K\pi\pi\pi$. Some preliminary results have been obtained. The invariant masses of $\gamma J/\psi$ are fitted with ee and $\mu\mu$ channels simultaneously, and the statistical significance is larger than 3.5σ. But from the distribution of $M(\gamma\psi(2S))$, there is no evident signal of $X(3872)$. And the ratio between $B(X(3872) \to \gamma\psi(2S))$ and $B(X(3872) \to \gamma J/\psi)$ is less than 0.59% at 90% C.L. The simultaneous

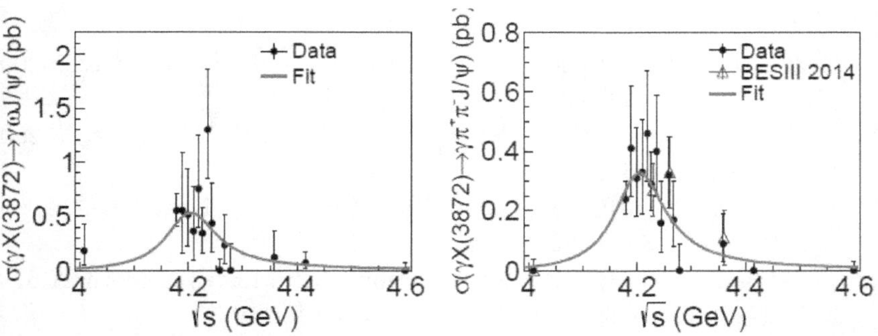

Fig. 4. Simultaneous fit to the cross sections of (left) $e^+e^- \to \gamma X(3872) \to \gamma\omega J/\psi$ and (right) $e^+e^- \to \gamma X(3872) \to \gamma\pi^+\pi^- J/\psi$.

fit to $M(DD^*)$ gives the significance is larger than 7.4σ with the numbers of signal are 25.5 ± 4.4 and 32.5 ± 5.5 for the $\gamma D\bar{D}$ and $\pi^0 D\bar{D}$ channels, respectively. And there is no evident signal for $X(3872) \to \gamma D^+D^-$.

The following table presents the ratios between the branching fractions of the measured channels and $X(3872) \to \pi^+\pi^- J/\psi$.

BESIII preliminary results

mode	$D^0\bar{D}^{*0} + c.c.$	$\gamma J/\psi$	$\gamma\psi(2S)$	γD^+D^-
ratio	14.8 ± 3.80	0.79 ± 0.28	< 0.42	< 0.99

3. Summary and outlook

Great progress on the studies of $X(3872)$ has been achieved at BESIII recently including: new decay mode $X(3872) \to \pi^0\chi_{c1}$ is observed, first firm observation of $X(3872) \to \omega J/\psi$, and more decays $D^0\bar{D}^{*0} + c.c.$, $\gamma J/\psi$, $\gamma\psi(2S)$, γD^+D^- are searched and measured. There are also hints of relation between $X(3872)$ and $Y(42xx)$ states, as well as evidence of $X(3915)$ and $X(3960)$ in $\omega J/\psi$ channel. These results are inputs for further analysis and theoretical interpretations. For example, the paper[14] uses results from BaBar, Belle, LHCb, and BESIII to do a global fit and provide the branching fractions of various $X(3872)$ decaying channels and B decaying into $X(3872)$, as well as the peak cross section of $e^+e^- \to \gamma X(3872)$ at 4.226 GeV. BESIII has collected about 3.8 fb^{-1} e^+e^- collision data between 4.13 and 4.44 GeV in year 2019, new results are on the way.

References

1. S. K. Choi *et al.* [Belle Collaboration], Phys. Rev. Lett. **91**, 262001 (2003) [hep-ex/0309032].
2. A. Abulencia *et al.* [CDF Collaboration], Phys. Rev. Lett. **98**, 132002 (2007) [hep-ex/0612053].
3. R. Aaij *et al.* [LHCb Collaboration], Phys. Rev. Lett. **110**, 222001 (2013) [arXiv:1302.6269 [hep-ex]].
4. M. Ablikim *et al.* [BESIII Collaboration], Nucl. Instrum. Meth. A **614**, 345 (2010) [arXiv:0911.4960 [physics.ins-det]].
5. S. Dubynskiy and M. B. Voloshin, Phys. Rev. D **77**, 014013 (2008) [arXiv:0709.4474 [hep-ph]].
6. T. Mehen, Phys. Rev. D **92**, no. 3, 034019 (2015) [arXiv:1503.02719 [hep-ph]].
7. M. Ablikim *et al.* [BESIII Collaboration], Phys. Rev. Lett. **122**, no. 20, 202001 (2019) [arXiv:1901.03992 [hep-ex]].
8. P. del Amo Sanchez *et al.* [BaBar Collaboration], Phys. Rev. D **82**, 011101 (2010) [arXiv:1005.5190 [hep-ex]].
9. K. Abe *et al.* [Belle Collaboration], hep-ex/0505037.
10. M. Ablikim *et al.* [BESIII Collaboration], Phys. Rev. Lett. **122**, no. 23, 232002 (2019) [arXiv:1903.04695 [hep-ex]].
11. B. Aubert *et al.* [BaBar Collaboration], Phys. Rev. Lett. **102**, 132001 (2009) [arXiv:0809.0042 [hep-ex]].
12. V. Bhardwaj *et al.* [Belle Collaboration], Phys. Rev. Lett. **107**, 091803 (2011) [arXiv:1105.0177 [hep-ex]].
13. R. Aaij *et al.* [LHCb Collaboration], Nucl. Phys. B **886**, 665 (2014) [arXiv:1404.0275 [hep-ex]].
14. C. Li and C. Z. Yuan, Phys. Rev. D **100**, no. 9, 094003 (2019) [arXiv:1907.09149 [hep-ex]].

Sensitivity to the $X(3872)$ total width at the Belle II experiment

Hikari Hirata, for the Belle II Collaboration

Graduate School of Science, Nagoya University, Nagoya, 464-8602, Japan
E-mail: hirata@hepl.phys.nagoya-u.ac.jp

The $X(3872)$ resonance is an exotic hadron candidate, and the measurement of its characteristics is fundamental to reveal its nature. In this paper, it is shown that the Belle II experiment at the SuperKEKB asymmetric energy e^+e^- collider is capable of measuring the total width down to 280 keV with more than 3σ significance with 50 ab^{-1} of data. We also report rediscovery of the $B \to K\psi(2S)$ $(\to J/\psi\pi^+\pi^-)$ decay, which is topologically similar to the doorway to the $B \to KX(3872)$ $(\to J/\psi\pi^+\pi^-)$ decay.

Keywords: Heavy quark; exotic hadron.

1. Introduction

The $X(3872)$ was discovered from the $J/\psi\pi^+\pi^-$ mass spectrum in B decays at the Belle experiment.[1] Since then, many properties have been measured in the Belle and other experiments. However, the internal structure has not yet been revealed. It is thus necessary to acquire further information about the $X(3872)$, for example its total width, for which only an upper limit of 1.2 MeV (at the 90% confidence level) has been obtained with $J/\psi\pi^+\pi^-$ decay.[2] The Belle II experiment at the SuperKEKB asymmetric-energy e^+e^- collider[3] is a substantial upgrade of the B factory facility at the Japanese KEK laboratory. Belle II aims to record 50 ab^{-1} of data, a factor of 50 more than its predecessor. It will provide unique opportunity for precise measurement of the $X(3872)$ total width.[4]

A milestone for the $X(3872)$ study with initial data is to discover control sample of the primary $X(3872)$ channel, $B \to K\psi(2S)$ $(\to J/\psi\pi^+\pi^-)$. We search for it using data collected by the Belle II detector[5,6] at a center-of-mass (CM) energy corresponding to the mass of the $\Upsilon(4S)$ resonance (10.58 GeV) from March to July 2019. The corresponding integrated luminosity is 5.15 fb^{-1}. In the following, it will be referred to as "early phase 3" data.

This proceeding describes the sensitivity study to the $X(3872)$ total width using simulation sample and the rediscovery of the $B \to K\psi(2S)$ decay with the early phase 3 data.

2. Sensitivity to the $X(3872)$ total width at Belle II

We propose an analysis with the $B^+ \to K^+(X(3872) \to D^0\bar{D}^0\pi^0)$ decay mode so as to improve the sensitivity to the $X(3872)$ total width. The advantage of the decay mode is the good $X(3872)$ mass resolution thanks to small Q-value. The sensitivity is estimated by using a Monte-Carlo (MC) simulation with the full Belle II setup. In this simulation, the signal mode is as follows.

$$B^+ \to K^+[X(3872) \to D^0\bar{D}^0\pi^0]$$
$$D^0 \to K^-\pi^+,\ K^-\pi^+\pi^0,\ K^-\pi^+\pi^-\pi^+$$

Here, the charge conjugated modes are always implied throughout this proceeding. Generic $B\bar{B}$ and $e^+e^- \to q\bar{q}$ events are used as background events.

Figure 1 (left) shows the $D^0\bar{D}^0\pi^0$ invariant mass distribution. Here, detailed conditions for reconstruction and selection are described in Ref. 4. As a result, signal yield expected with 1 ab^{-1} is 64.5 ± 23.9 and the mass resolution is 684 ± 8 keV. Compared with the $J/\psi\pi^+\pi^-$ mode,[2] the yield is around half, but the mass resolution is about 3 times better. The total width is extracted by fitting the mass spectrum with a function which is a sum of three components. The signal component is a Breit-Wigner function convoluted with the mass resolution. The component of crossfeed

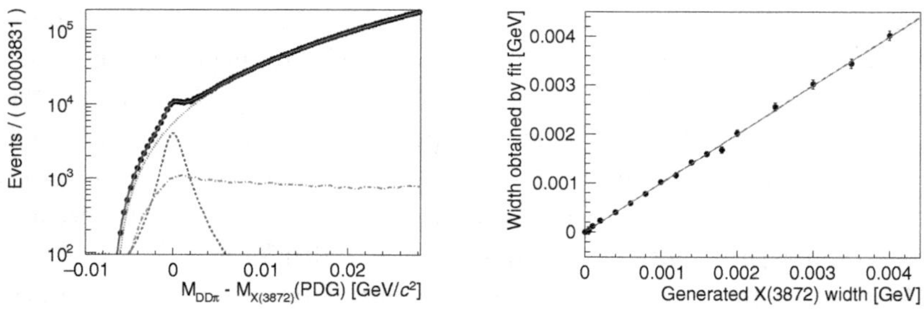

Fig. 1. (Left) The Distribution of difference between $D^0\bar{D}^0\pi^0$ reconstructed mass and $X(3872)$ mass and the fit result. Blue dashed, red dashdotted and green dotted lines are the signal, the crossfeed background and the general background components, respectively. (Right) The relation between the total width generated in simulating and that obtained by fit.

background, which comes from signal events reconstructed incorrectly, is generated by the MC simulation. The component of generic background is represented by a threshold function. A relation between the total width generated in the simulation and that obtained by fit is checked in Fig. 1 (right). The fit result of the relation with the linear function (red solid line) is consistent with $y = x$ (blue dashed line), which means that there is no bias in this fitting. Finally, obtained sensitivities as a function of the integrated luminosity are shown in Fig. 2. With 50 ab^{-1} data, the Belle II experiment is capable of measuring total widths with 3σ and 5σ significance with values down to 280 keV and 570 keV, respectively. Moreover, the upper limit is improved down to around 180 keV.

The total width provides us partial widths for each decay mode by combining branching fractions of the $X(3872)$.

$$\Gamma_{X(3872)\to f} = \Gamma_{X(3872)} \times \mathcal{BR}(X(3872) \to f), \tag{1}$$

where $\Gamma_{X(3872)}$ is the total width, and $\Gamma_{X(3872)\to f}$ and $\mathcal{BR}(X(3872) \to f)$ mean the partial width and the branching fraction for $X(3872) \to f$ decay, respectively. Here, f shows an arbitrary final state. $\mathcal{BR}(X(3872) \to f)$ can be obtained experimentaly as follows.

$$\mathcal{BR}(X(3872) \to f) = \frac{\mathcal{BR}(B^+ \to X(3872)K^+) \times \mathcal{BR}(X(3872) \to f)}{\mathcal{BR}(B^+ \to X(3872)K^+)} \tag{2}$$

The denominator is a so-called absolute branching fraction, and it can be measured by the missing mass technique at the B-factory experiments operating at the CM energy in the $\Upsilon(4S)$ resonance.[7] It plans to be measured precisely at Belle II.[6] If the partial width determined by $\mathcal{O}(100\text{ keV})$ or the

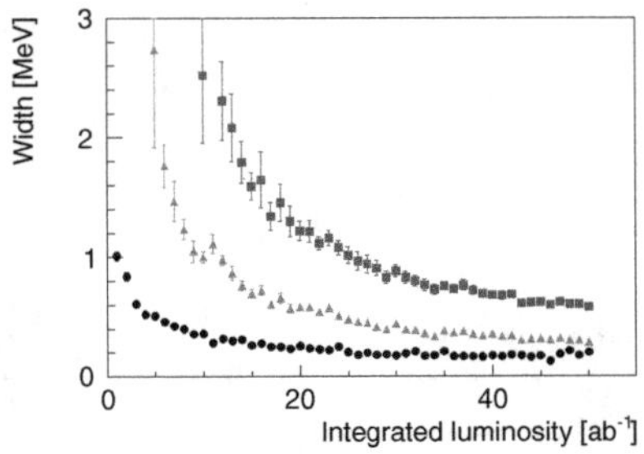

Fig. 2. The sensitivities of the total width as a function of the integrated luminosity. The black circles are the upper limits at the 90% confidence level. The red triangles (blue squares) are the measurable total widths with 3σ (5σ) significance.[4]

upper limit is improved at that level, it will have capability of testing the models.

3. Rediscovery of the $B \to K\psi(2S)$ decay

Rediscovering control modes for the $X(3872)$ with the initial data and evaluating the performance are important to prove an ability of measuring the $X(3872)$ charactristics. According to the previous studies at Belle, the $B \to KX(3872)$ ($\to J/\psi\pi^+\pi^-$) decay has the largest signal yield per unit of integrated luminosity. Therefore, we search for a topologically similar mode, $B \to K\psi(2S)$ ($\to J/\psi\pi^+\pi^-$) with the early phase 3 data at first. In this analysis, $B^+ \to \psi(2S)K^+$ and $B^0 \to \psi(2S)K_S$ are combined to maximize the signal significance.

Figure 3 (left) shows a $M_{J/\psi\pi^+\pi^-}$ distribution, which is defined as $M^{\rm recon}_{l^+l^-\pi^+\pi^-} - M^{\rm recon}_{l^+l^-} + m_{J/\psi}$.[1] Here, $M^{\rm recon}_{l^+l^-\pi^+\pi^-}$, $M^{\rm recon}_{l^+l^-}$ and $m_{J/\psi}$ are the reconstructed $l^+l^-\pi^+\pi^-$ invariant mass, the reconstructed l^+l^- invariant mass, and the J/ψ nominal mass,[8] respectively. In order to extract the number of $\psi(2S)$ from B decays as distinguished from the $\psi(2S)$ signal from non-B decays, we perform an Unbinned Extended Maximum-likelihood simultaneous fit to two $M_{J/\psi\pi^+\pi^-}$ distributions on a beam-energy constrained mass ($M_{\rm bc}$) signal region and on a $M_{\rm bc}$ sideband region. Here, $M_{\rm bc}$ is defined by $\sqrt{E^2_{\rm beam} - P^2_{B\rm cand}}$, where $E_{\rm beam}$ and $P_{B\rm cand}$ are the beam energy in the CM system and the CM momentum of the B candidate, respectively. The fit components for the $M_{\rm bc}$ sideband region are the $\psi(2S)$ signal from non-B decays and the combinatorial background. On the $M_{\rm bc}$ signal region, there are 3 components, which are the $\psi(2S)$ signal from B

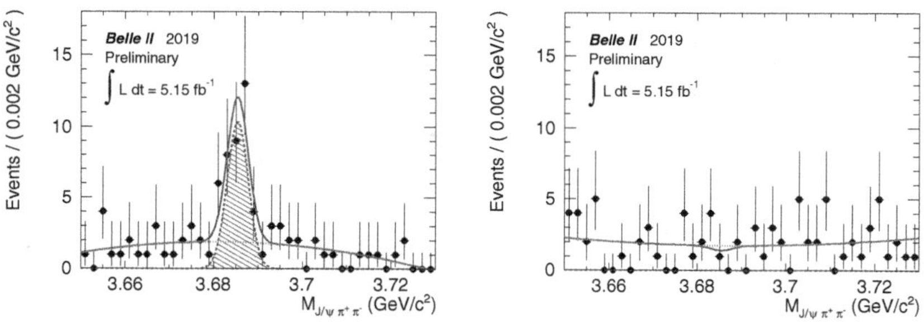

Fig. 3. $M_{J/\psi\pi^+\pi^-}$ distribution in the $M_{\rm bc}$ signal region for the data (left) and that in the $M_{\rm bc}$ sideband region (right) with signal hypothesis. The blue solid line shows the total fit result. The blue hatched area is the $\psi(2S)$ component from B decays, and red dashed lines show the combinatorial background.

decays in addition to the components of the sideband region. As a result of the simultaneous fit, the mean of the peak is obtained by 3685.39 ± 0.63 MeV/c^2, where the error is only statistical. It agrees well with the nominal $\psi(2S)$ mass. The statistical significance of the $\psi(2S)$ state is obtained by 5.0σ.

4. Conclusion

The Belle II experiment will improve the sensitivity to the X(3872) total width. By using the decay mode, $X(3872) \to D^0\bar{D}^0\pi^0$, Belle II will improve the upper limits of the total width at 90 % confidence level down to around 180 keV, and will measure the total widths with 3σ and 5σ significance with values down to 280 keV and 570 keV, respectively.

We also report rediscovery of the control mode of $B \to KX(3872)$ ($\to J/\psi\pi^+\pi^-$) decay, $B \to K\psi(2S)$ ($\to J/\psi\pi^+\pi^-$), with a significance of 5.0σ, using the early phase 3 data. The mean value of the peak agrees well with the nominal $\psi(2S)$ mass. Now, we are aiming to rediscover the $X(3872)$ to evaluate the measurement performance at Belle II with integrated luminosity of $\mathcal{O}(10\ \text{fb}^{-1})$.

Acknowledgments

We thank the conference organizers for hosting. This work is supported partly by Grant-in-Aid for Scientific Research (S) (No. 18H05226) and Grant-in-Aid for JSPS Fellows (No. 19J23314).

References

1. S.-K. Choi *et al.* (Belle Collaboration), Phys. Rev. Lett, **91**, 262001 (2003).
2. S.-K. Choi *et al.* (Belle Collaboration), Phys. Rev. D, **84** 052004 (2014).
3. Y. Ohnishi *et al.*, Prog. Theor. Exp. Phys. **2013**, 03A011.
4. H. Hirata *et al.* (Belle II Collaboration), JPS Conf. Proc **26**, 031008 (2019).
5. T. Abe *et al.* (Belle II Collaboration), KEK report 2010-1, arXiv:1011.0352.
6. E. Kou *et al.* (Belle II Collaboration), arXiv:1808.10567.
7. Y. Kato *et al.* (Belle Collaboration), Phys. Rev. D, **97** 012005 (2018).
8. M. Tanabashi *et al.* (Particle Data Group), Phys. Rev. D **98**, 030001 (2018).

Implications of spin symmetry for XYZ states

Q. Wang[1,2,*], V. Baru[3,4,5], E. Epelbaum[6], A. A. Filin[6], C. Hanhart[7], A.V. Nefediev[5,8], J.L. Wynen[7]

[1] *Institute of Quantum Matter, South China Normal University, Guangzhou 510006, China,*

[2] *Theoretical Physics Center for Science Facilities, Institute of High Energy Physics, Chinese Academy of Sciences, Beijing 100049, China,*

[3] *Helmholtz-Institut für Strahlen- und Kernphysik and Bethe Center for Theoretical Physics, Universität Bonn, D-53115 Bonn, Germany,*

[4] *Institute for Theoretical and Experimental Physics, B. Cheremushkinskaya 25, 117218 Moscow, Russia*

[5] *P.N. Lebedev Physical Institute of the Russian Academy of Sciences, 119991, Leninskiy Prospect 53, Moscow, Russia,*

[6] *Ruhr University Bochum, Faculty of Physics and Astronomy, Institute for Theoretical Physics II, D-44780 Bochum, Germany ,*

[7] *Institute for Advanced Simulation, Institut für Kernphysik and Jülich Center for Hadron Physics, Forschungszentrum Jülich, D-52425 Jülich, Germany,*

[8] *National Research Nuclear University MEPhI, 115409, Kashirskoe highway 31, Moscow, Russia*

[*] *E-mail: qianwang@m.scnu.edu.cn,*

Numerous exotic candidates containing a heavy quark and anti-quark (the so-called XYZ states) have been reported since the observation of the $X(3872)$ in 2003. For these systems a study of the implications of the heavy quark spin symmetry and its breaking is expected to provide useful guidance towards a better understanding of their nature. For instance, since the formation of the complete spin multiplets is described with the same parameter sets, in some cases the currently available experimental data on the XYZ states allows us to predict properties of spin partner states. To illustrate this point we extract the parameters of the two Z_b states by analyzing the most recent experimental data within an effective-field theory approach which treats both short-ranged contact interactions and the long-ranged one-pion/one-eta Goldstone boson exchanges (OPE/OEE) dynamically. The line shapes and pole positions of their spin partners are then predicted in a parameter-free way and await to be tested by future experimental data.

Keywords: Heavy quark spin symmetry; hadronic molecules; effective field theory.

1. Introduction

Heavy-quark spin symmetry (HQSS) states that the strong interaction is invariant under the rotation of a spin of a heavy quark when the heavy quark mass goes to infinity. Corrections scale as Λ_{QCD}/M_Q with $\Lambda_{QCD} \sim 200$ MeV the intrinsic mass scale of QCD and M_Q the heavy quark mass. Two important consequences follow: 1) in the heavy quark limit the heavy quark spin and the total angular momentum of the light degrees of freedom are conserved separately by the strong interactions. 2) A given heavy quark spin state should have spin partners with the same light cloud, but different heavy quark spins. It can be shown that the violations of the spin symmetry predictions differ strongly depending on the assumed structure of the states.[1] In this contribution, we will use the two Z_b states and their spin partners W_{bJ} $(J = 0, 1, 2)$ as an example to illustrate how the program works within the hadronic molecular picture, where the leading order spin symmetry violation enters through $B - B^*$ mass difference. In particular, an effective field theory (EFT), which incorporates both short-range and long-range OPE interactions at leading order, is proposed for the scattering processes involving (B, B^*) and $\left(\bar{B}, \bar{B}^*\right)$ doublets. The parameters are extracted from the data on the line shapes of the two Z_b states and used to predict the line shapes of their spin partners W_{bJ} as well as their pole positions.

2. Framework

As the bottom quark mass is much larger than the typical QCD scale, i.e. $M_b \gg \Lambda_{QCD}$, the heavy quark symmetry should be accurate for bottomonium-like systems, such as the $Z_b(10610)$, $Z_b(10650)$ and their spin partners. Both the contact, OPE and OEE potentials between the (B, B^*) and $\left(\bar{B}, \bar{B}^*\right)$ doublets were included within the EFT framework, as formulated in Refs. 2, 3. In addition, the inelastic channels, i.e. the $(\Upsilon(nS)\pi, \eta_b(nS)\pi)$ and $(\chi_{bJ}(mP)\pi, h_b(mP)\pi)$ channels with $n = 1, 2, 3$ and $m = 1, 2$ are also considered dynamically to satisfy the unitarity requirements. Inelastic contributions are taken into account via corrections to the potentials between the elastic channels, as the direct transitions between the inelastic channels can be neglected.[2–5] As the charge parities of the two Z_b and their spin partners are different, the negative C-parity states Z_b can be produced from the initial $\Upsilon(10860)$ through an emission of one pion $\Upsilon(10860) \to \pi Z_b^{(\prime)} \to \pi B^{(*)}\bar{B}^{(*)}$ while their positive C-parity partners, W_{bJ}, require an emission of the photon $\Upsilon(10860) \to \gamma W_{bJ} \to \gamma B^{(*)}\bar{B}^{(*)}$.

With the production amplitudes obtained by solving the partial-wave-decomposed coupled channel Lippmann-Schwinger equation numerically,[2,3] we obtained the invariant mass distributions for the Z_bs and their spin partners. By adjusting the unknown parameters to data in the Z_b channels and employing HQSS we predict the line shapes for the W_{bJ}'s.

3. Results and discussions

Table 1. The pole positions (on the sheet close to the physical one) and the residues in various S-wave $B^{(*)}\bar{B}^{(*)}$ channels for Scheme III. The energy $E_{\rm pole}$ is given relative to the nearest open-bottom threshold quoted in the third column. Uncertainties correspond to a 1σ deviation in the parameters allowed by the fit to the data in the Z_bs' channels. The poles are calculated for the cutoff $\Lambda = 1$ GeV.

J^{PC}	State	Threshold	$E_{\rm pole}$ w.r.t. threshold [MeV]	Residue at $E_{\rm pole}$
1^{+-}	Z_b	$B\bar{B}^*$	$(-2.3 \pm 0.5) - i(1.1 \pm 0.1)$	$(-1.2 \pm 0.2) + i(0.3 \pm 0.2)$
1^{+-}	Z'_b	$B^*\bar{B}^*$	$(1.8 \pm 2.0) - i(13.6 \pm 3.1)$	$(1.5 \pm 0.2) - i(0.6 \pm 0.3)$
0^{++}	W_{b0}	$B\bar{B}$	$(2.3 \pm 4.2) - i(16.0 \pm 2.6)$	$(1.7 \pm 0.6) - i(1.7 \pm 0.5)$
0^{++}	W'_{b0}	$B^*\bar{B}^*$	$(-1.3 \pm 0.4) - i(1.7 \pm 0.5)$	$(-0.9 \pm 0.3) - i(0.3 \pm 0.2)$
1^{++}	W_{b1}	$B\bar{B}^*$	$(10.2 \pm 2.5) - i(15.3 \pm 3.2)$	$(1.3 \pm 0.2) - i(0.4 \pm 0.2)$
2^{++}	W_{b2}	$B^*\bar{B}^*$	$(7.4 \pm 2.8) - i(9.9 \pm 2.2)$	$(0.7 \pm 0.1) - i(0.3 \pm 0.1)$

In what follows, we present three fitting schemes, namely **Scheme I**: purely S-wave momentum-independent contact interactions; **Scheme II**: full OPE potential as well as the $\mathcal{O}(Q^2)$ S-wave-to-D-wave counter term promoted to LO in addition to scheme I; **Scheme III**: the $\mathcal{O}(Q^2)$ S-wave-to-S-wave contact terms at NLO and the η-meson exchange potential on top of scheme II. The fit results of Scheme I agree with those of the parametrization proposed in Refs. 4, 5, which demonstrates that the two approaches are consistent with each other. As expected, the inclusion of the pion dynamics in Scheme II visibly improves the fit once the $\mathcal{O}(Q^2)$ S-wave-to-D-wave counter term is considered simultaneously, which is reflected in the reduction of χ^2/d.o.f. from 1.29 to 0.95, see also Fig. 1. The fit shows that HQSS violating the contact interactions are not required by the data. The transition to Scheme III represents a rather perturbative effect which leads to some further (although small) reduction of χ^2/d.o.f. to 0.83, however, the inclusion of the $\mathcal{O}(Q^2)$ S-wave-to-S-wave contact terms allows one to remove some higher-order regulator artefacts related with the iteration of the truncated potential within the integral equations (see Ref. 2 for the extended discussion).

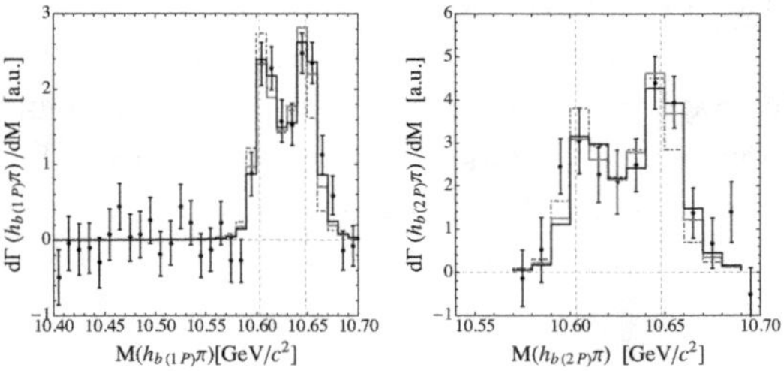

Fig. 1. The fitted line shapes in the 1^{+-} channel in the $h_b(1P)\pi$ and $h_b(2P)\pi$ channels for an illustration.[2] The line shapes of schemes I, II and III are shown by the blue dashed, red thick solid and black solid curves, respectively. The vertical dashed lines indicate the positions of the $B\bar{B}^*$ and $B^*\bar{B}^*$ thresholds. The experimental data are from Ref. 6

Using Scheme III as our final result we predict the line shapes (Fig. 2) and the pole positions (Table 1) of the W_{bJ}s in a parameter-free way. Our

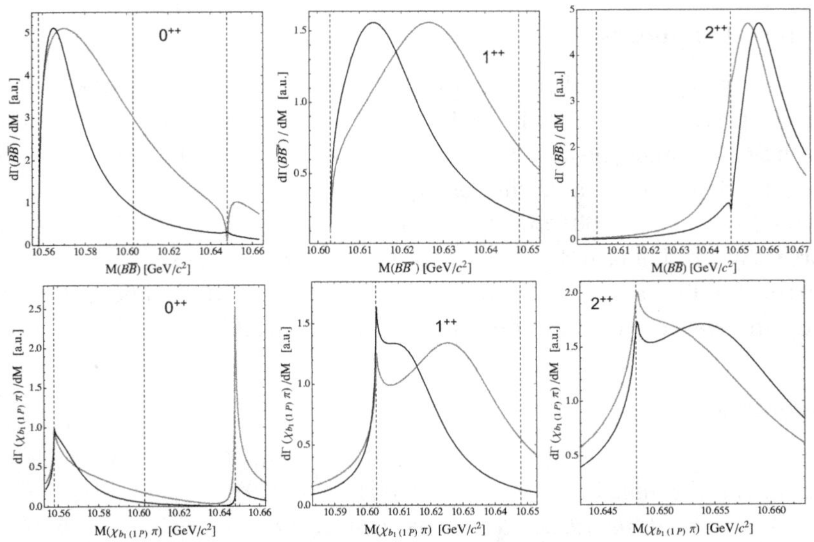

Fig. 2. Predicted line shapes for the 0^{++}, 1^{++} and 2^{++} channels in the corresponding lowest elastic channels and the $\chi_{b1}(1P)\pi$ channels. Upper panel: the line shapes of the $W_{b0}^{(\prime)}$, W_{b1} and W_{b2} in the $B\bar{B}$, $B\bar{B}^*$ and $B^*\bar{B}^*$ channels, respectively. Lower panel: the line shapes of the spin partners W_{bJ} states in the $\chi_{b1}(1P)\pi$ channel. The red and black lines show the results for schemes II and III, respectively, and the vertical dashed lines indicate the position of the $B\bar{B}, B\bar{B}^*$ and $B^*\bar{B}^*$ thresholds.

predictions for the line shapes are strongly asymmetric (either a bump just above threshold or a large sizable distortion at threshold) reflecting the assumed molecular nature of the states. Thus, an experimental conformation of those predictions would provide strong support for the hadronic molecule picture for both groups the W_b-states as well as the Z_b states.

4. Summary

A systematic EFT approach with respect to both chiral and heavy-quark spin symmetries is proposed for the two Z_b states. In our framework, both the short-range contact and the long-range OPE/OEE potentials are considered dynamically, as well as the inelastic channels. With the parameters extracted from the invariant mass distributions of the two Z_b states, the line shapes and pole positions of the spin partners W_{bJ}s are predicted in a parameter-free way. They can be confronted with the results from future high-luminosity and high-statistic experiments, such as Belle-II, to gain insights into the nature of these exotic candidates.

Acknowledgments

This work is supported in part by the National Natural Science Foundation of China (NSFC) and the Deutsche Forschungsgemeinschaft (DFG) through the funds provided to the Sino-German Collaborative Research Center "Symmetries and the Emergence of Structure in QCD" (NSFC Grant No. 11621131001 and DFG Grant No. TRR110). QW is also supported by the Thousand Talents Plan for Young Professionals and research startup funding at SCNU. Work of V.B. and A.N. was supported by the Russian Science Foundation (Grant No. 18-12-00226).

References

1. M. Cleven *et al.*, *Phys. Rev. D* **92**, 014005 (2015).
2. V. Baru, *et al.*, *Phys. Rev. D* **99**, 094013 (2019).
3. Q. Wang, *et al.*, *Phys. Rev. D* **98**, 074023 (2018).
4. F.-K. Guo, *et al.*, *Phys. Rev. D* **93**, 074031 (2016).
5. C. Hanhart, *et al.*, *Phys. Rev. Lett.* **115**, 202001 (2015).
6. A. Bondar *et al.* [Belle Collaboration], *Phys. Rev. Lett.* **108**, 122001 (2012); A. Garmash *et al.* [Belle Collaboration], *Phys. Rev. Lett.* **116**, 212001 (2016).

Nature of $X(3872)$ from the line shape

Xian-Wei Kang

Key Laboratory of Beam Technology of Ministry of Education, College of Nuclear Science and Technology, Beijing Normal University, Beijing 100875, China
Beijing Radiation Center, Beijing 100875, China
E-mail: 11112018023@bnu.edu.cn

J. A. Oller

Departamento de Física, Universidad de Murcia, E-30071 Murcia, Spain
E-mail: oller@um.es

Using a general parameterization of two-body scattering amplitude, we systematically analyze the corresponding data on $X(3872)$, more explicitly, the CDF data on inclusive $p\bar{p}$ scattering to $J/\psi\pi^+\pi^-$, and the Belle and BaBar data on B decays to $K\,J/\psi\pi^+\pi^-$ and $KD\bar{D}^{*0}$ around the $D^0\bar{D}^{*0}$ threshold. We achieve a good reproduction of data and find that the $X(3872)$ can be interpreted as a bound and/or virtual state, or even higher-order (double or triple) virtual sate pole. The latter point was not realized previously in the literature. As a result, the compositeness of the $X(3872)$ can vary largely from nearly 0 to 1. More higher-precision data is needed to discriminate its pole structure and nature.

Keywords: CDD pole; two-body scattering; effective range expansion; line shape, compositeness; bound and virtual state.

1. Introduction

A series of exotic hadron candidates, called XYZ particles because their properties are hard to accommodate or cannot be accounted for by conventional potential models, have been observed. A variety of theoretical interpretations have been proposed, e.g., kinematical effects, molecular states, quark-gluon hybrids, etc. We want to stress that, generally, different scenarios predict different constituents, and in practice, it may involve several mechanisms. In such circumstances the concept of compositeness is a valuable tool to obtain insight on the nature of the resonance. In most cases, those newly observed states stay close to threshold of hadron pairs[1,2] and the effective range expansion (ERE) may be applicable. We have such a good example in Ref. 3, where the scattering length and effective range of

$B\bar{B}^{(*)}$ are calculated by reproducing the pole location of Z_b states, and the molecular component of $B\bar{B}^{(*)}$ pairs are predicted. However, the ERE radius of convergence may be severely reduced in the presence of a Castillejo-Dalitz-Dyson (CDD) pole.[4]

2. Formalism and data analysis

Since the discovery of the $X(3872)$ by the Belle collaboration,[5] a voluminous amount of papers devoted to its study have appeared. Its quantum numbers have been well determined by now as $I^G J^{PC} = 0^+(1^{++})$.[6] Within the ERE analysis, Ref. 7 assigns the $X(3872)$ as a pure $D\bar{D}^*$ molecule, while in Ref. 8 the virtual-state nature is proposed.

A nearby CDD pole around threshold could spoil the applicability of ERE to study the $X(3872)$.[9] In the presence of a CDD pole, the scattering amplitude can be written as[10]

$$t(E) = \left(\frac{\lambda}{E - M_{\rm CDD}} + \beta - ik(E) \right)^{-1}, \tag{1}$$

where the CDD poles lies at $M_{\rm CDD}$, its residue is λ, β is the subtraction constant and $k(E) = \sqrt{2\mu(E + i\frac{\Gamma_*}{2})}$. Throughout the paper, energy E, $M_{\rm CDD}$ and pole position are defined with respect to the $D^0\bar{D}^{*0}$ threshold. The width of D^{*0} meson is $\Gamma_* \approx 65$ keV, and is taken into account in our formalism by employing a complex mass for the D^{*0}. This width may play an important role in a quantitative analysis, since the width of the $X(3872)$, Γ_X, is also very small (smaller than 1.2 MeV as determined by experiment[6]). Removing the extra zero due to the CDD pole, one ends with the function $d(E)$, which accounts for the final-state interactions,

$$d(E) = \frac{1}{1 + \frac{E - M_{\rm CDD}}{\lambda}(\beta - ik)}. \tag{2}$$

Equation (1) can be verified in the case in which cross-channel effect[11] can be disregarded, so that the partial-wave amplitude only has a right-hand cut.

By comparing Eq. (1) with the ERE

$$t(E) = \frac{1}{-1/a + 1/2rk^2 - ik}, \tag{3}$$

one obtains

$$1/a = \frac{\lambda}{M_{\rm CDD}}, \qquad r = -\frac{\lambda}{\mu M_{\rm CDD}^2}. \tag{4}$$

If $M_{\rm CDD}$ is far from threshold, r would have a natural size around 1 fm, and the two meson constituents dominate; otherwise when $M_{\rm CDD}$ is vanishing, so that the CDD pole lies very close to threshold, the compositeness will be a small number, which implies the non-molecule nature.

For the definition of compositeness (X) of a resonance, we adopt the one developed in Ref. 12, adapted to the non-relativistic near-threshold case in Ref. 3. Denoting by g_k^2 the residue in the variable k of the partial-wave amplitude for a pole, either a bound state or resonance, one has that $X = -ig_k^2$ for the former, while $X = |g_k^2|$ for a resonance if $M_R > 0$ (with M_R the mass of the resonance measured from the threshold).

We are now in position to obtain the theoretical formula for the event distributions, which are also the experimentally measured observables. We define the signal function as

$$\frac{d\hat{M}}{dE} = \frac{\Gamma_X |d(E)|^2}{2\pi|\alpha|^2}\,, \tag{5}$$

with α introduced such that in the vicinity of the pole position $d(E) \simeq \alpha/(E - E_p)$. As done in Ref. 7, the background term, in the energy region considered, is treated as a constant or a linear function. The event distribution $N(E)$ is then expressed by the incoherent sum of the signal part and the background. In all the cases, the overall normalization constants, the parameters characterizing the background and the parameters λ, β and $M_{\rm CDD}$, which determine the signal function, are free parameters, fitted to data. In practice, however, there are too many free parameters, which prevents us to extract any definite conclusion by reproducing the experimental data. To reduce the numbers of free parameters, we discuss several interesting scenarios that drive to different nature and pole structures for the $X(3872)$ (which may also represent feasible situations for other near-threshold states). As a result the number of free parameters in $d(E)$ for each case is 1 or 2, so that one can obtain for every scenario stable fits to data. The cases considered are:

Case 1: Using the ERE with only the scattering length, which predicts the $X(3872)$ as a pure composite bound state.[7]

Case 2: Imposing that $t(E)$ (for $\Gamma_* = 0$) has a double-virtual pole state as the limiting zero-width case of a resonance, one can express λ and β from E_R and $M_{\rm CDD}$. That is, there are two free parameters, denoted by Case 2.I (generating two virtual-state poles). Furthermore, if one imposes $M_{\rm CDD} = -3E_R$ then a triple virtual state pole happens for $\Gamma_* = 0$, and this is denoted as Case 2.II.

Case 3: Taking into account the coupled channel effects and imposing independence of the pole position with respect to the threshold of the charged or neutral $D\bar{D}^*$ pairs, so that the self-energy contribution due to re-scattering is necessarily suppressed. Then, the compositeness is very small at the level of only a few percent, which implies the "elementary" (or "bare") nature of this particle. Here, only E_R determines the shape of $t(E)$, but we encounter a quadratic equation with two solutions which give rise to Case 3.I and Case 3.II.

The results that follow by fitting simultaneously all the combined data are shown in Fig. 1. There one can see that all these curves describe data

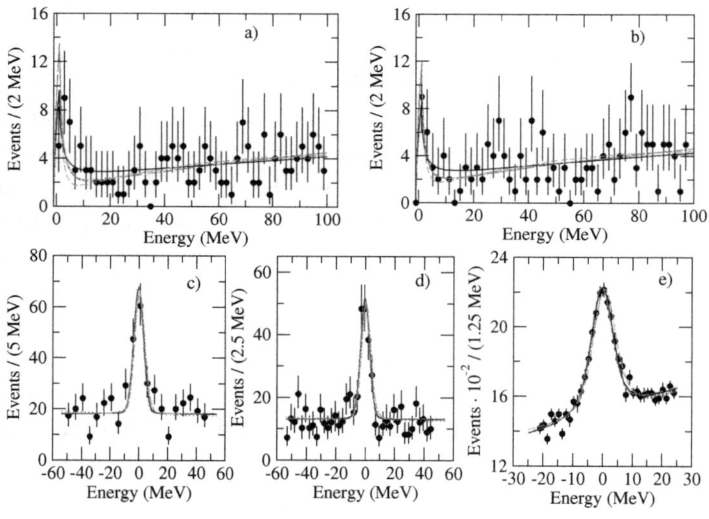

Fig. 1. Fit results for the different event distributions as a function of energy. Panels a) and b) denote the mode $D^0\bar{D}^{*0}$ from BaBar[13] and Belle,[16] respectively. Panels c), d) and e) denote the mode $J/\psi\pi^+\pi^-$ from BaBar,[14] Belle,[15] and CDF[17] collaborations, respectively. The black solid lines correspond to case 1. The CDD pole contribution is included in the expression of $d(E)$, and the rest of lines refer to cases making use of this parameterization. Case 2.I is shown by the red dotted lines, and case 2.II by the brown dashed lines, while cases 3.I and 3.II are given by the blue dash-double-dotted and green dash-dotted lines, in order. Part of the lines overlap so much that it is difficult to distinguish between them in the scale of the plots.

Table 1. Fits for all the cases including the pole positions and compositeness (X).

Case	Pole position [MeV]	X
1	$-0.19^{+0.01}_{-0.01} - i\,0.0325$	1.0
2.I	$-0.36^{+0.08}_{-0.10} - i\,0.18^{+0.01}_{-0.02}$	
	$-0.70^{+0.11}_{-0.13} + i\,0.17^{+0.02}_{-0.01}$	
2.II	$-0.33^{+0.04}_{-0.03} - i\,0.31^{+0.02}_{-0.01}$	
	$-0.84^{+0.07}_{-0.05} + i\,0.77^{+0.03}_{-0.04}$	
	$-1.67^{+0.10}_{-0.08} - i\,0.49^{+0.02}_{-0.02}$	
3.I	$-0.50^{+0.04}_{-0.03}$	$0.061^{+0.003}_{-0.002}$
	$-0.68^{+0.05}_{-0.03}$	
3.II	$-0.51^{+0.03}_{-0.01}$	$0.158^{+0.001}_{-0.001}$
	$-1.06^{+0.05}_{-0.02}$	

almost equally well. The resulting pole locations are given in Table 1, indicating the compositeness X for those cases in which we can calculate it, as discussed above.

3. Conclusion and discussion

We propose to use a more general parameterization of a two-body scattering amplitude around threshold. With this advanced tool, we derive the formula for the event distributions with the free parameters obtained by fitting to data, which is well reproduced. The pole structures are analyzed, and it turns out that the $X(3872)$ can be interpreted as either bound or virtual state, or even as an "elementary" state. That is, the current data based line shapes is not enough for determining the properties of the $X(3872)$. Let us notice especially the Case 2, which implies that the $X(3872)$ stems from a double- or triple-virtual state pole. More higher-precision data or other observables are highly desired to further discriminate between those scenarios.

References

1. L. Y. Dai, J. Haidenbauer and U. G. Meißner, Phys. Rev. D **96**, no. 11, 116001 (2017) [arXiv:1710.03142 [hep-ph]].
2. L. Y. Dai, M. Shi, G. Y. Tang and H. Q. Zheng, Phys. Rev. D **92**, 014020 (2015), [arXiv:1206.6911 [hep-ph]].
3. X. W. Kang, Z. H. Guo and J. A. Oller, Phys. Rev. D **94**, no. 1, 014012 (2016) [arXiv:1603.05546 [hep-ph]].
4. L. Castillejo, R. H. Dalitz and F. J. Dyson, Phys. Rev. **101**, 453 (1956).

5. S. K. Choi *et al.* [Belle Collaboration], Phys. Rev. Lett. **91**, 262001 (2003) [hep-ex/0309032].
6. C. Patrignani *et al.* (Particle Data Group), Chin. Phys. C **40**, 100001 (2016).
7. E. Braaten and J. Stapleton, Phys. Rev. D **81**, 014019 (2010) [arXiv:0907.3167 [hep-ph]].
8. C. Hanhart, Y. S. Kalashnikova, A. E. Kudryavtsev and A. V. Nefediev, Phys. Rev. D **76**, 034007 (2007) [arXiv:0704.0605 [hep-ph]].
9. C. Hanhart, Y. S. Kalashnikova and A. V. Nefediev, Eur. Phys. J. A **47**, 101 (2011) [arXiv:1106.1185 [hep-ph]].
10. X. W. Kang and J. A. Oller, Eur. Phys. J. C **77**, no. 6, 399 (2017) [arXiv:1612.08420 [hep-ph]].
11. J. A. Oller and E. Oset, Phys. Rev. D **60**, 074023 (1999) [hep-ph/9809337].
12. Z. H. Guo and J. A. Oller, Phys. Rev. D **93**, no. 9, 096001 (2016) [arXiv:1508.06400 [hep-ph]].
13. B. Aubert *et al.* [BaBar Collaboration], Phys. Rev. D **77**, 011102 (2008) [arXiv:0708.1565 [hep-ex]].
14. B. Aubert *et al.* [BaBar Collaboration], Phys. Rev. D **77**, 111101 (2008) [arXiv:0803.2838 [hep-ex]].
15. I. Adachi *et al.* [Belle Collaboration], arXiv:0809.1224 [hep-ex].
16. T. Aushev *et al.* [Belle Collaboration], Phys. Rev. D **81**, 031103 (2010) [arXiv:0810.0358 [hep-ex]].
17. T. Aaltonen *et al.* [CDF Collaboration], Phys. Rev. Lett. **103**, 152001 (2009) [arXiv:0906.5218 [hep-ex]].

Understanding $X(3872)$ and its decays in the extended Friedrichs scheme

Meng-Ting Yu and Zhi-Yong Zhou*

School of Physics, Southeast University, Nanjing 211189, People's Republic of China
**E-mail: zhouzhy@seu.edu.cn*

Zhiguang Xiao

Interdisciplinary Center for Theoretical Study, University of Science and Technology of China, Hefei, Anhui 230026, China
E-mail: xiaozg@ustc.edu.cn

We present that the $X(3872)$ could be represented as a dynamically generated state in the extended Friedrichs scheme, in which the ratio of "elementariness" and "compositeness" of the different components in the $X(3872)$ is about $Z_{c\bar{c}}$: $X_{\bar{D}^0D^{0*}}$: $X_{D^+D^{-*}}$: $X_{\bar{D}^*D^*}$ = 1 : (2.67 ∼ 8.85) : (0.45 ∼ 0.46) : 0.04. Furthermore, its decays to π^0 and a P-wave charmonium χ_{cJ} state with $J =$ 0, 1, or 2, $J/\psi\pi^+\pi^-$, and $J/\psi\pi^+\pi^-\pi^0$ could be calculated out with the help of Barnes-Swanson model. The isospin breaking effects is easily understood in this scheme. This calculation also shows that the decay rate of $X(3872)$ to $\chi_{c1}\pi^0$ is much smaller than its decay rate to $J/\psi\pi^+\pi^-$.

Keywords: Friedrichs model; exotic state; charmonium; molecular state.

1. Introduction

The $X(3872)$ state, first discovered by Belle,[1] has several enigmatic properties including its mass, narrow width, isospin breaking effects in $X(3872) \to J/\psi\pi^+\pi^-$ and $J/\psi\pi^+\pi^-\pi^0$,[2] as discussed in the literature. These properties of $X(3872)$ might be explained by describing it as a dynamically generated state due to the coupling of the bare $\chi_{c1}(2P)$ and $D\bar{D}^*$, $D^*\bar{D}^*$ continuum states in the extended Friedrichs scheme.[3,4] Recently, BESIII searched for the $X(3872)$ signals in $e^+e^- \to \gamma\chi_{cJ}\pi^0$ $(J = 0, 1, 2)$ and reported an observation of $X(3872) \to \chi_{c1}\pi^0$ with a ratio of branching fractions[5] $\frac{\mathcal{B}(X(3872)\to\chi_{c1}\pi^0)}{\mathcal{B}(X(3872)\to J/\psi\pi^+\pi^-)} = 0.88^{+0.33}_{-0.27} \pm 0.10$. Soon after, Belle made a search for $X(3872)$ in $B^+ \to \chi_{c1}\pi^0K^+$ but only reported an upper limit[6] $\frac{\mathcal{B}(X(3872)\to\chi_{c1}\pi^0)}{\mathcal{B}(X(3872)\to J/\psi\pi^+\pi^-)} < 0.97$ at 90% confidence level. We will present that by using the wave function of $X(3872)$ from the extended Friedrichs scheme and the Barnes-Swanson constituent interchange model,[7] the decays of $X(3872)$ to different hadronic final states could also be obtained.

2. Wave function of $X(3872)$ in the extended Friedrichs scheme

In 1948, Friedrichs proposed an exactly solvable model to understand an unstable state,[8] in which the free Hamiltonian H_0 has a bare continuous spectrum $[E_{th}, \infty)$, and a discrete eigenvalue E_0 imbedded in this continuous spectrum ($E_0 > E_{th}$). The interaction Hamiltonian V couples the bare continuous state and the discrete state of H_0 such that the discrete state is dissolved in the continuous state and a resonance is produced.

This model was developed to a general form with many discrete and continuous states,[9] and furthermore, some general properties of the Friedrichs model, such as the wave functions of resonance and the completeness relations, are obtained.[3,10] The free Hamiltonian H_0 can be expressed as $H = E_0|0\rangle\langle 0| + \sum_{n,S,L}\int_{E_{th,n}}^{\infty} \mathrm{d}E\, E|E;n,SL\rangle\langle E;n,SL|$ and the interaction Hamiltonian $V = \sum_{n,S,L}\int_{E_{th,n}}^{\infty} \mathrm{d}E f_{SL}^{n}(E)|0\rangle\langle E;n,SL| + h.c.$, where E_0 denotes the bare mass of the discrete state, n the "n" continuum state, $E_{th,n}$ the energy threshold of related continuum state, S and L the total spin and the angular momentum of the continuum states, f_{SL}^{n} the coupling functions between the bare state and the "n" continuum state with particular S, L quantum numbers. The eigenvalue problem of the full Hamiltonian $H = H_0 + V$ is exactly solvable as mentioned above. The eigenvalues of bound states, virtual states or resonant states could be found by solving $\eta(z) = 0$ on the complex energy plane where $\eta(z) = z - E_0 - \sum_n \int_{E_{th,n}}^{\infty} \frac{\sum_{S,L}|f_{SL}^{n}(E)|^2}{z-E}\mathrm{d}E$. The form factor f_{SL} is provided by some particular model, such as the quark pair creation model.[4,11] By using the results from the famous Godfrey-Isgur model[12] as the input, the $X(3872)$ state is dynamically generated by the coupling between the bare discrete $\chi_{c1}(2P)$ state and the continuum $D\bar{D}^*$ and $D^*\bar{D}^*$ states,[4] and its wave function could be explicitly written down as

$$|X\rangle = N_B\Big(|c\bar{c}\rangle + \int_{M_{00}}^{\infty} \mathrm{d}E \sum_{l,s} \frac{f_{ls}^{00}(E)}{z_X - E}(|E\rangle_{ls}^{D^0\bar{D}^{0*}} + C.C.)$$
$$+ \int_{M_{+-}}^{\infty} \mathrm{d}E \sum_{l,s} \frac{f_{ls}^{+-}(E)}{z_X - E}(|E\rangle_{ls}^{D^+D^{-*}} + C.C.) + \cdots\Big), \tag{1}$$

where $C.C.$ means the corresponding charge conjugate state, $|c\bar{c}\rangle$ denotes the bare $\chi_{c1}(2P)$ state and $|E\rangle^n_{ls} = \sqrt{\mu k}|k, j\sigma, ls\rangle$ denotes the two-particle "n" state with the reduced mass μ, the magnitude of one-particle three-momentum k in their $c.m.$ frame, total spin s, relative orbital angular momentum l, total angular momentum j, and its third component σ. Based on the probability explanation of the wave function, one could obtain the "elementariness" and "compositeness" of $X(3872)$ as $Z_{c\bar{c}} : X_{\bar{D}^0 D^{0*}} : X_{D^+ D^{-*}} : X_{\bar{D}^* D^*} = 1 : (2.67 \sim 8.85) : (0.45 \sim 0.46) : 0.04$, which means $\bar{D}^0 D^{0*}$ component is dominant in the $X(3872)$ state.

3. Decays of $X(3872)$

In calculating the decays of the $X(3872)$ to $\chi_{cJ}\pi^0$ for $J = 0, 1, 2$, the partial-wave decay amplitude reads

$$\begin{aligned} F_{l's'} &= {}_{l's'}\langle \chi_{cJ}\pi^0 | H_I | X(3872)\rangle = N_B \Big({}^{\chi_{cJ}\pi^0}_{l's'}\langle E' | H_I | c\bar{c}\rangle \\ &+ \int_{M_{00}}^{\infty} \mathrm{d}E \sum_{l,s} \frac{f^{00}_{ls}(E)}{z_X - E} \big({}^{\chi_{cJ}\pi^0}_{l's'}\langle E' | H_I | E\rangle^{D^0 \bar{D}^{0*}}_{ls} + C.C. \big) \\ &+ \int_{M_{+-}}^{\infty} \mathrm{d}E \sum_{l,s} \frac{f^{+-}_{ls}(E)}{z_X - E} \big({}^{\chi_{cJ}\pi^0}_{l's'}\langle E' | H_I | E\rangle^{D^+ D^{-*}}_{ls} + C.C. \big) + \cdots \Big). \end{aligned} \tag{2}$$

Once the matrix elements for $D\bar{D}^* \to \chi_{cJ}\pi^0$ are obtained, the partial decay widths and branching ratios are easily calculated directly.

The scattering amplitude of $D\bar{D}^* \to \chi_{cJ}\pi^0$ could be calculated by using the Barnes-Swanson model,[7,13] in which the meson meson scatterings are induced by the quark (antiquark)-quark (antiquark) interactions. Such interactions are supposed to come from the one gluon exchange (OGE) interaction and the confinement interaction. Four kinds of diagrams are considered, among which the quark-antiquark interactions are denoted as C_1, C_2, and the quark-quark(antiquark-antiquark) interactions are denoted as T_1, and T_2. The other four "post" diagrams are also considered similarly and averaged to reduce the so-called "prior-post" ambiguity.

The partial wave scattering amplitude for each diagram with only meson C being a P-wave state can be obtained by using

$$\begin{aligned} \mathcal{M}^1_{l'j_C, lj_B} &= \sqrt{\mu k \mu' k'} \sum_{mm'm_{l_C}} \langle j_B - m l m | 10\rangle \langle j_C - m' l' m' | 10\rangle \\ &\times \langle l_C m_{l_C} s_C (-m' - m_{l_C}) | j_C - m'\rangle \langle \phi_{14}\phi_{32} | \phi_{12}\phi_{34}\rangle \langle \omega_{14}\omega_{32} | H_C | \omega_{12}\omega_{34}\rangle \\ &\times \int d\Omega_k \int d\Omega_{k'} \langle \chi_C \chi_D | I^{m_{l_C}}_{Space}[\vec{k}, \vec{k}'] | \chi_A \chi_B\rangle Y^m_l(\hat{k}) Y^{m'*}_{l'}(\hat{k}') \end{aligned} \tag{3}$$

where $\langle\phi_{14}\phi_{32}|\phi_{12}\phi_{34}\rangle$ is the flavor factor, and $\langle\omega_{14}\omega_{32}|H_C|\omega_{12}\omega_{34}\rangle$ the color factor, which is $-4/9$ and $4/9$ for interactions of $q\bar{q}$ and qq respectively. χ_A represents the spin wave function of meson A. The spatial integral is

$$I^{m_{l_C}}_{Space}[\vec{k},\vec{k}'] = \int d^3p \int d^3q \psi^A_{000}(\vec{p}_A)\psi^B_{000}(\vec{p}_B)\psi^{C*}_{01m_{l_C}}(\vec{p}_C)\psi^{D*}_{000}(\vec{p}_D)T_{fi} \quad (4)$$

where $\psi_{n_rLm_L}(\vec{p}_r)$ is the wave function for the bare meson state, with n_r being the radial quantum number, L the relative angular momentum of the quark and anti-quark, m_L its third component, $\vec{p}_r$ the relative momentum of quark and antiquark in the meson. The quark interactions of spin-spin, color Coulomb, linear, OGE spin-orbit, linear spin-orbit, and OGE tensor terms are respectively expressed as

$$T_{fi} = \begin{cases} -\frac{8\pi\alpha_s}{3m_1m_2}[\vec{S}_1\cdot\vec{S}_2] \\ \frac{4\pi\alpha_s}{q^2}I \\ \frac{6\pi b}{q^4}I \\ \frac{4i\pi\alpha_s}{q^2}\{\vec{S}_1\cdot[\vec{q}\times(\frac{\vec{p}_1}{2m_1^2}-\frac{\vec{p}_2}{m_1m_2})]+\vec{S}_2\cdot[\vec{q}\times(\frac{\vec{p}_1}{m_1m_2}-\frac{\vec{p}_2}{2m_2^2})]\} \\ -\frac{3i\pi b}{q^4}[\frac{1}{m_1^2}\vec{S}_1\cdot(\vec{q}\times\vec{p}_1)-\frac{1}{m_2^2}\vec{S}_2\cdot(\vec{q}\times\vec{p}_2)] \\ \frac{4\pi\alpha_s}{m_1m_2q^2}[\vec{S}_1\cdot\vec{q}\vec{S}_2\cdot\vec{q}-\frac{1}{3}q^2\vec{S}_1\cdot\vec{S}_2] \end{cases} \quad (5)$$

where $\alpha_s = \sum_k \alpha_k e^{-\gamma_k q^2}$ is the same parametrization of the strong coupling as the in the GI model and m_1 and m_2 are the masses of the two interacting quarks.

The decay amplitude of $X(3872) \to J/\psi\rho$ and $J/\psi\omega$ is much simpler because there is only S-wave states involved in the scattering amplitudes.

Finally, we find that the decay rates of $X(3872)$ to $\chi_{cJ}\pi^0$ for $J = 0,1,2$ turn out to be one order of magnitude smaller than that to $J/\psi\pi^+\pi^-$. Our result is smaller than the central value measured by BESIII,[5] but we noticed that the result of BESIII has sizable uncertainties, and more data are needed to increase the statistics and reduce the error bar. The ratio $\frac{\mathcal{B}(X(3872)\to J/\psi\pi^+\pi^-\pi^0)}{\mathcal{B}(X(3872)\to J/\psi\pi^+\pi^-)}$ in our calculation is about 1.6,[14] which is consistent with our previous calculation[15] and the experiment measurements.[2]

Ackownledgment

This work is supported by China National Natural Science Foundation under contract No. 11975075, No. 11575177, and No. 11105138, and the Natural Science Foundation of Jiangsu Province of China under contract No. BK20171349.

References

1. S. K. Choi *et al.*, Observation of a narrow charmonium-like state in exclusive $B^{\pm} \to K^{\pm}\pi^{+}\pi^{-}J/\psi$ decays, *Phys. Rev. Lett.* **91**, 262001 (2003).
2. P. del Amo Sanchez *et al.*, Evidence for the decay X(3872) $\to J/\psi\omega$, *Phys. Rev.* **D82**, 011101 (2010).
3. Z. Xiao and Z.-Y. Zhou, On Friedrichs Model with Two Continuum States, *J. Math. Phys.* **58**, 062110 (2017).
4. Z.-Y. Zhou and Z. Xiao, Understanding $X(3862)$, $X(3872)$, and $X(3930)$ in a Friedrichs-model-like scheme, *Phys. Rev.* **D96**, 054031 (2017), [Erratum: Phys. Rev. D 96, 099905 (2017)].
5. M. Ablikim *et al.*, Observation of the decay $X(3872) \to \pi^0\chi_{c1}(1P)$, *Phys. Rev. Lett.* **122**, 202001 (2019).
6. V. Bhardwaj *et al.*, Search for X(3872) and X(3915) decay into $\chi_{c1}\pi^0$ in B decays at Belle, *Phys. Rev. D* **99**, 111101(R) (2019).
7. T. Barnes and E. S. Swanson, A Diagrammatic approach to meson meson scattering in the nonrelativistic quark potential model, *Phys. Rev.* **D 46**, 131 (1992).
8. K. O. Friedrichs, On the perturbation of continuous spectra, *Commun. Pure Appl. Math.* **1**, 361 (1948).
9. Z. Xiao and Z.-Y. Zhou, Partial Wave Decomposition in Friedrichs Model With Self-interacting Continua, *J. Math. Phys.* **58**, 072102 (2017).
10. Z. Xiao and Z.-Y. Zhou, Virtual states and the generalized completeness relation in the Friedrichs model, *Phys. Rev.* **D 94**, 076006 (2016).
11. L. Micu, Decay rates of meson resonances in a quark model, *Nucl. Phys.* **B10**, 521 (1969).
12. S. Godfrey and N. Isgur, Mesons in a Relativized Quark Model with Chromodynamics, *Phys. Rev.* **D 32**, 189 (1985).
13. T. Barnes, N. Black and E. S. Swanson, Meson meson scattering in the quark model: Spin dependence and exotic channels, *Phys. Rev.* **C63**, 025204 (2001).
14. Z.-Y. Zhou, M.-T. Yu and Z. Xiao, On decays of $X(3872)$ to $\chi_{cJ}\pi^0$ and $J/\psi\pi^+\pi^-$, *Phys. Rev.* **D100**, 094025 (2019).
15. Z.-Y. Zhou and Z. Xiao, Comprehending Isospin breaking effects of $X(3872)$ in a Friedrichs-model-like scheme, *Phys. Rev.* **D97**, 034011 (2018).

Exotica in QCD sum rules

Rui-Rui Dong
School of Physics, Beihang University, Beijing 100191, China

Hua-Xing Chen
School of Physics, Beihang University, Beijing 100191, China
School of Physics, Southeast University, Nanjing 210094, China
E-mail: hxchen@buaa.edu.cn

Wei Chen
School of Physics, Sun Yat-Sen University, Guangzhou 510275, China
E-mail: chenwei29@mail.sysu.edu.cn

Xiang Liu
School of Physical Science and Technology, Lanzhou University, Lanzhou 730000, China
E-mail: xiangliu@lzu.edu.cn

Shi-Lin Zhu
School of Physics and State Key Laboratory of Nuclear Physics and Technology, Peking University, Beijing 100871, China
E-mail: zhusl@pku.edu.cn

We use the method of QCD sum rules to study $ss\bar{s}\bar{s}$ tetraquark states of $J^{PC} = 1^{--}$ and $J^{PC} = 1^{+-}$. Our results suggest that the $Y(2175)$ can be interpreted as an $ss\bar{s}\bar{s}$ state of $J^{PC} = 1^{--}$, and it has a partner state, whose mass is $71^{+172}_{-\ 48}$ MeV larger.

Keywords: Exotic hadrons; QCD sum rules; interpolating currents.

1. Introduction

In recent years there have been many exotic hadrons observed in particle experiments,[1] which can not be explained in the traditional quark model. They are of particular importance to understand the low energy behaviours of Quantum Chromodynamics, and have attracted much attention from the hadron physics community.[2]

The $Y(2175)$ is one of them. It was first observed in 2006 by the BaBar Collaboration in the $\phi f_0(980)$ invariant mass spectrum,[3,4] and later confirmed in the Belle,[5] BESII,[6] and BESIII[7] experiments. There are many possible interpretations for the $Y(2175)$, such as a 2^3D_1 $s\bar{s}$ meson,[8,9] an $ss\bar{s}\bar{s}$ state of $J^{PC} = 1^{--}$,[10] a dynamically generated state in the $\phi K\bar{K}$ and $\phi\pi\pi$ systems,[11] and a hidden-strangeness baryon-antibaryon state,[12] etc.

Recently, the BESIII Collaboration reported their observation of another structure X in the $\phi\eta'$ mass spectrum of the $J/\psi \to \phi\eta\eta'$ decay in 2.0-2.1 GeV region.[13] Its spin-parity quantum number was determined to be either $J^P = 1^-$ or 1^+. One possible theoretical explanation is to interpret it as the second radial excitation of the $h_1(1380)$ with $I(J^P) = 0(1^+)$.[14]

The method of QCD sum rules has also been applied to study the above two structures,[15–18] which method has been widely and successfully applied to study hadron properties. When using this method to investigate a physical state, we need to construct the relevant interpolating current, but we still do not fully understand their relation. To better understand this, a good subject is to study the $Y(2175)$ of $J^{PC} = 1^{--}$.

Theoretically, the $ss\bar{s}\bar{s}$ interpolating currents of $J^{PC} = 1^{--}$ have been systematically constructed in Ref. 15, and there are only two independent ones. Experimentally, besides the $Y(2175)$, there might be another structure in the $\phi f_0(980)$ mass spectrum at around 2.4 GeV, whose evidences have been found in the BaBar,[4] Belle,[5] BESII,[6] and BESIII[7] experiments. Hence, it is quite interesting to study $ss\bar{s}\bar{s}$ tetraquark states of $J^{PC} = 1^{--}$. At the same time we shall also study $ss\bar{s}\bar{s}$ tetraquark states of $J^{PC} = 1^{+-}$.

2. $ss\bar{s}\bar{s}$ tetraquark states of $J^{PC} = 1^{--}$

There are two independent $[ss][\bar{s}\bar{s}]$ interpolating currents of $J^{PC} = 1^{--}$[15] (more discussions can be found in Ref. 19):

$$\begin{aligned}\eta_{1\mu} &= (s_a^T C\gamma_5 s_b)(\bar{s}_a\gamma_\mu\gamma_5 C\bar{s}_b^T) - (s_a^T C\gamma_\mu\gamma_5 s_b)(\bar{s}_a\gamma_5 C\bar{s}_b^T)\,,\\ \eta_{2\mu} &= (s_a^T C\gamma^\nu s_b)(\bar{s}_a\sigma_{\mu\nu} C\bar{s}_b^T) - (s_a^T C\sigma_{\mu\nu} s_b)(\bar{s}_a\gamma^\nu C\bar{s}_b^T)\,,\end{aligned} \tag{1}$$

where C is the charge-conjugation operator, a and b are color indices, and the sum over repeated indices is taken.

In Ref. 15 we separately use $\eta_{1\mu}$ and $\eta_{2\mu}$ to perform QCD sum rule analyses by calculating the diagonal terms:

$$\langle 0|T\eta_{1\mu}(x)\eta_{1\nu}^\dagger(0)|0\rangle \quad \text{and} \quad \langle 0|T\eta_{2\mu}(x)\eta_{2\nu}^\dagger(0)|0\rangle\,. \tag{2}$$

Although $\eta_{1\mu}$ and $\eta_{2\mu}$ are independent of each other, they may still correlate

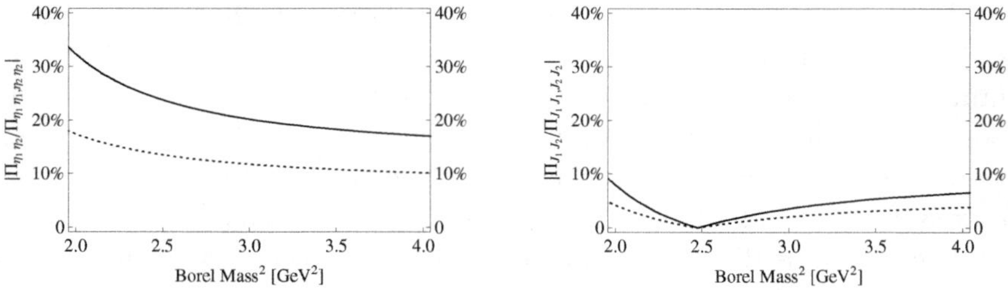

Fig. 1. $\left|\frac{\Pi_{\eta_1\eta_2}(M_B^2)}{\Pi_{\eta_1\eta_1}(M_B^2)}\right|$ (left-solid), $\left|\frac{\Pi_{\eta_1\eta_2}(M_B^2)}{\Pi_{\eta_2\eta_2}(M_B^2)}\right|$ (left-dotted), $\left|\frac{\Pi_{J_1J_2}(M_B^2)}{\Pi_{J_1J_1}(M_B^2)}\right|$ (right-solid), and $\left|\frac{\Pi_{J_1J_2}(M_B^2)}{\Pi_{J_2J_2}(M_B^2)}\right|$ (right-dotted), as functions of the Borel mass. Taken from Ref. 16.

with each other. To verify this, we further calculate the off-diagonal term:

$$\langle 0|T\eta_{1\mu}(x)\eta_{2\nu}^\dagger(0)|0\rangle \,, \tag{3}$$

and find it to be nonzero, as shown in the left panel of Fig. 1.[16]

By diagonalizing the matrix composed of Eqs. (2) and (3), we obtain two new currents $J_{1\mu}$ and $J_{2\mu}$:

$$\begin{aligned} J_{1\mu} &= \cos\theta\ \eta_{1\mu} + \sin\theta\ i\ \eta_{2\mu}\,, \\ J_{2\mu} &= \sin\theta\ \eta_{1\mu} + \cos\theta\ i\ \eta_{2\mu}\,. \end{aligned} \tag{4}$$

After choosing the mixing angle $\theta = -5.0^\circ$, they do not strongly correlate with each other, as shown in the right panel of Fig. 1. Hence, $J_{1\mu}$ and $J_{2\mu}$ can not strongly couple to the same physical state Y, and we assume

$$\langle 0|J_{1\mu}|Y_1\rangle = f_1\epsilon_\mu\,, \quad \text{and} \quad \langle 0|J_{2\mu}|Y_2\rangle = f_2\epsilon_\mu\,, \tag{5}$$

where Y_1 and Y_2 are two different states with $J^{PC} = 1^{--}$, and f_1 and f_2 are their relevant decay constants.

We separately use $J_{1\mu}$ and $J_{2\mu}$ to perform QCD sum rule analyses, and the masses are extracted to be

$$\begin{aligned} M_{Y_1} &= 2.41 \pm 0.25 \text{ GeV}\,, \\ M_{Y_2} &= 2.34 \pm 0.17 \text{ GeV}\,. \end{aligned} \tag{6}$$

3. $ss\bar{s}\bar{s}$ tetraquark states of $J^{PC} = 1^{+-}$

There are also two independent $[ss][\bar{s}\bar{s}]$ currents of $J^{PC} = 1^{+-}$:[17]

$$\begin{aligned} \xi_{1\mu} &= (s_a^T C s_b)(\bar{s}_a\gamma_\mu\gamma_5 C\bar{s}_b^T) - (s_a^T C\gamma_\mu\gamma_5 s_b)(\bar{s}_a C\bar{s}_b^T)\,, \\ \xi_{2\mu} &= (s_a^T C\gamma^\nu s_b)(\bar{s}_a\sigma_{\mu\nu}\gamma_5 C\bar{s}_b^T) - (s_a^T C\sigma_{\mu\nu}\gamma_5 s_b)(\bar{s}_a\gamma^\nu C\bar{s}_b^T)\,. \end{aligned} \tag{7}$$

We use them to perform QCD sum rule analyses, but only one of them, $\xi_{2\mu}$, leads to reliable sum rule results, and the mass is extracted to be

$$M_{\xi_2} = 2.00^{+0.10}_{-0.09} \text{ GeV}. \tag{8}$$

4. Summary and discussions

We use the method of QCD sum rules to study $ss\bar{s}\bar{s}$ tetraquark states of $J^{PC} = 1^{--}$ and $J^{PC} = 1^{+-}$.[15–17] There are two independent $ss\bar{s}\bar{s}$ currents of $J^{PC} = 1^{--}$. We calculate both their diagonal and off-diagonal correlation functions, and obtain two new currents which do not strongly correlate with each other: one of them couples to the $Y(2175)$, while the other may couple to another state at around 2.4 GeV, whose evidences can be found in the BaBar,[4] Belle,[5] BESII,[6] and BESIII[7] experiments.

There are also two independent $ss\bar{s}\bar{s}$ currents of $J^{PC} = 1^{+-}$. We use them to perform QCD sum rule analyses, but only one of them leads to reliable QCD sum rule results. The mass is extracted from this current to be $2.00^{+0.10}_{-0.09}$ GeV, suggesting that the structure X, newly observed by BESIII in the $\phi\eta'$ mass spectrum in 2.0-2.1 GeV region,[13] can be interpreted as an $ss\bar{s}\bar{s}$ tetraquark state of $J^{PC} = 1^{+-}$.

Acknowledgments

This project is supported by the National Natural Science Foundation of China under Grants No. 11722540, the China National Funds for Distinguished Young Scientists under Grant No. 11825503, the National Program for Support of Top-notch Young Professionals, and the Fundamental Research Funds for the Central Universities.

References

1. C. Patrignani *et al.* [Particle Data Group], Chin. Phys. C **40**, 100001 (2016).
2. Y. R. Liu, H. X. Chen, W. Chen, X. Liu and S. L. Zhu, Prog. Part. Nucl. Phys. **107**, 237 (2019); M. Karliner, J. L. Rosner and T. Skwarnicki, Ann. Rev. Nucl. Part. Sci. **68**, 17 (2018); F. K. Guo, C. Hanhart, U. G. Meissner, Q. Wang, Q. Zhao, and B. S. Zou, Rev. Mod. Phys. **90**, 015004 (2018); S. L. Olsen, T. Skwarnicki, and D. Zieminska, Rev. Mod. Phys. **90**, 015003 (2018); N. Brambilla, S. Eidelman, C. Hanhart, A. Nefediev, C. P. Shen, C. E. Thomas, A. Vairo and C. Z. Yuan, arXiv:1907.07583 [hep-ex].

3. B. Aubert *et al.* [BaBar Collaboration], Phys. Rev. D **74**, 091103 (2006).
4. B. Aubert *et al.* [BaBar Collaboration], Phys. Rev. D **76**, 012008 (2007).
5. C. P. Shen *et al.* [Belle Collaboration], Phys. Rev. D **80**, 031101 (2009).
6. M. Ablikim *et al.* [BES Collaboration], Phys. Rev. Lett. **100**, 102003 (2008).
7. M. Ablikim *et al.* [BESIII Collaboration], Phys. Rev. D **91**, 052017 (2015).
8. G. J. Ding and M. L. Yan, Phys. Lett. B **657**, 49 (2007).
9. C. Q. Pang, Phys. Rev. D **99**, 074015 (2019).
10. Q. F. Lü, K. L. Wang and Y. B. Dong, arXiv:1903.05007 [hep-ph].
11. A. Martinez Torres, K. P. Khemchandani, L. S. Geng, M. Napsuciale and E. Oset, Phys. Rev. D **78**, 074031 (2008); M. Napsuciale, E. Oset, K. Sasaki and C. A. Vaquera-Araujo, Phys. Rev. D **76**, 074012 (2007); L. Alvarez-Ruso, J. A. Oller and J. M. Alarcon, Phys. Rev. D **80**, 054011 (2009); S. Coito, G. Rupp and E. van Beveren, Phys. Rev. D **80**, 094011 (2009).
12. M. Abud, F. Buccella and F. Tramontano, Phys. Rev. D **81**, 074018 (2010); L. Zhao, N. Li, S. L. Zhu and B. S. Zou, Phys. Rev. D **87**, 054034 (2013).
13. M. Ablikim *et al.* [BESIII Collaboration], Phys. Rev. D **99**, 112008 (2019).
14. L. M. Wang, S. Q. Luo and X. Liu, arXiv:1901.00636 [hep-ph].
15. H. X. Chen, X. Liu, A. Hosaka and S. L. Zhu, Phys. Rev. D **78**, 034012 (2008).
16. H. X. Chen, C. P. Shen and S. L. Zhu, Phys. Rev. D **98**, 014011 (2018).
17. E. L. Cui, H. M. Yang, H. X. Chen, W. Chen and C. P. Shen, Eur. Phys. J. C **79**, 232 (2019).
18. Z. G. Wang, Nucl. Phys. **A 791**, 106 (2007); Z. G. Wang, arXiv:1901.04815 [hep-ph]; K. Azizi, S. S. Agaev and H. Sundu, arXiv:1906.04061 [hep-ph].
19. H. X. Chen, A. Hosaka and S. L. Zhu, Phys. Rev. D **74**, 054001 (2006); H. X. Chen, A. Hosaka and S. L. Zhu, Phys. Rev. D **76**, 094025 (2007); H. X. Chen, V. Dmitrasinovic, A. Hosaka, K. Nagata and S. L. Zhu, Phys. Rev. D **78**, 054021 (2008); H. X. Chen, A. Hosaka and S. L. Zhu, Phys. Rev. D **78**, 054017 (2008); H. X. Chen, Eur. Phys. J. C **73**, 2628 (2013); H. X. Chen, W. Chen, X. Liu and S. L. Zhu, Phys. Rept. **639**, 1 (2016).

Decoding the nature of the pentaquark states from LHCb

Meng-Lin Du,[1,*] Vadim Baru,[1,2,3] Feng-Kun Guo,[4,5] Christoph Hanhart,[6]
Ulf-G. Meißner,[1,6,7] José A. Oller,[8] and Qian Wang[9,10]

[1] *Helmholtz-Institut für Strahlen- und Kernphysik and Bethe Center for Theoretical Physics, Universität Bonn, D-53115 Bonn, Germany*
[2] *Institute for Theoretical and Experimental Physics, B. Cheremushkinskaya 25, 117218 Moscow, Russia*
[3] *P.N. Lebedev Physical Institute of the Russian Academy of Sciences, 119991, Leninskiy Prospect 53, Moscow, Russia*
[4] *CAS Key Laboratory of Theoretical Physics, Institute of Theoretical Physics, Chinese Academy of Sciences, Zhong Guan Cun East Street 55, Beijing 100190, China*
[5] *School of Physical Sciences, University of Chinese Academy of Sciences, Beijing 100049, China*
[6] *Institute for Advanced Simulation, Institut für Kernphysik and Jülich Center for Hadron Physics, Forschungszentrum Jülich, D-52425 Jülich, Germany*
[7] *Tbilisi State University, 0186 Tbilisi, Georgia*
[8] *Departamento de Física, Universidad de Murcia, E-30071 Murcia, Spain*
[9] *Institute of Quantum Matter, South China Normal University, Guangzhou 510006, China*
[10] *Theoretical Physics Center for Science Facilities, Institute of High Energy Physics, Chinese Academy of Sciences, Beijing 100049, China*
**E-mail: du@hiskp.uni-bonn.de*

The updated analysis of LHCb of the $\Lambda_b^0 \to J/\psi p K^-$ process with an order of magnitude larger data sample led to three narrow pentaquark states, i.e. $P_c(4312)$, $P_c(4440)$ and $P_c(4457)$. We analyze the measured $J/\psi p$ invariant mass distribution by solving the Lippmann-Schwinger equation with both the dynamical $\Sigma^{(*)}\bar{D}^{(*)}$ and $J/\psi p$ channels under the assumption that the crossed-channel effects provide a smooth incoherent background. We find that the data can be well described in the hadronic molecular picture, which predicts seven $\Sigma^{(*)}\bar{D}^{(*)}$ molecular states. In particular, we show that there is evidence in the data for an additional narrow $\Sigma_c^*\bar{D}$ bound state, which is different from the broad one reported by LHCb in 2015.

Keywords: Pentaquarks; hadronic molecules; heavy quark spin symmetry.

1. Introduction

Searching for multiquark exotic hadrons beyond the conventional quark model is one of the central tasks in modern particle physics. A first well-established signal of the pentaquark states was reported by LHCb[1] in 2015 in the $J/\psi p$ channel of $\Lambda_b^0 \to J/\psi K^- p$ decays. Two Breit-Wigner amplitudes, for a broad $P_c(4380)$ and a much narrower $P_c(4450)$, were introduced to obtain satisfactory fits of the structures in the $J/\psi p$ mass spectrum. In a more recent analysis[2] with an order of magnitude larger data sample, an additional narrow state, $P_c(4312)$, was observed and the $P_c(4450)$ structure split into two narrow structures called $P_c(4440)$ and $P_c(4457)$. At the same time, the broad $P_c(4380)$ lost its significance and a verification of its existence awaits a complete amplitude analysis.

Various theoretical interpretations were proposed to the nature of these pentaquark states, including hadronic molecules[3–11] (for a review on hadronic molecules see Ref. 12), compact pentaquark states[13,14] and hadrocharmonia.[15] The molecular interpretation attracts much attention as the three pentaquark states are close to the $\Sigma_c^{(*)}\bar{D}^{(*)}$ thresholds and can be explained simultaneously by employing the heavy quark spin symmetry (HQSS). By explicitly considering all the $\Sigma_c^{(*)}\bar{D}^{(*)}$ channels (here called elastic channels, since their thresholds are close to the P_c masses) and $J/\psi p$ channel (here called inelastic channel) dynamically, we solve the Lippmann-Schwinger equation to obtain the $J/\psi p$ invariant mass distribution. We show that the $J/\psi p$ distribution can be well described in the molecular scenario, with the $P_c(4312)$ a $J^P = \frac{1}{2}^-$ $\Sigma_c\bar{D}$, and $P_c(4440)$ and $P_c(4457)$ bound states of $\Sigma_c\bar{D}^{(*)}$ with quantum numbers $J^P = \frac{3}{2}^-$ and $\frac{1}{2}^-$, respectively. We also show that the HQSS requires a narrow $\Sigma_c^*\bar{D}$ bound state with a mass around 4.38 GeV, which is clearly visible in the data but different from the broad one reported by LHCb in 2015.

2. Framework

We expand the two-particle states in the basis of heavy-light spin degrees of freedom $|s_Q \otimes j_\ell\rangle$, with s_Q and j_ℓ the total spin of the heavy quarks and total angular momentum of the light degrees of freedom, respectively. The (Σ_c, Σ_c^*) and $(\bar{D}, \bar{D}^*)$ form the $|\frac{1}{2} \otimes 1\rangle$ and $|\frac{1}{2} \otimes \frac{1}{2}\rangle$ spin multiplets, respectively. One can rewrite the S-wave $\Sigma_c^{(*)}\bar{D}^{(*)}$ systems in terms of the $|s_Q \otimes j_\ell\rangle$ basis as[9]

$$\begin{pmatrix} |\Sigma_c\bar{D}\rangle \\ |\Sigma_c\bar{D}^*\rangle \\ |\Sigma_c^*\bar{D}^*\rangle \end{pmatrix}_{\frac{1}{2}} = \begin{pmatrix} \frac{1}{2} & \frac{-1}{2\sqrt{3}} & \sqrt{\frac{2}{3}} \\ \frac{-1}{2\sqrt{3}} & \frac{5}{6} & \frac{\sqrt{2}}{3} \\ \sqrt{\frac{2}{3}} & \frac{\sqrt{2}}{3} & -\frac{1}{3} \end{pmatrix} \begin{pmatrix} |0 \otimes \frac{1}{2}\rangle \\ |1 \otimes \frac{1}{2}\rangle \\ |1 \otimes \frac{3}{2}\rangle \end{pmatrix}, \tag{1}$$

$$\begin{pmatrix} |\Sigma_c\bar{D}^*\rangle \\ |\Sigma_c^*\bar{D}\rangle \\ |\Sigma_c^*\bar{D}^*\rangle \end{pmatrix}_{\frac{3}{2}} = \begin{pmatrix} \frac{-1}{\sqrt{3}} & \frac{1}{3} & \frac{\sqrt{5}}{3} \\ \frac{1}{2} & \frac{-1}{\sqrt{3}} & \frac{1}{2}\sqrt{\frac{5}{3}} \\ \frac{1}{2}\sqrt{\frac{5}{3}} & \frac{\sqrt{5}}{3} & \frac{1}{6} \end{pmatrix} \begin{pmatrix} |0\otimes\frac{3}{2}\rangle \\ |1\otimes\frac{1}{2}\rangle \\ |1\otimes\frac{3}{2}\rangle \end{pmatrix}, \tag{2}$$

$$|\Sigma_c^*\bar{D}^*\rangle_{\frac{5}{2}} = |1\otimes\frac{3}{2}\rangle, \tag{3}$$

where the subscripts on the left-hand side represent the total angular momentum $J = \frac{1}{2}, \frac{3}{2}$ and $\frac{5}{2}$. The rotation matrices will be denoted as R^J in the following. From the decomposition one can obtain the contact potentials

$$C^J_{\alpha\beta} = \sum_{i_J} R^J_{\alpha\, i_J}\; C_{j_\ell(i_J)}\; \left(R^J\right)^T_{i_J\, \beta}\,, \tag{4}$$

where $j_\ell(i_J)$ denotes the light-quark spin of the ith channel for a given J-multiplet, i.e.

$$C_{\frac{1}{2}} \equiv \langle s_Q\otimes\frac{1}{2}|\hat{\mathcal{H}}_I|s_Q\otimes\frac{1}{2}\rangle, \quad C_{\frac{3}{2}} \equiv \langle s_Q\otimes\frac{3}{2}|\hat{\mathcal{H}}_I|s_Q\otimes\frac{3}{2}\rangle.$$

The one-pion exchange (OPE) potential, V^J_{OPE}, can be obtained by making use of the effective Lagrangian, see e.g. in Ref. 16. The Feynman diagrams for the contact and OPE potentials are shown in Fig. 1. An important feature of the OPE is that it automatically mixes in D-waves which have to be included in the analysis.

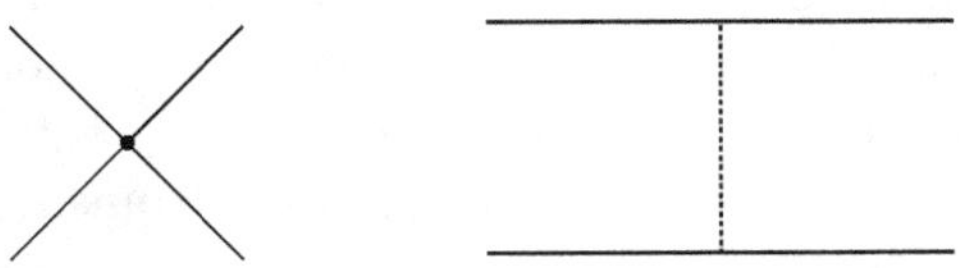

Fig. 1. Feynman diagrams of the elastic transitions for contact and OPE potentials.

Transitions between the elastic and inelastic channels can be related via HQSS as well. The $|1\otimes\frac{1}{2}\rangle$ and $|1\otimes\frac{3}{2}\rangle$ components in Eqs. (1)-(3) couple to $J/\psi p$ in the S and D wave, respectively. Then, the transition vertices $\mathcal{V}^J_{\alpha i}$ between the αth elastic and i^{th} inelastic channel, with $i = 1, 2$ denoting the S-wave and D-wave $J/\psi p$, respectively, can be parameterized as[16]

$$\begin{aligned} &\mathcal{V}^J_{\alpha 1} = g_S R^J_{\alpha 2}, && \mathcal{V}^J_{\alpha 2}(k) = g_D k^2 R^J_{\alpha 3}, && J = \frac{1}{2}, \frac{3}{2}, \\ &\mathcal{V}^J_{\alpha 1} = 0, && \mathcal{V}^J_{\alpha 2}(k) = g_D k^2, && J = \frac{5}{2} \end{aligned} \tag{5}$$

with k the magnitude of the J/ψ three-momentum in the c.m. frame of $J/\psi p$. The weak production amplitude for $\Lambda_b^0 \to K^- \Sigma_c^{(*)} \bar{D}^{(*)}$ can be parameterized in a similar way as

$$P_\alpha^J = \sum_n R_{\alpha n}^J \mathcal{F}_n^J \,, \tag{6}$$

with $\mathcal{F}_n^J = \langle \Lambda_b | \hat{\mathcal{H}}_W | K^- (s_Q \otimes j_\ell)_n^J \rangle$, where $(s_Q \otimes j_\ell)_n^J$ refers to the nth state in the $|s_Q \otimes j_\ell\rangle$ basis in Eqs. (1)-(3).

3. Numerical results

With the above ingredients, one can obtain the production amplitude for the $K^- J/\psi p$ system by solving the Lippmann-Schwinger equations,

$$U_\alpha^J(E,p) = P_\alpha^J(E,p) - \sum_\beta \int \frac{d^3\vec{q}}{(2\pi)^3} V_{\alpha\beta}^J(E,p,q) G_\beta(E,q) U_\beta^J(q),$$

$$U_i^J(E,k) = -\sum_\beta \int \frac{d^3\vec{q}}{(2\pi)^3} V_{\beta i}^J(k) G_\beta(E,q) U_\beta^J(q),$$

with U_α^J the production amplitude for the αth elastic channel and U_i^J for the ith $J/\psi p$ channel. The widths of the $\Sigma_c^{(*)}$ are taken into account by using complex masses in the two-body propagator $G_\beta(E,q)$. To fit the $J/\psi p$ invariant mass distribution, various incoherent smooth backgrounds, including the ones used in the experimental analysis,[2] are employed to model possible contributions from the crossed channels. The results are similar and will be included in the uncertainties. A typical behavior of the line shapes corresponding to the best fit with $\chi^2/\text{d.o.f.} = 0.98$ is shown in Fig. 2. It leads to the poles presented in Table 1. The $P_c(4312)$ couples dominantly to $\Sigma_c \bar{D}$ with $J^P = \frac{1}{2}^-$ and shows up as a (quasi) bound state. The latter feature is different from the findings in Ref. 17, where it is classified as a virtual state. The $P_c(4440)$ and $P_c(4457)$ couple dominantly to the $\Sigma_c \bar{D}^*$ with $J^P = \frac{3}{2}^-$ and $\frac{1}{2}^-$, respectively. Moreover, there is a narrow state located around 4380 MeV where the data show a peak, though less prominent than the well-known three pentaquark states. Its existence is a consequence of the HQSS of QCD.

The existence of the narrow $P_c(4380)$ and the other three predicted states with higher masses need to be checked using data with higher statistics and experiments with different reactions.

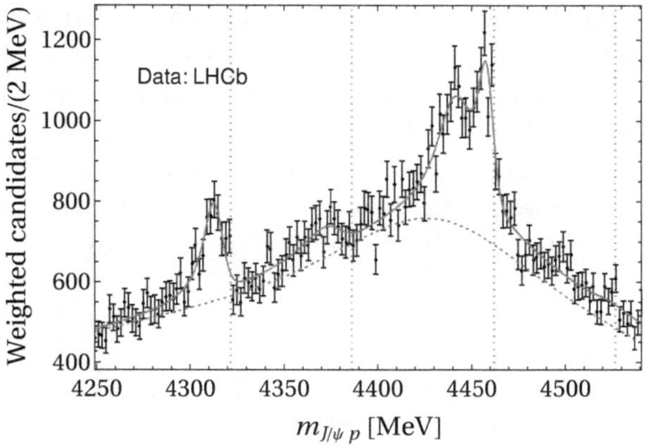

Fig. 2. The fitted invariant mass distribution versus the experimental data.[2] The background is shown as the dotted curve. The vertical dashed lines from left to right correspond to the $\Sigma_c\bar{D}$, $\Sigma_c^*\bar{D}$, $\Sigma_c\bar{D}^*$ and $\Sigma_c^*\bar{D}^*$ thresholds, respectively.

Table 1. The pole positions (on the sheets close to the physical one); the dominant channels (DCs) and their thresholds. The $P_c(4380)$ in boldface is the new state we advocate in this work.

	J^P	Pole [MeV]	DC (threshold [MeV])
$P_c(4312)$	$\frac{1}{2}^-$	$4314(2) - 5(2)i$	$\Sigma_c\bar{D}$ (4321.6)
$\mathbf{P_c(4380)}$	$\frac{3}{2}^-$	$4378(2) - 13(3)i$	$\Sigma_c^*\bar{D}$ (4386.2)
$P_c(4440)$	$\frac{3}{2}^-$	$4441(2) - 11(3)i$	$\Sigma_c\bar{D}^*$ (4462.1)
$P_c(4457)$	$\frac{1}{2}^-$	$4459(2) - 4(1)i$	$\Sigma_c\bar{D}^*$ (4462.1)
P_c	$\frac{1}{2}^-$	$4524(2) - 9(1)i$	$\Sigma_c^*\bar{D}^*$ (4526.7)
P_c	$\frac{3}{2}^-$	$4518(2) - 11(2)i$	$\Sigma_c^*\bar{D}^*$ (4526.7)
P_c	$\frac{5}{2}^-$	$4498(5) - 35(17)i$	$\Sigma_c^*\bar{D}^*$ (4526.7)

Acknowledgments

This work is supported in part by the National Natural Science Foundation of China (NSFC) and the Deutsche Forschungsgemeinschaft (DFG) through the funds provided to the Sino-German Collaborative Research Center "Symmetries and the Emergence of Structure in QCD" (NSFC Grant No. 11621131001 and DFG Grant No. TRR110), by the NSFC under Grants No. 11835015 and No. 11847612, by the Chinese Academy of Sciences (CAS) under Grants No. QYZDB-SSW-SYS013 and No. XDPB09, by the CAS Center for Excellence in Particle Physics (CCEPP), and by the Munich Institute for Astro- and Particle Physics (MIAPP) of the DFG cluster of excellence "Origin and Structure of the Universe". QW is also supported by the Thousand Talents Plan for Young Professionals and research startup funding at SCNU. The work of UGM is further supported by the CAS President's International Fellowship Initiative (PIFI) (Grant No. 2018DM0034) and by the VolkswagenStiftung (Grant No. 93562). The work of VB is supported by the Russian Science Foundation (Grant No. 18-12-00226). JAO would like to acknowledge partial financial support by the MINECO (Spain) and FEDER (EU) grant FPA2016-77313-P.

References

1. R. Aaij *et al.*, *Phys. Rev. Lett.* **115**, 072001 (2015).
2. R. Aaij *et al.*, *Phys. Rev. Lett.* **122**, 222001 (2019).
3. H.-X. Chen, W. Chen and S.-L. Zhu, *Phys. Rev.* **D100**, 051501 (2019).
4. R. Chen, Z.-F. Sun, X. Liu and S.-L. Zhu, *Phys. Rev.* **D100**, 011502 (2019).
5. F.-K. Guo, H.-J. Jing, U.-G. Meißner and S. Sakai, *Phys. Rev.* **D99**, 091501 (2019).
6. M.-Z. Liu, Y.-W. Pan, F.-Z. Peng, M. S. Sánchez, L.-S. Geng, A. Hosaka and M. P. Valderrama, *Phys. Rev. Lett.* **122**, 242001 (2019).
7. Z.-H. Guo and J. A. Oller, *Phys. Lett.* **B793**, 144 (2019).
8. C. W. Xiao, J. Nieves and E. Oset, Phys. Rev. D **100**, no. 1, 014021 (2019)
9. S. Sakai, H.-J. Jing and F.-K. Guo, *Phys. Rev.* **D100**, 074007 (2019).
10. L. Meng, B. Wang, G.-J. Wang and S.-L. Zhu, *Phys. Rev.* **D100**, 014031 (2019).
11. T. J. Burns and E. S. Swanson, arXiv:1908.03528 (2019).
12. F.-K. Guo, C. Hanhart, U.-G. Meißner, Q. Wang, Q. Zhao and B. S. Zou, Rev. Mod. Phys. **90**, no. 1, 015004 (2018).
13. A. Ali and A. Ya. Parkhomenko, *Phys. Lett.* **B793**, 365 (2019).
14. R. Zhu, X. Liu, H. Huang and C.-F. Qiao, *Phys. Lett.* **B797**, 134869 (2019).
15. M. I. Eides, V. Y. Petrov and M. V. Polyakov, arXiv:1904.11616 (2019).
16. M.-L. Du, V. Baru, F.-K. Guo, C. Hanhart, U.-G. Meißner, J. A. Oller and Q. Wang, arXiv:1910.11846 (2019).
17. C. Fernández-Ramírez *et al.* [JPAC Collaboration], Phys. Rev. Lett. **123**, 092001 (2019).

Three pentaquark states or more?

C. W. Xiao
School of Physics and Electronics, Central South University,
Changsha 410083, China
E-mail: xiaochw@csu.edu.cn

J. Nieves
IFIC, Centro Mixto Universitat de València-CSIC, Institutos de Investigación de Paterna,
Aptdo. 22085, 46071 Valencia, Spain
E-mail: jmnieves@ific.uv.es

E. Oset
Departamento de Física Teórica, Universitat de València,
and IFIC, Centro Mixto Universitat de València-CSIC, Institutos de Investigación de Paterna,
Aptdo. 22085, 46071 Valencia, Spain
E-mail: oset@ific.uv.es

We investigate the interaction of the coupled channels with the $\bar{D}^{(*)}\Sigma_c^{(*)}$ states by taking into account the heavy quark spin symmetry. By fitting only one parameter to the recent three pentaquark states reported by the LHCb collaboration, we can reproduce the three of them and identify them as molecular states of $1/2^-$ $\bar{D}\Sigma_c$, $1/2^-$ $\bar{D}^*\Sigma_c$, and $3/2^-$ $\bar{D}^*\Sigma_c$, and isospin $I = 1/2$. We also find another state around 4374 MeV, of $3/2^-$ $\bar{D}\Sigma_c^*$ structure, two other near degenerate states of $1/2^-$ $\bar{D}^*\Sigma_c^*$ and $3/2^-$ $\bar{D}^*\Sigma_c^*$ around 4520 MeV. In addition, a $5/2^-$ $\bar{D}^*\Sigma_c^*$ state at the same energy appears, which however does not couple to $J/\psi p$ in $S-$wave.

Keywords: Pentaqurark state; Coupled channel; Resonance; Molecular state.

1. Introduction

Some pentaquark signals were reported by the LHCb collaboration in 2015,[1,2] where two states were found, one at 4380 MeV and width $\Gamma \sim 205$ MeV and another one at 4450 MeV and width 40 MeV. Actually there had been several predictions for hidden charm molecular states in this region prior to the experimental discovery.[3–10] This discovery has drawn much theoretical attentions, see the review papers for references to all these works.[11–19]

With the advent of Run-2 data, the LHCb collaboration updated the results of Refs. 1, 2 reporting the observation of three clear narrow structures,[20] branded as

$$M_{P_{c1}} = (4311.9 \pm 0.7^{+6.8}_{-0.6})\,\text{MeV}, \quad \Gamma_{P_{c1}} = (9.8 \pm 2.7^{+3.7}_{-4.5})\,\text{MeV},$$
$$M_{P_{c2}} = (4440.3 \pm 1.3^{+4.1}_{-4.7})\,\text{MeV}, \quad \Gamma_{P_{c2}} = (20.6 \pm 4.9^{+8.7}_{-10.1})\,\text{MeV},$$
$$M_{P_{c3}} = (4457.3 \pm 0.6^{+4.1}_{-1.7})\,\text{MeV}, \quad \Gamma_{P_{c3}} = (6.4 \pm 2.0^{+5.7}_{-1.9})\,\text{MeV}.$$

One can see that the old peak at 4450 MeV is now splitting into two states at 4440 MeV and 4457 MeV, the last one very narrow, and a fluctuation observed in the old spectrum has given rise to a neat peak around 4312 MeV.

The new experimental findings have already had a reply from the theoretical community. In Ref. 21 sum rules are used that provide several scenarios to explain these states, the most favored ones being of $\Sigma_c^{(*)} D^{(*)}$ molecular nature. In Ref. 22 heavy quark spin symmetry (HQSS) is used with $\Sigma_c \bar{D}$, $\Sigma_c \bar{D}^*$, $\Sigma_c^* \bar{D}$, $\Sigma_c^* \bar{D}^*$ as single channels and seven bound states are found, three of which can be associated with the experimental states. One should mention that in that line there is previous work, including other coupled channels, and which also predicts seven states with isospin $I = 1/2$, and the widths of the states.[8]

In the works of Refs. 3, 7 $\bar{D}\Sigma_c$ and $\bar{D}^*\Sigma_c$, among other coupled channels, were used, but not $\bar{D}\Sigma_c^*$, $\bar{D}^*\Sigma_c^*$. HQSS relates the strength of the interaction of these channels and they were considered in Ref. 8. The advent of the LHCb data offers an opportunity to tune the regulator of the loops to adjust to some experimental data. This is the purpose of the present work. It is similar to the study of Ref. 22, but includes more channels than the $\Sigma_c^{(*)} D^{(*)}$ used in Ref. 22, and in addition we work with coupled channels rather than using single channels, which allows us to obtain also the widths.

2. Formalism

In Ref. 8 the Bethe-Salpeter equation (BSE) is used with the coupled channels in $I = 1/2$, $\eta_c N$, $J/\psi N$, $\bar{D}\Lambda_c$, $\bar{D}\Sigma_c$, $\bar{D}^*\Lambda_c$, $\bar{D}^*\Sigma_c$, $\bar{D}^*\Sigma_c^*$ for spin parity $J^P = 1/2^-$ and $J/\psi N$, $\bar{D}^*\Lambda_c$, $\bar{D}^*\Sigma_c$, $\bar{D}\Sigma_c^*$, $\bar{D}^*\Sigma_c^*$ for $J^P = 3/2^-$. In addition a single channel for $\bar{D}^*\Sigma_c^*$ in the $J^P = 5/2^-$ sector is also studied. The BSE in matrix form for the scattering matrix reads

$$T = [1 - V\,G]^{-1}\,V, \tag{1}$$

where G is the loop function of the meson-baryon intermediate states and the potential V, with the HQSS constraints, see more details in Refs. 8, 23.

3. Results

The novelty with respect to Ref. 8 is a different choice of the subtraction constant to regularize the meson-baryon loops (G) in dimensional regularization, since the position of the poles is tied to its value. A subtraction constant $a(\mu) = -2.3$ with $\mu = 1$ GeV was used in Ref. 8. The scheme produces seven states, three of which can be clearly associated to the recently found experimental resonances. The new information allows us to take a new value of $a(\mu = 1\,\text{GeV}) = -2.09$, such that the sum of masses of the three theoretical states matches the experimental results. With this constraint we fix the only free parameter of the model of Ref. 8. The results are reported in Tables 1 and 2 for $J^P = 1/2^-$ and $3/2^-$, respectively. In addition we get a mass of 4519.23 MeV and a zero width for the single channel $\bar{D}^*\Sigma_c^*$ with $J = 5/2^-$. This channel obviously does not couple to $J/\psi N$ so we should not see it in the $\Lambda_b \to J/\psi p K^-$ experiment. The states in Tables 1 and 2 all couple to $J/\psi N$ and in principle they could be seen in the experiment, although we cannot predict their strength in the spectrum. In Table 3, we show the results for the three resonances that we identify with the experimental states. The main channel is taken from the largest coupling. We find the last two states nearly degenerate, yet, the widths of the states force us to identify the $3/2^-$ state with the $P_c(4457)$. Note that the masses divert only in a few MeV from the experimental ones, and the three widths obtained are compatible with the experiment.

Table 1. The coupling constants of the ($I = 1/2, J^P = 1/2^-$) poles found in this work. The imaginary part of the energies corresponds to $\Gamma/2$.

	$\eta_c N$	$J/\psi N$	$\bar{D}\Lambda_c$	$\bar{D}\Sigma_c$	$\bar{D}^*\Lambda_c$	$\bar{D}^*\Sigma_c$	$\bar{D}^*\Sigma_c^*$
$(4306.38 + i7.62)$ MeV							
$\|g_i\|$	0.67	0.46	0.01	**2.09**	0.25	0.31	0.16
$(4452.96 + i11.72)$ MeV							
$\|g_i\|$	0.25	0.89	0.11	0.13	0.14	**2.03**	0.19
$(4520.45 + i11.12)$ MeV							
$\|g_i\|$	0.73	0.45	0.13	0.06	0.08	0.08	**1.92**

Table 2. Same as Table 1 for $J^P = 3/2^-$.

$(4374.33 + i6.87)$ MeV	$J/\psi N$	$\bar{D}^*\Lambda_c$	$\bar{D}^*\Sigma_c$	$\bar{D}\Sigma_c^*$	$\bar{D}^*\Sigma_c^*$
$\lvert g_i \rvert$	0.73	0.18	0.19	**1.94**	0.30
$(4452.48 + i1.49)$ MeV					
$\lvert g_i \rvert$	0.30	0.07	**1.82**	0.08	0.19
$(4519.01 + i6.86)$ MeV					
$\lvert g_i \rvert$	0.66	0.13	0.10	0.13	**1.82**

Table 3. Identification of the resonances found in this work with the experiments.

Mass [MeV]	Width [MeV]	Main channel	J^P	Experiments
4306.4	15.2	$\bar{D}\Sigma_c$	$1/2^-$	$P_c(4312)$
4453.0	23.4	$\bar{D}^*\Sigma_c$	$1/2^-$	$P_c(4440)$
4452.5	3.0	$\bar{D}^*\Sigma_c$	$3/2^-$	$P_c(4457)$

4. Conclusions

In summary, with the molecular picture in the coupled channels to $J/\psi p$ in S-wave, we find six states that couple to $J/\psi p$. Three of these resonances can be identified with the three states reported in Ref. 20 in base to their masses and widths. In addition, we provide a prediction of their J^P quantum numbers and of the nature of these states as basically $1/2^-$ $\bar{D}\Sigma_c$, $1/2^-$ $\bar{D}^*\Sigma_c$ and $3/2^-$ $\bar{D}^*\Sigma_c$. We find a fourth state, which couples mostly to $\bar{D}\Sigma_c^*$ with $3/2^-$, in a region where there is a small enhancement in the $J/\psi p$ spectrum of Ref. 20. The other two states, of $\bar{D}^*\Sigma_c^*$ nature, are around 4520 MeV, and although there are small peaks in that region in Ref. 20, one can only speculate at the present stage. They are also near degenerate with a $5/2^-$ state of the same nature, which however is not expected to show up in the LHCb experiment.

Acknowledgments

This research has been supported by the Spanish Ministerio de Ciencia, Innovación y Universidades and European FEDER funds under Contracts FIS2017-84038-C2-1-P, FIS2017-84038-C2-2-P and SEV-2014-0398 and by the EU STRONG-2020 project under the program H2020-INFRAIA-2018-1, grant agreement no. 824093.

References

1. R. Aaij *et al.* [LHCb Collaboration], Phys. Rev. Lett. **115**, 072001 (2015).
2. R. Aaij *et al.* [LHCb Collaboration], Chin. Phys. C **40**, 011001 (2016).
3. J. J. Wu, R. Molina, E. Oset and B. S. Zou, Phys. Rev. Lett. **105**, 232001 (2010).
4. W. L. Wang, F. Huang, Z. Y. Zhang and B. S. Zou, Phys. Rev. C **84**, 015203 (2011).
5. Z. C. Yang, Z. F. Sun, J. He, X. Liu and S. L. Zhu, Chin. Phys. C **36**, 6 (2012).
6. S. G. Yuan, K. W. Wei, J. He, H. S. Xu and B. S. Zou, Eur. Phys. J. A **48**, 61 (2012).
7. J. J. Wu, T.-S. H. Lee and B. S. Zou, Phys. Rev. C **85**, 044002 (2012).
8. C. W. Xiao, J. Nieves and E. Oset, Phys. Rev. D **88**, 056012 (2013).
9. T. Uchino, W. H. Liang and E. Oset, Eur. Phys. J. A **52**, 43 (2016).
10. M. Karliner and J. L. Rosner, Phys. Rev. Lett. **115**, 122001 (2015).
11. H. X. Chen, W. Chen, X. Liu and S. L. Zhu, Phys. Rept. **639**, 1 (2016).
12. R. F. Lebed, R. E. Mitchell and E. S. Swanson, Prog. Part. Nucl. Phys. **93**, 143 (2017).
13. A. Esposito, A. Pilloni and A. D. Polosa, Phys. Rept. **668**, 1 (2016).
14. F. K. Guo, C. Hanhart, U. G. Meißner, Q. Wang, Q. Zhao and B. S. Zou, Rev. Mod. Phys. **90**, 015004 (2018).
15. A. Ali, J. S. Lange and S. Stone, Prog. Part. Nucl. Phys. **97**, 123 (2017).
16. S. L. Olsen, T. Skwarnicki and D. Zieminska, Rev. Mod. Phys. **90**, 015003 (2018).
17. M. Karliner, J. L. Rosner and T. Skwarnicki, Ann. Rev. Nucl. Part. Sci. **68**, 17 (2018).
18. A. Cerri *et al.*, arXiv:1812.07638 [hep-ph].
19. Y. R. Liu, H. X. Chen, W. Chen, X. Liu and S. L. Zhu, Prog. Part. Nucl. Phys. **107**, 237 (2019).
20. R. Aaij *et al.* [LHCb Collaboration], Phys. Rev. Lett. **122**, 222001 (2019).
21. H. X. Chen, W. Chen and S. L. Zhu, Phys. Rev. D **100**, 051501 (2019).
22. M. Z. Liu, Y. W. Pan, F. Z. Peng, M. Sánchez Sánchez, L. S. Geng, A. Hosaka and M. Pavon Valderrama, Phys. Rev. Lett. **122**, 242001 (2019).
23. C. W. Xiao, J. Nieves and E. Oset, Phys. Rev. D **100**, 014021 (2019).

Vector charmonium-like states at BESIII

Jielei Zhang

College of Physics and Electronic Engineering, Xinyang Normal University,
Xinyang, 464000, China
E-mail: zhangjielei@ihep.ac.cn

The BESIII Experiment at the Beijing Electron Positron Collider has accumulated the large data samples for e^+e^- collisions with center-of-mass energies above 4 GeV. Using these data samples, some surprising discoveries are obtained about vector charmonium-like states, which contains a charm quark and anti-charm quark with quantum number $J^{PC} = 1^{--}$, but can not be simply described as $c\bar{c}$ bound states. In this proceeding, the new results on the vector charmonium-like states at BESIII will be presented.

Keywords: Charmonium; charmonium-like; BESIII.

1. Introduction

In recent years, charmonium physics has gained strong interest from both the theoretical and the experimental side, due to the observation of a dozen of charmonium-like states. These states do not fit in the conventional charmonium spectroscopy, and they are good candidates for exotic states beyond the quark model. Among them, the vector charmonium-like states, having quantum numbers $J^{PC} = 1^{--}$, are usually called Y-states, such as $Y(4260)$, $Y(4360)$ and $Y(4660)$. There are many theoretical models to explain these Y-states, such as hybrid charmonium, tetraquark state and the molecular state.[1] However, there is still no definite conclusion. The BESIII Experiment has accumulated the large data samples for e^+e^- collisions with center-of-mass energies above 4 GeV, which provides a good opportunity to study vector charmonium-like states. In this proceeding, we review the new results on the vector charmonium-like states at BESIII.

2. New results on vector charmonium-like states

Here the new results on the vector charmonium-like states at BESIII are presented, including decay channels $e^+e^- \to \pi^+\pi^- J/\psi$, $\pi^+\pi^- h_c$,

$\pi^+\pi^-\psi(3686)$, $\pi^+ D^0 D^{*-}$, $\omega\chi_{c0}$, $K\bar{K}J/\psi$, $\pi^+\pi^- D\bar{D}$, $\Lambda_c^+\bar{\Lambda}_c^-$, $K_s^0 K^\pm\pi^\mp\pi^0$, $K_s^0 K^\pm\pi^\mp\eta$ and $K_s^0 K^\pm\pi^\mp$.

BESIII reported a precise cross section measurement of $e^+e^- \to \pi^+\pi^- J/\psi$ from center-of-mass energies $\sqrt{s} = 3.77$ to 4.60 GeV.[2] Two resonant structures are observed, one with a mass of $(4222.0 \pm 3.1 \pm 1.4)$ MeV/c^2 and a width of $(44.1 \pm 4.3 \pm 2.0)$ MeV, and the other with a mass of $(4320.0 \pm 10.4 \pm 7.0)$ MeV/c^2 and a width of $(101.4^{+25.3}_{-19.7} \pm 10.2)$ MeV. The first resonance agrees with the $Y(4260)$ resonance. However, the measured width is much narrower than the $Y(4260)$ average width reported by previous experiments. The second resonance is observed for the first time and the statistical significance is estimated to be larger than 7.6σ. The second resonance has a mass and width comparable to the $Y(4360)$ resonance reported by *BABAR* and Belle in $e^+e^- \to \pi^+\pi^-\psi(3686)$. Additional, the existence of the $Y(4008)$ resonance can not be confirmed from the measurement.

BESIII measured the $e^+e^- \to \pi^+\pi^- h_c$ Born cross section using data from 3.896 to 4.600 GeV.[3] Two structures are observed in the cross section around 4.22 and 4.39 GeV/c^2, which are called $Y(4220)$ and $Y(4390)$, respectively. A fit with a coherent sum of two Breit-Wigner functions results in a mass of $(4218.4^{+5.5}_{-4.5} \pm 0.9)$ MeV/c^2 and a width of $(66.0^{+12.3}_{-8.3} \pm 0.4)$ MeV for the $Y(4220)$, and a mass of $(4391.5^{+6.3}_{-6.8} \pm 1.0)$ MeV/c^2 and a width of $(139.5^{+16.2}_{-20.6} \pm 0.6)$ MeV for the $Y(4390)$. The statistical significance of $Y(4220)$ and $Y(4390)$ is 10σ over one structure assumption. The parameters of the two structures are different from those of $Y(4260)$, $Y(4360)$, and $\psi(4415)$.

The Born cross section of $e^+e^- \to \pi^+\pi^-\psi(3686)$ is measured based on 5.1 fb^{-1} the e^+e^- collision data between 4.008 and 4.600 GeV.[4] The cross section is fit with three coherent Breit-Wigner functions. The parameters of two resonances, the $Y(4220)$ and $Y(4390)$, are determined from the fit to be $M = 4209.5 \pm 7.4 \pm 1.4$ MeV/c^2 and $\Gamma = 80.1 \pm 24.6 \pm 2.9$ MeV for the $Y(4220)$, and $M = 4383.8 \pm 4.2 \pm 0.8$ MeV/c^2 and $\Gamma = 84.2 \pm 12.5 \pm 2.1$ MeV for the $Y(4390)$. The resonance $Y(4220)$ is observed in the process $e^+e^- \to \pi^+\pi^-\psi(3686)$ for the first time with a significance of 5.8σ.

The Born cross section for the process $e^+e^- \to \pi^+ D^0 D^{*-}$ is precisely measured using the data samples collected at 84 energy points from 4.05 to 4.60 GeV with the BESIII detector.[5] Two enhancements are observed in the dressed cross sections around 4.23 and 4.40 GeV. The first resonant structure's parameters are measured to be $M = 4228.6 \pm 4.1 \pm 6.3$ MeV/c^2 and $\Gamma = 77.0 \pm 6.8 \pm 6.3$ MeV. The second resonant structure's parameters are strongly dependent on the model assumptions, necessitating further studies. This is the first experimental evidence for open-charm production associated with the Y states.

The process $e^+e^- \to \omega\chi_{c0}$ has been studied using 9 data samples collected at center-of-mass energies from $\sqrt{s} = 4.178$ to 4.278 GeV.[6] A clear enhancement is seen around $\sqrt{s} = 4.22$ GeV which confirms, and statistically improves upon, an earlier observation.[7] By fitting the $e^+e^- \to \omega\chi_{c0}$ cross section with a single resonance, the mass and width for the structure are determined to be $M = (4218.5\pm1.6\pm4.0)$ MeV/c^2 and $\Gamma = (28.2\pm3.9\pm1.6)$ MeV. The obtained resonance parameters are not compatible with the vector charmonium state $\psi(4160)$, ruling out its possible contribution to the structure. Figure 1 shows the measured mass and width of the $Y(4220)$ from the different processes. The masses are consistent with each other, while the widths are not. Further experimental studies with higher statistics are needed to draw a more reliable conclusion on the nature of this structure.

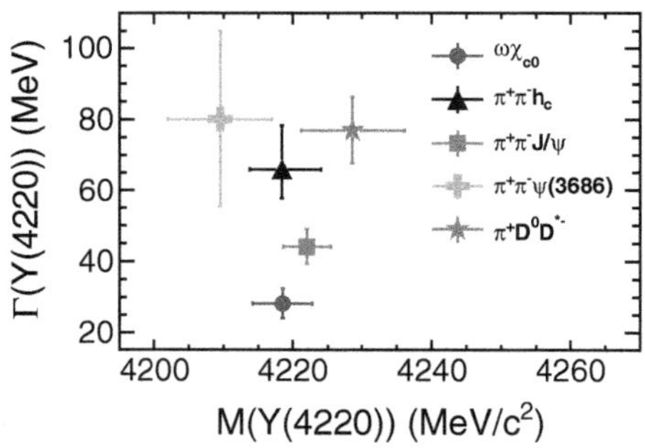

Fig. 1. Mass and width of the $Y(4220)$ obtained from the processes $e^+e^- \to \omega\chi_{c0}$, $\pi^+\pi^-h_c$, $\pi^+\pi^-J/\psi$, $\pi^+\pi^-\psi(3686)$ and $\pi^+D^0D^{*-}$ + c.c.

The process $e^+e^- \to K\bar{K}J/\psi$ is investigated at center-of-mass energies from 4.189 to 4.600 GeV.[8] The Born cross sections of $e^+e^- \to K^+K^-J/\psi$ and $K^0_SK^0_SJ/\psi$ are measured. The energy dependence of the cross section for $e^+e^- \to K^+K^-J/\psi$ is shown to differ from that for $\pi^+\pi^-J/\psi$ in the region around the $Y(4260)$. In addition, there is evidence for a structure around 4.5 GeV in the $e^+e^- \to K^+K^-J/\psi$ cross section that is not present in $\pi^+\pi^-J/\psi$. The ratio of cross sections for the reactions with neutral and charged kaons to be consistent with expectations from isospin conservation.

Several intermediate states of the reaction channels $e^+e^- \to \pi^+\pi^-D^0\bar{D}^0$ and $e^+e^- \to \pi^+\pi^-D^+D^-$ are studied using the data samples above 4.08 GeV.[9] For the first time in this final state, a $\psi(3770)$ signal is seen in the $D\bar{D}$ invariant mass spectrum, with a statistical significance of 5.2σ at $\sqrt{s} = 4.42$ GeV. Also, the process $e^+e^- \to D_1(2420)\bar{D} + c.c.$ are investigated. The neutral mode with $D_1(2420)^0 \to D^0\pi^+\pi^-$ is reported with statistical significance of 7.4σ at $\sqrt{s} = 4.42$ GeV for the first time. No evident signal for the process $e^+e^- \to D_1(2420)^0\bar{D}^0 + c.c., D_1(2420)^0 \to D^{*+}\pi^-$ is reported. Evidence for $e^+e^- \to D_1(2420)^+D^- + c.c., D_1(2420)^+ \to D^+\pi^+\pi^-$ is reported with statistical significance of 3.1σ and 3.0σ at $\sqrt{s} = 4.36$ and 4.42 GeV, respectively.

Using data collected at $\sqrt{s} = 4574.5, 4580.0, 4590.0, 4599.5$ MeV with the BESIII detector, the cross sections of $e^+e^- \to \Lambda_c^+\bar{\Lambda}_c^-$ have been measured with high precision.[10] The most precise cross section measurement is achieved so far at $\sqrt{s} = 4574.5$ MeV, which is only 1.6 MeV above the threshold. The measured value is $(236 \pm 11 \pm 46)$ pb, which highlights the enhanced cross section near threshold and indicates the complexity of production behavior of the Λ_c.

The Born cross sections for $e^+e^- \to K_s^0K^\pm\pi^\mp\pi^0$ and $K_s^0K^\pm\pi^\mp\eta$ are measured with data samples collected from 3.90 to 4.60 GeV.[11] Since no clear structure is observed, the upper limits of the product $\Gamma_{e^+e^-}\mathcal{B}(Y(4260) \to K_s^0K^\pm\pi^\mp\pi^0)$ at 90% C.L. are estimated to be less than 0.05 eV and that of $\Gamma_{e^+e^-}\mathcal{B}(Y(4260) \to K_s^0K^\pm\pi^\mp\eta)$ is estimated to be smaller than 0.19 eV. The $e^+e^- \to K_s^0K^\pm\pi^\mp$ Born cross sections also have been measured from 3.8 to 4.6 GeV.[12] The line shape of the Born cross section is consistent with only the continuum process, however a better fit is obtained by adding an additional resonance. The fit to the Born cross sections from this work, with $\psi(4160)[Y(4220)]$ added, is performed. Only evidence for the $\psi(4160)[Y(4220)]$ is observed with corresponding significance $2.5\sigma(2.2\sigma)$.

3. Summary

BESIII provide new access to charmonium-like states by e^+e^- annihilation and our understanding of these exotic states has been significantly improved. Here we review the recent results of vector charmonium-like states at BESIII experiment. Many decay channels' cross sections have been measured with high precision by BESIII. The $Y(4220)$ is observed in many hidden charm decay modes, i.e. $e^+e^- \to$

$\omega\chi_{c0}, \pi^+\pi^-h_c, \pi^+\pi^-J/\psi, \pi^+\pi^-\psi(3686)$, and also in one open charm decay mode $e^+e^- \to \pi^+D^0D^{*-}$. And no significant $Y(4220)$ signals are seen in light hadron final states. However, the nature of $Y(4220)$ is still not clear, which needs more efforts from experiment and theory. BESIII will collect more data for charmonium-like states study, and more exciting results from BESIII will come up soon.

Acknowledgments

This work is supported by the National Natural Science Foundation of China (NSFC) under Contract No. 11905179, Foundation of Henan Educational Committee under Contract No. 19A140015, and Nanhu Scholars Program for Young Scholars of Xinyang Normal University.

References

1. H. X. Chen, W. Chen, X. Liu and S. L. Zhu, Phys. Rept. **639**, 1 (2016).
2. M. Ablikim *et al.* [BESIII Collaboration], Phys. Rev. Lett. **118**, 092001 (2017).
3. M. Ablikim *et al.* [BESIII Collaboration], Phys. Rev. Lett. **118**, 092002 (2017).
4. M. Ablikim *et al.* [BESIII Collaboration], Phys. Rev. D **96**, 032004 (2017).
5. M. Ablikim *et al.* [BESIII Collaboration], Phys. Rev. Lett. **122**, 102002 (2019).
6. M. Ablikim *et al.* [BESIII Collaboration], Phys. Rev. D **99**, 091103 (2019).
7. M. Ablikim *et al.* [BESIII Collaboration], Phys. Rev. D **93**, 011102 (2016); Phys. Rev. Lett. **114**, 092003 (2015).
8. M. Ablikim *et al.* [BESIII Collaboration], Phys. Rev. D **97**, 071101 (2018).
9. M. Ablikim *et al.* [BESIII Collaboration], Phys. Rev. D **100**, 032005 (2019).
10. M. Ablikim *et al.* [BESIII Collaboration], Phys. Rev. Lett. **120**, 132001 (2018).
11. M. Ablikim *et al.* [BESIII Collaboration], Phys. Rev. D **99**, 012003 (2019).
12. M. Ablikim *et al.* [BESIII Collaboration], Phys. Rev. D **99**, 072005 (2019).

First evidence of $B \to h_c K$ and recent results on X and Y from Belle

R. Garg on behalf of the Belle Collaboration

Department of physics, Panjab University,
Chandigarh, Chandigarh 160014, India
E-mail: renu92garg@gmail.com

In this talk, we report the evidence of $B \to h_c K$ decay along with the first observation of the decay $\eta_c(2S) \to p\bar{p}\pi^+\pi^-$ at Belle based on the full data sample accumulated by the Belle experiment at KEKB asymmetric energy e^+e^- collider. We also present search for $B \to Y(4260)K$, $B \to X(3872)(\to \chi_{c1}\pi^0)K$ and $B \to X(3915)(\to \chi_{c1}\pi^0)K$ at Belle. As no signal is found, an upper limit on the product branching fractions is provided.

Keywords: Charmonium, Belle.

1. Introduction

Several new states have been discovered at B-factories. The B-factories provide a clean environment to search and study new states. The first charmonium like state $X(3872)$ has been observed by the Belle in 2003.[1] Although there are several measurements of its properties, the nature of this particle is unknown. It does not fit into the conventional charmonium spectrum. Theoretical models also fail to explain its nature. After that, the observation of $Y(4260)$ was also taken with great interest. In this report, we present the recent results of the charmonium (-like) state in B decays.

2. Search for the $B \to Y(4260)K$, $Y(4260) \to J/\psi\pi^+\pi^-$ decays[2]

In 2006, BaBar collaboration has measured the signal $B^+ \to Y(4260)(\to J/\psi\pi^+\pi^-)K^+$[3] based on the data sample of 211 fb^{-1} with statistical significance of 3.1σ. They set an upper limit on the product of branching fractions to be $\mathcal{B}(B^+ \to Y(4260)K^+) \times \mathcal{B}(Y(4260) \to J/\psi\pi^+\pi^-) < 2.9 \times 10^{-5}$. Based on the QCD sum rule approach,[4] it has been suggested that the product of branching fractions of $B^+ \to Y(4260)(\to J/\psi\pi^+\pi^-)K^+$ to be in the range $3.0 \times 10^{-8} - 1.8 \times 10^{-6}$. We aim to precisely measure the branching fractions.

In this analysis, $Y(4260)$ is reconstructed from $J/\psi\pi^+\pi^-$, where $J/\psi \to \ell^+\ell^-$ (ℓ is either e or μ). $B \to \psi' K$ and $B \to X(3872)K$ are used as control samples. An unbinned maximum likelihood (UML) fit is performed to the energy difference ($\Delta E = \sum E_i - E_{\rm beam}$, where $E_{\rm beam}$ and E_i are the beam energy and energy of $i^{\rm th}$ daughter in the center of mass frame) distribution. The statistical weight for each candidate to be a signal is determined from ${}_s\mathcal{P}lot$ technique[5] to subtract the combinatorial background from $M_{J/\psi\pi\pi}$. Signal yield is extracted from a UML fit to the weighted $M_{J/\psi\pi\pi}$ distribution. The fit results are shown in Fig. 1.

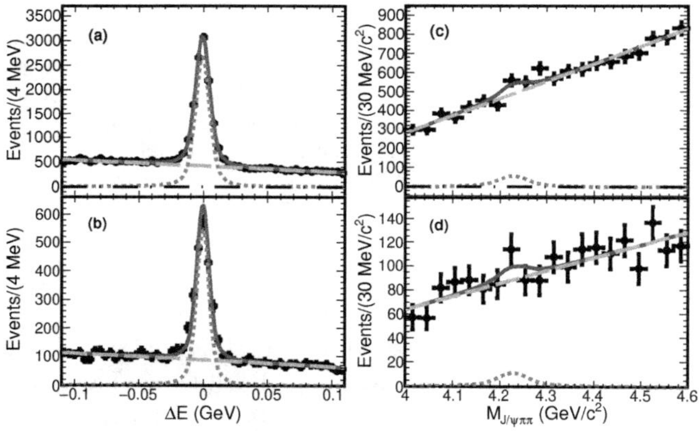

Fig. 1. Fit to the ΔE and ${}_s\mathcal{P}lot$ of $M_{J/\psi\pi\pi}$ distributions for $B^+ \to Y(4260)(\to J/\psi\pi^+\pi^-)K^+$.

The signal for charged and neutral $B \to Y(4260)(\to J/\psi\pi^+\pi^-)K$ is observed with signal significance 2.1 σ and 0.9 σ, respectively. In the absence of significant signal, we set an upper limit on the product of branching fractions to be 1.4×10^{-5} and 1.7×10^{-5} at 90% C.L. for the charged and neutral modes, respectively. The upper limit for the charged mode is consistent with BaBar results and for the neutral mode is given for the first time.

3. Search for $X(3872)$ and X(3915) decay into $\chi_{c1}\pi^0$ in B decays at Belle[6]

Recently, a new decay, $X(3872) \to \chi_{c1}\pi^0$ in $e^+e^- \to \chi_{c1}\pi^0\gamma$ has been reported by the BESIII.[7] If $X(3872)$ is dominated by a charmonium

$\chi_{c1}(2P)$ component, then the branching fraction of the pionic transition, $X(3872) \to \chi_{c1}\pi^0$, is expected to be very small due to isospin breaking by the light quark masses.[8] According to the BESIII measurement, $\mathcal{B}(X(3872) \to \chi_{c1}\pi^0)/\mathcal{B}(X(3872) \to J/\psi\pi^+\pi^-) = 0.88^{+0.33}_{-0.27} \pm 10$ is large as compared to $\mathcal{B}(B \to J/\psi\pi^0)/\mathcal{B}(B \to J/\psi\pi^+\pi^-) = 3.66 \times 10^{-3}$ and disfavors the $\chi_{c1}(2P)$ interpretation. The $X(3915)$ was first observed by the Belle in $B \to X(3915)(\to J/\psi\omega)K$ decay.[9] In $\chi_{c0}(2P)$ scenario, its width should be larger[10] and $J/\psi\omega$ is also suppressed by OZI.[11] However, the measured width $(20 \pm 5\ \mathrm{MeV/c^2})$[12] is smaller than expectations (> 100 MeV/c^2). If $X(3915)$ is a non-conventional $c\bar{c}$ state, single pionic transition maybe enhanced.

In this analysis, $X(3872/3915)$ is reconstructed from $\chi_{c1}\pi^0$. The χ_{c1} candidates are reconstructed from $J/\psi\gamma$, where $J/\psi \to \ell^+\ell^-$ (ℓ is either e or μ). $B \to \chi_{c1}K^{*+}$ events are vetoed by rejecting event with 791.8 $< M_{K^+\pi^0} <$ 991.8 MeV/c^2. The signal yield is extracted from a UML fit to the ΔE distribution and the signal yield of $X(3872/3915)$ is extracted from a UML fit to the weighted $M_{\chi_{c1}\pi^0}$ distribution produced by ${}_s\mathcal{P}lot$ technique. The fit results are shown in Fig. 2.

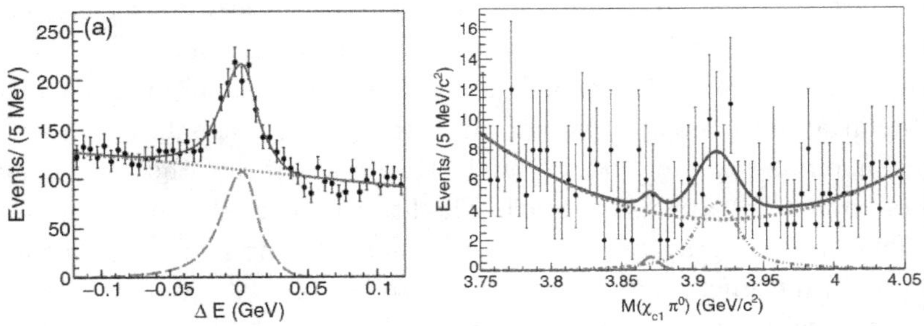

Fig. 2. Fit to the ΔE and ${}_s\mathcal{P}lot$ of $M_{\chi_{c1}\pi^0}$ distributions for $B^+ \to X(3872/3915)(\to \chi_{c1}\pi^0)K^+$.

The signal for $X(3872)$ and $X(3915)$ is observed with 0.31σ and 2.3σ signal significance, respectively. In the absence of significant signal, we set an upper limit (U.L.) on the product of branching fractions to be $\mathcal{B}(B \to X(3872)K) \times \mathcal{B}(X(3872) \to \chi_{c1}\pi^0) < 8.1 \times 10^{-6}$ and $\mathcal{B}(B \to X(3915)K) \times \mathcal{B}(X(3915) \to \chi_{c1}\pi^0) < 3.8 \times 10^{-5}$ at 90% C.L. which are compatible with $D^0D^{*0} + \chi_{c1}(2P)$ admixture scenario for $X(3872)$. We also estimate, $\mathcal{B}(X(3872) \to \chi_{c1}\pi^0/\mathcal{B}(X(3872) \to J/\psi\pi\pi) < 0.97$ at 90% C.L.

4. Evidence for $B \to h_c K$ and observation of $\eta_c(2S) \to p\bar{p}\pi^+\pi^-$ [13]

Since the current branching fraction of factorization suppressed[14] decay, $\mathcal{B}(B^+ \to \chi_{c0}K^+) = (1.49^{+0.15}_{-0.14}) \times 10^{-4}$ is smaller than the factorization allowed process $\mathcal{B}(B^+ \to \chi_{c1}K^+) = (4.84 \pm 0.23) \times 10^{-4}$, it is not strongly suppressed. Before the first experimental search, it was assumed that $\mathcal{B}(B^+ \to h_c K^+) \approx \mathcal{B}(B^+ \to \chi_{c0}K^+)$.[14] The upper limit on the branching fraction $\mathcal{B}(B^+ \to h_c K^+) < 3.8 \times 10^{-5}$ at 90% C.L. was obtained by the Belle in h_c search.[15] LHCb collaboration also set an upper limit on the product of branching fractions $\mathcal{B}(B^+ \to h_c K^+) \times \mathcal{B}(h_c \to p\bar{p}) < 6.4 \times 10^{-8}$ at 95% C.L.[16] Since $h_c \to p\bar{p}$ has never been observed and $\mathcal{B}(h_c \to p\bar{p}) < 1.5 \times 10^{-4}$ (90% C.L.), it does not strongly constrain the $B \to h_c K$. After the experimental upper limit, several theoretical predictions of $\mathcal{B}(B^+ \to h_c K^+)$ were given. In QCD factorization approach[17] the branching fraction has been calculated to be 2.7×10^{-5} and in the pQCD approach,[18] it is calculated to be 3.6×10^{-5}. Both results are slightly below the experimental results.

In this analysis, the h_c candidates are reconstructed from $\eta_c\gamma$ and $p\bar{p}\pi^+\pi^-$. The η_c candidates are reconstructed from ten different channels: $K^+K^0_S\pi^-$, $K^+K^-\pi^0$, $K^0_SK^0_S\pi^0$, $K^+K^-\eta$, $K^+K^-K^+K^-$, $\eta'(\to \eta\pi^+\pi^-)\pi^+\pi^-$, $p\bar{p}$, $p\bar{p}\pi^0$, $p\bar{p}\pi^+\pi^-$, and $\Lambda\bar{\Lambda}$. η candidates are reconstructed from $\gamma\gamma$ and $\pi^+\pi^-\pi^0$ channels. Multivariate analysis (MVA)[19] is used for each channel to separate signal from background. A simultaneous UML fit to the invariant mass of h_c is performed to $h_c \to \eta_c\gamma$ signal, $h_c \to p\bar{p}\pi^+\pi^-$ signal and $h_c \to p\bar{p}\pi^+\pi^-$ background. The decays of the other charmonium states to $p\bar{p}\pi^+\pi^-$ are also studied. The fit results are shown in Fig. 3.

The charged decay $B^+ \to h_c K^+$ is found with the signal significance of 4.8σ and branching fraction is measured to be $(3.7^{+1.0}_{-0.9} \pm 0.8) \times 10^{-5}$. It is consistent with exist upper limits. Since the neutral $B^0 \to h_c K^0_S$ decay is not found, an upper limit for branching fraction is set to be 1.4×10^{-5} at 90% C.L. and given for the first time. Further, the decay $\eta_c(2S) \to p\bar{p}\pi^+\pi^-$ is observed for the first time with 12.1σ significance in the invariant mass distribution of $p\bar{p}\pi^+\pi^-$ in $B \to p\bar{p}\pi^+\pi^- K$ channel. Other charmonium signals are consistent with current world average values.

5. Conclusion

We report recent results of the charmonium (-like) states at the Belle, the first observation of the decay $\eta_c(2S) \to p\bar{p}\pi^+\pi^-$, evidence of the decay $B^+ \to h_c K^+$ and the upper limits are set on the processes $B \to Y(4260)K$,

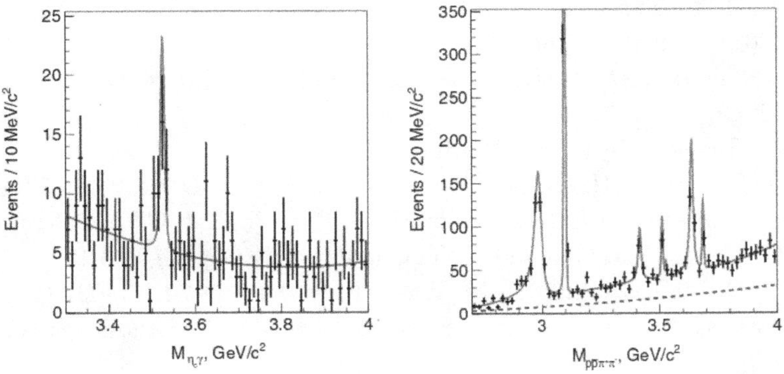

Fig. 3. Fit results in the $B^+ \to h_c K^+$ channel: $h_c \to \eta_c \gamma$ signal (left), $h_c \to p\bar{p}\pi^+\pi^-$ signal (right).

$B^0 \to h_c K_S^0$ and $X(3872/3915) \to \chi_{c1}\pi^0$. We expect with the 40 times of integrated luminosity in BelleII, a huge improvement in the measurements and will help to understand the exotic states.

References

1. S.-K. Choi *et al.* (Belle Collaboration), Phys. Rev. Lett. **91**, 262001 (2003).
2. R. Garg *et al.* (Belle Collaboration), Phys. Rev. D **99**, 071102(R) (2019).
3. B. Aubert *et al.* (BaBar Collaboration), Phys. Rev. D **73**, 011101(R) (2006).
4. R. M. Albuquerque, M. Nielsen, and C. M. Zanetti, Phys. Lett. B **747**, 83 (2015).
5. M. Pivk and F. R. Le Diberder, Nucl. Instrum. Methods Phys. Res. Sect. A **555**, 356 (2005).
6. V. Bhardwaj *et al.* (Belle Collaboration), Phys. Rev. D **99**, 111101(R) (2019).
7. M. Ablikim *et al.* (BESIII Collaboration), Phys. Rev. Lett. **122**, 202001 (2019).
8. S. Dubynskiy and M. B. Voloshin, Phys. Rev. D **77**, 014013 (2008).
9. S.-K. Choi *et al.* (Belle Collaboration), Phys. Rev. Lett. **94**, 182002 (2005).
10. F. K. Guo and Ulf-G. Meissner, Phys. Rev. D **86**, 091501(R) (2012).
11. S. L. Olsen, Phys. Rev. D **91**, 057501 (2015).

12. M. Tanabashi *et al.* (Particle Data Group), Phys. Rev. D **98**, 030001 (2018) and 2019 update.
13. K. Chilikin *et al.* (Belle Collaboration), Phys. Rev. D **100**, 012001 (2019).
14. M. Suzuki, Phys. Rev. D **66**, 037503 (2002).
15. F. Fang *et al.* (Belle Collaboration), Phys. Rev. D **74**, 012007 (2006).
16. R. Aaij *et al.* (LHCb Collaboration), Eur. Phys. J. C **73**, 2462 (2013).
17. C. Meng, Y. J. Gao and K. T. Chao, (2006), hep-ph/0607221.
18. X. Q. Li, X. Liu and Y. M. Wang, Phys. Rev. D **74**, 114029 (2006).
19. H. Voss, A. Hocker, J. Stelzer and F. Tegenfeldt, PoSACAT, 040 (2007).

Light-quark components analysis and the nature of the $Y(4260)$

Yun-Hua Chen

School of Mathematics and Physics, University of Science and Technology Beijing, Beijing 100083, China

Ling-Yun Dai

School of Physics and Electronics, Hunan University, Changsha 410082, China

Feng-Kun Guo

CAS Key Laboratory of Theoretical Physics, Institute of Theoretical Physics, Chinese Academy of Sciences, Beijing 100190, China
School of Physical Sciences, University of Chinese Academy of Sciences, Beijing 100049, China

Bastian Kubis

Helmholtz-Institut für Strahlen- und Kernphysik (Theorie) and Bethe Center for Theoretical Physics, Universität Bonn, 53115 Bonn, Germany

We study the processes $e^+e^- \to Y(4260) \to J/\psi\pi\pi(K\bar{K})$. The strong final-state interactions, especially the coupled-channel ($\pi\pi$ and $K\bar{K}$) final-state interaction in the S-wave are taken into account in a model-independent way using dispersion theory. It is found that the light-quark SU(3) octet state plays a significant role in these transitions, implying that the $Y(4260)$ contains a large light-quark component. Our findings suggest that the $Y(4260)$ is neither a hybrid nor a conventional charmonium state. Furthermore, through an analysis of the ratio of the light-quark SU(3) octet and singlet components, we show that the $Y(4260)$ does not behave like a pure $\bar{D}D_1$ hadronic molecule as well.

Keywords: Exotic states; dispersion theory; heavy quarkonium.

The nature of the $Y(4260)$ has remained controversial since its discovery by the *BABAR* Collaboration in the $J/\psi\pi^+\pi^-$ channel in the initial-state radiation process $e^+e^- \to \gamma_{\rm ISR}J/\psi\pi^+\pi^-$. Its mass is inconsistent with the naive quark model prediction for a normal vector charmonium,[1] and it does not show strong couplings to ground-state open-charm decay modes although it is above the $D\bar{D}$ threshold. Such peculiar properties have initiated a lot of theoretical and experimental studies, see Refs. 2, 3 for recent reviews. In this work, we study the $e^+e^- \to Y(4260) \to J/\psi\pi\pi(J/\psi K\bar{K})$ processes to extract information on possible light-quark components of the $Y(4260)$. If the $Y(4260)$ contains no light quarks (as in the hybrid-state or the conventional-charmonium scenarios), the light-quark source provided by the $Y(4260)$ has to be in the form of an SU(3) flavor singlet state. Thus the study of the different light-quark SU(3) flavor eigenstate components provided by the $Y(4260)$ in these processes may help to reveal its structure.

For the $Y(4260)J/\psi\pi\pi$ and $Y(4260)J/\psi K\bar{K}$ contact couplings, the leading chiral effective Lagrangian reads[4–7]

$$\begin{aligned}\mathscr{L}_{Y\psi\Phi\Phi} = {} & g_1\langle V_1^{\alpha}J_{\alpha}^{\dagger}\rangle\langle u_\mu u^\mu\rangle + h_1\langle V_1^{\alpha}J_{\alpha}^{\dagger}\rangle\langle u_\mu u_\nu\rangle v^\mu v^\nu \\ & + g_8\langle J_{\alpha}^{\dagger}\rangle\langle V_8^{\alpha}u_\mu u^\mu\rangle + h_8\langle J_{\alpha}^{\dagger}\rangle\langle V_8^{\alpha}u_\mu u_\nu\rangle v^\mu v^\nu + \text{H.c.}\,, \end{aligned} \tag{1}$$

where the parameters g_1 and h_1 correspond to the contributions from the SU(3) singlet component of the $Y(4260)$, and g_8 and h_8 are the corresponding parameters for the SU(3) octet component. The strong pion–pion final-state interactions (FSIs) as well as the $K\bar{K}$ coupled channel in the S-wave are taken into account model independently by using dispersion theory. In the dispersion theory, the left-hand-cut contributions are approximated by the sum of the $Z_c(3900)$-exchange mechanism and the triangle diagrams $Y(4260) \to \bar{D}D_1(2420) \to \bar{D}D^*\pi(\bar{D}D_s^*K) \to J/\psi\pi\pi(J/\psi K\bar{K})$.[8,9] The subtraction terms can be determined by matching to the chiral contact terms, since at low energies the amplitude should agree with the results required by chiral symmetry. To estimate the uncertainty due to the dispersive input for the $\pi\pi/K\bar{K}$ rescattering, we will use two different $T_0^0(s)$ matrices, the Dai–Pennington (DP)[10,11] and the Bern/Orsay (BO)[12,13] parametrizations, and compare the results. For details of the theoretical treatment, we refer to Ref. 14.

We fit to the experimental data of the $\pi\pi$ invariant mass spectra of $e^+e^- \to J/\psi\pi^+\pi^-$ and the ratios of the cross sections $\sigma(e^+e^- \to J/\psi K^+K^-)/\sigma(e^+e^- \to J/\psi\pi^+\pi^-)$ measured at $E = 4.26\,\text{GeV}$ by the BESIII Collaboration.[15,16] To illustrate the effect of the SU(3) octet component, we perform two fits. In Fit a we only consider the SU(3) singlet component, the Z_c-exchange terms, and the triangle diagrams, while in Fit b, the SU(3) octet component is taken into account in addition. The best fit results of the $\pi\pi$ mass spectra in $e^+e^- \to J/\psi\pi^+\pi^-$ are shown in Fig. 1. The fit results of the ratios of the cross sections $\sigma(e^+e^- \to J/\psi K^+K^-)/\sigma(e^+e^- \to J/\psi\pi^+\pi^-)$ are given in Table 1.

One observes that Fit a, in which the SU(3) octet chiral contact terms are not included, cannot describes the experimental data well, especially for

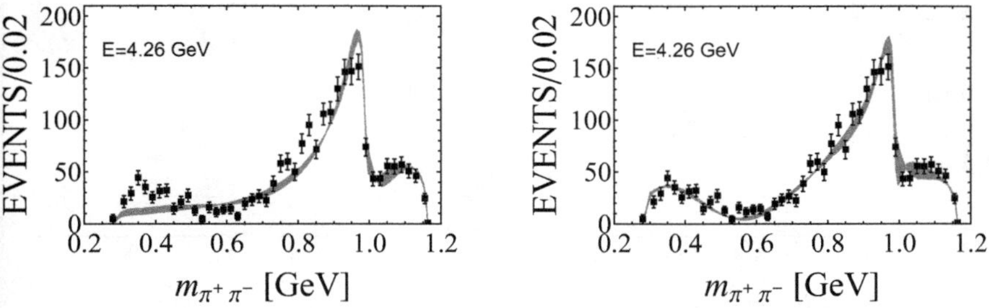

Fig. 1. Fit results of the $\pi\pi$ invariant mass spectra in $e^+e^- \to J/\psi\pi^+\pi^-$ for Fit a (left) and Fit b (right). The borders of the bands represent our best fit results using two different $T_0^0(s)$ matrices. The experimental data are taken from Ref. 15.

the broad peak below 0.6 GeV. In contrast, in Fit b with the SU(3) octet chiral contact terms added, the fit quality is improved significantly. Using the fit results, we can analyze the ratio of the parameters for the SU(3) octet component relative to those for the SU(3) singlet component. In the DP parametrization we find $g_8/g_1 = 1.2 \pm 0.2$ and $h_8/h_1 = 57 \pm 76$, while in the BO parametrization we obtain $g_8/g_1 = 1.1 \pm 0.1$ and $h_8/h_1 = 102 \pm 152$, which agree well with each other within errors. Note that in the $\bar{D}D_1$ hadronic molecule scenario of the $Y(4260)$, the light-quark component is in the form of $|u\bar{u} + d\bar{d}\rangle/\sqrt{2} = (\sqrt{2}V_1^{\text{light}} + V_8^{\text{light}})/\sqrt{3}$, where $V_1^{\text{light}} = \frac{1}{\sqrt{3}}(u\bar{u} + d\bar{d} + s\bar{s})$ and $V_8^{\text{light}} = \frac{1}{\sqrt{6}}(u\bar{u} + d\bar{d} - 2s\bar{s})$. They give the ratio of $1/\sqrt{2}$, which differs significantly from our results (given by the values of g_8/g_1). Therefore we conclude that the $Y(4260)$ contains a large light-quark SU(3) octet component, and the scenarios of a hybrid or conventional charmonium are disfavored since the light quarks have to be produced in an SU(3) singlet configuration inside such states. Also our study shows that the $Y(4260)$ cannot be a pure $\bar{D}D_1$ hadronic molecule as well.

In Fig. 2, we plot the moduli of the S- and D-wave amplitudes from the chiral contact terms, the Z_c-exchange terms, and the triangle diagrams for Fit b. One observes that the D-wave contribution is comparable to the S-wave contribution in almost the whole energy region. Such a large D-wave contribution again indicates that the $Y(4260)$ cannot be a conventional

Table 1. Experimental and theoretical values for the cross sections ratio $\sigma(e^+e^- \to J/\psi K^+K^-)/\sigma(e^+e^- \to J/\psi\pi^+\pi^-) \times 10^2$ at $E = 4.26$ GeV.

Experiment	Fit a, DP	Fit b, DP	Fit a, BO	Fit b, BO
4.99 ± 1.10	4.46 ± 0.82	4.67 ± 0.98	5.37 ± 1.03	5.38 ± 0.82

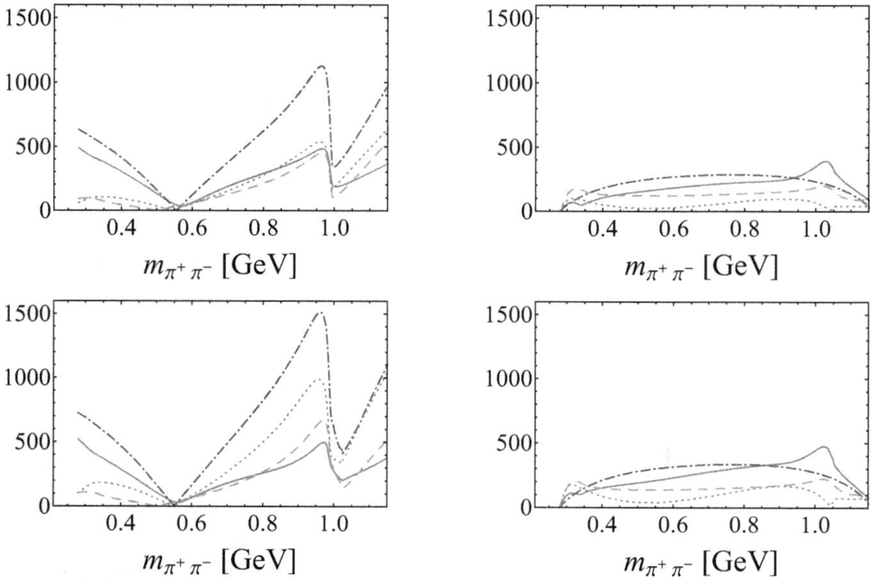

Fig. 2. The moduli of the S- (left) and D-wave (right) amplitudes for $e^+e^- \to J/\psi\pi^+\pi^-$ in Fit b, using the DP (top) or the BO (bottom) parametrizations. The red solid lines represent our best fit results, while the blue dot-dashed, darker green dashed, and magenta dotted lines correspond to the contributions from the chiral contact terms, Z_c-exchange, and the triangle diagrams, respectively.

charmonium state, for which the $\pi\pi$ S-wave should be dominant. Also note that in the $\bar{D}D_1$ hadronic molecule interpretation,[8,17] the $\pi\pi$ D-wave emerges naturally since the D_1 decays dominantly into D-wave $D^*\pi$.

At last, we anticipate a combined analysis of both the $Y(4260)$ and $Z_c(3900)$ data, which will be crucial to reveal the nature of both states.

Acknowledgments

This work is supported in part by the Fundamental Research Funds for the Central Universities under Grants No. 531107051122 and No. FRF-BR-19-001A, by the National Natural Science Foundation of China (NSFC) under Grants Nos. 11975028, 11805059, 11847612, and No. 11835015, by NSFC and Deutsche Forschungsgemeinschaft (DFG) through funds provided to the Sino–German Collaborative Research Center "Symmetries and the Emergence of Structure in QCD" (NSFC Grant No. 11621131001, DFG Grant No. TRR110), by the Thousand Talents Plan for Young Professionals, by the Chinese Academy of Sciences (CAS) Key Research Program of Frontier Sciences (Grant No. QYZDB-SSW-SYS013), by the CAS Key Research Program (Grant No. XDPB09), and by the CAS Center for Excellence in Particle Physics (CCEPP).

References

1. S. Godfrey and N. Isgur, Phys. Rev. D **32**, 189 (1985).
2. H.-X. Chen, W. Chen, X. Liu, and S.-L. Zhu, Phys. Rep. **639**, 1 (2016).
3. F.-K. Guo, C. Hanhart, U.-G. Meißner, Q. Wang, Q. Zhao, and B.-S. Zou, Rev. Mod. Phys. **90**, 015004 (2018).
4. T. Mannel and R. Urech, Z. Phys. C **73**, 541 (1997).
5. Y.-H. Chen, J. T. Daub, F.-K. Guo, B. Kubis, U.-G. Meißner, and B.-S. Zou, Phys. Rev. D **93**, 034030 (2016).
6. Y.-H. Chen, M. Cleven, J. T. Daub, F.-K. Guo, C. Hanhart, B. Kubis, U.-G. Meißner, and B.-S. Zou, Phys. Rev. D **95**, 034022 (2017).
7. Y.-H. Chen and F.-K. Guo, Phys. Rev. D **100**, 054035 (2019).
8. M. Cleven, Q. Wang, F.-K. Guo, C. Hanhart, U.-G. Meißner, and Q. Zhao, Phys. Rev. D **90**, 074039 (2014).
9. M. Albaladejo, F.-K. Guo, C. Hidalgo-Duque, and J. Nieves, Phys. Lett. B **755**, 337 (2016).
10. L.-Y. Dai and M. R. Pennington, Phys. Lett. B **736**, 11 (2014).
11. L.-Y. Dai and M. R. Pennington, Phys. Rev. D **94**, 116021 (2016).
12. I. Caprini, G. Colangelo, and H. Leutwyler, Eur. Phys. J. C **72**, 1860 (2012).
13. P. Büttiker, S. Descotes-Genon, and B. Moussallam, Eur. Phys. J. C **33**, 409 (2004).
14. Y.-H. Chen, L.-Y. Dai, F.-K. Guo, and B. Kubis, Phys. Rev. D **99**, 074016 (2019).
15. M. Ablikim *et al.* (BESIII Collaboration), Phys. Rev. Lett. **119**, 072001 (2017).
16. M. Ablikim *et al.* (BESIII Collaboration), Phys. Rev. D **97**, 071101 (2018).
17. Y. Lu, M. N. Anwar, and B.-S. Zou, Phys. Rev. D **96**, 114022 (2017).

Lifetime and dominant decay modes of the tetraquark with double beauty $bb\bar{u}\bar{d}$

J.-M. Richard

Université de Lyon, Institut de Physique des 2 Infinis de Lyon, IN2P3-CNRS–UCBL, 4 rue Enrico Fermi, 69622 Villeurbanne, France
E-mail: j-m.richard@ipnl.in2p3.fr

E. Hernández and A. Valcarce

Departamento de Física Fundamental e IUFFyM, Universidad de Salamanca, 37008 Salamanca, Spain

J. Vijande

Unidad Mixta de Investigación en Radiofísica e Instrumentación Nuclear en Medicina (IRIMED), Instituto de Investigación Sanitaria La Fe (IIS-La Fe)-Universitat de Valencia (UV) and IFIC (UV-CSIC) 46100 Valencia, Spain

We briefly review the stability of the $QQ\bar{q}\bar{q}$-type of tetraquarks with two heavy quarks and two light antiquarks. We present the first comprehensive estimate of the lifetime and leading decay modes of the exotic meson $bb\bar{u}\bar{d}$ with double beauty.

Keywords: Exotic hadron; lifetime; weak decay of beauty.

1. Introduction

There is an exciting activity dealing with exotic or crypto-exotic hadrons containing heavy quarks and/or antiquarks. Most experimental searches and results deal with hidden-flavor states such as the XYZ mesons or P_c baryons, which in some cases are remarkably narrow, but still unstable with respect to dissociation into lighter hadrons. There is also a renewed interest in flavor-exotic mesons which were already foreseen decades ago and seemingly will become accessible to the next generation of experiments. It is predicted that some $QQ\bar{q}\bar{q}$ states with two heavy quarks and two light antiquarks lie below their lowest dissociation threshold, and thus will provide the first evidence for *stable* multiquark states beyond the familiar nuclei. Some relevant reviews include Refs. 1–6.

The physics of weak interactions will be much stimulated by the occurrence of such states. The decay of strangeness has revealed the possibility of oscillations, the Cabibbo mixing scheme, and the dominance of nonleptonic modes for hyperon decays, due to the strong final state interaction. The decay of charmed hadrons has shown how the lifetimes and the semileptonic branching ratios are sensitive to the light constituents surrounding the charmed quark, with very different values for D^+ and D^0, and for the various charmed baryons. In contrast, the hadrons with single beauty, such as the B mesons and the Λ_b baryons have a remarkably similar lifetime $\tau \sim 1.5\,\text{ps}$, suggesting that the b quark, while decaying, ignores its environment.

Our aim is to review briefly the stability of the $T_{bb}^- \equiv bb\bar{u}\bar{d}$ isoscalar exotic meson in constituent models, and to study its decay modes. Naively, a hadron with two b quarks should decay about twice faster than a hadron with a single b. This is the result found for bbq baryons.[7] In the case of the tetraquark, two effects might act in the opposite direction. Firstly, for a typical decay $T_{bb} \to \bar{B}^{(*)} + D^{(*)} + X$, a fraction of the released energy is taken by the spectator $\bar{B}^{(*)}$, and less effective phase-space is left for the weak decay itself. It is known, from many examples in the study of nuclear β decays, how sensitive are the weak interactions upon the available phase space. Secondly, after the $b \to c$ transition, the hadronization is somewhat more laborious than in the case of B decay, because the antiquarks are further apart and not always in an optimal color state. This is why a detailed calculation is necessary, using realistic wave functions for both the tetraquark and the hadrons in the final state.[8]

2. Stability of doubly-heavy tetraquarks

The first theoretical indication for the stability of $QQ\bar{q}\bar{q}$ tetraquarks was made nearly four decades ago,[9] and confirmed by several groups, as reviewed, e.g., in Ref. 3. It can be understood in the following way. Let us denote by M and m the constituent masses, and start from the case where $M = m$, being first restricted to a spin-independent interaction, which is assumed to be *flavor independent.* If M is increased and m decreased with the average inverse mass kept constant, the energy of the tetraquark is lowered by the breaking of symmetry (namely charge conjugation), while the threshold energy remains constant. For a large enough mass ratio M/m, stability is reached. This is the same mechanism by which the hydrogen molecule is much more stable than the positronium molecule, when the bindings are scaled by their respective threshold energies.

Explicit four-body calculations show that this chromoelectric mechanism, though fully rigorous, it is not sufficient to provide binding, even for $bb\bar{q}\bar{q}$. Fortunately, there is another effect, due to chromomagnetic forces. In an isoscalar $QQ\bar{u}\bar{d}$ state with $J^P = 1^+$, the $\bar{u}\bar{d}$ pair is mainly in a spin $s_{ud} = 0$ and orbital momentum $\ell_{ud} = 0$ state, and benefits from a strong spin-spin attraction. In the threshold, the spin-spin effects are penalized by the mass of the heavy quark.

Note that we do not have to *assume* any diquark clustering beforehand. We simply note that the dynamics induces some correlation between the two heavy quarks, due to chromoelectricity, and between the two light antiquarks, due to chromomagnetism. Pairwise interactions based on color-octet exchange induce mixing between the $\bar{3}3$ and $6\bar{6}$ states in the $QQ - \bar{q}\bar{q}$ basis, enhancing the $\bar{3}3$ components for larger values of M due to the attractive chromoelectric interaction of the QQ pair that it is absent in the $Q\bar{q}$ threshold. This result is only valid in the bottom sector. In the charm sector, the binding mechanism is different: the $\bar{3}3$ and $6\bar{6}$ components have a similar probability and are mixed by the chromomagnetic interaction. A brute-force diquark approximation might be misleading.[10]

Within explicit quark models, the state of art of rigorous calculations is the following. The double-charm isoscalar tetraquark $cc\bar{u}\bar{d}$ with $J^P = 1^+$ is probably slightly bound below its formal threshold DD^*,[11] i.e., should be searched for as a peak in $DD\pi$ or $DD\gamma$. The double-beauty isoscalar T_{bb} is stable, lying approximately 150 MeV below the strong decay threshold $B^-\bar{B}^{*0}$ and 105 MeV below the electromagnetic decay threshold $B^-\bar{B}^0\gamma$. Thus it decays weakly.[8] The isoscalar $bc\bar{u}\bar{d}$ might be stable too, and if this is the case, can bear $J^P = 0^+$ or 1^+. While the $J^P = 0^+$ state would be strong and electromagnetic-interaction stable, the $J^P = 1^+$ would decay electromagnetically to $\bar{B}D\gamma$.[12,13]

3. T_{bb} weak decays

We have calculated the width of the isoscalar T_{bb} tetraquark due to all plausible semileptonic and nonleptonic decay modes.[8] The hadronic decays are calculated within the factorization approximation. Some of the corresponding processes are illustrated in Fig. 1. The largest partial widths are found to be of the order of 10^{-15} to 10^{-14} GeV. For the semileptonic modes, the corresponding decays are of the type $\bar{B}^* D^* \ell^- \bar{\nu}_\ell$, where $\ell = e$ or μ. Due to the large phase space available in all cases, the differences among the widths into the three lepton families are very small.

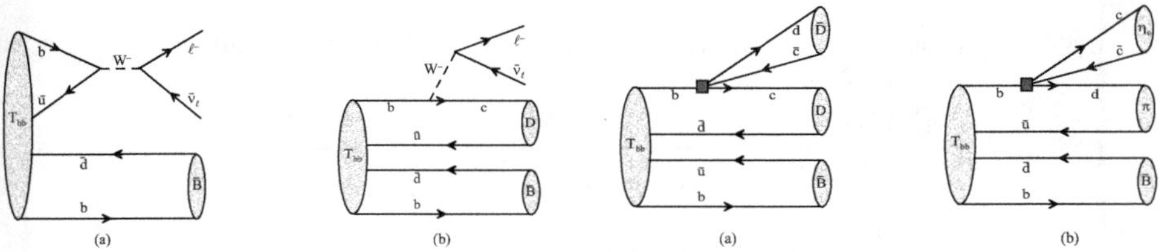

Fig. 1. Representative diagrams for semileptonic (left) and nonleptonic (right) decays of the T_{bb} tetraquark

For semileptonic decays involving two mesons in the final state, the processes involving a $b \to c$ vertex are favored compared to those involving a $b \to u$ vertex, due to the larger CKM matrix element. For the former, see Table 1, and, for the latter, a typical estimate $\Gamma(T_{bb} \to \bar{B}^{*0}\,\ell^-\,\bar{\nu}_\ell) \sim 0.04 \times 10^{-15}$ GeV. Besides, due to spin recoupling coefficients, the largest decay widths appear for vector mesons in the final state.

Table 1. Decay widths, in units of 10^{-15} GeV, of the leading semileptonic modes of T_{bb}

Final state	Γ	Final state	Γ
$B^{*-}\,D^{*+}\,\ell^-\,\bar{\nu}_\ell$ $\bar{B}^{*0}\,D^{*0}\,\ell^-\,\bar{\nu}_\ell$	9.02 ± 0.07	$B^{*-}\,D^{*+}\,\tau^-\,\bar{\nu}_\tau$ $\bar{B}^{*0}\,D^{*0}\,\tau^-\,\bar{\nu}_\tau$	1.55 ± 0.01
$B^{*-}\,D^{+}\,\ell^-\,\bar{\nu}_\ell$ $\bar{B}^{*0}\,D^{0}\,\ell^-\,\bar{\nu}_\ell$	3.59 ± 0.03	$B^{*-}\,D^{+}\,\tau^-\,\bar{\nu}_\tau$ $\bar{B}^{*0}\,D^{0}\,\tau^-\,\bar{\nu}_\tau$	0.727 ± 0.005
$B^{-}\,D^{*+}\,\ell^-\,\bar{\nu}_\ell$ $\bar{B}^{0}\,D^{*0}\,\ell^-\,\bar{\nu}_\ell$	4.63 ± 0.05	$B^{-}\,D^{*+}\,\tau^-\,\bar{\nu}_\tau$ $\bar{B}^{0}\,D^{*0}\,\tau^-\,\bar{\nu}_\tau$	0.86 ± 0.007
$B^{-}\,D^{+}\,l^-\,\bar{\nu}_l$ $\bar{B}^{0}\,D^{0}\,\ell^-\,\bar{\nu}_\ell$	1.92 ± 0.02	$B^{-}\,D^{+}\,\tau^-\,\bar{\nu}_\tau$ $\bar{B}^{0}\,D^{0}\,\tau^-\,\bar{\nu}_\tau$	0.409 ± 0.003

For the nonleptonic decays, the largest widths are for the decays of the type $\bar{B}^* D^* D_s{}^*$. All of them contain a $b \to c$ vertex and a D_s^* meson in the final state. See Table 2. Once again vector mesons are favored in the final state. Processes with D_s or a light meson final states arising from vacuum have decay widths comparable to the corresponding semileptonic decay.

In the event where an isoscalar T_{bc} tetraquark with $J^P = 0^+$ exists, then the decay $T_{bb}(1^+) \to T_{bc}(0^+)\ell^-\nu$ is about $3. \times 10^{-15}$ GeV. The semileptonic decay to the isoscalar $J^P = 0^+$ tetraquark T^0_{bc} is relevant but it is not found to be dominant.

We also estimated all plausible decay modes such as $\bar{B}^0 e^- \bar{\nu}_e$, or $B^{*-}D^+\pi^-$, etc. The total width turns out to be about 87×10^{-15} GeV, which gives an upper bound for the lifetime of about 7.6 ps.

Table 2. Decay widths, in units of 10^{-15} GeV, of the leading nonleptonic modes of T_{bb}

Final state	Γ	Final state	Γ
$B^{*-}\,D^{*+}\,D_s^-$	4.00 ± 0.06	$B^-\,D^{*+}\,D_s^{*-}$	3.15 ± 0.05
$\bar{B}^{*0}\,D^{*0}\,D_s^-$		$\bar{B}^0\,D^{*0}\,D_s^{*-}$	
$B^{*-}\,D^{*+}\,D_s^{*-}$	6.50 ± 0.09	$B^-\,D^+\,D_s^{*-}$	1.20 ± 0.02
$\bar{B}^{*0}\,D^{*0}\,D_s^{*-}$		$\bar{B}^0\,D^0\,D_s^{*-}$	
$B^{*-}\,D^+\,D_s^-$	2.57 ± 0.04	$B^{*-}\,D^{*+}\,\rho^-$	3.57 ± 0.09
$\bar{B}^{*0}\,D^0\,D_s^-$		$B^{*-}\,D^{*+}\,\pi^-$	1.28 ± 0.03
$B^{*-}\,D^+\,D_s^{*-}$	2.32 ± 0.03	$B^{*-}\,D^+\,\rho^-$	1.70 ± 0.04
$\bar{B}^{*0}\,D^0\,D_s^{*-}$		$B^{*-}\,D^+\,\pi^-$	0.70 ± 0.02
$B^-\,D^{*+}\,D_s^-$	2.78 ± 0.05	$B^-\,D^{*+}\,\rho^-$	2.01 ± 0.05
$\bar{B}^0\,D^{*0}\,D_s^-$		$B^-\,D^{*+}\,\pi^-$	0.77 ± 0.03

4. Outlook

We have examined the stability of $QQ\bar{q}\bar{q}$ tetraquarks, showing how the structure of the T_{QQ} state evolves from a molecular-like system to a compact-like structure when moving from the charm to the bottom sector, due to a competition between chromoelectric and chromomagnetic forces. We have also examined a variety of decay modes of T_{bb}, either semileptonic and nonleptonic. The leading modes involve a $\bar{B}^{(*)}$ and a $D^{(*)}$ in the final state. The experimental search might look at channels with a smaller branching ratio but an easier identification, as the final states pointed out in Refs. 14–16.

The most striking result is our estimate of the lifetime

$$\tau(T_{bb}) \sim 7.6\,\text{ps}\ , \tag{1}$$

to be compared with about 1.5 ps for hadrons with single beauty. For instance, if one searches for double-b hadrons by the method of displaced vertex,[17] the tetraquark will give a signature rather different from that of the baryons. See, also, the discussion in Refs. 15, 16.

References

1. H.-X. Chen, W. Chen, X. Liu and S.-L. Zhu, The hidden-charm pentaquark and tetraquark states, *Phys. Rept.* **639**, 1 (2016).
2. R. F. Lebed, R. E. Mitchell and E. S. Swanson, Heavy-Quark QCD Exotica, *Prog. Part. Nucl. Phys.* **93**, 143 (2017).
3. J.-M. Richard, Exotic hadrons: review and perspectives, *Few Body Syst.* **57**, 1185 (2016).

4. A. Ali, L. Maiani and A. D. Polosa, *Multiquark Hadrons* (Cambridge University Press, 2019).
5. N. Brambilla, S. Eidelman, C. Hanhart, A. Nefediev, C.-P. Shen, C. E. Thomas, A. Vairo and C.-Z. Yuan, The XYZ states: experimental and theoretical status and perspectives, *arXiv:1907.07583* (2019).
6. M. Tanabashi *et al.*, Review of Particle Physics, *Phys. Rev.* **D98**, 030001 (2018).
7. V. V. Kiselev and A. K. Likhoded, Baryons with two heavy quarks, *Phys. Usp.* **45**, 455 (2002), [Usp. Fiz. Nauk **172**, 497 (2002)].
8. E. Hernández, J. Vijande, A. Valcarce and J.-M. Richard, Spectroscopy, lifetime and decay modes of the T_{bb}^- tetraquark, *Phys. Lett.* **B800**, 135073 (2020).
9. J. P. Ader, J. M. Richard and P. Taxil, Do narrow heavy multiquark states exist?, *Phys. Rev.* **D25**, 2370 (1982).
10. J.-M. Richard, A. Valcarce and J. Vijande, Few-body quark dynamics for doubly heavy baryons and tetraquarks, *Phys. Rev.* **C97**, 035211 (2018).
11. D. Janc and M. Rosina, The $T_{cc} = DD^*$ molecular state, *Few Body Syst.* **35**, 175 (2004).
12. T. Caramés, J. Vijande and A. Valcarce, Exotic $bc\bar{q}\bar{q}$ four-quark states, *Phys. Rev.* **D99**, 014006 (2019).
13. A. Francis, R. Hudspith, R. Lewis and K. Maltman, Evidence for charm-bottom tetraquarks and the mass dependence of heavy-light tetraquark states from lattice QCD, *Phys. Rev.* **D99**, 054505 (2019).
14. Y. Xing and R. Zhu, Weak Decays of Stable Doubly Heavy Tetraquark States, *Phys. Rev.* **D98**, 053005 (2018).
15. A. Ali, Q. Qin and W. Wang, Discovery potential of stable and near-threshold doubly heavy tetraquarks at the LHC, *Phys. Lett.* **B785**, 605 (2018).
16. A. Ali, A. Y. Parkhomenko, Q. Qin and W. Wang, Prospects of discovering stable double-heavy tetraquarks at a Tera-Z factory, *Phys. Lett.* **B782**, 412 (2018).
17. T. Gershon and A. Poluektov, Displaced B_c^- mesons as an inclusive signature of weakly decaying double beauty hadrons, *JHEP* **01**, 019 (2019).

Spectrum of the fully-heavy tetraquark state $QQ\bar{Q}'\bar{Q}'$

Guang-Juan Wang*, Lu Meng and Shi-Lin Zhu

Center of High Energy Physics, Peking University, Beijing 100871, China
School of Physics and State Key Laboratory of Nuclear Physics and Technology, Peking University, Beijing 100871, China
**Email: wgj@pku.edu.cn*

We systematically study the mass spectrum of the S-wave fully-heavy tetraquark states $QQ\bar{Q}'\bar{Q}'$ $(Q, Q' = c, b)$ using two nonrelativistic quark models. We include the mixing effect between the $6_c - \bar{6}_c$ and $\bar{3}_c - 3_c$ color configurations and find that the $6_c - \bar{6}_c$ component may play a quite important role in the ground state. The obtained mass spectrum show that the ground states are located above the di-meson thresholds and may not support the existence of a bound tetraquark state $QQ\bar{Q}'\bar{Q}'$.

Keywords: Fully-heavy tetraquark states; quark model.

1. Introduction

Recently, the CMS Collaboration reported the observation of the $\Upsilon(1S)$ pair production.[1] The analysis with CMS data indicated a $bb\bar{b}\bar{b}$ signal around 18.4 GeV with a global significance of 3.6σ.[2] Later, the LHCb searched the $bb\bar{b}\bar{b}$ state in the $\Upsilon(1s)\mu^+\mu^-$ channel and did not find significant excess in the mass range $17.5 \sim 20.0$ GeV.[3] In the literature, there have been many theoretical studies about the fully-heavy tetraquark states with different models.[4–10] However, the existence of the fully-heavy tetraquark states is still in debating.

The fully-heavy tetraquark state $QQ\bar{Q}'\bar{Q}'$ is quite important. They are good candidates of the compact tetraquark states since the interactions between the inner heavy quarks are dominated by the short-range potentials. Then, we can use the nonrelativistic quark model with the short range one-gluon-exchange and a phenomenological linear confinement potentials to study the fully-heavy tetraquarks. In addition, the four inner quarks can form the color singlet representation in two ways, $6_c \otimes \bar{6}_c \to 1_c$ and $\bar{3}_c \otimes 3_c \to 1_c$. The study of the $QQ\bar{Q}'\bar{Q}'$ states may help us to investigate the color configurations in the multi-quark systems.

2. Formalism

For a $Q_1Q_2\bar{Q}_3\bar{Q}_4$ tetraquark state, the total Hamiltonian reads

$$H = \sum_{i=1}^{4} \frac{p_j^2}{2m_j} + \sum_{i<j} V_{ij} + \sum_i m_i = \frac{p^2}{2u} + V_I + h_{12} + h_{34}, \tag{1}$$

$$V_I = V_{13} + V_{14} + V_{23} + V_{24}, \;\; h_{ij} = \frac{p_{ij}^2}{2u_{ij}} + V_{ij} + m_i + m_j, \tag{2}$$

where h_{12}/h_{34} is the Hamiltonian of two quark/antiquark clusters and diagonal in the color-flavor-spin space. V_I denotes the potentials between the two clusters and may induce the mixing between different color-flavor-spin configurations. The non-relativistic quark model is a reasonable theoretical framework to depict interactions between the heavy quarks. In quark model I, the V_{ij} reads[11]

$$V_{ij}(r_{ij}) = \frac{\lambda_i}{2}\frac{\lambda_j}{2}\left(\frac{\alpha_s}{r_{ij}} - \frac{3b}{4}r_{ij} - \frac{8\pi\alpha_s}{3m_im_j}\mathbf{s}_i\cdot\mathbf{s}_j e^{-\sigma^2r^2}\frac{\sigma^3}{\pi^{3/2}} + V_{\text{cons}}\right), \tag{3}$$

with the running coupling constant

$$\alpha_s(Q^2) = \frac{12\pi}{(33-2N_f)\ln(A+Q^2/B^2)}. \tag{4}$$

where λ is the color matrix for a quark ($-\lambda^*$ for an antiquark). To investigate the model dependence, we also adopt another quark model to study the tetraquark state,[12]

$$\begin{aligned} V_{ij}(r_{ij}) = &-\frac{3}{16}\sum_{i<j}\lambda_i\lambda_j\Big(-\frac{\kappa(1-\exp(-r_{ij}/r_c))}{r_{ij}} + \lambda r_{ij}^p \\ &-\Lambda + \frac{8\pi}{3m_im_j}\kappa'(1-\exp(-r_{ij}/r_c))\frac{\exp(-r_{ij}^2/r_0^2)}{\pi^{3/2}r_0^3}\mathbf{s}_i\cdot\mathbf{s}_j\Big), \end{aligned} \tag{5}$$

with $r_0 = A(\frac{2m_im_j}{m_i+m_j})^{-B}$. Here, the hyperfine potential in model II contains all the mass information of the interacting quarks, and one expects it to play a more important role than that in model I. The parameters in the two models are determined by fitting the meson spectra.[13]

The wave function of a tetraquark state is described by three independent Jacobi coordinates as illustrated in Fig. 1. The wave function is

$$\psi_{JJ_z} = \sum\left[\varphi_{n_aJ_a}(\mathbf{r}_{12},\beta_a)\otimes\varphi_{n_bJ_b}(\mathbf{r}_{34},\beta_b)\otimes\phi_{NL_{ab}}(\mathbf{r},\beta)\right]_{JJ_z}, \tag{6}$$

$$\varphi_{n_aJ_aM_a} = \left[\phi_{n_al_a}(\mathbf{r}_{12},\beta_a)\chi_{s_a}\right]_{M_a}^{J_a}\chi_f\chi_{c_a}, \tag{7}$$

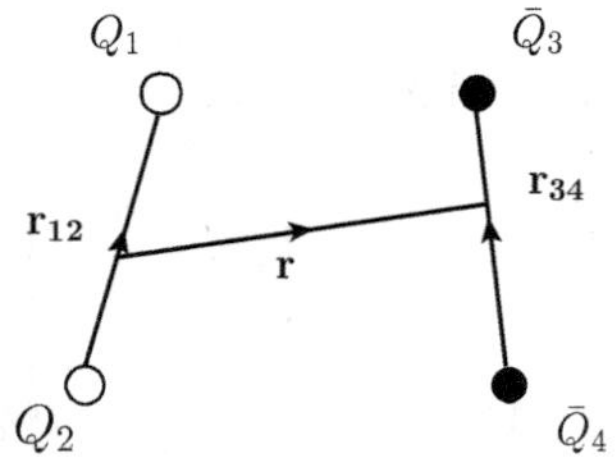

Fig. 1. The Jacobi coordinates in a tetraquark state.

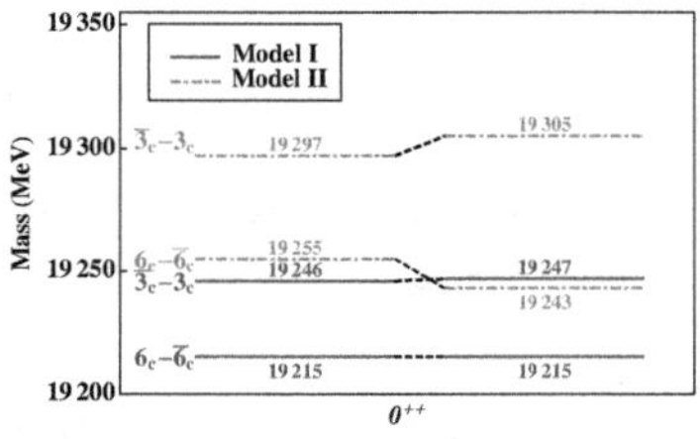

Fig. 2. The 0^{++} $bb\bar{b}\bar{b}$ states with and without color mixing.

where $\chi_{f,s,c}$ is the flavor, spin, and color wave functions. The spatial wave function ϕ is expressed by the Gaussian basis,

$$\phi_{n_a l_a m_a}(\mathbf{r}_{12},\beta_a) = i^{l_a} r_{12}^{l_a}\sqrt{\frac{4\pi}{(2l_a+1)!!}}\left(\frac{n_a\beta_a^2}{\pi}\right)^{3/4}(2n_a\beta_a^2)^{l_a/2}e^{-r_{12}^2\beta_a^2 n_a/2}Y_{l_a m_a}(\mathbf{\Omega}_{12}),$$

For the ground S-wave tetraquark states, we only consider the basis functions with the orbital angular momentum $l_a = l_b = L_{ab} = 0$ and neglect the contributions from the higher orbital excitations induced by the tensor or the spin-orbital potentials. Constrained by the Fermi statistic, the color-flavor-spin wave functions of the states with different J^{PC} reads

$$\begin{aligned}
0^{++} &: \chi_1 = \left[[QQ]^1_{\bar{3}_c}[\bar{Q}\bar{Q}]^1_{3_c}\right]^0_{1_c}, \quad \chi_2 = \left[[QQ]^0_{6_c}[\bar{Q}\bar{Q}]^0_{\bar{6}_c}\right]^0_{1_c}.\\
1^{+-} &: \chi_1 = \left[[QQ]^1_{\bar{3}_c}[\bar{Q}\bar{Q}]^1_{3_c}\right]^1_{1_c}.\\
2^{++} &: \chi_1 = \left[[QQ]^1_{\bar{3}_c}[\bar{Q}\bar{Q}]^1_{3_c}\right]^2_{1_c}.
\end{aligned} \tag{8}$$

With the Hamiltonian and the wave functions, we solve the Schrödinger equation with the variational method and obtain the mass spectrum.

3. Results and discussions

The tetraquark state with $J^{PC} = 0^{++}$ is an admixture of the $\bar{3}_c \otimes 3_c$ and $6_c \otimes \bar{6}_c$ configurations. With the coupled channel effects, we obtain the mass spectra of the tetraquark states as listed in Table 1. We present the 0^{++} $bb\bar{b}\bar{b}$ as an example in Table 2. We find that the two quark models reproduce the similar mass spectra up to tens of MeV.

The numerical results show that the $6_c \otimes \bar{6}_c$ is important even dominant in the ground state. The quark-quark interactions within the $\bar{3}_c$ and 6_c diquarks are attractive and repulsive, respectively. However, the interactions between 6_c diquark and $\bar{6}_c$ antidiquark are attractive, and much stronger than those between the $\bar{3}_c$ diquark and 3_c antidiquark. In the $6_c \otimes \bar{6}_c$ color configuration, the attractive interactions between the two quark clusters compete with the repulsive interactions within the clusters. If the attractive interactions win, there exist a $6_c \otimes \bar{6}_c$ tetraquark state. Moreover, if the attractive interactions are large enough, the $6_c \otimes \bar{6}_c$ state may be located below the $\bar{3}_c \otimes 3_c$ one.

In the S-wave tetraquark state with two identical quarks (antiquarks), the color interactions do not contribute to the mixing effects of different color configurations due to the orthogonality of the spin wave function. Only the hyperfine interaction contributes to the coupled channel effects. Then the hyperfine interaction in mode II, which is expected to be more important than that in model I, induces a much strong mixing effects between the $\bar{3}_c \otimes 3_c$ and $6_c \otimes 6_c$ components.

Table 1. The mass spectra and proportions of different color configurations in the fully-heavy tetraquark system with $J^{PC} = 0^{++}$ in two quark models.

	Model I	$\bar{3}_c \otimes 3_c$	$6_c \otimes \bar{6}_c$	Model II	$\bar{3}_c \otimes 3_c$	$6_c \otimes \bar{6}_c$
$cc\bar{c}\bar{c}$	6.377	11%	89%	6.371	43%	57%
	6.425	89%	11%	6.483	57%	43%
$bb\bar{b}\bar{b}$	19.215	1%	99%	19.243	17%	83%
	19.247	99%	1%	19.305	83%	17%
$bb\bar{c}\bar{c}$	12.847	14%	86%	12.886	53%	47%
	12.866	86%	14%	12.946	47%	53%

The tetraquarks with $J^{PC} = 1^{+-}$ and $J^{PC} = 2^{++}$ only contain the $\bar{3}_c \otimes 3_c$ color configuration. Their mass difference arises from the hyperfine potential. We summarize the mass spectra in Fig. 3. The results show that the lowest 0^{++} states are located about $300 \sim 450$ MeV above the corresponding scattering states. We also vary the mass of the constituent quark and display the mass dependence of the tetraquark state in Fig. 4. The results show that the ground tetraquark state is always located above the $\eta_Q \eta_Q$ threshold with increasing quark mass. This indicates that there is no bound $cc\bar{c}\bar{c}$, $bb\bar{b}\bar{b}$, and $bb\bar{c}\bar{c}$ ($cc\bar{b}\bar{b}$) tetraquark exist in the two quark models.

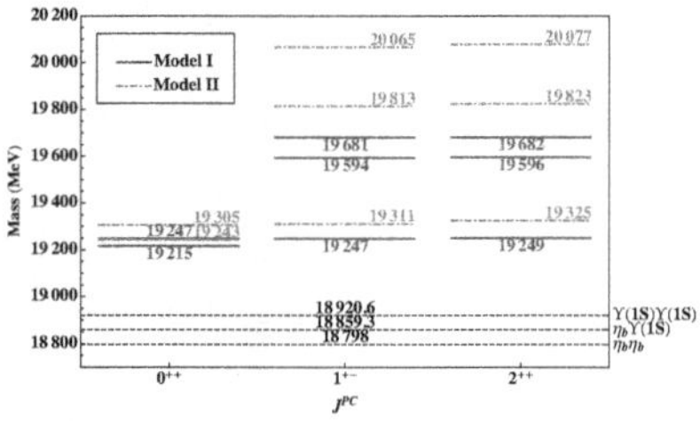

Fig. 3. The mass spectra of the ground $bb\bar{b}\bar{b}$ states.

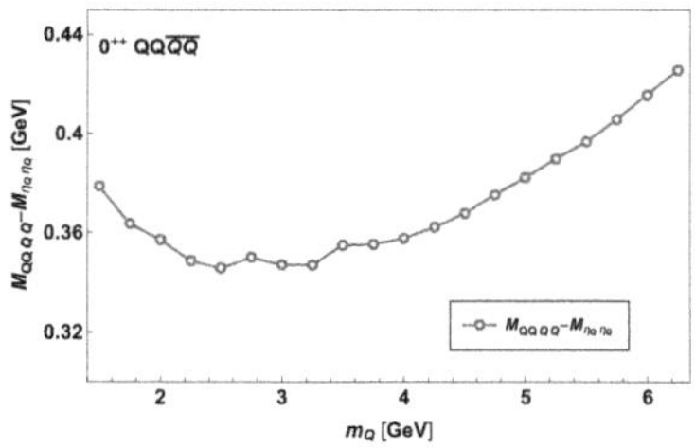

Fig. 4. The quark mass dependence of the 0^{++} tetraquark states in model II.

Acknowledgments

This project is supported by the National Natural Science Foundation of China under Grants 11575008, 11621131001. G. J. Wang is also supported by China Postdoctoral Science Foundation under Grant No. 2019M660279.

References

1. CMS Collab. (V. Khachatryan *et al.*), JHEP 05, 013 (2017).
2. S. Durgut, Search for Exotic Mesons at CMS,(2018), `https://meetings.aps.org/Meeting/APR18/Session/U09.6`.
3. LHCb Collab. (R. Aaij *et al.*), JHEP 10, 086 (2018).
4. Y. Bai, S. Lu, and J. Osborne, Phys. Lett. B **798**, 134930 (2019).
5. M. N. Anwar, J. Ferretti, F. K. Guo, E. Santopinto, and B. S. Zou, Eur. Phys. J. C **78**, 647 (2018).
6. W. Chen, H. X. Chen, X. Liu, T. G. Steele and S. L. Zhu, Phys. Lett. B **773**, 247 (2017).
7. W. Heupel, G. Eichmann, and C. S. Fischer, Phys. Lett. B **718**, 545 (2012).
8. J. Wu, Y. R. Liu, K. Chen, X. Liu, and S. L. Zhu, Phys. Rev. D **97**, 094015 (2018).
9. M. S. Liu, Q. F. L, X. H. Zhong, and Q. Zhao, Phys. Rev. D **100**, 016006 (2019).
10. C. Hughes, E. Eichten, and C. T. H. Davies, Phys. Rev. D **97**, 054505 (2018).
11. C. Y. Wong, E. S. Swanson, and T. Barnes, Phys. Rev. C **65**, 014903 (2002); **66**, 029901(E) (2002)].
12. B. Silvestre-Brac, Few Body Syst. **20**, 1 (1996).
13. G. J. Wang, L. Meng and S. L. Zhu, Phys. Rev. D **100**, 096013 (2019).

Hidden charm pentaquark states and $\Sigma_c^{(*)}\bar{D}^{(*)}$ interaction in chiral perturbation theory

Lu Meng*, Bo Wang, Guang-Juan Wang and Shi-Lin Zhu

School of Physics and State Key Laboratory of Nuclear Physics and Technology, Peking University, Beijing 100871, China
**E-mail: lmeng@pku.edu.cn*

We adopt the chiral perturbation theory to calculate the $\Sigma_c^{(*)}\bar{D}^{(*)}$ interaction to the next-to-leading order (NLO) and include the couple-channel effect in the loop diagrams. We reproduce the three P_c states in the molecular picture after including the $\Lambda_c\bar{D}^{(*)}$ intermediate states. We also discuss some novel observations arising from the loop diagrams.

Keywords: Pentaquark states; chiral perturbation theory; molecular states.

1. Introduction

In 2015, the LHCb Collaboration discovered two pentaquark candidates $P_c(4380)$ and $P_c(4450)$ in the $J/\psi p$ invariant mass spectrum of $\Lambda_b \to J/\psi K p$[1] (or see Ref. 2 for a review). Recently, the LHCb Collaboration updated these results.[3] The previously reported $P_c(4450)$ was resolved into two states $P_c(4440)$ and $P_c(4457)$, and a new state $P_c(4312)$ was observed with 7.3σ significance. These three states are narrow and their masses lie below the thresholds of $\Sigma_c\bar{D}^{(*)}$. Thus, they are good candidates of the $\Sigma_c\bar{D}^{(*)}$ molecules. The theoretical calculation in different approaches all supports their molecular nature.[4–7]

Chiral perturbation theory (ChPT) is the effective field theory of low energy QCD, which was used to build the modern theory of nuclear force.[8,9] In this work, we adopt the ChPT to study the $\Sigma_c^{(*)}\bar{D}^{(*)}$ interaction,[10,11] which is less model-dependent than other previous works. We take the loop diagrams into consideration, which give rise to some novel effects.

2. Formalism

The ChPT is performed to calculate the $\Sigma_c^{(*)}\bar{D}^{*}$ effective potentials to the next-to-leading order (NLO). The topological diagrams involved in the calculation are presented in Fig. 1. The leading order (LO) diagrams include

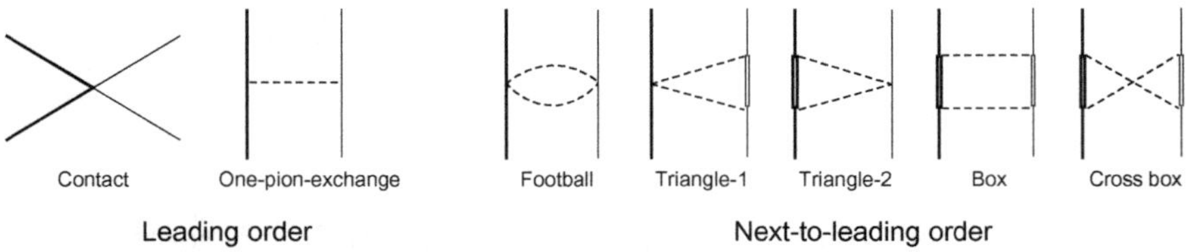

Fig. 1. Topological diagrams contributing to the LO and NLO $\Sigma_c^{(*)}\bar{D}^*$ interaction.

the tree level contact interaction and one-pion-exchange diagrams. The NLO diagrams are the two-pion-exchange ones. In our calculation, the intermediate matter fields (double lines in Fig. 1) could be Σ_c, Σ_c^*, $\bar{D}$ and $\bar{D}^*$. We keep the mass splittings between the heavy quark spin (HQS) partner states, thus our results could give the heavy quark spin symmetry (HQSS) violation effect. Since the $\Sigma_c^{(*)}\Lambda_c\pi$ couplings are competitive compared with the $\Sigma_c^{(*)}\Sigma_c^{(*)}\pi$ ones, we also include the Λ_c as the intermediate states in the loops.

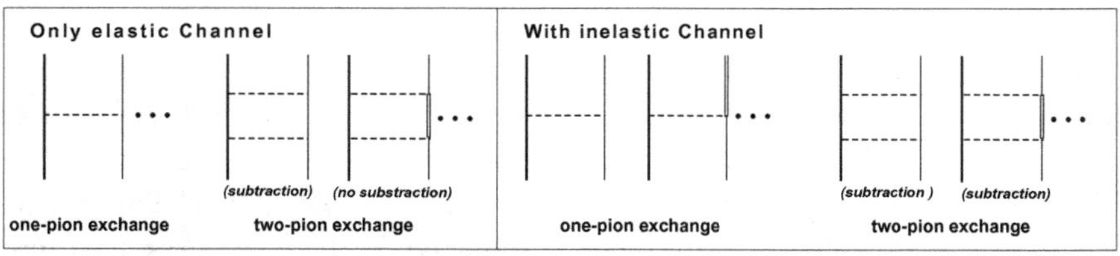

Fig. 2. Two approaches to couple-channel effect.

For the box diagrams, we could deal with the couple-channel effect through two approaches as illustrated in Fig. 2. From the view of time-ordered perturbation theory (TOPT), the box diagrams can be divided into the one-pion-iteration part and other part, see Fig. 3. For the one-pion-iteration part, the amplitude reads,

$$\begin{aligned}\mathcal{M} &\sim (E_{\text{inital}} - E_{\text{inter}})^{-1} \\ &\approx \left[\left(m_1 + m_2 + \frac{\boldsymbol{p}_1^2}{2m_1} + \frac{\boldsymbol{p}_2^2}{2m_2}\right) - \left(m_1' + m_2' + \frac{\boldsymbol{p}_1'^2}{2m_1'} + \frac{\boldsymbol{p}_2'^2}{2m_2'}\right)\right]^{-1}. \quad (1)\end{aligned}$$

When the intermediate states are the same as the external lines, i.e., $m_1 = m_2$ and $m_1' = m_2'$, like NN systems, the amplitude would be amplified to destroy the power counting. We adopt the Weinberg's formalism to handle this problem.[12] We subtract the one-pion-iteration part in the box diagrams and recover their contribution by solving the Lippmann-Schwinger or Schrödinger equation. For $\Sigma_c^{(*)}\bar{D}^{(*)}$ systems, the intermediate states could be different from the external lines. The power counting works well for these box diagrams. Therefore, we could either subtract the one-pion-iteration part and include inelastic channels in tree diagrams or keep the one-pion-iteration part and only include the elastic channels in tree diagrams. In this work, we adopt latter approaches as shown in the left panel of Fig. 2.

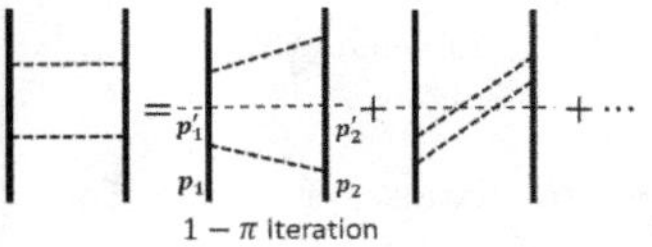

Fig. 3. Box diagrams in TOPT.

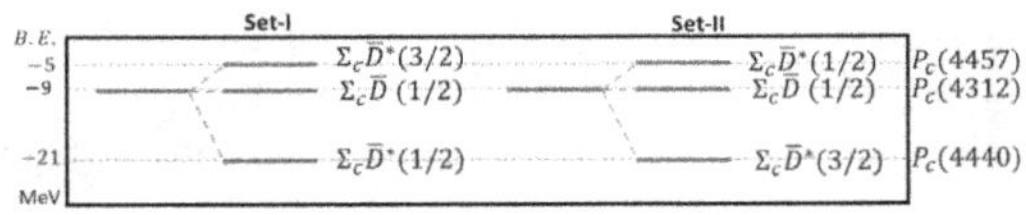

Fig. 4. Two possible spin assignments for P_c states.

3. Results and discussions

Our LO results keep the heavy quark symmetry, which read,

$$\mathcal{V}_{\Sigma_c\bar{D}} = V_c, \quad \mathcal{V}_{\Sigma_c\bar{D}^*} = V_c + V_{ss}\boldsymbol{S}_1 \cdot \boldsymbol{S}_2. \tag{2}$$

There are two possible spin assignments for the three P_c states as shown in Fig. 4. Three states have the same central potential and their binding energy splittings arise from the spin-spin interaction. If we treat spin-spin interaction as a perturbation, the ratio of the binding energy splitting is exactly the ratio of spin-spin interaction. Therefore, the spin Set-I in Fig. 4 is favored.

At the NLO, we include loop diagrams which bring some novel effects. The mass splittings between HQS partner states give rise to considerable HQSS violation for $\Sigma_c^{(*)}\bar{D}$ systems. Meanwhile, the new spin structure $(\boldsymbol{S}_1 \cdot \boldsymbol{S}_2)^2$ appears at the loop level for the $\Sigma_c^*\bar{D}^*$,[11] which is different from the two nucleon systems.

We have two unknown low energy constants (LECs), $\mathbb{D}_1$ and $\mathbb{D}_2$, to the NLO. We vary the two LECs and try to reproduce three states simultaneously. The result is presented in Fig. 5. Three bands correspond to the regions of parameters in which the bound states have the binding energy $[-30, 0]$ MeV. The three black lines represent the set of parameters corresponding to the experimental results. The left and right graphs in Fig. 5

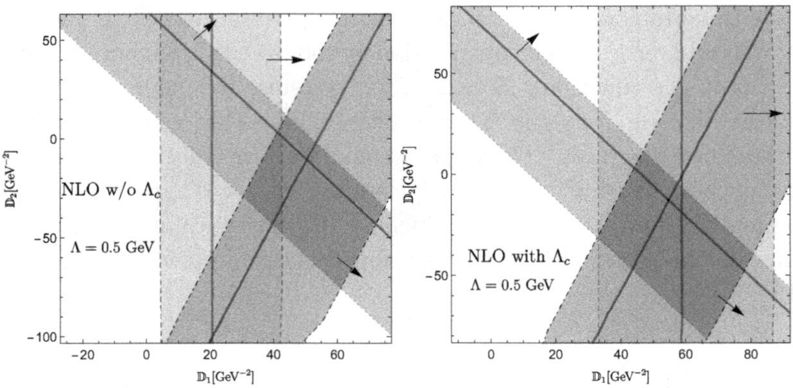

Fig. 5. The numerical results without and with Λ_c intermediate states to the NLO.

are the results without and with the Λ_c as intermediate states, respectively. The three P_c states can coexist only when the couple-channel effect from $\Lambda_c\bar{D}^{(*)}$ is considered.[11] With the $\mathbb{D}_1$ and $\mathbb{D}_2$ being fixed, there exist bound solutions for all seven S-wave $\Sigma_c^{(*)}\bar{D}$ channels as listed in Table 1. It is possible to find the predicted states in the future experiments.

Table 1. The binding energies ΔE of $I = \frac{1}{2}$ hidden-charm $\Sigma_c^{(*)}\bar{D}^{(*)}$ systems.

	$[\Sigma_c\bar{D}]_{\frac{1}{2}}$	$[\Sigma_c\bar{D}^*]_{\frac{1}{2}}$	$[\Sigma_c\bar{D}^*]_{\frac{3}{2}}$	$[\Sigma_c^*\bar{D}]_{\frac{3}{2}}$	$[\Sigma_c^*\bar{D}^*]_{\frac{1}{2}}$	$[\Sigma_c^*\bar{D}^*]_{\frac{3}{2}}$	$[\Sigma_c^*\bar{D}^*]_{\frac{5}{2}}$
ΔE	-4.6	-22.5	-3.2	-34.5	-14.3	-3.4	-0.3

In Table 2, we compare the $\Sigma_c^{(*)}\bar{D}^{(*)}$ systems with two nucleon systems. $\Sigma_c^{(*)}\bar{D}^{(*)}$ interactions have some interesting features which do not appear in N-N systems. For example, the chiral dynamics for $\Sigma_c^{(*)}\bar{D}^{(*)}$ systems is the interplay between the light-light diquark in the charmed baryons and the light quark in the charmed mesons. Thus, one should take all the heavy quark partner states of charmed hadrons into consideration to include the whole interacting elements.

Our analytical results are pion mass dependent. The lattice QCD simulations for the $\Sigma_c^{(*)}\bar{D}^{(*)}$ systems are called for. Our expressions could be used to extrapolate the lattice QCD results to the physical pion mass.[13]

Table 2. Comparisons of the $\Sigma_c^{(*)}\bar{D}^{(*)}$ and N-N systems.

	N-N	$\Sigma_c^{(*)}\bar{D}^{(*)}$
Chiral dynamics	$N-N$	(light diquark)-(light quark)
Inter. states	Δ	Λ_c
Inter. states	-	HQS partner states e.g., $c(qq)_{s=1}^{I=1} = \Sigma_c + \Sigma_c^*$
Heavy quark limit	-	Pinch singularity in elastic channel Including inelastic channel
M_{inter} vs M_{initial}	$M_\Delta > M_N$	e.g., $M_{\Lambda_c} + M_\pi < M_{\Sigma_c}$
Imaginary part	w/o	w/
Small scale expansion	works	fails in $\Sigma_c\bar{D} - \Lambda_c\bar{D}^* - \Sigma_c\bar{D}$
Spin structure	$\boldsymbol{\sigma_1}\cdot\boldsymbol{\sigma_2}$	$\boldsymbol{S_1}\cdot\boldsymbol{S_2}$, $(\boldsymbol{S_1}\cdot\boldsymbol{S_2})^2$...
Weinberg composite	$\lvert E_D\rvert \ll \frac{m_\pi^2}{2\mu}$	$\lvert E_{P_c}\rvert \sim \frac{m_\pi^2}{2\mu} \approx 9$ MeV

Acknowledgments

This project is supported by the National Natural Science Foundation of China under Grants 11575008, 11621131001.

References

1. LHCb Collab. (R. Aaij *et al.*), *Phys. Rev. Lett.* **115**, 072001 (2015).
2. H. X. Chen, W. Chen, X. Liu and S. L. Zhu, *Phys. Rept.* **639**, 1 (2016).
3. LHCb Collab. (R. Aaij *et al.*), *Phys. Rev. Lett.* **122**, 222001 (2019).
4. R.Chen, Z.F.Sun, X.Liu and S.L.Zhu, *Phys. Rev. D* **100**, 011502 (2019).
5. M. Z. Liu, Y. W. Pan, F. Z. Peng, M. Sánchez Sánchez, L. S. Geng, A. Hosaka and M. Pavon Valderrama, *Phys. Rev. Lett.* **122**, 242001 (2019).
6. F. K. Guo, H. J. Jing, U. G. Meißner and S. Sakai, *Phys. Rev. D* **99**, 091501 (2019).
7. M. L. Du, V. Baru, F. K. Guo, C. Hanhart, U. G. Meißner, J. A. Oller and Q. Wang, arXiv:1910.11846.
8. E. Epelbaum, H. W. Hammer and U. G. Meißner, *Rev. Mod. Phys.* **81**, 1773 (2009).
9. R. Machleidt and D. R. Entem, *Phys. Rept.* **503**, 1 (2011).
10. L. Meng, B. Wang, G. J. Wang and S. L. Zhu, *Phys. Rev. D* **100**, 014031 (2019).
11. B. Wang, L. Meng and S. L. Zhu, arXiv:1909.13054.
12. S. Weinberg, *Nucl. Phys. B* **363**, 3 (1991).
13. Z. W. Liu, J. M. M. Hall, D. B. Leinweber, A. W. Thomas and J. J. Wu, *Phys. Rev. D* ***95***, 014506 (2017).

Ω_c states with an extension of the local hidden gauge approach

E. Oset* and V. R. Debastiani
Departamento de Fisica Teorica and IFIC, University of Valencia-CSIC, Valencia, 46071, Spain
**E-mail: oset@ific.uv.es*

J. M. Dias
University of Sao Paulo, Sao Paulo, CEP 05508-090, Brazil

W. H. Liang
Department of Physics, Guangxi Normal University, Guilin, 541004, China

We briefly expose the method used to obtain Ω_c states of molecular type from the interaction of meson-baryon channels and how three of the states obtained are in very good agreement with experiment for masses and widths.

1. Introduction

In Ref. 1 the LHCb collaboration reported five new narrow Ω_c^0 states studying the $\Xi_c^+ K^-$ mass spectrum produced in high energy pp collisions: $\Omega_c(3000)$, $\Omega_c(3050)$, $\Omega_c(3066)$, $\Omega_c(3090)$ and $\Omega_c(3119)$. We present here some of the results obtained in Ref. 2 concerning the likely nature of these states as molecular states stemming from the interaction of meson-baryon in coupled channels. Predictions for such states and related ones had been done within quark models and others (see references in Ref. 2). Molecular states had also been used to make predictions in Refs. 3, 4 studying the interaction of coupled channels. A more updated study along these lines was done in Ref. 5, where predictions for charmed and strange baryons are done using an interaction based on SU(6) flavor-spin symmetry in the light quark sector and SU(2) spin symmetry in the heavy quark sector, extending the SU(3) Weinberg-Tomozawa interaction.

The experimental findings of Ref. 1 have brought a new wave of theoretical activity with many suggestions to explain the new states.

In the molecular picture, an update of the work of Ref. 4 was done in Ref. 6 using some information from the experimental spectrum to regularize the loops and then giving a description of the mass and width of two states of Ref. 1 as $J^P = 1/2^-$ meson-baryon molecular states.

In the present work we follow Refs. 5, 6 for the coupled channels and the unitarization procedure. We differ in the input for the interaction, which in our case is based on the local hidden gauge approach, exchanging vector mesons.[7]

In the present work the diagonal terms that we evaluate coincide with those of Ref. 6 where the model of Ref. 3 is used implementing also the exchange of vector mesons and SU(4) symmetry for mesons and baryons. We, instead, use explicit wave functions for the baryon states imposing flavor-spin symmetry on the light quark sector and singling out the heavy quarks. Hence, in the baryon sector we are not using SU(4) symmetry. For the diagonal terms we also show that one is exchanging light vectors and the heavy quarks are spectators. In this case we obtain the same matrix elements as in Ref. 6, but there are differences in the non diagonal ones. Since in the dominant terms we are exchanging only light vectors and the heavy quarks are spectators, the interaction automatically respects heavy quark symmetry.[8–10] The non diagonal terms which exchange heavy vectors do not fulfill heavy quark symmetry, but neither should them since these are terms of order $O(m_Q^{-2})$ in the heavy quark mass counting. In addition to the work of Ref. 6 we also include pseudoscalar-baryon($3/2^+$) components and we obtain two more states. We can identify two states of $J^P = 1/2^-$ and one of $J^P = 3/2^-$ with the states found in Ref. 1. We also look for vector-baryon states and find three states at higher energies.

2. Formalism

In the Tables below we show the coupled channels that we use in our study for states of $J^P = 1/2^-$ and $J^P = 3/2^-$, respectively, together with the threshold masses of the channels.

States	$\Xi_c \bar{K}$	$\Xi_c' \bar{K}$	ΞD	$\Omega_c \eta$	ΞD^*	$\Xi_c \bar{K}^*$	$\Xi_c' \bar{K}^*$
Threshold	2965	3074	3185	3243	3327	3363	3472

In order to get the interaction we exchange vector mesons between the meson and the baryon. One needs two ingredients, the upper meson-vector-

States	$\Xi_c^*\bar{K}$	$\Omega_c^*\eta$	ΞD^*	$\Xi_c\bar{K}^*$	$\Xi^* D$	$\Xi_c'\bar{K}^*$
Threshold	3142	3314	3327	3363	3401	3472

meson vertex and the lower baryon-vector-baryon vertex. In Ref. 11 we use wave functions to evaluate the upper vertex but we find that for practical purposes we can use the Lagrangian

$$\mathcal{L}_{VPP} = -ig \langle \, [P, \partial_\mu P] \, V^\mu \, \rangle \,, \tag{1}$$

with P and V, matrices representing $q\bar{q}$ in terms of the mesons (see Ref. 2).

For the lower vertex we use wave functions for the baryons where the heavy quark is singled out and the symmetry on the quarks is demanded on the flavor and spin wave functions for the light quarks (see Ref. 2 for details).

It becomes easy and systematic to evaluate all the matrix elements and we find

$$V_{ij} = D_{ij} \frac{1}{4f_\pi^2} (p^0 + p'^0) \tag{2}$$

where p^0, p'^0 are the energies of the mesons and f_π the pion decay constant. The D_{ij} coefficients are given in Ref. 2.

3. Results

We use the potential of Eq. (2) and the on-shell factorized Bethe-Salpeter equation[12,13]

$$T = [1 - VG]^{-1} \, V \,, \tag{3}$$

with G the meson-baryon loop function. We choose to regularize it with the cutoff method.

In the $J^P = 1/2^-$ sector and for $q_{max} = 650 MeV$ we find two states, whose couplings to the building coupled channels are given in Ref. 2. We find a state at 3054 MeV with 0.8 MeV width which couples mostly to $\Xi_c'\bar{K}$, and another one at 3091 MeV and width about 10 MeV which couples mostly to ΞD.

The strategy followed in these calculations is to fine tune the cutoff to adjust the pole position to some experimental data. We see that if we take q_{max} = 650 MeV the results agree well with the second and fourth resonances reported in Ref. 1, $\Omega_c(3050)$ and $\Omega_c(3090)$. It is interesting to note that cutoff values of this order are used in Ref. 14 for $\bar{K}N$ or in Ref. 15 for DN. Fitting one resonance is partly merit of fine tuning the cutoff, but then the second resonance and the widths are genuine predictions of the theory. Note that the widths are respectively 0.88 MeV and 10.24 MeV which agree remarkably well with the experiment, $0.8 \pm 0.2 \pm 0.1$ MeV and $8.7 \pm 1.0 \pm 0.8$ MeV, respectively.

Next we look for the states of $J^P = 3/2^-$ from the pseudoscalar-baryon($3/2^+$) interaction.

We find two states, one at 3125 MeV and the other one at 3290 MeV, both of them with zero or small width. The state at 3125 MeV couples strongly to $\Xi_c^* \bar{K}$ and $\Omega_c^* \eta$, more strongly to $\Xi_c^* \bar{K}$. The one at 3290 MeV couples mostly to $\Xi^* D$. In Ref. 2 we discuss more states that are degenerate in $1/2^-$ and $3/2^-$.

We can clearly associate the state found at 3125 MeV with the $\Omega_c(3119)$ observed experimentally, which has a width of $1.1 \pm 0.8 \pm 0.4$ MeV. The agreement is also remarkable.

4. Conclusions

We have studied Ω_c states which are dynamically generated from the interaction of meson-baryon in the charm sector. The interaction is obtained using an extension of the local hidden gauge approach with the exchange of vector mesons. We show that the dominant terms come from exchange of light vector mesons, leaving the heavy quarks as spectators. This has two good consequences: first we can map the interaction to what happens in SU(3) using chiral Lagrangians, and second, the fact that the heavy quarks are spectators in the interaction guarantee that the dominant terms in the $(1/m_Q)$ counting fulfill the rules of heavy quark symmetry.

We obtain two states with $J^P = 1/2^-$ which are remarkably close in mass and width to the experimental states $\Omega_c(3050)$, $\Omega_c(3090)$. In addition, we also obtain a $3/2^-$ state with zero width at 3125 MeV, which can be associated to the experimental $\Omega_c(3119)$ that also has a width of the order or smaller than 1 MeV.

Acknowledgments

V. R. Debastiani acknowledges the Programa Santiago Grisolia of Generalitat Valenciana (Exp. GRISOLIA/2015/005). J. M. Dias thanks the Fundação de Amparo à Pesquisa do Estado de São Paulo (FAPESP) for support by FAPESP grant 2016/22561-2. This work is partly supported by the Spanish Ministerio de Economia y Competitividad and European FEDER funds under the contract number FIS2014-57026-REDT, FIS2014-51948-C2- 1-P, and FIS2014-51948-C2-2-P, and the Generalitat Valenciana in the program Prometeo II-2014/068 (EO). This work is also partly supported by the National Natural Science Foundation of China under Grants No. 11975083, No. 11565007 and No. 11747307.

References

1. R. Aaij *et al.* [LHCb Collaboration], Phys. Rev. Lett. **118**, no. 18, 182001 (2017).
2. V. R. Debastiani, J. M. Dias, W. H. Liang and E. Oset, Phys. Rev. D **97**, no. 9, 094035 (2018).
3. J. Hofmann and M. F. M. Lutz, Nucl. Phys. A **763**, 90 (2005).
4. C. E. Jimenez-Tejero, A. Ramos and I. Vidana, Phys. Rev. C **80**, 055206 (2009).
5. O. Romanets, L. Tolos, C. Garcia-Recio, J. Nieves, L. L. Salcedo and R. G. E. Timmermans, Phys. Rev. D **85**, 114032 (2012).
6. G. Montaña, A. Feijoo and A. Ramos, Eur. Phys. J. A **54**, no. 4, 64 (2018).
7. M. Bando, T. Kugo, S. Uehara, K. Yamawaki and T. Yanagida, Phys. Rev. Lett. **54**, 1215 (1985).
8. N. Isgur and M. B. Wise, Phys. Lett. B **232**, 113 (1989).
9. M. Neubert, Phys. Rept. **245**, 259 (1994).
10. A. V. Manohar and M. B. Wise, Camb. Monogr. Part. Phys. Nucl. Phys. Cosmol. **10**, 1 (2000).
11. S. Sakai, L. Roca and E. Oset, Phys. Rev. D **96**, no. 5, 054023 (2017).
12. J. A. Oller and E. Oset, Phys. Rev. D **60**, 074023 (1999).
13. J. A. Oller and U. G. Meißner, Phys. Lett. B **500**, 263 (2001).
14. E. Oset and A. Ramos, Nucl. Phys. A **635**, 99 (1998).
15. W. H. Liang, T. Uchino, C. W. Xiao and E. Oset, Eur. Phys. J. A **51**, no. 2, 16 (2015).

Describing the charged charmonium like structures in the $e^+e^- \to \pi^+\pi^-\psi(3686)$ process based on the ISPE mechanism

Qi Huang

School of Physical Science and Technology, Lanzhou University, Lanzhou 730000, China
Research Center for Hadron and CSR Physics, Lanzhou University and Institute of Modern Physics of CAS, Lanzhou 730000, China
E-mail: huangqi@ucas.ac.cn

Dian-Yong Chen

School of Physics, Southeast University, Nanjing 210094, China
E-mail: chendy@seu.edu.cn

Xiang Liu

School of Physical Science and Technology, Lanzhou University, Lanzhou 730000, China
Research Center for Hadron and CSR Physics, Lanzhou University and Institute of Modern Physics of CAS, Lanzhou 730000, China
E-mail: xiangliu@lzu.edu.cn

Takayuki Matsuki

Tokyo Kasei University, 1-18-1 Kaga, Itabashi, Tokyo 173-8602, Japan
Theoretical Research Division, Nishina Center, RIKEN, Wako, Saitama 351-0198, Japan
E-mail: matsuki@tokyo-kasei.ac.jp

After performing a combined fit of the $\pi^\pm\psi(3686)$, dipion invariant mass spectra, and $e^+e^- \to \psi(3686)\pi^+\pi^-$ cross section to the experimental data for the process $e^+e^- \to \psi(3686)\pi^+\pi^-$, we find that the new charged charmonium-like structure observed by BESIII Collaboration in 2017 can be reproduced well based on the ISPE mechanism. This provides strong evidence to show that the ISPE mechanism can be an underlying mechanism resulting in such a novel phenomenon.

Keywords: BESIII; charged charmonium-like structure; ISPE mechanism.

1. Decay mechanisms for $e^+e^- \to \pi^+\pi^-\psi(3686)$

The decay mechanisms working in the process $e^+e^- \to \psi(3686)\pi^+\pi^-$ are shown in Fig. 1. Among these diagrams, the first one is the background

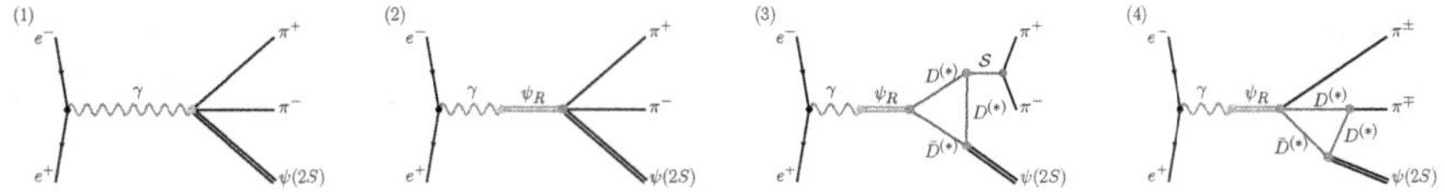

Fig. 1. (Color online). Feynman diagrams that depicting the decay mechanisms which give contributions to the process $e^+e^- \to \psi(3686)\pi^+\pi^-$.

contribution, while in the remaining diagrams, e^+e^- transforms into a ψ_R first, then ψ_R decays into the final $\psi(3686)\pi^+\pi^-$ by direct decay, cascade mode, and IPSE mechanism, respectively. As for the background contribution, its amplitude is expressed as

$$\mathcal{M}^{(1)}_{\rm NoR} = \bar{v}(k_2)\gamma^\mu u(k_1) g_{\rm NoR} e \frac{-g_{\mu\nu}}{s} \epsilon^{*\nu}_{\psi'} \mathcal{F}_{\rm NoR}(s), \tag{1}$$

with $\mathcal{F}_{\rm NoR}(s) = e^{-a_{\rm NoR}(\sqrt{s}-\Sigma m_f)^2}$ being a form factor.[1] While for diagrams in which an intermediate ψ_R participates, a general amplitude, with the help of vector meson dominance (VMD) ansatz,[2,3] is

$$\begin{aligned}\mathcal{M}^{(i)}_R &= \bar{v}(k_2)\gamma^\alpha u(k_1) e \frac{-g_{\alpha\beta}}{s} \frac{e m_R^2}{f_R} \frac{1}{s - m_R^2 + i m_R \Gamma_R} \\ &\quad \times \left(-g^{\beta\mu} + \frac{p_0^\beta p_0^\mu}{m_R^2}\right) \mathcal{A}^{(i)}_{\mu\zeta} \epsilon^*_{\psi'}(p_3)^\zeta \mathcal{F}_R(s),\end{aligned} \tag{2}$$

where $\mathcal{F}_R(s) = e^{-a_R[\sqrt{s}-m_R]}$ is a form factor introduced to balance the divergence coming from phase space, and $\mathcal{A}_{\mu\nu}$ expresses the decay amplitude of $\psi_R \to \pi^+\pi^-\psi(3686)$ as

$$\begin{aligned}\mathcal{A}^{(2)}_{\mu\zeta} &= \frac{F_{\rm Dir}}{f_\pi^2} g_{\mu\zeta} \Big\{ \Big[m_{\pi\pi}^2 - \kappa_R (\Delta M)^2 \Big(1 + \frac{2m_\pi^2}{m_{\pi\pi}^2}\Big)\Big]_{\rm S-wave} \\ &\quad + \Big[\frac{3}{2}\kappa_{\rm Dir}\big((\Delta M)^2 - m_{\pi\pi}^2\big)\Big(1 - \frac{4m_\pi^2}{m_{\pi\pi}^2}\Big)\Big(\cos\alpha^2 - \frac{1}{3}\Big)\Big]_{\rm D-wave} \Big\},\end{aligned} \tag{3}$$

$$\mathcal{A}^{(3)}_{\mu\zeta} = \frac{1}{m_{\pi\pi}^2 - m_\sigma^2 + i m_\sigma \Gamma_\sigma(m_{\pi\pi})} \Big(f_R g_{\mu\zeta} + e^{i\varphi_R} g_R p_{0\zeta} p_{3\mu} \Big), \tag{4}$$

$$\begin{aligned}\mathcal{A}^{(4)}_{\mu\zeta} &= C_{00} g_{\mu\zeta} + C_{12} p_{1\mu} p_{2\zeta} + C_{13} p_{1\mu} p_{3\zeta} + C_{22} p_{2\mu} p_{2\zeta} + C_{23} p_{2\mu} p_{3\zeta} \\ &\quad + C_{32} p_{3\mu} p_{2\zeta} + C_{33} p_{3\mu} p_{3\zeta}.\end{aligned} \tag{5}$$

In Eqs. (3-5), $\Delta M = m_R - m_{\psi'}$ is the mass splitting of ψ_R and $\psi(3686)$, α is the angle between ψ_R and π^- in the $\pi^+\pi^-$ rest frame, φ_R is the phase angle between the S and D wave coupling, and the coefficients C_{ij}, $\{(i,j) = (0,1,2,3)\}$ can be obtained by the loop integrals of the ISPE amplitudes. We want to emphasize here that since the mass and width of σ meson are comparable, we adopt a momentum dependent width.

2. Numerical results and discussion

Accroding to Ref. 4, apart from $\psi(4160)$ and $\psi(4415)$, in 4.1-4.6 GeV there exist another two vector charmonia states. One is the well established $Y(4220)$, which can be assigned as a state of 4S-3D mixing, and the other is the partner of $Y(4220)$, which is predicted as $\psi(4380)$. Ref. 4 finds that when the predicted $\psi(4380)$ is included, the cross section of $e^+e^- \to \psi(2S)\pi^+\pi^-$ can be described better. Hence, to test their conclusions, we do both 4R and 3R fits in which $\psi(4380)$ participates or not, respectively. With the parameters listed in Tables 1 and 2, the results of our fit of the total cross section, the $\pi^\pm\psi(3686)$, and dipion invariant mass spectra of $e^+e^- \to \psi(3686)\pi^+\pi^-$ process are presented in Figs. 2 and 3. Then, using the parameters in Tables 1 and 2, we extract the branching ratios $\mathcal{B}[\psi_R \to \psi(3686)\pi^+\pi^-]$ in Table 3.

Table 1. The fitted values of the parameters in three-charmonium scenario.

Prameters	$\psi(4160)$	$\psi(4220)$	$\psi(4415)$
$\phi_{\rm Dir}$	-1.631	-1.506	1.762
$F_{\rm Dir}$	-3.477	5.881	0.982
κ_{Dir}	-1.829	-0.025	-0.918
ϕ_σ	1.130	2.009	2.241
f_σ (GeV^2)	7.562	5.785	-3.971
g_σ	77.498	21.803	-1.920
φ_σ	1.186	-2.426	-1.923
$\phi_{\rm ISPE}$	-1.320	2.980	-2.813
$g_{RD\bar{D}\pi}$ (GeV^{-3})	1.717	1.574	0.002
$g_{RD^*\bar{D}\pi}$	-1.834	-0.670	0.025
$g_{RD^*\bar{D}^*\pi}$ (GeV^{-1})	0.120	-0.096	0.105
a_R (GeV^{-1})	6.0	5.0	4.0
$g_{\rm NoR}$=18.940 GeV		$a_{\rm NoR}$=4.702 GeV^{-2}	
$\chi^2/d.o.f. = 3.704$			

Table 2. The same as Table 1 but for four-charmonium scenario.

Parameter	$\psi(4160)$	$\psi(4220)$	$\psi(4415)$	$\psi(4380)$
$\phi_{\rm Dir}$	-0.146	1.661	-0.423	-0.345
$F_{\rm Dir}$	-12.243	3.361	0.504	1.523
$\kappa_{\rm Dir}$	-0.151	-0.215	-1.190	-0.463
ϕ_σ	0.517	-0.950	2.910	-0.328
f_σ (GeV^2)	11.953	1.813	1.832	-5.940
φ_σ	-1.680	0.287	-0.838	-1.455
g_σ	58.717	-6.612	-7.864	-25.906
$\phi_{\rm ISPE}$	-1.759	2.875	1.396	-5.475
$g_{RD\bar{D}\pi}$ (GeV^{-3})	-2.223	-1.736	-0.257	0.276
$g_{RD^*\bar{D}\pi}$	1.646	0.353	-0.470	1.249
$g_{RD^*\bar{D}^*\pi}$ (GeV^{-1})	0.323	0.131	0.133	-0.022
a_R (GeV^{-1})	5.5	4.0	3.3	3.0
$g_{\rm NoR}$=-4.847 GeV			$a_{\rm NoR}$=2.058 GeV^{-2}	
$\chi^2/d.o.f. = 2.767$				

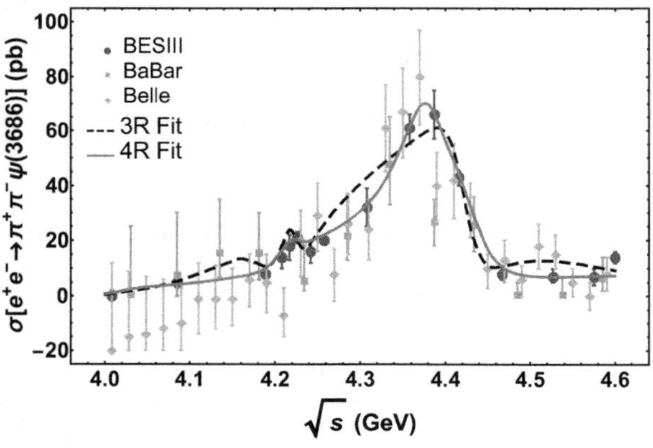

Fig. 2. (Color online). A combined fit on the cross section of $e^+e^- \to \psi(2S)\pi^+\pi^-$ given by BESIII in Ref. 5.

Table 3. Branching ratio of $\psi_R \to \pi^+\pi^-\psi(3686)$ obtained from the present fit in unit of 10^{-3}.

	$\psi(4160)$	$\psi(4220)$	$\psi(4415)$	$\psi(4380)$
3R Fit	15.711	8.101	9.65	-
4R Fit	4.389	7.370	5.368	17.285

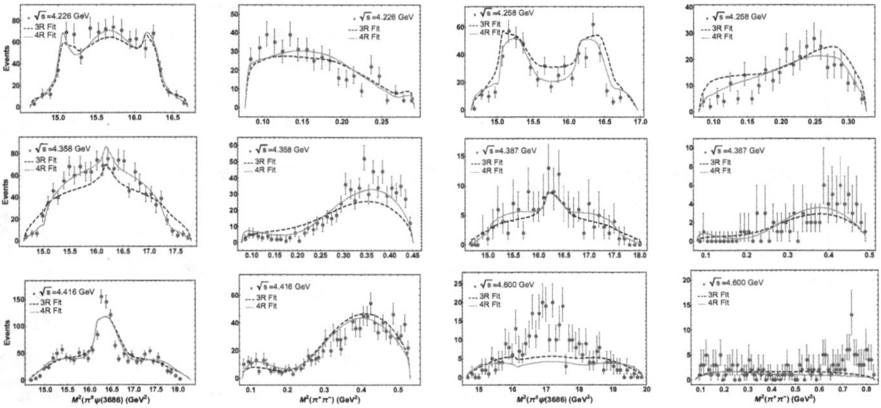

Fig. 3. (Color online). Our combined fit on the $\psi(2S)\pi^{\pm}$ and $\pi^+\pi^-$ invariant mass spectrums that lying on 6 different center-of-mass energies that reported by BESIII in Ref. 5. Here, the black dashed curve and red solid curve are the fitting results under three-charmonium and four-charmonium scenarios, respectively.

3. Conclusion

From Figs. 2 and 3, we can see that the cross section and 12 invariant mass spectra of $e^+e^- \to \psi(2S)\pi^+\pi^-$ can be well reproduced by just one set of parameters. Also, the structure around 4.032 GeV, which lies on the $\psi(2S)\pi^{\pm}$ invariant mass spectra at energies 4.258 and 4.416 GeV, nicely shows up. In addition, the introduction of a new $\psi(3D)$ in 4R fit makes not only the cross section, but also the invariant mass spectra, much better than the 3R fit. Besides, when it comes to Table 3, we can easily see that when it is 3R fit, the extracted branching ratios of $\psi_R \to \psi(3686)\pi^+\pi^-$ are all near 10^{-2}, and while it is 4R fit, most of them are around 10^{-3}, which is closer to the experimental measurements.[5]

References

1. D. Y. Chen, X. Liu and T. Matsuki, Phys. Rev. D **88**, no. 3, 036008 (2013).
2. Q. Y. Lin, X. Liu and H. S. Xu, Phys. Rev. D **88**, 114009 (2013).
3. H. Xu, J. J. Xie and X. Liu, Eur. Phys. J. C **76**, no. 4, 192 (2016).
4. J. Z. Wang, Z. F. Sun, X. Liu and T. Matsuki, Eur. Phys. J. C **78**, no. 11, 915 (2018).
5. M. Ablikim *et al.* [BESIII Collaboration], Phys. Rev. D **96**, no. 3, 032004 (2017).

Multiple charm and hidden charm mesons with strangeness

X. L. Ren

Ruhr-Universität Bochum, Fakultät für Physik und Astronomie, Institut für Theoretische Physik II, D-44780 Bochum, Germany.

Brenda B. Malabarba

Universidade de Sao Paulo, Instituto de Fisica, C.P. 05389-970, Sao Paulo, Brazil.

L. S. Geng

School of Physics & Beijing Key Laboratory of Advanced Nuclear Materials and Physics, Beihang University, Beijing 100191, China.

Beijing Advanced Innovation Center for Big Date-based Precision Medicine, Beihang University, Beijing100191, China.

K. P. Khemchandani

Universidade Federal de Sao Paulo, C.P. 01302-907, Sao Paulo, Brazil.

A. Martínez Torres

Universidade de Sao Paulo, Instituto de Fisica, C.P. 05389-970, Sao Paulo, Brazil.
E-mail: amartine@if.usp.br

In this talk we summarize our latest results on the study of three-body systems with explicit/hidden charm and with strangeness. In particular, we focus on the KDD and $KD\bar{D}^*$ systems, where a charm +2, isospin 1/2 and strangeness +1 state is found with a mass around 4140 MeV in the former and a K^* state, with hidden charm, at a mass around 4307 MeV is obtained in the latter. Both states are predictions of our model and the experimental confirmation of them would be important to understand the properties of the strong interaction in the presence of heavy quarks.

1. Introduction

In the present understanding of Quantum Chromodynamics, nature permits the existence of mesons and baryons with several units of charm quantum number. Some examples are the doubly charmed mesons and baryons T_{cc}, Ξ^+_{cc}, Ω^+_{cc} and triply charmed baryons arising from the dynamics involved in systems like $\Xi_{cc}D$, $\Xi_{cc}D^*$, $\Xi_{cc}\Lambda_c$, $\Xi_{cc}\Sigma_c$.[1–4] In the last years, the interest of investigating systems with charm and bottom quantum numbers has been extended to three-body systems, like $BD\bar{D}$, BDD, BBB^*, DDK.[5–7] However, an experimental investigation of these states is necessary to corroborate the available information on their properties from theoretical works. And such a research proposal is presently missing in the projects of the different experimental facilities. The search for this kind of states would definitely shed light on the working of the strong interaction in the presence of heavy quarks and can open new research lines, like the search of excited states of kaons and/or K^*'s at energies around 4-5 GeV with hidden charm and/or bottom quantum numbers.

In this talk we show our results for two kind of three-body systems with strangeness, the KDD system, where we find a state with mass around 4140 MeV, in agreement with the findings of Ref. 7, based on a different formalism, and the $KD\bar{D}^*$ system, where we find the existence of a K^* vector meson with hidden charm and mass around 4307 MeV, as also found in Ref. 6.

2. Formalism and Results

To study the formation of states in three-body systems like KDD and $KD\bar{D}^*$ we solve the Faddeev equations using as kernel two-body t-matrices obtained from the resolution of the Bethe-Salpeter equation in its on-shell factorization form.[8,9] The advantage of the latter is that the Bethe-Salpeter equation, which is given by

$$t = t + vgt, \tag{1}$$

where v is the kernel and g is a loop function of two hadrons, becomes an algebraic equation, which is solved by considering two-body coupled channels. Using these t-matrices, we solve the Faddeev equations[10]

$$\begin{aligned} T^1 &= t^1 + t^1 G[T^2 + T^3], \\ T^2 &= t^2 + t^2 G[T^1 + T^2], \\ T^3 &= t^3 + t^3 G[T^1 + T^2], \end{aligned} \tag{2}$$

within the approach developed in Refs. 11–13 for the case of the DDK system, and that developed in Refs. 14–16 for the $KD\bar{D}^*$ system.

To determine the kernel v in Eq. (1) we have used effective Lagrangians based on the chiral and heavy quark symmetries of the strong interaction.[17–22] It is now widely accepted that the non-perturbative coupled channel dynamics generates the state $D^*_{s0}(2317)$ in the KD system,[23–27] while in the $D\bar{D}^*$ system we have the formation of the isospin 0 state $X(3872)$ and of the isospin 1 state $Z_c(3900)$.[28–32]

With these ingredients, the solution of Eq. (2) for the DDK, $DD\pi$ and $DD_s\eta$ coupled channel system reveals the formation of a bound state with isospin 1/2, charm +2, strangeness +1 and mass around 4140 MeV.[33] The state is found to arise when one of the DK subsystems, together with its coupled channels, generate the $D^*_{s0}(2317)$. This result is similar to the one found in Ref. 7 in which the system $D - D^*_{s0}(2317)$ was studied by means of an effective two-body potential. We have also estimated the size of such an exotic state to know whether it is compact or not. To do this, we have treated the state found as a $D - D^*_{s0}(2317)$ state and estimate the root mean square (RMS) distance among the constituent hadrons[33] by constructing the wave function of the state, for which we follow Ref. 34. We find the RMS$\sim 1.0 - 1.4$ fm, which is larger than that of $D^*_{s0}(2317)$.[26,33] This result has been also confirmed by the authors of Ref. 35. The state found in Refs. 7, 33 is a three-body bound state, thus, it has a zero width. However, a state with these properties can decay via triangular loops to two-body channels like D_sD, D_sD^* and DD^*_s producing a small width. We have also studied such decay mechanisms in Ref. 36, finding a total width for the state to be around 2-3 MeV.

In case of the $KD\bar{D}^*$ system, the resolution of Eq. (2) shows the generation of a K^* meson with mass around 4307 MeV when the KD system in isospin 0 forms the $D^*_{s0}(2317)$ and the interaction in the $D\bar{D}^*$ system give rise to $X(3872)$ in isospin 0 and $Z_c(3900)$ in isospin 1.[37] This three-body state, as in case of the one found in the DDK system, appears below the $KD\bar{D}^*$ threshold, thus, it is a bound state and has zero width. However, the $Z_c(3900)$, considering it to be a state generated from the $D\bar{D}^*$ and $J/\psi\pi$ coupled channels, has a width of around 30 MeV from its decay to $J/\psi\pi$.[38] When considering this width into the calculation, the three-body state found in the $KD\bar{D}^*$ system gains a width of around 18 MeV. This three-body state can also decay to two-body channels, like $J/\psi K^*(892)$, $\bar{D}D_s$, $\bar{D}D^*_s$ and $\bar{D}^*D^*_s$ via triangular loops. In Ref. 39 we have obtained these decay widths, finding the values ~ 7 MeV for $J/\psi K^*(892)$, ~ 0.5 for $\bar{D}D^*_s$, $\bar{D}^*D^*_s$ and ~ 1 MeV for $\bar{D}D_s$. A three-body state like this, with a dominant $KZ_c(3900)$ component, can also easily decay to a state formed by $J/\psi\pi K$, thus, the reconstruction on the $J/\psi\pi K$ invariant mass in reactions involving these particles in their final states can be a way of finding this $K^*(4307)$ experimentally.

3. Conclussions

We have presented in this talk our latest results for the KDD and $KD\bar{D}^*$ systems and the formation of exotic states in them whose properties can be understood as molecular kind of states. In particular, in the KDD system we find a state around 4140 MeV and in the $KD\bar{D}^*$ system we obtain a K^* vector meson with mass around 4307 MeV.

Acknowledgments

This work was partly supported the National Natural Science Foundation of China (NSFC) under Grants Nos. 11735003 and 11522539, by DFG and NSFC through funds provided to the Sino-German CRC 110 "Symmetries and the Emergence of Structure in QCD" (Grant No. TRR110), and Conselho Nacional de Desenvolvimento Científico e Tecnológico (CNPq) under Grant Nos. 310759/2016-1 and 311524/2016-8. A. M. T gratefully acknowledges the support from the program "Mobilidade Santander - Docentes" (Edital PRPG-11/2019).

References

1. M. Mattson *et al.* [SELEX Collaboration], *Phys. Rev. Lett.* **89**, 112001 (2002).
2. S. Zouzou, B. Silvestre-Brac, C. Gignoux and J. M. Richard, *Z. Phys. C* **30**, 457 (1986).
3. R. Chen, A. Hosaka and X. Liu, *Phys. Rev. D* **96**, no. 11, 114030 (2017).
4. R. Chen, F. L. Wang, A. Hosaka and X. Liu, Phys. Rev. D **97**, no. 11, 114011 (2018).
5. J. M. Dias, V. R. Debastiani, L. Roca, S. Sakai and E. Oset, Phys. Rev. D **96**, no. 9, 094007 (2017).
6. L. Ma, Q. Wang and U.-G. Meißner, *Chin. Phys. C* **43**, no. 1, 014102 (2019).
7. M. Sanchez Sanchez, L. S. Geng, J. X. Lu, T. Hyodo and M. P. Valderrama, *Phys. Rev. D* **98**, no. 5, 054001 (2018).
8. J. A. Oller and E. Oset, *Nucl. Phys. A* **620**, 438 (1997) Erratum: [*Nucl. Phys. A* **652**, 407 (1999)].
9. E. Oset and A. Ramos, *Nucl. Phys. A* **635**, 99 (1998).
10. L. D. Faddeev, *Sov. Phys. JETP* **12**, 1014 (1961) [*Zh. Eksp. Teor. Fiz.* **39**, 1459 (1960)].
11. A. Martinez Torres, K. P. Khemchandani and E. Oset, *Phys. Rev. C* **77**, 042203 (2008).

12. K. P. Khemchandani, A. Martinez Torres and E. Oset, *Eur. Phys. J. A* **37**, 233 (2008).
13. A. Martinez Torres and D. Jido, *Phys. Rev. C* **82**, 038202 (2010).
14. J. J. Xie, A. Martinez Torres and E. Oset, *Phys. Rev. C* **83**, 065207 (2011).
15. J. J. Xie *et. al.*, *Phys. Rev. C* **83**, 055204 (2011).
16. A. Martinez Torres, E. J. Garzon, E. Oset and L. R. Dai, *Phys. Rev. D* **83**, 116002 (2011).
17. J. Gasser and H. Leutwyler, *Phys. Rept.* **87**, 77 (1982).
18. J. Gasser and H. Leutwyler, *Nucl. Phys. B* **250**, 465 (1985).
19. H. Georgi, *Phys. Lett. B* **240**, 447 (1990).
20. N. Isgur and M. B. Wise, *Phys. Rev. Lett.* **66**, 1130 (1991).
21. M. B. Wise, *Phys. Rev. D* **45**, no. 7, R2188 (1992).
22. G. Burdman and J. F. Donoghue, *Phys. Lett. B* **280**, 287 (1992).
23. E. E. Kolomeitsev and M. F. M. Lutz, *Phys. Lett. B* **582**, 39 (2004).
24. F. K. Guo *et. al.*, *Phys. Lett. B* **641**, 278 (2006).
25. D. Gamermann, E. Oset, D. Strottman and M. J. Vicente Vacas, *Phys. Rev. D* **76**, 074016 (2007).
26. A. Martinez Torres, E. Oset, S. Prelovsek and A. Ramos, *JHEP* **1505**, 153 (2015).
27. M. Altenbuchinger, L.-S. Geng and W. Weise, *Phys. Rev. D* **89**, no. 1, 014026 (2014)
28. F. E. Close and P. R. Page, *Phys. Lett. B* **578**, 119 (2004).
29. C. Y. Wong, *Phys. Rev. C* **69**, 055202 (2004).
30. E. S. Swanson, *Phys. Lett. B* **588**, 189 (2004).
31. D. Gamermann and E. Oset, *Eur. Phys. J. A* **33**, 119 (2007).
32. F. Aceti *et. al.*, *Phys. Rev. D* **90**, no. 1, 016003 (2014).
33. A. Martinez Torres, K. P. Khemchandani and L. S. Geng, *Phys. Rev. D* **99**, no. 7, 076017 (2019).
34. D. Gamermann, J. Nieves, E. Oset and E. Ruiz Arriola, *Phys. Rev. D* **81**, 014029 (2010).
35. T. W. Wu *et. al.*, *Phys. Rev. D* **100**, no. 3, 034029 (2019).
36. Y. Huang, M. Z. Liu, Y. W. Pan, L. S. Geng, A. Martínez Torres and K. P. Khemchandani, arXiv:1909.09021 [hep-ph].
37. X. L. Ren, B. B. Malabarba, L. S. Geng, K. P. Khemchandani and A. Martinez Torres, *Phys. Lett. B* **785**, 112 (2018).
38. M. Tanabashi *et al.*, *Phys. Rev. D* **98**, no. 3, 030001 (2018).
39. X. L. Ren, B. B. Malabarba, K. P. Khemchandani and A. Martinez Torres, *JHEP* **1905**, 103 (2019).

Strange pentaquarks with a hidden heavy quark-antiquark pair

Sachiko Takeuchi[1,2,3,*], Alessandro Giachino[4,5], Makoto Takizawa[6,2,7], Elena Santopinto[4] and Makoto Oka[8]

[1] *Japan College of Social Work, Kiyose, Tokyo 204-8555, Japan*
**E-mail: s.takeuchi@jcsw.ac.jp*

[2] *Theoretical Research Division, Nishina Center, RIKEN, Hirosawa, Wako, Saitama 351-0198, Japan*

[3] *Research Center for Nuclear Physics (RCNP), Osaka University, Ibaraki, Osaka 567-0047, Japan*

[4] *Istituto Nazionale di Fisica Nucleare (INFN), Sezione di Genova, via Dodecaneso 33, 16146 Genova, Italy*

[5] *Dipartimento di Fisica dell'Università di Genova, via Dodecaneso 33, 16146 Genova, Italy*

[6] *Showa Pharmaceutical University, Machida, Tokyo 194-8543, Japan*

[7] *J-PARC Branch, KEK Theory Center, Institute for Particle and Nuclear Studies, KEK, Tokai, Ibaraki 319-1106, Japan*

[8] *Advanced Science Research Center, Japan Atomic Energy Agency, Tokai, Ibaraki 319-1195, Japan*

The strange pentaquarks with hidden heavy quark pair ($q^3c\bar{c}$ and $q^3b\bar{b}$) are investigated by the coupled-channel quark cluster model. Two types of the q^3 color-octet configurations are found to provide the attraction, which makes bound states, sharp resonances, and cusps in the baryon meson scattering. A resonance appears at around 4500 MeV in the strange hidden charm sector. Such structures are more clearly seen in the hidden bottom systems.

Keywords: Exotic hadron; pentaquark; flavor singlet baryon; quark model.

1. Introduction

Recently, the $P_c^+(4312)$ state and the two narrow peaks, $P_c^+(4440)$ and $P_c^+(4457)$, have been found in $\Lambda_b^0 \to J/\psi K^- p$ decay by LHCb.[1,2] Many theoretical works have been proposed to investigate the hidden charm systems (*e.g.* see Ref. 3). In Ref. 4, it has been reported that the quark cluster model gives a bound state, resonances, and a cusp in the NJ/ψ or $Y_c\bar{D}$ scattering. Together with the pion exchange, the model seems to reproduce the observed peaks.[5,6] Those structures appear because of the three light quarks in the color-octet isospin-1/2 spin-3/2 compact configuration. Such a configuration appears when the baryon (q^2Q) and the meson ($q\bar{Q}$) come close to each other, and produces a channel-dependent short-range interaction between the hadrons.

Here we discuss the strange pentaquarks with the hidden heavy quark pair, $udsQ\bar{Q}$. Now the three light quarks can also be color-octet and flavor-singlet. Note that this flavor-singlet configuration for the S-wave three quarks can form only when the quarks are color-octet. In this work we show that this configuration also causes resonances. They are more clearly seen in the hidden bottom sector, whereas the one in the charm sector may be observed, *e.g.*, in $\Lambda_b^0 \to J/\psi\Lambda\phi$ decay or in $\Xi_b^- \to J/\psi\Lambda K^-$ decay.

2. Method

Let us first classify possible configurations of the three light quarks in a $q^3Q\bar{Q}$ system. Since the whole system is color-singlet and the hidden heavy quark part ($Q\bar{Q}$) is color-singlet or octet, the remaining light quarks (q^3) are also color-singlet or octet. A color-singlet q^3 belongs to the usual flavor spin $\mathbf{56}_{f\sigma}$ multiplet, which consists of the flavor-octet and the flavor-decuplet baryons. A color-octet q^3 belongs to the $\mathbf{70}_{f\sigma}$ multiplet, which consists of flavor-singlet spin-1/2, the flavor-octet spin-1/2, the flavor-octet spin-3/2, and flavor-decuplet spin-1/2 configurations[7]:

$$\mathbf{56}_{f\sigma} = \mathbf{8}_f \times \mathbf{2}_\sigma + \mathbf{10}_f \times \mathbf{4}_\sigma \tag{1}$$

$$\mathbf{70}_{f\sigma} = \mathbf{1}_f \times \mathbf{2}_\sigma + \mathbf{8}_f \times \mathbf{2}_\sigma + \mathbf{8}_f \times \mathbf{4}_\sigma + \mathbf{10}_f \times \mathbf{2}_\sigma\ . \tag{2}$$

The color spin interaction for the quarks contributes differently to the two terms in rhs of Eq. (1), which causes the hyperfine splitting (HFS) of the baryon masses. It also contributes to the terms in Eq. (2), which gives the HFS of the pentaquarks, $q^3Q\bar{Q}$, at the heavy quark limit (CS):

$$\mathrm{CS} = -\langle q^3[fs]_c | \sum (\lambda\cdot\lambda)(\sigma\cdot\sigma) | q^3[fs]_c \rangle, \tag{3}$$

where $q^3[fs]_c$ stands for the q^3 configuration with the flavor f, spin s, and color c. The difference between CS and the one that contributes to the threshold energy (CS_T) shows whether the color-spin term of the Hamiltonian works as an attractive force ($\mathrm{CS}-\mathrm{CS}_T < 0$) or not ($> 0$). For the

Table 1. The classification of the isospin $\frac{1}{2}$ strangeness 0 and isospin 0 strangeness -1 $q^3Q\bar{Q}$ states. $[fs]_c$ stands for q^3 with flavor, f, spin s, and color c. $S_{Q\bar{Q}}$ stands for the $Q\bar{Q}$ spin, and J for the total spin of $uudQ\bar{Q}$. The lowest S-wave threshold for $Q = c$, CS, and CS_T are also listed.

$[fs]_c$	CS	$S_{Q\bar{Q}}$	J	the lowest S-wave threshold		CS_T	$\mathrm{CS}-\mathrm{CS}_T$
$[8\frac{1}{2}]_1$	-8	0	$\frac{1}{2}$	$N\eta_c$	$\Lambda\eta_c$	-8	0
		1	$\frac{1}{2}, \frac{3}{2}$	NJ/ψ	$\Lambda J/\psi$		
$[1\frac{1}{2}]_8$	-14	0	$\frac{1}{2}$		$\Lambda_c D_s$	-8	-6
		1	$\frac{1}{2}, \frac{3}{2}$		$\Lambda_c D_s{}^{(*)}$		
$[8\frac{1}{2}]_8$	-2	0	$\frac{1}{2}$	$\Lambda_c\overline{D}$	$\Lambda_c D_s$	-8	6
		1	$\frac{1}{2}, \frac{3}{2}$	$\Lambda_c\overline{D}^{(*)}$	$\Lambda_c D_s{}^{(*)}$		
$[8\frac{3}{2}]_8$	2	0	$\frac{3}{2}$	$\Sigma_c^*\overline{D}$	$\Xi_c'\bar{D}$	$\frac{8}{3}$	$-\frac{2}{3}$
		1	$\frac{1}{2}, \frac{3}{2}, \frac{5}{2}$	$\Sigma_c{}^{(*)}\overline{D}^{(*)}$	$\Xi_c'\bar{D}^{(*)}$, $\Xi_c^*\bar{D}^*$		

isospin-0 $udsQ\bar{Q}$, the terms with $\mathbf{1}_f$ or $\mathbf{8}_f$ are relevant. In Table 1, for each $[fs]_c$ and the spin of $Q\bar{Q}$, $S_{Q\bar{Q}}$, we list possible total spin, J, CS, the relevant lowest S-wave threshold with CS_T. It is found that the configurations of $[fs]_c$=$[1\frac{1}{2}]_8$ and $[8\frac{3}{2}]_8$ provide attraction.

In order to understand if such an attraction is strong enough to produce resonances or bound states, we investigate the situation dynamically by employing the coupled-channel quark cluster model for the $uudQ\overline{Q}$ $I(J^P) = \frac{1}{2}(\frac{1}{2}^-, \frac{3}{2}^-, \frac{5}{2}^-)$ and the $udsQ\overline{Q}$ $I(J^P) = 0(\frac{1}{2}^-, \frac{3}{2}^-, \frac{5}{2}^-)$ systems with $Q = c$ and b. The model is essentially the same as used in Ref. 4 with the difference that now we are considering also the strange and bottom quarks. The Hamiltonian for quarks, H_q, consists of the kinetic term, K, the confinement term, V_{conf}, the color Coulomb term, V_{Coul}, and the color spin term, V_{CS}:

$$H_q = K + V_{\mathrm{conf}} + V_{\mathrm{Coul}} + V_{\mathrm{CS}} \ . \tag{4}$$

Both of V_{Coul} and V_{CS} are considered to come from the effective one-gluon exchange interaction between quarks. The s and b quark masses used in our model and the obtained hadron masses are listed in Table 2. The hadron masses are reproduced well, but to use the observed threshold energies, we introduce the mass correction to the kinetic term. In the long range region, this model becomes essentially a free baryon-meson model. In the short range region, an interaction between the baryon and the meson arises from the quark degrees of freedom and from the interaction between quarks.

Table 2. The quark masses used in the model and the obtained hadron masses (in MeV). The numbers in parentheses are the isospin-averaged experimentally observed masses.[8]

$m_u(=m_d)$	m_s	m_c	m_b		
300	510	1741.5	5110.9		
N	Λ	Σ	Σ^*	Ξ	Ξ^*
922(939)	1110(1116)	1188(1193)	1394(1385)	1328(1318)	1523(1533)
Λ_c	Σ_c	Σ_c^*	Ξ_c	Ξ_c'	Ξ_c^*
2292(2286)	2454(2454)	2516(2518)	2495(2469)	2578(2579)	2630(2646)
Λ_b	Σ_b	Σ_b^*	Ξ_b	Ξ_b'	Ξ_b^*
5640(5620)	5826(5813)	5847(5833)	5827(5794)	5936(5935)	5952(5954)
η_c	J/ψ	$\overline{D}$	$\overline{D}^*$	D_s	D_s^*
2981(2984)	3101(3097)	1863(1867)	2005(2009)	1969(1968)	2112(2112)
η_b	Υ	B	B^*	B_s	B_s^*
9395(9399)	9460(9460)	5276(5279)	5331(5325)	5356(5367)	5417(5415)

Table 3. The number of structures in the baryon meson scattering with the number of the relevant q^3 configurations which provide the attraction.

S	J	$[8\frac{3}{2}]_8$	$uudb\bar{b}$	$uudc\bar{c}$
0	$\frac{1}{2}$	1	1	1
	$\frac{3}{2}$	2	2	1
	$\frac{5}{2}$	1	1	1

S	J	$[1\frac{1}{2}]_8$	$[8\frac{3}{2}]_8$	$udsb\bar{b}$	$udsc\bar{c}$
−1	$\frac{1}{2}$	2	1	3	1
	$\frac{3}{2}$	1	2	3	1
	$\frac{5}{2}$	0	1	1	1

3. Results

In Table 3, we summarize the number of the bound states and resonances that appear in the quark cluster model calculation. Resonances we count here are the ones that the diagonal phase shift rises more than $\pi/2$. The number of the bound states and resonances roughly corresponds to the number of quark configurations that provide the attraction.

Correspondence between the attractive color-octet configurations and the structures in the scattering is more prominent in the $q^3b\bar{b}$ systems. As shown in Fig. 1, for example, the $uudb\bar{b}$ $J = 3/2$ system has two resonances, which corresponds to the two configurations of $[8\frac{3}{2}]_8$. In $udsb\bar{b}$ $J = 3/2$ system has one more resonance at the $\Lambda_b B_s^*$ threshold, which corresponds to $[1\frac{1}{2}]_8$. There are one bound state in each of the $q^3Q\bar{Q}$ $J = 5/2$ systems, where the three light quark configuration is $[8\frac{3}{2}]_8$.

The resonances can still be seen in the $q^3c\bar{c}$ systems. As for the spin-3/2 $udsc\bar{c}$ systems, there is a resonance at around 4500 MeV, in which both of

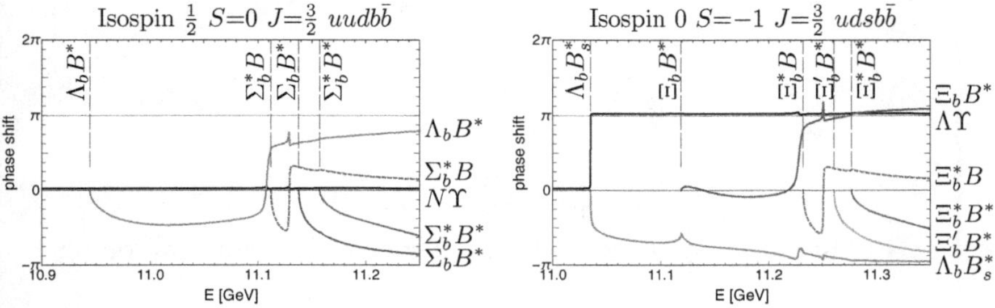

Fig. 1. Hidden bottom baryon meson scattering phase shifts of the isospin 1/2 (left) and the strangeness −1 isospin 0 systems (right). Total angular momentum is 3/2.

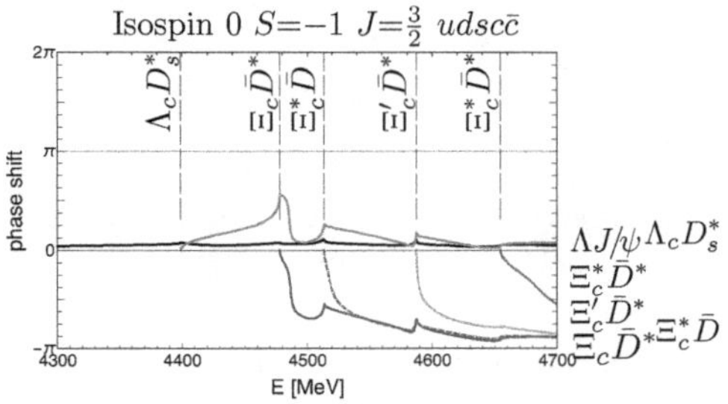

Fig. 2. Hidden charm baryon meson scattering phase shifts of the strangeness −1 isospin 0 systems. Total angular momentum is 3/2.

the $[1\frac{1}{2}]_8$ and $[8\frac{3}{2}]_8$ configurations contribute (Fig. 2). Energetically, it can be seen in $\Lambda_b^0 \to J/\psi\Lambda\phi$ decay or in $\Xi_b^- \to J/\psi\Lambda K^-$ decay.

The energies and channels in which the resonances appear may vary by the choice of the potential parameters. The existence of the peaks, however, seems to be robust. Investigating the pentaquark systems will show us properties of the color non-singlet configuration, and will give us information to understand features of the low energy non perturbative QCD.

This work is supported in part by by JSPS KAKENHI No. 16K05361.

References

1. R. Aaij *et al.*, *Phys. Rev. Lett.* **115**, 072001 (2015).
2. R. Aaij *et al.*, *Phys. Rev. Lett.* **122**, 222001 (2019).
3. H.-X. Chen, W. Chen, X. Liu and S.-L. Zhu, *Phys. Rept.* **639**, 1 (2016).
4. S. Takeuchi and M. Takizawa, *Phys. Lett.* **B764**, 254 (2017).
5. Y. Yamaguchi, A. Giachino, A. Hosaka, E. Santopinto, S. Takeuchi and M. Takizawa, *Phys. Rev.* **D96**, 114031 (2017).
6. Y. Yamaguchi, *et al.*, arXiv:1907.04684 [hep-ph].
7. Y. Irie, M. Oka and S. Yasui, Phys. Rev. D **97** 034006 (2018).
8. M. Tanabashi *et al.*, *Phys. Rev.* **D98**, 030001 (2018).

Hidden-charm pentaquark resonances in $J/\psi p$ scattering

Hongxia Huang*, Jun He and Jialun Ping

Department of Physics, Nanjing Normal University
Nanjing, Jiangsu 210023, China
**E-mail: hxhuang@njnu.edu.cn*
www.njnu.edu.cn

Fan Wang

Department of Physics, Nanjing University
Nanjing, Jiangsu 210093, China

Recently, the LHCb Collaboration reported their new results about the hidden-charm pentaquarks, $P_c(4312)$, $P_c(4440)$ and $P_c(4457)$, observed in $J/\psi p$ invariant mass spectrum. Based on a calculation of the cross section of $J/\psi p$ scattering in quark delocalization color screening model, $P_c(4312)$ can be identified as a hidden-charm molecular state $\Sigma_c D$ with $J^P = \frac{1}{2}^-$. The two-peak structure can be reproduced at about 4450 MeV, which corresponds to the $P_c(4440)$ and $P_c(4457)$. They are the resonances molecular states $\Sigma_c D^*$ of $J^P = \frac{1}{2}^-$ and $J^P = \frac{3}{2}^-$. All these heavy pentaquarks are worth searching in the future experiments.

Keywords: Hidden-charm pentaquarks; resonances; two-peak structure; scattering process.

1. Introduction

In 2015, the report of two hidden-charm pentaquark states $P_c(4380)$ and $P_c(4450)$ by the LHCb Collaboration[1] attracted people's interest in the pentaquarks with heavy quarks. Four years later, the LHCb Collaboration reported the observation of three new pentaquarks, named as $P_c(4312)$, $P_c(4440)$, and $P_c(4457)$.[2] All of them are observed in the $J/\psi p$ invariant mass spectrum from the $\Lambda_b \to J/\psi p K^-$ decay, corresponding to three peaks in the $J/\psi p$ invariant mass distributions. All peaks are all very sharp, and a two-peak structure near the mass of 4450 MeV is obvious. Immediately after the report of the LHCb Collaboration, several theoretical work have been done to study the mass spectrum of these states.[3–6]. Ref. 7 studied the isospin breaking decays of the molecular structure of the $P_c(4457)$.

Searching for the existence of multiquark states is an important issue of the hadron physics. To provide the necessary information for experiments, mass spectrum calculation alone is not enough. The study of hadron-hadron scattering, as well as the main production process of multiquark states, is indispensable. Recently, we investigated the hidden-charm and hidden-bottom pentaquarks in the framework of the quark delocalization color screening model(QDCSM),[8,9] and a few narrow pentaquark resonances with hidden-charm above 4.2 GeV, and some narrow pentaquark resonances with hidden-bottom above 11 GeV are found from corresponding scattering processes. In this work, we continue to study the cross section of the $J/\psi p$ scattering to investigate the three peaks in the the $J/\psi p$ scattering.

2. The quark delocalization color screening model

The detail of QDCSM used in the present work can be found in the references.[10] In the following, only the Hamiltonian and parameters are given.

$$
\begin{aligned}
H &= \sum_{i=1}^{5}\left(m_i + \frac{p_i^2}{2m_i}\right) - T_c + \sum_{i<j} V_{ij}, \qquad (1)\\
V_{ij} &= V^G(r_{ij}) + V^\chi(r_{ij}) + V^C(r_{ij}),\\
V^G(r_{ij}) &= \frac{1}{4}\alpha_s \boldsymbol{\lambda}_i \cdot \boldsymbol{\lambda}_j \left[\frac{1}{r_{ij}} - \frac{\pi}{2}\left(\frac{1}{m_i^2} + \frac{1}{m_j^2} + \frac{4\boldsymbol{\sigma}_i\cdot\boldsymbol{\sigma}_j}{3m_im_j}\right)\delta(r_{ij}) - \frac{3}{4m_im_jr_{ij}^3}S_{ij}\right],\\
V^\chi(r_{ij}) &= \frac{\alpha_{ch}}{3}\frac{\Lambda^2}{\Lambda^2 - m_\chi^2}m_\chi \left\{\left[Y(m_\chi r_{ij}) - \frac{\Lambda^3}{m_\chi^3}Y(\Lambda r_{ij})\right]\boldsymbol{\sigma}_i\cdot\boldsymbol{\sigma}_j\right.\\
&\quad \left. + \left[H(m_\chi r_{ij}) - \frac{\Lambda^3}{m_\chi^3}H(\Lambda r_{ij})\right]S_{ij}\right\}\boldsymbol{\lambda}_i^F\cdot\boldsymbol{\lambda}_j^F, \quad \chi = \pi, K, \eta
\end{aligned}
$$

$$
\begin{aligned}
V^C(r_{ij}) &= -a_c\boldsymbol{\lambda}_i\cdot\boldsymbol{\lambda}_j[f(r_{ij}) + V_0],\\
f(r_{ij}) &= \begin{cases} r_{ij}^2 & \text{if } i,j \text{ occur in the same baryon orbit}\\ \frac{1-e^{-\mu_{ij}r_{ij}^2}}{\mu_{ij}} & \text{if } i,j \text{ occur in different baryon orbits}\end{cases}\\
S_{ij} &= \frac{(\boldsymbol{\sigma}_i\cdot\boldsymbol{r}_{ij})(\boldsymbol{\sigma}_j\cdot\mathbf{r}_{ij})}{r_{ij}^2} - \frac{1}{3}\boldsymbol{\sigma}_i\cdot\boldsymbol{\sigma}_j.
\end{aligned}
$$

All symbols have their usual meanings, and all parameters are also taken from our previous work.[8]

3. The results and discussions

The well developed resonating group method (RGM)[11] is applied to calculate the phase shifts and the cross section of the baryon-meson scattering. The wave function of the baryon-meson system and the detail of this method can be found in Ref. 9. The channels involved are listed in Table 1.

Table 1. The channels involved in the calculation.

$S = 1/2$	$J/\psi p$	$\Sigma_c D$	$\Sigma_c D^*$	$\Sigma_c^* D^*$
$S = 3/2$	$J/\psi p$	$\Sigma_c D^*$	$\Sigma_c^* D$	$\Sigma_c^* D^*$

The cross section of the $J/\psi p$ is shown in Fig. 1. Obviously, there are four sharp peaks in the cross section of the $J/\psi p$ scattering, which indicates that there are four narrow resonance states. The first one locates at the mass of 4307.9 MeV with a very narrow partial width of about 1.2 MeV. The mass is under the threshold of the $\Sigma_c D$, so it is the the $\Sigma_c D$ molecular pentaquark. Moreover, both the mass and decay width are close to the experimental value of the $P_c(4312)$, and this peak is just corresponding to the peak appearing in the experimental $J/\psi p$ invariant mass distributions, which indicates that the newly reported $P_c(4312)$ state could be identified as the $\Sigma_c D$ molecular pentaquark with $J^P = \frac{1}{2}^-$ in our model calculation. Particularly, there are two sharp peaks in the cross section of the $J/\psi p$ channel near the mass of 4450 MeV. The one is $\Sigma_c D^*$ with $J^P = \frac{1}{2}^-$ shown in Fig. 1(a), and another one is $\Sigma_c D^*$ with $J^P = \frac{3}{2}^-$ shown in Fig. 1(b). The masses and the partial decay widths can be read from Fig. 1, they are: $\Sigma_c D^*$ of $J^P = \frac{1}{2}^-$, 4459.7 MeV and 3.9 MeV; $\Sigma_c D^*$ of $J^P = \frac{3}{2}^-$, 4445.7 MeV and 1.5 MeV. It is obvious that these two peaks are very close to each other and have maximum cross sections of about 2 fm^2, which can reproduce a two-peak structure at about 4450 MeV as the one shown in the experimental $J/\psi p$ invariant mass distributions. Compared with experimental data, the $P_c(4440)$ is more possible to be the molecular pentaquark $\Sigma_c D^*$ of $J^P = \frac{3}{2}^-$, and the $P_c(4457)$ can be explained as the molecular pentaquark $\Sigma_c D^*$ of $J^P = \frac{1}{2}^-$.

Besides the three resonances discussed above, one may see the fourth peak showed in Fig. 1 (b). It is the molecular pentaquark $\Sigma_c^* D$ with $J^P = \frac{3}{2}^-$. The mass of this resonance is 4376.4 MeV, which is very close to the reported $P_c(4380)$. However, the two body decay width (with heavy hadron in the final state) is only 2.4 MeV, much smaller than the experimental value

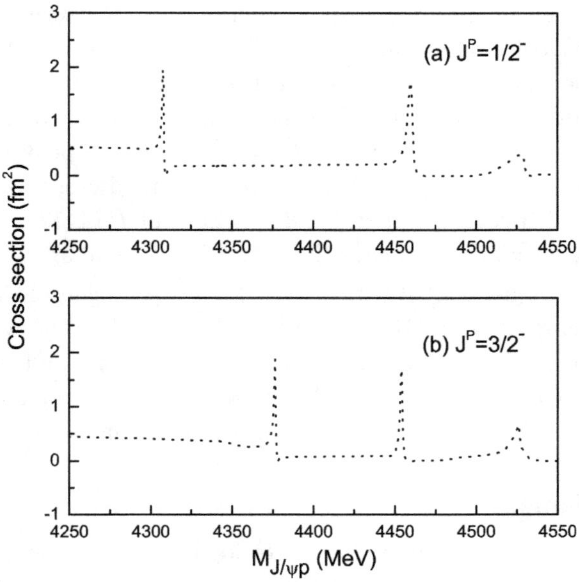

Fig. 1. The cross section of the $J/\psi p$ channel with $J^P = \frac{1}{2}^-$ and $J^P = \frac{3}{2}^-$ respectively.

of $P_c(4380)$ state. We suggest the experiment to find whether there is any narrow resonance near the $P_c(4380)$.

In addition, in Fig. 1(a), we also find there is a cusp near the mass of 4527 MeV, which is near the threshold of the $\Sigma_c^* D^*$. Our previous calculation showed that the single channel $\Sigma_c^* D^*$ of $J^P = \frac{1}{2}^-$ was bound, coupling with $\eta_c p$ scattering channel, the state becomes a resonance with mass 4525.8 MeV and width 4.0 MeV.[9] However, it does not show up in the scattering channels $J/\psi p$, $\Sigma_c D$, and $\Sigma_c D^*$, and it only appears as a cusp in the cross section of the $J/\psi p$.[9] This is consistent with the experimental result, there is no distinct signal near the threshold of the $\Sigma_c^* D^*$ in the experimental data. Besides, the $\Sigma_c^* D^*$ state of $J^P = \frac{3}{2}^-$ has the similar situation. It appears as a resonance with mass of 4523.0 MeV and width of 1.0 MeV in the $\Lambda_c D^*$ scattering process, but it only shown up as a cusp in the cross section of the $J/\psi p$ (see Fig. 1(b)).[9]

4. Summary

To summarize, inspired by the three new narrow pentaquarks $P_c(4312)$, $P_c(4440)$, and $P_c(4457)$ observed in the process of $\Lambda_b \to J/\psi pK$ reported by LHCb, we investigate the cross section of $J/\psi p$ scattering in the quark delocalization color screening model. Four peaks are found in the cross section. The mass and width of the first one is close to the $P_c(4312)$, which indicates that the $P_c(4312)$ can be interpreted as the hidden-charm molecular pentaquarks $\Sigma_c D$ with $J^P = \frac{1}{2}^-$. The two-peak structure can be reproduced at about 4450 MeV in our theoretical cross section of $J/\psi p$ scattering, which correspond to the $P_c(4440)$ and $P_c(4457)$. They are the resonances molecular pentaquark $\Sigma_c D^*$ of $J^P = \frac{1}{2}^-$ and $J^P = \frac{3}{2}^-$. Another molecular pentaquark $\Sigma_c^* D$ with $J^P = \frac{3}{2}^-$ is also suggested in our calculation, the mass of which is close to the $P_c(4380)$, but the width is much smaller than it. Besides, the $\Sigma_c^* D^*$ of both $J^P = \frac{1}{2}^-$ and $J^P = \frac{3}{2}^-$ are possible molecular pentaquarks. All these narrow pentaquarks are worth searching for or being confirmed in future experiments.

Acknowledgment

This work is supported partly by the National Natural Science Foundation of China under Contract Nos. 11675080, 11775118 and 11535005, 11675228.

References

1. R. Aaij, *et al.* (LHCb Collaboration), Phys. Rev. Lett. **115**, 072001 (2015).
2. R. Aaij, *et al.* (LHCb Collaboration), Phys. Rev. Lett. **122**, 222001 (2019).
3. H. X. Chen, W. Chen and S. L. Zhu, Phys. Rev. D **100**, 051501(R) (2019).
4. R. Chen, X. Liu, Z. F. Sun and S. L. Zhu, Phys. Rev. D **100**, 011502(R) (2019).
5. M. Z. Liu, Y. W. Pan, F. Z. Peng, M. S. Sanchez, L. S. Geng, A. Hosaka and P. Valderrama, Phys. Rev. Lett. **122**, 242001 (2019).
6. J. He, Eur. Phys. J. C **79**, 393 (2019).
7. F. K. Guo, H. J. Jing, Ulf-G. Meissner and S. Sakai, Phys. Rev. D **99**, 091501(R) (2019).
8. H. X. Huang, C. R. Deng, J. L. Ping and F. Wang, Eur. Phys. J. C **76**, 624 (2016).
9. H. X. Huang and J. L. Ping, Phys. Rev. D **99**, 014010 (2019).
10. F. Wang, G. H. Wu, L. J. Teng and T. Goldman, Phys. Rev. Lett. **69**, 2901 (1992).
11. M. Kamimura, Supp. Prog. Theo. Phys. **62**, 236 (1977).

Tri-meson bound states with heavy flavor

Li Ma

Department of physics, Beijing Jiaotong University,
Beijing 100044, China
E-mail: ma.li@bjtu.edu.cn

Through the Born-Oppenheimer approximation and One-Pion-Exchange model, We have explored four doubly heavy tri-meson systems DD^*K, $D\bar{D}^*K$, $BB^*\bar{K}$ and $B\bar{B}^*\bar{K}$. After solving the three-body Schrödinger equation with considering the coupled-channel effects and S-D wave mixing, we find loosely bound molecular states for the four doubly heavy tri-meson systems. For instance, With an estimated error, the mass of the DD^*K or $BB^*\bar{K}$ molecule is $4317.92^{+3.66}_{-4.32}$ MeV or $11013.65^{+8.68}_{-9.02}$ MeV. We also extend our investigation to the triply heavy tri-meson system BBB^* via the Born-Oppenheimer Potential method. By sharing one virtual pion, the BBB^* system has a loosely bound state as long as the molecular states of its two-body subsystem BB^* exist.

Keywords: Tri-meson bound state; One-Pion-Exchange model; Born-Oppenheimer approximation.

1. Introduction

The numerous exotic states which beyond the conventional quark model has grown dramatically in recent years. Some of them sit near the open flavor thresholds with narrow widths could be regarded as the analogous of the deuteron. Similarly, the existence of the triton inspires us to ask whether a bound state formed by three hadrons could exist or not. The studies on the NDK, $\bar{K}DN$, $ND\bar{D}$, $N\bar{K}K$, DKK and $DK\bar{K}$ systems were publicised.[1–5] Similar explorations on the $BD\bar{D}$ and BDD systems were claimed.[6] E. Wilbring *et al.* had studied the molecule-meson scattering amplitudes for the $Z_b(10610)$ or $Z_b(10650)$ in the molecule picture.[7] They had predicted that the universal bound states for three B and B^* mesons had been ruled out. The system $BB^*B^* - B^*B^*B^*$ has been calculated recently based on the colored interaction.[8] Discussions on the ΩNN and $\Omega\Omega N$ systems were claimed in Ref. 9.

2. Doubly heavy tri-hadron system

In this section we perform our calculations on the four doubly heavy tri-meson systems DD^*K, $D\bar{D}^*K$, $BB^*\bar{K}$ and $B\bar{B}^*\bar{K}$. Since the mass of K or $\bar{K}$ is much smaller than the charm meson or bottom meson, we can employ the Born-Oppenheimer Approximation to simplify our calculations. The Lagrangians with the SU(2) chiral symmetry and C parity conservation read $\mathcal{L}_P = -i\frac{2g}{f_\pi}\bar{M}P_b^{*\mu}\partial_\mu\phi_{ba}P_a^\dagger + i\frac{2g}{f_\pi}\bar{M}P_b\partial_\mu\phi_{ba}P_a^{*\mu\dagger}$, where the heavy flavor meson fields $P^{(*)}$ represent $P^{(*)} = (D^{(*)0}, D^{(*)+})$, $(B^{(*)-}, \bar{B}^{(*)0})$ or $(K^{(*)-}, \bar{K}^{(*)0})$. The ϕ in the above denotes the exchanged pion matrix. There are three coupling coefficients should be determined. One is the pion decay constant $f_\pi = 132$ MeV, determined by chiral symmetry. One is the pionic coupling constant $g = 0.59$ fixed by the width of D^{*+}.[10] The other one is the coupling constant g'. Here, we take the experimental result $g' = 0.90$ which was obtained from the full widths of K^* in the PDG.[10] For suppressing the contribution from the ultraviolet region of the exchanged momentum, we take the conventional form of form factor $F(q) = \frac{\Lambda^2 - m_\pi^2}{\Lambda^2 - q^2}$ in each vertex. In order to fix the parameter Λ in the form factor, we introduce the assumption that the $D_{s1}(2460)$ reported in the experiment is dominant by a loosely bound D^*K molecule,[11] then the parameter Λ could be fixed definitely at 803.20 MeV. For the bottom sector, we also adopt the speculations proposed in Ref. 12, that the $D_{s1}(2460)$ has a counter partner state in bottom meson sector which might be a good candidate of the $B^*\bar{K}$ molecular state. Thus the parameter Λ could be fixed at 1451.00 MeV.

In the framework of the Born-Oppenheimer Approximation (BOA), there are two degree of freedoms in the DD^*K system as an example, the motion of the K related to the D and D^*, and the relative motion between D and D^*. Based on the main idea of BOA, both of the two degrees of freedom are independent, and they can be treated separately. After solving the Schrödinger equation as well as the treatment of S-D wave mixing, we find out the loosely bound states for the DD^*K, $D\bar{D}^*K$, $BB^*\bar{K}$ and $B\bar{B}^*\bar{K}$ with the isospin $\frac{1}{2}$. The reduced three-body binding energy for the DD^*K ($D\bar{D}^*K$) is 8.29 MeV. The reduced three-body binding energy for the $BB^*\bar{K}$ ($B\bar{B}^*\bar{K}$) is 41.76 MeV.

We also plot the dependence of the reduced three body binding energy $|E_3|$ on the two body binding energy $|E_b|$ for the DD^*K and $D\bar{D}^*K$ in Figure 1. The dependence for the $BB^*\bar{K}$ and $B\bar{B}^*\bar{K}$ also plotted in Figure 1. Where the red and green bands are from PDG2016 and PLB647,133, respectively. The reduced three-body binding energies are

$E_{I=1/2}^{DD^*K} = 8.29^{+4.32}_{-3.66}$ MeV, $E_{I=1/2}^{BB^*\bar{K}} = 41.76^{+8.49}_{-8.84}$ MeV. The error band is estimated as $m_K/(2m_u)$ with m_u the reduced mass of the two heavy particles. As the long-distance DD^* potential is related to that of the $D\bar{D}^*$ potential by G-parity,[13] there could also exist a three-body $D\bar{D}^*K$ bound state, but with additional uncertainty $m_\pi^2/(2m_u)$ (which characterizes the natural energy scale of OPE[14]) from the unknown short-distance interaction. Thus, for the $D\bar{D}^*K$ and $B\bar{B}^*\bar{K}$ system, the three-body binding energies are $E_{I=1/2}^{D\bar{D}^*K} = 8.29^{+6.55}_{-6.13}$ MeV, $E_{I=1/2}^{B\bar{B}^*\bar{K}} = 41.76^{+8.68}_{-9.02}$ MeV. With the additional uncertainty from the missing short-distance interaction. Those correspond to two bound states with masses $4317.92^{+6.13}_{-6.55}$ MeV and $11013.65^{+8.68}_{-9.02}$ MeV.

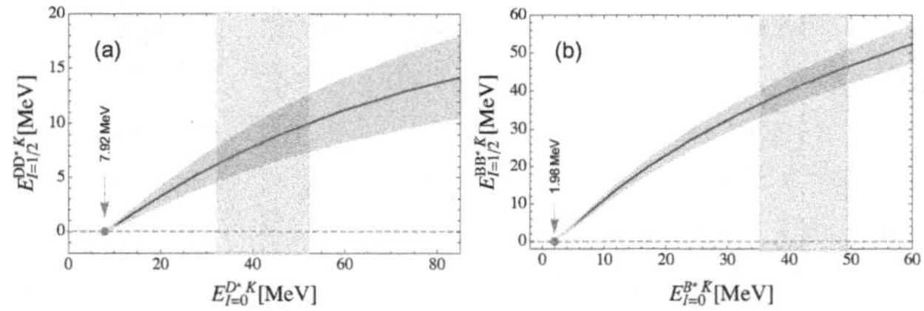

Fig. 1. The dependence of the three-body binding energy $|E_3|$ on the two-body binding energy $|E_b|$. (a) is for the charm sector. (b) is for the bottom sector.

3. Triply heavy tri-hadron system

In this section we focus on the system BBB^*. For simplicity, we use a, b and c to label the three constituents. If we assume that the meson b and c are much heavier than the meson a, then we can still use the Born-Oppenheimer approximation to separate the degree of freedom of a from the three-body system. We can fix the meson b and c with the distance r_{bc} when discussing the dynamics for the meson a. Thus the three-body system could degenerate into a two-body system consisting of mesons b and c but with an additional BO potential created by the meson a. As we know, only the meson a is much lighter than the other ones can the separation in this way be a good approximation for this system. Based on this separation, we obtain the wave functions $\psi(\vec{r}_{ab}, \vec{r}_{ac})$ for the meson a. The remaining degree of freedom is the relative motion between meson b and c which can be described by a wave function $\Phi(\vec{r}_{bc})$ that can be determined from the Schrödinger equation. Then the total wave function of the system has the form $\Psi_T = \Phi(\vec{r}_{bc})\psi(\vec{r}_{ab}, \vec{r}_{ac})$. Nevertheless, the truth is that the three constituents in the $B_a^{(*)}B_b^{(*)}B_c^{(*)}$ have little mass difference. Every meson has the priority to be the light meson and separated from the three-body system. Thus we have three different kinds of separation schemes. That is we can divide the system $B_a^{(*)}B_b^{(*)}B_c^{(*)}$ into three kinds of two-body subsystems, i.e., $B_a^{(*)}B_b^{(*)}$ with the BO potential created by the meson c, $B_b^{(*)}B_c^{(*)}$ with the BO potential created by the meson a and $B_a^{(*)}B_c^{(*)}$ with the BO potential created by the meson b. The three simplification schemes constitute three kinds of basic configurations. A simple approximation for the eigenstates of the three-body system could be combinations of them.

In order to verify the approximate is good enough, we apply it to a three-nucleon system. which has sufficient experimental data. Once the parameter $\Lambda = 899.60$ MeV is fixed by the binding energy of deuteron, the reduced three-body binding energy and the total binding energy relative to the three free nucleons are 5.38 MeV and 7.62 MeV, respectively. The later is comparable with the empirical binding energies of the triton (8.48 MeV) and helium-3 (7.80 MeV) nuclei.

After careful calculations, we find no critical value for the two-body binding energy, which indicates the lower limit of the required binding energy of their subsystem BB^* to form a three-body bound state. That is to say, no matter how small the two-body binding energy is, as long as the two-body subsystem BB^* has a loosely bound state, the three-body system BBB^* is most likely to have a loosely bound state too. To show this conclusion explicitly, we plot the dependence of the three-body binding energy on the variety of two-body binding energy in Fig. 2, where $E^{BBB^*}_{I=3/2(1/2)}$ and $E^{BB^*}_{I=1(0)}$ denote the reduced three-body binding energy and two-body binding energy, respectively.

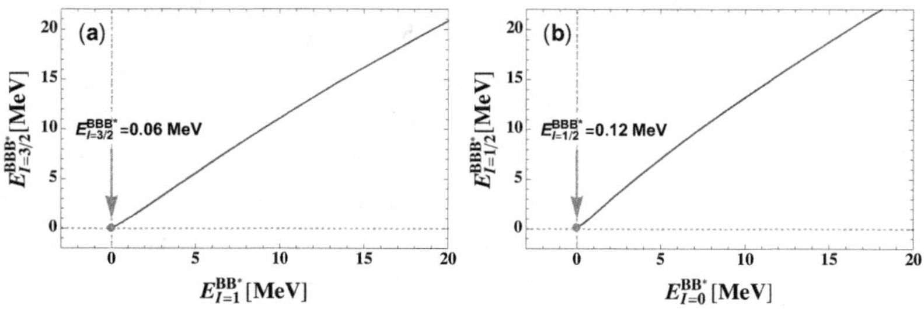

Fig. 2. Dependence of the reduced three-body binding energy on the two-body binding energy of its subsystem BB^*. (a) corresponds to the isospin states $|1, \frac{3}{2}, \pm\frac{1}{2}(\pm\frac{3}{2})\rangle$ and $|1, \frac{1}{2}, \pm\frac{1}{2}\rangle$ cases, while (b) corresponds to the isospin state $|0, \frac{1}{2}, \pm\frac{1}{2}\rangle$ case.

4. Summary

The large number of exotic states inspire us exploring the scenario of hadronic molecules. Analogous of the triton leaves us an open issue on the existence of tri-hadron molecules. In the framework of OPE model and BOA, we have claimed four tri-meson molecules with doubly heavy flavor, i.e. DD^*K, $D\bar{D}^*K$, $BB^*\bar{K}$ and $B\bar{B}^*\bar{K}$. For instance, With the estimated error, the mass of the DD^*K or $BB^*\bar{K}$ molecule is $4317.92^{+3.66}_{-4.32}$ MeV or $11013.65^{+8.68}_{-9.02}$ MeV. We also extend our discussions to triply heavy system BBB^* and obtain a conclusion that the BBB^* system has a loosely bound state as long as the molecular states of its two-body subsystem BB^* exist. Wish further experiment on B mesons decay and pp collision can unveil the existence of tri-hadron molecules.

References

1. C. W. Xiao, M. Bayar and E. Oset, Phys. Rev. D **84**, 034037 (2011).
2. J. J. Xie, A. Martinez Torres and E. Oset, Phys. Rev. C **83**, 065207 (2011).
3. V. R. Debastiani, J. M. Dias and E. Oset, Phys. Rev. D **96**, no. 1, 016014 (2017).
4. D. Jido and Y. Kanada-En'yo, Phys. Rev. C **78**, 035203 (2008).
5. A. Martinez Torres, K. P. Khemchandani and E. Oset, Phys. Rev. C **79**, 065207 (2009).
6. J. M. Dias, V. R. Debastiani, L. Roca, S. Sakai and E. Oset, arXiv:1709.01372 [hep-ph].
7. E. Wilbring, H.-W. Hammer and U. G. Meiner, arXiv:1705.06176 [hep-ph].
8. H. Garcilazo and A. Valcarce, Phys. Lett. B **784**, 169 (2018) doi:10.1016/j.physletb.2018.07.055 [arXiv:1808.00226 [hep-ph]].
9. H. Garcilazo and A. Valcarce, Phys. Rev. C **99**, no. 1, 014001 (2019) doi:10.1103/PhysRevC.99.014001 [arXiv:1901.05678 [hep-ph]].
10. C. Patrignani *et al.* [Particle Data Group], Chin. Phys. C **40**, no. 10, 100001 (2016).
11. A. Martnez Torres, E. Oset, S. Prelovsek and A. Ramos, JHEP **1505**, 153 (2015).
12. F. K. Guo, P. N. Shen and H. C. Chiang, Phys. Lett. B **647**, 133 (2007).
13. E. Klempt, F. Bradamante, A. Martin and J. M. Richard, Phys. Rept. **368**, 119 (2002).
14. E. Braaten and M. Lu, Phys. Rev. D **76**, 094028 (2007).

Session 4

Hadron Decays, Production and Interactions

Recent results on CP violation in charm sector by LHCb

M. Saur
On behalf of the LHCb Collaboration

University of Chinese Academy of Sciences,
Beijing, China
E-mail: miroslav.saur@cern.ch

The charm sector is an unique probe to possible new physics coupling to up-type quarks and offers a complimentary approach to direct searches for particles originating from processes not described by the Standard Model of Particle physics (SM). Contributions beyond the SM could originate from the higher-order loop effects and assuming generic beyond SM scenarios, much higher mass scales are accessible in comparison with direct searches. Measurements of Charge-Parity (*CP*) violation and mixing in charm decays are promising method to probe the aforementioned effects. The violation of *CP* symmetry in charm was observed for the first time by the LHCb experiment in 2019.[1]

These proceedings present a measurement of the mass difference between neutral charm meson mass eigenstates, a search for time-dependent *CP* violation in neutral charm meson decays and the first observation of *CP* violation in charm hadrons decays.

Keywords: LHCb; proton-proton collisions; charm physics; *CPV*.

1. Introduction

Charge-Parity violation (*CPV*) was postulated as one of three conditions for baryon asymmetry of the Universe.[2] While in the SM, *CPV* is introduced via the Cabibbo-Kobayashi-Maskawa (CKM) mechanism, the amount of *CPV* generated is too small to explain observed matter-antimatter asymmetry. The violation of *CP* symmetry has been successfully measured in strange and bottom hadrons decays and also for the first time in the decays of D^0 meson by the LHCb experiment in 2019.[1] In comparison to the strange or the bottom sector, the charm sector is unique in providing an access to the SM and possible beyond the SM couplings to up-type quarks. Theoretical predictions for *CP* violation in the charm sector vary in range $\mathcal{O}(10^{-3} - 10^{-4})$. However, theoretical predictions in the charm sector are difficult to compute reliably. Possible beyond the SM effects could affect the size of predicted *CP* violating effects in charm decays and therefore searches for *CP* asymmetries can serve as a additional probe of dynamics beyond the SM.

D^0 mesons are produced as a flavour eigenstates, but decay as mass eigenstates labelled as D_1 and D_2. The relation between flavour and mass eigenstates is given by:

$$|D_{1,2}\rangle \equiv p|D^0\rangle \pm q|\overline{D}^0\rangle, \tag{1}$$

where p and q are complex parameters. Related mixing parameters characterising time-dependent oscillations are:

$$x \equiv \frac{(m_1 - m_2)}{\mathbf{\Gamma}},\ y \equiv \frac{(\mathbf{\Gamma}_1 - \mathcal{W}_2)}{2\mathbf{\Gamma}}, \tag{2}$$

where $m_{1,2}$ and $\mathbf{\Gamma}_{1,2}$ are the masses and widths of the mass eigenstates $D_{1,2}$ and $\mathbf{\Gamma} \equiv (\mathbf{\Gamma}_1 + \mathbf{\Gamma}_2)/2$. Although $D^0 - \overline{D}^0$ mixing is a well measured experimental fact,[3] most of those measurements provide limited direct information on mixing parameters x and y

Defining Γ as the decay width and f as an allowed final state, we can distinguish two main types of *CP* violation within the scope of the SM: direct ($\Gamma(D^0 \to f) \neq \Gamma(\overline{D}^0 \to \bar{f})$) and indirect *CPV*. Indirect *CPV* can be also divided into *CP* violation in mixing ($\Gamma(D^0 \to \overline{D}^0) \neq \Gamma(\overline{D}^0 \to D^0)$) and *CP* violation in the interference between mixing and decay ($\Gamma(D^0 \to \overline{D}^0 \to f, t) \neq \Gamma(\overline{D}^0 \to D^0 \to f, t)$).

The Large Hadron Collider beauty experiment (LHCb) is a dedicated heavy-flavour experiment with a broad physics program and with a special focus on *CPV* measurements. Being a single-arm forward spectrometer with excellent vertex, primary vertex and momentum resolution in combination with highly efficient particle-identification systems, and very large $c\bar{c}$ production cross-section, LHCb can study charm physics with higher precision than previous dedicated b-factory experiments. In the period from 2010 to 2018, LHCb has collected 9 fb^{-1} of data, roughly corresponding to the order of 10^{10} D^0 decays.

These proceedings report three recent *CPV* measurements done by LHCb collaboration: measurement of the mass difference between neutral charm-meson eigenstates,[4] a search for time-dependent *CPV* in D^0 decays[5] and the first observation of *CP* violation in charm decays.[1]

2. Measurement of the mass difference between neutral charm-meson eigenstates

In order to extract mixing parameters the usage of decay channels with a rich resonance spectrum is convenient. The analysis of $D^0 \to K^0_S\pi^+\pi^-$ decays is reported, using the data-sample corresponding to 3 fb^{-1} recorded in Run 1 of the LHC. For enhanced sensitivity, this analysis uses the model-independent bin-flip method[6] which is optimised for determination of the mixing parameter x. The dataset is binned in the Dalitz plane to keep strong phases approximately constant and also divided into 20 decay-time bins. The bin-flip method relies on the ratio of yields in opposite Dalitz plot bins across the bisection for acceptance effects cancellation and further enhance sensitivity to x.

A simultaneous least-squares fit has been done to the ratios of yields in the symmetric decay time bins, where the fits results are shown in Fig. 1. The resulting mixing and *CPV* parameters are

$$\begin{aligned} x_{CP} &= [2.7 \pm 1.6\,(\text{stat.}) \pm 0.4\,(\text{syst.})] \times 10^{-3}, \\ y_{CP} &= [7.4 \pm 3.6\,(\text{stat.}) \pm 1.1\,(\text{syst.})] \times 10^{-3}, \\ \Delta x &= [-0.53 \pm 0.70\,(\text{stat.}) \pm 0.22\,(\text{syst.})] \times 10^{-3}, \\ \Delta y &= [0.6 \pm 1.6\,(\text{stat.}) \pm 0.3\,(\text{syst.})] \times 10^{-3}. \end{aligned}$$

These results are the most precise determination of the parameter x_{CP} obtained by single experiment. The current world average, including re-

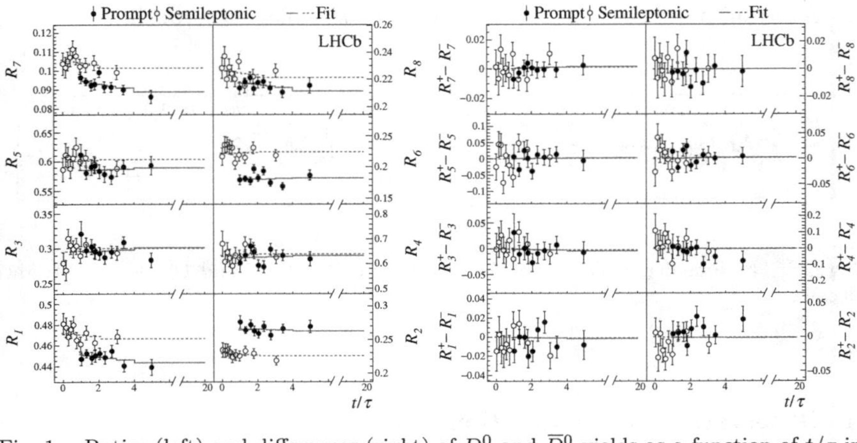

Fig. 1. Ratios (left) and differences (right) of D^0 and $\overline{D}^0$ yields as a function of t/τ in each Dalitz plane bin and fit projections for prompt (closed circles and solid line) and semileptonic (open circles and dashed line) decays.

ported results, yields $x>0$ with the significance above 3 standard deviations,[3] which is a first evidence of non-zero mass difference between neutral charm-meson eigenstates.

3. Search for time-dependent *CP* violation in $D^0 \to K^+K^-$ and $D^0 \to \pi^+\pi^-$

A search for time-dependent *CP* violation parameter A_Γ in $D^0 \to K^+K^-$ and $D^0 \to \pi^+\pi^-$ decays has been performed using data from 2015 and 2016, corresponding to 1.9 fb^{-1}. The time dependent asymmetry $\mathcal{A}_{CP}$ can be written as:

$$\mathcal{A}_{CP}(f,t) = \mathcal{A}_{CP}^{decay}(f) - A_\Gamma \frac{t}{\tau(D^0)}, \tag{3}$$

where $\mathcal{A}_{CP}^{decay}(f)$ is direct *CPV* in decay and $\tau(D^0)$ is the lifetime of the D^0 meson. The parameter A_Γ can be extracted a via linear fit to the time-dependent asymmetries in $D^0 \to K^+K^-$ and $D^0 \to \pi^+\pi^-$ decays, as shown in Fig. 2. The measured values are:

$$A_\Gamma(D^0 \to K^+K^-) = [1.3 \pm 3.5\,(\text{stat.}) \pm 0.7\,(\text{syst.})] \times 10^{-4}$$
$$A_\Gamma(D^0 \to \pi^+\pi^-) = [11.3 \pm 6.9\,(\text{stat.}) \pm 0.8\,(\text{syst.})] \times 10^{-4}.$$

Combining these results with previous LHCb measurements[7] the obtained value of A_Γ is:

$$A_\Gamma = [0.9 \pm 2.1\,(\text{stat.}) \pm 0.7\,(\text{syst.})] \times 10^{-4}.$$

All results are compatible with the no *CP* violation hypothesis.

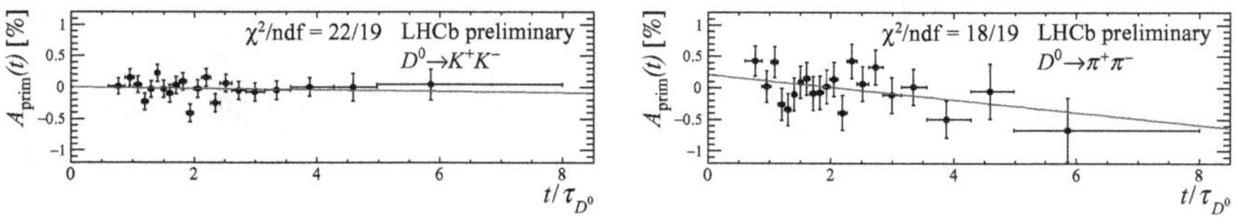

Fig. 2. The time-dependent fits of A_Γ to $D^0 \to K^+K^-$ (left) and $D^0 \to \pi^+\pi^-$ (right) decays.

4. Observation of *CP* violation in charm decays

The recently reported first observation of *CP* violation in charm hadrons decays by the LHCb collaboration[1] is utilising around 44 (9) $\times 10^6$ and

14 (3) $\times 10^6$ $D^0 \to K^+K^-$ and $D^0 \to \pi^+\pi^-$ prompt (semileptonic) decays, respectively. This dataset, roughly corresponding to 6 fb^{-1}, was recorded from 2015 to 2018 at collision energy 13 TeV.

For the case of this measurement, *CP* violation parameter $\mathcal{A}_{CP}$ is defined as:

$$\mathcal{A}_{CP}(f) \approx a_{CP}^{dir}(f) - A_\Gamma(f)\frac{\langle t(f)\rangle}{\tau(D^0)}, \tag{4}$$

where $\langle t(f)\rangle$ denotes the mean decay time of $D^0 \to f$ influenced by the experimental efficiency, $a_{CP}^{dir}(f)$ is the direct *CP* asymmetry, $\tau(D^0)$ is the D^0 lifetime and A_Γ the asymmetry between the $D^0 \to f$ and $\overline{D}^0 \to f$ effective decay widths.

However, $\mathcal{A}_{CP}$ values as defined above are not accessible directly by an experimental methods and must be extracted from the data. Directly measurable value is difference between raw yields, A_{raw}, of $D^0 \to K^+K^-$ and $\overline{D}^0 \to K^-K^+$ decays or between $D^0 \to \pi^+\pi^-$ and $D^0 \to \pi^-\pi^+$, respectively. The raw asymmetry A_{raw} can be very well approximated, up to the order $\mathcal{O}(10^{-6})$, as a linear combination of physical *CP* asymmetry $\mathcal{A}_{CP}$, detection asymmetry of D^0 which is equal to zero due to charge conjugated final states, mother particle production asymmetry and detection asymmetry of the tagging particle. These detection and production asymmetries are cancelled by equalising kinematics distributions of K^+K^- and $\pi^+\pi^-$ decay modes and then taking the difference. This difference can be written as:

$$\begin{aligned}\Delta\mathcal{A}_{CP} \equiv \mathcal{A}_{CP}(K^+K^-) - \mathcal{A}_{CP}(\pi^+\pi^-) = \\ A_{raw}^{equalised}(K^+K^-) - A_{raw}^{equalised}(\pi^+\pi^-).\end{aligned} \tag{5}$$

The $\Delta\mathcal{A}_{CP}$ observable is mainly sensitive to direct *CP* violation.

The final results for prompt and semileptonic (SL) decays are

$$\begin{aligned}\Delta\mathcal{A}_{CP}^{prompt} &= [-18.2 \pm 3.2\,(\text{stat.}) \pm 0.9\,(\text{syst.})] \times 10^{-4}, \\ \Delta\mathcal{A}_{CP}^{SL} &= [-9 \pm 8\,(\text{stat.}) \pm 5\,(\text{syst.})] \times 10^{-4}.\end{aligned}$$

These results are in good agreement with previous LHCb measurements.[8,9] When all these results are combined together, the resulting resulting values for $\Delta\mathcal{A}_{CP}$ is equal to

$$\Delta\mathcal{A}_{CP}^{combined} = (-15.4 \pm 2.9) \times 10^{-4},$$

where the uncertainty includes statistical and systematic contributions. This result deviates from zero *CP* asymmetry hypothesis by 5.3 standard deviation. For additional information, the LHCb average of A_Γ[10,11] is combined with the obtained $\Delta A_{CP}^{combined}$ result and using equation 4 it is possible to derive

$$\Delta a_{CP}^{dir} = (-15.7 \pm 2.9) \times 10^{-4},$$

which shows sensitivity to direct *CP* violation. Finally, the combined fit of Heavy Flavour Averaging Groups shown on Fig. 3, clearly shows current world average with excluded no *CPV* hypothesis at the level 5.4 standard deviations.

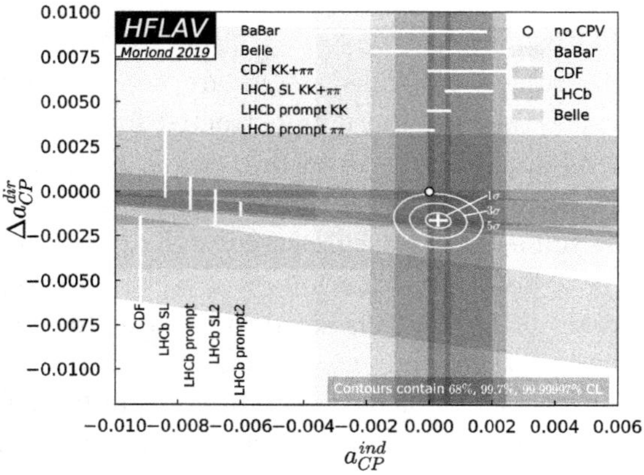

Fig. 3. World average of the direct and indirect *CP* violation parameters. Courtesy of HFLAV.[3]

5. Conclusion

The LHCb experiment has provided several new measurements of *CP* violation in charm sector. The most important result is the observation of *CPV* in D^0 decays for the first time. The measurements of the first hint of mass difference in neutral charm-meson eigenstates and searches for time-dependent *CP* violation in two-body D^0 decays are also reported. The LHCb detector is currently ongoing substantial upgrade[12] with planned collected luminosity around 30 fb^{-1} by the end of Run III.

References

1. R. Aaij *et al.*, Observation of *CP* violation in charm decays, *Phys. Rev. Lett.* **122**, 211803 (2019).

2. A. D. Sakharov, Violation of CP Invariance, C asymmetry, and baryon asymmetry of the universe, *Pisma Zh. Eksp. Teor. Fiz.* **5**, 32 (1967).
3. Y. S. Amhis *et al.*, Averages of b-hadron, c-hadron, and τ-lepton properties as of 2018 (2019).
4. R. Aaij *et al.*, Measurement of the mass difference between neutral charm-meson eigenstates, *Phys. Rev. Lett.* **122**, 231802 (2019).
5. T. Pajero *et al.*, Search for time-dependent *CP* violation in $D^0 \to K^+K^-$ and $D^0 \to \pi^+\pi^-$ decays, *LHCb CONF note.*
6. A. Di Canto, J. Garra Tic, T. Gershon, N. Jurik, M. Martinelli, T. Pilar, S. Stahl and D. Tonelli, Novel method for measuring charm-mixing parameters using multibody decays, *Phys. Rev.* **D99**, 012007 (2019).
7. R. Aaij *et al.*, Measurement of the *CP* violation parameter A_Γ in $D^0 \to K^+K^-$ and $D^0 \to \pi^+\pi^-$ decays, *Phys. Rev. Lett.* **118**, 261803 (2017).
8. R. Aaij *et al.*, Measurement of *CP* asymmetry in $D^0 \to K^-K^+$ and $D^0 \to \pi^-\pi^+$ decays, *JHEP* **07**, 041 (2014).
9. R. Aaij *et al.*, Measurement of the difference of time-integrated *CP* asymmetries in $D^0 \to K^-K^+$ and $D^0 \to \pi^-\pi^+$ decays, *Phys. Rev. Lett.* **116**, 191601 (2016).
10. R. Aaij *et al.*, Measurement of indirect *CP* asymmetries in $D^0 \to K^-K^+$ and $D^0 \to \pi^-\pi^+$ decays, *JHEP* **04**, 043 (2015).
11. R. Aaij *et al.*, Search for *CP* violation in the phase space of $D^0 \to \pi^+\pi^-\pi^+\pi^-$ decays, *Phys. Lett.* **B769**, 345 (2017).
12. Framework TDR for the LHCb Upgrade: Technical Design Report, CERN-LHCC-2012-007, (2012).

Curious link of exclusive and inclusive CPV in charmless 3-body B^+ decays

George W.-S. Hou
Department of Physics, National Taiwan University, Taipei 10617, Taiwan
E-mail: wshou@phys.ntu.edu.tw

The LHCb experiment has measured CP violation (CPV) across the Dalitz plot of charmless decays of B+ mesons to 3 charged tracks, namely in $K\pi\pi$, KKK, $\pi\pi\pi$ and πKK final states, with strikingly large CPV that vary strongly with Dalitz variables. Identifying these processes with $b \to sq\bar{q}$, $s\bar{s}s$ and $b \to dq\bar{q}$, $d\bar{s}s$, where q = u, d, then the "sum rule" that requires two-loop absorptive parts by unitarity works well for inclusive $b \to s$ CPV, but less well for inclusive $b \to d$ case. We elucidate the situation, and argue that the 30 year old unitarity/CPT argument is valid to this day, while affirming quark-hadron duality.

Keywords: CPV; 3-body; charmless; unitarity/*CPT*; sum rule.

1. The Striking Plot, and a Curiosity

The Bander-Silverman-Soni (BSS) mechanism[1] induces direct CPV (DCPV) in B^+ decay via the $b \to sq\bar{q}$ transition: CPV phase in V_{ub} via the Tree (T) diagram, the Penguin (P) diagram gives absorptive part via $b \to si\bar{i}$ (i = u, c) then on-shell $i\bar{i} \to q\bar{q}$ rescattering. Since the B factory discovery of charmless 3-body B decays, the LHCb experiment has now measured inclusive CPV for $B^\pm$, namely[2]

$$A_{CP}(B^\pm \to K^\pm\pi^+\pi^-) = +0.025 \pm 0.004 \pm 0.004 \pm 0.007 \quad [\text{PDG: } +0.027], \tag{1}$$

$$A_{CP}(B^\pm \to K^\pm K^+K^-) = -0.036 \pm 0.004 \pm 0.002 \pm 0.007 \quad [\text{PDG: } -0.033], \tag{2}$$

$$A_{CP}(B^\pm \to \pi^\pm\pi^+\pi^-) = +0.058 \pm 0.008 \pm 0.009 \pm 0.007 \quad [\text{PDG: } +0.057], \tag{3}$$

$$A_{CP}(B^\pm \to \pi^\pm K^+K^-) = -0.123 \pm 0.017 \pm 0.012 \pm 0.007 \quad [\text{PDG: } -0.122], \tag{4}$$

and across the Dalitz plot, e.g. in Fig. 1 for $B^\pm \to \pi^\pm\pi^+\pi^-$, the "artist's palette". Strikingly large CPV vary strongly across the Dalitz plot, reflecting the presence of hadron level rescattering and resonant interference. The asymmetries in Fig. 1 are in bins with same number of events, where one has[2] roughly 181k and 109k events in $K\pi\pi$ and KKK modes,[a] and 25k and 6k in $\pi\pi\pi$ and πKK.

[a] Note that LHCb has excellent K/π separation or PID, while pions suffer more background.

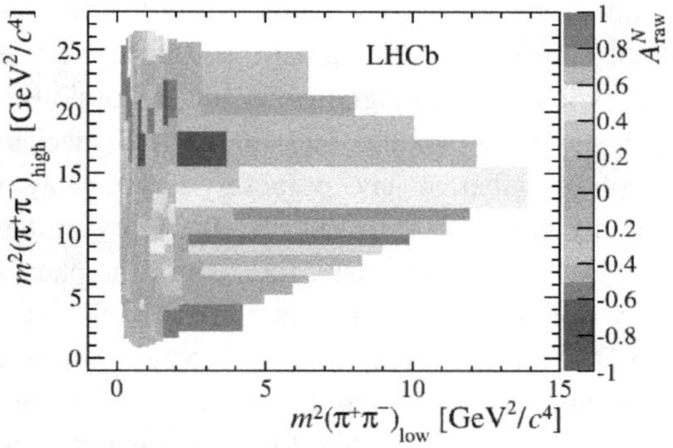

Fig. 1: CPV across the Dalitz plot[2] for $B^\pm \to \pi^+\pi^+\pi^-$ decay, an "artist's palette".

The curiosity is then, when weighted by PDG values of ~ 5.1, 3.4, 1.5, 0.5 ($\times 10^{-5}$) for the respective decay branching ratios, and asymmetries given in [...] in the equations without error, combining Eqs. (1) and (2) one finds 0.3%, and combining Eqs. (3) and (4) one finds 1.1%, for CPV in $b \to s$ and $b \to d$ transitions, respectively. From large DCPV approaching $\pm O(1)$ locally in the Dalitz plot, to a few % to O(10%) for the inclusive charmless mesonic 3-body final states, the DCPV for inclusive charmless $B^\pm$ decay to even or odd number of kaons in the final state is actually sub-percent! This is reminiscent of a "prediction" made 30 years ago,[3,4] where quark level inclusive $b \to sq\bar{q}$ and $b \to dq\bar{q}$ decays were found at the sub-percent level (a "sum rule"), when taking into account an additional T-P interference, with extra absorptive part from off-shell gluon propagator in P as demanded by unitarity and *CPT* arguments.

In the following, since *CP*-invariant phase is needed for CPV, we first discuss Final State Interactions (FSI), progressing from elastic to inelastic, and from soft (hadronic) to hard, such as the annihilation of $c\bar{c}$ into charmless $q\bar{q}$ (q = u, d, s) from the leading $b \to sc\bar{c}$ decay, crucial for the penguin process.

2. FSI: elastic vs inelastic

In summer 2001, Belle observed[5,b] the color-suppressed $\bar{B}^0 \to D^0\pi^0$ decay at 9.3σ, and with the $D^{*0}\pi^0$, $D^0\eta^0$ and $D^0\omega^0$ modes above 4σ. The observed rates were much larger than theory predictions of the time, indicating the presence of FSI,[6] or elastic $D^{(*)}\pi \to D^{(*)}\pi$ rescattering, while the $D^{(*)0}\eta$ and $D^{(*)0}\omega$ final states imply SU(3) symmetry. Before long, the formalism was extended[7] to the more lucrative charmless two body B decays, such as $B \to K\pi$, to pursue DCPV. This now involved meson loops, with D or light mesons running in the loop,

[b] If half a year later, the leading authors would have been K.-F. Chen, H.-C. Huang and R.-S. Lu. This is one of two first discovery papers by the young NTU-Belle group.

which raises the questions: How does it work when the mesons in the loop are far off-shell? What does one do for inelastic scattering vertices when far off-shell hadrons are involved? The diagramatics and discussions, followed to this day,[8] are really just cherry-picking, lack in control mechanism, and quite arbitrary. It is an approach that we have purposely stayed away from.

But we are certainly not adverse to FSI. Pre-B-factory, in the CLEO era, we had already pursued[9,10,11] FSI vigorously, for sake of the phases it may provide that are necessary for DCPV. And when elastic FSI was experimentally observed in $\bar{B}^0 \to D^0\pi^0$, we extended the work[6] to suggest something similar in $B \to K\pi$, i.e. experimental indications for presence of elastic $K\pi \to K\pi$ FSI scattering.[12] This work was actually quite fine, but was not viewed favorably by PRL reviewers. But the consolation soon came with Belle observation[13,c] of DCPV in $B^0 \to K^+\pi^-$ decay, which was again the effort of NTU-Belle.

We note that the original expectation in the late 1990s of QCD factorization in charmless two body B decays did not come to pass. For ourselves, when the experimentally firm DCPV difference[14] between $B^0 \to K^+\pi^-$ and $B^+ \to K^+\pi^0$ was not corroborated by a large and negative $\sin\phi_s$, suggesting instead "enhanced color-suppressed" amplitude C as culprit, we no longer trusted perturbative QCD estimates of strong phases in B decay to hadronic final states.

3. FSI: soft vs hard

In a 2015 LHCb talk[15] presented in Brazil (and in other LHCb papers), attempt was given to elucidate the volatility of DCPV across the Dalitz plot via $\pi^+\pi^- \leftrightarrow K^+K^-$ rescattering. Invoking *CPT* invariance, it was assumed that "all the inelasticity of the $\pi\pi$ interaction goes into the KK channel" in the rescattering region of (1, 2.2) GeV^2, which was rooted in an earlier discussion[16] that "*CPT* symmetry constrains hadron rescattering so that the sum of the partial decay widths of all channels <u>with the same final-state quantum numbers related by the scattering matrix</u> must equal that of their charge-conjugated decays".[2] We scrutinize this notion: Although the statement is not incorrect, the underlined part (by us) is often not sufficiently respected, i.e. treated in some cavalier way. We then broaden our criticism before turning to quark level language.

3.1. *CPT and inelastic hadron scattering*

The upper plots[15] in Fig. 2 zooms in on the CPV in Dalitz plot of lower $\pi^+\pi^-$, K^+K^- mass regions for $B^\pm \to K^\pm\pi^+\pi^-$, $K^\pm K^+K^-$ decays, where the positive tendency for $B^\pm \to K^\pm\pi^+\pi^-$ seems to counterbalance the negative tendency for $B^\pm$

---

[c] Strictly speaking, "observation" was achieved by combining with BaBar result.

$\rightarrow K^{\pm}K^{+}K^{-}$ in the $m^2_{\pi\pi}$, m^2_{KK} regions of (1, 2.2) GeV^2 (as marked), lending credence to $\pi^+\pi^- \leftrightarrow K^+K^-$ rescattering[2,15,16] as argued by *CPT*. However, the scattering matrix would connect $\pi^+\pi^- \rightarrow K^+K^- + (n\pi)^0$ that have "the same final-state quantum numbers" for n = 0–3, where $(n\pi)^0$ is charge neutral. These thresholds can be drawn for $B^{\pm} \rightarrow K^{\pm}\pi^+\pi^-$ at $m^2_{\pi\pi} \approx 1.28$, 1.62, 1.99 GeV^2, respectively, and much denser for $K^+K^- \rightarrow$ multi-pion final states for $B^{\pm} \rightarrow K^{\pm}K^{+}K^{-}$. The scattering matrix is thus rather large and inelastic, and thinking only in terms of $\pi^+\pi^- \leftrightarrow K^+K^-$ rescattering as guided by *CPT* is too simplistic.

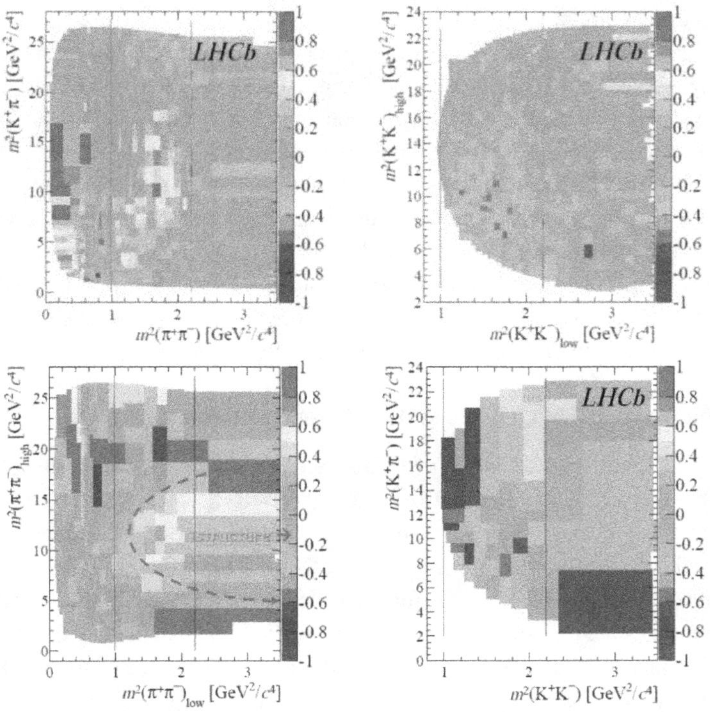

Fig. 2: CPV in Dalitz plot for [upper] $B^{\pm} \rightarrow K^{\pm}\pi^+\pi^-$, $K^{\pm}K^{+}K^{-}$ and [lower] $B^{\pm} \rightarrow \pi^{\pm}\pi^+\pi^-$, $\pi^{\pm}K^+K^-$ decays, with (1, 2.2) GeV^2 as marked.[15]

A similar argument applied to $B^{\pm} \rightarrow \pi^{\pm}\pi^+\pi^-$, $\pi^{\pm}K^+K^-$ decays as illustrated in lower plots of Fig. 2 makes the case further, where even the LHCb discussion[15] admits to "other mechanisms in action" for the $\pi^{\pm}\pi^+\pi^-$ final state.

3.2. *a Critique*

The lower-left plot in Fig. 2 actually zooms into the lower $m^2_{\pi\pi}$ region of the "artist's palette", the most colorful Fig. 1 for $B^{\pm} \rightarrow \pi^{\pm}\pi^+\pi^-$ decay, where the latter gives a broader picture. LHCb has recently conducted an amplitude analysis[17] of the upper-left corner of rather low mass $m_{\pi\pi} \sim m_\rho$, together with the high mass $m_{\pi\pi}$ above 4 GeV or so. The analysis went beyond S-P wave interference,[2,15] and observed DCPV also in D-wave, but things are still not fully settled. Note that we have marked in the lower-left plot of Fig. 2, that there is some starting "structure" towards the right as outlined by the (magenta) dashed curve, which extends into the brilliant and bright yellow-to-red region of the full Dalitz plot of Fig. 1. The lower boundary of this structure is around the $c\bar{c}$ threshold, namely an underlying c-anti-c to q-anti-q annihilation effect, where q is a light quark (u, d, s). This $c\bar{c} \to q\bar{q}$ annihilation amounts to $D\bar{D} + X \to$ charmless rescattering,[4] as stated in the original BSS mechanism at quark level.

We see that there are "soft" FSI, which amounts to $K\pi\pi \leftrightarrow KKK$ and $\pi\pi\pi \leftrightarrow \pi KK$ rescattering, where an $n\pi$ system can be thrown in easily on either side to add inelasticity. The complexity seems rather difficult to address. But one also has hard FSI, where $c\bar{c} \to q\bar{q}$, or equivalently a double charm (but with no overall charm quantum number — *CPT* connection!) $D\bar{D} + X$ rescattering into 3-body charmless mesonic final states from the leading $B \to D\bar{D} + X_s$ decay process, as discussed in our work 30 years ago, which brings about subtleties.[3,4]

We now turn to discuss *CPT* and unitarity at the quark level.

4. Unitarity/CPT at Quark Level: a 30-year "Sum Rule"

In 1988 we published a study[18] of inclusive $b \to sg^*$ decay, where the g^* could be light-like (on-shell gluon), time-like (g^* splits to $q\bar{q}$,[d] which is the usual Penguin for B decays), and space-like (the gluon attaches to the spectator). We corrected the $b \to sg$ formula and showed that the time-like process dominates, in contrast to the usual Penguin for kaon decay, where one cannot distinguish between time-like and space-like processes, while light-like is ill-defined.

The quark level $b \to sq\bar{q}$ process would match to inclusive charmless $B \to K + X$ decays by quark-hadron duality, where X has no net strange quantum number. Having done the inclusive rate, and being a student of Amarjit Soni, it was natural to explore inclusive CPV via the BSS mechanism.[1] At MPI Munich, we paired up with Jean-Marc Gérard, dividing efforts between inclusive vs exclusive processes. We soon ran into a paradox: the direct application of BSS mechanism did not seem to respect unitarity, nor *CPT*. We will not go into the details of the discussion[3,4] 30 years ago (including a debate with Lincoln Wolfenstein[19]), but just give a brief account and come back to the comparison with charmless B^+ decays to 3-body mesonic final states measured by LHCb.

---

[d] The treatment[3,18] of $g^* \to gg$ was later shown to be absent (see PRD article[4] for references).

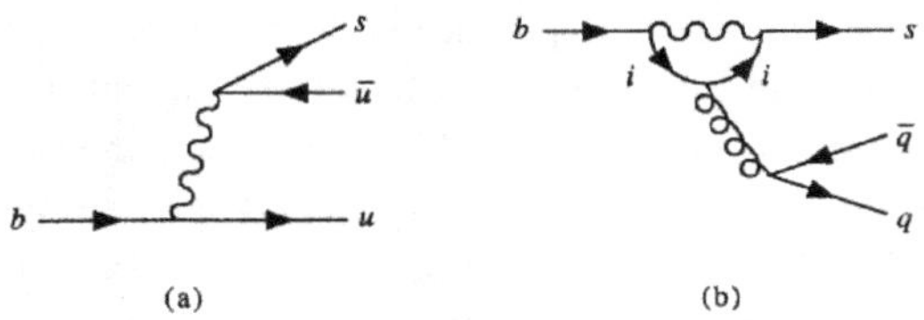

Fig. 3: BSS mechanism[1] for DCPV in $b \to sq\bar{q}$ decay (taken from Ref. 4) via T-P interference.

4.1. *Unitarity/CPT at Quark Level*

Fig. 3 corresponds to the BSS mechanism[1] of T-P interference, where CPV phase is in T amplitude, while the absorptive part arises from the i = c cut of the Penguin loop. Although the mechanism was intended for T-P interference such as for $b \to su\bar{u}$ process, with the pure Penguin process at hand, we purposed to study CPV in pure Penguin process, where one can imagine P-P interference, with one dispersive P-amplitude providing CPV phase, while the other P-amplitude is absorptive. But we soon found that this procedure for T-P interference in $b \to su\bar{u}$ process, extended to include P-P interference for pure penguin $b \to su\bar{u}$, $sd\bar{d}$ and $ss\bar{s}$ processes, did not respect unitarity, nor *CPT*. The situation was the same for the corresponding $b \to du\bar{u}$, $dd\bar{d}$ and $ds\bar{s}$ processes. We refer to the PRD article[4] for a detailed exposition.

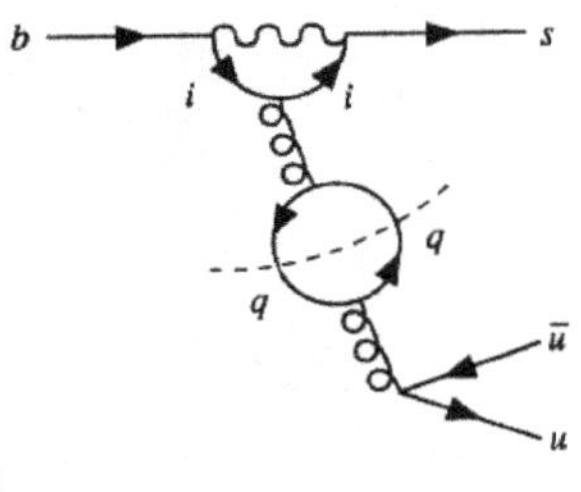

Fig. 4: Diagram (taken from Ref. 4) needed to interfere with Fig. 3(a) to restore unitarity/*CPT*.

It turns out that, to maintain unitarity and *CPT*, one needs to include an additional two-loop diagram that interferes with Fig. 3(a): the "double-Penguin" displayed in Fig. 4! The gluon propagator carries a "bubble" or $q\bar{q}$ loop (the q = u, d, s cut illustrated by dashed line) which possesses a rather large absorptive part (see footnote d). Interfering with only $b \to su\bar{u}$ of Fig. 3(a), it tends to cancel the CPV in this mode and restore unitarity/*CPT*.

One can visualize this unitarity/*CPT* argument. In Fig. 5 we make[4] a term-by-term amplitude-squared analysis of the BSS mechanism. Fig. 5(a) is T-T interference in $b \to su\bar{u}$, which conserves *CP* because there is no "strong phase", while one has the CKM factor $|V_{ub}|^2$ hence no CPV phase. Fig. 5(b) is the usual BSS mechanism, or T-P interference. However, as explained, we purposed to study DCPV in pure penguins, which is Fig. 5(c), where the i = u and c cuts of P-P interference illustrates two possibilities for quark level strong phases.[e] It was found that, stopping at this level, where pure Penguin decays such as $b \to sd\bar{d}$, $ss\bar{s}$ would also possess DCPV, one would not respect unitarity nor *CPT* (see PRD article[4] for detailed discussion). One needs to include Fig. 5(d).

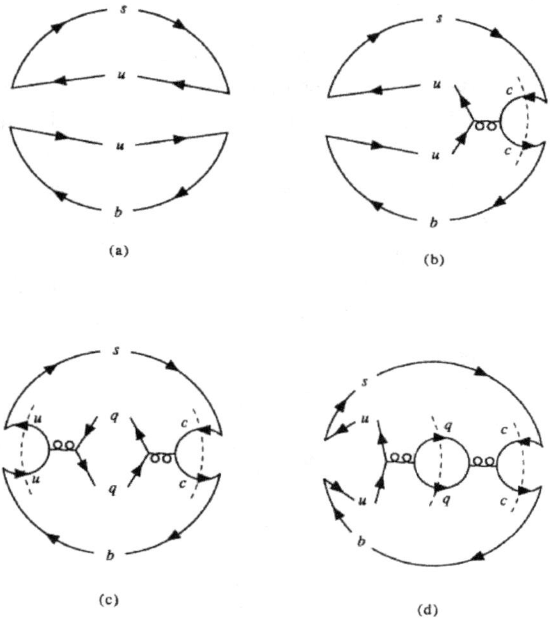

Fig. 5: (a) T-T interference ($b \to su\bar{u}$); (b) T-P interference ($b \to su\bar{u}$), or BSS mechanism; (c) P-P interference (pure Penguin $b \to sq\bar{q}$); (d) T–double-P interference. Figure taken from Ref. 4.

Comparing Figs. 5(d) and 5(c), one can understand how they are connected. The amplitude-squared in Fig. 5(c) also represents the $b \to sq\bar{q}$ decay rate (q = u, d, s) to charmless final state. But for the u-quark cut in the Penguin, it could also correspond to the $b \to su\bar{u}$ process illustrated in Fig. 5(d), with the q = u, d, s final states now correspond to the cut of the Penguin bubble, or absorptive part of a time-like gluon propagator. The two diagrams are therefore related by unitarity (the "same" diagram), hence ignoring Fig. 5(d) amounts to a fallacy. As for *CPT*, note that on-shell color-octet $q\bar{q}$ and $c\bar{c}$ have <u>the same final-state</u>

[e] The dispersive top-loop provides a subtraction to the u- and c-loops by GIM mechanism.

quantum numbers related by the scattering matrix, via one gluon exchange. At the quark level, there are no issues as discussed in Sec. 3 with hadronic processes. Thus, these final states feed each other,[4] governed by *CPT*. Such arguments guided us 30 years ago to identify the interference of T (Fig. 3(a)) and "double-P" (Fig. 4) that restored unitarity/*CPT*.

TABLE I. (Semi-)inclusive branching ratio B and asymmetries for $b \to s$ and $b \to d$ processes for $\rho = -0.5$, $\eta = 0.15$, and $m_t = M_W$. a_0 is with Figs. 1(a) and 1(b) only, while a is the result with Fig. 1(c) taken into account. The entry "0.0" stands for a very small positive number.

	B (%)	a_0 (%)	a (%)
$b \to su\bar{u}$	0.46	1.2	0.0
$b \to sd\bar{d} + ss\bar{s}$	0.54	0.5	0.5
Total $b \to s$ (no charm)	1.19	0.7	0.2
$b \to du\bar{u}$	0.71	−0.7	−0.0
$b \to dd\bar{d} + ds\bar{s}$	0.07	−4.2	−4.2
Total $b \to d$ (no charm)	0.80	−1.0	−0.4

Fig. 6: Table I, or numerical results, from Ref. 4. Note $\rho \approx -0.12$, $\eta \approx 0.35$ at present.

4.2. *a 30-year "Sum Rule"*

To illustrate the numerics, let us first display Table I from the PRD article[4] in Fig. 6. This corresponds to $\rho = -0.5$, $\eta = 0.15$, ca. 1990, for the Wolfenstein parameters governing V_{ub}. The current values are $\rho \approx -0.12$, $\eta \approx 0.35$ are rather different, so we are only concerned with gross features.[f] But by including Fig. 4, i.e. from a_0 to a in Fig. 6 (Table I), the asymmetries for $b \to su\bar{u}$ and $b \to du\bar{u}$ drop to 0, while the inclusive $b \to sq\bar{q}$ asymmetry drop from percent level (noting that η is now 2x larger) to 0.2%. This is what we mentioned as a "prediction" from almost 30 years ago. We will comment on $b \to dq\bar{q}$ later.

It is heuristic to match $b \to sq\bar{q}$ to charmless $B^\pm$ decay to odd number of kaons in 3-body mesonic final state. But one may wish to match $b \to su\bar{u}$ with $B^\pm \to K^\pm\pi^+\pi^-$, and $b \to ss\bar{s}$ with $B^\pm \to K^\pm K^+K^-$, which may also seem plausible at first sight. However, if one takes $B^\pm \to K^\pm$ as matching to $b \to s$, then $[K^-\pi^+\pi^-]$ and $[K^-K^+K^-]$ can be represented as $[s\, u\bar{d}\, \bar{u}d]$ and $[s\, u\bar{s}\, \bar{u}s]$, respectively. Thus, both $b \to su\bar{u}$ and $b \to sd\bar{d}$ feed $B^\pm \to K^\pm\pi^+\pi^-$, while both $b \to ss\bar{s}$ and $b \to su\bar{u}$ feed $B^\pm \to K^\pm K^+K^-$ (though $b \to ss\bar{s}$ should dominate over $b \to su\bar{u}$), so the matching of quark level processes to the mesonic level ones are not trivial. Besides the

---

[f] Given the numerical accuracy, we do not distinguish between ρ, η and the "barred" counterparts.

wide difference in ρ, η values ca. 1990, this makes the comparison of Eqs. (1) and (2) with $b \to su\bar{u}$ and $b \to sd\bar{d} + ss\bar{s}$ entries difficult.

With the heuristic argument that $b \to ds\bar{s}$ should dominate over $b \to du\bar{u}$ in generating $B^{\pm} \to \pi^{\pm}K^{+}K^{-}$, and that η is now a little more than 2x larger than ca. 1990, it is intriguing that ~ −12% from Eq. (4) seems to match ~ −4% of Table I quite well. What is troubling is the +1.1% asymmetry for inclusive $B^{\pm}$ decay to even number of kaons in 3-body mesonic final state, which does not match the negative sign in Table I. But $\rho \approx -0.5$ ca. 1990 is rather different from the current value of −0.12. The trouble is, to update the 30 year old result, one needs to fully update the effective field theory, including the on-shell rescattering as outlined. We shy away from it at this stage given the current level of precision. This also implies that further experimental updates are desired, to eliminate statistical fluctuations as another possible cause of disagreement.

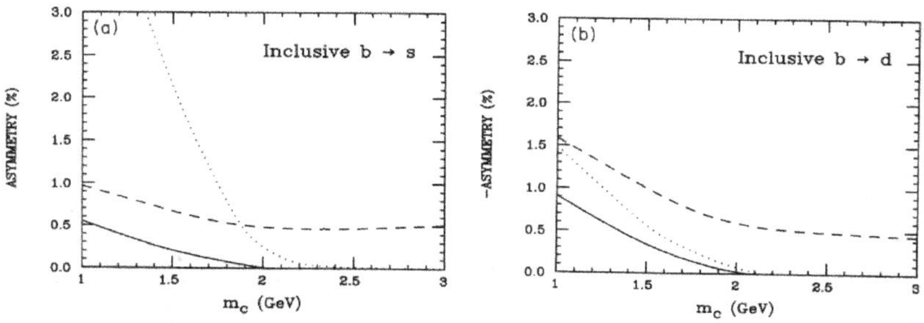

Fig. 7: Inclusive (a) $b \to s$ and (b) $b \to d$ asymmetries vs m_c,[4] where dotted line is Fig. 5(b) alone, dashed line includes Fig. 5(c), and solid line includes Fig. 5(d). *CPT* requires inclusive asymmetry to vanish when $2m_c > m_b$,[3,4] which is not respected by the dashed line, but restored when Fig. 5(d) is included. The smallness of the inclusive asymmetry is in part due to $c\bar{c}$ threshold suppression.

Before commenting further on how to update the theory, we address a question of curiosity: if pure Penguin DCPV, as well as the T–double-P interference are at $O(\alpha_s^2)$ hence higher order, why is it numerically so important? The issue turns out to be the relative proximity of the $c\bar{c}$ threshold to m_b, and the factor of 3 from q = u, d, s for pure Penguin final state. In Fig. 7 we give the inclusive $b \to s$ and $b \to d$ asymmetries from our PRD article.[4] As explained in the caption, the inclusion of T–double-P interference, Fig. 5(d), restores *CPT*, and one sees threshold suppression as behind the smallness of the inclusive asymmetry. But at the deeper level, if the $c\bar{c}$ cut on the P-amplitude suffers threshold suppression, the double-Penguin, where the time-like gluon absorptive part brings in $q\bar{q}$ cut which does not suffer threshold suppression, it brings in further a factor of 3 enhancement due to three light quarks. Thus, the impact of $O(\alpha_s^2)$ process over

$O(\alpha_s)$ process for charmless B decays is indeed a subtle one. It is quite satisfying that it is now reflected in charmless 3-body mesonic decays measured by LHCb.

Of course, one should keep in mind that the smallness of $|V_{ub}|$ elevates the importance of the Penguin.

4.3. *Improving the Inclusive DCPV Theory*

As stated, we have not attempted to update the numerics from 30 years ago.

- Interested theorists can try to improve things, but it's a little muddied and complicated, as one needs to update to an effective field theory language, not just updating ρ, η and m_b, m_c, as we have already commented.
- One needs to take into account the light-like penguin $b \to sg$, with g on-shell. This process has been updated[20,21] and estimated at O(0.5%), hence not small compared with time-like penguin $b \to sq\bar{q}$. But "on-shell" gluons do not give much absorptive parts, so how much does it contribute to DCPV? Further, in what part of "$h'^{\pm}h^-h^+$" ($h = K, \pi$) Dalitz plot does $b \to sg$ interfere with $b \to sq\bar{q}$? These would not be easy questions to address, and would have to be incorporated in the effective field theory.
- The good news is that the near cancellation of inclusive asymmetries is guaranteed by the strength of the absorptive part of the double-Penguin. But does it hold true for ρ that is considerably weaker than what was considered around 1990? And, is there any way to access the individual quark level processes such as $b \to su\bar{u}$, $sd\bar{d}$, $ss\bar{s}$ separately? In any case, it would be interesting to see further experimental (and theoretical) development.

5. Conclusion

Thirty years ago,[3,4] "*we found the surprising result that $O(\alpha_s^2)$ contributions are as important as $O(\alpha_s)$ effects for individual semi-inclusive modes such as* $b \to su\bar{u}$. *These unexpected results were uncovered because we chose to study the rates and asymmetries of semi-inclusive charmless* B *decays, and we therefore had to pay better attention to general conditions such as unitarity and CPT.*" And "*it is indeed subtle*".[4]

> "*The upshot of our results is that the total inclusive charmless* $b \to s$ *and* $b \to d$ *decay asymmetries are rather suppressed.*"[4]

In 2014, the LHCb experiment measured[2] DCPV in charmless $B^{\pm} \to h'^{\pm}h^-h^+$ decays, with rather large asymmetries that vary strongly across the Dalitz plot. Such hadronic level asymmetries are not precluded by the above statement, but data does support sub-percent CPV in charmless inclusive 3-body decays, especially for B^+ decay to an odd number of charged kaons, corresponding to charmless inclusive $b \to$

$sq\bar{q}$. This affirms the rather small inclusive charmless asymmetry, or "sum rule", stated above, and also affirms quark-hadron duality holds well in such decays. For charmless 3-body B^+ decay to an even number of charged kaons that correspond to charmless inclusive $b \to dq\bar{q}$, the inclusive asymmetry seems just above percent level, which should be further clarified.

Acknowledgment

This work is supported by grants MOST 106-2112-M-002-015-MY3 and NTU 108L104019.

References

1. M. Bander, D. Silverman and A. Soni, *CP* Noninvariance in the Decays of Heavy Charged Quark Systems, *Phys. Rev. Lett.* **43**, 242 (1979).
2. R. Aaij *et al.* (LHCb Collaboration), Measurements of *CP* violation in the three-body phase space of charmless $B^{\pm}$ decays, *Phys. Rev. D* **90**, 112004 (2014).
3. J.-M. Gérard and W.-S. Hou, *CP* Nonconservation and *CPT*: Reassessment of Loop Effects in Charmless *B* Decays, *Phys. Rev. Lett.* **62**, 855 (1989).
4. J.-M. Gérard and W.-S. Hou, *CP* violation in inclusive and exclusive charmless *B* Decays, *Phys. Rev. D* **43**, 2909 (1991).
5. K. Abe *et al.* (Belle Collaboration), Observation of the Color-Suppressed Decay $\bar{B}^0 \to D^0\pi^0$, *Phys. Rev. Lett.* **88**, 052002 (2002).
6. C.-K. Chua, W.-S. Hou and K.-C. Yang, Final state rescattering and color-suppressed $\bar{B}^0 \to D^{(*)0}h^0$ decays, *Phys. Rev. D* **65**, 096007 (2002).
7. H.-Y. Cheng, C.K. Chua and A. Soni, Final state interactions in hadronic *B* decays, *Phys. Rev. D* **71**, 014030 (2005).
8. See e.g. I. Bediaga and T. Frederico and P.C. Magalhães, Charm Penguin in $B^{\pm} \to K^{\pm}K^{+}K^{-}$: Partonic and hadronic loops, *Phys. Lett. B* **785**, 581 (2018); and plenary talk by P.C. Magalhães at this conference.
9. X.-G. He, W.-S. Hou and K.-C. Yang, Prospects for Direct *CP* Violation in Exclusive and Inclusive Charmless *B* Decays, *Phys. Rev. Lett.* **81**, 5738 (1998).
10. N.G. Deshpande, X.-G. He, W.-S. Hou and S. Pakvasa, Electroweak Penguins, Final State Interaction Phases, and *CP* Violation in $B \to K\pi$ Decays, *Phys. Rev. Lett.* **82**, 2240 (1999).
11. W.-S. Hou and K.-C. Yang, Possibility of Large Final State Interaction Phases in Light of $B \to K\pi$ and $\pi\pi$ Data, *Phys. Rev. Lett.* **84**, 4806 (2000).

12. C.-K. Chua, W.-S. Hou and K.-C. Yang, Indication for Large Rescatterings in Charmless Rare B Decays, *Mod. Phys. Lett. A* **18**, 1763 (2003).
13. Y. Chao, P. Chang *et al.* (Belle Collaboration), Evidence for Direct *CP* Violation in $B^0 \to K^+\pi^-$ Decay, *Phys. Rev. Lett.* **93**, 191802 (2004).
14. S.-W. Lin, Y. Unno, W.-S. Hou, P. Chang *et al.* (Belle Collaboration), *Nature* **452**, 332 (2008).
15. A.C. dos Reis, CP violation in charmless $B^{\pm}$ three-body decays: where Flavour Physics meets Hadron Physics, talk at *XIII International Workshop on Hadron Physics* (Rio de Janeiro, Brazil, March 2015), *J. Phys. Conf. Ser.* **706**, 042001 (2016).
16. H.-Y. Cheng and C.K. Chua, Branching fractions and direct *CP* violation in charmless three-body decays of B mesons, *Phys. Rev. D* **88**, 114014 (2013).
17. R. Aaij *et al.* (LHCb Collaboration), Amplitude analysis of the $B^+ \to \pi^+\pi^+\pi^-$ decay, arXiv:1909.05212 [hep-ex], submitted to *Phys. Rev. D*.
18. W.-S. Hou, QCD induced charmless B decays, *Nucl. Phys. B* **308**, 561 (1988).
19. L. Wolfenstein, Final state interactions and *CP* violation in weak decays, *Phys. Rev. D* **43**, 151 (1991).
20. C. Greub and P. Liniger, The rare decay $b \to sg$ beyond leading logarithms, *Phys. Lett. B* **494**, 237 (2000).
21. C. Greub and P. Liniger, Calculation of next-to-leading QCD corrections to $b \to sg$, *Phys. Rev. D* **63**, 054025 (2001).

Semileptonic and leptonic charm decays at BESIII

Yue Wang
University of Science and Technology of China,
Hefei, China
E-mail: wy0108@mail.ustc.edu.cn

Using the data sample taken at the center-of-mass energies 3.773, 4.18 and 4.6 GeV by the BESIII detector, BESIII colloboration studied the semileptonic and leptonic charmed hadron decays very well in recent years. The studies mainly contains the precision measurements of decay constants, form factors and CKM matrix elements. The properties of $a_0(980)$, $f_0(500)$, $f_0(980)$ and $K_1(1270)^0$ have also been extracted. BESIII also test the LFU at a very high precision.

Keywords: Charm physics; semileptonic decays; leptonic decays; lepton flavor universality; CKM matrix.

1. Introduction

Semileptonic and leponic decays of pseudoscalar mesons are well described in the Standard Model (SM). The decay amplitudes are proportional to the product of the hadronic current and the leptonic current. The hadronic current, which describes the strong interaction in the bound state for leptonic decays and in the hadronic transition in the semileptonic decays, can not be given directly. Instead, it can be written in terms of decay constants and form factors. The charm mesons and baryons may be not heavy enough for applying perturbation method and not heavy enough for heavy quark theory, so it is important to measure the decay constants and form factors in experiment.

Meanwhile, some recent results[1–8] in the measurements of semileptonic B decays have shown hints of violation of lepton flavor universality (LFU), which requires that different generations of leptons have the same coupling strength with gauge bosons. For the leptonic D decays and the semileptonic D decays to pseudoscalar mesons, the hadronic currents are almost cancelled out when we calculate the ratio of decay rates to different generations of leptons. With these precise SM predictions, we can test for LFU in the charm sector at a very good precision, which is important to understand the potential mechanism in the B anomaly.

Measurements of the leptonic and semileptonic D decays at BESIII mainly used the 2.93 fb^{-1} and 3.19 fb^{-1} data collected at the center mass energies 3.773 GeV and 4.178 GeV, respectively. At these energies, the D mesons are produced in pair. If we first fully reconstruct a D meson using hadronic decays, the other D meson is then guaranteed to exist in the recoiling system.

2. Leptonic decays

Leptonic D decays give us direct access to the corresponding decay constants and CKM matrix elements. BESIII has recently published the measurement of the BF of $D_s^+ \to \mu^+\nu_\mu$.[9] The BF is measured to be $\mathcal{B}(D_s^+ \to \mu^+\nu_\mu) = (5.49\pm0.16\pm0.15)\times10^{-3}$. Given the lifetime of D_s^+ meson,[10] we obtain $f_{D_s^+}|V_{cs}| = 246.2\pm3.6\pm3.5$ MeV. If we input the CKM matrix element $|V_{cs}|$ from a global fit,[10] we have $f_{D_s^+} = 252.9\pm3.7\pm3.5$ MeV. In contrast, inputting $f_{D_s^+}$ from LQCD calculation[11,12] gives $|V_{cs}| = 0.985\pm0.014\pm0.014$. In any case, these are the most precise measurements to date. Averaging this measurement of $\mathcal{B}(D_s^+ \to \mu^+\nu_\mu)$ with the previous measurements[13–16] and combining the world average of $\mathcal{B}(D_s^+ \to \tau^+\nu_\tau)$,[10] we find $\mathcal{B}_{D_s^+\to\tau^+\nu_\tau}\bar{\mathcal{B}}_{D_s^+\to\mu^+\nu_\mu} = 9.98 \pm 0.52$, which is consistent with the SM prediction 9.75.

BESIII has also report the first observation of $D^+ \to \tau^+\nu_\tau$ with a signicicance of 5.1σ.[17] The data sample is divided according to the energy deposited in the electromagnetic calorimeter into pion and muon dominated samples to better constrain the background from $D^+ \to \mu^+\nu_\mu$. We have $\mathcal{B}(D^+ \to \tau^+\nu_\tau) = (1.20\pm0.24\pm0.12)\times10^{-3}$. If we input the CKM matrix element $|V_{cd}|$ from a global fit,[10] we have $f_{D^+} = 224.5\pm22.8\pm11.3\pm0.9$ MeV, where the last uncertainty is due to external input parameters. In contrast, inputting f_{D^+} from[18,19] gives $|V_{cd}| = 0.237\pm0.024\pm0.012\pm0.001$. Taking the world average BF of $D^+ \to \mu^+\nu_\mu$,[10] we obtain $\mathcal{B}_{D^+\to\tau^+\nu_\tau}/\mathcal{B}_{D^+\to\mu^+\nu_\mu} = 3.21\pm0.64\pm0.43$, which is also consistent with the SM prediction 2.66.

3. Semileptonic decays

Semileptonic D decays to pseudoscalar mesons can be fully characterized by the single kinematic variable q^2. The form factors, which describe the hadronic transition, are usually parametrized as the function of q^2 with respect to its value $q^2 = 0$. The most widely used parametrizations include the Single Pole Model,[20] the Modified Pole Model,[21] the ISGW2 Model[22] and the series expansion parametrization.[23] In experiment, one usually divide the q^2 distribution into several intervals and measure the partial decay rate in each interval. The measured partial decay rates are then fitted using the theoretically expected form to extract the form factor at $q^2 = 0$.

In recent years, $D^0 \to K^- e^+ \nu_e$, $D \to \pi e^+ \nu_e$, $D^0 \to K^- \mu^+ \nu_\mu$, and $D \to \eta^{(\prime)} e^+ \nu_e$, etc[24–29] were well studied by BESIII. The measured BFs and the product of form factors at $q^2 = 0$ and CKM matrix elements are summarized in Table 1. Here the results are extracted using the 2-parameter series expansion parametrization. Most of the measurements

Table 1. The measured BFs and product of form factors at $q^2 = 0$ and CKM matrix elements for some semileptonic D decays to psuedoscalar mesons.

$\mathcal{B}(D^0 \to K^- e^+ \nu_e)$	$(3.505 \pm 0.014 \pm 0.033)\%$	$f_+^{D^0 \to K^-}(0)\|V_{cs}\|$	$0.7172 \pm 0.0025 \pm 0.0035$
$\mathcal{B}(D^0 \to K^- \mu^+ \nu_\mu)$	$(3.431 \pm 0.019 \pm 0.035)\%$	$f_+^{D^0 \to K^-}(0)\|V_{cs}\|$	$0.7133 \pm 0.0038 \pm 0.0030$
$\mathcal{B}(D^+ \to \bar{K}^0 e^+ \nu_e)$	$(8.60 \pm 0.06 \pm 0.015)\%$	$f_+^{D^+ \to \bar{K}^0}(0)\|V_{cs}\|$	$0.7053 \pm 0.0040 \pm 0.0112$
$\mathcal{B}(D_s^+ \to K^0 e^+ \nu_e)$	$(3.25 \pm 0.38 \pm 0.16) \times 10^{-3}$	$f_+^{D_s^+ \to K^0}(0)\|V_{cd}\|$	$0.162 \pm 0.019 \pm 0.003$
$\mathcal{B}(D^0 \to \pi^- e^+ \nu_e)$	$(2.95 \pm 0.04 \pm 0.03) \times 10^{-3}$	$f_+^{D^0 \to \pi^-}(0)\|V_{cd}\|$	$0.1435 \pm 0.0018 \pm 0.0009$
$\mathcal{B}(D^+ \to \pi^0 e^+ \nu_e)$	$(3.63 \pm 0.08 \pm 0.05) \times 10^{-3}$	$f_+^{D^+ \to \pi^0}(0)\|V_{cd}\|$	$0.1400 \pm 0.0026 \pm 0.0007$
$\mathcal{B}(D^+ \to \eta e^+ \nu_e)$	$(10.74 \pm 0.81 \pm 0.51) \times 10^{-4}$	$f_+^{D^+ \to \eta}(0)\|V_{cd}\|$	$0.0786 \pm 0.0064 \pm 0.0021$
$\mathcal{B}(D_s^+ \to \eta e^+ \nu_e)$	$(2.323 \pm 0.063 \pm 0.063)\%$	$f_+^{D_s^+ \to \eta}(0)\|V_{cs}\|$	$0.4455 \pm 0.0053 \pm 0.0044$
$\mathcal{B}(D_s^+ \to \eta' e^+ \nu_e)$	$(0.824 \pm 0.073 \pm 0.027)\%$	$f_+^{D_s^+ \to \eta'}(0)\|V_{cs}\|$	$0.477 \pm 0.049 \pm 0.011$

are consistent with the LQCD calculations.[30–33] We can also find that $f_+^{D_s^+ \to K^0}(0)/f_+^{D^+ \to \pi^0}(0) = 1.16 \pm 0.14 \pm 0.02$, which is consistent with U-spin symmetry.

As introduced before, the ratio of the decay rates for the semileptonic D decays to pseudoscalar mesons with muon or with electron can be precisely determined in the SM theory, which makes it an ideal place for test of LFU. Such studies have been performed at BESIII, yielding[25,34,35] $\Gamma(D^0 \to K^- \mu^+ \nu_\mu)/\Gamma(D^0 \to K^- e^+ \nu_e) = 0.974 \pm 0.014$, $\Gamma(D^+ \to \bar{K}^0 \mu^+ \nu_\mu)/\Gamma(D^+ \to \bar{K}^0 e^+ \nu_e) = 0.988 \pm 0.033$, $\Gamma(D^0 \to \pi^- \mu^+ \nu_\mu)/\Gamma(D^0 \to \pi^- e^+ \nu_e) = 0.922 \pm 0.037$, $\Gamma(D^+ \to \pi^0 \mu^+ \nu_\mu)/\Gamma(D^+ \to \pi^0 e^+ \nu_e) = 0.964 \pm 0.045$. These are all consistent with the SM prediction within 2σ.[36]

The situation is more complex in the case of semileptonic D decays to vector mesons with the presence of extra polarization vectors. The decay rate is described by three extra angular variables in addition to q^2. Meanwhile, we now have four form factors $A_{0,1,2}(q^2)$ and $V(q^2)$ ($A_3(q^2)$ is not independent), where $A_0(q^2)$ is often ignored. Due to the limited statistics, these form factors are usually modeled using the Single Pole Model. since

$A_1(q^2)$ appears in every helicity amplitude, we often measure the form factor ratios $r_V = V(0)/A_1(0)$ and $r_2 = A_2(0)/A_1(0)$, as they do not require additional input like D meson lifetime and CKM matrix elements. The results from $D^0 \to \bar{K}^0\pi^- e^+\nu_e$,[37] $D^+ \to \omega e^+\nu_e$,[38] $D \to \pi\pi e^+\nu_e$[39] and $D_s^+ \to K^{*0}e^+\nu_e$[29] at BESIII are summarized in Table 2. We also notice that significant S-wave contribution is observed for $f_0(500)$ in $D^+ \to \pi^-\pi^+e^+\nu_e$ with $\mathcal{B}(D^+ \to f_0(500)e^+\nu_e, f_0(500) \to \pi^+\pi^-) = (0.630 \pm 0.043 \pm 0.032) \times 10^{-3}$, while no evidence for $D^+ \to f_0(980)e^+\nu_e$ is observed.

Table 2. The measured BFs and form factor ratios for some of the semileptonic D decays to vector mesons.

Decay	BF	r_V	r_2
$D^0 \to K^{*-}e^+\nu_e$	$(1.36 \pm 0.03 \pm 0.03)\%$	$1.46 \pm 0.07 \pm 0.02$	$0.67 \pm 0.06 \pm 0.01$
$D^+ \to \omega e^+\nu_e$	$(1.63 \pm 0.11 \pm 0.08) \times 10^{-3}$	$1.24 \pm 0.09 \pm 0.06$	$1.06 \pm 0.15 \pm 0.05$
$D^0 \to \rho^- e^+\nu_e$	$(1.45 \pm 0.05 \pm 0.04) \times 10^{-3}$	$1.70 \pm 0.08 \pm 0.05$	$0.85 \pm 0.06 \pm 0.04$
$D^+ \to \rho^0 e^+\nu_e$	$(1.86 \pm 0.07 \pm 0.06) \times 10^{-3}$	$1.70 \pm 0.08 \pm 0.05$	$0.85 \pm 0.06 \pm 0.04$
$D_s^+ \to K^{*0}e^+\nu_e$	$(2.37 \pm 0.26 \pm 0.20) \times 10^{-3}$	$1.67 \pm 0.34 \pm 0.16$	$0.77 \pm 0.28 \pm 0.07$

It is also interesting to search for semileptonic D decays to scalar mesons which may help us understand the internal structure of the light scalar mesons. BESIII has recently searched for $D \to a_0(980)e^+\nu_e$,[40] where $D^0 \to a_0(980)^- e^+\nu_e$ is observed for the first with 6.4σ significance. The BF is measured to be $\mathcal{B}(D^0 \to a_0(980)^- e^+\nu_e, a_0(980)^- \to \eta\pi^-) = (1.33^{+0.33}_{-0.29} \pm 0.09) \times 10^{-4}$. The significance for $D^+ \to a_0(980)^0 e^+\nu_e$ is 2.9σ with $\mathcal{B}(D^+ \to a_0(980)^0 e^+\nu_e, a_0(980)^0 \to \eta\pi^0) = (1.66^{+0.81}_{-0.66} \pm 0.11) \times 10^{-4}$, and less than 3.0×10^{-4} at 90% confidence level. Ref. 41 proposed a model-independent method to study the nature of light scalar mesons with $R = [\mathcal{B}(D^+ \to f_0(980)e^+\nu_e) + \mathcal{B}(D^+ \to f_0(500)e^+\nu_e)]/\mathcal{B}(D^+ \to a_0(980)^0 e^+\nu_e)$. R is estimated to be 1.0 ± 0.3 for two-quark description of these scalar mesons, and 3.0 ± 0.9 for tetraquark description. With BESIII's results we have $R > 2.7$ at 90% confidence level, which favors the tetraquark description.

BESIII report the first observation of $D^+ \to \bar{K}_1(1270)^0 e^+\nu_e$ with a statistical significance greater than 10σ.[42] Its decay BF is determined to be $\mathcal{B}[D^+ \to \bar{K}_1(1270)^0 e^+\nu_e] = (2.30 \pm 0.26^{+0.18}_{-0.21} \pm 0.25) \times 10^{-3}$, the last uncertainty originates from the input BF of $\bar{K}_1(1270)^0 \to K^-\pi^+\pi^0$. This result agrees with the CLFQM and LCSR predictions when $\theta_{K_1} \approx 33^\circ$ or 57°,[43] and clearly rules out the predictions when setting θ_{K_1} negative.[44] Making use of the measured value for the BF of $D^0 \to K_1(1270)^- e^+\nu_e$[45]

and the world-average lifetimes of the D^0 and D^+ mesons,[10] we determine the partial decay width ratio $\Gamma[D^+ \to \bar{K}_1(1270)^0 e^+\nu_e]/\Gamma[D^0 \to K_1(1270)^- e^+\nu_e] = 1.2^{+0.7}_{-0.5}$, which is consistent with unity as predicted by isospin conservation.

The first measurements of the absolute BFs of semileptonic Λ_c^+ decays are also performed at BESIII using 567 fb^{-1} data taken at center mass energy $\sqrt{s} = 4.6$ GeV.[46,47] We find $\mathcal{B}(\Lambda_c^+ \to \Lambda e^+\nu_e) = (3.63\pm0.38\pm0.20)\%$, $\mathcal{B}(\Lambda_c^+ \to \Lambda\mu^+\nu_\mu) = (3.49 \pm 0.46 \pm 0.26)\%$ and $\Gamma(\Lambda_c^+ \to \Lambda\mu^+\nu_\mu)/\Gamma(\Lambda_c^+ \to \Lambda e^+\nu_e) = 0.96 \pm 0.16 \pm 0.04$. These are consistent with a recent LQCD calculation.[48]

4. Summary

To summary, the precision of decay constants, form factors and CKM matrix elements in the charm sector has been well improved by BESIII. Meanwhile, no significant evidence of LFU violation was found at BESIII. Search for semileptonic D decays to scalar mesons were performed at BESIII and the current results are in favor of the tetraquark description of $a_0(980)$, $f_0(500)$ and $f_0(980)$. Our measurement of $D^+ \to \bar{K}_1(1270)^0 e^+\nu_e$ clearly rules out the predictions when setting θ_{K_1} negative.

References

1. J. P. Lees *et al.* (BaBar Collaboration), Phys. Rev. Lett. **109**, 101802 (2012).
2. J. P. Lees *et al.* (BaBar Collaboration), Phys. Rev. D**88**, 072012 (2013).
3. R. Aaij *et al.* (LHCb Collaboration), Phys. Rev. Lett. **115**, 111803 (2015).
4. M. Huschle *et al.* (Belle Collaboration), Phys. Rev. D**92**, 072014 (2015).
5. Y. Sato *et al.* (Belle Collaboration), Phys. Rev. D**94**, 072007 (2016).
6. R. Aaij *et al.* (LHCb Collaboration), Phys. Rev. Lett. **113**, 151601 (2014).
7. R. Aaij *et al.* (LHCb Collaboration), JHEP **08**, 055 (2017).
8. A. Abdesselam *et al.* (Belle Collaboration), arXiv:1904.08794.
9. M. Ablikim *et al.* (BESIII Collaboration), Phys. Rev. Lett. **122**, 071802 (2019).
10. M. Tanabashi *et al.* (Particle Data Group), Phys. Rev. D**98**, 030001 (2018).
11. A. Bazavov *et al.* (Fermilab Lattice and MILC Collaborations), Phys. Rev. D**98**, 074512 (2018).
12. N. Carrasco *et al.* (ETM Collaboration), Phys. Rev. D**91**, 054507 (2015).
13. J. P. Alexander *et al.* (CLEO Collaboration), Phys. Rev. D**79**, 052001 (2009).
14. P. del Amo Sanchez *et al.* (BaBar Collaboration), Phys. Rev. D**82**, 091103 (2010).
15. A. Zupanc *et al.* (Belle Collaboration), JHEP **09**, 139 (2013).
16. M. Ablikim *et al.* (BESIII Collaboration), Phys. Rev. D**94**, 072004 (2016).
17. M. Ablikim *et al.* (BESIII Collaboration), Phys. Rev. Lett. **123**, 211802 (2019).
18. N. Carrasco *et al.* (ETM Collaboration), Phys. Rev. D**91**, 054507 (2015).
19. A. Bazavov *et al.* (Fermilab Lattice and MILC Collaboration), Phys. Rev. D**98**, 074512 (2018).
20. M. Wirbel, B. Stech, and M. Bauer, Z. Phys. C**29**, 637 (1985).

21. D. Becirevic and A. B. Kaidalov, Phys. Lett. B**478**, 417 (2000).
22. D. Scora and N. Isgur, Phys. Rev. D**52**, 2783 (1995).
23. T. Becher and R. J. Hill, Phys. Lett. B**633**, 61 (2006).
24. M. Ablikim *et al.* (BESIII Collaboration), Phys. Rev. D**92**, 072012 (2015).
25. M. Ablikim *et al.* (BESIII Collaboration), Phys. Rev. Lett. **122**, 011804 (2019).
26. M. Ablikim *et al.* (BESIII Collaboration), Phys. Rev. D**96**, 012002 (2017).
27. M. Ablikim *et al.* (BESIII Collaboration), Phys. Rev. D**97**, 092009 (2018).
28. M. Ablikim *et al.* (BESIII Collaboration), Phys. Rev. Lett. **122**, 121801 (2019).
29. M. Ablikim *et al.* (BESIII Collaboration), Phys. Rev. Lett. **122**, 061801 (2019).
30. H. Na *et al.* (HPQCD Collaboration), Phys. Rev. D**82**, 114506 (2010).
31. H. Na *et al.* (HPQCD Collaboration), Phys. Rev. D**84**, 114505 (2011).
32. V. Lubicz *et al.* (ETM Collaboration), Phys. Rev. D**96**, 054514 (2017).
33. R. Li *et al.* (Fermilab Lattice and MILC Collaborations), arXiv:1901.08989.
34. M. Ablikim *et al.* (BESIII Collaboration), Eur. Phys. J. C**76**, 369 (2016).
35. M. Ablikim *et al.* (BESIII Collaboration), Phys. Rev. Lett. **121**, 171803 (2018).
36. L. Riggio, G. Salerno and S. Simula, Eur. Phys. J. C**78**, 501 (2018).
37. M. Ablikim *et al.* (BESIII Collaboration), Phys. Rev. D**99**, 011103 (2019).
38. M. Ablikim *et al.* (BESIII Collaboration), Phys. Rev. D**92**, 071101 (2015).
39. M. Ablikim *et al.* (BESIII Collaboration), Phys. Rev. Lett. **122**, 062001 (2019).
40. M. Ablikim *et al.* (BESIII Collaboration), Phys. Rev. Lett. **121**, 081802 (2018).
41. W. Wang and C.-D. Lu, Phys. Rev. D**82**, 034016 (2010).
42. M. Ablikim *et al.* (BESIII Collaboration), arXiv:1907.11370.
43. H. Y. Cheng and X. W. Kang, Eur. Phys. J. C **77**, 587 (2017).
44. S. Momeni and R. Khosravi, J. Phys. G: Nucl. Part. Phys. **46**, 105006(2019).
45. M. Artuso *et al.* (CLEO Collaboration), Phys. Rev. Lett. **99**, 191801 (2007).
46. M. Ablikim *et al.* (BESIII Collaboration), Phys. Rev. Lett. **115**, 221805 (2015).
47. M. Ablikim *et al.* (BESIII Collaboration), Phys. Lett. B**767**, 42 (2017).
48. S. Meinel, Phys. Rev. Lett. **118**, 082001 (2017).

A novel approach on semileptonic decays and its application to helicity amplitudes

L. R. Dai*, Y. N. Zhou, Z. Y. Liu

Department of Physics, Liaoning Normal University, Dalian, 116029, China
**E-mail: dailr@lnnu.edu.cn*
www.lnnu.edu.cn

E. Oset

Departamento de Física Teórica and IFIC, Centro Mixto
Universidad de Valencia-CSIC, Institutos de Investigacíon de Paterna,
Aptdo. 22085, 46071 Valencia, Spain

X. Zhang

Institute of modern physics, Chinese Academy of Sciences, Lanzhou 730000, China
University of Chinese Academy of Sciences, Beijing 101408, China

A novel approach is presented to study the semileptonic decays of mesons and some interesting application to helicity amplitudes. The results show that the differential widths are very different for the different components $M = \pm 1, 0$. In particular, we suggest the magnitude $\frac{d\Gamma}{dM^{(\nu l)}_{\rm inv}}|_{M=-1} - \frac{d\Gamma}{dM^{(\nu l)}_{\rm inv}}|_{M=+1}$ to test different models beyond the standard model.

Keywords: Semileptonic decay; helicity amplitude; angular momentum algebra.

1. Introduction

The semileptonic decay of $\bar{B} \to D\bar{\nu}l^-$ and $D \to \bar{K}\nu l^+$ and the related reactions with B^* or D^* offer a good ground to study heavy flavor symmetry,[1–4] but how to evaluate the amplitudes of the processes?

In the conventional approaches the amplitudes of the processes are conveniently parameterized in terms of certain structures and their associated form factors, and some information is taken from experiment, such as in the works of quark models[5] and QCD sum rules.[6]

Actually the relevant form factors would be the same for D or D^* and B or B^*. Yet, the structures can be very different for the produced pseudoscalar or vector meson states.

Now a novel approach is presented to study the semileptonic decays of mesons[7] and some interesting application to helicity amplitudes[8] within the standard model (SM) and also beyond the standard model (BSM) to search for evidence of new physics. In our novel approach, we obtained an analytical amplitudes for each reaction with no any free parameter by utilizing heavy quark symmetry and elaborate angular momentum algebra to evaluate explicitly the weak matrix elements. In this process we relate all the processes and only one experimental value of branching ratio is needed for the B and D sectors, the rest of the results are all predictions. Then we extended the formalism allowing a $(\gamma^\mu - \alpha\gamma^\mu\gamma_5)$ structure for the weak hadronic vertex which makes it easy to make predictions for different values of α that could occur in different models BSM ($\alpha = 1$ for the SM). We find that some observables are very sensitive to the value of α, which should stimulate experimental work to investigate possible physics BSM.

2. Formalism

We want to study the $B \to D^*\bar{\nu}l$ decay, which is depicted in Fig. 1 for $B^- \to D^{*0}\bar{\nu}_l l^-$ process, where $(J'M')JM$ are the modulus and third com-

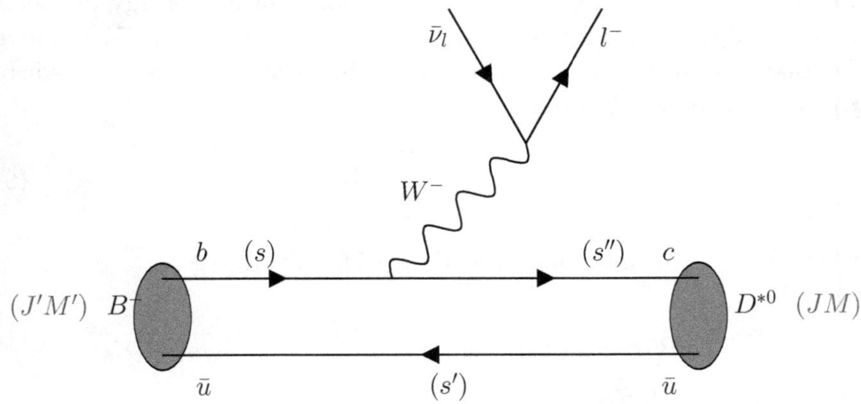

Fig. 1. Diagrammatic representation of $B^- \to D^{*0}\bar{\nu}_l l^-$ at the quark level.

ponent of the initial (final) meson spin, and the rates for the different J', J cases were evaluated. The sum over the M' and M components was done in [7] and keep track of the individual M' and M contributions in [8].

The Hamiltonian of the weak interaction is given by

$$H = \mathcal{C} L^{\alpha} Q_{\alpha} \,, \tag{1}$$

where the $\mathcal{C}$ contains the couplings of the weak interaction. The constant $\mathcal{C}$ plays no role in our study because we are only concerned about ratios of rates. The leptonic current is given by

$$L^{\alpha} = \langle \bar{u}_l | \gamma^{\alpha}(1-\gamma_5) | v_{\nu} \rangle \,, \tag{2}$$

and the quark current by

$$Q^{\alpha} = \langle \bar{u}_c | \gamma^{\alpha}(1-\gamma_5) | u_b \rangle \,. \tag{3}$$

In evaluating the quark current, we use the ordinary spinors

$$u_r = \widetilde{A} \begin{pmatrix} \chi_r \\ \widetilde{B} \boldsymbol{\sigma} \cdot \boldsymbol{p} \, \chi_r \end{pmatrix} ; \widetilde{A} = \sqrt{\frac{E_p + m}{2m}} \, ; \widetilde{B} = \frac{1}{E_p + m} \,, \tag{4}$$

where χ_r are the pauli bispinors and m, p and E_p are the mass, momentum and energy of the quark. We take for heavy quark symmetry

$$\frac{p_b}{m_b} = \frac{p_B}{m_B} \,; \qquad \frac{E_b}{m_b} = \frac{E_B}{m_B} \,, \tag{5}$$

where m_B, p_B, and E_B are the mass, momentum and energy of the B meson, and the same for the c quark related to the D^* meson. We evaluate the matrix elements in the frame where the $\bar{\nu} l$ system is at rest, where $\boldsymbol{p}_B = \boldsymbol{p}_{D^*} = \boldsymbol{p}$, with p given by

$$p = \frac{\lambda^{1/2}(m_B^2, M_{\rm inv}^{2(\nu l)}, m_{D^*}^2)}{2 M_{\rm inv}^{(\nu l)}} \,, \tag{6}$$

where $M_{\rm inv}^{(\nu l)}$ is the invariant mass of the νl pair, and

$$u_r = A' \begin{pmatrix} \chi_r \\ B' \boldsymbol{\sigma} \cdot \boldsymbol{p}_B \, \chi_r \end{pmatrix} ; A' = \sqrt{\frac{E_B + m_B}{2 m_B}} \, ; B' = \frac{1}{m_B + E_B} \tag{7}$$

and similarly, A, B would be defined for the D^* meson.

For $J' = 0, J = 1$, the different helicity contributions are given by

1) $M = 0$

$$\begin{aligned} \sum |t|^2 = {} & \frac{m_l^2}{m_\nu m_l} \frac{M_{\rm inv}^{2(\nu l)} - m_l^2}{M_{\rm inv}^{2(\nu l)}} \left\{ A A' (B + B') p \right\}^2 \alpha^2 \\ & + \frac{2}{m_\nu m_l} \left(\widetilde{E}_\nu \widetilde{E}_l + \frac{1}{3} \widetilde{p}_\nu^2 \right) \left\{ A A' (1 + B B' p^2) \right\}^2 \alpha^2 \,, \end{aligned} \tag{8}$$

2) $M = 1$

$$\sum |t|^2 = \frac{2}{m_\nu m_l}\left(\widetilde{E}_\nu \widetilde{E}_l + \frac{1}{3}\widetilde{p}_\nu^2\right)\left\{AA'[(1-BB'p^2)\alpha + (B'p - Bp)]\right\}^2, \quad (9)$$

3) $M = -1$

$$\sum |t|^2 = \frac{2}{m_\nu m_l}\left(\widetilde{E}_\nu \widetilde{E}_l + \frac{1}{3}\widetilde{p}_\nu^2\right)\left\{AA'[(1-BB'p^2)\alpha - (B'p - Bp)]\right\}^2. \quad (10)$$

3. Results of helicity amplitudes

Now we study $B^- \to D^{*0}\bar{\nu}_l l^-$ decay, the final differential width is given by

$$\frac{d\Gamma}{dM_{\text{inv}}^{(\nu l)}} = \frac{2m_\nu 2m_l}{(2\pi)^3}\frac{1}{4M_B^2}p'_{D^*}\widetilde{p}_\nu \sum |t|^2, \quad (11)$$

where p'_{D^*} is the D^* momentum in the B rest frame and $\widetilde{p}_\nu$ the $\bar{\nu}$ momentum in the νl rest frame. The total differential width is given by

$$R = \frac{d\Gamma}{dM_{\text{inv}}^{(\nu l)}}|_{M=0} + \frac{d\Gamma}{dM_{\text{inv}}^{(\nu l)}}|_{M=-1} + \frac{d\Gamma}{dM_{\text{inv}}^{(\nu l)}}|_{M=+1}. \quad (12)$$

It is interesting to see what happens for the ratio $\frac{1}{R}(\frac{d\Gamma}{dM_{\text{inv}}^{(\nu l)}}|_{M=-1} - \frac{d\Gamma}{dM_{\text{inv}}^{(\nu l)}}|_{M=+1})$. In Fig. 2 we can see that this magnitude keeps rising up as α goes from 1.5 to about 0.2. For $\alpha = 0.1$ the shape changes drastically, and as α goes to zero it changes very fast. This is because for $\alpha \to -\alpha$ the magnitude changes sign. We also show the value of the magnitude for $\alpha = -0.5$, which indeed is symmetric of the one for $\alpha = 0.5$ with respect to the $M_{\text{inv}}^{(\nu l)}$ axis. It is clear that magnitudes like this, which can change sign from one model to another, should be very useful in the search of contributions BSM.

Acknowledgments

LRD acknowledge supports from the National Natural Science Foundation of China (Grant Nos. 11975009, 11575076) and Spanish Ministerio de Economia y Competitividad and European FEDER funds under Contracts No. FIS2017-84038-C2-1-P B and No. FIS2017-84038-C2-2-P B, and the Generalitat Valenciana in the program Prometeo II-2014/068, and the project Severo Ochoa of IFIC, SEV-2014-0398 (EO).

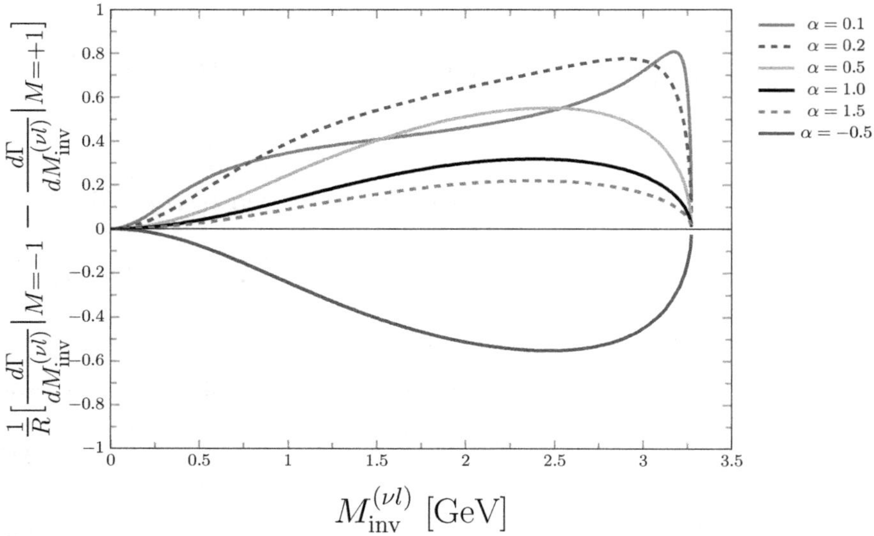

Fig. 2. The difference of differential width $\left(\frac{d\Gamma}{dM_{\rm inv}^{(\nu l)}}|_{M'=-1} - \frac{d\Gamma}{dM_{\rm inv}^{(\nu l)}}|_{M'=+1}\right)$ divided by the total width R of Eq. (12).

References

1. D. Bigi, P. Gambino and S. Schacht, *JHEP* **1711**, 061 (2017).
2. M. Antonelli *et al.*, *Phys. Rept.* **494**, 197 (2010).
3. S. Fajfer, J.F. Kamenik, and I. Nišandžić, *Phys. Rev. D* **85**, 094025 (2012).
4. X. G. He, G. Valencia, *Phys. Lett. B* **779**, 52 (2018).
5. C. Albertus, E. Hernández, J. Nieves, and J. M. Verde-Velasco, *Phys. Rev. D.* **71**, 113006 (2005).
6. N. Ghahramany, R. Khosravi and K. Azizi, *Phys. Rev. D* **78**, 116009 (2008).
7. L. R. Dai, X. Zhang and E. Oset, *Phys. Rev. D* **98**, 036004 (2018).
8. L. R. Dai and E. Oset, *Eur. Phys. J. C* **78**, 951 (2018).

Radiative corrections for Dalitz decays of π^0, $\eta^{(\prime)}$ and Σ^0

Tomáš Husek

IFIC, Universitat de València – CSIC,
Apt. Correus 22085, E-46071 València, Spain
E-mail: thusek@ific.uv.es

We briefly summarize current experimental and theoretical results on the π^0 Dalitz decay, including the new value for the ratio $R = \Gamma(\pi^0 \to e^+e^-\gamma(\gamma))/\Gamma(\pi^0 \to \gamma\gamma) = 11.978(6) \times 10^{-3}$, which is by two orders of magnitude more precise than the current PDG average. Furthermore, we discuss radiative corrections for the Dalitz decays $\eta^{(\prime)} \to \ell^+\ell^-\gamma$ beyond the soft-photon approximation. The corrections inevitably depend on the $\eta^{(\prime)} \to \gamma^*\gamma^{(*)}$ transition form factors. Finally, we present (inclusive) radiative corrections for the $\Sigma^0 \to \Lambda e^+e^-$ decay, evaluated as well beyond the soft-photon approximation, i.e., over the whole range of the Dalitz plot and with no restrictions on the energy of the radiative photon. We calculate explicitly the 1γIR contribution and the correction to the $\Sigma^0\Lambda\gamma$ vertex and confirm that these can be neglected.

Keywords: Meson and hyperon decays; electromagnetic corrections; form factors of hadrons.

1. The neutral-pion Dalitz decay

The neutral-pion Dalitz decay $\pi^0 \to e^+e^-\gamma$, the second most significant decay channel of π^0, allows us to access information about the pion transition form factor. Moreover, it is used as the normalization channel in the rare-pion-decay ($\pi^0 \to e^+e^-$) search, and its eventual inaccurate knowledge is a limiting factor for rare-kaon-decay measurements, such as $K^+ \to \pi^+e^+e^-$ or $K^\pm \to \pi^\pm\pi^0e^+e^-$. In general, to provide relevant experimental results, the correct and consistent incorporation of properly calculated radiative corrections is crucial. Radiative corrections to the total decay rate of the π^0 Dalitz decay were first (numerically) addressed in Ref. 1. A pioneering study of the corrections to the differential decay rate was published by Lautrup and Smith,[2] although only in the soft-photon approximation, and extended later in the work of Mikaelian and Smith.[3]

A reinvestigation[4] of this old topic was motivated by the needs of NA48/NA62 experiments at CERN. Among other goals, the slope a_π of the singly-virtual electromagnetic transition form factor $\mathcal{F}_{\pi^0\gamma^*\gamma^*}(0,q^2)$ was expected to be directly measured with unprecedented precision. Compared to Ref. 3, we took into account the one-photon-irreducible (1γIR) contribution. Finally, we developed a C code, which returns the radiative correction for any given kinematically allowed point and which propagated to the Monte Carlo event generator within the NA62 experiment. Afterwards, 1.1×10^6 fully reconstructed π^0 Dalitz decays were analyzed with the result $a_\pi^{\rm NA62} = 3.68(57)\,\%$.[5]

Having at hand the complete set of the next-to-leading-order (NLO) QED radiative corrections $\delta(x, y)$ and a relatively good knowledge of the form-factor shape (in the neutral-pion case represented by the form-factor slope a_π), we can determine very precisely (and at the same time reliably) the ratio $R = \Gamma(\pi^0 \to e^+e^-\gamma(\gamma))/\Gamma(\pi^0 \to \gamma\gamma)$. With a rather conservative choice $a_\pi^{\rm univ} \equiv 3.55(70)\,\%$ for the slope we find $R = 11.978(5)(3) \times 10^{-38}$; the former uncertainty stands for the form-factor effects and the latter for neglecting the higher-order corrections. Note that the value $a_\pi^{\rm univ}$ covers a whole interval of numerical values suggested by various theoretical predictions and experiments. The new result for R represents a significant improvement (by two orders of magnitude) to the current PDG-based value $R^{\rm PDG} = 11.88(35) \times 10^{-3}$. It should thus be used in future theoretical predictions or experimental analyses. Since the sum of all the branching ratios of the neutral-pion decays should be equal to 1, R translates into the branching ratios of the two main π^0 decay modes: $\mathcal{B}(\pi^0 \to \gamma\gamma) = 98.8131(6)\,\%$ and $\mathcal{B}(\pi^0 \to e^+e^-\gamma(\gamma)) = 1.1836(6)\,\%$.[8]

2. Dalitz decays of $\eta^{(\prime)}$ mesons

Taking the pion case as a starting point, let us shortly point out the subtleties and difficulties one encounters when investigating the radiative corrections for the Dalitz decays of $\eta^{(\prime)}$ mesons. The mass of the η meson lies above the muon-pair-production threshold, and η' is even more massive than the lowest-lying resonances ρ and ω. In this way, the calculation becomes unavoidably sensitive to the width of the broad ρ resonance. The form-factor effects are thus not negligible and a particular form-factor model needs to be taken into account.

Naive radiative corrections for the $\eta \to e^+e^-\gamma$ process were published in Ref. 9, where only the numerical value of the mass of the decaying pseudoscalar was changed with respect to the earlier work 3. In Ref. 10 the list of the NLO corrections in the QED sector is completed and the previous approach[9] is improved: We take into account muon loops and hadronic corrections as a part of the vacuum-polarization contribution, 1γIR contribution at one-loop level (using a vector-meson-dominance-inspired model incorporating the η-η' mixing), higher-order final-state-lepton-mass corrections and form-factor effects. Moreover, we systematically study additional processes including η' decays: $\eta \to \mu^+\mu^-\gamma$, $\eta' \to e^+e^-\gamma$ and $\eta' \to \mu^+\mu^-\gamma$. Tables of values suitable for interpolation and further use can be found together with Ref. 10. The resulting corrections are most significant for the $\eta' \to e^+e^-\gamma$ decay. The form-factor effects are considerable and change the shape of the resonance peaks.

3. The Dalitz decay of Σ^0

So far we discussed radiative corrections to electromagnetic decays in the mesonic sector. In general, performing experiments with hyperons is more complicated since they are very unstable. The (direct as well as transition) electromagnetic form factors can be accessed at high energies, using electron–positron annihilation into hyperon–antihyperon pairs. At low energies, we would like to make use of a Dalitz decay (of a baryon into another baryon and the electron–positron pair). Indeed, in this sector there is an interesting decay to be studied in greater detail: $\Sigma^0 \to \Lambda e^+e^-$.

The Dalitz decay of Σ^0 allows us to study the singly-virtual transition form factor within the small window of virtualities up to $M_{\Sigma^0} - M_\Lambda \approx 77\,\text{MeV}$. According to theoretical predictions,[11,12] the transition radii are very small and of the same order as the NLO QED radiative corrections. Consequently, from the experimental point of view, not only the extraction of the form factor parameters will be challenging and require a high-precision measurement, but the good knowledge of the radiative corrections is necessary.

Radiative corrections to the decay rate as well as, within the soft-photon approximation, to the differential decay width were studied in Ref. 13. In our recent work[14] we study the inclusive corrections *beyond* the soft-photon limit, taking into account the complete set of contributions at NLO, i.e. including in addition the 1γIR contribution and QED corrections to the $\Sigma^0\Lambda\gamma$ vertex. By the loop-momenta-power counting, form factors need to be taken into account to regulate the UV behavior of the loop integrals of the two latter mentioned contributions. For the 1γIR contribution, the UV convergence is achieved already for a simple model with constant electric and magnetic form factors: $G_\text{E}(q^2) = 0$ and $G_\text{M}(q^2) = \kappa$, where κ is related to the magnetic moment. For the triangle diagram, a stronger UV sup-

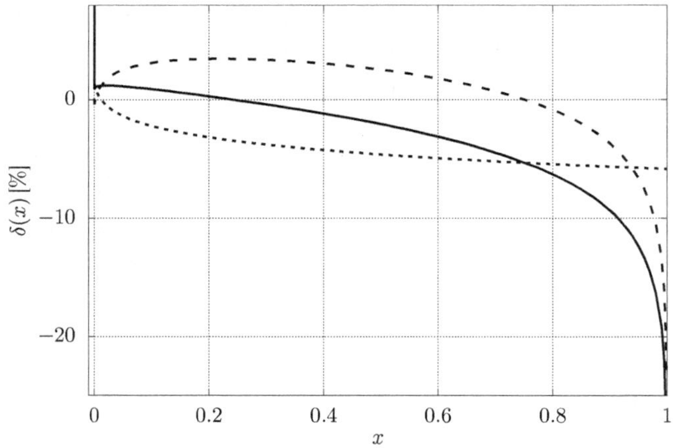

Fig. 1. The inclusive NLO QED radiative correction $\delta(x)$ (solid line) in comparison to its constituents for the decay $\Sigma^0 \to \Lambda e^+e^-$. The virtual corrections are depicted as the dotted line, the bremsstrahlung as the dashed line.

pression is necessary. The effect of these contributions was confirmed to be negligible independently of the chosen model. Numerically, the main difference between our results[14] and the work[13] thus lies in the bremsstrahlung contribution. Regarding this contribution, a low-energy expansion of the form factor can be used and the model dependence can be conveniently scaled out to a large extent in the final correction.

The radiative corrections to the one-fold differential decay width are shown in Fig. 1. Integrating the two-fold differential decay width (taken up to NLO) over the whole Dalitz plot reveals

$$R_{\Sigma^0_{\mathrm{D}}} \equiv \frac{\Gamma(\Sigma^0 \to \Lambda e^+e^-)}{\Gamma(\Sigma^0 \to \Lambda\gamma)} = 5.544(2) \times 10^{-3}\,, \tag{1}$$

which is perfectly consistent with the NLO result for the rate given in Ref. 13 and obtained using a different method: $R^{\mathrm{S\&S}}_{\Sigma^0_{\mathrm{D}}} = (5.532 + 0.627a) \times 10^{-3}$. Here, a is related to the magnetic radius; we used $a \equiv \frac{1}{6}\langle r^2_{\mathrm{M}}\rangle(M_{\Sigma^0} - M_\Lambda)^2 \approx 0.0183(26)$.[11] The result (1) can be again translated into branching ratios: $\mathcal{B}(\Sigma^0 \to \Lambda\gamma) = 99.4486(5)\,\%$ and $\mathcal{B}(\Sigma^0 \to \Lambda e^+e^-) = 0.5514(5)\,\%$.

Finally, from Fig. 1 we can estimate the size of the correction to the (magnetic) form-factor slope by taking half of the slope of the curve in the low-x region, however farther from the threshold:

$$\Delta a \equiv a_{(+\mathrm{QED})} - a \simeq \frac{1}{2}\frac{\mathrm{d}\delta(x)}{\mathrm{d}x}\bigg|_{x=x_0}\,, \tag{2}$$

with $\nu^2 \ll x_0 \ll 1$. We find $\frac{1}{2}\frac{\mathrm{d}\delta(x)}{\mathrm{d}x}\Big|_{x=x_0} \approx -3.5\,\%$. Since this value is bigger than the estimate on a itself, it is indeed essential to consider radiative corrections when extracting the magnetic radius from experiment.

The complete sets of NLO radiative corrections in the QED sector for the discussed decays—the Dalitz decays of π^0,[4] η'[10] and Σ^0[14]—are available and their use in future experimental analyses should be essential. Finally, note that in Refs. 4,10,14 we study fully inclusive radiative corrections: no momentum or angular cuts on additional photon(s) are applied.

Acknowledgments

The presented work is based on collaboration with K. Kampf, J. Novotný, S. Leupold and E. Goudzovski. This work has been supported in part by Grants No. FPA2017-84445-P and SEV-2014-0398 (AEI/ERDF, EU) and by PROMETEO/2017/053 (GV).

References

1. D. Joseph, *Nuovo Cim.* **16**, p. 997 (1960).
2. B. Lautrup and J. Smith, *Phys. Rev. D* **3**, 1122 (1971).
3. K. O. Mikaelian and J. Smith, *Phys. Rev. D* **5**, 1763 (1972).
4. T. Husek *et al.*, *Phys. Rev. D* **92**, p. 054027 (2015).
5. C. Lazzeroni *et al.*, *Phys. Lett. B* **768**, 38 (2017).
6. M. Tanabashi *et al.*, *Phys. Rev. D* **98**, p. 030001 (2018).
7. H. J. Behrend *et al.*, *Z. Phys. C* **49**, 401 (1991).
8. T. Husek *et al.*, *Phys. Rev. Lett.* **122**, p. 022003 (2019).
9. K. O. Mikaelian and J. Smith, *Phys. Rev. D* **5**, 2890 (1972).
10. T. Husek *et al.*, *Phys. Rev. D* **97**, p. 096013 (2018).
11. B. Kubis and U. G. Meißner, *Eur. Phys. J. C* **18**, 747 (2001).
12. C. Granados *et al.*, *Eur. Phys. J. A* **53**, p. 117 (2017).
13. D. P. Sidhu and J. Smith, *Phys. Rev. D* **4**, 3344 (1971).
14. T. Husek and S. Leupold, [arXiv:1911.02571] (2019).

Status of quarkonium production

H. S. Chung

Physik-Department, Technische Universität München,
85748 Garching, Germany
E-mail: heesok.chung@tum.de

We review the status of heavy quarkonium production phenomenology based on the nonrelativistic QCD effective field theory approach. We show the relation between the description of the heavy quarkonium state in nonrelativistic QCD and production processes of heavy quarkonia. We examine exclusive electromagnetic production and inclusive production processes and assess the current understanding of the quarkonium production mechanism.

Keywords: Heavy quarkonium; nonrelativistic QCD.

1. Introduction

Heavy quarkonia are considered good laboratories to study the interplay between perturbative and nonperturbative physics because they encompass not only the heavy quark mass scale but also scales comparable or below $\Lambda_{\rm QCD}$ through the bound state dynamics. The separation of the heavy quark mass scale from smaller scales allow us to employ nonrelativistic effective field theory approaches to heavy quarkonia.

The nonrelativistic QCD (NRQCD) effective field theory[1,2] is obtained by integrating out the the heavy quark mass scale. The scales that appear in NRQCD are mv and mv^2, where m is the heavy quark mass and v is the typical velocity of the heavy quark inside the quarkonium at rest. In NRQCD, a heavy quarkonium state is described by a Fock state expansion, where the dominant Fock state is the $|Q\bar{Q}\rangle$ with the $Q\bar{Q}$ in the color singlet state, and the higher Fock states such as $|Q\bar{Q}g\rangle$, where g is a dynamical gluon in the Coulomb gauge. The higher Fock states can involve $Q\bar{Q}$ in color octet states, and the contribution from those higher Fock states are suppressed by powers of v.

One way to probe the leading and subleading Fock states of heavy quarkonia is to use the production cross sections. In general, the contribution at leading order in v comes from the leading Fock state, while higher order corrections in v can involve contributions from higher Fock states. Hence, heavy quarkonium production processes provide important tests of the NRQCD formalism.

In this paper, we review the status of heavy quarkonium production phenomenology. In Sec. 2, we examine the status of exclusive electromagnetic production, and we discuss inclusive production in Sec. 3. We conclude in Sec. 4.

2. Exclusive Electromagnetic Production of Quarkonia

In NRQCD, exclusive electromagnetic production amplitudes of a quarkonium H factorize into nonperturbative NRQCD matrix elements and perturbative short-distance coefficients:

$$i\mathcal{M} = \sum_n c_n \langle H|\mathcal{O}_n|0\rangle, \tag{1}$$

where $|0\rangle$ is the QCD vacuum, and $\mathcal{O}_n$ are NRQCD operators. The long-distance matrix elements (LDMEs) $\langle H|\mathcal{O}_n|0\rangle$ have known scalings in v, which are determined from the dimensions of the operators $\mathcal{O}_n$ and the probability of the $Q\bar{Q}$ Fock state they create. Hence, the sum is organized in powers of v, and in practice, is truncated to desired accuracy. The c_n are short-distance coefficients (SDCs) that can be computed in perturbation theory. In the following sections, we give examples of phenomenological studies of exclusive quarkonium production based on Eq. (1).

2.1. *Exclusive production of P-wave charmonia*

We first consider the production of P-wave charmonia, which include χ_{cJ} and h_c, where $J = 0, 1$, and 2. The production of χ_{cJ} are of particular interest, because they have well-known exclusive decay modes into $\gamma\gamma$ (for $J = 0$ and 2) and $J/\psi + \gamma$. For the production of χ_{cJ}, contributions from color-octet Fock states appear from relative order v^2 accuracy.[3]

The χ_{cJ} can be produced in lepton colliders through the process $e^+e^- \to \chi_{cJ} + \gamma$. This process was proposed in Ref. 4 as a way to probe heavy quarkonia with $C = +1$, where the first calculation at leading orders in α_s and v was presented. Order-α_s and color-singlet order-v^2 corrections,[5–9] and the color-octet order-v^2 corrections[3] have been computed. Belle[10] reported the measured cross section for $J = 1$, which agree well with theoretical predictions within uncertainties. One obstacle in reducing theoretical uncertainties is our poor knowledge of the LDMEs, especially for the color octet LDME.

2.2. *Exclusive production of S-wave quarkonia*

Now we discuss the production of S-wave quarkonia, which include the J/ψ, η_c, $\Upsilon(nS)$, and the η_b. At leading order in v, one LDME appears in Eq. (1), which is sensitive to the color-singlet S-wave $Q\bar{Q}$ Fock state of the S-wave quarkonium. This LDME can be related to the quarkonium wavefunction at the origin. For S-wave quarkonia, color-octet LDMEs appear from relative order v^3 accuracy, so that the higher Fock state contributions do not appear at relative order v^2 accuracy.[11,12]

One way to produce η_c in lepton colliders is through the process $e^+e^- \to \eta_c + \gamma$. This process was proposed in Ref. 4 as a way to probe heavy quarkonia with $C = +1$, where the first calculation at leading orders in α_s and v appeared. The order-$\alpha_s v^0$, $\alpha_s^0 v^2$, $\alpha_s v^2$, and $\alpha_s^2 v^0$ corrections have also been computed.[5,6,9,13–16] Recently, resummation of the logarithms in m^2/s to next-to-leading logarithmic accuracy have been computed,[17] based on the pioneering work in Ref. 18. Interestingly, the theoretical prediction is 4σ above the measured upper limit from Belle.[10]

An interesting case of exclusive J/ψ production is Higgs boson decay into $J/\psi+\gamma$. This process was first proposed as a way to measure the Higgs-charm coupling.[19] The decay rate is sensitive to both the size and the sign of the coupling, owing to quantum interference. In the Standard Model, the branching ratio of this process is about 3×10^{-6}. Order-v^2 corrections and resummation of leading logarithms,[20] resummation of next-to-leading logarithms,[21,22] and order-v^4 corrections that are sensitive to the color-octet Fock states[23] have been computed to reduce the theoretical uncertainty to about 3%.

Finally, we discuss the double quarkonium production process $e^+e^- \to J/\psi + \eta_c$. At leading orders in α_s and v, the theoretical prediction is much smaller than the the measured cross section at B factories.[25–28] It was later found that the order-v^2 and order-α_s corrections are large and positive, and the discrepancy between the theory and measurement seem to have been resolved.[28–31] However, it was later revealed that the large order-α_s correction is dominated by the double logarithms of the form $\alpha_s \log^2(m^2/s)$, which may spoil the convergence of perturbation series.[32] Since the process $e^+e^- \to J/\psi + \eta_c$ occurs at next-to-leading power at $m/\sqrt{s}$, standard methods of resumming logarithms do not apply to this process. The double logarithms have been identified as endpoint logarithms, but an all-orders treatment that may lead to the resummation of logarithms is still out of reach.[33]

3. Inclusive Production of Quarkonia

In the NRQCD factorization formalism, the inclusive production cross section of a heavy quarkonium is given by[2]

$$\sigma_H = \sum_n \sigma_{Q\bar{Q}(n)} \langle 0|\mathcal{O}_n^H|0\rangle, \tag{2}$$

where n denote the angular momentum and color states of the $Q\bar{Q}$. The $\sigma_{Q\bar{Q}(n)}$ are the short-distance cross sections that correspond to the perturbative cross section of $Q\bar{Q}$ in the state n, and the LDMEs $\langle 0|\mathcal{O}_n^H|0\rangle$ describe the evolution of a $Q\bar{Q}$ pair into a heavy quarkonium $H + X$, where X includes any light degrees of freedom of scales mv and below. The production LDMEs can be computed if they correspond to the leading Fock state of the quarkonium H; in such case, the color-singlet LDME at leading order in v is given in terms of the quarkonium wavefunction at the origin. On the other hand, it is not known how to compute the color-octet LDMEs. Hence, the color-octet LDMEs are usually extracted from data.

In case of inclusive production of J/ψ, at leading order in v, the color-singlet ${}^3S_1$ channel contributes to the right-hand side of Eq. (2). To relative order v^4 accuracy, color-octet ${}^1S_0$, ${}^3S_1$, and ${}^3P_J$ channels appear. In many inclusive production processes, including hadroproduction, the short-distance cross sections in the color octet channels are enhanced compared to that in the color singlet channel,[34] to the extent that the color octet channels dominate the cross section. Hence, it is necessary to determine the three unknown color-octet LDMEs to make predictions of inclusive J/ψ cross sections.

The determinations of J/ψ color-octet LDMEs depend strongly on the choice of experimental data.[35–38] Especially, the relative signs of the color-octet ${}^3S_1$ and ${}^3P_J$ LDMEs differ between determinations based on large p_T hadroproduction data[35,37,38] and determinations from global fits[36] that include small p_T data. It has been pointed out that the determinations based on hadroproduction data tend to overestimate the measured $ep \to J/\psi + X$ and $e^+e^- \to J/\psi + X$ cross sections.[39] On the other hand, global fits are in tension with LHC measurements of J/ψ polarization[40] and the momentum distribution of J/ψ in jet.[41]

4. Summary and Discussion

In this paper, we reviewed the status of quarkonium production phenomenology. Exclusive production of heavy quarkonium provide good tests of the nonrelativistic EFT formalism, as well as resummation and evolution in perturbative QCD. In order to obtain precise theoretical predictions, nonperturbative matrix elements, including ones sensitive to color octet Fock states, need to be determined accurately. Measurements of exclusive production and decay processes are invaluable in constraining such nonperturbative quantities. On the other hand, a satisfactory theoretical description of inclusive production mechanism of heavy quarkonium is still missing. Ongoing experiments call for theoretical effort towards QCD-based understanding of the quarkonium inclusive production mechanism.

References

1. W. E. Caswell and G. P. Lepage, Phys. Lett. **167B**, 437 (1986).
2. G. T. Bodwin, E. Braaten and G. P. Lepage, Phys. Rev. D **51**, 1125 (1995) Erratum: [Phys. Rev. D **55**, 5853 (1997)].
3. N. Brambilla, W. Chen, Y. Jia, V. Shtabovenko and A. Vairo, Phys. Rev. D **97**, no. 9, 096001 (2018).
4. H. S. Chung, J. Lee and C. Yu, Phys. Rev. D **78**, 074022 (2008).
5. W. L. Sang and Y. Q. Chen, Phys. Rev. D **81**, 034028 (2010).
6. D. Li, Z. G. He and K. T. Chao, Phys. Rev. D **80**, 114014 (2009).
7. Y. J. Li, G. Z. Xu, K. Y. Liu and Y. J. Zhang, JHEP **1401**, 022 (2014).
8. K. T. Chao, Z. G. He, D. Li and C. Meng, arXiv:1310.8597 [hep-ph].
9. G. Z. Xu, Y. J. Li, K. Y. Liu and Y. J. Zhang, JHEP **1410**, 071 (2014).
10. S. Jia *et al.* [Belle Collaboration], Phys. Rev. D **98**, no. 9, 092015 (2018).
11. G. T. Bodwin and A. Petrelli, Phys. Rev. D **66**, 094011 (2002) Erratum: [Phys. Rev. D **87**, no. 3, 039902 (2013)].
12. N. Brambilla, E. Mereghetti and A. Vairo, JHEP **0608**, 039 (2006) Erratum: [JHEP **1104**, 058 (2011)].
13. V. V. Braguta, Phys. Rev. D **82**, 074009 (2010).
14. F. Feng, Y. Jia and W. L. Sang, Phys. Rev. Lett. **115**, no. 22, 222001 (2015).
15. L. B. Chen, Y. Liang and C. F. Qiao, JHEP **1801**, 091 (2018).
16. Y. Fan, J. Lee and C. Yu, Phys. Rev. D **87**, no. 9, 094032 (2013).
17. H. S. Chung, J. H. Ee, D. Kang, U. R. Kim, J. Lee and X. P. Wang, JHEP **1910**, 162 (2019).
18. Y. Jia and D. Yang, Nucl. Phys. B **814**, 217 (2009).
19. G. T. Bodwin, F. Petriello, S. Stoynev and M. Velasco, Phys. Rev. D **88**, no. 5, 053003 (2013).

20. G. T. Bodwin, H. S. Chung, J. H. Ee, J. Lee and F. Petriello, Phys. Rev. D **90**, no. 11, 113010 (2014).
21. G. T. Bodwin, H. S. Chung, J. H. Ee and J. Lee, Phys. Rev. D **95**, no. 5, 054018 (2017).
22. G. T. Bodwin, H. S. Chung, J. H. Ee and J. Lee, Phys. Rev. D **96**, 116014 (2017).
23. N. Brambilla, H. S. Chung, W. K. Lai, V. Shtabovenko and A. Vairo, Phys. Rev. D **100**, no. 5, 054038 (2019).
24. E. Braaten and J. Lee, Phys. Rev. D **67**, 054007 (2003) Erratum: [Phys. Rev. D **72**, 099901 (2005)].
25. K. Y. Liu, Z. G. He and K. T. Chao, Phys. Lett. B **557**, 45 (2003).
26. K. Abe *et al.* [Belle Collaboration], Phys. Rev. D **70**, 071102 (2004).
27. B. Aubert *et al.* [BaBar Collaboration], Phys. Rev. D **72**, 031101 (2005).
28. E. Braaten and J. Lee, Phys. Rev. D **67**, 054007 (2003) Erratum: [Phys. Rev. D **72**, 099901 (2005)].
29. Y. J. Zhang, Y. j. Gao and K. T. Chao, Phys. Rev. Lett. **96**, 092001 (2006).
30. Z. G. He, Y. Fan and K. T. Chao, Phys. Rev. D **75**, 074011 (2007).
31. G. T. Bodwin, J. Lee and C. Yu, Phys. Rev. D **77**, 094018 (2008).
32. Y. Jia, J. X. Wang and D. Yang, JHEP **1110**, 105 (2011).
33. G. T. Bodwin, H. S. Chung and J. Lee, Phys. Rev. D **90**, no. 7, 074028 (2014).
34. P. L. Cho and A. K. Leibovich, Phys. Rev. D **53**, 6203 (1996).
35. B. Gong, L.-P. Wan, J.-X. Wang, and H.-F. Zhang, Phys. Rev. Lett. **110**, 042002 (2013).
36. M. Butenschoen and B. A. Kniehl, Phys. Rev. D **84**, 051501 (2011).
37. H. S. Shao, H. Han, Y. Q. Ma, C. Meng, Y. J. Zhang and K. T. Chao, JHEP **1505**, 103 (2015).
38. G. T. Bodwin, K. T. Chao, H. S. Chung, U. R. Kim, J. Lee and Y. Q. Ma, Phys. Rev. D **93**, no. 3, 034041 (2016).
39. M. Butenschoen and B. A. Kniehl, Mod. Phys. Lett. A **28**, 1350027 (2013).
40. S. Chatrchyan *et al.* [CMS Collaboration], Phys. Lett. B **727**, 381 (2013).
41. R. Bain, L. Dai, A. Leibovich, Y. Makris and T. Mehen, Phys. Rev. Lett. **119**, no. 3, 032002 (2017).

Associated quarkonium production at ATLAS

T. Zakareishvili
on behalf of the ATLAS collaboration
High Evenry Physics Institute of Tbilisi State University
Tbilisi, Georgia
E-mail: Tamar.Zakareishvili@cern.ch

The associated production of a vector boson with quarkonia is a key new observable for understanding the quarkonium production mechanisms, including the separation of single and double parton scattering components. Using the 2011 ATLAS dataset of 4.5 fb^{-1} of $\sqrt{s}$ =7 TeV pp collisions at the LHC, the first observations were made of the production of $W^{\pm}$ + prompt J/ψ events in hadronic collisions, and using 20.3 fb^{-1} of data collected with the ATLAS experiment, in pp collisions at $\sqrt{s}$ = 8 TeV, the first measurements of associated Z + J/ψ production were made for both prompt and non-prompt J/ψ production mechanisms. The cross section ratio of prompt J/ψ production in association with a vector boson ($W^{\pm}$ and Z^0) to the inclusive vector boson production is measured, and the double parton scattering contribution to the cross section is estimated for both analyses. Normalised differential production cross-section ratios are also determined as functions of J/ψ transverse momentum.

Keywords: Quarkonium; associated production; ATLAS; LHC.

1. Introduction

Quarkonium is formed from a quark-antiquark pair of the same flavor and should represent one of the simplest systems described by QCD theory. However, the mechanisms responsible for the production of quarkonia in hadronic collisions are not fully understood. Study of the production of a prompt J/ψ meson in association with a vector boson offers new tests of Quantum Chromodynamics (QCD) at the perturbative/non-perturbative boundary, offers tests to try to separate single and double parton scattering components, and also helps develop the framework for future probes of the Higgs sector and beyond-the-standard-model searches in such final states. The ATLAS experiment[1] at the Large Hadron Collider (LHC) aims to investigate the above mentioned points using a quarkonium system as a probe.

2. Production of prompt J/Ψ mesons in association with a $W^{\pm}$ boson

Using the 2011 ATLAS dataset of 4.5 fb^{-1} of $\sqrt{s} = 7$ TeV pp collisions at the LHC, the first observation is made of the production of $W^{\pm}+$ prompt J/ψ events in hadronic collisions, using $W^{\pm} \rightarrow \mu\nu_{\mu}$ and $J/\psi \rightarrow \mu^{+}\mu^{-}$ channels.[2]

To be considered for this analysis, events are required to have at least three identified muons. One pair of oppositely charged muons is required to form a J/ψ candidate, while an additional muon must combine with the events missing transverse momentum $E_{\mathrm{T}}^{\mathrm{miss}}$ to form a $W^{\pm}$ candidate.

In the Standard Model, a single partonparton interaction can produce a J/ψ meson in association with a vector boson either through a prompt QCD subprocess, or through the production of a vector boson with a b-quark and its subsequent decay into a J/ψ (non-prompt production).

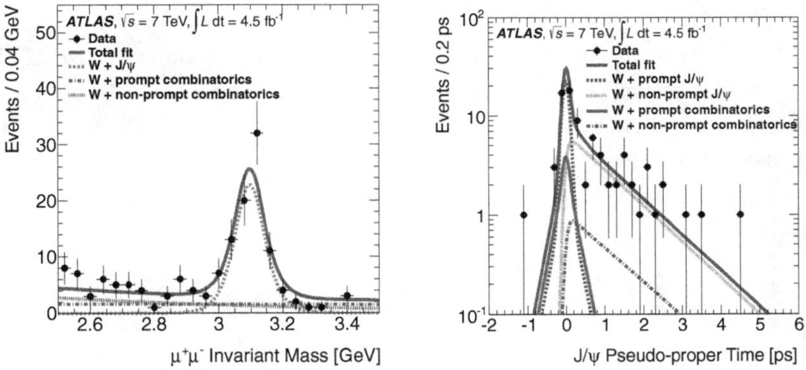

(a) (b)

Fig. 1. Projections in (a) invariant mass and (b) pseudo-proper time of the two-dimensional masspseudo-proper time fit used to extract the prompt J/ψ candidates.[2]

Prompt J/ψ candidates are distinguished from those originating from b-hadron decays through the spatial separation of the primary collision vertex and the J/ψ decay vertex. For this the pseudo-proper time

$$\tau \equiv \frac{\vec{L} \cdot \vec{p}_{\mathrm{T}}^{J/\psi}}{p_{\mathrm{T}}^{J/\psi}} \cdot \frac{m_{\mu^{+}\mu^{-}}}{p_{\mathrm{T}}^{J/\psi}}, \tag{1}$$

is used, where $\vec{L}$ is the separation vector from the primary vertex to the J/ψ decay vertex and $m_{\mu^+\mu^-}$ is the invariant mass of the J/ψ candidate. Prompt J/ψ decays have a pseudo-proper time consistent with zero within resolution. Non-prompt J/ψ decay vertices are displaced from the primary vertex and have positive pseudo-proper time on average apart from resolution effects.

An unbinned maximum likelihood fit in J/ψ candidate invariant mass and pseudo-proper time is applied to obtain yields (using the sPlot procedure[4]) of prompt J/ψ, non-prompt J/ψ, and prompt/non-prompt combinatoric backgrounds, see Figure 1.

It is possible for the $W^\pm$ and J/ψ to originate from two different parton interactions in the same protonproton collision, in a double parton scattering (DPS) process. The standard ansatz[5] is adopted that, for a collision in which a hard process occurs, the probability of an additional process is parameterized as

$$P_{J/\psi|W^\pm} = \frac{\sigma_{J/\psi}}{\sigma_{\text{eff}}}. \tag{2}$$

The effective area parameter σ_{eff} accounts for the geometric size of the proton and transverse parton correlations, and is assumed to be independent of the scattering process. It is taken to be 15 ± 3 (stat.)$^{+5}_{-3}$ (syst.) mb as measured using $W^\pm \to l\nu_l + 2$ jet events.[11]

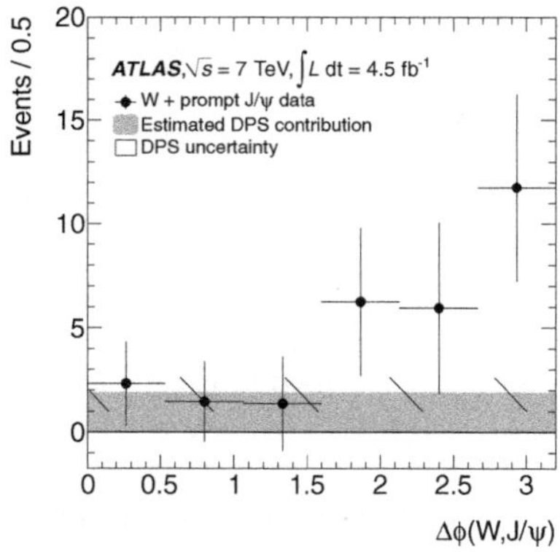

Fig. 2. The sPlot-weighted azimuthal angle between the $W^\pm$ and the J/ψ is shown for $W^\pm$ + prompt J/ψ candidates.[2]

A uniform distribution in the azimuthal angle between the $W^\pm$ and J/ψ momenta is expected from DPS, under the assumption that the two interactions are independent. For the Single Parton Scattering (SPS) contribution simulations in both the colour-singlet and colour-octet models predict a distribution strongly peaked near $\Delta\phi = \pi$. The sPlot-weighted distribution of this variable for the data has a peak near π and a tail extending towards zero, as shown in Figure 2. This suggests that the observed $W^{\pm}$ + prompt J/ψ candidates include both SPS and DPS events.

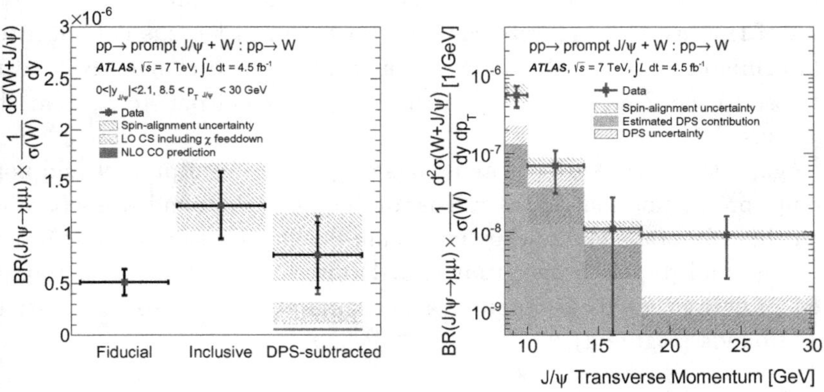

(a) (b)

Fig. 3. (a) Fiducial, inclusive and DPS-subtracted cross-section ratio of $W^{\pm}$ + prompt J/ψ to inclusive $W^{\pm}$ production; (b) The inclusive (SPS+DPS) cross-section ratio $dR^{\rm incl}_{J/\psi}/dp_{\rm T}$ as a function of J/ψ transverse momentum.[2]

The fiducial cross-section ratio of $W^{\pm}$ + prompt J/ψ production to inclusive $W^{\pm}$ production in the fiducial region was measured. Then, after correcting for J/ψ acceptance, the inclusive cross-section ratio was calculated, and after subtracting the double parton scattering component, the DPS-subtracted cross-section ratio was obtained (Figure 3(a)). The differential cross-section ratio $dR^{\rm incl}_{J/\psi}/dp_{\rm T}$ as a function of $p_{\rm T}^{J/\psi}$ was also determined (Figure 3(b)), with an estimate of the differential DPS contribution. The data suggest that SPS is the dominant contribution to the total rate at low J/ψ transverse momenta. The distribution of the azimuthal angle between the produced $W^{\pm}$ and J/ψ suggests the presence of both SPS and DPS contributions, while the DPS estimate accounts for a large fraction of the observed signal. The predictions from both LO CS[8] and NLO CO[9] theoretical models lie below the measured values.

3. Production of prompt J/ψ mesons in association with a Z^0 boson

Using 20.3 fb^{-1} of data collected by the ATLAS experiment at the LHC in pp collisions at $\sqrt{s}$ =8 TeV, the first measurement of associated $Z + J/\psi$ production was carried out for both prompt and non-prompt mechanisms of J/ψ production.[10]

The final state of this measurement is $Z(\rightarrow l^+l^-) + J/\psi(\rightarrow \mu^+\mu^-)$, where $l = \mu, e$, and therefore candidate events are required to have two pairs of leptons with opposite charges. Each pair of leptons is then fitted to a common vertex, with the invariant mass of the first pair required to be close to the Z boson mass and that of the second pair to be near the J/ψ mass.

Again, the $J/\psi \rightarrow \mu^+\mu^-$ candidates originating from prompt and non-prompt production sources, as well as continuum background, are separated with the help of a simultaneous maximum likelihood fit of the J/ψ invariant mass and pseudo-proper time (Eq. 1) distribution. The sPlot tool[4] is used to extract yields for prompt signal, non-prompt signal, and various backgrounds (Figure 4).

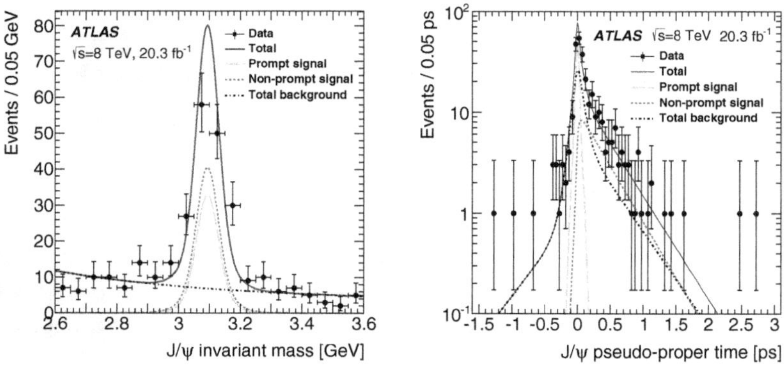

(a) (b)

Fig. 4. Projections of the unbinned mass and pseudo-proper time maximum-likelihood fit in (a) invariant mass and (b) pseudo-proper time of the associated-production sample. The fit is used to extract the prompt and non-prompt signal fractions.[10]

Figure 5 shows the distribution in the azimuthal angle $\Delta\phi$ between the Z boson and the J/ψ momentum vectors.

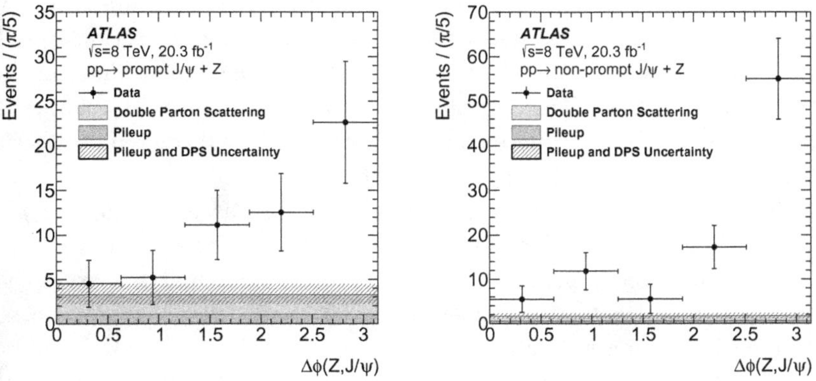

(a) (b)

Fig. 5. Azimuthal angle between the Z boson and the J/ψ meson after the application of the sPlot weights to separate the prompt (left) and non-prompt (right) yield from background contributions.[10]

By conservatively assuming that all observed signal in the first bin is due to DPS, the data uncertainties and uncertainties inherent in the DPS estimate allow a lower limit $\sigma_{\text{eff}} > 5.3$ mb (3.7 mb) at 68% (95%) confidence level to be extracted from the Z + prompt J/ψ data.

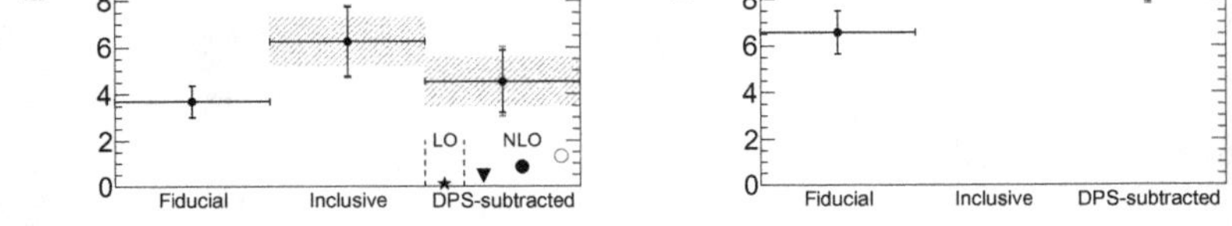

(a) (b)

Fig. 6. Production cross-section ratios of J/ψ in association with a Z boson, relative to inclusive Z production, for prompt and non-prompt J/ψ production.[10]

The fiducial cross-section ratio of Z^0 + prompt J/ψ production to inclusive Z^0 production was measured, without applying corrections for the incomplete geometric acceptance for the J/ψ decay muons, nor for the Z boson acceptance. The acceptance-corrected inclusive production cross-section ratio was also calculated, and finally the DPS-subtracted production cross-section ratio was obtained (Figure 6).

The differential cross-section ratio $\mathrm{d}R^{\mathrm{incl}}_{J/\psi}/dp_{\mathrm{T}}$ as a function of $p_{\mathrm{T}}^{J/\psi}$ was also determined (Figure 7), with an estimate of the differential DPS contribution. Theory predicts that colour-octet contributions exceed the production rate from color-singlet processes, with colour-octet processes becoming increasingly dominant for higher p_{T} of the J/ψ. The combination of DPS and NLO NRQCD contributions tends to underestimate the production rate observed in data, with the discrepancy increasing with transverse momentum and reaching a factor of 4 5 for $p_{\mathrm{T}}^{J/\psi} > 18$ GeV.

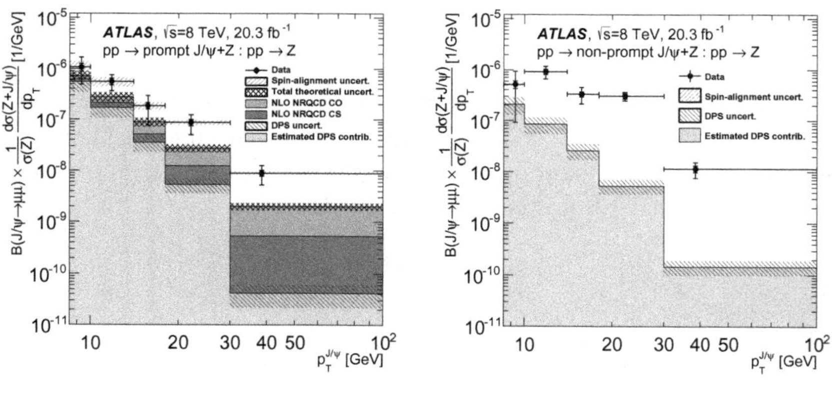

(a) (b)

Fig. 7. Production cross-section of J/ψ in association with a Z boson as a function of the p_{T} of prompt J/ψ, and non-prompt J/ψ, normalised to the inclusive Z cross-section.[10]

4. Summary and conclusions

The ATLAS Collaboration has observed $W^{\pm}$ + prompt J/ψ production at 5.1σ significance in 4.5 fb^{-1} of $\sqrt{s} = 7$ TeV pp collisions at the LHC. W + prompt J/ψ candidates include both SPS and DPS events. SPS is the dominant contribution to the total rate at low p_{T} of the J/ψ. The ATLAS Collaboration has also observed Z^0 + prompt J/ψ production at

5σ significance and Z^0 + non-prompt J/ψ production at 9σ significance in 20.3 fb^{-1} of $\sqrt{s} = 8$ TeV pp collisions at the LHC. Z^0 + prompt J/ψ candidates include both SPS and DPS events. A lower limit has been set on $\sigma_{\text{eff}}(> 5.3$ mb). A higher production rate is predicted through CO transitions than through CS processes. The expected production rate from the sum of singlet and octet contributions is lower than the data by a factor of 2 to 5.

In a J/ψ pair production study,[15] the effective cross-section of DPS is measured to be $\sigma_{\text{eff}} = 6.3 \pm 1.6$ (stat.) ± 1.0 (syst.) mb, which is lower than for other final states,[16] but is in agreement with the lower limit shown above.

Acknowledgments

The author was supported by Shota Rustaveli National Science Foundation of Georgia through the grant projects № FR 17_184 and № 04/48, and The Volkswagen Foundation *Ref.*: 93 562.

References

1. ATLAS Collaboration, JINST 3 (2008) S08003.
2. ATLAS Collaboration, JHEP 04 (2014) 172.
3. J.P. Lansberg, et al, Phys.Lett. B781 (2018) 485-491.
4. M. Pivk, et al., Nucl.Instrum.Meth. A555: 356-369, 2005.
5. N. Paver, et al, Nuovo Cim. A70 (1982) 215 SISSA-7/82/EP.
6. J.P. Lansberg, el al., Phys.Lett. B726: 218-222, 2013.
7. R. Gavin, et al., Comput.Phys.Commun. 184 (2013) 208-214.
8. J.P. Lansberg, et al., Phys.Lett. B726 (2013) 218-222, Erratum: Phys.Lett. B738 (2014) 529-529.
9. G. Li, et al., Phys. Rev. D 83 (2011) 014001.
10. ATLAS Collaboration, Eur.Phys.J. C75 (2015) 229.
11. ATLAS Collaboration, New J.Phys. 15 (2013) 033038.
12. ATLAS Collaboration, Nucl.Phys. B 850 (2011) 387-444.
13. B. Gong, et al., JHEP 1303 (2013) 115.
14. S. Mao, et al., JHEP 1102:071,2011.
15. ATLAS Collaboration, Eur.Phys.J. C77 (2017) 76.
16. ATLAS Collaboration, Phys.Lett. 790 (2019) 595.

Three-body open flavor decays of higher vector charmonium and bottomonium

Xin-Zhen Weng[1,*], Li-Ye Xiao[2,†],
Wei-Zhen Deng[1,‡], Xiao-Lin Chen[1,§], and Shi-Lin Zhu[1,3,4,||]
[1] *School of Physics and State Key Laboratory of Nuclear Physics and Technology, Peking University, Beijing 100871, China*
[2] *University of Science and Technology Beijing, Beijing 100083, China*
[3] *Center of High Energy Physics, Peking University, Beijing 100871, China*
[4] *Collaborative Innovation Center of Quantum Matter, Beijing 100871, China*
E-mail: [*] *xzhweng@pku.edu.cn* [†] *lyxiao@ustb.edu.cn*
[‡] *dwz@pku.edu.cn* [§] *chenxl@pku.edu.cn* [||] *zhusl@pku.edu.cn*

With an extended quark pair creation model we systematically study the OZI-allowed three-body open flavor decays of higher vector charmonium and bottomonium states. We obtain that the $BB^*\pi$ and $B^*B^*\pi$ partial decay widths of $\Upsilon(10860)$ are consistent with experiment, and the corresponding partial decay widths of $\Upsilon(11020)$ can reach up to 2∼3 MeV. Meanwhile the partial widths of $DD^*\pi$ and $D^*D^*\pi$ modes for most higher vector charmonium states can reach up to several MeV.

Keywords: Quark pair creation model; three-body decays; charmonium; bottomonium.

1. Introduction

After the observation of $X(3872)$ by the Belle Collaboration in 2003,[1] lots of charmonium-like states have been found in experiment.[2] Among them, the 1^{--} states are of special importance because they can be easily produced in the e^+e^- annihilation. To reveal the nature of charmonium-like states, it is important to have a better understanding of the normal charmonium states. Considering the strong decay properties playing a pivotal role in uncovering the nature of hadrons, in the present work we focus on the strong decays of charmonium. Besides the two-body open flavor decays, three-body open flavor decays are also important to the study of charmonium states with the experimental process. In Ref. 3, we extended the quark pair creation ($^3P_0$) model to second order and studied the $\psi(4660)\to\Lambda_c\bar{\Lambda}_c$ process. In this work, we further use the extended $^3P_0$ model to study the three-body open flavor decays of higher vector charmonium and bottomonium states.

2. The $^3P_0$ Model

In the $^3P_0$ model, a light quark pair is created from the vacuum, and then regroups with the quarks within the initial hadron to produce two daughter hadrons.[4–6] The corresponding interaction Hamiltonian reads[7]

$$H_{q\bar{q}} = \gamma \sum_f 2m_f \int \mathrm{d}^3x \bar{\psi}_f \psi_f, \tag{1}$$

where γ is the quark pair creation strength parameter, and ψ_f stands for the Dirac field of light quark f with constituent quark mass m_f. To study the

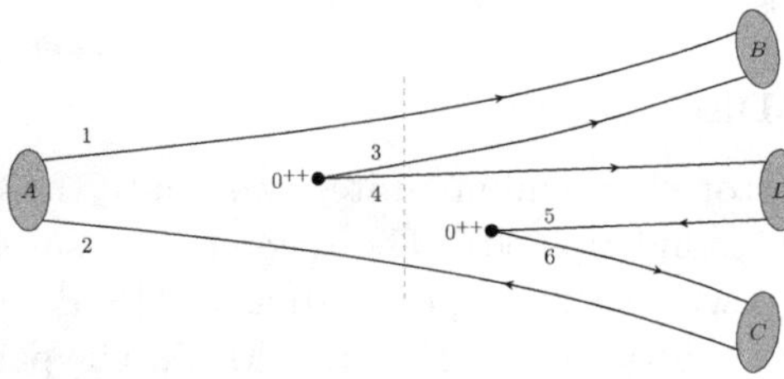

Fig. 1. The quarkonium (A) decays into three mesons ($B + C + D$). The intermediate state is marked by a red dashed line.

decays of higher heavy quarkonium into two light-heavy mesons plus a light meson (see Fig. 1), we extend the $^3P_0$ model to second order, where two light $q\bar{q}$ pairs are created from the vacuum.[3,8] The corresponding helicity amplitude $\mathcal{M}^{M_{J_A} M_{J_B} M_{J_C} M_{J_D}}$ is defined as

$$\begin{aligned} &\delta^3 \left(\mathbf{p}_A - \mathbf{p}_B - \mathbf{p}_C - \mathbf{p}_D \right) \mathcal{M}^{M_{J_A} M_{J_B} M_{J_C} M_{J_D}} \\ &\quad = \sum_k \frac{\langle BCD | H_{q\bar{q}} | k \rangle \langle k | H_{q\bar{q}} | A \rangle}{E_k - E_A} \end{aligned} \tag{2}$$

where $|k\rangle$ is the intermediate state; E_A (E_k) is energy of the initial (intermediate) state; $\mathbf{p}_i$'s stand for the momenta of the hadrons. We apply the quark-hadron duality[9] to simplify the calculation. Since the intermediate state differs from the initial state by a created quark pair at the quark level, we assume[3,8]

$$E_k - E_A \approx 2m_q. \tag{3}$$

Under this approximation, we can rewrite the Eq. (2) as

$$\delta^3 \left(\mathbf{p}_A - \mathbf{p}_B - \mathbf{p}_C - \mathbf{p}_D \right) \mathcal{M}^{M_{J_A} M_{J_B} M_{J_C} M_{J_D}} \approx \frac{\langle BCD | H_{q\bar{q}} H_{q\bar{q}} | A \rangle}{2m_q}. \tag{4}$$

It should be mentioned that the above assumption is reasonable at the hadron level as well. Taking charmonium for example, the possible intermediate states with quantum number $J^{PC} = 1^{--}$ are $D\bar{D}_1$, $D^*\bar{D}_0$, $D^*\bar{D}_1$, $D^*\bar{D}_2$, $J/\psi f_0(500)$, $h_c(1P)\pi$, $h_c(1P)\eta$, $\chi_{c0}(1P)\omega$, $\chi_{c2}(1P)\omega$ and tetraquark states[10] and so on. Their masses are approximately $(4.0 \sim 4.1)$ GeV. Thus it is justness to take $E_k - E_A$ as a constant for the higher mass states like $\psi(4360)$, $\psi(4415)$, and $\psi(4660)$. But for the lower mass state, such as $\psi(4040)$ or $\psi(4160)$, the value of $E_k - E_A$ is sensitive to the masses of intermediate state due to its smallness. In this case, taking $E_k - E_A$ as a constant will result in large uncertainties in the calculation. We will focus on the higher mass states in the following.

3. Results and Discussions

For the higher vector charmonium states, we study the decay properties of $\psi(4360)$, $\psi(4415)$, and $\psi(4660)$. The theoretical results are collected in Tables 1–3. It is shown that taking $\psi(4360)$ as $\psi(4^3S_1)$[11] or $\psi(3^3D_1)$,[12] the $DD^*\pi$ partial decay width can reach up to 1 MeV. The partial decay width is large enough to be observed in future experiments if $\psi(4360)$ indeed turns out to be the charmonium state $\psi(4^3S_1)$ or $\psi(3^3D_1)$.

Table 1. Partial decay widths (MeV) of the vector charmonium with a mass of 4368 MeV.

State	$\psi(4^3S_1)$	$\psi(3^3D_1)$
$\Gamma_{DD\pi}$	0.27	0.14
$\Gamma_{DD^*\pi}$	1.40	1.21
$\Gamma_{D^*D^*\pi}$	0.60	0.25
$\Gamma_{DD\eta}$	0.6 keV	0.3 keV

Table 2. Partial decay widths (MeV) of the vector charmonium with a mass of 4421 MeV.

State	$\psi\,(4^3S_1)$	$\psi\,(5^3S_1)$	$\psi\,(3^3D_1)$
$\Gamma_{DD\pi}$	0.38	0.11	0.21
$\Gamma_{DD^*\pi}$	2.01	0.96	1.84
$\Gamma_{D^*D^*\pi}$	1.07	0.59	0.52
$\Gamma_{DD\eta}$	5.4 keV	1.7 keV	2.9 keV

For the $\psi(4415)$ state, when we take it as the $\psi(4^3S_1)$ assignment, the decay widths of the $DD^*\pi$ and $D^*D^*\pi$ modes are larger than 1 MeV, while the $DD\pi$ mode is also sizable, as shown in Table 2. The predicted branching ratios of the $DD\pi$ and $DD^*\pi$ channels are consistent with the upper limits (2.2% and 11% respectively) measured by the Belle Collaboration.[13,14] We also consider the possibility of $\psi(5^3S_1)$ or $\psi(3^3D_1)$, which gives similar results. More experiments are needed to determine the inner structure of the $\psi(4415)$ state.

As to $\psi(4660)$ we have studied its decays into $\Lambda_c\bar{\Lambda}_c$ in Ref. 3. This state has many possible assignments, namely $\psi(4S, 5S, 6S)$ and $\psi(3D, 4D, 5D)$.[3]

With those possible assignments, we calculate its decay properties, as list in Table 3. From Table 3, we see that the $DD^*\pi$ and $D^*D^*\pi$ channels are always the dominant channels, and the corresponding widths can reach up to several MeV. The decay widths of $DD\rho$ and $DD\omega$ modes are also quite large. Moreover, if $\psi(4660)$ is $\psi(3D)$ state, the $DD\rho$ decay width can reach up to 1.86 MeV.

Table 3. The partial decay width (MeV) of the vector charmonium with a mass of 4643 MeV.

State	$\psi(4^3S_1)$	$\psi(5^3S_1)$	$\psi(6^3S_1)$	$\psi(3^3D_1)$	$\psi(4^3D_1)$	$\psi(5^3D_1)$
$\Gamma_{DD\pi}$	1.14	0.31	0.09	0.63	0.17	0.05
$\Gamma_{DD^*\pi}$	6.65	2.83	1.10	6.99	2.99	1.16
$\Gamma_{D^*D^*\pi}$	5.97	2.68	1.13	4.12	2.11	0.96
$\Gamma_{DD\rho}$	0.85	0.41	0.16	1.86	0.64	0.22
$\Gamma_{DD\omega}$	0.24	0.12	0.05	0.59	0.20	0.07
$\Gamma_{DD\eta}$	53.2 keV	15.3 keV	4.2 keV	29.1 keV	8.2 keV	2.2 keV
$\Gamma_{DD^*\eta}$	0.25	0.12	0.05	0.20	0.11	0.05
$\Gamma_{D^*D^*\eta}$	58.2 keV	38.7 keV	19.1 keV	8.5 keV	9.9 keV	7.5 keV
$\Gamma_{D_sD_s\eta}$	3.0 keV	0.8 keV	0.2 keV	1.6 keV	0.4 keV	0.1 keV
$\Gamma_{D_sD_s^*\eta}$	1.9 keV	1.3 keV	0.6 keV	12 eV	11 eV	7 eV

Besides the vector charmonium states, we also discuss the decay properties of $\Upsilon(10860)$ and $\Upsilon(11020)$, which are above the $BB\pi$ threshold. The predictions are presented in Table 4. For $\Upsilon(10860)$, the partial decay ratios of the $BB\pi$ and $B^*B^*\pi$ modes are consistent with the experimental data, and the partial ratio of the $BB^*\pi$ mode is also very close to the experimental data.[15] For $\Upsilon(11020)$, the decay widths of $BB\pi$, $BB^*\pi$, and $B^*B^*\pi$ are 0.34, 3.17 and 2.69 MeV, respectively. The latter two are quite large, and have a good potential to be observed by the Belle II Collaboration in the near future.

Table 4. The $B^{(*)}\bar{B}^{(*)}\pi$ partial decay widths of the vector bottomonium (MeV). $\mathcal{B}_{\rm exp}$ represents the branching ratio for each corresponding channel.

Meson	State	Mode	$\Gamma_{\rm th}$	$\mathcal{B}_i$	$\Gamma_{\rm th.}$[16]	$\mathcal{B}_{\rm exp.}$[15]
$\Upsilon(10860)$	5^3S_1	$BB\pi$	0.20	0.4%		$(0.0 \pm 1.2)\%$
		$BB^*\pi$	1.22	2.4%	(23–30) keV	$(7.3 \pm 2.3)\%$
		$B^*B^*\pi$	0.61	1.2%	(5–6.6) keV	$(1.0 \pm 1.4)\%$
$\Upsilon(11020)$	6^3S_1	$BB\pi$	0.34	0.7%		
		$BB^*\pi$	3.17	6.5%		
		$B^*B^*\pi$	2.69	5.5%		

Acknowledgments

This project is supported by the National Natural Science Foundation of China under Grants No. 11575008 and 11621131001 and National Key Basic Research Program of China (Grants No. 2015CB856700). This project is also in part supported by China Postdoctoral Science Foundation under Grant No. 2017M620492.

References

1. S. K. Choi *et al.*, *Phys. Rev. Lett.* **91**, 262001 (2003).
2. H.-X. Chen, W. Chen, X. Liu and S.-L. Zhu, *Phys. Rept.* **639**, 1 (2016).
3. L.-Y. Xiao, X.-Z. Weng, Q.-F. Lü, X.-H. Zhong and S.-L. Zhu, *Eur. Phys. J.* **C78**, 605 (2018).
4. L. Micu, *Nucl. Phys. B* **10**, 521 (1969).
5. A. Le Yaouanc, L. Oliver, O. Pene and J. C. Raynal, *Phys. Rev.* **D8**, 2223 (1973).
6. A. Le Yaouanc, L. Oliver, O. Pene and J. C. Raynal, *Phys. Rev.* **D9**, 1415 (1974).
7. E. S. Ackleh, T. Barnes and E. S. Swanson, *Phys. Rev.* **D54**, 6811 (1996).
8. X.-Z. Weng, L.-Y. Xiao, W.-Z. Deng, X.-L. Chen and S.-L. Zhu, *Phys. Rev.* **D99**, 094001 (2019).
9. M. A. Shifman, Quark hadron duality arXiv:hep-ph/0009131.
10. Y. Cui, X.-L. Chen, W.-Z. Deng and S.-L. Zhu, *HEPNP* **31**, 7 (2007).
11. J. Segovia, D. R. Entem, F. Fernandez and E. Hernandez, *Int. J. Mod. Phys.* **E22**, 1330026 (2013).
12. B.-Q. Li and K.-T. Chao, *Phys. Rev.* **D79**, 094004 (2009).
13. G. Pakhlova *et al.*, *Phys. Rev. Lett.* **100**, 062001 (2008).
14. G. Pakhlova *et al.*, *Phys. Rev.* **D80**, 091101 (2009).
15. M. Tanabashi *et al.*, *Phys. Rev.* **D98**, 030001 (2018).
16. Yu. A. Simonov and A. I. Veselov, *JETP Lett.* **88**, 5 (2008).

η decay program at GlueX

A. Somov, for the GlueX Collaboration
Jefferson Lab, Newport News, VA, USA
E-mail: somov@jlab.org

The GlueX detector in the experimental Hall D at Jefferson Lab provides a unique capability to perform a precision measurement of the $\eta \to \gamma\gamma$ decay width via the Primakoff effect and study rare decays of η mesons. Measurement of the η decay width is essential for the determination of fundamental properties, such as the ratios of the light quark masses and the η-η' mixing angle, and will provide an important test of chiral symmetry breaking in QCD. Physics of rare η decays ranges from critical tests of the chiral perturbation theory to the search for lepto-phobic dark matter candidates. We will give an overview of the ongoing PrimEx η experiment and physics program of rare η decays.

Keywords: Radiative decay width of η meson; rare η decays.

1. Introduction

The GlueX detector in the experimental Hall D at Jefferson Lab was originally designed to search for gluonic excitations in the spectra of light mesons using photon beams. The detector provides good reconstruction of both neutral and charged particles and allows for identifications of multi-particle final states. The detector started collecting data in the Spring of 2016 and has acquired a data sample of about 300 fb^{-1} since then. Two new experiments to study various decays of eta mesons with the GlueX detector have been recently approved: an experiment to perform a precision measurement of the $\eta \to \gamma\gamma$ decays width via the Primakoff effect (the PrimEx η experiment), and an experiment to study η rare decays, the Jefferson Eta Factory (JEF). These experiments will be described below.

2. PrimEx η Experiment

The PrimEx η experiment will measure the cross section of the $\eta \to \gamma\gamma$ decays as a function of the η production angle. The resonant production of mesons in the Coulomb field of a nucleus by photons is known as the

Primakoff effect. The cross section of this process on a zero-spin nucleus is related to the η decay width, $\Gamma_{\gamma\gamma}$. The two-photon decay width will be extracted from the cross section measurement with a precision of 3.2%. The PrimEx η results will have a significant impact on various physics topics:

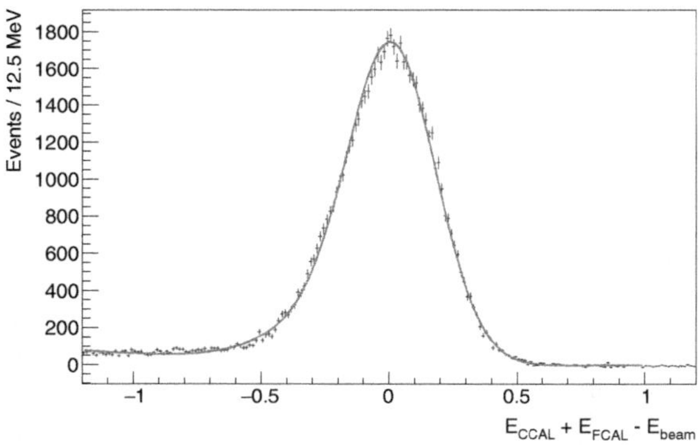

Fig. 1. Elasticity distribution of reconstructed Compton candidates.

- The measurement will significantly improve the world's average value of $\Gamma_{\gamma\gamma}$[1] and will help to understand some discrepancies between $\Gamma_{\gamma\gamma}$ obtained from e^+e^- collider experiments and the Primakoff experiment at Cornell.[2]
- The measurement will improve partial widths of all other η decays, which are determined using the measured $\Gamma_{\gamma\gamma}$ and the corresponding branching ratios. The branching ratios are typically known with a relatively good precision.
- Extraction of the partial width of $\eta \to \pi^0\pi^0\pi^0$ and $\eta \to \pi^+\pi^-\pi^0$ decays is of particular interest. These decays proceed through the isospin breaking generated by the mass difference $m_d - m_u$ and can be used for the determination of the light-quark mass ratio $(m_u - m_d)/m_s$.[3]
- The SU(3) symmetry breaking induced by m_s gives rise to the mixing of η and η'. The precision measurement of the η radiative decay width will provide an important input for the determination of the $\eta - \eta'$ mixing angle.

The PrimEx η experiment started collecting data in Spring of 2019 using a He target and has acquired about 30% of the required statistics. The cross section will be normalized using the Compton process, which will also be used to monitor the luminosity and control the detector stability during the run. In order to reconstruct electrons and photons at small angles originating from Compton events, the GlueX detector was instrumented with a small (24 cm x 24 cm) lead tungstate ($PbWO_4$) calorimeter, positioned downstream of the GlueX forward calorimeter (FCAL). The elasticity distribution, defined as the energy of two reconstructed clusters minus the beam energy, for Compton candidates is presented in Fig. 1. The Compton calorimeter was also the prototype of a large-scale calorimeter, which will be used to upgrade the FCAL. This upgrade is required for the JEF experiment, described in Section 3.

3. JEF Experiment

The primary goal of the JEF experiment is to perform precision measurements of various η and η' decays mainly reconstructed in neutral modes. The experiment requires an upgrade of the inner part (80 cm x 80 cm) of the GlueX lead glass forward calorimeter with high-granularity, high-resolution $PbWO_4$ crystals. The calorimeter will improve the separation of clusters in the forward direction and the energy resolution of reconstructed photons by about a factor of two. The experiment is tentatively scheduled to start data taking in 2023 and will run concurrently with the GlueX spectroscopy program. The main physics topics considered for JEF are listed in Table 1 and described below:

Table 1. The η decays highlighted in the JEF experiment.

Mode	Branching Ratio	Physics Highlight	Photons
$\gamma + B'$	beyond SM	dark vector boson	4
$\pi^0 + \phi'$	beyond SM	dark scalar boson	4
$\pi^0 2\gamma$	$(2.7 \pm 0.5) \times 10^{-4}$	χPTh at $\mathcal{O}(p^6)$	4
$3\pi^0$	$(32.7 \pm 0.2)\%$	$m_u - m_d$	6
$\pi^+\pi^-\pi^0$	$(22.9 \pm 0.3)\%$	$m_u - m_d$, CV	2
3γ	$< 1.6 \times 10^{-5}$	CV, CPV	3
ancillary:			
4γ	$< 2.8 \times 10^{-4}$	$< 10^{-11}$	4
$2\pi^0$	$< 3.5 \times 10^{-4}$	CPV, PV	4
$2\pi^0\gamma$	$< 5 \times 10^{-4}$	CV, CPV	5
$3\pi^0\gamma$	$< 6 \times 10^{-5}$	CV, CPV	7
$4\pi^0$	$< 6.9 \times 10^{-7}$	CPV, PV	8
$\pi^0\gamma$	$< 9 \times 10^{-5}$	CV, Ang. mom. viol.	3

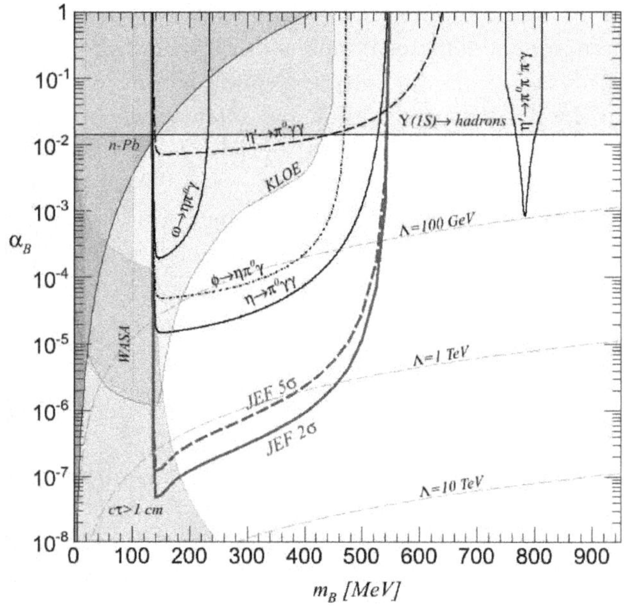

Fig. 2. Limits on leptophobic gauge B-boson coupling α_B as a function of the mass $m_{\rm B}$, described in Ref. 5. Projected JEF search regions are superimposed on the plot.

- A search of the leptophobic dark B boson, which couples predominantly to quarks. The search will be performed using $\eta \to \gamma + B(\to \gamma + \pi^0)$ channel,[4,5] which will cover the B boson mass range 0.14 - 0.54 GeV. The JEF sensitivity to the baryonic fine structure constant $\alpha_{\rm B}$ is about two orders of magnitude better than the existing bound and corresponds to 10^{-7} (see Fig. 2). In the η' mass range, the B-boson will be searched using $\eta' \to B\gamma(\to \pi^+\pi^-\pi^0)$ decays.
- A search of the electrophobic dark scalar[6] ϕ' in the decays $\eta \to \pi^0 + \phi'(\to \gamma + \gamma)$ in the mass range between 0.01 GeV and 0.4 GeV.
- Perform measurements of the Dalitz distribution of the $\eta \to \pi^0\gamma\gamma$ decays. The measurements will allow to determine two low-energy constants in the $O(p^6)$ chiral Lagrangian[7] and study the contribution of scalar resonances in this channel. The highly boosted $\eta's$ produced in Hall D is expected to have significantly smaller background compared with the previous measurements by low-energy experiments. The projected invariant mass distribution of the reconstructed 4γ events is shown in Fig. 3.

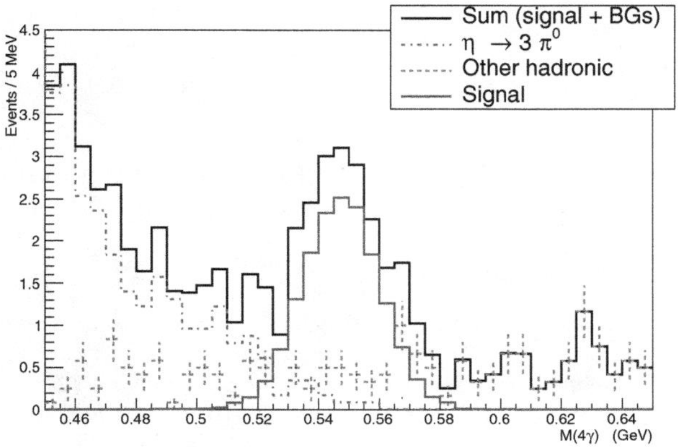

Fig. 3. Four photon invariant mass distributions for the signal $\eta \to \pi^0\gamma\gamma$ events (blue solid curve), $\eta \to 3\pi^0$ background (dash-dotted curve), and the hadronic background predicted by PYTHIA (red dashed curves). The black solid curves correspond to the sum of the signal and backgrounds. All yields are normalized to 1 day of taking data with GlueX.

Acknowledgment

This material is based upon work supported by the U.S. Department of Energy, Office of Science, Office of Nuclear Physics under contract DE-AC05-06OR23177.

References

1. M. Tanabashi *et al.*, Phys. Rev. D **98**, no. 3, 030001 (2018).
2. A. Browman, J. DeWire, B. Gittelman, K. M. Hanson, E. Loh and R. Lewis, Phys. Rev. Lett. **32**, 1067 (1974).
3. H. Leutwyler, Phys. Lett. B **374**, 181 (1996). A. V. Anisovich and H. Leutwyler, Phys. Lett. B **375**, 335 (1996). H. Leutwyler, Phys. Lett. B **378**, 313 (1996).
4. A. E. Nelson and N. Tetradis, Phys. Lett. B **221**, 80 (1989).
5. S. Tulin, Phys. Rev. D **89**, no. 11, 114008 (2014).
6. Y. S. Liu, D. McKeen and G. A. Miller, Phys. Rev. Lett. **117**, no. 10, 101801 (2016). Y. S. Liu, I. C. Clot and G. A. Miller, Nucl. Phys. B, 114638 (2019).
7. L. Ametller, *et al*, Phys. Lett. B **276**, 185 (1992).

Photoproduction of the S-, P- and D-wave resonances on protons in the $\pi^+\pi^-$ channel

Ł. Bibrzycki

Institute of Computer Science, Pedagogical University of Kraków, 30-084 Kraków, Poland
Physics Department, Indiana University, Bloomington, IN 47405, USA,
Theory Center, Thomas Jefferson National Accelerator Facility, USA
E-mail: lukasz.bibrzycki@up.krakow.pl

P. Bydžovský

Nuclear Physics Institute, CAS, 25068 Řež, Czech Republic

R. Kamiński

Institute of Nuclear Physics, Polish Academy of Sciences, 31-342 Kraków, Poland

A. P. Szczepaniak

Physics Department, Indiana University, Bloomington, IN 47405, USA
Center for Exploration of Energy and Matter, Indiana University, Bloomington, IN 47403, USA
Theory Center, Thomas Jefferson National Accelerator Facility, USA

We show that the photoproduction of resonances in the $\pi^+\pi^-$ system can be described by a combination of two production mechanisms: the long range production due to one pion exchange known in the literature as the Deck mechanism and the short range production related to exchanges of heavier particles and quark/gluon processes. The model with these components very well reproduces experimental mass distributions for partial waves S, P and D.

Keywords: Photoproduction; partial wave analysis; resonances.

1. Introduction

The understanding of spectrum of resonances and the mechanisms of their production are in direct relation to the fundamental properties of QCD like the confinement. Photoproduction processes provide a favourable environment to study hadron spectroscopy with particular stress put on production of exotic states.[1] Reliable photoproduction models are also badly needed in order to describe a wealth of data to be expected in near future from experiments at JLab, ELSA, MAMI, BESIII and SPring-8. Here we concentrate on the $\gamma p \to \pi^+\pi^- p$ reaction in the context of previous[2] and ongoing photoproduction experiments conducted at JLab.

2. Model description

To define the basic notation for our process of interest

$$\gamma(q,\lambda) + p(p_1,\lambda_1) \to p(p_2,\lambda_2) + \pi^+(k_1) + \pi^-(k_2) \tag{1}$$

we write the relation between invariant amplitude and the S-matrix as

$$S_{fi} = \delta_{fi} + i(2\pi)^4\,\delta^4(p_2 + k_1 + k_2 - p_1 - q)\,\mathcal{T}_{fi} \tag{2}$$

where λ's denote particle helicities. The invariant double-differential cross section expressed in terms of $\pi\pi$ partial waves is given by

$$\frac{d^2\sigma}{d|t|\,d\sqrt{s_{\pi\pi}}} = \frac{1}{64(2\pi)^4}\frac{|\mathbf{k}|}{(s-m^2)^2}\sum_{lm}\sum_{\lambda_2\lambda\lambda_1}|\mathcal{T}^{lm}|^2\,, \tag{3}$$

where $|\mathbf{k}| = \sqrt{s_{\pi\pi}/4 - m_\pi^2}$ is the magnitude of pion momenta in the $\pi\pi$ rest frame and the partial wave expansion is performed in the helicity frame.[3,4] The q, p_1, p_2, k_1 and k_2 denote respectively the 4-momenta of photon, initial proton, final proton, positive pion and negative pion. The partial wave amplitudes used in Eq.(3) include both long range (pion exchange) and short range $\pi^+\pi^-$ production amplitudes which play the role of Born terms as well as the final state $\pi\pi$ rescattering term. The one pion exchange photoproduction amplitude does not conserve the electromagnetic current so the respective Deck amplitude includes pion pole modification (third term in Eq.(4)), so that gauge invariance is properly imposed.[5]

$$M_{\rm Deck}^{\lambda_2\lambda\lambda_1} = e\left[\frac{\epsilon_\lambda\cdot k_2}{q\cdot k_2}T^-_{\lambda_1\lambda_2} - \frac{\epsilon_\lambda\cdot k_1}{q\cdot k_1}T^+_{\lambda_1\lambda_2} + \frac{\epsilon_\lambda\cdot(p_1+p_2)}{q\cdot(p_1+p_2)}(T^+_{\lambda_1\lambda_2} - T^-_{\lambda_1\lambda_2})\right] \tag{4}$$

In Eq.(4) e is the electric charge, ϵ_λ is the photon polarization vector and $T^+_{\lambda_1\lambda_2}$ and $T^-_{\lambda_1\lambda_2}$ are $\pi^\pm N$ scattering amplitudes (for this purpose we have used the SAID parametrisation[6]). As shown in[7] the final state interactions can be included in a partial wave projected photoproduction amplitude in a unitary way if the long range (Deck) and short range contributions are combined as

$$\mathcal{T}^{lm} = M^{lm}_{\rm Deck}e^{i\delta_l}\cos\delta_l + M^{lm}_{\rm SR}e^{i\delta_l}\sin\delta_l, \tag{5}$$

where we have adopted the phase shifts δ_l from[8] and the short range contribution for each partial wave is expressed by

$$M_{SR} = A_l + B_l\,s_{\pi\pi} \tag{6}$$

(helicity indices have been dropped for brevity). Such a simple form of M_{SR} is justified since the short range exchanges correspond to remote singularities which do not produce strong $s_{\pi\pi}$ dependence in the physical region.

3. Numerical results

In order to compare our model with the mass distributions for low partial waves determined by the CLAS collaboration,[2] we fitted the A and B parameters used in Eq.(6) to the data. In what follows, the full model containing both the short range and long range contributions along with the $\pi\pi$ final state interactions is labeled as "Deck+FSI+short range". In Fig. 1

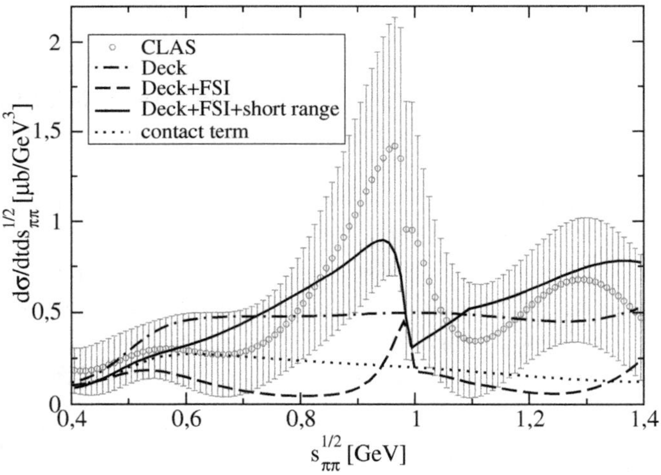

Fig. 1. $S-$wave double differential cross section at E_γ=3.3 GeV and $-t$=0.55 GeV^2. Dash-dotted line - pure Deck model; dashed line - Deck model with final state $\pi\pi$ interactions; solid line - Deck model with FSI and the short range term; dotted line - contribution of the contact term; red points - CLAS fit to the experimental data. The error band shows the total uncertainty that combines the systematic and statistical uncertainties (color online).

we show the model curve for the $S-$wave mass distribution and compare it with the CLAS fit.[2] The model's behaviour slightly differs from CLAS data above 1 GeV which can be attributed to the absence of the $K\bar{K}$ channel. The P-wave mass distribution provided by the model (Deck+FSI+ short

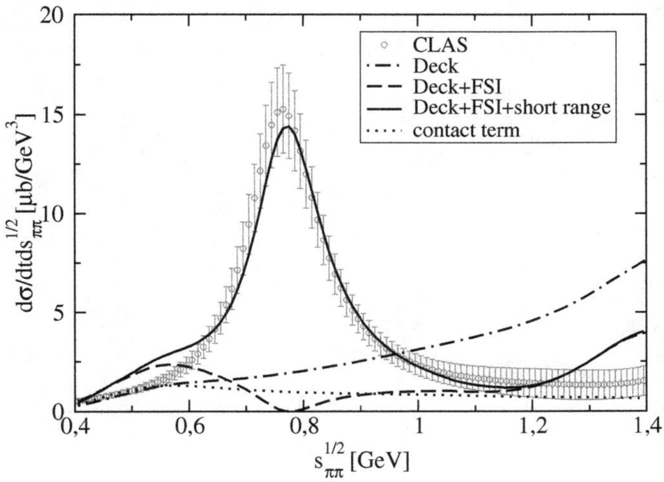

Fig. 2. P−wave double differential cross section at E_γ=3.3 GeV and $-t$=0.55 GeV^2. Dash-dotted line - pure Deck model; dashed line - Deck model with final state $\pi\pi$ interactions; solid line - Deck model with FSI and the short range term; dotted line - contribution of the contact term; red points - CLAS fit to the experimental data. The band shows the total uncertainty of the fit (color online).

range), especially in the resonance region, is in agreement with data, as shown in Fig. 2. The agreement is mainly due to the short range production which implies, as expected, that the P-wave resonance $\rho(770)$ is the standard $q\bar{q}$ state.

Fig. 3 shows that inclusion of the final state interactions in the Deck D-wave amplitude, similarly as in the $P-$wave, results in developing the minimum rather than the maximum for the invariant masses around the $f_2(1270)$. The minimum in the D wave is due to the $\pi\pi$ phase shift passing $\pi/2$ at about 1.25 GeV. This observation is compatible with previous analyses.[3] The model develops the maximum in the D-wave if the short range component is included (see[4] for values of A and B parameters).

4. Summary and outlook

The photoproduction of $\pi^+\pi^-$ resonances in low partial waves can be treated as a superposition of two production modes where the production region is either diffuse or compact. Such description is successful in

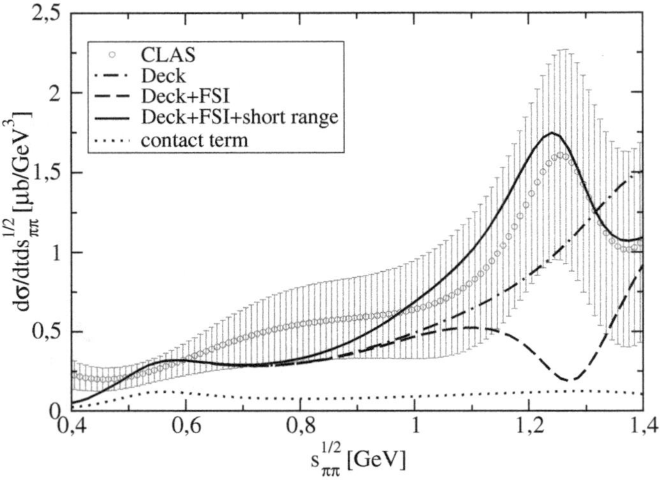

Fig. 3. $D-$wave double differential cross section at E_γ=3.3 GeV and $-t$=0.55 GeV^2 with $M \leq 1$. Dash-dotted line - pure Deck model; dashed line - Deck model with final state $\pi\pi$ interactions; solid line - Deck model with FSI and the short range term; dotted line - contribution of the contact term; red points - CLAS fit to the experimental data. The band shows the total uncertainty of the fit (color online).

description of mass distributions in the S, P and D waves which are dominated by $f_0(980)$, $\rho(770)$ and $f_2(1270)$ resonances respectively. The large relative magnitude of the short range components of P and D partial wave amplitudes agrees with the expectation that resonances in these waves are conventional $q\bar{q}$ states. The compact source component of the $f_0(980)$ photoproduction amplitde is smaller which implies that this resonance is rather a weakly bound tetraquark state. Authors' further theoretical effort is directed towards describing the short range production component in a more microscopically motivated way.

Acknowledgments

This work was supported by the Polish Science Center (NCN) Grant No. 2018/29/B/ST2/02576.

References

1. A.P. Szczepaniak, M. Swat, *Phys.Lett. B* **516**, 72 (2001).

2. M. Battaglieri *et al.* (CLAS collaboration), *Phys. Rev. D* **80**, 072005 (2009).
3. Ł. Bibrzycki, R. Kamiński, *Phys. Rev. D* **87**, 114010 (2013).
4. Ł. Bibrzycki, P. Bydžovský, R. Kamiński and A. P. Szczepaniak, *Phys. Lett. B* **789**, 287 (2019).
5. J. Pumplin, *Phys. Rev. D* **2** 1859, (1970).
6. R.L. Workman, R.A. Arndt, W.J. Briscoe, M.W. Paris, I.I. Strakovsky, *Phys. Rev. C* **86** 035202, (2012); (SAID) http://gwdac.phys.gwu.edu/.
7. I.J. Aithchison, M.G. Bowler, *J. Phys. G* 3 1503 (1977).
8. P. Bydžovský, R. Kamiński and V. Nazari, *Phys. Rev. D* **94**, 116013 (2016).

Baryon-baryon interactions in chiral perturbation theory

X.-L. Ren

Ruhr University Bochum, Faculty of Physics and Astronomy, Institute for Theoretical Physics II, D-44870 Bochum, Germany
E-mail: xiulei.ren@rub.de

We report on a study of baryon-baryon scattering by applying time-ordered perturbation theory to the covariant chiral perturbation theory. Through working out the rules of time-ordered diagrams, we built up a systematic framework to investigate the chiral potentials in the SU(3) sector. At leading order, we found that the baryon-baryon potentials are perturbatively renormalizable, and the corresponding integral equations have unique solutions in all partial waves. We also discussed the issue of additional finite subtractions required, e.g., in the $^3P_0$ partial waves to improve the ultraviolet convergence of loop integrals for the nucleon-nucleon and lambda-nucleon scatterings.

Keywords: Nucleon-nucleon interactions; chiral Lagrangians; renormalization.

1. Introduction

Nuclear systems, especially with non-vanishing strangeness, provide a necessary connection between nuclear, particle and astrophysics. Thus, as the fundamental interaction, the nucleon-nucleon (NN), hyperon-nucleon (YN), and hyperon-hyperon (YY) forces are important and have been extensively studied.

Chiral perturbation theory (ChPT), as an effective field theory of low-energy QCD, provides a model-independent framework to investigate the baryon-baryon (BB) interactions, as shown in the pioneering work by Weinberg.[1,2] Currently, the chiral NN interaction has been formulated up to high orders.[3–5] Due to the scarce of the experimental data, the studies of YN interactions have been performed only up to next-to-leading order (NLO).[6] In 2012, a modified Weinberg approach to NN scattering has been proposed in Ref. 7. This novel framework employs time-ordered perturbation theory (TOPT) and relies on the manifestly Lorentz-invariant effective Lagrangian. It has already been explored for the NN and YN scattering.[8–10,a]

[a]An alternative power counting by keeping the full form of nucleon spinor has been

However, a systematic approach to the SU(3) sector using TOPT, which would allow a straightforward generalization beyond the leading order, has not been formulated. In Ref. 14, we filled this gap and worked out in details the TOPT diagrammatic rules. As a simple application, the phase shifts of NN scattering were presented. Most recently, we have also extended the application to the lambda-nucleon scattering in Ref. 15.

2. Baryon-baryon interactions at leading order

According to the Weinberg power counting, we can calculate the chiral potential, which is defined as a sum of two-baryon irreducible contributions of time-ordered diagrams. The corresponding scattering amplitude can be obtained by solving the coupled-channel generalization of the Kadyshevsky equation, $T = V + VGT$, which is self-consistently derived in TOPT.

At leading order, the BB interactions contain the contact terms, which are identical to those of the non-relativistic approach, see Refs. 16, 17, and one-meson-exchange potential (OMEP), which reads as

$$
\begin{aligned}
V_{\mathrm{LO},M_P}^{IJ,KL} &= -\frac{f_{IKP} f_{JLP} \mathcal{I}_{IJ,KL}}{2\,\omega(\vec{q})} \left[\frac{1}{\omega(\vec{q}) + \omega(\vec{p}_K) + \omega(\vec{p}_J) - E - i\,\epsilon} \right. \\
&+ \left. \frac{1}{\omega(\vec{q}) + \omega(\vec{p}_L) + \omega(\vec{p}_I) - E - i\,\epsilon} \right] \frac{(m_I + m_K)\,(m_J + m_L)}{\sqrt{m_I m_J m_K m_L}} \\
&\times \frac{(m_K \vec{\sigma}_1 \cdot \vec{p}_I - m_I \vec{\sigma}_1 \cdot \vec{p}_K)\,(m_L \vec{\sigma}_2 \cdot \vec{p}_J - m_J \vec{\sigma}_2 \cdot \vec{p}_L)}{\sqrt{\omega\,(\vec{p}_I) + m_I}\,\sqrt{\omega\,(\vec{p}_J) + m_J}\,\sqrt{\omega\,(\vec{p}_K) + m_K}\,\sqrt{\omega\,(\vec{p}_L) + m_L}},
\end{aligned}
\tag{1}
$$

for $B_I(\vec{p}_I) + B_J(\vec{p}_J) \to B_K(\vec{p}_K) + B_L(\vec{p}_L)$ scattering process with the momentum shift $\vec{q}$. The isospin factors $\mathcal{I}_{IJ,KL}$ and the values of f_{IKP} can be found in Ref. 16. The distinctive feature of our OMEP as compared to its non-relativistic analog is that its iterations within the integral equation lead to ultraviolet finite diagrams. As a consequence, the ultraviolet cutoff can be safely removed at LO. Furthermore, our OMEP also results in the unique solutions of the BB scattering amplitudes for all partial waves.

3. Phase shifts of nucleon-nucleon scattering

After performing the partial wave decomposition of the potential and the integral equation, we can obtain the phase shifts. Here, we would like to

---

suggested in Ref. 11 and applied to the YN and YY systems.[12,13].

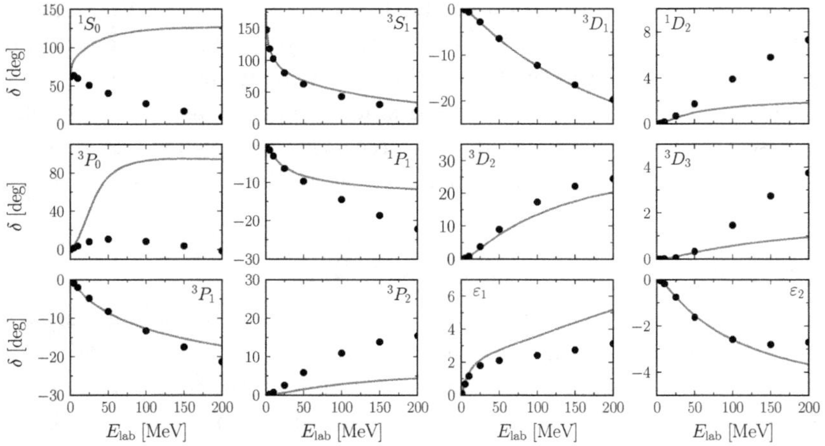

Fig. 1. Nucleon-nucleon phase shifts and mixing angles at LO. Lines correspond to our results while solid dots refer to the Nijmegen partial wave analysis.[18]

take NN scattering as an example to present our results. The study of hyperon-nucleon interactions, e.g. ΛN system, is given in Ref. 15.

In Fig. 1, the phase shifts of NN scattering are presented. For ${}^1S_0$ and ${}^3S_1$ waves, the contact terms are fixed by reproducing the scattering lengths, respectively. Our LO calculation provides a reasonable description of the empirical phase shifts, except the cases of the ${}^1\mathrm{S}_0$ and ${}^3\mathrm{P}_0$ partial waves. These large discrepancies suggest that at least some part of the subleading corrections must be treated non-perturbatively.

To improve the description of the ${}^3P_0$ phase shift, for example, we include the lowest-order contact term in the LO potential,

$$V_{\mathrm{LO}}^{{}^3P_0}(p',p) = C\,p'p + V_\pi \equiv V_C + V_\pi, \tag{2}$$

where V_π stands for the OPEP of Eq. (1) in the ${}^3P_0$ wave. Since V_C is separable, the scattering amplitude can be written as

$$T = T_\pi + (1 + T_\pi\, G)\,\bar{\xi}\,\mathcal{X}\,\xi(1 + G\,T_\pi), \tag{3}$$

where $\mathcal{X} = \left[C^{-1} - \xi\, G\,\bar{\xi} - \xi\, G\, T_\pi G\,\bar{\xi}\right]^{-1}$, $\xi(p) = p$, $\bar{\xi}(p') = p'$, and the amplitude T_π satisfies the equation $T_\pi = V_\pi + V_\pi\, G\, T_\pi$.

Next, we use the subtractive renormalization to obtain the renormalized amplitude,

$$T(p',p) = T_\pi(p',p) + \frac{\bar{\Xi}(p')\,\Xi(p)}{\frac{1}{C_R} - (\xi\, G\,\bar{\xi})^R - (\xi\, G\, T_\pi G\,\bar{\xi} - \alpha)}, \tag{4}$$

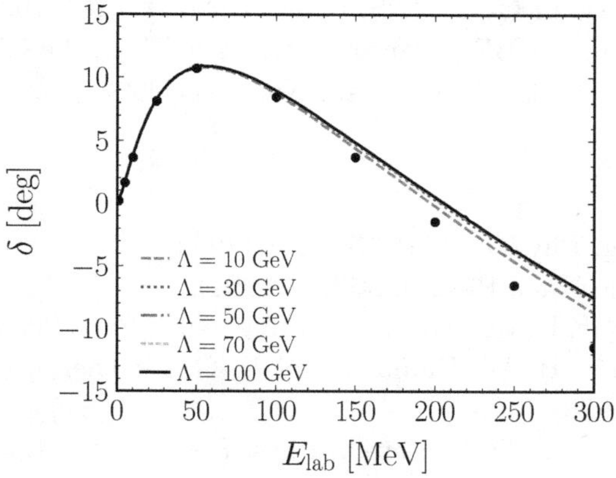

Fig. 2. Cutoff-dependence of the calculated $^3P_0$ phase shifts.

with $\bar{\Xi}(p') = (1 + T_\pi\, G)\,\bar{\xi}$ and $\Xi(p) = \xi(1 + G\,T_\pi)$. The α subtracts the overall divergence of $\xi\, G\, T_\pi G\,\bar{\xi}$. In numerical evaluation, we fix the bare coupling $1/C = 1/C_R - \alpha$ via the low-energy phase shift of $^3P_0$ varying the momentum cutoff from 10 GeV to 100 GeV. The resulting phase shift in the $^3P_0$ channel is plotted in Fig. 2 and shows a very good description of the data. The cutoff independent behavior indicates the achievement of the renormalized amplitude. The similar studies of NN $^1S_0$ partial wave and the related partial waves in the lambda-nucleon scattering can be found in Refs. 9, 15.

4. Conclusion

In this talk we presented the leading order studies of nucleon-nucleon and hyperon-nucleon scatterings by applying time-ordered perturbation theory to the Lorenz-invariant chiral effective Lagrangians. We found that the chiral potentials at leading order are renormalizable and the scattering equations have unique solutions for all partial waves. The established formalism can be used to study baryon-baryon scatterings based on a renormalizable formulation with the corrections beyond LO treated perturbatively.

Acknowledgment

The author acknowledges the financial support from the Hadron 2019 conference. This work was supported in part by DFG and NSFC through funds provided to the Sino-German CRC 110 “Symmetries and the Emergence of Structure in QCD" (NSFC Grant No. 11621131001, DFG Grant No. TRR110), the BMBF (Grant No. 05P18PCFP1), the GSRNSF (Grant No. FR17-354), and the RSF (Grant No. 18-12-00226).

References

1. S. Weinberg, Phys. Lett. B **251**, 288 (1990).
2. S. Weinberg, Nucl. Phys. **B363**, 3 (1991).
3. P. F. Bedaque, U. van Kolck, Ann. Rev. Nucl. Part. Sci. **52**, 339 (2002).
4. E. Epelbaum, H. W. Hammer and U.-G. Meißner, Rev. Mod. Phys. **81**, 1773 (2009).
5. R. Machleidt and D. R. Entem, Phys. Rept. **503**, 1 (2011).
6. J. Haidenbauer, S. Petschauer, N. Kaiser, U.-G. Meißner, A. Nogga and W. Weise, Nucl. Phys. A **915**, 24 (2013).
7. E. Epelbaum and J. Gegelia, Phys. Lett. B **716**, 338 (2012).
8. E. Epelbaum, A. M. Gasparyan, J. Gegelia and M. R. Schindler, Eur. Phys. J. A **50**, 51 (2014).
9. E. Epelbaum, A. M. Gasparyan, J. Gegelia and H. Krebs, Eur. Phys. J. A **51**, 71 (2015).
10. K. W. Li, X. L. Ren, L. S. Geng and B. Long, Phys. Rev. D **94**, 014029 (2016).
11. X.-L. Ren, K. W. Li, L. S. Geng, B. W. Long, P. Ring and J. Meng, Chin. Phys. C **42**, 014103 (2018).
12. K. W. Li, X.-L. Ren, L. S. Geng and B. W. Long, Chin. Phys. C **42**, 014105 (2018).
13. K. W. Li, T. Hyodo and L. S. Geng, Phys. Rev. C **98**,065203 (2018)
14. V. Baru, E. Epelbaum, J. Gegelia and X.-L. Ren, Phys. Lett. B **798**, 134987 (2019).
15. X.-L. Ren, E. Epelbaum and J. Gegelia, arXiv:1911.05616 [nucl-th].
16. H. Polinder, J. Haidenbauer and U.-G. Meißner, Nucl. Phys. A **779**, 244 (2006).
17. H. Polinder, J. Haidenbauer and U.-G. Meißner, Phys. Lett. B **653**, 29 (2007).
18. V. G. J. Stoks, R. A. M. Klomp, M. C. M. Rentmeester and J. J. de Swart, Phys. Rev. C **48**, 792 (1993).

Simulation study of the $\bar{p}p \to \bar{\Sigma}^0\Lambda$ reaction at $\bar{\text{P}}$ANDA at FAIR

G. Perez-Andrade[1*] , Karin Schönning[2], W.I. Andersson[2], Jenny Regina[2] and Michael Papenbrock[2] on behalf of the $\bar{\text{P}}$ANDA Collaboration

[1] *IKP, Forschungszentrum Jülich,, Jülich, Germany Jülich, 52428, Germany*

[2] *Dept. of Physics and Astronomy, Uppsala University, Uppsala, Sweden Uppsala, 75237, Sweden*

[*] *E-mail: g.perez@fz-juelich.de*

The $\bar{\text{P}}$ANDA[a] experiment is one of the pillars of the new Facility for Antiproton and Ion Research,[1] currently under construction in Darmstadt, Germany. $\bar{\text{P}}$ANDA will be a fixed-target experiment which will study non-perturbative strong interaction phenomena in antiproton-proton collisions in the beam momentum range of 1.5 - 15 GeV/c. Strangeness production will be addressed through $\bar{p}p \to$ *Hyperon-Antihyperon* processes. The results of a feasibility study to measure the $\bar{p}p \to \bar{\Sigma}^0\Lambda$ reaction, at antiproton beam momenta of 1.771 GeV/c and 6 GeV/c are presented and discussed in this paper.

Keywords: Hyperon; non-perturbative QCD; PANDA experiment.

1. Introduction

To achieve a better understanding of the strong interaction mechanism, a coherent explanation of the strong coupling (α_s) over a wide range of momentum transfer (Q^2) is needed. At high Q^2, α_s is weak enabling the use of perturbative Quantum Chromodynamics (pQCD). At intermediate and low Q^2 where quarks are confined into hadrons, α_s strongly grows so that pQCD is no longer effective. The fact that different strategies must be used at certain Q^2 ranges, in the non-perturbative (confinement) domain,[2] remains as one of the open topics in modern physics.

1.1. *Hyperons*

Hyperons/antihyperons ($Y/\bar{Y}$) are baryons containing at least one strange quark ($s/\bar{s}$). Strangeness production is controlled by $m_s \sim 100$ MeV which

[a] antiProton ANnhilation at DArmstadt.

is close to the QCD cut-off ($\Lambda_{QCD} \sim 200$ MeV) and therefore, reactions containing hyperons test the confinement scale. Moreover, most ground state hyperons decay weakly, emitting their daughter particles in accordance with the mother hyperon spin vector. As a consequence, measuring the decay angular distributions gives access to some hyperon spin observables, which contain information on the production process.

Single-strangeness hyperon reactions of the form $\bar{p}p \rightarrow \bar{Y}Y$ were explored by the PS185 collaboration at CERN[3–6] up to $\sim$ 6 GeV/c. The results constituted a basis for the development and testing of different theoretical models for strangeness production based on: i) meson exchange (MEX),[7] ii) quark-gluon degrees of freedom (QG)[8] and iii) the combination of i) and ii).[9] These models describe the total and differential cross-sections, however they have different degrees of freedom, mainly describe the single-strangeness sector and lack a complete spin dynamics description. These reasons motivate new experimental measurements in the single and multi-strangeness sector in a wider beam momenta range.

2. The $\bar{\text{P}}$ANDA experiment

The High Energy Storage Ring (HESR) will provide luminosities of $\mathcal{L} \sim 10^{31}$ cm^{-2}s^{-1} in the starting stage (Day-One), until it achieves the desired $\mathcal{L} \sim 2{\times}10^{32}$ cm^{-2}s^{-1} at a later stage. The $\bar{\text{P}}$ANDA layout includes a *Target* and a *Forward Spectrometer* for a nearly 4π-solid angle coverage.[10] Each spectrometer consists of state-of-the-art detector systems for particle tracking and momentum reconstruction, particle identification (PID), calorimetry and muon detection. The $\bar{\text{P}}$ANDA physics program includes a wide range of hyperon topics including hyperon production, spectroscopy and structure.[10]

Feasibility simulation studies of exclusive event reconstruction for single and multi-strangeness channels are currently being performed for $\bar{\text{P}}$ANDA, using a dedicated simulation and analysis framework with features such as ideal pattern recognition and PID. Cross section distributions based on experimental data have been used wherever available for the data generation. Model predictions are used otherwise.

3. Analysis strategy

The reaction $\bar{p}p \rightarrow \bar{\Sigma}^0\Lambda$ was studied at $p_{beam} = 1.771$ GeV/c and 6 GeV/c. These beam momenta were chosen to allow a comparison with previous measurements and other $\bar{\text{P}}$ANDA simulation studies. Each sample contained ~10k events and the simulations were performed using decay models based on the experimental results from PS185.[4,6] The analysis strategy is described here and details can be found in Ref. 11

We study $\bar{\Sigma}^0\Lambda$ where $\bar{\Sigma}^0 \to \bar{\Lambda}\gamma$ and subsequently $\bar{\Lambda} \to \bar{p}\pi^+$, whereas $\Lambda \to \pi^- p$. At both beam momenta, the event *pre-selection* consists of i) the final state particles identification, ii) a photon selection based on the reaction kinematics and iii) the combination of the $\bar{p}\pi^+/p\pi^-$ and a vertex fit of the decay tracks to obtain the $\bar{\Lambda}/\Lambda$ candidates. The only significant background source is false combinations.

In the *final selection* at p_{beam} = 1.771 GeV/c, the $\bar{\Sigma}^0$ candidates are identified and combined with pre-selected Λ candidates. Then, conservation of the initial $\bar{p}p$ system four-momentum is required for the $\bar{\Sigma}^0\Lambda$ system. At p_{beam} = 6 GeV/c, a $\bar{\Sigma}^0$ candidates selection prior to their combination with the Λ candidates is performed. Such is based on the photons energy boosted to their corresponding $\bar{\Sigma}^0$ candidate rest frame. In this case, four-momentum conservation is also imposed as the final step.

At both beam momenta, a background sample ($\bar{p}p \to$ *anything*) of 10^7 events was obtained with the DPM generator.[12] Due to the large $\bar{\Lambda}\Lambda$ channel cross-section[3] at p_{beam} = 1.771 GeV/c, the analysis strategy was also tested on an independent background sample of 10^5 $\bar{\Lambda}\Lambda$ events, generated with a decay model based on experimental data at this beam momentum. At p_{beam} = 6 GeV/c the reaction threshold for higher mass $\bar{Y}Y$ states is surpassed,[3] all including a $\bar{\Lambda}\Lambda$ in the final state. To evaluate how the selection criteria works in reconstructing such channels, all events containing $\bar{\Lambda}/\Lambda$ hyperons were removed from the full DPM sample in a second background sample hereby called *DPM-filtered.* For both beam momenta, the background sources motivated additional mass window cuts, and at p_{beam} = 6 GeV/c a cut in the $\bar{\Lambda}$ decay vertex position was added.

4. Results

The final number of events, the signal to background ratio (S/B) and reconstruction efficiencies (ϵ) for signal and background samples are summarized in Table 1 for p_{beam} = 1.771 GeV/c (left) and p_{beam} = 6 GeV/c (right). In cases where no events survive the selection, a limit of events given by the 90% confidence level is used to account for statistical fluctuations.

At p_{beam} = 1.771 GeV/c the DPM sample is fully removed while only a few $\bar{\Lambda}\Lambda$ events remain, showing that the applied criteria is successful in selecting signal events. At p_{beam} = 6 GeV/c the DPM sample was fully suppressed, while a few DPM-filtered (DPM-filt) events survived the selection. The latter highlights the importance of testing this strategy in other independent hyperon channel samples. At both p_{beam}, high S/B ratios (purity) were obtained. The latter is important given that pure samples are required for spin observables reconstruction. False combinations (Combi) are well suppressed in both beam momenta cases, but are the highest contribution to the total background. This is attributed to neutrons being wrongly identified as photons, which could be solved with an additional selection based on the photons recorded time and position.[13]

Table 1. Reconstruction efficiencies for samples at p_{beam} =1.771 and 6 GeV/c.

	p_{beam} = 1.771 GeV/c			p_{beam} = 6 GeV/c		
Channel	Events	S/B	$\epsilon(\%)$	Events	S/B	$\epsilon(\%)$
$\bar{\Sigma}^0\Lambda$	526 ± 23	—	5.3 ± 0.2	614 ± 25	—	6.1 ± 0.3
Combi	38	14	0.38	111	5.5	1.1
DPM	< 50.6	> 11	$< 5.1 \times 10^{-4}$	30	20.7	3.0×10^{-6}
$\bar{\Lambda}\Lambda$	4	120	4.0×10^{-5}	—	—	—
DPM-filt	—	—	—	< 18	>34.7	$< 3.9 \times 10^{-6}$

In addition to this study, other $\bar{\text{P}}$ANDA feasibility studies for different hyperon channels have been performed. In particular, the $\bar{p}p \to \bar{\Lambda}\Lambda$ and $\bar{p}p \to \bar{\Xi}^+\Xi^-$ channels reconstruction have been studied in, Refs. 14, 15 where different selection criteria were used. In general, the $\bar{p}p \to \bar{\Lambda}\Lambda$ channel reconstruction is based on vertex fits, mass window selection and decay vertex position cuts. In $\bar{p}p \to \bar{\Xi}^+\Xi^-$ a Decay Chain Fitting tool was used for the full channel reconstruction. In Table 2, the estimated $\bar{p}p \to \bar{\Sigma}^0\Lambda$ reconstruction rates obtained in this work together with the obtained in Refs. 14, 15 are summarized. All reconstruction rates were calculated considering *Day-One* HESR luminosity, *i.e.* $\mathcal{L} \sim 10^{31}$ cm^{-2}s^{-1}. The calculated rates show that it will be possible to start collecting large and clean exclusive hyperon samples already during the first phase of the experiment.[16]

Table 2. Reconstruction rates at *Day-One* luminosity.

p_{beam} [GeV/c]	Reaction	σ [μb]	ϵ [%]	Rate [s^{-1}]
1.64	$\bar{p}p \to \bar{\Lambda}\Lambda$	64	16	10
1.77	$\bar{p}p \to \bar{\Sigma}^0\Lambda$	11	5.3	0.6
6.0	$\bar{p}p \to \bar{\Sigma}^0\Lambda$	20	6.1	1.6
4.6	$\bar{p}p \to \bar{\Xi}^+\Xi^-$	~ 1	8.2	0.1
7.0	$\bar{p}p \to \bar{\Xi}^+\Xi^-$	~ 0.3	7.9	0.03

5. Summary

Hyperons provide a testing tool for the strong interaction in the Q^2 region where pQCD is not applicable. At the $\bar{\text{P}}$ANDA experiment, high single and multi-strange hyperon production rates are expected from Day-one and these will increase when the full luminosity is achieved.

References

1. FAIR webpage `https://fair-center.eu/`.
2. A. Deur *et al.* 2016, *Progress in Particle and Nuclear Physics* **90** 1–74.
3. T. Johansson 2003, *Proc. Int. AIP Conf. Proc. 8th Int. Conf. on Low Energy Antiproton Physics (Italy)* p 95.
4. P. D. Barnes *et al.* 1996 *Phys. Rev.* C **54** 2831.
5. P. D. Barnes *et al.* 1990 *Phys. Lett.* B **246** 273.
6. H. Becker *et al.* 1978 *Nucl. Phys.* B **141** 48.
7. F. Tabakin and R. A. Eisenstein 1985, *Phys. Rev.* C **31** 1857.
8. H. R. Rubinstein and H. Snellman 1985 *Phys. Lett.* B**165** 187.
9. P. G. Ortega *et al.* 2011 *Phys. Lett.* B **696** 352.
10. Erni W. *et al.* 2009 *arXiv preprint* arXiv: 0903.3905.
11. G. Pérez Andrade, 2019 , Master thesis, Uppsala University. Available from `http://urn.kb.se/resolve?urn=urn:nbn:se:uu:diva-375552`.
12. K. Götzen 2014, $\bar{p}p$ Cross-section estimator. Found in `https://panda.gsi.de/pbarx/index.php?&mode=1`.
13. Albrecht Gillitzer, 2018 *Private communication.*
14. W. Ikegami-Andersson, 2019, Proc. 6th International Workshop for Next Generation Scientists (FAIRNESS 2019).
15. W. Ikegami-Andersson, 2020, PhD thesis Uppsala Univeristy (In preparation).
16. Jennifer Pütz *et al* 2016, *J. Phys.: Conf. Ser.* **742** 012028.

Recent results from SND experiment at the VEPP-2000 collider

L. V. Kardapoltsev[ab*], M. N. Achasov[ab], A. Yu. Barnyakov[ab], A. A. Baykov[ab], K. I. Beloborodov[ab], A. V. Berdyugin[ab], A. G. Bogdanchikov[a], A. A. Botov[a], T. V. Dimova[ab], V. P. Druzhinin[ab], V. B. Golubev[ab], A. G. Kharlamov[ab], A. A. Korol[ab], D. P. Kovrizhin[ab], A. S. Kupich[a], R. A. Litvinov[a], K. A. Martin[a], N. A. Melnikova[ab] A. E. Obrazovsky[a], E. V. Pakhtusova[a], K. V. Pugachev[ab], Y. S. Savchenko[ab], S. I. Serednyakov[ab], D. A. Shtol[ab], Z. K. Silagadze[ab], I. K. Surin[ab], Yu. V. Usov[ab], V. N. Zabin[a] and V. V. Zulanov[a,b]

[a] *Budker Institute of Nuclear Physics, Novosibirsk, 630090, Russia*
[b] *Novosibirsk State University, Novosibirsk, 630090, Russia*
* *E-mail: l.v.kardapoltsev@inp.nsk.su*

We present recent results on study of exclusive processes of e^+e^- annihilation into hadrons with the SND detector. Data were collected at the VEPP-2000 e^+e^- collider in the center-of-mass energy range from 0.3 to 2 GeV. In particular, we have measured the cross sections for $e^+e^- \to \pi^+\pi^-$ and $n\bar{n}$, studied the radiative decay of excited light vector mesons to $\eta\pi^0\gamma$, and performed a search for the rare processes of e^+e^- annihilation into C-even resonances η and $f_1(1285)$.

Keywords: e^+e^-; VEPP-2000.

1. Introduction

SND[1–4] is the universal nonmagnetic detector consisting of a nine-layer drift chamber, aerogel Cherenkov counters, a three-layer spherical electromagnetic calorimeter with 1640 NaI(Tl) crystals, and a muon system.

SND collected data at e^+e^- collider VEPP-2000[5] starting from 2010 in the energy range 0.3–2.0 GeV. During the 2010–2013 data taking period the luminosity was limited by the deficit of positrons. From 2014 the VEPP-2000 accelerator complex was under reconstruction. The experiments were restarted by the end of 2016 with a new 10 times more intensive positron source. The data sample with an integrated luminosity of about 70 pb^{-1} has been collected during 2010-2013 and about 200 pb^{-1} in 2017–2019. The achieved luminosity near 2 GeV is about 4×10^{31} cm^{-2}s^{-1}.

Main physics task of the SND experiment is the study of all possible processes of e^+e^- annihilation into hadrons below 2 GeV. In particular, these measurements are used to obtain the total hadronic cross section needed for Standard Model calculation of the anomalous magnetic moment of muon and running electromagnetic coupling constant. A detailed study of dynamic of exclusive processes is also performed. In this report we present SND results on the processes $e^+e^- \to \pi^+\pi^-$, $e^+e^- \to n\bar{n}$, $e^+e^- \to \eta\pi^0\gamma$, $e^+e^- \to \eta$, and $e^+e^- \to f_1(1285)$.

2. Measurement of the $e^+e^- \to \pi^+\pi^-$ cross section

The process $e^+e^- \to \pi^+\pi^-$ is very important for calculation of the hadronic contribution to the $(g-2)$ of muon. There are many measurements of this process, some of them have systematic uncertainty less than 1%.

Our preliminary measurement is based on 4.7 pb^{-1} data, 10% of the full SND data set in the energy region under study. The event selection is based on excellent e/π separation provided by the three layer SND calorimeter. Figure 1 represents the measured $e^+e^- \to \pi^+\pi^-$ cross section. The currently achieved systematic uncertainty is 0.8%.

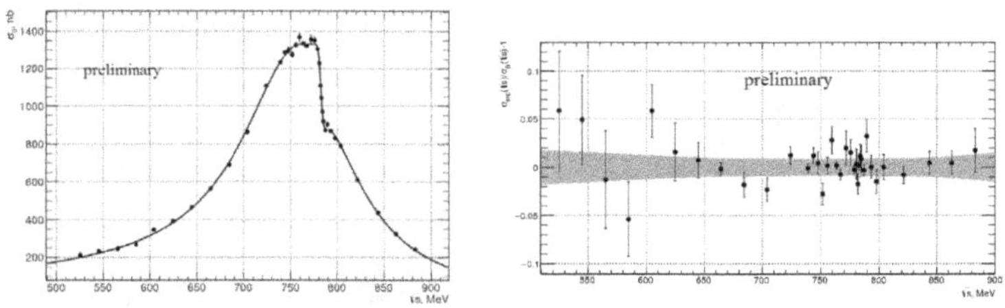

Fig. 1. Left panel: The $e^+e^- \to \pi^+\pi^-$ cross section measured by SND. The curve is the result of the VMD fit. Right panel: The relative difference between the SND $e^+e^- \to \pi^+\pi^-$ data and the fit to the data. The band represents the statistical and systematic uncertainty of the fit combined in quadrature. .

The difference between the currently most accurate BABAR[6] and KLOE[7,8] $e^+e^- \to \pi^+\pi^-$ data and the SND fit is shown in Fig. 2, where the green band represents the statistical and systematic uncertainties of the SND fit combined in quadrature. The systematic difference is observed between the SND and BABAR data below 0.7 GeV and between the SND and KLOE data above 0.7 GeV.

We also calculate the contribution to the muon anomalous magnetic moment from the $e^+e^- \to \pi^+\pi^-$ channel in the energy region 0.53–0.88 GeV using the new SND data. It is equal to $(411.8 \pm 1.0 \pm 3.7) \times 10^{-10}$.

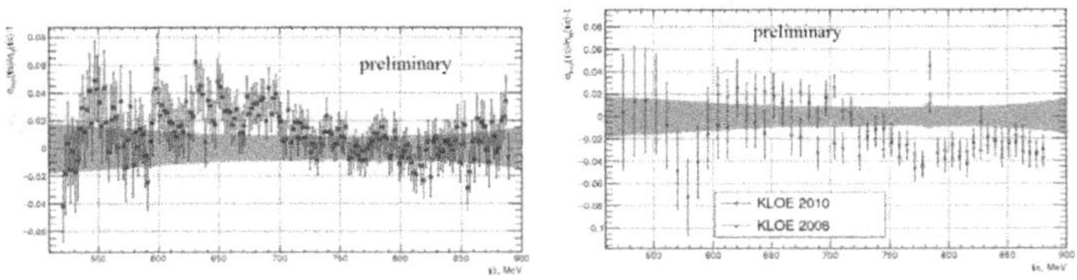

Fig. 2. The relative difference between the BABAR[6] (left panel) and KLOE[7,8] (right panel) $e^+e^- \to \pi^+\pi^-$ data and the SND fit. The band represents the statistical and systematic uncertainty of the SND fit combined in quadrature. .

This value is in good agreement with the previous SND result[?] $(408.9 \pm 1.3 \pm 5.3) \times 10^{-10}$ and the BABAR result[6] $(414.9 \pm 0.3 \pm 2.1) \times 10^{-10}$.

3. Study of the process $\mathbf{e^+e^- \to n\bar{n}}$

The process $e^+e^- \to n\bar{n}$ was previously measured by FENICE[9] and SND[10] using the 2011-2012 data set. The new SND measurement is based on 2017 data and uses a different method of signal-background separation compared with Ref. 10.

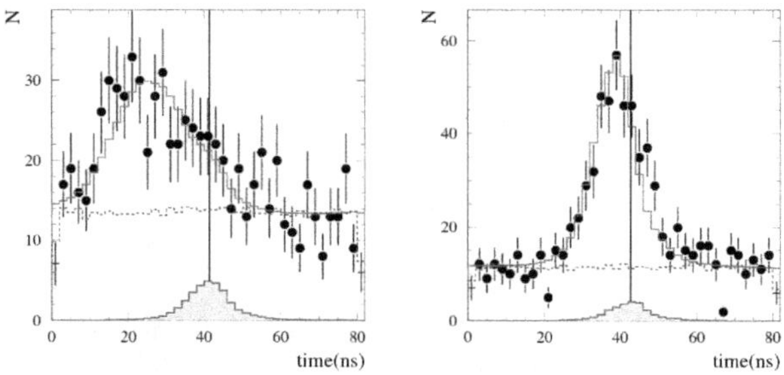

Fig. 3. The time distribution of the calorimeter trigger signal for $n\bar{n}$ candidates in data (points with error bars) for the energy regions: 1.880–1.884 GeV (left panel) and 1.90–1.94 GeV (right panel). The solid histogram represents the result of the fit with a sum of signal and background distributions. The dashed histogram shows cosmic background, while the shaded histogram represents e^+e^- annihilation background. .

The time distribution of the calorimeter trigger signal is analyzed to separate signal from background. This distribution is shown in Fig. 3 for two energy regions. It is fitted by a sum of distributions for signal, cosmic background, and beam plus e^+e^- annihilation background. The latter is estimated from data recorded below the $n\bar{n}$ threshold. The $n\bar{n}$ time distribution is shifted relative to other e^+e^- annihilation events due to small antineutron velocity.

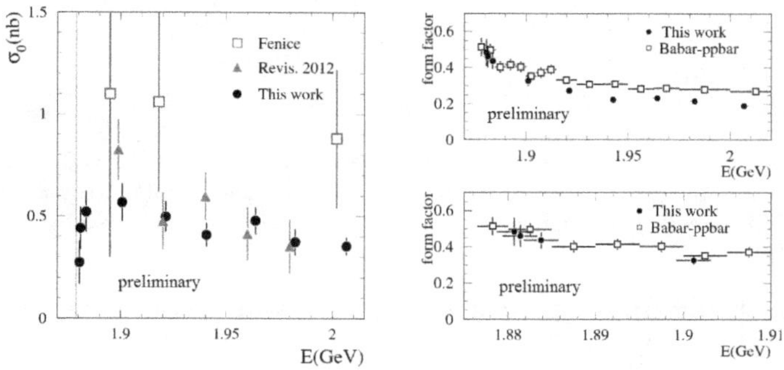

Fig. 4. Left panel: The $e^+e^- \to n\bar{n}$ cross section measured by SND using the 2012 (triangles) and 2017 (circles) data set in comparison with the FENICE results[9] (squares). Right panel: The neutron effective form factor measured by SND (circles) in comparison with the proton effective form factor measured by BABAR[11] (squares). .

The measured cross section is shown in Fig. 4 (left). The statistical accuracy of the measurement is significantly improved compared with the previous SND measurement.[10] However the new SND result is lower than the previous one by about 30% at 1.9 GeV and by two times near 2 GeV. The main reasons are underestimated beam background and not quite correct MC simulation. The triangles in the plot represent the revised 2012 data, which are in agreement with the new measurement.

The systematic uncertainty on the cross section is estimated to be about 20%, mainly due to MC simulation. Currently we are working to decrease it.

4. Radiative process $e^+e^- \to \eta\pi^0\gamma$

The process $e^+e^- \to \eta\pi^0\gamma$ above 1.05 GeV is studied for the first time. The five-photon final state is used for the analysis. In this process, there is a significant contribution of the $\omega\eta$ intermediate state, which is seen as a peak at ω mass in the $\pi^0\gamma$ mass distribution. The non-$\omega\eta$ signal is also observed. It may arise from the processes $e^+e^- \to a_0(1450)\gamma$ and $a_2(1320)\gamma$. Figure 5 represents the measured $e^+e^- \to \omega\eta$ cross section in comparison with the SND and CMD-3 measurements in the decay mode $\omega \to 3\pi$ and the first measurement of the non-$\omega\eta$ cross section in the energy range 1.05-2.00 GeV.

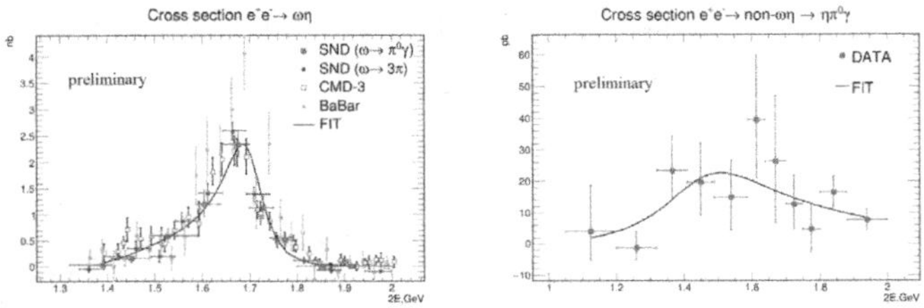

Fig. 5. Left panel: The $e^+e^- \to \omega\eta$ cross section measured by SND in the $\eta\pi^0\gamma$ final state in comparison with the SND[12] and CMD-3[13] measurements in the $\pi^+\pi^-\pi^0\eta$ final state. Right panel: The non-$\omega\eta$ $e^+e^- \to \eta\pi^0\gamma$ cross section measured by SND. .

5. Search for direct production of C-even resonances in e^+e^- annihilation

The direct production of C-even resonances in e^+e^- annihilation is a rare process suppressed by α^2 compared with single-photon annihilation. For spin-0 resonances there is additional helicity suppression.

The search of the process $e^+e^- \to \eta$ is performed using a special 650 nb^{-1} data sample recorded in 2018 at the center of mass energy equal to the η mass. The visible η line shape is determined by the beam energy spread. The beam energy and spread were measured with a special system based on Compton backscattering of laser photons off the electron beam. The decay mode $\eta \to 3\pi^0$ is used for analysis, in which the single photon annihilation background is absent. Zero signal events have been selected. Since the $e^+e^- \to \eta$ is proportional to the $\eta \to e^+e^-$ branching fraction, the 90% confidence level upper limit $B(\eta \to e^+e^-) < 7 \times 10^{-7}$ is set,[14] which improves the previous limit of the HADES Collaboration[15] by a factor of 3. The theoretical prediction for this decay is about 5×10^{-9} (see[16] and references therein).

The predicted branching fraction for the axial vector resonance f_1 decay to e^+e^- is $(3.5 \pm 1.8) \times 10^{-9}$.[17] The corresponding f_1 production cross section is relatively large, about 30 pb.

We analyze the 15 pb^{-1} data sample recorded in the energy range 1.2–1.4 GeV. About 4 pb^{-1} were collected at the f_1 maximum. The decay mode $f_1 \to \eta\pi^0\pi^0 \to 6\gamma$, which is not produced in single photon annihilation, is used for the analysis.

After applying selection and background suppression cuts, two events are observed at the f_1 resonance peak and zero events beyond the resonance. The expected background from the processes $e^+e^- \to \omega\pi^0$, $\eta\gamma$, and $\omega\pi^0\pi^0$ does not exhibit the resonance behavior and is about 0.25 events at the f_1

peak. The observed two events correspond to the cross section of 43^{+32}_{-23} pb and the $f_1 \to e^+e^-$ branching fraction $(4.9^{+3.6}_{-2.6}) \times 10^{-9}$.[18] The significance of the f_1 signal is 2.4σ.

Acknowledgment

This work is supported in part by the RFBR grant 18-02-00147.

References

1. M. N. Achasov *et al.*, Nucl. Instrum. Methods Phys. Res., Sect. A **598**, 31 (2009).
2. V. M. Aulchenko *et al.*, Nucl. Instrum. Methods Phys. Res., Sect. A **598**, 102 (2009).
3. A. Y. Barnyakov *et al.*, JINST **9**, C09023 (2014).
4. V. M. Aulchenko *et al.*, Nucl. Instrum. Methods Phys. Res., Sect. A **598**, 340 (2009).
5. A. Romanov *et al.*, in *Proceedings of Particle Accelerator Conference PAC 2013, Pasadena, CA USA, 2013*, p. 14.
6. J. P. Lees *et al.* (BaBar Collaboration), Phys. Rev. D **86**, 032013 (2012).
7. F. Ambrosino *et al.* (KLOE Collaboration), Phys. Lett. B **670**, 285 (2009).
8. F. Ambrosino *et al.* (KLOE Collaboration), Phys. Lett. B **700**, 102 (2011).
9. A. Antonelli *et al.* (FENICE Collaboration), Nucl. Phys. B **517**, 3 (1998).
10. M. N. Achasov *et al.* (SND Collaboration), Phys. Rev. D **90**, no. 11, 112007 (2014)
11. J. P. Lees *et al.* (BaBar Collaboration), Phys. Rev. D **87**, 092005 (2013).
12. M. N. Achasov *et al.* (SND Collaboration), Phys. Rev. D **99**, 112004 (2019).
13. R. R. Akhmetshin *et al.* (CMD-2 Collaboration), Phys. Lett. B **773**, 150 (2017).
14. M. N. Achasov *et al.* (SND Collaboration), Phys. Rev. D **98**, 052007 (2018).
15. G. Agakishiev *et al.* (HADES Collaboration), Phys. Lett. B **731**, 265 (2014).

16. P. Sanchez-Puertas, arXiv:1709.04792 [hep-ph].
17. A. I. Milstein and A. S. Rudenko, arXiv:1909.07938 [hep-ph].
18. M. N. Achasov *et al.* (SND Collaboration), Phys. Lett. B **800**, 135074 (2020), arXiv:1906.03838 [hep-ex].

Charmonium resonances and Fano line shapes

Xu Cao

Institute of Modern Physics, Chinese Academy of Sciences
Lanzhou, 730000, P. R. China
and University of Chinese Academy of Sciences
Beijing, 100049, P. R. China
E-mail: caoxu@impcas.ac.cn

H. Lenske

Institut für Theoretische Physik, Universität Gießen
Gießen, D-35392, Germany
and GSI Helmholtzzentrum für Schwerionenforschung
Darmstadt, D-64291, Germany
E-mail: Horst.Lenske@theo.physik.uni-giessen.de

Anomalous line shapes of quarkonia are explained naturally as an interference effect of a $c\bar{c}$ confined closed channel with the surrounding continua, well established in other fields of physics as Fano-resonances. We discuss a quark model coupled-channel analysis describing quarkonium as a mixing of closed $Q\bar{Q}$ and molecular-like $D\bar{D}$ open channels. The asymmetric line shapes observed in $\psi(3770)$ production cross sections in e^+e^- annihilation to $D^0\bar{D}^0$ and D^+D^-, respectively, are described very well. The method allows to extract directly from the data the amount of $Q\bar{Q} \leftrightarrow D\bar{D}$ configuration mixing.

Keywords: Charmonium; line shapes; continua; coupled-channel.

1. Introduction

The observations of a multitude of unexpected states in the charmonium region is puzzling. Those sharp spectral structures are close resemblance to the phenomenon of bound states embedded into the continuum, $BSEC$ or BIC. In fact, long-lived states observed as sharp resonances above particle emission thresholds are an ubiquitous phenomenon in quantum systems of any scale. The existence of such states was postulated already at the very beginning of modern quantum physics by von Neumann and Wigner.[1] Fano revived that idea[2] and established the effect in atomic physics explaining self-ionizing states as a quantum interference phenomenon of closed channel two-electron configurations with energetically degenerate single electron scattering states. Nowadays, modern laser techniques allow to simulate and manipulate such configuration mixing effects in atoms by a suitable choice of atto-second laser pulses.[3] BIC are observed abundantly in molecular spectra. Zhang et al.[4] found the BIC phenomenon in the Hubbard model. Our own studies are showing that Fano-states do also exist in nuclei,[5] caused by the interference of closed channel multi-particle hole configurations with nucleon-core continuum states. A defining BIC-feature is the unusual small width-to-centroid energy ratio in the order of 10^{-3} or less.

We point out the relevance of the BIC-phenomenon for quarkonium physics and keep the presentation of the formalism general. However, charmonium and bottonium states are of particular interest because of the appearance of sharp resonances with a width smaller by orders of magnitudes than the centroid mass. The steadily increasing number of unexplained so-called XYZ-states in the charmonium region indicates the need for extensions of the theoretical tool box. Coupling among open channels is a well studied problem in many areas of hadron physics, e.g. in meson production on the nucleon.[6] The mixing of closed and open channels, however, is a less well established phenomenon in hadron physics.

2. Ingredients and Discussion

The description of quarkonia including configuration mixing of confined $Q\bar{Q}$ heavy quark states and asymptotically emerging $D\bar{D}$ scattering states is in fact a demanding task. A vast amount of work, however, is centered around the Cornell quark model,[7] giving us a concrete physical model at hand for the spectroscopy of $Q\bar{Q}$ charmonium. Herein the $Q\bar{Q}$ spectra are obtained by a variational approach, using spherical harmonic oscillator wave functions as trial functions. Those obtained states are considered as bare charmonium states without continuum coupling.

The interacting system is described in terms of the $Q\bar{Q}$ and the $D\bar{D}$ configurations as effective degrees of freedom. Rather than describing $D\bar{D}$ propagation as a four body problem we approximate the motion by an effective two-body meson configuration with the light quarks as spectators. This allows to reformulate the problem in terms of $Q\bar{Q}$ closed channels coupled to $Q\bar{Q}$ open channels which, however, are dressed by light quarks. The intrinsic D-meson structure is fully taken into account in the evaluation of matrix elements. In a partial wave representation the state vector is

$$\Psi = \sum_{A} z_A |A; P, q\rangle + \sum_{BC} \int dk k^2 z_{BC}(k) |BC; P, k\rangle \tag{1}$$

given in terms of quarkonium states $|A; P, q\rangle$ and $D\bar{D}$ channel states $|BC; P, k\rangle$. The total four-momentum is $P^2 = (p_B + p_C)^2 = s$, where s is the Mandelstam center-of-mass (*c.m.*) energy. Relative motion in the $Q\bar{Q}$ and the $D\bar{D}$ is described by the momenta q and k, respectively. The quarkonium states and the $D\bar{D}$ open channel states are interacting through an interaction V, taken from the $^3P_0$-model Lagrangian.[8] For simplicity, we consider the coupling of a single closed, confined $Q\bar{Q}$ channel to a set of open $D\bar{D}$ channels. The configuration coefficients are determined by the linear system, which have to be solved as a partial wave boundary condition problem. How to construct such a solution is found e.g. in Fano's original paper.[2] The essential, but not surprising result is that the channel coupling induces dispersive self-energies. Quarkonium states, for example, gain the dynamical self-energy $\Sigma_A(s) = \Delta_A(s) - \frac{i}{2}\Gamma_A(s)$ leading to a mass shift Δ_A, a width Γ_A and resulting in the finite life time $\tau_A \sim \frac{1}{\Gamma_A}$.

In the open channels the coupling introduces an additional configuration phase shift, given by $\cot\delta_c(s) = -s + m_A^2(s)/\sqrt{s}\Gamma_A(s)$, where $m_A^2(s) = s_A^2 + 2\sqrt{s_A}\Delta_A(s)$. The $Q\bar{Q}$ channel is now dressed by (virtual) admixtures of $D\bar{D}$ components, while the open channel components are contributing to an incoherent smooth background.[2] In atomic physics, those states are well known as *dark states.* Here, we concentrate on the so-called *bright* channel, including the resonance.

As an example we investigate the production of $\psi(3770)$ in leptonic $|\alpha\rangle = |e^+e^-\rangle$ annihilation reactions, well studied experimentally and analysed by various theoretical approaches.[9–12] We define the (in general complex) Fano-line shape parameter q_0 and production cross section becomes

$$\sigma_{\alpha\beta} = \sigma^{(BC)}_{\alpha\beta} \frac{|q_{\alpha\beta} - \cot\delta_c|^2}{1 + \cot^2\delta_c} \tag{2}$$

The Fano-form factor, multiplying the open channel production cross section $\sigma^{(BC)}_{\alpha\beta}$, is, in fact, a scale-independent quantity: Irrespective of the scale of the physical system the same line shape will be observed for all systems with the same q_0, producing a line shape with an interference dip at energy $s = s_0$ where $cot\delta_c(s_0) = q_0$. By resolving such structures experimentally, important spectral information on the population of the (dressed) confined relative to the open channel components will be derived. The dependence of the Fano-parameter depends on the producing reaction, in principle provides the possibility to disentangle reaction dynamics and configuration dynamics by producing the same state by different probes.

Eq.(2) implies a separation of coupled channels effects from the intrinsic dynamics of the $Q\bar{Q}$ and the $D\bar{D}$ subsystems: $Q\bar{Q}$ and coupled channels

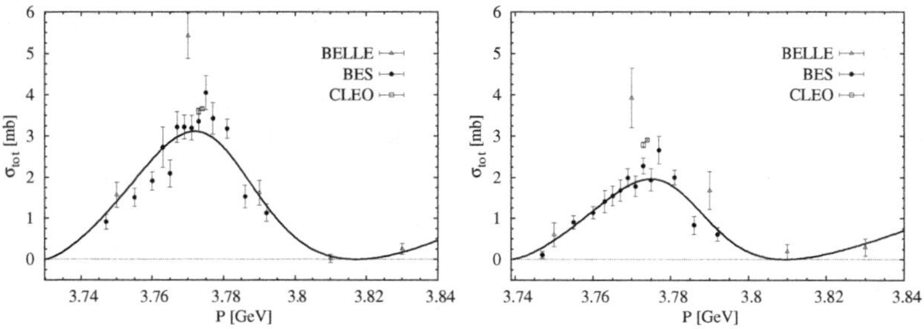

Fig. 1. Total cross sections of $e^+e^- \to D^0\bar{D}^0$ (right) and $e^+e^- \to D^+D^-$ (left) reactions. Theoretical curves are compared to the data which summarized in Ref. 10

dynamics are attached to the line shape parameter $q_{\alpha\beta}$ while $D\bar{D}$ dynamics is contained in $\sigma^{(BC)}_{\alpha\beta}$. An important conclusion is that the line shape seen in a production cross section must not be identical to the line shape observed in (hypothetical) $D\bar{D}$ elastic scattering. In order to estimate contributions of $D\bar{D}$ interactions we introduce one boson exchange (OBE) interactions.

With FSI the $D\bar{D}$ production amplitude is given by a Lippmann-Schwinger type equation accounting for the rescattering processes. Within the quark model the direct $Q\bar{Q}$ production amplitudes are easily calculated.[7] The ratio $\frac{\Gamma_{e^+e^-}(^3D_1)}{\Gamma_{e^+e^-}(^3S_1)} \simeq 1.5 \times 10^{-3}$, which is determined by the wave functions and their derivatives at the origin, indicates a strong suppression of the direct $^3D_1$ production. The observed widths, however, do not show that behaviour. Either the $^3D_1$ states contain an unexpected large amount of $^3S_1$ configurations or they contain a substantial admixtures of $D\bar{D}$ components. The $D\bar{D}$ vertex is not affected by that suppression. The $\gamma^* \to D\bar{D}$ vertex can be resolved into $\gamma^* \to Q\bar{Q}$ loops which subsequently decay into the final $D\bar{D}$ configuration. At the energies considered here, it is justified to replace the loops by a sum over the $^3S_1$, $^3D_1$ quark model states and the $D\bar{D}$ production will proceed mainly through $^3S_1$ intermediate states (or $^3S_1$ admixtures to $^3D_1$ states, respectively). For $\psi(3770)$ this means that the $D\bar{D}$ component is mainly produced through the nearby $\psi(3686)$ $^3S_1$ sub-threshold state. In Fig. 1, our final result for the production of $\psi(3770)$ are summarized. Treating for comparison the Fano-parameter as a free parameter leads to $q_0 = -0.4$ for $e^+e^- \to D^0\bar{D}^0$ and $q_0 = -0.3$ for D^+D^- reactions, respectively. The different q_0 values in the neutral and charged $D\bar{D}$ channels are plausible considering the impact of $D^0(D^+)$ mass gap and coulomb forces to the amplitudes.

3. Summary

In summary, we have proposed a novel view on charmonium line shapes observed in production cross sections by considering charmonium states as Fano resonances embedded into the $D\bar{D}$ continuum. The quark model was taken as a guidance to the internal structure of charmonium and open charm mesons. As an example we analysed the $e^+e^- \to D\bar{D}$ reaction populating the $\psi(3770)$ state. More details can be found in Ref. 13. Quarkonium decay studies should take into account the coherent superposition of confined $Q\bar{Q}$ components and open charm (or bottom) continuum configuration providing a natural explanation for variations of line shapes in different production and decay channels. It is intriguing to apply the approach to other hadronic states, especially with respect of the newly found and hardly understood XYZ states. A parameterization originated from Eq. (2) has applied to other interesting channels in e^+e^- annihilation.[14,15]

Acknowledgments

This work was supported by the Deutsche Forschungsgemeinschaft (DFG) (CRC16, Grants No. B7), DFG project No. Le439/7 and in part by I3HP SPHERE, the LOEWE and the National Natural Science Foundation of China (Grant Nos. 11347146 and 11405222).

References

1. J. von Neumann and E. Wigner, Phys. Z. **30**, 465 (1929).
2. U. Fano, Phys. Rev. **124**, 1866 (1961).
3. C. Ott *et al.*, Science **340**, 716 (2013).
4. J. M. Zhang *et al.*, Phys. Rev. Lett. **109**, 116405 (2012).
5. S. E. A. Orrigo *et al.*, Phys. Lett. **B633**, 469 (2006).
6. X. Cao, V. Shklyar, and H. Lenske, Phys. Rev. C **88**, 055204 (2013).
7. E. Eichten *et al.*, Phys. Rev. D **17**, 3090 (1978); **21**, 203 (1980).
8. L. Micu, Nucl. Phys. **B10**, 521 (1969).
9. H. B. Li *et al*, Phys. Rev. D **81**(R), 011501 (2010).
10. Y. J. Zhang and Q. Zhao, Phys. Rev. D **81**, 034011 (2010).
11. Y. R. Liu *et al.*, Phys. Rev. D **82**, 014011 (2010).
12. G. Y. Chen and Q. Zhao, Phys. Lett. **B718**, 1369 (2013).
13. X. Cao and H. Lenske, arXiv:1408.5600; arXiv:1410.1375.
14. X. Cao, J. P. Dai and Y. P. Xie, Phys. Rev. D **98**, no. 9, 094006 (2018).
15. R. Wang, X. Cao and X. Chen, Phys. Lett. B **747**, 321 (2015).

$\pi\pi$ and $K\pi$ scattering amplitudes from lattice QCD

L. Leskovec

Thomas Jefferson National Accelerator Facility, 12000 Jefferson Avenue, Newport News, Virginia 23606, USA
E-mail:leskovec@jlab.org

S. Meinel and G. Rendon

Department of Physics, University of Arizona, Tucson, AZ 85721, USA
E-mail:smeinel@email.arizona.edu,jgrs@email.arizona.edu

J. W. Negele and A. Pochinsky

Center for Theoretical Physics, Massachusetts Institute of Technology, Cambridge, Massachusetts 02139, USA
E-mail:negele@mit.edu,avp@mit.edu

M. Petschlies

Helmholtz-Institut für Strahlen- und Kernphysik,
Rheinische Friedrich-Wilhelms-Universität Bonn, Nußallee 14-16, D-53115 Bonn, Germany
E-mail:marcus.petschlies@hiskp.uni-bonn.de

S. Paul

Institute of Nuclear Physics, Johannes Gutenberg Universität,
Johann-Joachim-Becher-Weg 45, 55099 Mainz
E-mail:srijitpaul@gmail.com

S. Silvi

Institute for Advanced Simulation, Forschungszentrum Jülich, 52425 Germany
E-mail:g.silvi@fz-juelich.de

S. Syritsyn

RIKEN BNL Research Center, Brookhaven National Laboratory, Upton, New York 11973, USA, and Department of Physics and Astronomy, Stony Brook University, Stony Brook, New York 11794, USA
E-mail:ssyritsyn@quark.phy.bnl.gov

We report on our study of low-lying resonances in $\pi\pi$ and $K\pi$ scattering from lattice QCD. Based on ab-initio multi-hadron spectroscopy and the Lüscher finite-volume method, we have investigated the elastic scattering amplitudes for $\pi\pi$ P-wave, $K\pi$ S-wave and P-wave partial waves. A particular focus of the discussion will be the parametrizaton of the amplitudes with respect to the identification of resonance parameters. The shown results are obtained for two gauge field ensembles at pion masses 317 MeV and 178 MeV.

Keywords: Lattice QCD; scattering amplitude; finite volume; Lüscher method.

1. Introduction

The elastic scattering in the $\pi\pi$ isospin-1 as well as the $K\pi$ isospin-1/2 channel are of great phenomenological interest, as they both feature resonant excitations in the $\ell = 1$ partial wave: the ρ and K^*, respectively. In addition, precise knowledge of the scattering phase shift versus center of mass energy of the 2-hadron system is essential input for the lattice investigation of many $2 \to 1$ transitions via an intermediate resonance, such as $\pi\gamma^* \to (\rho \to \pi\pi)$. An equally sought-after excitation is the $K_0^*(700)$ with $I\,(J^P) = 1/2\,(0^+)$ in the $K\pi$ S-wave scattering, also designated $K_0^*(800)$ or κ. The precise nature of this state has been difficult to ascertain in experiment.[1–3] Lattice QCD has matured to a state where tracking the pole position with pion mass towards the physical point has become feasible with increasing control over statistical and systematic uncertainties.[4,5]

2. $\pi\pi$ scattering analysis

To determine the $\pi\pi$ $I\,(J^P) = 1\,(1^-)$ elastic scattering parameters with the Lüscher finite-volume method we extract the lattice spectrum from solving the generalized eigenvalue problem,[6] for correlation matrices built from a basis of lattice-irreducible single- and two-hadron interpolators. Our analysis is performed on the $N_f = 2 + 1$ isotropic, clover-improved Wilson ensemble "C13"[a] with parameters as listed in Table 1. For the single open channel case with a partial wave cut-off $\ell_{\text{max}} = 1$, the Lüscher

Table 1. Parameters for gauge field ensembles C13, D6

Label	a/ fm	L/ fm	m_π/ MeV	m_K/ MeV	$m_\pi L$
C13	0.11403 (77)	3.6	317	530	5.9
D6	0.08766(79)	4.2	178	515	3.8

[a]Provided by K. Orginos, based on XSEDE resources.

quantization condition maps one-to-one finite-volume lattice energy levels and infinite-volume elastic scattering amplitude or phase shift $\delta_{\ell=1}$ values, as shown for our spectrum data by the discrete data points in the left plot of Fig. 1. Parametrizations of Breit-Wigner type lead to resonance mass $m_\rho = 797.6\,(2.8)_{\rm stat}\,(2.4)_{\rm sys}\,(5.4)_a$ MeV and effective coupling $g_{\rho\pi\pi} = 5.69\,(13)_{\rm stat}\,(16)_{\rm sys}$ at $m_\pi = 317$ MeV.[7] In summary we show a compilation of available lattice results for the ρ resonance mass based on $N_f = 2{+}1$ (degenerate up and down quark plus strange) and $N_f = 2{+}1{+}1$ (including charm) dynamical quarks in the right plot of Fig. 1.[b]

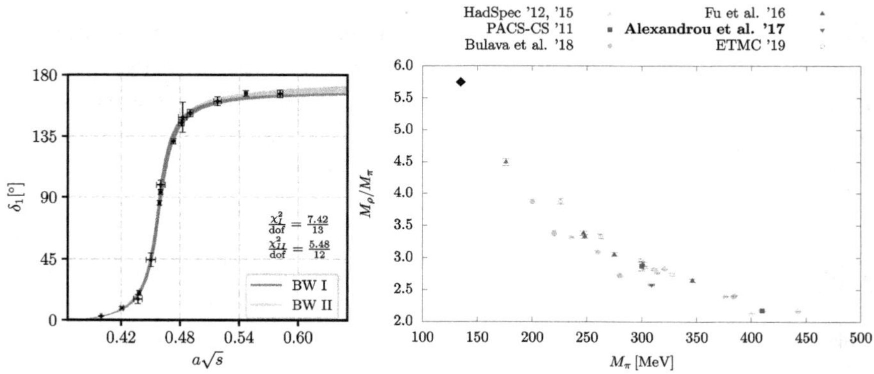

Fig. 1. Left: solutions for Lüscher quantization condition $(\sqrt{s},\, \delta_{\ell=1})$ and overlaid Breit-Wigner fits of the energy dependence.[7] Right: comparison of contemporary results for ρ resonance mass from lattice simulations with $N_f = 2+1$ and $N_f = 2+1+1$ dynamical quark flavors.

3. $K\pi$ scattering analysis

The analysis for the $K\pi$ isospin $I = 1/2$ channel, technically similar to the $\pi\pi$ case, is complicated by S- and P-wave mixing due to non-equal masses and thus reduced symmetry in the $K\pi$ system.[8,9] We perform the inverse Lüscher analysis and use our spectrum data to constrain the scattering K-matrix elements via the quantization condition. In a combined S- and P-wave analysis[10] we draw from a variety of K-matrix parametrizations and fit the predicted to the observed finite-volume spectrum. We here restrict

[b]Results from simulations with same lattice action but different lattice spacing values are not distinguished.

the display to results for two specific models: Bugg's parametrization[11] and the conformal mapping by Peláez and Rodas.[12] Our preliminary results for the K-matrix fits on ensembles `C13` and `D6` and pion mass of 317 MeV and 178 MeV, respectively, in terms of the S-wave phase shift is shown in Fig. 2 and the pole parameters are listed in Table 2. In the figure we also compare to experimental data.[13]

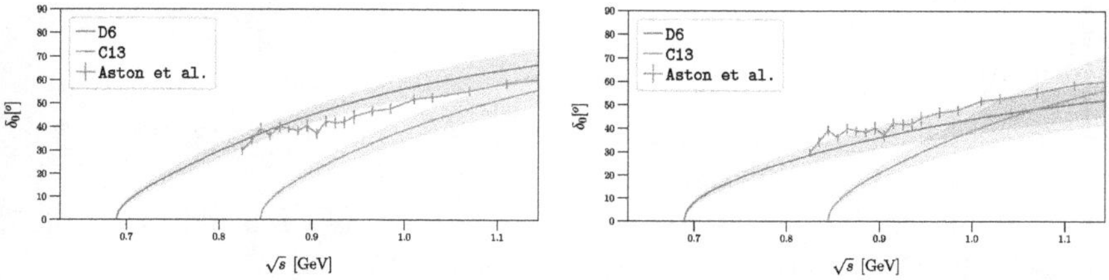

Fig. 2. Results from the K-matrix fits for the S-wave phase shift for Bugg's model (upper panel) and the conformal mapping.

Table 2. Results for S-, P-wave poles for Bugg's parametrization and conformal mapping.

m_π [MeV]	Model	T-matrix Poles [GeV]	χ^2/dof
317	Conformal Mapping	S-wave: $(0.841, -0.329) \pm (0.149, 0.081)$ P-wave: $(0.895, -0.00250) \pm (0.006, 0.00021)$	0.758
	Bugg's parametrization	S-wave: $(0.840, -0.342) \pm (0.077, 0.044)$ P-wave: $(0.895, -0.00250) \pm (0.006, 0.00021)$	0.757
178	Conformal mapping	S-wave: $(0.712, -0.440) \pm (0.040, 0.081)$ P-wave: $(0.872, -0.013) \pm (0.008, 0.001)$	1.01
	Bugg's parametrization	S-wave: $(0.711, -0.297) \pm (0.074, 0.029)$ P-wave: $(0.872, -0.013) \pm (0.008, 0.001)$	1.39

4. Outlook

We find our preliminary results for $K\pi$ K-matrix fits and K_0^* pole analysis at 178 and 317 MeV are encouraging. The $\pi\pi$ P-wave scattering analysis on ensemble at 178 MeV pion mass, as well as the inclusion of the $K\bar{K}$ channel for $\pi\pi$ and the $K\eta$ channel for $K\pi$ are under way. Additional data for fine lattice spacing, but at pion mass of 280 MeV is in production. A publication with the details of the finalized analysis, of which preliminary results were demonstrated here, is in preparation.

Acknowledgments

M.P. gratefully acknowledges support by the Sino-German collaborative research center CRC-110. The computational resources for this project have been granted by NERSC, a U.S. Department of Energy Office of Science User Facility operated under Contract No. DE-AC02-05CH11231, and by the LRZ Germany under the GCS project "Weak decays of beauty and charm mesons to multi-hadron final states on the lattice".

References

1. S. N. Cherry and M. R. Pennington, There is no kappa(900), *Nucl. Phys.* **A688**, 823 (2001).
2. D. V. Bugg, The Kappa in $J/\psi \to K^+\pi^-K^-\pi^+$, *Eur. Phys. J.* **A25**, 107 (2005), [Erratum: Eur. Phys. J. A **26**, 151(2005)].
3. D. V. Bugg, The Kappa in E791 data for $D \to K_{\pi\pi}$, *Phys. Lett.* **B632**, 471 (2006).
4. D. J. Wilson et al., The quark-mass dependence of elastic πK scattering from QCD, *Phys. Rev. Lett.* **123**, 042002 (2019).
5. R. Brett et al., Determination of s- and p-wave $I = 1/2$ $K\pi$ scattering amplitudes in $N_\mathrm{f} = 2 + 1$ lattice QCD, *Nucl. Phys.* **B932**, 29 (2018).
6. M. Lüscher and U. Wolff, How to Calculate the Elastic Scattering Matrix in Two-dimensional Quantum Field Theories by Numerical Simulation, *Nucl. Phys.* **B339**, 222 (1990).
7. C. Alexandrou et al., P-wave $\pi\pi$ scattering and the ρ resonance from lattice QCD, *Phys. Rev.* **D96**, 034525 (2017).
8. L. Leskovec and S. Prelovsek, Scattering phase shifts for two particles of different mass and non-zero total momentum in lattice QCD, *Phys. Rev.* **D85**, 114507 (2012).
9. G. Rendon, L. Leskovec, S. Meinel, J. Negele, S. Paul, M. Petschlies, A. Pochinsky, G. Silvi and S. Syritsyn, $K\pi$ scattering and the $K^*(892)$ resonance in 2+1 flavor QCD, *PoS* **LATTICE2018**, 073 (2019).
10. G. Rendon, Talk presented at Lattice 2019.
11. D. V. Bugg, An Update on the Kappa, *Phys. Rev.* **D81**, 014002 (2010).
12. J. R. Pelaez and A. Rodas, Pion-kaon scattering amplitude constrained with forward dispersion relations up to 1.6 GeV, *Phys. Rev.* **D93**, 074025 (2016).
13. D. Aston *et al.*, A Study of $K^-\pi^+$ Scattering in the Reaction $K^-p \to K^-\pi^+n$ at 11-GeV/c, *Nucl. Phys.* **B296**, 493 (1988).

Exploratory study of the coupled-channel scattering in finite volume and the $Z_c(3900)$ resonance

Mei-Zhu Yan, Dan Zhou* and Zhi-Hui Guo♯

Department of Physics and Hebei Advanced Thin Films Laboratory, Hebei Normal University, Shijiazhuang 050024, China

*E-mail: danzhou@hebtu.edu.cn

♯E-mail: zhguo@hebtu.edu.cn

In this note we first introduce the theoretical formalism of the unitarization of the two-body scattering both in the infinite and finite volumes. Then we apply this formalism to the study of the $Z_c(3900)$ resonance in the $J/\psi\,\pi$ and $D\bar{D}^*$ coupled-channel scattering. The recent lattice finite-volume spectra are confronted with our predictions.

Keywords: Finite-volume spectrum; exotic charmed mesons.

1. Introduction

Important progresses on the study of unstable hadrons in the two-body scattering have been made recently in the lattice QCD computation. The rich and precise finite-volume spectra from the lattice QCD simulations provide valuable information of the underlying hadron interactions, which also give complementary inputs for the phenomenological analyses, compared to the experimental measurements. One powerful way to calculate the physical observables from the finite-volume discrete spectra is the celebrated Lüscher method,[1] which indeed has been used by many lattice analyses, such as in Refs. 2, 3. The finite-volume unitarized chiral perturbation theory (χPT) approach was proposed in Ref. 4, which also has been demonstrated to be useful in the description of the lattice energy levels.[5–7] In Refs. 5, 6, we have successfully applied the unitarization χPT approach to describe the lattice energy levels of the $\pi\eta, K\bar{K}, \pi\eta'$ and $D\pi, D\eta, D_s\bar{K}$ coupled-channel scattering, respectively. For the $\pi\eta$ scattering, the $a_0(980)$ resonance properties and the physical $\pi\eta$ phase shifts and inelasticities are determined.[5] After the successful fits to the lattice energy levels of the $D\pi, D\eta, D_s\bar{K}$ scattering in Ref. 3, we predict the physical $D\pi$ scattering amplitudes and determine the properties of the $D_0^*(2400)$ resonance.[6] In this note we first recapitulate the theoretical formalism of the finite-volume unitarized χPT approach. Then we apply the latter approach to the study of the $J/\psi\,\pi$ and $D\bar{D}^*$ coupled-channel scattering and confront our results with the recent lattice energy levels in Refs. 8, 9.

2. Unitarized amplitudes in finite volume

Following the basic idea of the N/D method, the on-shell two-body scattering amplitudes that satisfy the unitarity condition can be written as[10]

$$T(s) = [1 + N(s) \cdot G(s)]^{-1} \cdot N(s)\,, \tag{1}$$

where by construction the function $G(s)$ only contains the right-hand unitarity cut and the function $N(s)$ only incorporates the crossed-channel dynamics, including the possible local contact terms. One way to write the explicit expression of the function $G(s)$ is

$$G(s) = i \int \frac{\mathrm{d}^4 k}{(2\pi)^4} \frac{1}{(k^2 - m_1^2 + i\epsilon)[(p-k)^2 - m_2^2 + i\epsilon]}\,, \tag{2}$$

with $s = p^2$. Dimensional regularization can be used to calculate the above integral and the explicit result reads[10]

$$\begin{aligned} x_\pm &= \frac{s + m_1^2 - m_2^2}{2s} \pm \frac{q(s)}{\sqrt{s}}\,, \\ G(s)^{\mathrm{DR}} &= \frac{1}{16\pi^2}\left[a(\mu^2) + \log\frac{m_2^2}{\mu^2} - x_+ \log\frac{x_+ - 1}{x_+} - x_- \log\frac{x_- - 1}{x_-}\right], \end{aligned} \tag{3}$$

being μ the regularization scale. In the dispersive representation of the function $G(s)$, $a(\mu^2)$ in the above equation corresponds to the subtraction constant, which has to be determined by the experimental or lattice inputs. There are also other ways to express the function $G(s)$, such as to include the form factors or to introduce sharp momentum cutoff to regularize the divergent integral in Eq. (2).[11,12] The merit of the representation of the $G(s)$ function in Eq. (3) is that it has the correct analytical behaviors. In the multiple-channel scattering case, $T(s)$ and $N(s)$ in Eq. (1) become matrices spanned in the channel space and $G(s)$ corresponds to the diagonal matrix with non-vanishing elements given by Eq. (2).

In order to study the lattice discrete energy levels, one has to introduce the finite-volume effects. This can be done by calculating the function $G(s)$ of Eq. (2) in the finite cubic box of length L with periodic boundary conditions. In the center of mass (CM) frame, the resulting finite-volume correction ΔG to the function $G(s)$ in the infinite volume reads[4]

$$\Delta G = \frac{1}{L^3} \sum_{\vec{n}}^{|\vec{q}|<q_{\max}} I(|\vec{q}|) - \int^{|\vec{q}|<q_{\max}} \frac{\mathrm{d}^3\vec{q}}{(2\pi)^3} I(|\vec{q}|) \,, \tag{4}$$

with

$$\vec{q} = \frac{2\pi}{L}\vec{n}, \; (\vec{n} \in \mathbb{Z}^3) \,, \; I(|\vec{q}|) = \frac{w_1 + w_2}{2w_1 w_2 \left[s - (w_1 + w_2)^2\right]} \,, \; w_i = \sqrt{|\vec{q}|^2 + m_i^2} \,.$$

The three-momentum cutoff $q_{\max}$ is introduced to regularize separately the two terms in Eq. (4) and it is demonstrated that ΔG is rather weakly dependent on $q_{\max}$,[5] due to the cancellation of its dependences of the two terms in that equation. The function $G(s)$ used in the finite-volume study is then given by

$$\widetilde{G}^{\mathrm{DR}} = G^{\mathrm{DR}} + \Delta G \,. \tag{5}$$

The finite-volume spectra correspond to the solutions

$$\det\left[1 + N(s) \cdot \widetilde{G}^{\mathrm{DR}}\right] = 0 \,. \tag{6}$$

To proceed the study of the $J/\psi\pi$ and $D\bar{D}^*$ coupled-channel scattering, labeled as 1 and 2 in order, one needs to parameterize the matrix elements of the $N(s)$ in Eq. (1). The $J/\psi\pi \to J/\psi\pi$ amplitude has been found to be tiny[13] and we will simply fix the matrix element $N_{11} = 0$. The transition amplitude N_{12} for the $J/\psi\pi \to D\bar{D}^*$ will be approximated as a constant. According to the discussion in Ref. 12 and references therein, there is no resonance if one also takes the constant approximation for N_{22}. At least one next-to-leading order term in the momentum expansion should be kept in order to generate the resonance in the coupled-channel scattering. Therefore we will take one non-vanishing energy dependent term in the N_{22}. The explicit expression for the $N(s)$ is given by

$$N(s) = \begin{pmatrix} N_{11} & N_{12} \\ N_{12} & N_{22} \end{pmatrix} = \begin{pmatrix} 0 & c_{12} \\ c_{12} & c_{22}^0 + \frac{c_{22}^1}{m_D}(\sqrt{s} - m_D - m_D^*) \end{pmatrix} \,. \tag{7}$$

3. Lattice finite-volume spectra and relevant resonances

Several lattice finite-volume energy levels from the $J/\psi\pi$ and $D\bar{D}^*$ scattering are recently calculated in Refs. 8, 9. Both studies in the two former references claim that a standard narrow resonance state around the $D\bar{D}^*$ threshold is disfavored, although more precise energy levels with different volumes and lighter pion masses are needed to reach definite conclusions.

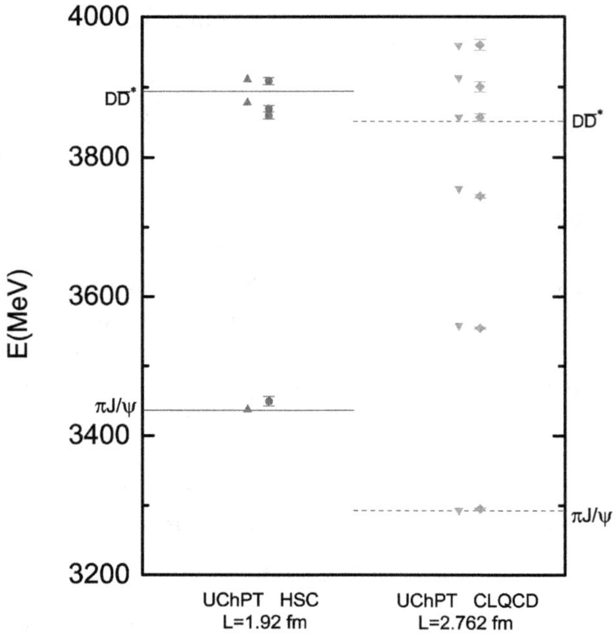

Fig. 1. The comparisons of our predictions and the lattice energy levels from Ref. 8 (HSC) and Ref. 9 (CLQCD). Notice that different meson masses are used in the two lattice simulations.

Here we take an exploratory study to use the formalism in Eq. (6) to confront the lattice spectra in Refs. 8, 9. To fix the unknown parameters $c_{12}, c_{22}^0, c_{22}^1$ in Eq. (7) and the subtraction constants in Eq. (3), which are assumed to be independent on the quark masses, we will use the $D\bar{D}^*$ amplitudes from an upcoming $J/\psi\pi$ and $D\bar{D}^*$ coupled-channel study in Ref. 14 to constrain the free parameters. We leave the details of the coupled-channel analyses in a future work.[14] Our primary aim here is to check whether it is possible to find a solution to reasonably describe the lattice spectra in Refs. 8, 9, while in the meantime there is a Z_c resonant state.

The comparisons of our predictions and the lattice spectra are given in Fig. 1. It is clear that our theoretical predictions are compatible with the lattice results within uncertainties. With the same parameters that give the energy levels in Fig. 1, we do find in the complex energy plane the Z_c resonance poles, among which the most relevant one is $(3887.2 - i1.5)$ MeV in the second Riemann sheet, although other shadow poles are also found in different Riemann sheets. However it is pointed out that none of the Z_c

poles correspond to the conventional narrow-width resonance, since they do not appear in the proper Riemann sheets and the phase shifts around the $D\bar{D}^*$ threshold do not pass ninety degrees as the conventional resonance.

Acknowledgments

This work is partially supported by the NSFC under Grants No. 11975090, No. 11575052, and the Natural Science Foundation of Hebei Province under Contract No. A2015205205.

References

1. M. Luscher, Nucl. Phys. B **354**, 531 (1991).
2. J. J. Dudek *et al.* [Hadron Spectrum Collaboration], Phys. Rev. **D93**, no. 9, 094506 (2016).
3. G. Moir, M. Peardon, S. M. Ryan, C. E. Thomas and D. J. Wilson, JHEP **1610**, 011 (2016).
4. M. Doring, U. G. Meissner, E. Oset and A. Rusetsky, Eur. Phys. J. A **47**, 139 (2011); Eur. Phys. J. A **48**, 114 (2012).
5. Z. H. Guo, L. Liu, U. G. Meissner, J. A. Oller and A. Rusetsky, Phys. Rev. D **95**, no. 5, 054004 (2017).
6. Z. H. Guo, L. Liu, U. G. Meissner, J. A. Oller and A. Rusetsky, Eur. Phys. J. C **79**, no. 1, 13 (2019).
7. J. J. Xie and E. Oset, Eur. Phys. J. A **48**, 146 (2012); H. X. Chen and E. Oset, Phys. Rev. D **87**, no. 1, 016014 (2013); D. Zhou, E. L. Cui, H. X. Chen, L. S. Geng and L. H. Zhu, Phys. Rev. D **91**, no. 9, 094505 (2015); J. X. Lu, X. L. Ren and L. S. Geng, Eur. Phys. J. C **77**, no. 2, 94 (2017).
8. G. K. C. Cheung *et al.* [Hadron Spectrum Collaboration], JHEP **1711**, 033 (2017).
9. T. Chen *et al.* [CLQCD Collaboration], Chin. Phys. C **43**, no. 10, 103103 (2019).
10. J. A. Oller and E. Oset, Phys. Rev. D **60**, 074023 (1999).
11. J. A. Oller and E. Oset, Nucl. Phys. A **620**, 438 (1997) Erratum: [Nucl. Phys. A **652**, 407 (1999)].
12. M. Albaladejo, P. Fernandez-Soler, F. K. Guo and J. Nieves, Phys. Lett. B **767**, 465 (2017).
13. X. H. Liu, F. K. Guo and E. Epelbaum, Eur. Phys. J. C **73**, no. 1, 2284 (2013).
14. Zhi-Hui Guo, et al., in preparation.

On the quark-mass dependence of meson masses and decay constants

Xiao-Yu Guo and Matthias F. M. Lutz

GSI Helmholtzzentrum für Schwerionenforschung GmbH,
Planck Str. 1, 64291 Darmstadt, Germany

We study the dependence of meson masses and decay constants on the up, down and strange quark masses in QCD in terms of an effective field theory approach. The role of dynamical vector meson degrees of freedom is scrutinized by a one-loop approximation to the self-energies of the Goldstone bosons and vector mesons as well as the decay constants of the Goldstone bosons. Our results are confronted with QCD lattice data from various lattice groups. From a successful global fit to such lattice data we are lead to the conclusion that dynamical vector mesons imply significant changes in the Gasser and Leutwylers low-energy constants.

Keywords: Vector meson; chiral extrapolation; lattice QCD.

1. Introduction

Chiral perturbation theory (ChPT) is explaining a plethora of strong interaction phenomena at momentum transfers well below 1 GeV. As an effective field theory, ChPT is formulated with Goldstone boson (π, K and η) fields which are identified as the most relevant degrees of freedom. Are there further degrees of freedom that are of dynamical importance in the study of hadron physics at momentum transfers below 1 GeV? The seminal work by Sakurai[1] illustrated how the light vector mesons may play an important role. How to combine the vector meson (ρ, ω, K^* and ϕ) fields and the Goldstone boson fields consistently into an effective theory poses a considerable challenge of QCD.[2–6]

In a series of papers, a possible construction path for such an effective field theory approach was suggested in.[5–7] It is conjectured that chiral QCD shows a significant mass gap at large number of colors (N_c). Below some hard scale $\Lambda_{HG} \sim 2$ GeV only the Goldstone and light-vector mesons are relevant in this particular limit. This hadrogenesis conjecture leads to a systematic effective field theory approach with the light pseudoscalar and vector mesons as effective degrees of freedom only. The leading order chiral Lagrangian was constructed in Refs. 6, 7.

In this proceeding we report on our next-to-leading order computation of the light pseudoscalar and vector meson masses as well as the pseudoscalar meson decay constants. Our approach is confronted with a large set of QCD lattice data.

2. Meson masses and decay constants at the one-loop level

The hadrogensis Lagrangian relevant for our current study is given in Refs. 6, 7. Such a Lagrangian is an extension of the Gasser-Leutwyler SU(3) chiral Lagrangian[8] by incorporating vector mesons as flavor-SU(3) nonet matter fields.

At the one-loop-level the meson polarization tensors Π_H are evaluated in a finite box (see Ref. 9), where H stands for either a Goldstone boson $H = P$ or a vector meson $H = V$. The corrections are composed of tree, bubble and tadpole type contributions specified in terms of the LECs.[7] The polarization tensors take the form

$$M_H^2 - \Pi_H^{\text{tree-level}} - \Pi_H^{\text{tadpole}} - \Pi_H^{\text{bubble}}/Z_H = 0\,, \tag{1}$$

where the wave-function factor Z_H is taken into account such that the residuum of the propagator pole is always normalized to one

$$Z_H = \Big(1 + \frac{\partial}{\partial s}\,\Pi_H^{\text{bubble}}\Big)\Big/\Big(1 - \frac{\partial}{\partial s}\,\Pi_H^{\text{tree-level}}\Big)\,. \tag{2}$$

For a given meson of type H the bubble-loop function may depend on any of the pseudo-scalar or vector meson on-shell masses. As a consequence we have to find solutions for a set of coupled and nonlinear equations.

The bubble-loop polarizations to Goldstone bosons Π_P^{bubble} also drives the Goldstone-boson decay constants away from the conventional ChPT formula $f_P^{\chi-\text{PT}}$,

$$f_P = f_P^{\chi-\text{PT}} + \Big(\sqrt{Z_P^{\text{bubble}}} - 1\Big) f - \frac{f}{\sqrt{Z_P^{\text{bubble}}}}\,\frac{\Pi_P^{\text{bubble}}}{m_P^2}\,,$$

$$\text{with}\quad Z_P^{\text{bubble}} = 1 + \frac{\partial}{\partial\, m_P^2}\,\Pi_P^{\text{bubble}}(m_P^2)\,, \tag{3}$$

Given the subtraction schemes as established in Ref. 7, the chiral-limit vector-meson mass M and Goldstone-boson decay constant f are not renormalized by our subtracted one-loop contributions.

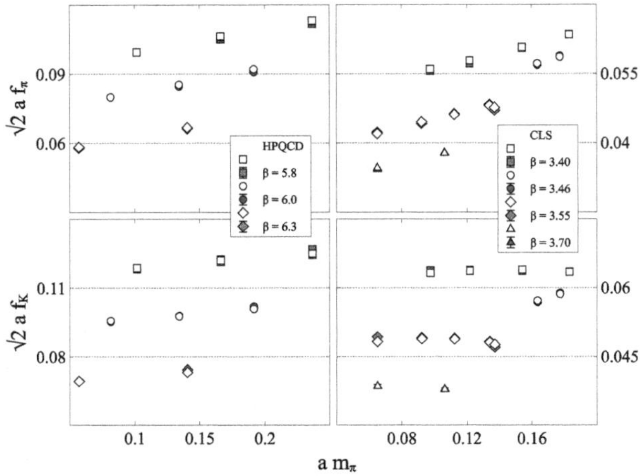

Fig. 1. Our results from Fit 2 for pion and kaon decay constants in lattice units. The lattice results are given by filled symbols in different colors, shown with statistical errors. They are compared to the chiral extrapolation results in open symbols.

Order-by-order renormalizability requires that the scale dependence from loops must be compensated by the counter terms at the same order. This requirement imposes constraints on the effective couplings.[7]

At subleading order, the non-vanishing polarization tensor $\Pi_{\omega\phi}$ leads to a mixing of ω and ϕ fields. The mixing is characterized by a mixing angle, ϵ, with

$$\omega = \omega' \cos\epsilon + \phi' \sin\epsilon \,, \qquad \phi = \phi' \cos\epsilon - \omega' \sin\epsilon \,,$$
$$\text{with} \qquad \Pi_{\omega\phi} = \frac{1}{2}\Big(\Pi_\phi - \Pi_\omega\Big)\,\tan(2\,\epsilon)\,, \tag{4}$$

where ω' and ϕ' are the physical field generating the corresponding mass eigenstates. The mixing angle ϵ depends on the squared mass s. Two physically relevant evaluation points of ϵ are $\epsilon_\omega \equiv \epsilon|_{s=m_\omega^2}$ and $\epsilon_\phi \equiv \epsilon|_{s=m_\phi^2}$. The ϵ_ϕ can be determined from the radiative decay $\phi \to \pi_0\gamma$, as $|\epsilon_\phi| \simeq 3.32^\circ$ from Ref. 10.

3. Fit to QCD lattice data

The meson mass and decay-constant formula (1),(3) involve a number of LECs including the ones of the Gasser-Leutwyler Lagrangian. Three LECs have been estimated by various decays and electromagnetic processes in Refs. 5, 6. There are 11 LECs left to be determined. From a fit to the

Table 1. Gasser-Leutwyler low-energy parameters from the three fit scenarios of Refs. 19 and 20, given at the renormalization scale $\mu = 0.77$ GeV.

	Fit 1	Fit 2	Fit 3	two-loop χPT[21]
f [MeV]	73	71	68	64 - 71
$L_4 \times 10^3$	-0.04	0.05	0.13	0.3 - 0.76
$L_5 \times 10^3$	-0.01	0.02	0.04	0.50 - 1.01
$L_6 \times 10^3$	-0.07	0.02	0.04	0.14 - 0.49
$L_7 \times 10^3$	-0.14	-0.13	-0.11	(-0.34) - (-0.19)
$L_8 \times 10^3$	0.75	0.67	0.65	0.17 - 0.47

lattice data of light meson masses and decay constants at various unphysical quark masses as well as their PDG values, we are able to determine these parameters.

For vector meson masses, we consider the energy levels in a finite box as computed by PACS-CS, QCDSF-UKQCD and HSC.[11–15] We perform a least chi square fit with an asymmetric global systematical error. We arrive at $\chi^2/N \sim 1$ with a systematical upper and lower error of 10 MeV and 20 MeV respectively.

Our fits well reproduced the lattice energy levels obtained by $q\bar{q}$ interpolating fields, but also the levels involving meson-meson interpolating fields given by HSC. For the decay constants, we fit to lattice results of pion and kaon decay constants on HPQCD, CLS and ETMC ensembles[16–18] assigning a global systematical uncertainty to be around 0.9MeV.

On each ensemble, we solve the mass formula self-consistently together with the mixing angles ϵ_ω and ϵ_ϕ and request the polarization tensor $\Pi_{\omega'\phi'}$ to vanish at $s = m_\omega^2$ and $s = m_\phi^2$. At physical quark masses, we get $\epsilon_\omega = 19.9^\circ - 21.9^\circ$. We perform global fits to the lattice data within 3 different scenarios (for details see Refs. 19, 20). Fit 1 relies on lattice data for the light meson masses only. Fit 2 considers both the light meson masses as the pion and kaon decay constants from HPQCD and CLS. Fit 3 considers in addition the pion decay constants from ETMC. Fig. 1 shows the fitting quality of f_π and f_K fitting to HPQCD and CLS results from Fit 2. The results from either of our three scenarios are in excellent agreement with the lattice data.

In Table 1 we list the Gasser-Leutwyler LECs from our fits compared with a two-loop calculation in conventional ChPT.[21] While most of the LECs from our fits are consistent with two-loop ChPT estimations, we found the inclusion of vector meson degrees of freedom significantly changed the value of L_5.

4. Summary

We studied the dynamical role of light vector mesons given the hadrogenesis Lagrangian. The quark-mass dependence of meson masses and decay constants were evaluated at the one-loop level. We reported on an excellent reproduction of the lattice data at different unphysical quark masses from 6 lattice groups. Our analysis predicts an unexpected quark-mass dependence of the $\omega - \phi$ mixing angles. A most striking implication from dynamical vector mesons is a significant reduction of Gasser-Leutwyler LEC, L_5, as compared to the results of 2-loop calculations in conventional ChPT.

References

1. J. J. Sakurai, *Currents and mesons* (University of Chicago Press, Chicago, USA, 1969).
2. M. Harada and K. Yamawaki, *Phys. Rept.* **381**, 1 (2003).
3. U. G. Meißner, *Phys. Rept.* **161**, 213 (1988).
4. J. Bijnens and E. Pallante, *Mod. Phys. Lett.* **A11** (1996).
5. M. F. M. Lutz and S. Leupold, *Nucl. Phys.* **A813**, 96 (2008).
6. C. Terschlüsen, S. Leupold and M. F. M. Lutz, *Eur. Phys. J.* **A48**, 190 (2012).
7. R. Bavontaweepanya, X.-Y. Guo and M. F. M. Lutz, *Phys. Rev.* **D98**, 056005 (2018).
8. J. Gasser and H. Leutwyler, *Nucl. Phys.* **B250**, 465 (1985).
9. M. F. M. Lutz, R. Bavontaweepanya, C. Kobdaj and K. Schwarz, *Phys. Rev.* **D90**, 054505 (2014).
10. F. Klingl, N. Kaiser and W. Weise, *Z. Phys.* **A356**, 193 (1996).
11. S. Aoki *et al.*, *Phys. Rev.* **D79**, 034503 (2009).
12. W. Bietenholz *et al.*, *Phys. Rev.* **D84**, 054509 (2011).
13. H.-W. Lin *et al.*, *Phys. Rev.* **D79**, 034502 (2009).
14. J. J. Dudek *et al.*, *Phys. Rev.* **D88**, 094505 (2013).
15. R. Brett *et al.*, *Nucl. Phys.* **B932**, 29 (2018).
16. R. J. Dowdall *et al.*, *Phys. Rev.* **D88**, 074504 (2013).
17. M. Bruno, T. Korzec and S. Schaefer, *Phys. Rev.* **D95**, 074504 (2017).
18. K. Ottnad and C. Urbach, *Phys. Rev.* **D97**, 054508 (2018).
19. X.-Y. Guo and M. F. M. Lutz, *Nucl. Phys.* **A988**, 48 (2019).
20. X.-Y. Guo and M. F. M. Lutz, *Nucl. Phys.* **A988**, 36 (2019).
21. J. Bijnens and G. Ecker, *Ann. Rev. Nucl. Part. Sci.* **64**, 149 (2014).

Interactions between two heavy mesons within heavy meson chiral effective field theory

Zhan-Wei Liu[1,2,*], Bo Wang[3], Hao Xu[4], Ning Li[5], Xiang Liu[1,2] and Shi-Lin Zhu[3,6,7]

[1]*School of Physical Science and Technology, Lanzhou University, Lanzhou 730000, China*

[2]*Research Center for Hadron and CSR Physics, Lanzhou University & Institute of Modern Physics of CAS, Lanzhou 730000, China*

[3]*Center of High Energy Physics, Peking University, Beijing 100871, China*

[4]*Department of Applied Physics, School of Science, Northwestern Polytechnical University, Xi'an 710072, China*

[5]*Facility for Rare Isotope Beams and Department of Physics and Astronomy, Michigan State University, East Lansing, MI 48824, USA*

[6]*School of Physics and State Key Laboratory of Nuclear Physics and Technology, Peking University, Beijing 100871, China*

[7]*Collaborative Innovation Center of Quantum Matter, Beijing 100871, China*

[*]*E-mail: liuzhanwei@lzu.edu.cn*

We have studied the interactions between two heavy mesons ($D^{(*)}$-$D^{(*)}$, $\bar{D}^{(*)}$-$\bar{D}^{(*)}$, $B^{(*)}$-$B^{(*)}$, or $\bar{B}^{(*)}$-$\bar{B}^{(*)}$) within heavy meson chiral effective field theory and investigated possible molecular states. The effective potentials are obtained with Weinberg's scheme up to one-loop level. At the leading order, four body contact interactions and one pion exchange contributions are considered. In addition to two pion exchange diagrams, we include the one-loop chiral corrections to contact terms and one pion exchange diagrams at the next-to-leading order. The effective potentials both in momentum space and coordinate space are investigated and discussed extensively. The possible molecular states are also studied and the binding energies are provided by solving the Schrodinger equation. The results will be helpful for the experimental search for the doubly-heavy molecular states.

Keywords: Doubly-heavy molecular states; effective potential; interaction between two heavy mesons; chiral perturbation theory.

1. Introduction

Hadron structures have still not been clearly disclosed since the first hadron, proton, was discovered in the beginning of last century. Although the quark model can deal with most hadron phenomenology, more and more exotic behaviors have been noticed in hadron physics in experiment. The structures are closely related to the hadron interactions at low energies, and the relevant studies would also help us to understand the non-pertubative properties of QCD.

The low-lying baryons, $N(1440)$, $N(1535)$, and $\Lambda(1405)$, cannot be interpreted well even only for their masses within conventional quark model. After considering the hadron interactions between πN, $\bar{K}N$, and so on, we can obtain the reasonable mass poles for them within Hamiltonian effective field theory, and moreover, both the scattering data at experiment and the lattice QCD simulations can studied well with such a scheme.[1–4] In the heavy flavor sector, the low-mass puzzle for $\Lambda_c(2940)$ can be resolved with the interaction of D^*N included.[5]

The exotic hadrons totally without the bare core of conventional quark model are more sensitive to the hadron interactions. Recently discovered Z_b and P_c states are probably such molecular states.[6–8] $B^{(*)}$ and $\bar{B}^{(*)}$ are attracted by each other and can form hidden-bottom bound states, probably $Z_b(10610)$ and $Z_b(10650)$. In this proceeding, we focus on doubly-heavy molecules.

In this proceeding, we systematically investigate the interaction between two heavy mesons, $D^{(*)}$-$D^{(*)}$, $\bar{D}^{(*)}$-$\bar{D}^{(*)}$, $B^{(*)}$-$B^{(*)}$, or $\bar{B}^{(*)}$-$\bar{B}^{(*)}$, within chiral effective theory. After deriving the effective potentials for the two interacting heavy mesons in Sec. 2, we solve the Schrödinger equation to study the doubly-heavy molecular states in Sec. 3. Since doubly-charmed baryon Ξ_{cc}^{++} has been observed at LHCb,[9] we believe such doubly-heavy molecules could be detected in experiment soon.

2. Interactions between two heavy mesons

Chiral effective field theory can systematically deal with the low-energy processes order by order. The results are expanded in power series of small momenta, small masses or residual masses, and so on, rather than the strong coupling α_s. At first we need to construct the Lagrangians up to the next-to-leading order. The details can be found in Refs. 10–12. Here we only list the Lagrangians at leading order.

$$\begin{aligned}\mathcal{L}_{4H}^{(0)} &= D_a \mathrm{Tr}[H\gamma_\mu \bar{H}]\mathrm{Tr}[H\gamma^\mu \bar{H}] + D_b \mathrm{Tr}[H\gamma_\mu\gamma_5 \bar{H}]\mathrm{Tr}[H\gamma^\mu\gamma_5 \bar{H}] \\ &\quad + E_a \mathrm{Tr}[H\gamma_\mu \lambda^a \bar{H}]\mathrm{Tr}[H\gamma^\mu \lambda_a \bar{H}] + E_b \mathrm{Tr}[H\gamma_\mu\gamma_5\lambda^a \bar{H}]\mathrm{Tr}[H\gamma^\mu\gamma_5\lambda_a \bar{H}], \quad (1)\end{aligned}$$

$$\mathcal{L}_{H\phi}^{(1)} = -\langle (iv\cdot\partial H)\bar{H}\rangle + \langle Hv\cdot\Gamma\bar{H}\rangle + g\langle H \not{u}\gamma_5\bar{H}\rangle - \frac{1}{8}\Delta\langle H\sigma^{\mu\nu}\bar{H}\sigma_{\mu\nu}\rangle. \quad (2)$$

Here H and $\bar{H}$ are related to the $B^{(*)}$ field, Γ and u contain the pseudoscalar mesons. The first Lagrangian in Eq. (1) can describe the contact interactions $\bar{B}^{(*)}$-$\bar{B}^{(*)}$-$\bar{B}^{(*)}$-$\bar{B}^{(*)}$, and the second in Eq. (2) contains the vertice like $\bar{B}^{(*)}$-$\bar{B}^{(*)}$-π, $\bar{B}^{(*)}$-$\bar{B}^{(*)}$-π-π.

After constructing the Lagrangians, we can derive the effective potentials in momentum space from calculating Feynman diagrams in Weinberg scheme.[13] At leading order, there are contact and one-pion-exchange diagrams in which the vertice are from the Lagrangians at leading order. At next to leading order, there are four types of diagrams. The first is from two-pion-exchange diagrams shown in Fig. 1. The second and third are loop corrections to the contact terms and the one-pion-exchange terms. The fourth is from the tree diagrams with vertice in the Lagrangians at the next-to-leading order, which can absorb the ultraviolet divergences from the loop diagrams.

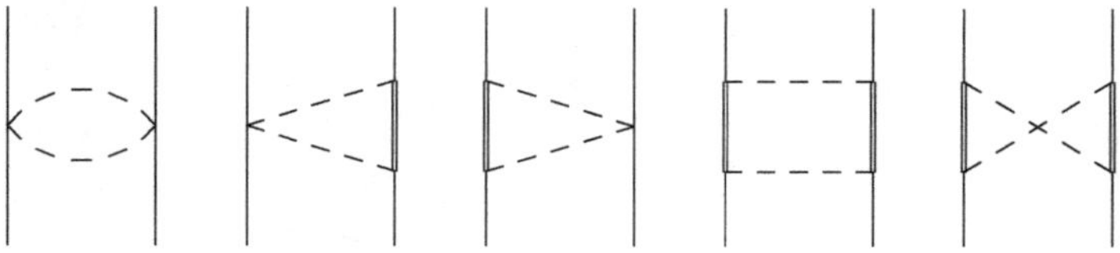

Fig. 1. The 2π-exchange diagrams including the football diagram (F.1), triangle diagrams (T.1 and T.2), box diagram (B.1), and crossed diagram (R.1).

We calculate these diagrams using dimensional regularization and modified minimal subtraction scheme. Before providing the numerical potentials, there are still some unknown low-energy constants which cannot be determined with chiral effective field theory itself. We try to estimate them via combining the resonance saturation method, SU(4) approximation, non-analytic approximation, fitting the lattice QCD results, and so on.

Now we can obtain the effective potentials in momentum space. We show the potentials for the $\bar{B}^*\bar{B}^*$ with $I(J^P) = 0(1^+)$ in the left diagram of Fig. 3 as an example. From the figure, one notices that the corrections at the next-to-leading order cannot be neglected.

3. Doubly-heavy molecular states

With the effective potentials obtained within chiral effective field theory, we can investigate the doubly-heavy molecular states made of $\bar{B}^{(*)}$-$\bar{B}^{(*)}$ or

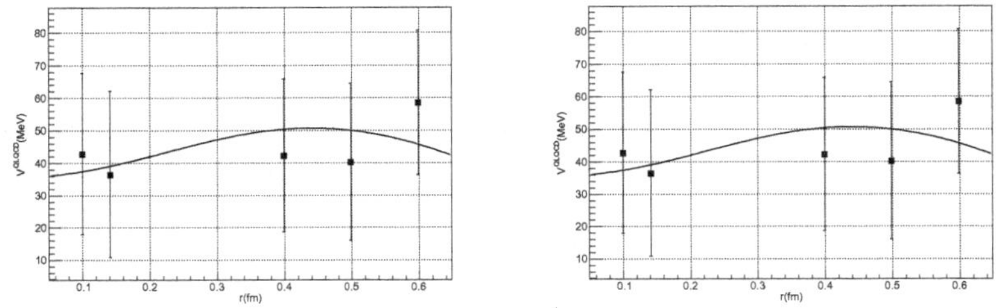

(a) $I = 0$ (b) $I = 1$

Fig. 2. Fitting the $\bar{B}\bar{B}$ potentials. The data of the quenched lattice QCD are derived from Ref. 14 where $m_\pi = 402.5 \pm 6.7$ MeV.

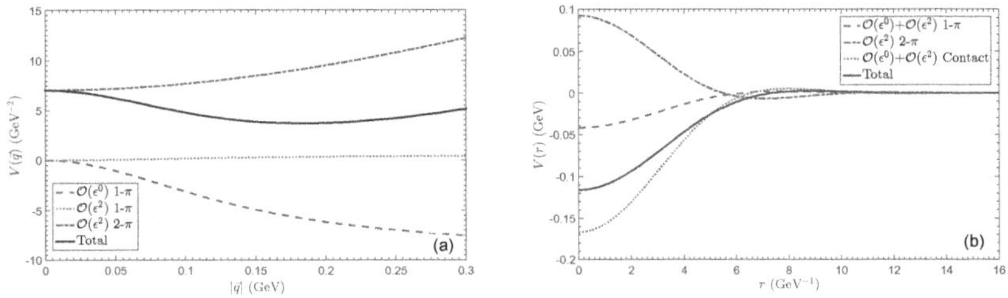

(a) momentum space (b) coordinate space

Fig. 3. Effective potentials for the $\bar{B}^*\bar{B}^*$ with $I(J^P) = 0(1^+)$

$D^{(*)}$-$D^{(*)}$. First, we need to get the effective potentials in coordinate space from those in momentum space in Sec. 2 via Fourier transformation. As an example, we present the potentials in coordinate space for the $\bar{B}^*\bar{B}^*$ with $I(J^P) = 0(1^+)$ in the right diagram of Fig. 3. One can study the properties of the molecular states via solving the Schrödinger equation in coordinate space.

The possible molecular states are listed in Refs. 11, 12. Among them, there are two bound states in the channels of $\bar{B}\bar{B}^*$ and $\bar{B}^*\bar{B}^*$ with $I(J^P) = 0(1^+)$ with the masses being $m_{\bar{B}\bar{B}^*} \simeq 10591.4^{+9.2}_{-12.9}$ MeV, $m_{\bar{B}^*\bar{B}^*} \simeq 10625.5^{+16.3}_{-21.5}$ MeV. They are below the threshold of $\bar{B}\bar{B}\pi$, and thus the strong decay is forbidden. These two molecules are stable under the strong interactions and can be searched for in the electromagnetic channels such as $\bar{B}\bar{B}\gamma$, $\bar{B}\bar{B}\gamma\gamma$, and so on.

This project is supported by the National Natural Science Foundation of China under Grants No. 11705072 and 11965016, 11575008, 11621131001, and the China National Funds for Distinguished Young Scientists under Grants No. 11825503.

References

1. Z.-W. Liu, et al., Hamiltonian effective field theory study of the **N***(**1535**) resonance in lattice QCD, Phys. Rev. Lett. 116 (2016) 082004.
2. Z.-W. Liu, et al., Structure of the $\mathbf{\Lambda(1405)}$ from Hamiltonian effective field theory, Phys. Rev. D95 (2017) 014506.
3. Z.-W. Liu, et al., Hamiltonian effective field theory study of the **N***(**1440**) resonance in lattice QCD, Phys. Rev. D95 (2017) 034034.
4. J.-j. Wu, D. B. Leinweber, Z.-w. Liu, and A. W. Thomas, Structure of the Roper Resonance from Lattice QCD Constraints, Phys. Rev. D97 (2018) 094509.
5. S.-Q. Luo, B. Chen, Z.-W. Liu, and X. Liu, Resolving the low mass puzzle of $\Lambda_c(2940)^+$, (2019) arXiv: 1910.14545.
6. I. Adachi: in Flavor physics and CP violation. Proceedings, 9th International Conference, FPCP 2011, Maale HaChamisha, Israel, May 23-27, 2011 (2011) arXiv: 1105.4583.
7. R. Aaij et al. (LHCb), Observation of a narrow pentaquark state, $P_c(4312)^+$, and of two-peak structure of the $P_c(4450)^+$, Phys. Rev. Lett. 122 (2019) 222001.
8. R. Aaij et al. (LHCb), Observation of $J/\psi p$ Resonances Consistent with Pentaquark States in $\Lambda_b^0 \to J/\psi K^- p$ Decays, Phys. Rev. Lett. 115 (2015) 072001.
9. R. Aaij et al. (LHCb), Observation of the doubly charmed baryon Ξ_{cc}^{++}, Phys. Rev. Lett. 119 (2017) 112001.
10. Z.-W. Liu, N. Li, and S.-L. Zhu, Chiral perturbation theory and the $\bar{B}\bar{B}$ strong interaction, Phys. Rev. D89 (2014) 074015.
11. B. Wang, Z.-W. Liu, and X. Liu, $\bar{B}^{(*)}\bar{B}^{(*)}$ interactions in chiral effective field theory, Phys. Rev. D99 (2019) 036007.
12. H. Xu, B. Wang, Z.-W. Liu, and X. Liu, DD^* potentials in chiral perturbation theory and possible molecular states, Phys. Rev. D99 (2019) 014027.
13. S. Weinberg, Nuclear forces from chiral Lagrangians, Phys. Lett. B251 (1990) 288.
14. W. Detmold, K. Orginos, and M. J. Savage, BB Potentials in Quenched Lattice QCD, Phys. Rev. D76 (2007) 114503.

Nonleptonic decays of heavy hadrons producing pesudoscalar and vector mesons

E. Oset

Departamento de Fisica Teorica and IFIC, University of Valencia-CSIC, Valencia, 46071, Spain
E-mail: oset@ific.uv.es

W. H. Liang

Department of Physics, Guangxi Normal University, Guilin, 541004, China

We have addressed the study of non-leptonic weak decays of heavy hadrons (Λ_b, Λ_c, B and D), with external and internal emission to give two final hadrons, taking into account the spin-angular momentum structure of the mesons and baryons produced. The agreement found with data is quite good in general.

1. Introduction

Non-leptonic and semi-leptonic weak decays of heavy hadrons have become an important source of information on hadron dynamics. A recent review on the subject can be seen in Ref. 1. In non-leptonic decays, the most typical situations appear in external emission, which we depict in Fig. 1(a) for meson decay, and internal emission, depicted in Fig. 1(b).[2–5]

In external emission, a $q\bar{q}$ state from the W decay vertex can lead to a pseudoscalar meson (P) or a vector meson (V), and then the other two final q and $\bar{q}$ states again can produce a pseudoscalar or a vector meson. We have thus four possibilities PP, PV, VP and VV for production. In internal emission, a q state from the first decay vertex and a $\bar{q}$ from the second decay vertex merge to produce either a pseudoscalar or a vector, and the remaining $q\bar{q}$ pair can again produce a pseudoscalar or a vector. Once again we have four possibilities, PP, PV, VP and VV for production.

There is a very large amount of decays of this type measured, and tabulated in the PDG,[6] including B, B_s, D, D_s decays, and the internal or external emission modes. There are also data in baryon decays as Λ_b or Λ_c, both in internal or external emission. A correlation of all these data from a theoretical perspective is worth in itself and this is the purpose of the present work.

We identify reactions that have the same quarks in the initial and final states, and the same decay topology, and only differ by the spin rearrangements in the mesons. We then assume the radial matrix elements to be similar in these reactions and carry out the nontrivial Racah algebra on the weak Hamiltonian to describe the reactions and relate them.

Another aspect of our approach is that it allows us to establish a relationship with approaches based on heavy quark symmetry[7,8] and improve upon them, in particular in the $B \to$ VP reactions, where the strict heavy quark symmetry gives zero for the matrix element.

Our approach leads to predictions in fair agreement with experimental data, in particular for final states in the charm sector, which is not so well studied.

An added value to the present work is the prediction of decay rates for Λ_b and Λ_c baryons into one baryon and a meson, which unlike B decays, have not been much studied theoretically.

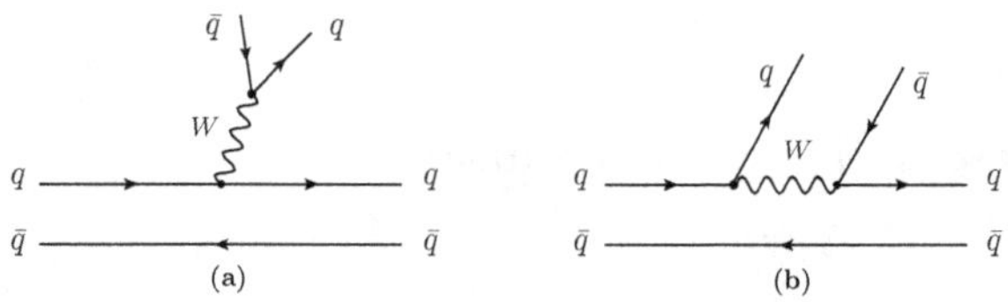

Fig. 1. Meson decay with external emission (a) or internal emission (b).

2. Formalism

We briefly summarize the cases studied in Ref. 9 and the numerical results can be found there.

2.1. Λ_b *external decay*

We look at the process $\Lambda_b \to D_s^- \Lambda_c$ which is depicted in Fig. 2.

The weak Hamiltonian is of the type $\gamma^\mu(1-\gamma_5)$ in each of the weak vertices and then we have an amplitude

$$t = \langle c|\gamma^\mu(1-\gamma_5)|b\rangle \, \langle s|\gamma_\mu(1-\gamma_5)|c'\rangle, \tag{1}$$

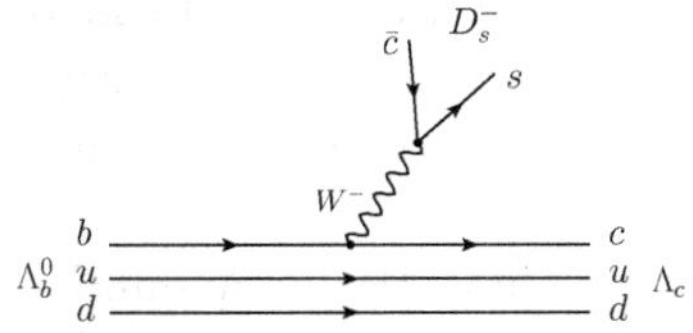

Fig. 2. Quark description in $\Lambda_b \to D_s^- \Lambda_c$ decay with external emission.

where c' corresponds to the $\bar{c}$ state that forms the D_s^-.

In order to evaluate these matrix elements, we choose for convenience a reference frame where the D_s^- is produced at rest.

Furthermore, neglecting the internal momentum of the quark versus p, which is of the order of 5000 MeV, we can write

$$\frac{p_c^\mu}{m_c} = \frac{p_{\Lambda_c}^\mu}{M_{\Lambda_c}}, \qquad \frac{p_b^\mu}{m_b} = \frac{p_{\Lambda_b}^\mu}{M_{\Lambda_b}}, \tag{2}$$

since these ratios are related to just the velocity of Λ_c or Λ_b. This approximation is discussed in detail in Ref. 10 and shown to be exact in the limit of infinite hadron mass, which means that the rules of heavy quark symetry are satisfied.

3. Cases studied

3.1. Λ_b, Λ_c *decays in external emission*

(1) $\Lambda_b \to D_s^- \Lambda_c,\ D_s^{*-} \Lambda_c$;
(2) $\Lambda_b \to \pi^- \Lambda_c,\ \rho^- \Lambda_c$;
(3) $\Lambda_b \to D^- \Lambda_c,\ D^{*-} \Lambda_c$; (Cabibbo suppressed)
(4) $\Lambda_b \to K^- \Lambda_c,\ K^{*-} \Lambda_c$; (Cabibbo suppressed)
(5) $\Lambda_c \to \pi^+ \Lambda,\ \rho^+ \Lambda$;
(6) $\Lambda_c \to K^+ \Lambda,\ K^{*+} \Lambda$. (Cabibbo suppressed)

3.2. Λ_b, Λ_c *decays in internal emission*

(1) $\Lambda_b \to J/\psi \Lambda,\ \eta_c \Lambda$;
(2) $\Lambda_b \to D^0 \Lambda,\ D^{*0} \Lambda$. (Cabibbo suppressed)

3.3. *B decays in external emission*

(1) $\bar{B}^0 \to D_s^- D^+,\ D_s^{*-} D^+,\ D_s^- D^{*+},\ D_s^{*-} D^{*+}$;

(2) $\bar{B}^0 \to \pi^- D^+, \quad \rho^- D^+, \quad \pi^- D^{*+}, \quad \rho^- D^{*+};$
(3) $B^- \to D_s^- D^0, \quad D_s^{*-} D^0, \quad D_s^- D^{*0}, \quad D_s^{*-} D^{*0};$
(4) $B^- \to \pi^- D^0, \quad \rho^- D^0, \quad \pi^- D^{*0}, \quad \rho^- D^{*0};$
(5) $\bar{B}_s^0 \to D_s^- D_s^+, \quad D_s^{*-} D_s^+, \quad D_s^- D_s^{*+}, \quad D_s^{*-} D_s^{*+};$
(6) $\bar{B}_s^0 \to \pi^- D_s^+, \quad \rho^- D_s^+, \quad \pi^- D_s^{*+}, \quad \rho^- D_s^{*+};$
(7) $B_c^- \to D_s^- \eta_c, \quad D_s^{*-} \eta_c, \quad D_s^- J/\psi, \quad D_s^{*-} J/\psi;$
(8) $B_c^- \to \pi^- \eta_c, \quad \rho^- \eta_c, \quad \pi^- J/\psi, \quad \rho^- J/\psi.$

3.4. *B decays in internal emission*

(1) $\bar{B}^0 \to \eta_c \bar{K}^0, \quad J/\psi \bar{K}^0, \quad \eta_c \bar{K}^{*0}, \quad J/\psi \bar{K}^{*0} \;;$
(2) $\bar{B}^0 \to \psi(2S) \bar{K}^0, \quad \psi(2S) \bar{K}^{*0} \;;$
(3) $\bar{B}^0 \to D^0 \pi^0, \quad D^{*0} \pi^0, \quad D^0 \rho^0, \quad D^{*0} \rho^0 \;;$
(4) $B^- \to \eta_c K^-, \quad J/\psi K^-, \quad \eta_c K^{*-}, \quad J/\psi K^{*-} \;;$
(5) $B^- \to \psi(2S) K^-, \quad \psi(2S) K^{*-} \;;$
(6) $B^- \to D^0 \pi^-, \quad D^{*0} \pi^-, \quad D^0 \rho^-, \quad D^{*0} \rho^- \;;$
(7) $\bar{B}_s \to \eta_c \eta, \quad J/\psi \eta, \quad \eta_c \phi, \quad J/\psi \phi \;;$
(8) $\bar{B}_s \to \eta_c \eta', \quad J/\psi \eta' \;;$
(9) $\bar{B}_s \to D^0 K^0, \quad D^{*0} K^0, \quad D^0 K^{*0}, \quad D^{*0} K^{*0} \;;$
(10) $B_c^- \to \eta_c D_s^-, \quad J/\psi D_s^-, \quad \eta_c D_s^{*-}, \quad J/\psi D_s^{*-} \;;$
(11) $B_c^- \to D^0 D^-, \quad D^{*0} D^-, \quad D^0 D^{*-}, \quad D^{*0} D^{*-} \;.$

We have also studied the D decays in internal and external emission.

4. Conclusions

We have made a study of the properties of internal and external emission in the weak decay of heavy hadrons from the point of view of the spin-angular momentum structure, differentiating among the vector and pseudoscalar decay modes. The rest of the structure is given by intrinsic form factors related to the spatial wave functions of the quark states, which do not differentiate the spin of the mesons formed. In this sense, for similar masses of the decay products, like $\eta_c, J/\psi$ or D, D^*, we can obtain rates of decays up to a global factor. Although the approach satifies rules of heavy quark symmetry, we obtain finite terms of order $(\frac{p}{m_Q})^2$, and actually we show that the $B \to$ PV, VP decay modes are proportional to $(\frac{p}{m_Q})^2$, with m_Q the heavy quark mass, and would be neglected in a strict heavy quark symmetry counting. We show that these modes have a similar strength than the $B \to$ PP, VV modes which survive in the heavy quark limit, and these predictions are corroborated by experiment.

The derivation of the final formulas requires a good deal of angular momentum algebra, but the final formulas are rather easy.

We applied the formulas to correlate a large amount of data in Λ_b, Λ_c decays and B or D decays that involve more than 100 reactions. We have taken a given datum for a certain decay rate and then have made predictions for the related reactions. The agreement in general is quite good and the discrepancies are systematic, appearing essentially when there are light particles in the final state.

Acknowledgments

This work is partly supported by the Spanish Ministerio de Economia y Competitividad and European FEDER funds under the contract number FIS2014-57026-REDT, FIS2014-51948-C2- 1-P, and FIS2014-51948-C2-2-P, and the Generalitat Valenciana in the program Prometeo II-2014/068 (EO). This work is also partly supported by the National Natural Science Foundation of China under Grants No. 11975083, No. 11565007 and No. 11747307.

References

1. E. Oset *et al.*, Int. J. Mod. Phys. E **25**, 1630001 (2016).
2. H. Muramatsu *et al.* [CLEO Collaboration], Phys. Rev. Lett. **89**, 251802 (2002); Erratum: [Phys. Rev. Lett. **90**, 059901 (2003)].
3. L. L. Chau, Phys. Rept. **95**, 1 (1983).
4. L. L. Chau and H. Y. Cheng, Phys. Rev. D **36**, 137 (1987); Addendum: [Phys. Rev. D **39**, 2788 (1989)].
5. H. Y. Cheng and C. W. Chiang, Phys. Rev. D **81**, 074031 (2010).
6. C. Patrignani *et al.* (Particle Data Group), Chin. Phys. C **40**, 100001 (2016) and 2017 update.
7. E. E. Jenkins, M. E. Luke, A. V. Manohar and M. J. Savage, Nucl. Phys. B **390**, 463 (1993).
8. G. Kramer, T. Mannel and W. F. Palmer, Z. Phys. C **55**, 497 (1992).
9. W. H. Liang and E. Oset, Eur. Phys. J. C **78**, no. 6, 528 (2018)
10. L. R. Dai and E. Oset, Eur. Phys. J. C **78**, no. 11, 951 (2018)

Session 5

Analysis Tools

Freed-isobar analysis of light mesons at COMPASS

F. Krinner for the COMPASS Collaboration

Max Planck Institut for Physics,
80805 Munich, Bavaria, Germany
E-mail: fkrinner@mpp.mpg/de

Modern hadron-spectroscopy experiments such as COMPASS collect data samples of unprecedented size, so that novel analysis techniques become possible and necessary. One such technique is the freed-isobar partial-wave analysis (PWA). In this approach, fixed parametrizations for the amplitudes of intermediate states commonly modeled using Breit-Wigner shapes are replaced by sets of step-like functions that are determined from the data. This approach not only reduces the model dependence of partial-wave analyses, but also allows us to study the amplitudes of the intermediate states and their dependence on the parent system.

We will also present results of a freed-isobar PWA performed on the large data set on diffractive production of three charged pions collected by the COMPASS experiment, which consists of 46×10^6 exclusive events. We will focus on results for the wave with spin-exotic quantum numbers $J^{PC} = 1^{-+}$, in particular on its decay into $\rho(770)\pi$. Here, the freed-isobar PWA method provides insight into the interplay of three- and two-particle dynamics and confirms the decay of the spin-exotic $\pi_1(1600)$ resonance to $\rho(770)\pi$ in a model-independent way.

Keywords: Exotic; model independent; partial-wave analysis; PWA; freed-isobar.

1. Diffractive 3π production

The data set presented here was collected in 2008 by the COMPASS experiment located at CERN's North Area using a 190 GeV/c π^- beam impinging on liquid hydrogen serving as a proton target. Using the two-stage COMPASS spectrometer, the process

$$\pi^-_{\text{beam}} + p_{\text{target}} \to \pi^-\pi^+\pi^- + p_{\text{recoil}} \tag{1}$$

is selected resulting in a sample of 46×10^6 exclusive events. These events have been analyzed in an extensive PWA[1] and the resonance parameters of eleven isovector 3π resonances have been extracted.[2]

2. Freed-isobar PWA method

The measured intensity distribution $\mathcal{I}(\vec{\tau})$ for a given $m_{3\pi}$ and t' bin is a function of the five phase space variables $\vec{\tau}$ and is modeled as the modulus square of a coherent sum over partial-wave amplitudes:

$$\mathcal{I}(\vec{\tau}) = \left| \sum_{\text{waves}} \mathcal{T}_{\text{wave}} \mathcal{A}_{\text{wave}}(\vec{\tau}) \right|^2 . \tag{2}$$

Here, the complex-valued transition amplitudes $\mathcal{T}_{\text{wave}}$ encode the strengths and phases with which single partial waves contribute and the decay amplitudes $\mathcal{A}_{\text{wave}}(\vec{\tau})$ describe the 3π decay and encode the $\vec{\tau}$-dependence of the partial waves.

The conventional PWA model uses a set of 88 partial waves that differ in spin quantum numbers and 2π resonance content.[1] The decay amplitudes are split up under the assumption that production and decay of appearing 2π and 3π resonances factorize and the process can therefore be described as two subsequent two-particle decays. This assumption is known as the isobar model:

$$\mathcal{A}_{\text{wave}}(\vec{\tau}) = \psi_{\text{wave}}(\vec{\tau}) \Delta_{\text{wave}}(m_{\pi\pi}) + \text{Bose symm.}, \tag{3}$$

where $\psi_{\text{wave}}(\vec{\tau})$ describes the dependence of the partial wave on the decay angles which is fully determined from first principles by the appearing spin quantum numbers. The dynamic isobar amplitudes $\Delta_{\text{wave}}(m_{\pi\pi})$ in contrast describe the 2π resonance content—or isobar—of the partial waves and have to be known beforehand. Bose symmetrization is necessary due to the two identical final-state π^-.

The choice of 2π resonance content, its parameterizations and its parameters introduces possible model bias to the partial wave model and neglects possible contributions from low-intensity resonances or final-state interactions of the isobar with the third pion. To overcome this, we re-analyzed the data set using the freed-isobar approach, in which we replace the fixed dynamic isobar amplitudes by sets of indicator functions spanning the kinematically allowed $m_{\pi\pi}$ range:

$$\Delta_{\text{wave}}(m_{\pi\pi}) \to \sum_{\text{bins}} \Delta^{\text{bin}}_{\text{wave}}(m_{\pi\pi}) \quad \text{with} \quad \Delta^{\text{bin}}_{\text{wave}}(m_{\pi\pi}) = \begin{cases} 1 & m_{\pi\pi} \in \text{bin}, \\ 0 & \text{otherwise.} \end{cases} \tag{4}$$

This replacement alleviates the necessity for fixed dynamic isobar amplitudes from the partial wave model, resulting in a much higher number of parameters, that can lead to mathematical ambiguities within the model—so-called zero modes—that have to be identified and resolved. The detailed origin of these ambiguities in general and for the data presented here is discussed in Refs. 3–5.

We replaced the dynamic isobar parameterizations of 12 of the 88 waves in the PWA model of Ref. 1 following Eq. (4). Eleven of these waves represent the waves describing the largest intensities in the model, so that 75% of the total intensity is described by the freed waves. This minimizes possible effects of imperfections in the fixed isobar parameterizations of the remaining waves. The twelfth wave is the spin-exotic $1^{-+}1^{+}[\pi\pi]_{1^{--}}\pi$P wave,[a] which will be presented in Sec. 3. We use an $m_{\pi\pi}$ bin width of $40\,\mathrm{MeV}/c^2$, and a region of narrower bins of $20\,\mathrm{MeV}/c^2$ around the $\rho(770)$ resonance. The analysis was performed in 50 equidistant bins in the mass range $0.5 < m_{3\pi} < 2.5\mathrm{GeV}/c^2$ and four non-equidistant bins in the squared four-momentum transfer t', resulting in 200 independently fitted kinematic cells.

3. Results for the spin-exotic wave

The result of this analysis is shown in Fig. 2. As expected, the resulting two-dimensional intensity distribution on the right side is dominated by a clear peak corresponding to the decay $\pi_1(1600) \to \rho(770)\pi$. This confirms the existence of this decay without any assumption on the dynamic amplitude of the $[\pi\pi]_{1^{--}}$ system. The left plot shows the coherent sum of all $m_{\pi\pi}$ bins, compared to the result from the conventional PWA. Both distributions are dominated by a similar peak corresponding to the $\pi_1(1600)$ resonance. The freed-isobar result has a higher over-all intensity, since it also can pick up intensity away from the $\rho(770)$ peak.

Fig. 1 shows the results of the analysis of the data in the highest t' bin for three $m_{3\pi}$ bins, below, on, and above the nominal mass of the $\pi_1(1600)$ resonance. The left column shows the resulting intensity distributions and the right column shows the same data as Argand diagrams. In all three $m_{3\pi}$ bins, the fixed Breit-Wigner amplitude from the conventional PWA agrees nicely with the result of the freed-isobar analysis.

The intensity distribution assumes its highest value in the $m_{3\pi}$ at the $\pi_1(1600)$ mass and the Argand diagrams rotate counter-clockwise moving

[a]The wave name is given by $J_{3\pi}^{PC} M^{\varepsilon} [\pi\pi]_{J_{\pi\pi}^{PC}} \pi$L, where J^{PC} are the quantum number of the 3π and $\pi\pi$ systems, M^{ε} is the spin-projection and reflectivity quantum number of the 3π system and L is the orbital angular between the $\pi\pi$ system and the third pion.

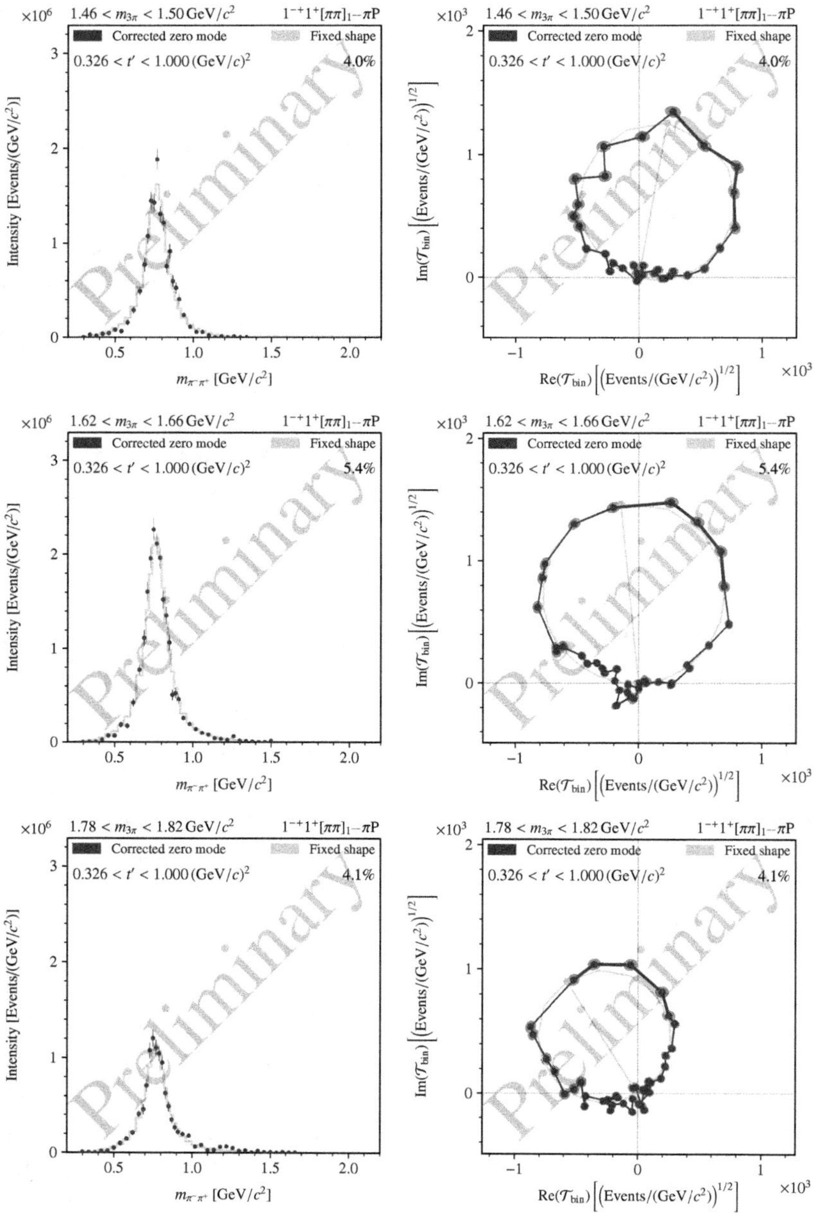

Fig. 1. Results of the freed-isobar PWA for three $m_{3\pi}$ bins (rows) in the highest t' bin for the $1^{-+}1^{+}[\pi\pi]_{1^{--}}\pi P$ wave (blue). The Breit-Wigner amplitude for the $\rho(770)$ as used in the conventional PWA is overlaid in gray. Left: Intensity distributions. Right: Argand diagrams; the orange dot indicates the nominal resonance position of the gray curve.

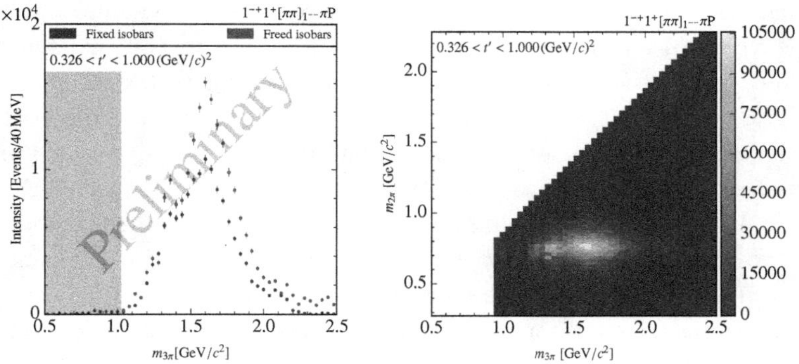

Fig. 2. Results of the freed-isobar analysis for the $1^{-+}1^{+}[\pi\pi]_{1^{--}}\pi P$ wave in the highest t' bin. Left: coherent sum over all $m_{\pi\pi}$ bins (red) compared to the corresponding result of the conventional PWA (blue). Right: Two-dimensional intensity distribution as function of $m_{3\pi}$ and $m_{\pi\pi}$.

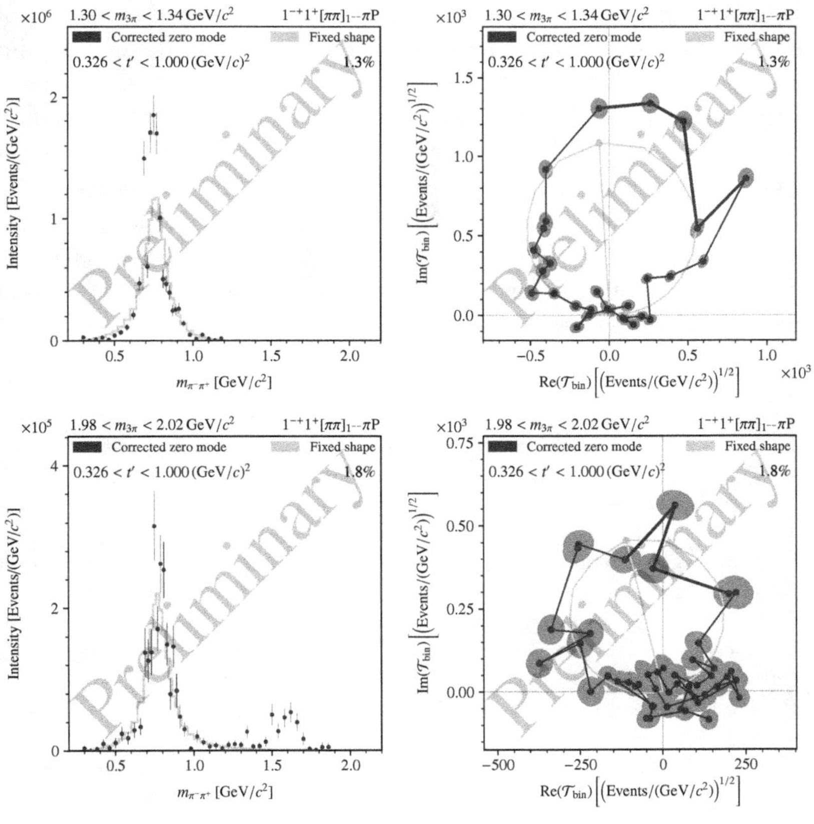

Fig. 3. As Fig. 1 for two additional $m_{3\pi}$ bins far away from the $\pi_1(1600)$ region.

through the $\pi_1(1600)$ resonance, which both reflects the dynamic amplitude of the $\pi_1(1600)$ mother resonance, which further confirms its existence.

Since the fixed Breit-Wigner amplitude is adequate for all $m_{3\pi}$ slices in Fig. 1, we show two additional bins in Fig. 3 where the shapes do not match equally well. At low $m_{3\pi}$ the dynamic isobar amplitude, exhibits a sharper peak than the fixed Breit-Wigner, i.e. a slightly deformed Argand circle. At high $m_{3\pi}$, it exhibits a second peak in the intensity spectrum at a two-pion mass of $1.6\,\mathrm{GeV}/c^2$, which could correspond to an excited ρ' and is unaccounted for in the conventional approach. Whether this peak actually stems from a resonance, requires a subsequent resonance-model fit.

4. Conlusion

We showed results for the dynamic isobar amplitudes of the spin-exotic $1^{-+}1^{+}[\pi\pi]_{1^{--}}\pi\mathrm{P}$ wave extracted from COMPASS data on diffractively produced 3π using the freed-isobar approach.[3–5] The extracted dynamic isobar amplitudes are in general in agreement with the assumption of a dominant presence of the $\rho(770)$ resonance in the $\pi\pi$ P-wave wave as assumed in the conventional PWA approach.[1] In addition, the $\pi_1(1600)$ is confirmed to appear in this partial wave. Nevertheless, there are deviations from the Breit-Wigner amplitude that may originate from distortions of the dynamic isobar amplitude due to non-resonant contributions, such as the Deck effect[6] or re-scattering effects, or additional excited isobar resonances. To clarify the origin of these deviations, a resonance-model fit to the presented data is work in progress.

References

1. C. Adolph *et al.*, Resonance Production and $\pi\pi$ S-wave in $\pi^- + p \to \pi^-\pi^-\pi^+ + p_{\mathrm{recoil}}$ at $190\,\mathrm{GeV}/c$, *Phys. Rev.* **D95**, 032004 (2017).
2. M. Aghasyan *et al.*, Light isovector resonances in $\pi^- p \to \pi^-\pi^-\pi^+ p$ at 190 GeV/c, *Phys. Rev.* **D98**, 092003 (2018).
3. F. Krinner, D. Greenwald, D. Ryabchikov, B. Grube and S. Paul, Ambiguities in model-independent partial-wave analysis, *Phys. Rev.* **D97**, 114008 (2018).
4. F. Krinner, Recent progress in the partial-wave analysis of the diffractively produced $\pi^-\pi^+\pi^-$ final state at COMPASS, *EPJ Web Conf.* **199**, 02003 (2019).
5. F. Krinner, Freed-Isobar Partial-Wave Analysis, PhD thesis, Technische Universität München (2018).
6. R. T. Deck, Kinematical interpretation of the first π–ρ resonance, *Phys. Rev. Lett.* **13**, 169 (1964).

Wave-selection techniques for partial-wave analysis in light-meson spectroscopy

F. M. Kaspar*, B. Grube, F. Krinner†, S. Paul and S. Wallner

Physics Department, Technical University of Munich,
Garching bei München, Bavaria 85748, Germany
†Max Planck Institut for Physics,
80805 Munich, Bavaria, Germany
**E-mail: florian.kaspar@tum.de*

The light-meson spectrum can be studied by analyzing data from diffractive dissociation of pion or kaon beams. The contributions of the various states that are produced in these reactions are disentangled by the means of partial-wave analysis. A challenge in these analyses is that the partial-wave expansion has to be truncated, i.e. that only a finite subset of the infinitely many partial-wave amplitudes can be inferred from the data. In recent years, different groups have applied regularization techniques in order to determine the contributing waves from the data. However, to obtain meaningful results the choice of the regularization term is crucial. We present our recent developments of wave-selection methods for partial-wave analyses based on simulated data for diffractively produced three-pion events.

Keywords: Partial-wave analysis; meson spectroscopy; model selection; HADRON 2019; analysis tools; regularization.

1. Introduction

The light-meson spectrum can be studied by analyzing data from diffractive dissociation of pion or kaon beams. The intensity distribution of the particles is usually modeled by the isobar model and the partial-wave analysis is performed using log-likelihood fits in bins of the final-state mass. A detailed description can be found in Ref. 1. The model parameters are a set of complex numbers $\{T_i\}$, each corresponding to the transition amplitude of a partial wave i. The number of partial waves in the decomposition is in principle infinite. However, only a finite subset can be fitted to data. Historically, this subset has been selected manually. However, this procedure may introduce hard-to-diagnose biases. It is therefore desirable to apply an algorithm to select relevant waves. In Ref. 2, the selection was performed using the so-called LASSO method, described in Ref. 3, which has also been applied to Baryon Spectroscopy, e.g., in Ref. 4. The method regularizes the fit to the data by adding a penalty term to the log-likelihood function and thereby forces the intensities $\{|T_i|^2\}$ of insignificant waves to zero in order to exclude them from the model. In Ref. 5, a similar method has been used, albeit with a different penalty term. We compare both penalty terms on simulated data for the process $\pi^- + p \to \pi^-\pi^-\pi^+ + p$ and suggest extensions that combine the advantages of both. We address strategies for tuning free parameters and the issue of multiple fit solutions.

2. Model Selection

We demonstrate the effects of different regularization procedures on a Monte Carlo data set of approximately 37×10^3 events in the region $1.80\,\text{GeV} < m_{3\pi} < 1.82\,\text{GeV}$. The data have been generated according to a reference wave set of 126 waves. The intensities of the individual waves vary over several orders of magnitude in order to imitate real data.

The selection procedure starts from a large, systematically constructed wave pool where the spin and angular momentum quantum numbers are limited. In this case, there are 753 waves in the pool. An unregularized fit of the whole wave pool is not able to reproduce the reference model. Figure 1 shows the partial-wave intensities of the fitted model (black) in descending order with the reference fit intensities (red) sorted accordingly. For many waves, the intensities of the fitted waves do not match the ones of the reference model. Figure 2 demonstrates the effect of the LASSO penalty, i.e.

$$\log \mathcal{L} - \frac{1}{\Gamma} \sum_i |T_i|, \tag{1}$$

where $\Gamma = 0.3$ in this case. See section 3 for our plans on parameter tuning.

To avoid discontinuities of the derivatives of $\mathcal{L}$ close to $T_i = 0$, we use the approximation $|T_i| \approx \sqrt{|T_i|^2 + \epsilon}$ with $\epsilon = 10^{-5}$. We observe a smooth drop in intensity down to the scale set by ϵ. For most large-intensity waves, we correctly recover the reference values. However, we observe a bias towards smaller intensities in the regularized fit.

Figure 3 demonstrates the effect of the "Biggest Conceivable Model" method (BCM), that has been developed in Ref. 5 and has first been applied to 3π real data in Ref. 6. The penalty term is the logarithm of a Cauchy distribution, i.e.

$$\log \mathcal{L} - \sum_i \log\left(1 + |T_i|^2 / \Gamma^2\right), \tag{2}$$

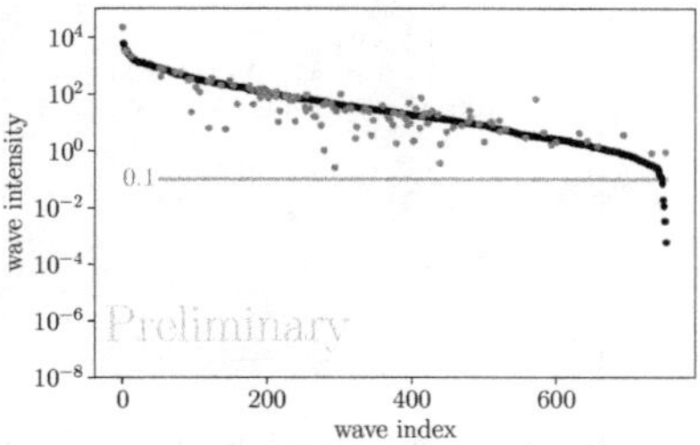

Fig. 1. Fit without regularization.

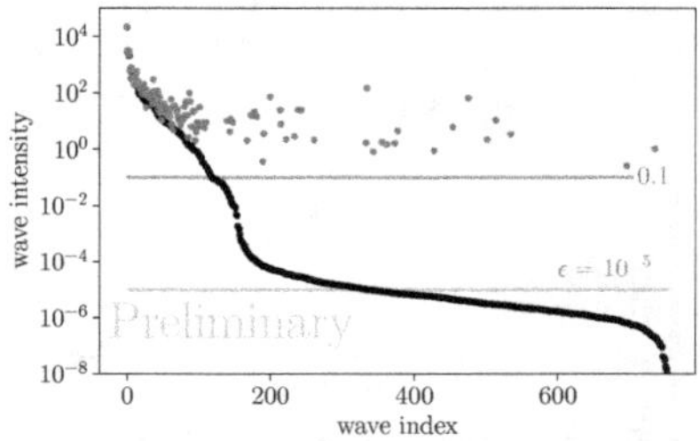

Fig. 2. Fit with LASSO regularization.

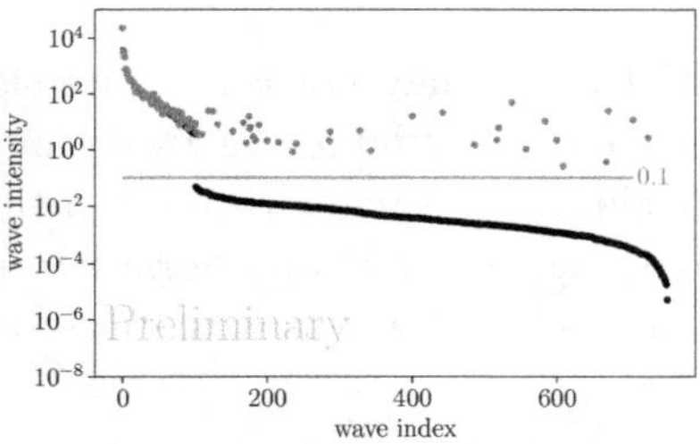

Fig. 3. Fit with BCM regularization.

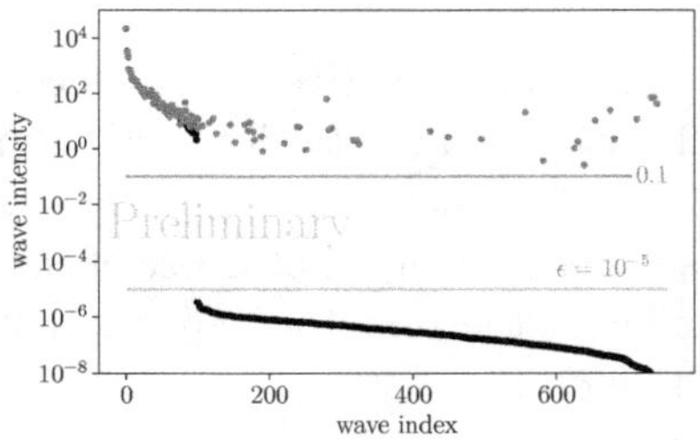

Fig. 4. Fit gen. Pareto regularization.[a]

where $\Gamma = 0.2$ for Fig. 3. The BCM method yields a discontinuous drop in intensity indicating the selection of waves above the drop. This behavior appears for sufficiently small value of Γ. In contrast to the LASSO, the bias towards smaller intensities is significantly reduced. This is because the penalty term rises only logarithmically. However, the deselected waves are not pushed all the way to zero (or ϵ) intensity, which may lead to a bias.

We also investigate the usefulness of other penalty terms that combine advantages of both of the aforementioned ones. In Ref. 7, the generalized Pareto distribution, i.e.

$$\log \mathcal{L} - \frac{1}{\zeta} \sum_i \log\left(1 + \zeta \left|T_i\right| / \Gamma\right), \tag{3}$$

has been proposed for selection. Its behavior for $\Gamma = 0.1$ and $\zeta = 0.5$ shown in Figure 4 is very similar to the BCM method for large intensities and to the LASSO for small intensities. The deselected waves have almost zero intensity, which further reduces their influence on the selected waves. For $\zeta \to 0$ the penalty approximates the LASSO.

---

[a] This plot has been added after the conference to demonstrate this penalty on the same data set.

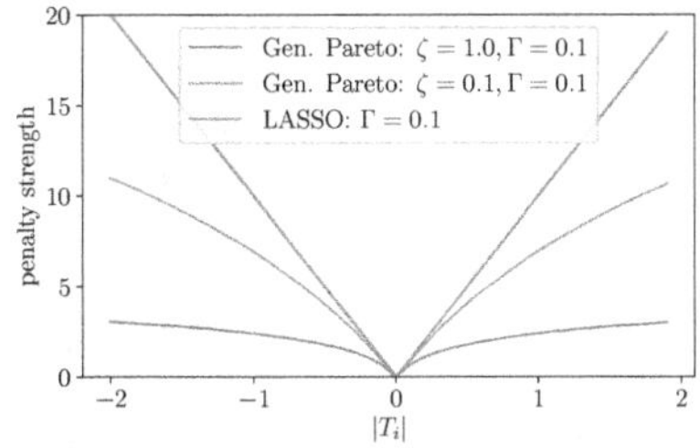

Fig. 5. Comparison of the LASSO with generalized Pareto.

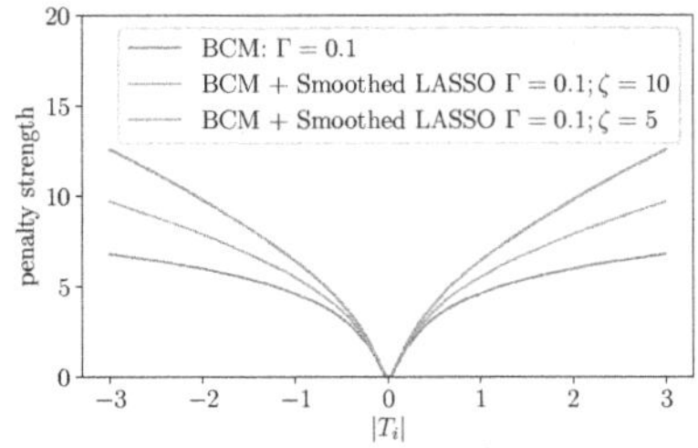

Fig. 6. Comparison of BCM with and without additional LASSO term.

Due to destructive interference, partial waves may not contribute significantly to the total intensity although being very large. A logarithmic penalty becomes too flat in order to provide a useful regularization in this case. Introducing a linear rise of the penalty by adding an appropriate term similar to the smoothed approximation of the LASSO (but with larger effective ϵ), i.e.

$$\log \mathcal{L} - \sum_i \left[\log\left(1 + |T_i|^2 / \Gamma^2\right) + \frac{1}{\zeta}\sqrt{|T_i|^2 / \Gamma^2 + 1} - \frac{1}{\zeta} \right], \tag{4}$$

can provide a trade-off between bias on the intensity and the introduction of destructive interference. The second free parameter ζ adjusts the relative strengths of the two terms.

A comparison of the different penalties is shown in Fig. 5 and Fig. 6.

3. Parameter Tuning

All penalty terms have free parameters and, a priori, it is not clear how to choose them. If the regularization is too strong, the fit will not describe the data, if it is too weak, the fit will describe fluctuations and a similar situation like in Fig. 1 will arise. In Ref. 2, the authors rely on the information criteria AIC[8] and BIC.[9] However, it is not clear whether the assumptions required for their applicability are fulfilled in our case (see references for details). Therefore, we are currently investigating a cross-validation procedure similar to Ref. 4.

4. Coping with Multiple Fit Solutions

The model-selection fits suffer from many local minima of the negative log-likelihood function. This causes a strong dependence of the fit result on the start parameter values. For every fit, we sample multiple random

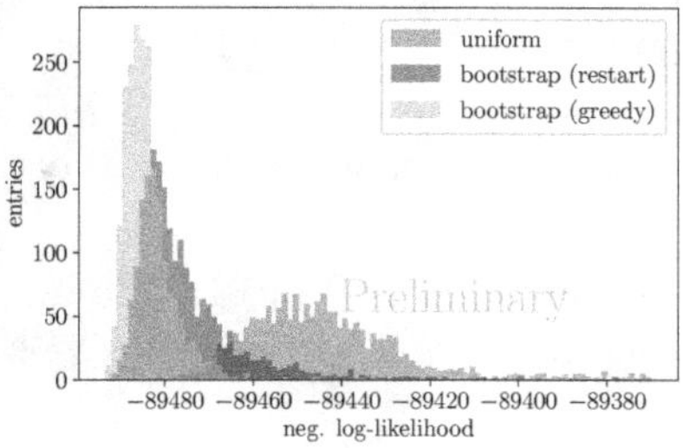

Fig. 7. Distribution of negative log-likelihood values for different start-parameter generation methods.

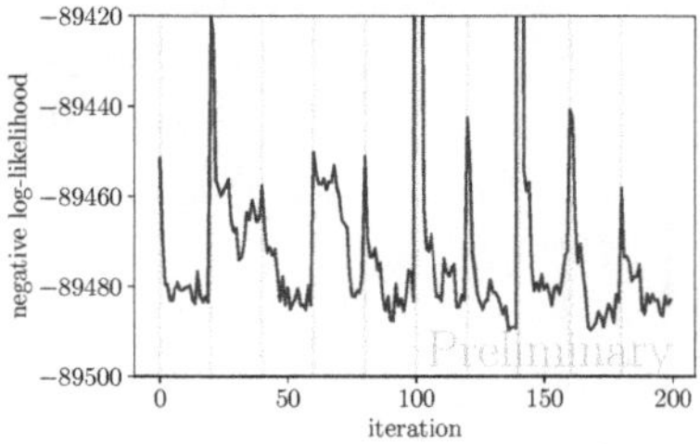

Fig. 8. Negative log-likelihood for boostrap start-parameter generation. Gray lines mark hard restarts.

start values and perform the optimization for each of them. We use the parameters with the smallest negative log-likelihood as our best estimate. We investigate the bootstrap restarting method described in Ref. 10 as a more efficient alternative to uniform start parameter sampling. The method uses parameter estimates of fits on a perturbed data set as a start-parameter proposal. We also consider a mixed approach, where a uniform draw of the start parameters is performed on every 20th step (hard restart) as opposed to only using the bootstrap ansatz (greedy). Figure 7 shows the concentration of the negative log-likelihood towards smaller values for both bootstrap methods in contrast to the spread out distribution for uniform sampling. In Figure 8 the value of the negative log-likelihood is shown over the number of draws for the reset method. The uniform restarts are marked by a light gray line every 20th step. How well this heuristic approach performs on our problem in practice is currently under investigation.

Acknowledgment

This work was supported by the BMBF, the Excellence Cluster ORIGINS which is funded by the Deutsche Forschungsgemeinschaft (DFG, German Research Foundation) under Germany's Excellence Strategy EXC-2094 390783311, and the Maier-Leibnitz-Laboratorium der Universitt und der Technischen Universität München.

References

1. C. Adolph *et al.*, Resonance production and $\pi\pi$ S-wave in $\pi^- + p \to \pi^-\pi^-\pi^+ + p_{\text{recoil}}$ at 190 GeV/c, *Phys. Rev. D* **95**, 032004 (2017).

2. B. Guegan, J. Hardin, J. Stevens and M. Williams, Model selection for amplitude analysis, *JINST* **10**, P09002 (2015).
3. R. Tibshirani, Regression Shrinkage and Selection via the Lasso, *Journal of the Royal Statistical Society. Series B (Methodological)* **58**, 267 (1996).
4. J. Landay, M. Dring, C. Fernández-Ramírez, B. Hu and R. Molina, Model selection for pion photoproduction, *Physical Review C* **95** (2017).
5. Karl Bicker, Model Selection for and Partial-Wave Analysis of a Five-Pion Final State at the COMPASS Experiment at CERN, PhD Thesis, Technische Universität München, (BY, Germany, 2016).
6. Oliver Drotleff, Model Selection of $\pi^- + p \to \pi^- \pi^+ \pi^- + p$ at the COMPASS Experiment at CERN, Diploma Thesis, Technische Universität München (2015).
7. A. Armagan, D. B. Dunson and J. Lee, Generalized double Pareto shrinkage, *Statistica Sinica* **23**, 119 (2013).
8. H. Akaike, A new look at the statistical model identification, *IEEE Transactions on Automatic Control* **19**, 716 (Dec 1974).
9. G. Schwarz, Estimating the dimension of a model, *Ann. Statist.* **6**, 461 (03 1978).
10. S. N. Wood, Minimizing model fitting objectives that contain spurious local minima by bootstrap restarting, *Biometrics* **57**, 240 (2001).

Δ(1232) Contribution in the π-photoproduction on nucleons in covariant chiral perturbation theory

G. H. Guerrero Navarro

Departamento de Física Teorica and Instituto de Fisica Corpuscular (IFIC), Centro Mixto UVEG-CSIC,
Valencia E-46071, Spain
E-mail: gusguena@ific.uv.es

We study the effects of the $\Delta(1232)$ resonance as an effective degree of freedom for charged and neutral pion photo-production on nucleons. Different observables have been calculated for these processes by using relativistic chiral perturbation theory up to $\mathcal{O}(p^3)$ in the δ counting, thus, including pion loops. We compare our model with a large database containing the available experimental data and constrain some unknown low energy constants.

1. Introduction

The pion photo-production processes on nucleons can be well described at low energies by using only nucleons and pions as the effective degrees of freedom. Baryon chiral perturbation theory (BChPT) reproduces fairly well these processes already at order $\mathcal{O}(p^4)$ both in heavy baryon and covariant approaches.[1] Despite the relative success of these calculations there are still some disagreements with data. Here, it will be shown that the effects of the $\Delta(1232)$ resonance mechanisms play a fundamental role and are essential to correctly reproduce the energy dependence of various observables.[2,3]

In this work, we consider nucleons, pions and the lowest lying resonance, $\Delta(1232)$, as the relevant degrees of freedom. The calculation is carried out up to and including $\mathcal{O}(p^3)$ in the δ power counting rule.[4] Furthermore, we use the extended on mass shell scheme (EOMS),[5] a covariant approach to BChPT.

This framework has been quite successful in the analysis of several processes, such as Compton scattering,[6,7] pion nucleon scattering[8–10] and weak pion production off nucleons.[11,12] Here, we focus on the different contributions of the $\Delta(1232)$ in the charged and neutral pion photo-production on nucleons at low energies and we compare the results of our model with the Δ-less approach.

2. Theoretical model

The terms of the chiral Lagrangian contributing up to $\mathcal{O}(p^3)$ to the amplitude for $\gamma N \to \pi N'$ in the isospin limit can be classified according to their chiral order and the particles involved as

$$\mathcal{L} = \mathcal{L}_N^{(1)} + \mathcal{L}_N^{(2)} + \mathcal{L}_{\pi\pi}^{(2)} + \mathcal{L}_N^{(3)} + \mathcal{L}_{\pi\pi}^{(4)} + \left(\mathcal{L}_{\Delta\pi N}^{(1)} + \mathcal{L}_{\Delta\gamma N}^{(2)}\right), \tag{1}$$

where $\mathcal{L}^{(k)}$ indicates Lagrangian terms of $\mathcal{O}(p^k)$. For detailed definitions of $\mathcal{L}_{\pi\pi}^{(2,4)}$ see Refs. 13, 14. On the other hand, for the nucleonic terms, $\mathcal{L}_N^{(1,2,3)}$, we follow the conventions given in Ref. 15 In this work, the relevant LECs are the chiral limit axial coupling, the nucleon mass and the pion decay constant, $\{g, m, F_0\}$ in $\mathcal{L}_N^{(1)}$, as well as $\{c_1, c_6, c_7\}$ in $\mathcal{L}_N^{(2)}$ and the set $\{d_8, d_9, d_{16}, d_{18}, d_{20}, d_{21}, d_{22}\}$ in $\mathcal{L}_N^{(3)}$. Finally, we have the $\Delta(1232)$ pieces, $\mathcal{L}_{\Delta\pi N}^{(1)}(h_A)$ and $\mathcal{L}_{\Delta\gamma N}^{(1)}(g_M)$, where $\{h_A, g_M\}$ are respectively related to the strong and electromagnetic decay constants, Γ_Δ^{EM} and Γ_Δ^{strong}, of the $\Delta(1232)$. See Refs. 16, 17 for details. The diagrams up to $\mathcal{O}(p^3)$ that contribute to the $\gamma N \to \pi N'$ reaction are the direct and crossed terms with a Δ as an intermediate state and leading to an $\mathcal{O}(p^{(5/2)})$ amplitude in the δ counting, as explained in Ref. 18.

For the pion photoproduction process the Lagrangian pieces shown in Eq. (1) generate different contributions at tree and one-loop level. The amputated amplitude order by order reads

$$\begin{aligned}\hat{\mathcal{M}} = \mathcal{M}_{\text{tree}}^{(1)}(\widetilde{g}, \widetilde{m}, F_0, M) + \mathcal{M}_{\text{tree}}^{(2)}(\widetilde{c}_1, \widetilde{c}_6, \widetilde{c}_7) + \mathcal{M}_{\text{tree}}^{(5/2)}(h_A, g_M) \\ + \mathcal{M}_{\text{tree}}^{(3)}(l_3, l_4, d_8, d_9, d_{16}, d_{18}, d_{20}, d_{21}, d_{22}) + \widetilde{\mathcal{M}}_{\text{loop}}^{(3)},\end{aligned} \tag{2}$$

with M the chiral limit pion mass from $\mathcal{L}_{\pi\pi}^{(2)}$, and $\widetilde{\mathcal{M}}_{\text{loop}}^{(3)} = \mathcal{M}_{\text{loop}}^{(3)} - \text{PCBT}$ the EOMS regularized loop amplitude. PCBT are the power counting breaking terms which are substracted through a redefinition of LECs at the lower order amplitudes, i.e., $\{\widetilde{g}, \widetilde{m}, \widetilde{c}_1, \widetilde{c}_6, \widetilde{c}_7\}$ are the EOMS shifted LECs used to regularize the loop diagrams.[18]

The loops appearing in the external legs of the Feynman diagrams are introduced by the Lehman-Symanzik-Zimmerman reduction formula.[19] In particular, we use the EOMS regularized wave function renormalizations $\mathcal{Z}_N = 1 + \widetilde{\delta}_N^{(2)}$ and $\mathcal{Z}_\pi = 1 - 2l_4 m_\pi^2 + \widetilde{\delta}_{Z_\pi}^{(2)}$ as in Ref. 11. The UV divergences from the loop calculations are subtracted following the $\widetilde{MS} = \overline{MS} - 1$ renormalization scheme. After the full renormalization procedure we implement further simplifications by expanding the chiral limit quantities in terms of its physical partners. Specifically, we take

$\{\widetilde{g}(g_A, d_{16}), \widetilde{m}(m_N, \widetilde{c}_1), F_0(F_\pi, l_4), M(m_\pi, l_3, l_4)\}$. In this way, we are able to write the renormalized amplitude $\mathcal{M} = \sqrt{\mathcal{Z}_\pi}\mathcal{Z}_N\hat{\mathcal{M}}$ as

$$\mathcal{M} = \mathcal{M}^{(1)}_{\text{tree}}(g_A, m_N, F_\pi, m_\pi) + \mathcal{M}^{(2)}_{\text{tree}}(\widetilde{c}_6, \widetilde{c}_7) + \mathcal{M}^{(5/2)}_{\text{tree}}(h_A, g_M)$$
$$+ \mathcal{M}^{(3)}_{\text{tree}}(d_8, d_9, d_{18}, d_{20}, 2d_{21} - d_{22}) + \widetilde{\mathcal{M}}^{(3)}_{\text{loop}} + (\widetilde{\delta}^{(2)}_{Z_N} + \frac{1}{2}\widetilde{\delta}^{(2)}_{Z_\pi})\mathcal{M}^{(1)}_{\text{tree}}. \quad (3)$$

Using this expression for the amplitude, we can compare with the experimental data currently available,[18] for various observables and constrain some of the unknown third order LECs.[18]

3. Results and discussion

Here, we present the comparison among two different models at $\mathcal{O}(p^3)$, one with the explicit inclusion of the Δ resonance and the other one without it. For the experimental database, we take into account a range of photon energies $E_\gamma^{lab} = 155$ MeV - 215 MeV, including different observables such as the angular and total cross section, as well as the beam polarizabilities.[18] In order to test the different models we minimize the χ squared taking the values for the unknown LECs as the fitting parameters.

In fact, many of the LECs are known from the analysis of other processes, in particular at $\mathcal{O}(p^3)$ we can set the values $\widetilde{c}_6 = 5.07(15)$,[12,20] $\widetilde{c}_7 = -2.68(08)$,[12,20] $d_{18} = [-0.20(80)\ \text{GeV}^{-2}]$,[8] $d_{22} = [5.20(02)\text{GeV}^{-2}]$[21] and $h_A = 2.87(03)$.[22] Consequently, the only fitting parameters are those given in Table 1 for both models.

According to the results in Table 1, the fit corresponding to the inclusion of Δ mechanisms results in a much better agreement with data than for the Δ-less case. This result is mostly driven by the π^0 photoproduction channel, where χ^2 is drastically reduced by the inclusion of the Δ at $\mathcal{O}(p^{5/2})$. On the other hand for $\pi^\pm$ channels the models show a similar agreement with or without the Δ inclusion. This is due to the fact that these channels are more sensitive to the lower orders. For that reason, considering higher order corrections does not improve much the agreement. In Fig. 1, we show the quality of the fits as a function of the energy. It can be clearly seen the superior convergence of the model with the Δ resonance at the studied energies.

This research was supported by MINECO (Spain) and the ERDF (European Commission) grant No. FIS2017-84038-C2-2-P, and SEV-2014-0398.

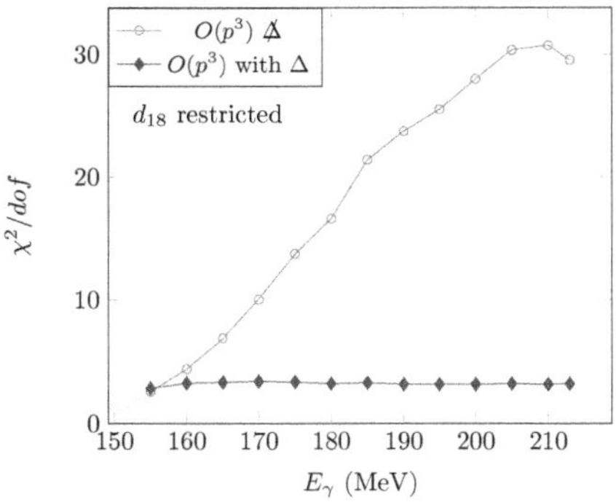

Fig. 1. χ^2 per degree of freedom as a function of the maximum photon energy of the data for the different studied models.

Table 1. The values of the LECs are dimensionless for g_M and in units of GeV^{-2} for ds. Fit refers to the standard setting, fit-$\not\Delta$ removes Δ mechanisms. Here d_{18} is restricted to $-0.20(80)$, and therefore shown in boldface, while g_M is left free

LEC	Fit	Fit - $\not\Delta$
$d_8 + d_9$	1.16(01)	3.53(01)
$d_8 - d_9$	1.09(18)	5.31(24)
d_{18}	**0.60**	**−1.00**
d_{20}	−0.74(17)	−3.81(19)
d_{21}	4.32(14)	6.98(15)
g_M	2.90(01)	...
$\chi^2_{TOT}/d.o.f.$	**3.22**	**29.5**
$\chi^2_{\pi^0}/d.o.f.$	3.58	37.2
$\chi^2_{\pi^+}/d.o.f.$	1.89	1.79
$\chi^2_{\pi^-}/d.o.f.$	1.99	1.90

References

1. M. Hilt, S. Scherer and L. Tiator, Threshold π^0 photoproduction in relativistic chiral perturbation theory, *Phys. Rev.* **C87**, 045204 (2013).
2. A. N. Hiller Blin, T. Ledwig and M. J. Vicente Vacas, Chiral dynamics in the $\vec{\gamma}p \to p\pi^0$ reaction, *Phys. Lett.* **B747**, 217 (2015).
3. A. N. Hiller Blin, T. Ledwig and M. J. Vicente Vacas, $\Delta(1232)$ resonance in the $\vec{\gamma}p \to p\pi^0$ reaction at threshold, *Phys. Rev.* **D93**, 094018 (2016).
4. V. Pascalutsa and D. R. Phillips, Effective theory of the delta(1232) in Compton scattering off the nucleon, *Phys. Rev.* **C67**, 055202 (2003).
5. T. Fuchs, J. Gegelia, G. Japaridze and S. Scherer, Renormalization of relativistic baryon chiral perturbation theory and power counting, *Phys. Rev.* **D68**, 056005 (2003).
6. V. Lensky and V. Pascalutsa, Predictive powers of chiral perturbation theory in Compton scattering off protons, *Eur. Phys. J.* **C65**, 195 (2010).
7. A. Hiller Blin, T. Gutsche, T. Ledwig and V. E. Lyubovitskij, Hyperon forward spin polarizability γ_0 in baryon chiral perturbation theory, *Phys. Rev.* **D92**, 096004 (2015).
8. J. M. Alarcon, J. Martin Camalich and J. A. Oller, Improved description of the πN-scattering phenomenology in covariant baryon chiral perturbation theory, *Annals Phys.* **336**, 413 (2013).

9. D.-L. Yao, D. Siemens, V. Bernard, E. Epelbaum, A. M. Gasparyan, J. Gegelia, H. Krebs and U.-G. Meißner, Pion-nucleon scattering in covariant baryon chiral perturbation theory with explicit Delta resonances, *JHEP* **05**, 038 (2016).
10. D. Siemens, V. Bernard, E. Epelbaum, A. Gasparyan, H. Krebs and U.-G. Meißner, Elastic pion-nucleon scattering in chiral perturbation theory: A fresh look, *Phys. Rev.* **C94**, 014620 (2016).
11. D.-L. Yao, L. Alvarez-Ruso, A. N. Hiller Blin and M. J. Vicente Vacas, Weak pion production off the nucleon in covariant chiral perturbation theory, *Phys. Rev.* **D98**, 076004 (2018).
12. D.-L. Yao, L. Alvarez-Ruso and M. J. Vicente Vacas, Neutral-current weak pion production off the nucleon in covariant chiral perturbation theory, *Phys. Lett.* **B794**, 109 (2019).
13. J. Gasser and H. Leutwyler, Chiral Perturbation Theory to One Loop, *Annals Phys.* **158**, 142 (1984).
14. J. Gasser, M. E. Sainio and A. Svarc, Nucleons with Chiral Loops, *Nucl. Phys.* **B307**, 779 (1988).
15. N. Fettes, U.-G. Meißner, M. Mojzis and S. Steininger, The Chiral effective pion nucleon Lagrangian of order p^4, *Annals Phys.* **283**, 273 (2000), [Erratum: Annals Phys.288,249(2001)].
16. V. Pascalutsa, The Delta(1232) Resonance in Chiral Effective Field Theory, *Prog. Part. Nucl. Phys.* **61**, 27 (2008).
17. V. Pascalutsa, M. Vanderhaeghen and S. N. Yang, Electromagnetic excitation of the Delta(1232)-resonance, *Phys. Rept.* **437**, 125 (2007).
18. G. H. Guerrero Navarro, M. J. Vicente Vacas, A. N. H. Blin and D.-L. Yao, Pion photoproduction off nucleons in covariant chiral perturbation theory, *Phys. Rev.* **D100**, 094021 (2019).
19. H. Lehmann, K. Symanzik and W. Zimmermann, On the formulation of quantized field theories, *Nuovo Cim.* **1**, 205 (1955).
20. T. Bauer, J. C. Bernauer and S. Scherer, Electromagnetic form factors of the nucleon in effective field theory, *Phys. Rev.* **C86**, 065206 (2012).
21. D.-L. Yao, L. Alvarez-Ruso and M. J. Vicente-Vacas, Extraction of nucleon axial charge and radius from lattice QCD results using baryon chiral perturbation theory, *Phys. Rev.* **D96**, 116022 (2017).
22. V. Bernard, E. Epelbaum, H. Krebs and U.-G. Meißner, New insights into the spin structure of the nucleon, *Phys. Rev.* **D87**, 054032 (2013).

Partial wave analysis using the PAWIAN software package

M. Albrecht*, B. Kopf†, X. Qin and U. Wiedner

Ruhr-Universität Bochum
Universitätsstr. 150, 44801 Bochum, Germany
*E-mail: *malte@ep1.rub.de, †bertram@ep1.rub.de*

PAWIAN is a powerful, user-friendly and highly modular partial wave analysis software package with the aim to support analyses for a multitude of different cases in the field of hadron physics. Real data originating from the $\bar{p}p$ annihilation process and from e^+e^- reactions are currently under investigation with PAWIAN. The software is written in C++ and follows an object-oriented approach with a wide range of flexibility. The software supports parallelization of the calculations and coupled channel analyses. The code allows to be easily extended for further decay models, complete amplitudes or other descriptions for the dynamics.

After an overview of the general features and capabilities of PAWIAN, some recent improvements concerning the treatment of analyticity as well as the extraction of pole parameters are discussed.

Keywords: Partial wave analysis; amplitude analysis; hadron spectroscopy.

1. Introduction

The development of the PAWIAN (PArtial Wave Interactive ANalysis) software package for partial wave analysis (PWA) is ongoing since a few years and with the complexity of its features also the number of supported physics cases increases steadily. Modern hadron physics experiments like the BESIII experiment located at the Beijing Electron Positron Collider II in China, or the $\overline{\text{P}}$ANDA experiment currently under construction at the Facility for Antiproton and Ion Research (FAIR) in Germany now and in the future provide high statistics data sets to perform precision spectroscopy of hadronic states. A highlight of these experiments is the investigation of possibly exotic states, which includes the search for glueballs, hybrids or multiquark states. To identify these states and map out their spectrum, spin-parity quantum numbers, pole positions or masses and widths as well as coupling strengths have to be determined. For this purpose partial wave analyses of the experimental data have to be performed, which requires a flexible and powerful software package.

2. Main Features of the Software Package

The main feature of the PAWIAN software is to provide the machinery necessary to perform unbinned maximum likelihood fits as well as binned fits to experimental data. The multi-dimensional, complex fit function is given by an expression for the transition amplitude of the reaction under study. This expression is dynamically generated from configuration parameters defined by the user. PAWIAN is written in C++ and licensed under the GNU General Public Licence v3. The main features and some results from earlier analyses are reported in Ref. 1.

PAWIAN relies on few external packages, keeping the dependencies as simple as possible: The package qft++[2] is used to numerically evaluate quantum field theory expressions. For the unbinned maximum likelihood fits the minimizer MINUIT2[3] is utilized, providing a gradient descent algorithm. However, the software allows for an easy replacement of the minimizer package. Data from different experiments and therefore different production processes can be analyzed. So far, $\overline{p}p$ and e^+e^- annihilation, as well as two-photon production and two-body scattering processes (e.g. $\pi\pi$ scattering) are supported. The production and decay amplitudes can be described with different spin formalisms (canonical, helicity, Rarita-Schwinger formalism). Similarly, also different options for the description of the dynamics of contributing resonances (Breit Wigner, K matrix, ...) are supported. All necessary options to set up a fit are controlled by the users and fed to the program using plain text configuration files with a human readable syntax. This includes the definition of the complete decay chain, spin-formalism, dynamics for each resonance as well as parameters needed to steer the fit procedure and network communication. Furthermore, the software supports the simultaneous analysis of multiple data sets at once. This includes especially the combined analysis of data from different production processes and / or resulting in different final states. When using this *coupled-channel mode*, PAWIAN generates one common global parameter list, of which many parameters may be shared among the individual reactions in the fit. This powerful approach provides valuable physical constraints and therefore allows to properly respect principles as e.g. unitarity. To facilitate such coupled channel fits, proper parallelization is an indispensible feature of the software. PAWIAN features full parallelization on the event level using a server-client network architecture. Since individual events are statistically independent from each other, this approach to parallelization has proven to scale performantly up to a level of several hundreds of clients per fit and does not require any communication between the individual client processes. Only the server process communicates directly with the minimizer and periodically saves a set of current parameters. The communication between the clients and the server consists of lists of parameters and likelihood values only, amounting to a network load of only a few kb/s per server process.

Apart from performing the fit, the software features a simple event generator as well as quality assurance tools. The generator can use an amplitude model with a given parameter set to generate events according to the desired model. Histogramming can be performed using the external ROOT package, which is however not needed during the fit. Also this interface to the user is kept flexible and histogramming may also be performed with other packages (e.g. via python scripts, ...). Different analyses have already been performed using the features listed above,[4,5] including a sophisticated coupled-channel analysis utilizing different $\bar{p}p$ data sets as well as scattering data.[6]

3. Description of Resonances

When performing fits to experimental data, one crucial point that shall be highlighted here is the description of resonance decay dynamics. It is well known that the commonly used Breit-Wigner parameterization is only valid for relatively narrow, isolated resonances, far away from thresholds and with a strong coupling to only one channel. In most cases, especially in the light hadron regime, all of these prerequisites are not met and broad, overlapping states of the same quantum numbers that decay into multiple final states are observed. Although PAWIAN features the description of resonances using relativistic Breit-Wigner parameterizations, a more sophisticated approach is needed in most cases. One possibility is to make use of the K matrix approximation for the description of resonances in 2-body scattering processes, which can be extended in the so-called P vector approach[7] to describe the production of resonances (e.g. in $\bar{p}p$ or e^+e^- annihilation). Both can be used (also in conjunction) within PAWIAN. This is an indispensible feature to perform coupled-channel analyses and provides strong physical constraints. In general for a scattering process the S matrix can be written as $S = I + 2i\sqrt{\rho}T\sqrt{\rho}$, where I is the identity matrix, ρ is the diagonal phase space matrix and T stands for the Lorentz-invariant transition matrix. The T matrix can now be expressed in terms of a real, symmetric matrix K as $T = (I - iK\rho)^{-1}K$. In our implementation the elements of this K matrix are in a simplified version given by

$$K_{ij}(s) = \sum_{\alpha} \frac{g_{\alpha i} g_{\alpha j}}{m_{\alpha}^2 - s} + \sum_{k} c_{kij} s^k, \tag{1}$$

where the indices i and j are the different reaction channels and α represents the poles of the K matrix. The parameters $g_{\alpha i}$ and m_α are the coupling strength of a resonance α to the channel i and its mass in the K matrix representation. Without violating unitarity, it is furthermore allowed to add polynomial terms of order k in s to the K matrix elements, sometimes called background terms. Their coefficients are denoted here with c_{kij}. In general the phase space factors ρ appearing in the formulae above are given by $\rho = \sqrt{\left[1 - \left(\frac{m_a+m_b}{s}\right)^2\right] \cdot \left[1 - \left(\frac{m_a-m_b}{s}\right)^2\right]}$. This expression is without further restrictions usable if s is larger than the threshold given by $m_a + m_b$, which is always the case in single channel descriptions, where all resonances are also situated above threshold. When dealing with coupled channel fits however, resonances may be below threshold for specific channels and the given function for ρ clearly violates constraints from analyticity due to unphysical behavior below threshold on the real axis (pole at $s = 0$ and unphysical cuts for $m_a \neq m_b$). To overcome these restrictions, we replace the standard phase-space factors with a Chew-Mandelstam matrix whose elements are calculated by the functions as defined in Ref. 8. Thus, $T = (1 - iK\rho)^{-1}K$ is replaced by $T = (1 - KC(s))^{-1}K$ with $C(s)$ being the Chew-Mandelstam matrix. The connection between this matrix and the phase space factors is given by $Im(C_{ii}(s)) = \rho_{ii}(s)$. Figure 1 shows a comparison of ρ and the Chew-Mandelstam function for a specific decay channel. The approach discussed so far is only valid for the case of vanishing widths of scattering particles. A recipe to extend the formalism to particles with

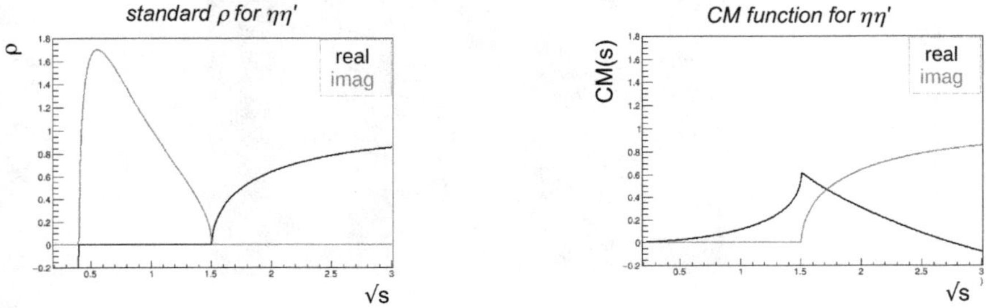

Fig. 1. Comparison of the standard phase space factor with the Chew-Mandelstam function for an exemplary case ($\eta\eta'$ channel)

non-negligible decay widths (e.g. ρ, K^*,...) is given in Ref. 8 as well and is implemented in PAWIAN via the usage of pre-calculated lookup tables. Configuration parameters for a K matrix (g-factors, masses, phase space factors, Adler-0 terms,...) are again controlled by the user in dedidcated configuration files.

4. Extraction of Pole Parameters

While masses and widths of resonances described by Breit-Wigner functions can be determined directly in the fit, the parameters of resonances described with the K matrix approach need to be determined from the pole positions in the complex energy plane of the T matrix on the Riemann sheet closest to the physical sheet. Apart from the fit itself, the Chew-Mandelstam function is also used for the extraction of these resonance properties. A tool is provided with the PAWIAN package which allows to scan the complex energy plane of the T matrix for poles for a given set of parameters. An example of the complex energy plane is shown in Figure 2. Besides the mass and width given by the position of the pole in the complex plane, also the coupling of a resonance to a specific channel is accessible. This quantity is related to the partial decay width and may be accessed through the residue of the respective pole of the T matrix with a projection to the relevant decay channel k on the sheet closest to the physical one. In general the residues are determined by calculating the integral along a closed contour around the pole α as $Res^{\alpha}_{k\to k} = \frac{1}{2\pi i}\oint \sqrt{\rho_k}T_{k\to k}(z)\sqrt{\rho_k}dz$. In PAWIAN we make use of a numerical estimation of this integral by means of a Laurent expansion as described e.g. in Ref. 9. This way no integration is needed and the calculation is very fast.

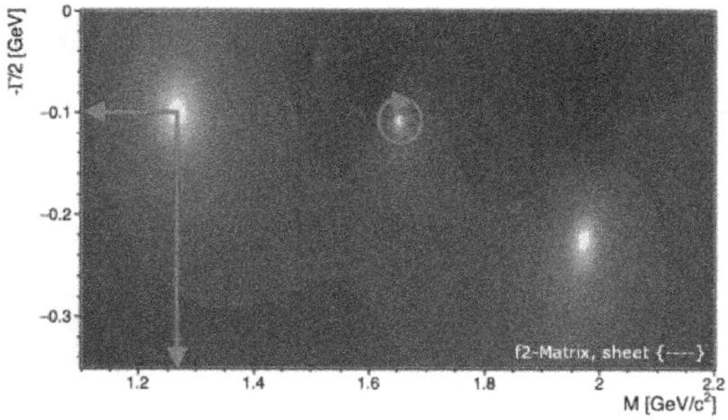

Fig. 2. Example for poles in the complex energy plane of the T matrix. Poles on one specific Riemann sheet (− − −−) for resonances in the D wave can be seen.

References

1. B. Kopf, H. Koch, J. Pychy and U. Wiedner, Partial wave analysis for $\bar{p}p$ and e^+e^- annihilation processes, *Hyperfine Interact.* **229**, 69 (2014).
2. M. Williams, Numerical Object Oriented Quantum Field Theory Calculations, *Comput. Phys. Commun.* **180**, 1847 (2009).
3. F. James and M. Roos, Minuit: A System for Function Minimization and Analysis of the Parameter Errors and Correlations, *Comput. Phys. Commun.* **10**, 343 (1975).
4. C. Amsler, F. H. Heinsius, H. Koch, B. Kopf, U. Kurilla, C. A. Meyer, K. Peters, J. Pychy, M. Steinke and U. Wiedner, Spin Density Matrix of the ω in the Reaction $\bar{p}p \to \omega\pi^0$, *Eur. Phys. J.* **C75**, 124 (2015).
5. M. Ablikim *et al.*, Observation of $\eta_c \to \omega\omega$ in $J/\psi \to \gamma\omega\omega$, *Phys. Rev.* **D100**, 052012 (2019).
6. M. Albrecht *et al.*, Coupled Channel Analysis of $\bar{p}p \to \pi^0\pi^0\eta$, $\pi^0\eta\eta$ and $K^+K^-\pi^0$ at 900 MeV/c and of $\pi\pi$-Scattering Data (2019).
7. I. J. R. Aitchison, K-MATRIX FORMALISM FOR OVERLAPPING RESONANCES, *Nucl. Phys.* **A189**, 417 (1972).
8. J. L. Basdevant and E. L. Berger, Unstable Particle Scattering and an Analytic Quasiunitary Isobar Model, *Phys. Rev.* **D19**, 239 (1979).
9. M. Doring, C. Hanhart, F. Huang, S. Krewald, U. G. Meissner and D. Ronchen, The reaction $\pi^+p \to K^+\Sigma^+$ in a unitary coupled-channels model, *Nucl. Phys.* **A851**, 58 (2011).

Partial wave mixing in Hamiltonian effective field theory

Yan Li[*,a], Jia-jun Wu[a], Curtis D. Abell[b], Derek B. Leinweber[b]
and Anthony W. Thomas[b,c]

[a] *School of Physical Sciences, University of Chinese Academy of Sciences (UCAS), Beijing 100049, China*
[b] *Special Research Center for the Subatomic Structure of Matter (CSSM), Department of Physics, University of Adelaide, Adelaide 5005, Australia*
[c] *ARC Centre of Excellence for Particle Physics at the Terascale (CoEPP), Department of Physics, University of Adelaide, Adelaide, South Australia 5005, Australia*
[*] *E-mail: liyan175@mails.ucas.edu.cn*

We explore partial-wave mixing in the finite volume based on HEFT, and provide the P-Matrix to show the degree of partial-wave mixing. An example of isospin-2 $\pi\pi$ scattering is used to check the consistency between HEFT and Lüscher's method.

Keywords: Finite-volume effect; partial wave mixing; lattice QCD.

1. Introduction

The infinite-volume phase shift observed in experiment and the finite-volume spectrum obtained from Lattice QCD proved to be related by the model independent Lüscher's formula.[1–3] Another equivalent approach connecting them is Hamiltonian effective field theory (HEFT).[4–7] In HEFT, one has the infinite- and finite-volume Hamiltonians parameterized by a common set of parameters. The phase shift can be predicted through the Hamiltonian constrained by the Lattice QCD spectrum.

As a consequence of the violation of spherical symmetry in the finite volume, the partial wave quantum numbers (l, m) fail to be good ones. That different partial waves are mixed in the finite volume is called partial-wave mixing. Here we explore partial-wave mixing in HEFT, and provide the P-Matrix defined in Eq. (4) to reflect the degree of partial-wave mixing. Finally, an example of isospin-2 $\pi\pi$ scattering is used to examine the consistency between this method and Lüscher's method.

2. Partial Wave Mixing in HEFT

In the infinite volume, to utilize the spherical symmetry, one combines the plane wave states $|\mathbf{k}\rangle$ with spherical harmonics $Y_{l,m}(\hat{\mathbf{p}})$ to define

$$|k;l,m\rangle := \int d\Omega_{\hat{\mathbf{k}}}\, Y_{lm}(\hat{\mathbf{k}})\; |\mathbf{k}\rangle\ . \tag{1}$$

Then spherical symmetry only allows interactions of the following form

$$\hat{V} = \int \frac{k'^2\, dk'}{(2\pi)^3} \int \frac{k^2\, dk}{(2\pi)^3} \sum_{l,m} v_l(k',k)\; |k';l,m\rangle\, \langle k;l,m|\ . \tag{2}$$

In the finite volume, only the periodic plane wave states denoted by $|\mathbf{n}\rangle$ will survive. Therefore the integration in Eq. (1) should be replaced by a summation of discrete momentum. Furthermore, one can define new states in the finite volume as follows

$$|N;l,m\rangle = \sum_{|\mathbf{n}|^2=N} \sqrt{4\pi}\, Y_{lm}(\hat{\mathbf{n}})\, |\mathbf{n}\rangle\ . \tag{3}$$

That $|N;l,m\rangle$ are in general not orthogonal to each other reflects the partial-wave mixing. One can define the inner product matrix (named as P-Matrix)

$$[P_N]_{l',m';l,m} := \langle N;l',m'|N;l,m\rangle = 4\pi \sum_{|\mathbf{n}|^2=N} Y^*_{l'm'}(\hat{\mathbf{n}}) Y_{lm}(\hat{\mathbf{n}}) \tag{4}$$

to show the degree of partial-wave mixing. If one considers a finite but non-zero momentum $2\pi\sqrt{N}/L$, in the infinite-volume limit $L \to \infty$, one can see the recovery of the spherical symmetry as follows

$$[P_N]_{l',m';l,m} \xrightarrow{N\to\infty} C_3(N)\, \delta_{l',l}\, \delta_{m',m}\ , \tag{5}$$

where $C_3(N)$ counts the $\mathbf{n}$ satisfying $|\mathbf{n}|^2 = N$. Here we provide a few examples of the numerical values for $P_N/C_3(N)$ with $N = 1$ and 941 in Fig. 1, and it is indeed approaching to an identity matrix.

To make use of the finite-volume symmetry, one combines $|N;l,m\rangle$ into states that respect the cubic symmetry

$$|N,l;\Gamma,f,\alpha\rangle = \sum_m [C_l]_{\Gamma,f,\alpha;m}\, |N;l,m\rangle\ , \tag{6}$$

where (Γ, f, α) represents the α-th vector of the f-th occurrence (in a given l) of the irreducible representation Γ, and $[C_l]_{\Gamma,f,\alpha;m}$ are group theoretic constants independent of N. The inner products between $|N,l;\Gamma,f,\alpha\rangle$

relate to the P-Matrix directly from the definition. The states after orthonormalization will be the final basis.

Working with $|\mathbf{n}\rangle$, one uses a momentum cutoff $N_{\rm cut}$ to make the finite-volume Hamiltonian of finite dimension. With $|N;l,m\rangle$, one then uses a partial wave cut $l_{\rm cut}$ to decouple the potential from high partial waves. With $|N,l;\Gamma,f,\alpha\rangle$, different irreducible representations of the cubic group are also decoupled. Finally, the final dimension of the finite-volume Hamiltonian will be significantly reduced.

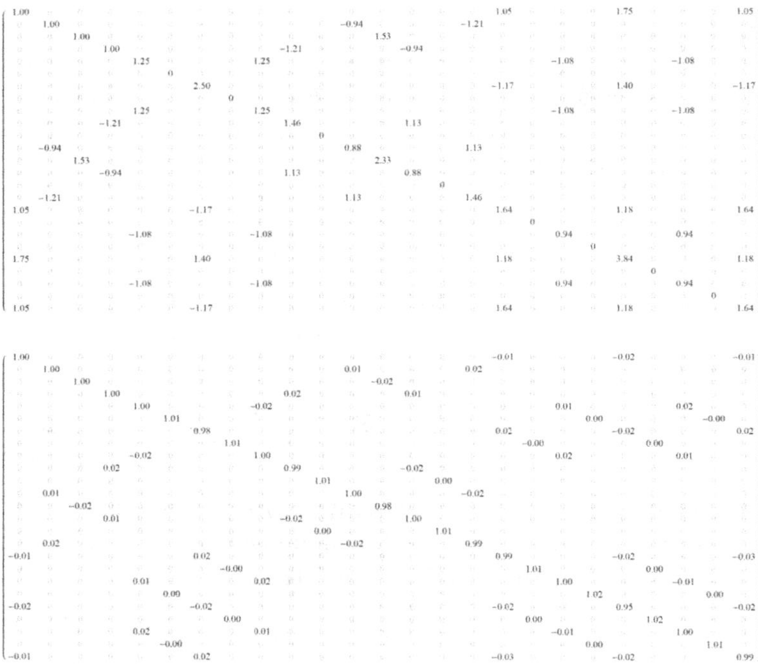

Fig. 1. $P_N/C_3(N)$ with $N = 1,941$ and $C_3(N) = 6,552$ respectively: The 25×25 matrix ordered as $(l,m) = (0,0)$, $(1,-1)$, $(1,0)$, $(1,1)$, $\cdots$, $(4,4)$.

3. Example of Isospin-2 $\pi\pi$ Scattering

The parametrization for the potential is chosen to be (measured in units of lattice spacings)

$$v_l(p,k) = f_l(p)\, G_l\, f_l(k)\,, \qquad f_l(k) = \frac{(d_l\, k)^l}{\left(1 + (d_l\, k)^2\right)^{l/2+2}}\,, \tag{7}$$

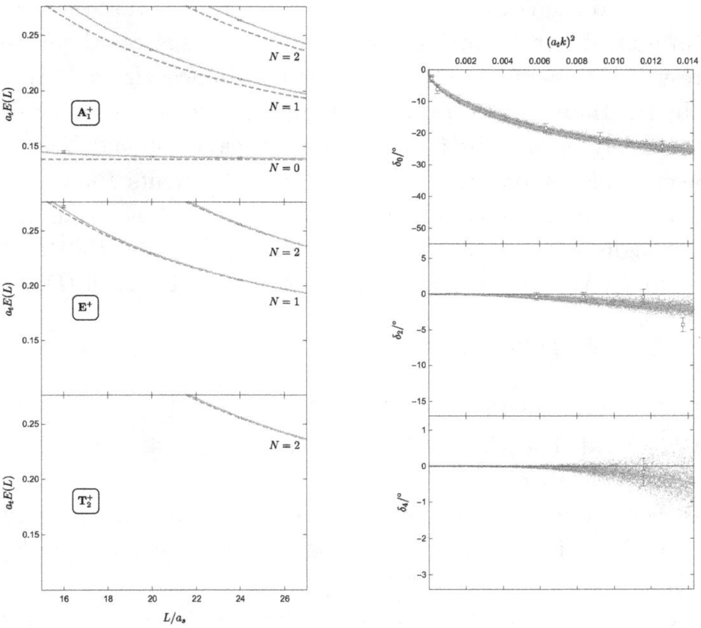

Fig. 2. Left: Data points describe the lattice spectrum for irreps $\mathbf{A}_1^+$, $\mathbf{E}^+$, $\mathbf{T}_2^+$ – from Ref. 8. Dashed curves represent the non-interacting rest-frame pion-pair energies $2\sqrt{m_\pi^2 + k_N^2}$. Solid curves represent the HEFT prediction of the volume dependent spectrum using the fitted parameters. Right: Data points are phase shifts predicted by Lüscher's method – from Ref. 8. Solid curves represent the HEFT prediction of the *s*- (top), *d*- (middle) and *g*- (bottom) wave phase shifts and the scattered points describe the uncertainty.

where there are 2 parameters G_l and d_l for each partial wave. The partial wave cut $l_{\text{cut}} = 4$ and Bose symmetry only allow *s*-, *d*- and *g*-waves. The lattice QCD results from Ref. 8 and the HEFT results are shown in Fig. 2, in which HEFT and Lüscher's method are indeed consistent.

4. Summary

We have explored partial-wave mixing in the finite volume based on HEFT, and provided the P-Matrix defined in Eq. (4) to show the degree of partial-wave mixing. Then an example of isospin-2 $\pi\pi$ scattering was used to check the consistency between HEFT and Lüscher's method.

Acknowledgments

It is a pleasure to thank Stephen Sharpe for interesting discussions on the research presented herein during his visit as a George Southgate Fellow. The finite-volume energy levels and their covariances from Ref. 8 were provided by the Hadron Spectrum Collaboration – no endorsement on their part of the analysis presented in the current paper should be assumed. This project is also supported by the Thousand Talents Plan for Young Professionals. This research was supported by the Australian Research Council through ARC Discovery Project Grants Nos. DP150103101 and DP180100497 (A.W.T.) and DP150103164 and DP190102215 (D.B.L.).

References

1. M. Lüscher, Volume Dependence of the Energy Spectrum in Massive Quantum Field Theories. 1. Stable Particle States, *Commun. Math. Phys.* **104**, 177 (1986).
2. M. Lüscher, Volume Dependence of the Energy Spectrum in Massive Quantum Field Theories. 2. Scattering States, *Commun. Math. Phys.* **105**, 153 (1986).
3. M. Lüscher, Two particle states on a torus and their relation to the scattering matrix, *Nucl. Phys.* **B354**, 531 (1991).
4. J. M. M. Hall, A. C. P. Hsu, D. B. Leinweber, A. W. Thomas and R. D. Young, Finite-volume matrix Hamiltonian model for a $\Delta \to N\pi$ system, *Phys. Rev.* **D87**, 094510 (2013).
5. J. M. M. Hall, W. Kamleh, D. B. Leinweber, B. J. Menadue, B. J. Owen, A. W. Thomas and R. D. Young, Lattice QCD Evidence that the $\Lambda(1405)$ Resonance is an Antikaon-Nucleon Molecule, *Phys. Rev. Lett.* **114**, 132002 (2015).
6. Z.-W. Liu, W. Kamleh, D. B. Leinweber, F. M. Stokes, A. W. Thomas and J.-J. Wu, Hamiltonian effective field theory study of the $\mathbf{N^*(1535)}$ resonance in lattice QCD, *Phys. Rev. Lett.* **116**, 082004 (2016).
7. J.-j. Wu, D. B. Leinweber, Z.-w. Liu and A. W. Thomas, Structure of the Roper Resonance from Lattice QCD Constraints, *Phys. Rev.* **D97**, 094509 (2018).
8. J. J. Dudek, R. G. Edwards and C. E. Thomas, S and D-wave phase shifts in isospin-2 pi pi scattering from lattice QCD, *Phys. Rev.* **D86**, 034031 (2012).

Three-particle system in a finite volume

Jin-Yi Pang*
College of Science, University of Shanghai for Science and Technology, Shanghai, 200093, China
Helmholtz-Institut für Strahlen- und Kernphysik and Bethe Center for Theoretical Physics, Universität Bonn, Bonn, 53115, Germany
**E-mail: pang@hiskp.uni-bonn.de*
www.hiskp.uni-bonn.edu

Lattice QCD simulation for hadron spectroscopy has reached three-body threshold. However, it is difficult to go further without an appropriate method to analyze the finite volume effect in three-body system. We derive the three-body quantization condition in a finite volume using an effective field theory in the particle-dimer picture. It is a powerful and transparent method to interpret three-body physical observables from lattice simulations. We review the whole method including formalism, quantization condition and the application for both 3-body bound states and scattering states.

Keywords: Lattice QCD; three-body system; finite volume; dimer formalism.

1. Introduction

The analysis of the lattice data in the three-particle sector has attracted most interest[1–12] (for the recent review, see Ref. 13). This is because lattice simulation has touched the three-body kinematics region. Our works[7,8,10,12] has showed its transparent and powerful to the problem via the non-relativistic effective theory and the particle-dimer picture in a finite volume. It is demonstrated that effective field theory is very important to connect lattice simulation results with physical experiment data.[8] Three-body force, the critical point of the problem is fit to the lattice spectrum.[7]

The paper is arranged as followed. We show the formalism describing 3-body system in Sec. 2. The formalism is put into a finite volume and quantization condition is derived in Sec. 3. We apply the quantization condition to analyze the 3-body bound states and scattering states in Sec. 4 and 5. Sec. 6 gives the conclusion and outlook.

2. Particle-dimer formalism

We introduce particle-dimer formalism to make the 3-body effective theory more compact and clear. Dimer field is an auxiliary degree of freedom which describes a 2-body subsystem. For simplicity, only the scalar dimer is considered.[a]

The particle-dimer scattering equation is written down as,

$$\mathcal{M}(\mathbf{p},\mathbf{q};E) = Z(\mathbf{p},\mathbf{q};E) + 8\pi \int^{\Lambda} \frac{d^3k}{(2\pi)^3} Z(\mathbf{p},\mathbf{k};E)\tau(\mathbf{k};E)\mathcal{M}(\mathbf{k},\mathbf{q};E). \quad (1)$$

As long as 3-body force is determined, we can solve the scattering equation. The analytic structure of particle-dimer scattering amplitude $\mathcal{M}$ then encodes all the 3-body dynamics.

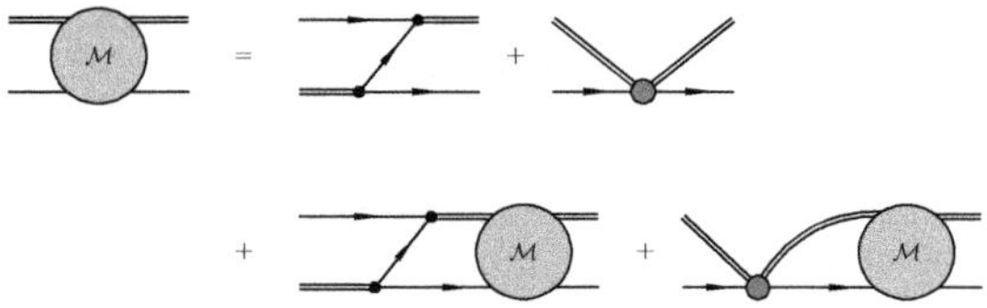

Fig. 1. Particle-dimer scattering equation

In the infinite volume, rotational symmetry is respected. So the scattering equation can be projected to the specific angular momentum quantum number.

3. Projected quantization condition

3.1. *Particle-dimer scattering equation in a finite volume*

Imposing periodical boundary condition for lattice configuration makes the momentum of the system discrete. Whereas the local LECs is non-sensitive to the volume effect. Therefore both 2-body input and 3-body force are preserved. The finite volume effects are reflected by changing loop integrals to loop sums. There are two corrections in the finite volume scattering

[a]The dimer with arbitrary spin can be checked in our paper.[8]

equation, iteration part and dimer propagator.

$$\mathcal{M}_L(\mathbf{p},\mathbf{q};E) = Z(\mathbf{p},\mathbf{q};E) + \frac{8\pi}{L^3}\sum_{\mathbf{k}}^{\Lambda} Z(\mathbf{p},\mathbf{k};E)\tau_L(\mathbf{k};E)\mathcal{M}_L(\mathbf{k},\mathbf{q};E)\,, \tag{2}$$

where $\mathbf{k} = \frac{2\pi}{L}\mathbf{n}\,,\ \mathbf{n} \in \mathbb{Z}^3$ is a quantized three-momentum in a finite volume

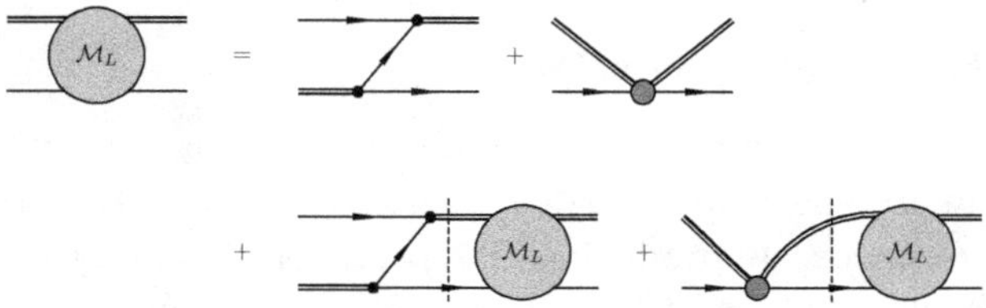

Fig. 2. Finite volume scattering equation

3.2. *Quantization condition*

In a finite volume, the spectrum is discrete both below and above threshold. All the energy levels are determined by the poles of finite volume scattering amplitude $\mathcal{M}_L$. Based on the finite volume scattering equation, Eq. (2), these poles are obtained from quantization condition,

$$\det\left(\tau_L^{-1} - Z\right) = 0\,. \tag{3}$$

Now we have written down the 3-body quantization condition. It gives the finite volume spectrum of 3-body system. Inversely, 3-body force can be fixed by plugging lattice simulation results into quantization condition. With 3-body force and pertinent 2-body input, we can solve the scattering equation in the infinite volume, *i.e.*, Eq. (1). Particle-dimer scattering amplitude $\mathcal{M}$ encodes all the 3-body dynamics information.

4. Three-body bound states

We apply 3-body quantization condition on the system where Efimov states exist. The 3-body force plays an essential role for the forming of Efimov states which are the typical 3-body bound states.

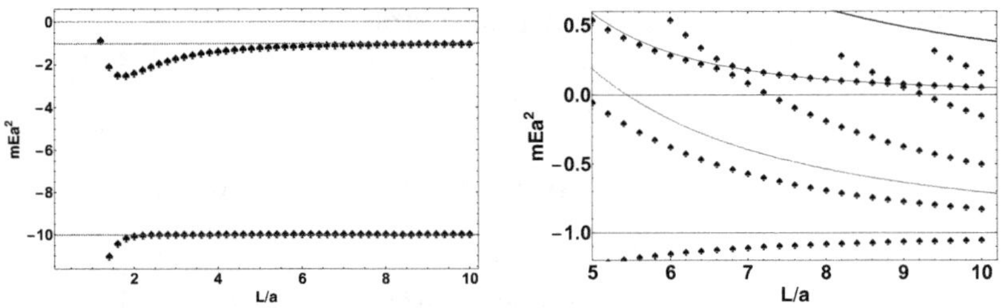

Fig. 3. The two Efimov states in a finite volume and the 3-body lattice spectrum.

5. Three-body scattering states

Avoided level crossing happens between the two kinds of scattering states. In 2-body lattice spectrum, avoided level crossing usually implies the existence of resonance. Reproducing this behavior in 3-body spectrum is a good sign for us to interpret 3-body resonance in the future simulation.

6. Conclusion

We have reviewed the effective field theory describing 3-body system. Particle-dimer formalism shows its powerful to analyze the 3-body problems both in the infinite volume and a finite volume. 3-body quantization condition is derived. The 3-body bound states and scattering discrete energy levels are obtained in a finite volume.

Our formalism needs further development including introducing twisted boundary condition, relativistic kinematics and so on. The application on realistic system, such as triton and Roper resonance is on-going as well.

Acknowledgments

The work is done in collaboration with H.-W. Hammer and A. Rusetsky. The authors would like to thank M. Döring, M. Mai, Ulf.-G. Meißner and J.-J. Wu for the collaboration on the most of works. We would also like to thank S. Beane, W. Detmold, M. Hansen, F. Romero-López, M. Savage, S. Sharpe, C. Thomas and C. Urbach for interesting discussions. JYP acknowledges the support from the DFG (CRC 110 "Symmetries and the Emergence of Structure in QCD").

References

1. K. Polejaeva and A. Rusetsky, Eur. Phys. J. A **48** (2012) 67, arXiv:1203.1241 [hep-lat].

2. U.-G. Meißner, G.-Rios and A. Rusetsky, Phys. Rev. Lett. **114** (2015) 091602 Erratum: Phys. Rev. Lett. **117** (2016) 069902, arXiv:1412.4969 [hep-lat].
3. M. T. Hansen and S. R. Sharpe, Phys. Rev. D **90** (2014) 116003, arXiv:1408.5933 [hep-lat].
4. M. T. Hansen and S. R. Sharpe, Phys. Rev. D **92** (2015) 114509, arXiv:1504.04248 [hep-lat].
5. S. Kreuzer and H.-W. Hammer, Phys. Lett. B **694** (2011) 424, arXiv:1008.4499 [hep-lat].
6. S. Kreuzer and H.-W. Hammer, Eur. Phys. J. A **43** (2010) 229, arXiv:0910.2191 [nucl-th].
7. H.-W. Hammer, J.-Y. Pang and A. Rusetsky, JHEP **1709** (2017) 109, arXiv:1706.07700 [hep-lat].
8. H.-W. Hammer, J.-Y. Pang and A. Rusetsky, JHEP **1710** (2017) 115, arXiv:1707.02176 [hep-lat].
9. Y. Meng, C. Liu, U.-G. Meißner and A. Rusetsky, arXiv:1712.08464 [hep-lat].
10. M. Döring, H.-W. Hammer, M. Mai, J.-Y. Pang, A. Rusetsky and J. Wu, Phys. Rev. D **97** (2018) no.11, 114508, arXiv:1802.03362 [hep-lat].
11. T. D. Blanton, F. Romero-López and S. R. Sharpe, arXiv:1901.07095 [hep-lat].
12. J.-Y. Pang, J.J. Wu, H.-W. Hammer, Ulf.-G. Meißner and A. Rusetsky Phys. Rev. D **99** (2019) no.7, 074513, arXiv:1902.01111 [hep-lat].
13. M. T. Hansen and S. R. Sharpe, arXiv:1901.00483 [hep-lat].

The Schmid theorem revisited

E. Oset*, S. Sakai and V. R. Debastiani

Departamento de Fisica Teorica and IFIC, University of Valencia-CSIC, Valencia, 46071, Spain
**E-mail: oset@ific.uv.es*

We investigate the Schmid theorem, which states that if one has a tree level mechanism with a particle decaying to two particles and one of them decaying posteriorly to two other particles, the possible triangle singularity developed by the mechanism of elastic rescattering of two of the three decay particles does not change the cross section provided by the tree level. We investigate the process in terms of the width of the unstable particle produced in the first decay and determine the limits of validity and violation of the theorem.

Keywords: Style file; LATEX; Proceedings; World Scientific Publishing.

1. Introduction

Let us look at a process proceeding at tree level, depicted in Fig. 1, in which particle A decays into R and 1, and R further decays into particles 2 and 3. Triangle singularities stem from the related reaction mechanism, which is depicted in Fig. 2, where the particle A decays into R and 1, posteriorly R decays into 2 and 3 and later 1 and 2 fuse to give a new particle or simply rescatter to give the same state (or another one if there are inelasticities). Diagrammatically the process is depicted in Fig. 2 where there is a loop containing particles 1, 2 and R as intermediate states. This loop function can lead to some singularities (for the limit of zero width of the R particle) when all the three intermediate particles are placed on shell, particle 1 and 3 are antiparallel and furthermore a condition is fulfilled which in classical terms can be stated as corresponding to the decay of A at rest into R and 1, assuming R moves forward and 1 backward, then R decays into 2 and 3, and letting 3 go forward, 2 eventually goes backward, which means in the same direction of 1. If 2 moves faster than 1 it can catch it and either rescatter or fuse. These conditions are the essence of Coleman-Norton theorem.[1] Early studies of triangle singularities were done in Ref. 2 but the systematic study was done by Landau.[3] A modern and practical approach is given in Ref. 4.

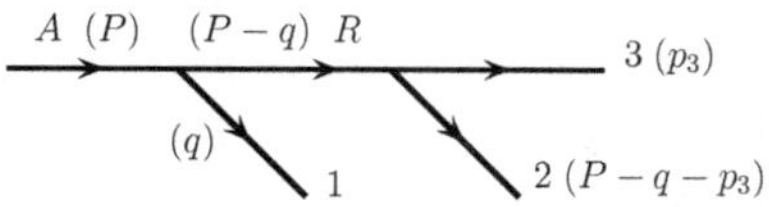

Fig. 1. Tree level diagram for the process $A \to 1 + 2 + 3$ mediated by a resonance R that decays into particles 2 and 3. In brackets the momenta of the particles.

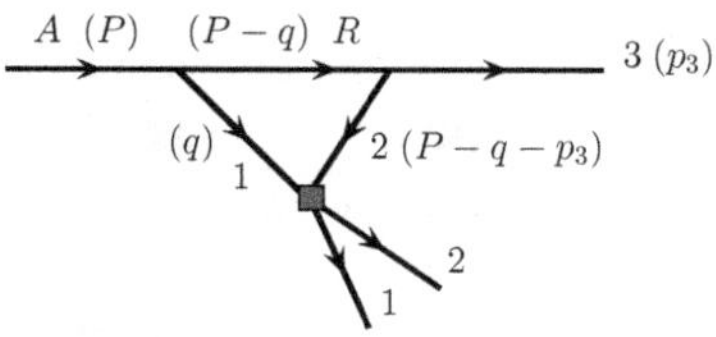

Fig. 2. Triangle mechanism emerging from the mechanism of Fig. 1, with final state interaction of particles 1 and 2. In brackets the momenta of the particles.

A thorough discussion of this early work was done by Schmid in a clarifying article.[5] There a surprise appeared, known nowadays as Schmid theorem that states that if the rescattering of particles 1 and 2 occurs, going to the same state $1 + 2$, the triangle singularity does not lead to any observable effect in magnitudes like cross sections or differential widths. It is simply reabsorbed by the S-wave of the tree level amplitude (the same mechanism without rescattering, see Fig. 1 modifying only the phase of this partial amplitude. In angle-integrated cross sections the effect of the triangle singularity disappears.

Some debate originated on the limits to the Schmid theorem and its range of validity.[6–9] In Ref. 6, for example, it was shown that if the scattering amplitude of $1 + 2 \to 1 + 2$ contains inelasticities then the Schmid theorem does not hold.

Along the literature on this topic it is customary to find the statement that due to the Schmid theorem, whenever one has rescattering in the loop to go to the same states as inside the loop, there is no need to evaluate the triangle diagram because its contribution is reabsorbed by the tree level. The purpose of the present paper is to get an insight on the theorem and see where it holds exactly, when it fails and when it is just an approximation and how good or bad can it be. For that, a new derivation is carried out, and a study in terms of the width of the intermediate state R is done. The failure of the theorem when the $t_{12,12}$ scattering matrix has inelasticities is also shown in detail, providing the quantitative amount of the breaking of the theorem. Apart from the limit of small Γ_R, where the theorem strictly holds, we study in a particular case what happens for a realistic width, and a typical $1+2 \to 1+2$ scattering amplitude, which serves as a guide on how much to trust the theorem to eventually neglect the contribution of the triangle loop.

2. Formalism

Let us study the decay process of a particle A into two particles 1 and R, with a posterior decay of R into particles 2 and 3. This is depicted in Fig. 1.

We shall also assume for simplicity that all vertices are scalar, and can be represented by just one coupling in each vertex. The conclusions that we will reach are the same for more elaborate couplings, with spin or momentum dependence.

The next step is to allow particles 1 and 2 to undergo final state interaction, as depicted in Fig. 2.

The vertex with particles 1 and 2 symbolizes the $1+2 \to 1+2$ scattering matrix. One peculiar property of triangle diagrams is that they can develop a singularity when all the particles inside the loop in Fig. 2 are placed on shell in the integration and the particles R and 3 go parallel in the A rest frame.[3] The conditions for this to occur, relating the invariant mass of particles 1, 2 with the mass of A, can be seen in a pedagogical description of the process in Ref. 4.

One should note that in the situation where the R resonance is placed on shell in the triangle loop, as well as particles 1 and 2, the tree level mechanism of Fig. 1 will also have a singularity in the limit of zero width for the resonance R, since the amplitude goes as $(M_{\rm inv}(R) - M_R + i\Gamma_R/2)^{-1}$. However, the tree diagram does not have the restriction that particles R and 3 should be parallel and hence the region where R can be placed on shell is much wider than for the triangle singularity.

Let $t_t^{(0)}$ be the S-wave projection of the tree level amplitude, t_t, of the diagram of Fig. 1, evaluated in the rest frame of $1+2$, referred to the angle between the particles 1 and 3. Let t_L be the amplitude corresponding to the triangle loop of Fig. 2. The Schmid theorem states that

$$t_t^{(0)} + t_L = t_t^{(0)} e^{2i\delta} \tag{1}$$

where δ is the S-wave phase shift of the scattering amplitude $1+2 \to 1+2$ (assume also for simplicity that this is the only partial wave in $1+2 \to 1+2$). The formula holds for the case that there is only elastic scattering $1+2 \to 1+2$. In the case that there can be inelastic channels the formula is modified and the Schmid theorem does not strictly hold.

The interesting thing about Eq. (1) is that

$$\left|t_t^{(0)} + t_L\right|^2 = \left|t_t^{(0)} e^{2i\delta}\right|^2 = \left|t_t^{(0)}\right|^2 \tag{2}$$

and consequently, since $t_t^{(\ell \neq 0)}$ and $t_t^{(0)}$ do not interfere in the angle integration, the angle integrated cross section is just given by the tree level alone.

3. Results and conclusions

In Ref. 10 we have done a thorough analytical and numerical study with many results that we summarize here.

We have done a new derivation of the Schmid theorem and have studied the results as a function of Γ_R, the width of the intermediate state in the triangle loop that decays into an external particle and an internal one. The Schmid theorem holds strictly in the limit when $\Gamma_R \to 0$. We show this again and illustrate it with a numerical example.

The first thing that we find is that when $\Gamma_R \to 0$ the relative weight of the triangle singularity versus the tree level contribution goes to zero. This means that, in a strict sense, the Schmid theorem would be irrelevant because where it holds and shows that the singularity changes only the phase of the S-wave part of the tree level amplitude, $t_t^{(0)}$, the strength of the triangle diagram or of $t_t^{(0)}$ is very small compared with the whole contribution of the full tree level amplitude.

We conducted some tests to see what happens when we have Γ_R finite and also when the scattering amplitude of the two particles that interact in the loop sit on top of a resonance.

In general lines we see that for finite Γ_R widths the Schmid theorem does not strictly hold but the $A \to 1+2+3$ process still has some memory of the theorem in the sense that the coherent sum of tree level and triangle singularity gives rise to a differential width where the interference is much smaller than usual, with final results even smaller than the incoherent sum of the two mechanisms.

We also saw that in the case of a $1+2 \to 1+2$ scattering amplitude with inelasticities the Schmid theorem does not hold and we could quantize the contribution of the triangle singularity, both analytically and numerically in the example we discussed.

The biggest contribution of the triangle singularity appears when the $1+2 \to 1+2$ scattering amplitude seats on top of a resonance. In this case, and for finite Γ_R, the contribution of the triangle singularity shows up clearly and can be even dominant over the tree level in some cases.

The message in general is that for each particular case one has to calculate both the tree level and the triangle mechanism and sum them coherently to see what comes out. Invoking the Schmid theorem to rule out the triangle mechanism in the case that the rescattering of particles $1, 2$ goes to the same channel should not be done.

Acknowledgments

V. R. Debastiani acknowledges the Programa Santiago Grisolia of Generalitat Valenciana (Exp. GRISOLIA/2015/005). S. Sakai acknowledges the support by NSFC and DFG through funds provided to the Sino-German CRC110 "Symmetries and the Emergence of Structure in QCD" (NSFC Grant No. 11621131001), by the NSFC (Grant No. 11747601), by the CAS Key Research Program of Frontier Sciences (Grant No. QYZDB-SSW-SYS013) and by the CAS Key Research Program (Grant No. XDPB09). This work is also partly supported by the Spanish Ministerio de Economia y Competitividad and European FEDER funds under the contract number FIS2014-57026-REDT, FIS2014-51948-C2-1-P, and FIS2014-51948-C2-2-P, and the Generalitat Valenciana in the program Prometeo II-2014/068.

References

1. S. Coleman and R. E. Norton, Nuovo Cim. **38**, 438 (1965).
2. R. Karplus, C. M. Sommerfield and E. H. Wichmann, Phys. Rev. **111**, 1187 (1958).
3. L. D. Landau, Nucl. Phys. **13**, 181 (1959).
4. M. Bayar, F. Aceti, F. K. Guo and E. Oset, Phys. Rev. D **94**, no. 7, 074039 (2016).
5. C. Schmid, Phys. Rev. **154**, 1363 (1967).
6. A. V. Anisovich and V. V. Anisovich, Phys. Lett. B **345**, 321 (1995).
7. I. J. R. Aitchison and C. Kacser, Phys. Rev. **173**, 1700 (1968).
8. I. J. R. Aitchison and C. Kacser, Phys. Rev. **142**, 1104 (1966).
9. A. P. Szczepaniak, Phys. Lett. B **757**, 61 (2016).
10. V. R. Debastiani, S. Sakai and E. Oset, Eur. Phys. J. C **79**, no. 1, 69 (2019).

Triangle singularity in $J/\psi \to \eta\pi^0\phi$

Hao-Jie Jing[1,2], Feng-Kun Guo[1,2], Shuntaro Sakai[1] and Bing-Song Zou[1,2,3]

[1] *CAS Key Laboratory of Theoretical Physics, Institute of Theoretical Physics, Chinese Academy of Sciences, Zhong Guan Cun East Street 55, Beijing 100190, China*
[2] *School of Physical Sciences, University of Chinese Academy of Sciences, Beijing 100049, China*
[3] *Central South University, Hunan 410083, China*

Recently, the observation of the $a_0(980)^0 - f_0(980)$ mixing in the isospin breaking decay $J/\psi \to \eta\pi^0\phi$ was reported by the BESIII Collaboration. In the Dalitz plot for that decay with the η reconstructed from two photons, there is a band around 1.4 GeV on the $\pi^0\phi$ distribution. Here, we study the effects of a set of $K^*\bar{K}K$ triangle diagrams, and show that due to triangle singularities such diagrams can lead to a peak around 1.4 GeV in the $\pi^0\phi$ invariant mass distribution. To examine whether the band is due to triangle singularities or due to a resonance, more experimental data are needed.

Keywords: Effective Lagrangian; hadronic decay; triangle singularity.

1. Introduction

In the Dalitz plot of the $J/\psi \to \eta\pi^0\phi$ process measured by the BESIII Collaboration (Fig. 1 of Ref. 1), there is a clear peak near the $K\bar{K}$ threshold on the $\eta\pi^0$ distribution which can be understood as the $a_0(980)^0 - f_0(980)$ mixing. In addition, there is also a clear accumulation of events for the $\pi^0\phi$ invariant mass being around 1.4 GeV, which would be a peak if the Dalitz plot is projected to the $\pi^0\phi$ invariant mass distribution. Such a peak can originate from the isovector $C(1480)$ claimed in Ref. 2 or the isoscalar $h_1(1380)$ whose mass is about 1.41 GeV because the S-wave $\pi^0\phi$ has $J^{PC} = 1^{+-}$ and this process is forbidden by the isospin symmetry. On the other hand, because the peak position in the $\pi^0\phi$ invariant mass distribution, $m_{\pi^0\phi} \sim 1.4$ GeV, is close to the position of the leading singularity of the triangle-loop diagrams composed of intermediate $K^*\bar{K}K$, such diagrams can lead to a peak and a significant isospin breaking enhancement here. Thus we consider the triangle diagrams shown in Fig. 1. The triangle singularity (TS) for each of these diagrams is at the physical boundary when the kinematics is such that the following processes happen (for a

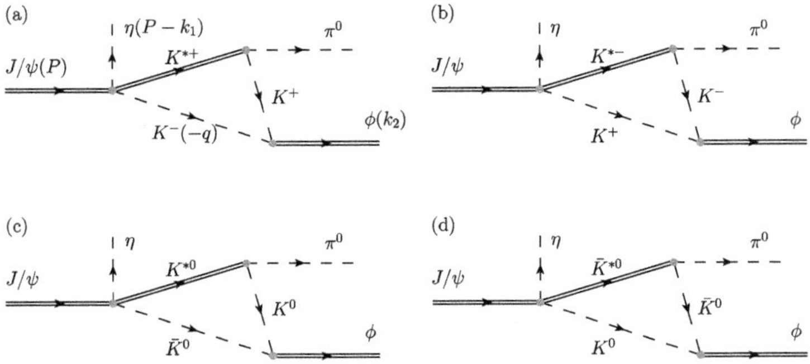

Fig. 1. Triangle diagrams in the J/ψ decay into $\eta\pi^0\phi$ that can produce a peak at $m_{\pi^0\phi} \simeq 1.4$ GeV without a resonance.

review of TSs, see Ref. 3): first, the J/ψ decays into $\eta K^*\bar{K}$, and the K^* decays into πK subsequently; then the K moves in the same direction with the $\bar{K}$ and catches up with it, and the $K\bar{K}$ pair finally forms the ϕ. The results reported in this talk are based on the published article, Ref. 4.

2. Formalism

The diagrams we consider for the $J/\psi \to \eta\pi^0\phi$ are shown in Fig. 1. To obtain the decay amplitude, we use the following effective Lagrangians[5–10]:

$$\mathcal{L}_{JVPP} = g_1 J_\mu \langle V^\mu PP \rangle, \qquad \mathcal{L}_{VPP} = -ig_2 \langle V^\mu [P, \partial_\mu P] \rangle, \tag{1}$$

where g_1 and g_2 are coupling constants, and J_μ is the field operator of the J/ψ, which as a charmonium is a light-flavor SU(3) singlet state. The light pseudoscalar meson octet and vector meson nonet matrices are denoted by P and V_μ, respectively,

$$P = \frac{1}{\sqrt{2}} \begin{pmatrix} \pi^0 + \frac{1}{\sqrt{3}}\eta & \sqrt{2}\pi^+ & \sqrt{2}K^+ \\ \sqrt{2}\pi^- & -\pi^0 + \frac{1}{\sqrt{3}}\eta & \sqrt{2}K^0 \\ \sqrt{2}K^- & \sqrt{2}\bar{K}^0 & -\frac{2}{\sqrt{3}}\eta \end{pmatrix}, \quad V_\mu = \frac{1}{\sqrt{2}} \begin{pmatrix} \rho^0 + \omega & \sqrt{2}\rho^+ & \sqrt{2}K^{*+} \\ \sqrt{2}\rho^- & -\rho^0 + \omega & \sqrt{2}K^{*0} \\ \sqrt{2}K^{*-} & \sqrt{2}\bar{K}^{*0} & \sqrt{2}\phi \end{pmatrix}_\mu . \tag{2}$$

In order to compare our results with the experimetenal data directly, we take into account the $\phi \to K^+K^-$ decay, i.e., $\eta\pi^0K^+K^-$ are in the final state. In this case, tree diagrams also contribute to the process in addition to the triangle loop ones. The diagrams considered here are shown in Fig. 2. The triangle diagrams involve the $K\bar{K} \to K\bar{K}$ final state interaction (FSI). We consider the P-wave $K\bar{K}$ scattering amplitude with quantum numbers $(I, J) = (0, 1)$ and $(1, 1)$. The amplitudes for the P-wave $K\bar{K}$ FSI vertices

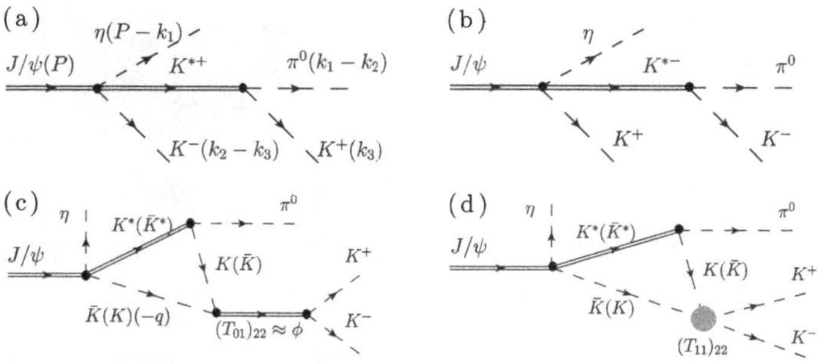

Fig. 2. Feynman diagrams for the J/ψ decay into $\eta\pi^0 K^+ K^-$ considered in this work. In the triangle diagrams, both the charged and neutral K and K^* mesons are taken into account as in Fig. 1. In the second line, $(T_{01})_{22}$ and $(T_{11})_{22}$ refer to the P-wave $K\bar{K} \to K\bar{K}$ scattering amplitudes with isospin $I = 0$ and $I = 1$, respectively.

are as follows:

$$-it_{(I,1)} = -i\tilde{t}_{(I,1)}(p_K^{\text{in}} - p_{\bar{K}}^{\text{in}})^\mu (p_K^{\text{out}} - p_{\bar{K}}^{\text{out}})_\mu, \tag{3}$$

$$-i\tilde{t}_{(0,1)} = i\frac{g_2^2}{m_{K\bar{K}}^2 - m_\phi^2 + im_\phi\Gamma_\phi}, \quad -i\tilde{t}_{(1,1)} = -i\frac{3(T_{11})_{22}}{4q_{\text{cm}}^{\text{in}} q_{\text{cm}}^{\text{out}}}, \tag{4}$$

where the $K(\bar{K})$ momenta in the loop and final state are denoted by p_K^{in} and p_K^{out} ($p_{\bar{K}}^{\text{in}}$ and $p_{\bar{K}}^{\text{out}}$), respectively, and $q_{\text{cm}}^{\text{in}}, q_{\text{cm}}^{\text{out}}$ are the magnitudes of the corresponding center-of-mass (cm) momenta. T_{11} is a component of the $\pi\pi$ and $K\bar{K}$ coupled-channel T matrix with the quantum numbers $(I = 1, J = 1)$ (see Ref. 11 for the specific form). Because the TS appears when the internal particles are on-shell, the off-shell contribution that is neglected here just gives non-singular contributions that are smooth in invariant mass distributions. Then, the structure around 1.4 GeV would not be changed by approximating the momentum q^{in} in the denominator of Eq. (4) for the $K\bar{K}$ scattering vertex with the on-shell momentum $q_{\text{cm}} = q^{\text{out}} = \sqrt{m_{K\bar{K}}^2/4 - m_K^2}$.

3. Results

Since we only care about the shape and relative size of the invariant mass distributions, the value of g_1 and g_2 are irrelevant for the characteristic feature of the nontrivial structures in question. Here, we set $g_1 = 1$ and set $g_2 \simeq 4.5$ from reproducing the observed ϕ and K^* widths. And we use dimensional regularization with the $\overline{\text{MS}}$ subtraction scheme to regularize the UV divergence. It is noteworthy that the TS is an infrared singularity

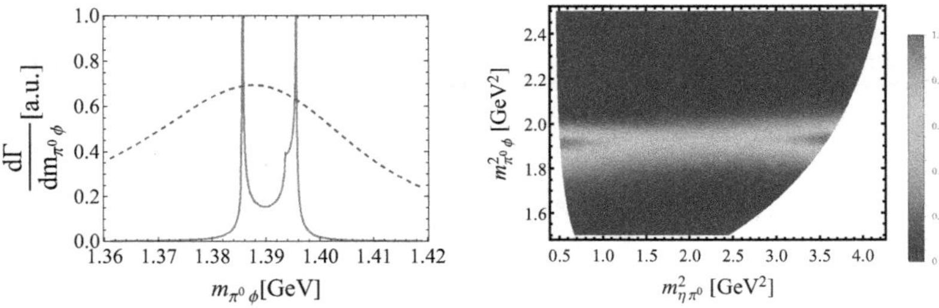

Fig. 3. Left panel: $\pi^0\phi$ invariant mass distribution in arbitrary units (a.u.) for the decay $J/\psi \to \eta\pi^0\phi$ from the mechanism shown in Fig. 1. The red-solid line shows the differential width on the $\pi^0\phi$ distribution obtained by switching off the K^* width, and the blue-dashed curve (500 times larger than the original value) corresponds to the result including the width of K^*. Right panel: The part of Dalitz plot where the TSs occur for the decay $J/\psi \to \eta\pi^0\phi$.

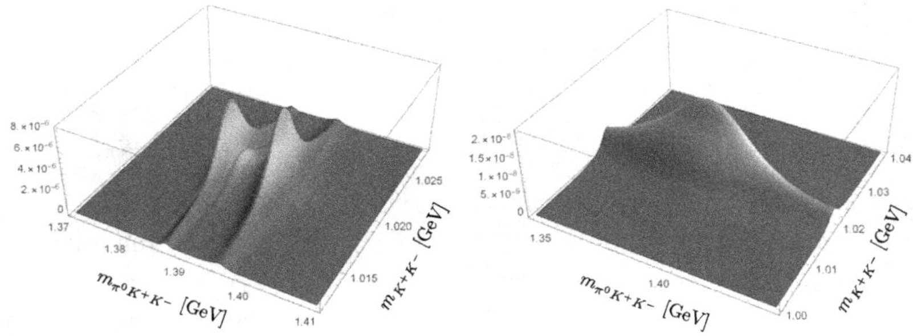

Fig. 4. The $J/\psi \to \eta\pi^0 K\bar{K}$ differential width as a function of invariant masses of $K\bar{K}$ and $\pi^0 K\bar{K}$ in arbitrary units, by considering only triangle loop diagrams of Fig. 2(c). The K^* width is considered in the right panel, but not in the left one.

and thus it is not sensitive to renormalization scale μ. In our calculation, we set $\mu = 1.2$ GeV.

The effect of a TS is very sensitive to the kinematical variables. In this case it appears on the physical boundary only when the $m_{K\bar{K}} \in [0.9874, 1.0259]$ and $[0.9952, 1.0337]$ GeV for the charged and neutral intermediate states, respectively. If the band around 1.4 GeV in the $\pi^0\phi$ distribution observed by the BESIII Collaboration is due to a resonance, its signal would only depend on $m_{\pi^0K^+K^-}$ irrespective of $m_{K^+K^-}$.

As shown in Fig. 3 for the $\pi^0\phi$ mass distribution of $J/\psi \to \eta\pi^0\phi$, the $K^*\bar{K}K$ triangle diagrams can produce a band in the Dalitz plot which is located at the $m_{\pi^0\phi}$ distribution around 1.4 GeV by virtue of TSs. From Fig. 4 and Fig. 5, the effect is most evident with the $m_{K^+K^-}$ in the ϕ

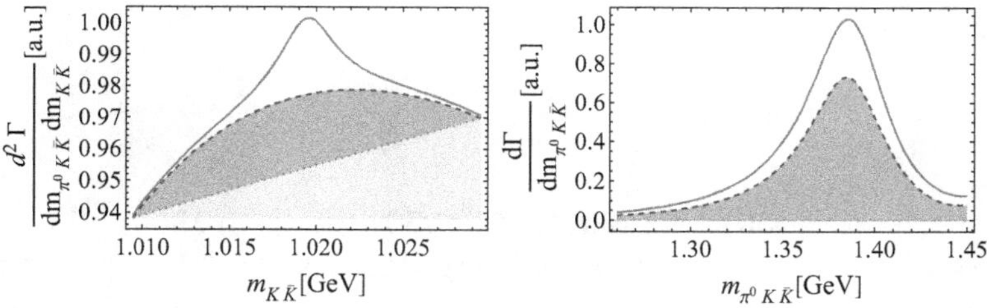

Fig. 5. Left panel: the K^+K^- invariant mass distribution in the ϕ window with $m_{\pi^0 K\bar{K}} = 1.39$ GeV for $J/\psi \to \eta\pi^0 K\bar{K}$. All the diagrams in Fig. 2 are included for the solid line, and the tree and $I = 0$ $K\bar{K}$ FSI diagrams are included for the dashed one. The dotted line is a linear interpolation between the two ends of the ϕ-resonance window, $m_{K\bar{K}} \in [1.01, 1.03]$ GeV. Right panel: the $\pi^0 K^+K^-$ energy spectrum of the same reaction with $m_{K\bar{K}}$ integrated in the ϕ window, i.e., the part below the straight dotted line in the left panel is subtracted. The K^* width is considered in these plots.

resonance mass region. Thus, by taking events with $m_{K^+K^-}$ away from that region, we expect the band at $m_{\pi^0 K^+K^-} \sim 1.4$ GeV in the Dalitz plot would become weaker and eventually invisible if $m_{K^+K^-}$ is sufficiently away from the ϕ mass region. In order to reveal the physics behind the band at $m_{\pi^0\phi} \sim 1.4$ GeV observed by the BESIII Collaboration, more data for the $J/\psi \to \eta\pi^0\phi$ are needed to make the Dalitz plot more clear, and to check whether the peak structure on the $\pi^0\phi$ distribution around 1.4 GeV persists in other K^+K^- invariant mass regions.

This work is supported in part by the National Natural Science Foundation of China (NSFC) under Grants No. 11847612, No. 11835015 and No. 11621131001.

References

1. M. Ablikim *et al.*, *Phys. Rev. Lett.* **121**, 022001 (2018).
2. S. I. Bityukov *et al.*, *Phys. Lett.* B **188**, 383 (1987).
3. F.-K. Guo, X.-H. Liu and S. Sakai, arXiv:1912.07030 [hep-ph].
4. H. J. Jing *et al.*, *Phys. Rev. D* **100**, 114010 (2019).
5. M. Bando *et al.*, *Phys. Rev. Lett.* **54** 1215, (1985).
6. U.-G. Meißner, *Phys. Rept.* **161** 213, (1988).
7. F. Klingl *et al.*, *Z. Phys. A* **356**, 193 (1996).
8. U.-G. Meißner and J. A. Oller, *Nucl. Phys. A* **679**, 671 (2001).
9. L. Roca *et al.*, *Nucl. Phys. A* **744**, 127 (2004).
10. W. H. Liang *et al.*, *Phys. Rev. D* **99**, 094020 (2019).
11. J. A. Oller *et al.*, *Phys. Rev. D* **59**, 074001 (1999).

$Z_c(4430)$, $Z_c(4200)$, $Z_1(4050)$, and $Z_2(4250)$ as triangle singularities

Satoshi X. Nakamura

University of Science and Technology of China, Hefei 230026, China
E-mail: satoshi@ustc.edu.cn

$Z_c(4430)$ discovered in $\bar{B}^0 \to \psi(2S)K^-\pi^+$, $Z_c(4200)$ found in $\bar{B}^0 \to J/\psi K^-\pi^+$, and $Z_1(4050)$ and $Z_2(4250)$ observed in $\bar{B}^0 \to \chi_{c1}K^-\pi^+$ are candidates of charged charmonium-like states. All surviving theoretical models interpreted these candidates as four-quark states, until we recently identified a compelling alternative. We discuss that kinematical singularities in triangle loop diagrams induce a resonance-like behavior that can consistently explain the properties (spin-parity, mass, width, and Argand plot) of $Z_c(4430)$, $Z_c(4200)$, $Z_1(4050)$ and $Z_2(4250)$ from experiments. In terms of the triangle singularities, we can also naturally understand interesting experimental findings such as the appearance (absence) of $Z_c(4200)(Z_c(4430))$-like contribution in $\Lambda_b^0 \to J/\psi p\pi^-$, and the highly asymmetric spectrum bump for $Z_1(4050)$; other theoretical models have not successfully addressed these points.

Keywords: Hadron spectroscopy; exotic hadrons.

1. Introduction

A current trend of the hadron spectroscopy is to establish the existence and the internal structure of exotic hadrons which are not accommodated by the conventional $q\bar{q}$ and qqq structures. Such exotic hadrons could be tetraquark, pentaquark, hadron molecule, or hybrid states. Possible experimental signatures of the exotic states are: (i) the mass does not fit a quark-model prediction; (ii) the state matches a state for which Lattice QCD predicts a high gluon content; (iii) the state has a peculiar decay pattern; and so on. But the listed signatures may seem model-dependent criteria, and one may wonder if there is a more unambiguous signature.

The discoveries of charged quarkonium-like state candidates, Z_c and Z_b, are encouraging. For example, $Z_c(4430)$ was discovered in the $\psi(2S)\pi^+$ invariant mass distribution of $\bar{B}^0 \to \psi(2S)K^-\pi^+$,[1–3] while $Z_c(4200)$ in the $J/\psi\pi^+$ distribution of $\bar{B}^0 \to J/\psi K^-\pi^+$.[4] $Z_1(4050)$ and $Z_2(4250)$ are also reported in the analysis of $\bar{B}^0 \to \chi_{c1}K^-\pi^+$.[5] If these spectrum bumps are really associated with the existence of resonances, the quark content of these states minimally need four quarks, a clear signature of exotics. Among the charged charmonium-like states, $Z_c(4430)$ has been an outstanding exotic candidate,[6] and all the surviving theoretical interpretations of $Z_c(4430)$ considered it to be a genuine four-quark state (including hadron-molecule interpretations) until the present work. Theoretical interpretations of $Z_c(4200)$, $Z_1(4050)$ and $Z_2(4250)$ are similar, although these exotic candidates are reported only by the Belle experiment, and yet to be confirmed by an independent experiment.

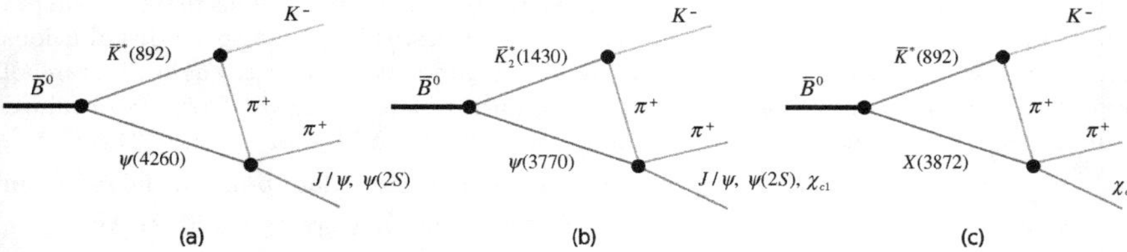

Fig. 1. Triangle diagrams contributing to $\bar{B}^0 \to \psi_f K^- \pi^+$; $\psi_f = J/\psi, \psi(2S), \chi_{c1}$. A $Z_c(4430)$ [$Z_c(4200)$]-like bump in the $\psi_f\pi$ invariant mass distribution is generated by TS from the diagram (a) [(b)], while a $Z_1(4050)$ [$Z_2(4250)$]-like bump in the $\chi_{c1}\pi^+$ distribution by the TS from the diagram (c) [(b)]. Figures taken from Refs. 7,8. Copyright (2019) APS.

We propose a completely different scenario based on the triangle singularity (TS) to interpret these exotic candidates.[7,8] The TS arise in triangle diagrams as shown in Fig. 1 when the processes are kinematically allowed to occur even at the classical level. The TS is a kinematical effect and its existence and location are completely determined by the particle masses involved in the process in the zero-width limit of unstable particles. Although the unstable particles have finite widths in reality and thus the TS are somewhat relaxed, the TS still significantly enhance the amplitudes. In what follows, we demonstrate that the TS arising from Fig. 1 induce a resonance-like behaviors, with which we can consistently understand the experimentally determined properties of $Z_c(4430)$, $Z_c(4200)$, $Z_1(4050)$ and $Z_2(4250)$ such as spin-parity, mass, width, and Argand plot.

2. Results for $Z_c(4430)$ and $Z_c(4200)$

The $\psi(2S)\pi$ [$J/\psi\pi$] invariant mass distribution for $\bar{B}^0 \to \psi(2S)K^-\pi^+$ [$\bar{B}^0 \to J/\psi K^-\pi^+$] is shown in Fig. 2(a) [2(b)]. The triangle diagrams of Figs. 1(a) and 1(b) generate the red solid curves in Figs. 2(a) and 2(b),

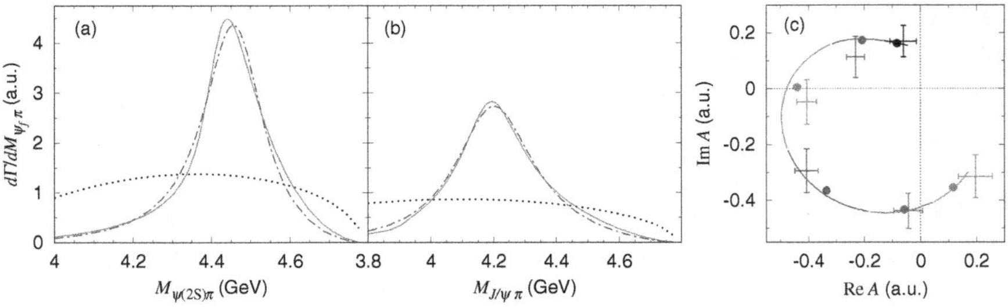

Fig. 2. The $\psi_f\pi$ ($\psi_f = J/\psi, \psi(2S)$) invariant mass distributions of $\bar{B}^0 \to \psi(2S)K^-\pi^+$ (a) and $\bar{B}^0 \to J/\psi K^-\pi^+$ (b). Triangle diagrams Figs. 1(a) and 1(b) generate the red solid curves in panels (a) and (b), respectively. Breit-Wigner amplitudes fitted to the red solid curves are shown by the green dash-dotted curves. The phase-space distributions are given by the dotted curves. (c) $Z_c(4430)$ Argand plot. The triangle diagram of Fig. 1(a) generates six curved segments. Six data points of the LHCb[3] are obtained from fitting the data in six bins that equally separate 18 GeV$^2 \leq m^2_{\psi(2S)\pi} \leq 21.5$ GeV^2; $m^2_{\psi(2S)\pi}$ increases counterclockwise. A curved segment and a data point in the same bin are indicated by the same color. Figure taken from Ref. 7. Copyright (2019) APS.

respectively. The phase-space distributions are also shown by the black dotted curves for comparison. Because of the presence of the TS in the triangle diagram, a resonance-like peak clearly shows up at $m_{\psi(2S)\pi} \sim 4.45$ GeV in Fig. 2(a) ($m_{J/\psi\pi} \sim 4.2$ GeV in Fig. 2(b)).

We simulate the peaks from the TS in terms of Z_c^+ excitations. We use a model that goes as $\bar{B}^0 \to Z_c^+ K^-$ followed by $Z_c^+ \to \psi_f \pi^+$ to fit the Dalitz plot distributions generated by the triangle diagrams of Figs. 1(a) and 1(b). The Breit-Wigner form is used to model the Z_c propagation. The kinematical region included in the fit have the Dalitz plot distribution larger than 10% of the peak height. As shown by the green dash-dotted curves in Figs. 2(a) and 2(b), the Breit-Wigner form can fit the peaks very well. The resulting Breit-Wigner parameters agree with the experimentally determined ones remarkably well.

Now we show the triangle amplitude for $Z_c(4430)$ in the form of the Argand plot so that we can compare it with the counterpart from the LHCb.[3] As seen in Fig. 2(c), the LHCb's $Z_c(4430)$ Argand plot is consistently described by the triangle diagram of Fig. 1(a). Thus the counterclockwise behavior of the LHCb's $Z_c(4430)$ Argand plot[3] is not necessarily pointing to the existence of a resonance.

According to the LHCb analysis,[9] a significantly better description of the $\Lambda_b^0 \to J/\psi p\pi^-$ data is obtained by including the $Z_c(4200)$ amplitude. We point out that triangle diagrams can contribute to this process with TS in the $Z_c(4200)$-region. The LHCb analysis[9] also found that their description of the $\Lambda_b^0 \to J/\psi p\pi^-$ data is hardly improved by including a $Z_c(4430)$ contribution. This can be understood in terms of the TS because, within experimentally observed hadrons, there is no triangle diagram to cause a TS at the $Z_c(4430)$ position for the case of $\Lambda_b^0 \to J/\psi p\pi^-$.

3. Results for $Z_1(4050)$ and $Z_2(4250)$

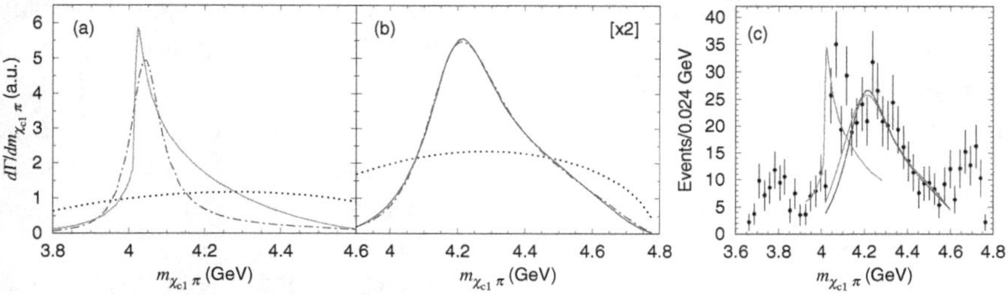

Fig. 3. Distributions of the $\chi_{c1}\pi^+$ invariant mass for $\bar{B}^0 \to \chi_{c1}K^-\pi^+$. The red solid curve in the panel (a) is generated by the triangle diagram of Fig. 1(c). The blue solid curve in the panel (b) is from the diagram Fig. 1(b). The other features are the same as those in Fig. 2. (c) The red and blue curves in the panels (a) and (b) are superimposed on the Belle data (Fig. 14 of Ref. 5). Figure taken from Ref. 8. Copyright (2019) APS.

The $\chi_{c1}\pi^+$ invariant mass distributions for $\bar{B}^0 \to \chi_{c1}K^-\pi^+$ are shown in Fig. 3. The triangle diagram of Fig. 1(c) gives the red solid curve in Fig. 3(a). The diagram Fig. 1(b) generates the blue solid curve in Fig. 3(b). Clear resonance-like peaks are induced by the triangle singularities at $m_{\chi_{c1}\pi} \sim 4.025$ GeV in Fig. 3(a) and $m_{\chi_{c1}\pi} \sim 4.22$ GeV in Figs. 3(b). A characteristic feature of the bump in Fig. 3(a) is that it has a significantly asymmetric shape.

We can again simulate the TS-induced bumps with the fake Z_c-excitation mechanisms. The $\bar{B}^0 \to \chi_{c1}K^-\pi^+$ Dalitz plot distribution from the triangle diagram of Fig. 1(c) [1(b)] is fitted with the mechanism of $\bar{B}^0 \to ZK^-$ followed by $Z \to \chi_{c1}\pi^+$ by adjusting their Breit-Wigner mass and width. We include the kinematical region where the Dalitz plot distribution is larger than 10% of the peak height. The green dash-dotted curves in Fig. 3 are showing the quality of the fits. The Breit-Wigner form cannot fit well the red solid curve with the asymmetric bump in Fig. 3(a). Meanwhile, the bump in Figs. 3(b) is reasonably well fitted. The resulting Breit-Wigner parameters are in good agreement with those from the Belle analysis[5] on $Z_1(4050)$ and $Z_2(4250)$.

The asymmetry of the bump shape generated by the triangle diagram Fig. 1(c) seems important to reproduce the Belle data in the $Z_1(4050)$-region.[5] To make this statement clear, as in Fig. 3(c), we superimpose the spectra from the triangle diagrams of Figs. 1(c) and 1(b) on the Belle data (Fig. 14 of Ref. 5). Although this is a qualitative comparison where any interferences among different mechanisms are not taken into account, the spectrum bumps from the triangle diagrams reasonably capture the features of the data. Particularly, the data has a very sharp rise and a moderate fall-off at the $Z_1(4050)$-region, and the asymmetric bump shape from the triangle diagram of Fig. 1(c) reproduces it well. The Belle analysis[5] was not able to fit well this sharp peak of the data as seen in Fig. 14 of Ref. 5, probably because they used the Breit-Wigner form to simulate this bump. The data seem to disfavor the Breit-Wigner shape. As seen in Fig. 3(a), the triangle diagram of Fig. 1(c) generates the spectrum bump, the shape of which is significantly different from the Breit-Wigner.

The asymmetric bump shape associated with $Z_1(4050)$ is interesting because it could sensitively discriminate different theoretical interpretations of $Z_1(4050)$. A successful model should explain this characteristic spectrum shape of $Z_1(4050)$, in addition to the mass, width, and J^P. So far, this question has been successfully addressed by our model only. Higher statistics data is also highly hoped to establish the spectrum shape because the error bars are still rather large in the Belle data.

Acknowledgment

This work is in part supported by National Natural Science Foundation of China (NSFC) under contracts 11625523.

References

1. S.K. Choi et al. (Belle), *Phys. Rev. Lett.* **100**, 142001 (2008).
2. K. Chilikin et al. (Belle), *Phys. Rev. D* **88**, 074026 (2013).
3. R. Aaij et al. (LHCb), *Phys. Rev. Lett.* **112**, 222002 (2014).
4. K. Chilikin et al. (Belle), *Phys. Rev. D* **90**, 112009 (2014).
5. R. Mizuk et al. (Belle), *Phys. Rev. D* **78**, 072004 (2008).
6. https://physics.aps.org/synopsis-for/10.1103/PhysRevLett.112.222002.
7. S.X. Nakamura and K. Tsushima, *Phys. Rev. D* **100**, 051502(R) (2019).
8. S.X. Nakamura, *Phys.Rev. D* **100**, 011504(R) (2019).
9. R. Aaij et al. (LHCb), *Phys. Rev. Lett.* **117**, 082003 (2016).

Triangle singularity in the $J/\psi \to K^+K^-f_0(980)$ decay

Niu Su

School of Physics, Beihang University, Beijing 100191, China

Hua-Xing Chen

School of Physics, Beihang University, Beijing 100191, China
School of Physics, Southeast University, Nanjing 210094, China
E-mail: hxchen@buaa.edu.cn

Wei-Hong Liang

Department of Physics, Guangxi Normal University, Guilin 541004, China
E-mail: liangwh@gxnu.edu.cn

Eulogio Oset

Departamento de Física Teórica and IFIC,
Centro Mixto Universidad de Valencia-CSIC Institutos de Investigación de Paterna,
Aptdo. 22085, 46071 Valencia, Spain
E-mail: eulogio.oset@ific.uv.es

En Wang

School of Physics and Microelectronics, Zhengzhou University,
Zhengzhou, Henan 450001, China
E-mail: wangen@zzu.edu.cn

We study the $J/\psi \to K^+K^-f_0(980)$ decay and find that the mechanism to produce this decay develops a triangle singularity around $M_{\rm inv}(K^-f_0) \approx$ 1515 MeV. Its branching ratio is of the order of 10^{-5}, accessible in present facilities.

Keywords: Triangle singularity; light scalar mesons; chiral unitary model.

1. Introduction

Discussed already in Ref. 1, the triangle singularities were fashionable in the sixties. The conditions for this are made specific by Coleman-Norton,[2] whose analytical expressions can be found in Ref. 3 and in a simpler form in Ref. 4. In recent years this field has experienced a revival, due to the spark raised by the puzzle of the anomalously large isospin breaking in the

$\eta(1405) \to \pi^0 f_0(980)$ reaction,[5] which was explained in Ref. 6 in terms of the $K\bar{K}K^*$ triangle singularity. More examples can be found in Ref. 7.

In this paper we shall study the $J/\psi \to K^+K^- f_0(980)$ decay,[8] where the $f_0(980)$ resonance is not directly produced but come from the rescattering of the $K\bar{K}$ components.[9] This reaction involves the $K\bar{K}\phi$ intermediate states, as depicted in Fig. 1, which have not been discussed previously. The triangle singularity point of this reaction is given by the equation[4]:

$$q_{\rm on} - q_{a-} = 0\,, \tag{1}$$

where $q_{\rm on}$ is the on-shell momentum of the ϕ, and q_{a-} the on-shell K^- momentum in the loop, antiparallel to the ϕ. Eq. (1) is fulfilled only in a very narrow window of K^+K^- energies between 987 MeV and 993 MeV, where the $f_0(980)$ peaks. This reaction contains relevant information concerning the nature of the $f_0(980)$ resonance.[10]

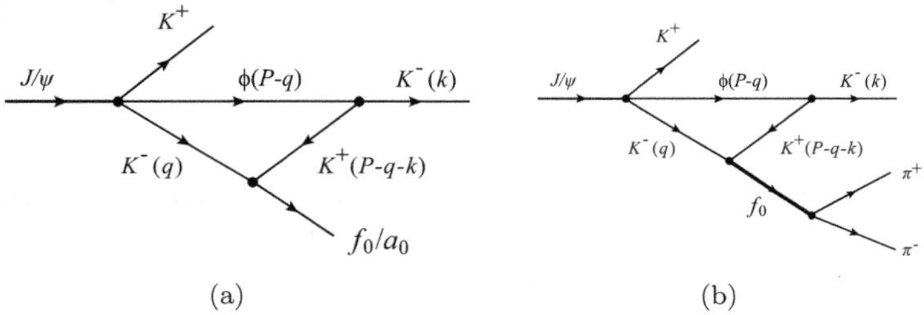

Fig. 1. Triangle diagrams for (a) the $J/\psi \to K^+K^- f_0(980)$ decay and (b) the $J/\psi \to K^+K^- f_0(980) \to K^+K^-\pi^+\pi^-$ reaction. Taken from Ref. 8.

2. Formalism

Our mechanism for the $J/\psi \to K^+K^- f_0(980)$ decay is depicted in Fig. 1. There is a primary decay of $J/\psi \to K^+K^-\phi$, a second decay of $\phi \to K^+K^-$, and the posterior fusion of K^+K^- to produce the $f_0(980)$.

The $J/\psi \to K^+K^-\phi$ decay. The $J/\psi \to K^+K^-\phi$ decay can proceed via S-wave,[11,12] whose amplitude is written as:

$$t_{J/\psi\to K^+K^-\phi} = \mathcal{A}\; \vec{\epsilon}(J/\psi)\cdot\vec{\epsilon}(\phi)\,. \tag{2}$$

The $J/\psi \to K^+K^- f_0(980)$ decay. The $J/\psi \to K^+K^- f_0(980)$ decay can be produced via the triangle diagram depicted in Fig. 1(a). We follow

Refs. 4 and 13 and write its amplitude as:

$$t = \mathcal{A}\, g_V\; g_{f_0,K^+K^-}\; \vec{\epsilon}(J/\psi) \cdot \vec{k}\; t_T\,, \tag{3}$$

where

$$\begin{aligned} t_T = & \int \frac{d^3q}{(2\pi)^3} \frac{1}{8\omega_{K^+}\omega_{K^-}\omega_\phi} \frac{1}{k^0 - \omega_{K^+} - \omega_\phi + i\Gamma_\phi/2} \\ & \times \Big(2M_{\text{inv}}(K^- f_0)\omega_{K^-} + 2k^0\omega_{K^+} - 2(\omega_{K^-} + \omega_{K^+})(\omega_{K^-} + \omega_{K^+} + \omega_\phi) \Big) \\ & \times \frac{1}{M_{\text{inv}}(K^- f_0) + \omega_{K^-} + \omega_{K^+} - k^0} \frac{1}{M_{\text{inv}}(K^- f_0) - \omega_\phi - \omega_{K^-} + i\Gamma_\phi/2} \\ & \times \frac{1}{M_{\text{inv}}(K^- f_0) - \omega_{K^-} - \omega_{K^+} - k^0 + i\epsilon} \times \left(2 + \frac{\vec{q}\cdot\vec{k}}{\vec{k}^2} \right) . \end{aligned} \tag{4}$$

We refer to Ref. 8 for detailed discussions. Using Eq. (3), we can write the differential mass distribution for the $J/\psi \to K^+K^- f_0(980)$ decay as

$$\frac{d\Gamma_{J/\psi \to K^+K^- f_0(980)}}{dM_{\text{inv}}(K^- f_0)} = \frac{1}{(2\pi)^3} \frac{1}{4M^2_{J/\psi}} p'_{K^+} \tilde{p}'_{K^-} \overline{\sum} \sum |t|^2 . \tag{5}$$

The $J/\psi \to K^+K^- f_0(980) \to K^+K^-\pi^+\pi^-$ reaction. We further consider that the $f_0(980)$ will be seen in the $\pi^+\pi^-$ mass distribution, as depicted in Fig. 1(b). Its amplitude t' is very similar to t given in Eq. (3), just with g_{f_0,K^+K^-} replaced by the transition amplitude $t_{K^+K^-,\pi^+\pi^-}$.[14] We follow Ref. 11 and write the double differential mass distribution for the $J/\psi \to K^+K^- f_0(980) \to K^+K^-\pi^+\pi^-$ reaction as:

$$\frac{d^2\Gamma_{J/\psi \to K^+K^- f_0(980) \to K^+K^-\pi^+\pi^-}}{dM_{\text{inv}}(K^- f_0)\, dM_{\text{inv}}(\pi^+\pi^-)} = \frac{p''_{K^+} \tilde{p}''_{K^-} \tilde{p}''_{\pi^+}}{(2\pi)^5 \cdot 4M^2_{J/\psi}} \overline{\sum} \sum |t'|^2 . \tag{6}$$

3. Results

First we show t_T in Fig. 2(a), that is the contribution of the triangle loop defined in Eq. (4). There is a very narrow window of the f_0 mass where the triangle singularity condition is exactly fulfilled, *i.e.*, from 987 MeV to 993 MeV. In view of this we plot t_T for $m_{f_0} = 989$ MeV. We observe a neat peak around $M_{\text{inv}}(K^- f_0) \approx 1515$ MeV.

Then we show $\frac{1}{\Gamma_{J/\psi}} \frac{d\Gamma_{J/\psi \to K^+K^- f_0}}{dM_{\text{inv}}(K^- f_0)}$ as a function of $M_{\text{inv}}(K^- f_0)$ in Fig. 2(b), that is the differential branching ratio of the $J/\psi \to K^+K^- f_0(980)$ decay defined in Eq. (5). Again we observe a clear peak around $M_{\text{inv}}(K^- f_0) \approx 1515$ MeV.

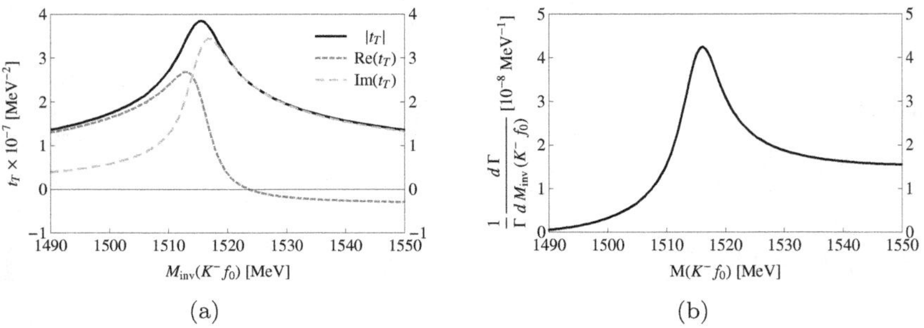

Fig. 2. (a) the triangle amplitude t_T and (b) the differential branching ratio $\frac{1}{\Gamma_{J/\psi}}\frac{d\Gamma_{J/\psi\to K^+K^-f_0}}{dM_{\rm inv}(K^-f_0)}$, as functions of $M_{\rm inv}(K^-f_0)$ for $m_{f_0} = 989$ MeV.

Finally, we take into account the mass distribution of the $f_0(980)$ reflected by the $t_{K^+K^-,\pi^+\pi^-}$ amplitude, and integrate Eq. (6) to obtain

$$\mathcal{B}(J/\psi \to K^+K^-f_0(980) \to K^+K^-\pi^+\pi^-) = 7.6 \times 10^{-6}\,. \qquad (7)$$

This rate should be multiplied by two to account for the triangle diagram of Fig. 1(b) and another diagram with K^+ and K^- interchanged.

4. Conclusion

We have studied the $J/\psi \to K^+K^-f_0(980)$ decay and have seen that they are driven by a triangle singularity, peaking at $M_{\rm inv}(K^-f_0) \approx 1515$ MeV. This singularity is tied to the nature of the $f_0(980)$ resonance as dynamically generated from the interaction of pseudoscalar mesons, and its experimental observation should bring valuable information on the important issue of the nature of the low-lying scalar mesons. We find the branching ratio of the order of 10^{-5}, which are accessible in present facilities, where many J/ψ branching ratios of the order of $10^{-6} \sim 10^{-7}$ have already been measured.

Acknowledgments

This work is supported by the National Natural Science Foundation of China under Grants Nos. 11505158, 11565007, 11722540, and 11847317, the Fundamental Research Funds for the Central Universities, the Academic Improvement Project of Zhengzhou University, the Spanish Ministerio de Economia y Competitividad, and European FEDER funds under the contract number FIS2011-28853-C02-01, FIS2011-28853-C02-02, FIS2014-57026-REDT, FIS2014-51948-C2-1-P, and FIS2014-51948-C2-2-P.

References

1. R. Karplus, C. M. Sommerfield and E. H. Wichmann, Phys. Rev. **111**, 1187 (1958); L. D. Landau, Nucl. Phys. **13**, 181 (1959).
2. S. Coleman and R. E. Norton, Nuovo Cim. **38**, 438 (1965).
3. X. H. Liu, M. Oka and Q. Zhao, Phys. Lett. B **753**, 297 (2016).
4. M. Bayar, F. Aceti, F. K. Guo and E. Oset, Phys. Rev. D **94**, 074039 (2016).
5. M. Ablikim *et al.* [BESIII Collaboration], Phys. Rev. Lett. **108**, 182001 (2012).
6. J. J. Wu, X. H. Liu, Q. Zhao and B. S. Zou, Phys. Rev. Lett. **108**, 081803 (2012); F. Aceti, W. H. Liang, E. Oset, J. J. Wu and B. S. Zou, Phys. Rev. D **86**, 114007 (2012).
7. N. N. Achasov, A. A. Kozhevnikov and G. N. Shestakov, Phys. Rev. D **92**, 036003 (2015); M. Mikhasenko, B. Ketzer and A. Sarantsev, Phys. Rev. D **91**, 094015 (2015); J. J. Xie, L. S. Geng and E. Oset, Phys. Rev. D **95**, 034004 (2017); X. H. Liu and U. G. Meissner, Eur. Phys. J. C **77**, 816 (2017); E. Oset and L. Roca, Phys. Lett. B **782**, 332 (2018); Z. Cao and Q. Zhao, Phys. Rev. D **99**, 014016 (2019); F. K. Guo, Phys. Rev. Lett. **122**, 202002 (2019); L. R. Dai, Q. X. Yu and E. Oset, Phys. Rev. D **99**, 016021 (2019).
8. W. H. Liang, H. X. Chen, E. Oset and E. Wang, Eur. Phys. J. C **79**, 411 (2019).
9. J. A. Oller and E. Oset, Nucl. Phys. A **620**, 438 (1997) Erratum: [Nucl. Phys. A **652**, 407 (1999)]; N. Kaiser, Eur. Phys. J. A **3**, 307 (1998); M. P. Locher, V. E. Markushin and H. Q. Zheng, Eur. Phys. J. C **4**, 317 (1998); J. Nieves and E. Ruiz Arriola, Nucl. Phys. A **679**, 57 (2000).
10. E. Oset *et al.*, Int. J. Mod. Phys. E **25**, 1630001 (2016); J. R. Pelaez, Phys. Rept. **658**, 1 (2016); H. X. Chen, A. Hosaka and S. L. Zhu, Phys. Lett. B **650**, 369 (2007); H. X. Chen, A. Hosaka and S. L. Zhu, Phys. Rev. D **76**, 094025 (2007).
11. R. Pavao, S. Sakai and E. Oset, Eur. Phys. J. C **77**, 599 (2017).
12. S. Sakai, E. Oset and A. Ramos, Eur. Phys. J. A **54**, 10 (2018).
13. F. Aceti, J. M. Dias and E. Oset, Eur. Phys. J. A **51**, 48 (2015).
14. W. H. Liang and E. Oset, Phys. Lett. B **737**, 70 (2014); J. J. Xie, L. R. Dai and E. Oset, Phys. Lett. B **742**, 363 (2015).

Identification of a visible narrow cusp structure in $\Lambda_c^+ \to pK^-\pi^+$

Xiao-Hai Liu

Center for Joint Quantum Studies and Department of Physics, School of Science, Tianjin University, Tianjin 300350, China
E-mail: xiaohai.liu@tju.edu.cn

Gang Li

School of Physics and Engineering, Qufu Normal University, Shandong 273165, China
E-mail: gli@qfnu.edu.cn

Ju-Jun Xie

Institute of Modern Physics, Chinese Academy of Sciences, Lanzhou 730000, China
E-mail: xiejujun@impcas.ac.cn

Qiang Zhao

Institute of High Energy Physics and Theoretical Physics Center for Science Facilities, Chinese Academy of Sciences, Beijing 100049, China
E-mail: zhaoq@ihep.ac.cn

A resonance-like structure as narrow as 10 MeV is observed in the K^-p invariant mass distributions in $\Lambda_c^+ \to pK^-\pi^+$. This precise measurement is based on a data sample of about 1.5 million events, and the bin width of K^-p invariant mass is only 1 MeV. The narrow peak precisely lies on the $\Lambda\eta$ threshold, because of which it is natural to identify it as a threshold cusp. Being different from the common two-body unitary cusp, we find that the narrowness of this cusp can be induced by a nearby triangle singularity of the $\Lambda a_0(980)$ or $\eta\Sigma(1660)$ rescattering process.

Keywords: Cusp; rescattering process; triangle singularity.

1. Introduction

The rapid variation of the scattering amplitude at some energy threshold will lead to the cusp phenomenon in the corresponding energy spectrum. Although the cusp phenomena have been noticed and discussed by people for many years, it is very difficult to observe these phenomena experimentally. One of the few clearly observed cusps was reported in 2005 by the

NA48/2 collaboration in $K^{\pm} \to \pi^{\pm}\pi^{0}\pi^{0}$ decays,[1] which results from the charge-exchange rescattering process $\pi^{+}\pi^{-} \to \pi^{0}\pi^{0}$.[2–4] This measurement was based on a data sample of 23 million $K^{\pm} \to \pi^{\pm}\pi^{0}\pi^{0}$ decays. The measurement on the cusp in $K \to 3\pi$ decays provides a precise method to determine the $\pi\pi$ scattering lengths.[2–4] The similar proposals have been extended to some other precise experiments.[5,6]

Recently in the $\Lambda_c^{+} \to pK^{-}\pi^{+}$decays, the Belle collaboration reported the observation of a narrow structure in the $K^{-}p$ invariant mass spectrum, of which the mass is 1663 MeV, and the width is 10 MeV.[7] The signal yields of $\Lambda_c^{+} \to pK^{-}\pi^{+}$decays at Belle is about 1.452×10^{6} and the bin width of $K^{-}p$ invariant mass is only 1 MeV, which means this is a very precise measurement. There are several hyperon resonances whose masses are close to 1663 MeV, such as $\Lambda(1670)$, $\Lambda(1690)$, $\Sigma(1660)$ and $\Sigma(1670)$, but all of their widths are much larger than 10 MeV. Therefore none of those established hyperons can account for this narrow structure. Interestingly, the peak position of this structure is just coincide with the $\Lambda\eta$ threshold$\sim$ 1663.5 MeV. It is therefore natural to expect that this peak may be correlated to the $\Lambda\eta$ threshold cusp. However, the two-body unitary cut usually cannot lead to such a narrow peak at the threshold. There should be some other effects which narrows the common two-body threshold cusp.

2. The Model

We notice that the three-body decays $\Lambda_c^{+} \to pK^{-}\pi^{+}$could receive contributions from the rescattering processes shown in Fig. 1.[8] The weak decays $\Lambda_c^{+} \to \Lambda a_0(980)^{+}$ and $\Lambda_c^{+} \to \eta\Sigma^{*+}$ are Cabibbo-favored processes, with Σ^{*+} representing an excited hyperon with $I = 1$,[a] and all of the relevant strong interaction couplings are sizable. Therefore it can be expected that these rescttering contributions may play a role in $\Lambda_c^{+} \to pK^{-}\pi^{+}$. More importantly, these triangle diagrams process some intriguing singularities which affect the $K^{-}p$ invariant mass distribution. The amplitude $T(s)$ corresponding to Fig. 1(a)/(b) has a threshold singularity at $s_{\rm th} = (m_{\Lambda}+m_{\eta})^{2}$ with $s \equiv (p_{K^-}+p_p)^{2} \equiv M_{K^-p}^{2}$. The right-hand unitary cut stating at s_{th} in the complex s-plane leads to a cusp in the $K^{-}p$ spectrum. If all of the three internal particles of Fig. 1(a)/(b) can be on-shell simultaneously, the triangle singularity (TS) of the rescattering diagram can be located on the physical boundary. However, such kinematic conditions cannot be perfectly satis-

[a]We only focus on the discussion of $\Sigma(1660)$ in this work.

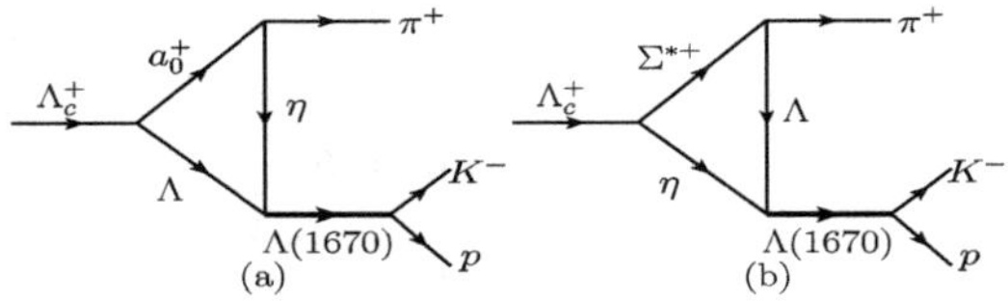

Fig. 1. Rescattering processes contributing to $\Lambda_c^+ \to pK^-\pi^+$. Kinematic conventions: (a) $\Lambda(q_1, m_1)a_0(q_2, m_2)\eta(q_3, m_3)$; (b) $\eta(q_1, m_1)\Sigma^*(q_2, m_2)\Lambda(q_3, m_3)$.

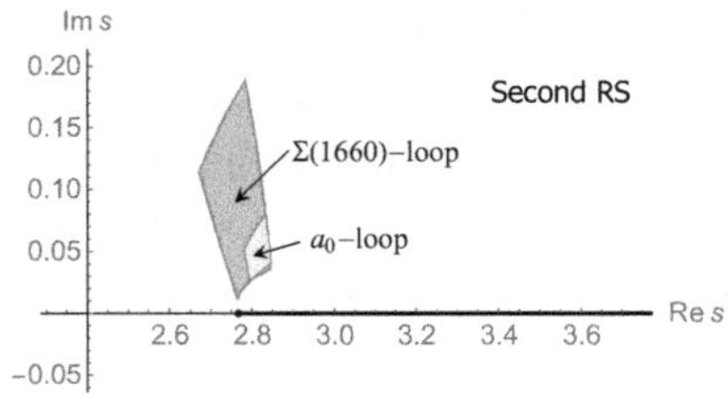

Fig. 2. The TS location of $T(s, m_2^2)$ in the complex s-plane. The thick line on the real axis represents the unitary cut starting from s_{th}.

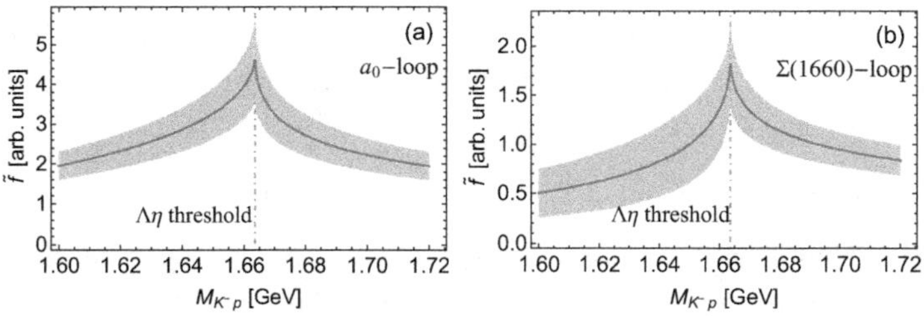

Fig. 3. Modified invariant mass distribution of K^-p. The bands are obtained by taking into account uncertainties of the mass and width of $a_0(980)/\Sigma(1660)$.

fied for the a_0-loop and $\Sigma(1660)$-loop. By setting $m_2 = M_{a_0,\Sigma^*} - i\Gamma_{a_0,\Sigma^*}/2$, the area where the TSs can appear is illustrated by the two colored blocks in Fig. 2, which are obtained by varying M_{a_0} in the range of $960 \sim 1000$ MeV, Γ_{a_0} in $50 \sim 100$ MeV, and M_{Σ^*} in $1630 \sim 1690$ MeV, Γ_{Σ^*} in $40 \sim 200$ MeV, respectively.[8–10] As shown in Fig. 2, the TSs of both the a_0-loop and $\Sigma(1660)$-loop are located on the upper half plane of the second Riemann sheet (RS). Usually the singularities on the upper half plane of the second RS are supposed to be far from the physical region, and their influence to the physical amplitude is negligible. However, we notice that the TSs of a_0- and $\Sigma(1660)$-loop are still very close to the threshold in the sense of distance traveled in the complex plane, which means they can still affect the amplitude in the vicinity of $s_{\rm th}$. The threshold $s_{\rm th}$ is the closest point in the physical region to such TSs, and it can be expected that the cusp structure at $s_{\rm th}$ could be enhanced by the nearby TS.

The rescattering amplitude of $\Lambda_c^+ \to pK^-\pi^+$via the rescattering process is given by

$$T = \frac{1}{s - M^2_{\Lambda(1670)} + iM_{\Lambda(1670)}\Gamma_{\Lambda(1670)}} \times \int \frac{d^4q_1}{(2\pi)^4} \frac{\mathcal{A}}{(q_1^2 - m_1^2)(q_2^2 - m_2^2)(q_3^2 - m_3^2)}, \tag{1}$$

with $\mathcal{A}$ the production of amplitudes $\mathcal{M}(\Lambda_c^+ \to \Lambda a_0^+)\mathcal{M}(a_0^+ \to \eta\pi^+)\mathcal{M}(\eta\Lambda \to \Lambda(1670))\mathcal{M}(\Lambda(1670) \to K^-p)$ and $\mathcal{M}(\Lambda_c^+ \to \eta\Sigma^{*+})\mathcal{M}(\Sigma^{*+} \to \Lambda\pi^+)\mathcal{M}(\eta\Lambda \to \Lambda(1670))\mathcal{M}(\Lambda(1670) \to K^-p)$ for a_0-loop and Σ^*-loop, respectively. The $\eta\Lambda \to K^-p$ reaction is dominated by the $\Lambda(1670)$ ($J^P = 1/2^-$) pole around the energy region we are interested in. Besides, $M_{\Lambda(1670)}$ is close to the $\Lambda\eta$ threshold, and the threshold cusp just grows up on the $\Lambda(1670)$ resonance bump in the M_{K^-p} spectrum. Therefore the line-shape of the $\Lambda(1670)$ pole may disturb the identification of the threshold cusp. In order to eliminate the influence of the $\Lambda(1670)$ pole in identifying the cusp, we define a new distribution function

$$\tilde{f}(M_{K^-p}) = \left|\frac{s - M^2_{\Lambda^*} + iM_{\Lambda^*}\Gamma_{\Lambda^*}}{M_{\Lambda^*}\Gamma_{\Lambda^*}}\right|^2 \times \frac{d\Gamma}{dM_{K^-p}}. \tag{2}$$

The modified distribution curves are shown in Figs. 3(a) and (b). A narrow peak just staying at the $\Lambda\eta$ threshold–1663 MeV can be clearly seen in both of the two figures. This is the cusp structure enhanced and narrowed by the nearby TSs in the unphysical region. We then conclude that even without introducing a genuine hyperon state, the threshold cusp enhanced by the nearby TS can still simulate the narrow resonance-like peak observed by Belle in $\Lambda_c^+ \to pK^-\pi^+$decays.

3. Summary

In summary, we find that the narrow resoance-like structure in the K^-p invariant mass spectrum can be identified as the cusp phenomenon resulted by the $\Lambda a_0(980)$ and $\eta\Sigma(1660)$ rescatterings. The observation reported by Belle is based on a huge data sample of $\Lambda_c^+ \to pK^-\pi^+$decays, and the narrow peak is found to be just located on the $\Lambda\eta$ threshold in this precise measurement. There are no established narrow hyperons whose masses just lying on this threshold. Therefore identifying this peak as a cusp phenomenon is a reasonable and natural explanation. But the two-body unitary cusp is usually not prominent over the background. Our analysis indicates that the cusp correlated with the two-body unitary cut is further enhanced and narrowed by the nearby TS on the second RS of the rescattering amplitudes. As a direct prediction, one may expect to observe the similar phenomenon in $\Lambda_c \to \Sigma\pi\pi$ decays due to the $\eta\Lambda \to \pi\Sigma$ reaction.

Acknowledgments

This work is supported by the National Natural Science Foundation of China (NSFC) under Grants No. 11975165, No. 11735003, No. 11835015, No. 11521505 and No. 11675091.

References

1. J. R. Batley *et al.* [NA48/2 Collaboration], Observation of a cusp-like structure in the $\pi^0\pi^0$ invariant mass distribution from $K^\pm \to \pi^\pm\pi^0\pi^0$ decay and determination of the $\pi\pi$ scattering lengths, *Phys. Lett. B* **633**, 173 (2006).
2. P. Budini and L. Fonda, Pion-Pion Interaction from Threshold Anomalies in K^+ Decay, *Phys. Rev. Lett.* **6**, 419 (1961).
3. N. Cabibbo, Determination of the $a_0 - a_2$ pion scattering length from $K^+ \to \pi^+\pi^0\pi^0$ decay, *Phys. Rev. Lett.* **93**, 121801 (2004).
4. G. Colangelo, J. Gasser, B. Kubis and A. Rusetsky, Cusps in $K \to 3\pi$ decays, *Phys. Lett. B* **638**, 187 (2006).
5. B. Kubis and S. P. Schneider, The Cusp effect in $\eta' \to \eta\pi\pi$ decays, *Eur. Phys. J. C* **62**, 511 (2009).
6. X. H. Liu, F. K. Guo and E. Epelbaum, Extracting $\pi\pi$ S-wave scattering lengths from cusp effect in heavy quarkonium dipion transitions, *Eur. Phys. J. C* **73**, 2284 (2013).
7. C. P. Shen (on behalf of Belle collaboration), Recent Belle results on singly charmed baryon and Belle II prospect of baryons, in *Workshop on Singly and Doubly Charmed Baryons*, (Paris, France, 2018).
8. X. H. Liu, G. Li, J. J. Xie and Q. Zhao, Visible narrow cusp structure in $\Lambda_c^+ \to pK^-\pi^+$ enhanced by triangle singularity, *Phys. Rev. D* **100**, 054006 (2019).
9. X. H. Liu, M. Oka and Q. Zhao, Searching for observable effects induced by anomalous triangle singularities, *Phys. Lett. B* **753**, 297 (2016).
10. X. H. Liu, Q. Wang and Q. Zhao, Understanding the newly observed heavy pentaquark candidates, *Phys. Lett. B* **757**, 231 (2016).

The first observation of narrow peak and isospin-violating $\Lambda(1405)$ production

L. R. Dai*, Y. Hu, X. J. Cai

Department of Physics, Liaoning Normal University, Dalian, 116029, China
**E-mail: dailr@lnnu.edu.cn*
www.lnnu.edu.cn

E. Oset, R. Pavao and S. Sakai

Departamento de Física Teórica and IFIC, Centro Mixto
Universidad de Valencia-CSIC, Institutos de Investigacíon de Paterna,
Aptdo. 22085, 46071 Valencia, Spain

The decay of Λ_c^+ into $\pi^+\pi^0\Lambda(1405)$ with the $\Lambda(1405)$ decaying into $\pi^0\Sigma^0$ through a triangle mechanism is studied. This process is prohibited by the isospin symmetry, but enhanced by the contribution of the triangle diagram. We find a narrow peak in the $\pi^0\Sigma^0$ invariant mass distribution. The observation of the peak in the isospin-violating production would provide further support for hadronic molecular picture of the $\Lambda(1405)$.

Keywords: Isospin-violating; triangle diagram; molecular state.

1. Introduction

The decay of heavy hadrons containing a charm or bottom quark is an exciting field in hadron physics.[1]

In our recent work we found that the "$a_1(1420)$" peak in the $\pi f_0(980)$ invariant mass distribution observed by COMPASS experiment[2] can be understood as a peak of a triangle singularity (TS).[3] The triangle diagram is formed by the $a_1(1260)$ decaying into $K\bar{K}^*$ ($K^*\bar{K}$) with a subsequent $\bar{K}^* \to \pi\bar{K}$ ($K^* \to \pi K$) decay and merging $K\bar{K}$ to form $f_0(980)$. The $a_1(1260)$ and the $f_0(980)$ have large couplings to the $K\bar{K}^* + c.c.$ and the $K\bar{K}$ channels, since they are dynamically generated through the coupled-channel effect of hadrons.[4–7] The singularity would be smeared by the width of hadrons and appear as a broad peak in actual reactions. This peak purely comes from a kinematical effect and we cannot associate this peak with a resonant state.

Recently many hadronic molecular states have been summarized in the work of [8]. The observation of the peak from the triangle singularity would provide further clues to clarify the nature of the hadronic molecules.

In the present work we further extend meson decay to baryon decay. It is interesting to see what will happen. In fact, the molecule nature of $\Lambda(1405)$ resonance was predicted in [9, 10, 11, 12]. This is the first time to study the anomalous enhancement of the isospin-violating baryon $\Lambda(1405)$ production by a triangle singularity in Λ_c decays into $\pi^+\pi^0\pi^0\Sigma^0$ process to test the molecule nature of $\Lambda(1405)$ resonance.

2. Formalism

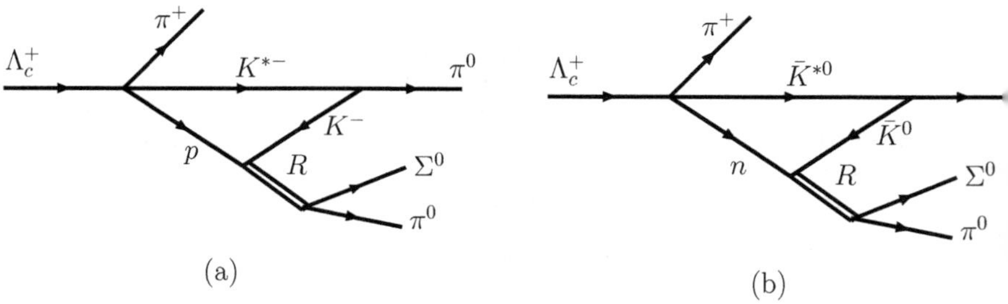

Fig. 1. Diagram for the decay of $\Lambda_c^+ \to \pi^+\pi^0\pi^0\Sigma^0$.

In the present study, we investigate the $\Lambda_c^+ \to \pi^+\pi^0\pi^0\Sigma^0$ decays via $\Lambda(1405)$ formation. Fig. 1 (a) shows the process of $\Lambda_c^+ \to \pi^+K^{*-}p$ followed by the K^{*-} decay into π^0K^- and the merging of the K^-p into $\Lambda(1405)$; and Fig. 1 (b) for the decay of $\Lambda_c^+ \to \pi^+\bar{K}^{*0}n$ followed by the $\bar{K}^{*0}$ decay into $\pi^0\bar{K}^0$ and the merging of the $\bar{K}^0n$ into $\Lambda(1405)$ generate a singularity.

We will see a signal for the $\Lambda(1405)$ around 1420 MeV because it comes from $\bar{K}N$ which couples to the second pole at 1420 MeV in the invariant mass of $\pi^0\Sigma^0$. In the study of Ref. 11, the $\Lambda(1405)$ appears as the dynamically generated state of K^-p, $\bar{K}^0n$, $\pi^0\Lambda$, $\pi^0\Sigma^0$, $\eta\Lambda$ $\eta\Sigma^0$, $\pi^+\Sigma^-$, $\pi^-\Sigma^+$, $K^+\Xi^-$ and $K^0\Xi^0$ in the coupled-channels calculation, in which we can obtain the scattering amplitudes for $t_{K^-p\to\pi^0\Sigma^0}$ and $t_{\bar{K}^0n\to\pi^0\Sigma^0}$.

The final triangle amplitude in Fig. 1 (a) decay is given by

$$t_T(m_{K^{*-}}, M_p, m_{K^-}) = \int \frac{d^3q}{(2\pi)^3} \frac{2M_p}{8\omega_{K^{*-}}\omega_p\omega_{K^-}} \frac{1}{k^0-\omega_{K^-}-\omega_{K^{*-}}+i\frac{\Gamma_{K^{*-}}}{2}} \tag{1}$$
$$\times \left(2+\frac{\vec{q}\cdot\vec{k}}{\vec{k}^2}\right) \frac{1}{P^0+\omega_p+\omega_{K^-}-k^0} \frac{1}{P^0-\omega_p-\omega_{K^-}-k^0+i\epsilon}$$
$$\times \frac{2P^0\omega_p+2k^0\omega_{K^-}-2(\omega_p+\omega_{K^-})(\omega_p+\omega_{K^-}+\omega_{K^{*-}})}{P^0-\omega_{K^{*-}}-\omega_p+i\frac{\Gamma_{K^{*-}}}{2}},$$

with $P^0 = M_{\text{inv}}(\pi^0\Lambda(1405))$, $\omega_p = \sqrt{\vec{q}^{\,2}+M_p^2}$, $\omega_{K^-} = \sqrt{(\vec{q}+\vec{k})^2+m_{K^-}^2}$, $\omega_{K^{*-}} = \sqrt{\vec{q}^{\,2}+m_{K^{*-}}^2}$. The energy k^0 and momentum $|\vec{k}|$ of π^0 emitted from $\bar{K}^*$ are given in [13], and we obtain the final differential distribution for four particles in the final state for Fig. 1 (a).

It is very interesting to find that the contributions from Fig. 1 (a) and 1 (b) will cancel each other if the same masses is used for $\bar{K}N$ and K^*, which means the isospin conservation. Then an important thing is to investigate the isospin-breaking effect. That means, for the first time, we will precisely look at the $\Lambda(1405)$ formation in an isospin forbidden mode. We expect that the formation will be driven by a triangle singularity and the shape will be narrower than usual, because it will be tied to the different masses of $\bar{K}N$. In the following we consider the isospin-breaking effect using the different masses for K^-p or $\bar{K}^0n$, and also for K^{*-} and $\bar{K}^{*0}$.

Finally we get the final differential distributions,

$$\frac{1}{\Gamma_{\Lambda_c^+}} \frac{d^2\Gamma}{dM_{\text{inv}}(\pi^0\Lambda(1405))dM_{\text{inv}}(\pi^0\Sigma^0)} = \frac{1}{(2\pi)^5} \frac{M_{\Sigma^0}}{M_{\Lambda_c^+}} \widetilde{p}_{\pi^+}\widetilde{q}_{\Sigma^0}\widetilde{p}_{\pi^0}\frac{1}{2}g^2\frac{A^2}{\Gamma_{\Lambda_c^+}}|\vec{k}|^3 \tag{2}$$
$$\times \left|t_T(m_{K^{*-}}, M_p, m_{K^-})t_{K^-p\to\pi^0\Sigma^0} - t_T(m_{\bar{K}^{*0}}, M_n, m_{\bar{K}^0})t_{\bar{K}^0n\to\pi^0\Sigma^0}\right|^2,$$

with

$$\widetilde{p}_{\pi^+} = \frac{\lambda^{1/2}(M_{\Lambda_c^+}^2, M_{\text{inv}}^2(\pi^0\Lambda(1405)), m_{\pi^+}^2)}{2M_{\Lambda_c^+}}, \quad \widetilde{q}_{\Sigma^0} = \frac{\lambda^{1/2}(M_{\text{inv}}^2(\pi^0\Sigma^0), m_{\pi^0}^2, M_{\Sigma^0}^2)}{2M_{\text{inv}}(\pi^0\Sigma^0)},$$
$$\widetilde{p}_{\pi^0} = |\vec{k}| = \frac{\lambda^{1/2}(M_{\text{inv}}^2(\pi^0\Lambda(1405)), m_{\pi^0}^2, M_{\text{inv}}^2(\pi^0\Sigma^0))}{2M_{\text{inv}}(\pi^0\Lambda(1405))}. \tag{3}$$

From the experimental branching ratio of $Br(\Lambda_c^+ \to \pi^+K^{*-}p) = (1.5 \pm 0.5) \times 10^{-2}$ decay,[14] we can determine the value of the constant $|A|$.

3. Results

In Fig. 2 we plot the contribution of the triangle loop defined in Eq. (1) fixed at $M_{\text{inv}}(R) = 1420$ MeV. The peak of the real part is related to the

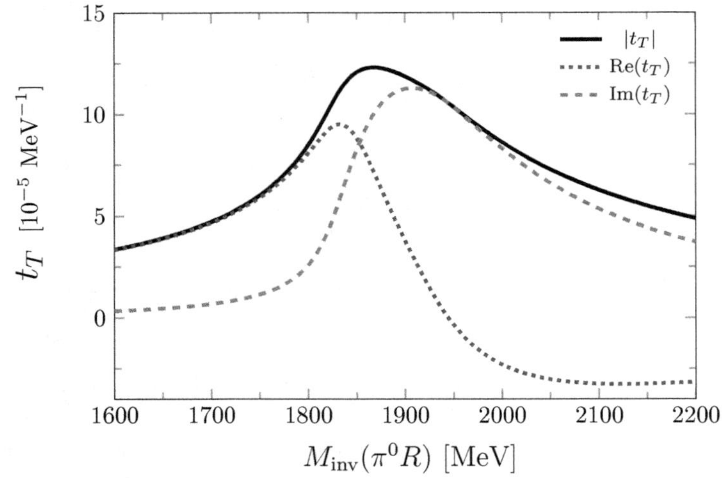

Fig. 2. Triangle amplitude t_T for the decay of $\Lambda_c^+ \to \pi^+ K^{*-} p$ followed by the K^{*-} decay into $\pi^0 K^-$ and the merging of the $K^- p$ into $\Lambda(1405)$, taking $M_{\mathrm{inv}}(R)$=1420 MeV.

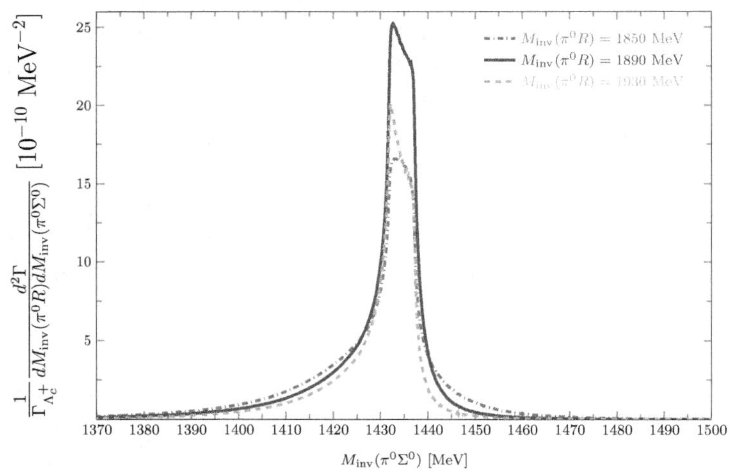

Fig. 3. Double differential width of $\Lambda_c^+ \to \pi^+ \pi^0 \pi^0 \Sigma^0$, keeping $M_{\mathrm{inv}}(\pi^0 R)$ ($R \equiv \pi^0\Sigma^0$).

$K^{*-}p$ threshold and the one of the imaginary part, that dominates for the larger $\pi^0 R$ invariant masses, to the triangle singularity.

In Fig. 3 it is seen that the distribution with largest strength is near $M_{\mathrm{inv}}(\pi^0 R)$=1890 MeV. and we also see a strong peak around 1432 MeV for the three different masses of $M_{\rm inv}(\pi^0 R)$. Further integrating, we obtain the differential width $\frac{1}{\Gamma_{\Lambda_c^+}} \frac{d\Gamma}{dM_{\rm inv}(\pi^0 R)}$ where a clear peak is seen around 1880 MeV for $\Lambda(1405)$ production.[13] Finally we obtain the branching fraction

$$Br(\Lambda_c^+ \to \pi^+\pi^0\Lambda(1405);\ \ \Lambda(1405) \to \pi^0\Sigma^0) = (4.17 \pm 1.39) \times 10^{-6}. \quad (4)$$

This number is within measurable range. The errors come from the experimental errors in the branching ratio of $Br(\Lambda_c^+ \to \pi^+ K^{*-} p)$.

Acknowledgments

LRD acknowledge supports from the National Natural Science Foundation of China (Grant Nos. 11975009, 11575076) and Spanish Ministerio de Economia y Competitividad and European FEDER funds under Contracts No. FIS2017-84038-C2-1-P B and No. FIS2017-84038-C2-2-P B, and the Generalitat Valenciana in the program Prometeo II-2014/068, and the project Severo Ochoa of IFIC, SEV-2014-0398 (EO).

References

1. E. Oset *et al.*, *Int. J. Mod. Phys. E* **25**, 1630001 (2016).
2. C. Adolph *et al.* [COMPASS Collaboration], *Phys. Rev. Lett.* **115**, 082001 (2015).
3. F. Aceti, L. R. Dai and E. Oset, *Phys. Rev. D* **94**, 096015 (2016).
4. L. Roca, E. Oset and J. Singh, *Phys. Rev. D* **72**, 014002 (2005).
5. J. A. Oller and E. Oset, *Nucl. Phys. A* **620**, 438 (1997); Erratum: [*Nucl. Phys. A* **652**, 407 (1999)]
6. D. Gamermann, E. Oset, D. Strottman and M. J. Vicente Vacas, *Phys. Rev. D* **76**, 074016 (2007).
7. J. J. Xie, L. R. Dai and E. Oset, *Phys. Lett. B* **742**, 363 (2015).
8. F. K. Guo, C. Hanhart, U. G. Meißner, Q. Wang, Q. Zhao and B. S. Zou, *Rev. Mod. Phys.* **90**, 015004 (2018).
9. R. H. Dalitz and S. F. Tuan, *Phys. Rev. Lett.* **2**, 425 (1959).
10. N. Kaiser, P. B. Siegel and W. Weise, *Nucl. Phys. A* **594**, 325 (1995).
11. E. Oset and A. Ramos, *Nucl. Phys. A* **635**, 99 (1998).
12. J. A. Oller and U. G. Meissner, *Phys. Lett. B* **500**, 263 (2001).
13. L. R. Dai, R. Pavao, S. Sakai and E. Oset, *Phys. Rev. D* **97**, 116004 (2018)
14. M. Tanabashi *et al.* [Particle Data Group], *Phys. Rev. D* **98**, 030001 (2018).

Session 6

QCD and Hadron Structure

Study of baryon form factors at BESIII

X. R. Zhou, for the BESIII Collaboration

Modern Physics Department, University of Science and Technology of China, Hefei, 230026, China
E-mail: zxrong@ustc.edu.cn
www.ustc.edu.cn

The nucleon is the main building blocks of our visible universe. The electromagnetic form factors serve as the structure observables as well as the test ground for strong interaction. The hyperons, on the other hand, provide a new perspective to dissect the structure of neuclon by replacing one or two quarks into stranges. In this proceeding, I will introduce a series of results obtained from BESIII experiment for the baryon form factors in time-like including both nucleon and hyperon. Furthermore, the prospect of the studies for baryon form factors at BESIII is presented.

Keywords: Form factor; nucleon; hyperon; time-like.

1. Introduction

Proton and neutron are a bound system of three quarks with a structure deviated from a point like particle. The nucleon electromagnetic form factors (EMFFs) are used to parametrize this internal structure. They originiate from the electromagnetic currents in the nucleon. One can access the EMFFs via the electromagnetic probe for space like momentum transfer or time like momentum transfer q^2. The electric G_E and magnetic G_M are used to paramterize the internal structure and dynamics of the nucleon. In time-like, the EMFFs can be accessed through a measurement of the Born cross section in electron-positron annihilation, as described[1]:

$$\frac{d\sigma}{d\Omega} = \frac{\alpha^2\beta C}{4q^2}(|G_M(q^2)|^2(1+\cos^2\theta) + \frac{4M_B^2}{q^2}|G_E(q^2)|^2\sin^2\theta) \qquad (1)$$

where α is the fine structure constant, $\beta = \sqrt{1-1/\tau}$ is the velocity of the final state baryon, $\tau = q^2/4M_B^2$, M_B is the mass of baryon, and C is the S-wave Sommerfeld-Gamow factor,[2] which equals to 1 for neutral baryons and is $y/(1-e^{-y})$ with $y = \pi\alpha\sqrt{1-\beta^2}/\beta$ for charged baryons. As a consequence of the optical theorem, time-like EMFFs above the two-pion threshold have a non-zero imaginary part.[3] This means that if G_E and G_M are different, they have a non-zero relative phase. This phase must be zero at the kinematic threshold, where by definition, the electric and the magnetic FFs are equal.

Experimentally, there have been many studies on proton FFs. In the time-like region,[4–12] measurements have been performed on the direct production channel $e^+e^- \to p\bar{p}$, in the radiative return channel $e^+e^- \to p\bar{p}\gamma_{ISR}$, where γ_{ISR} refers to a photon emitted by initial state radiation (ISR), and in $p\bar{p} \to e^+e^-$ experiments. In most cases these measurements only extracted the effective FF or the ratio of $|G_E|$ and $|G_M|$ with uncertainties above 10%. For the neutron FFs on the contrary, the experimental results are very rare. The first result[6] was reported in 1998 and the measurement appeared to be in disagreement with simple quark model prediction extensions to large q^2. Recent results[13] came from the VEPP experiment. However, all results of neutron FF have precision with over 20% uncertainty and are lacking of the knowledge about $|G_E/G_M|$ ratio due to the limiting statistics.

Besides, for both proton and neutron pair production cross section,unusual behavior in the near-threshold region has been observed,[8,13] which provide us the nautrual extension to study the hyperon pair production near threshold. Experimentally, the BABAR experiment measured the hyperon final states with larger uncertainties compared to the proton case.[14] However, the cross section of $e^+e^- \to \Lambda\bar{\Lambda}$ in a wide c.m. energy region from threshold to $\sqrt{s} = 2.27$ GeV was measured to be $204 \pm 60 \pm 20$ pb, which indicates a possible nonvanishing cross section at threshold.

2. The BESIII Experiment

The Beijing Electron-Positron Collider II(BEPCII) is a symmetric electron-positron collider, creating data on or off resonances in the energy region between the center-of-mass energy $\sqrt{s} = 2.0$ to 4.6 GeV. The Beijing Spectrometer III (BESIII) experiment is located at the BEPCII and has been optimized for the reconstruction of charged particles and photons using an inner tracking for the momentum measurement in the Main Drift Chamber (MDC) and energy deposition in an electromagnetic calorimeter. A time-of-flight (TOF) system outside of the drift chamber can measure the flight time of charged particles. A muon system made by RPC is used to identify muons.

The BESIII started taking data since 2009 for physics. It has collected a data set with an integrated luminosity of about 0.6 pb^{-1} between $\sqrt{s} = 2.0$ to 3.08 GeV, which is outside of a resonance. Besides, BESIII accumulated large data sample at energy region of $\sqrt{s} = 3.773$ to 4.60 GeV, which enables the measurement of the FFs using the ISR technique. BESIII is an excellent laboratory for the measurement of baryon time-like FFs, with both ISR and scan methods can be performed, and the kinematical threshold for different baryon pair production is covered by the energy range of BEPCII.

3. Nucleon FFs at BESIII

The process of $e^+e^- \to p\bar{p}$ is studied at 22 center-of-mass energy points from 2.0 to 3.08 GeV.[15] The Born cross section for this process is measured with the energy-scan technique and it is found to be consistent with the previously published data, with best accuracy achieving 3%, dominanted by systematic. In addition, the EMFF ratio $|G_E/G_M|$ and the absolute value of the electric and magnetic FFs are measured by studying the helicity angle of the proton at 16 energy points. $|G_E/G_M|$ and $|G_M|$ are determined with high accuracy, providing uncertainties comparable to data in the space-like region, with best precision of $|G_E/G_M|$ to be 3.4% dominated by statistical. The data points and results of $|G_E/G_M|$ are shown in Figure 1. This measurement strongly favors the result of Babar over that of PS170.

Besides the energy-scan technique, BESIII also measured the Born cross section for the process $e^+e^- \to p\bar{p}$ using the ISR technique with an undetected photon.[16] The collected events at the 6 energy points for $\sqrt{s} > 4$ GeV with an integrated luminosity of 7.5 fb^{-1} are analyzed in the 30 $M_{p\bar{p}}$ intervals between 2.0 and 3.8 GeV/c^2. The best precision of the Born cross section is 8%. The proton angular distriutions have been also analyzed to determine the proton FF ratio in 3 $M_{p\bar{p}}$ intervals between 2.0 and 3.0 GeV/c^2. The uncertainty on the measured proton FF ratio is dominanted by the statistical uncertainty due to the limited range of the proton angular distribution. In addition, the branching fractions of the J/ψ, $\psi(3686)$ to $p\bar{p}$ decays are also measured with good agreements with the world average values.

The reconstruction of $e^+e^- \to n\bar{n}$ is challenging due to the reconstruction of two neutral hadrons in the final state in absence of a hadronic calorimeter. To improve the detection efficiency, a method for the flight time measurement for neutral particles has been developed for this analysis. Besides, to maximize the efficiency, the data is classified into three sub-sets depending on the detector response of the signal event. With the scan data taken at BESIII from energy points of 2.0 to 3.08 GeV, the process $e^+e^- \to n\bar{n}$ is expected to be studied with better precision than previous measurement.

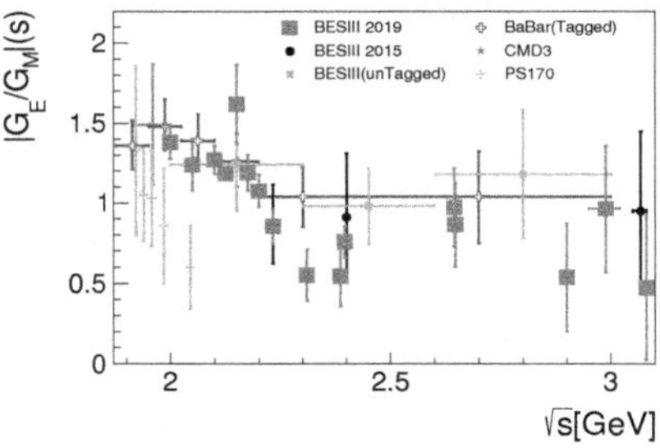

Fig. 1. Result from the ratio $|G_E/G_M$ of the proton with the scan technique, compared with previous experimental results.

4. Hyperon FFs at BESIII

As introduced before, time-like FFs can be complex and therefore the there are phase difference between G_E and G_M. In the weak decay of hyperon to a baryon and pseudoscalar meson, the parity is violated in S wave while conserved in P wave, the polarization can be measured by the angular distribution of daughter baryon. Owing to the self-analysing decay of the hyperons, by baryon production and decay can be fully described by the EMFF ratio, phase difference $\Delta\Phi$ and decay asymmetry.

The process $e^+e^- \to \Lambda\bar{\Lambda} \to p\pi^-\bar{p}\pi^+$ is studied with 66.9 pb^{-1} of data collected at 2.396 GeV.[17] The cross section and the effective form factor are obtained to be $\sigma = 118.7 \pm 5.3 \pm 5.1$ pb and $|G| = 0.123 \pm 0.003 \pm 0.003$. The effective FFs is about one half of that of the proton at the corresponding excess energy.[15] A multi-dimensional analysis with a complete decomposition of the spin structure of the reaction enables a determination of FFs. The ratio $|G_E/G_M|$ is determined with unprecedented precision to be $0.96 \pm 0.14 \pm 0.02$. The relative phase between G_E and $G_M|$ is determined for the first time to be $\Delta\Phi = 37° \pm 12° \pm 6°$. The first uncertainties in the above values are statistical while the second systematic. These results are obtained using the recently established and most precise value of the asymmetry parameter $\alpha_\Lambda = 0.750 \pm 0.010$ Measured by BESIII.

Based on an integrated luminosity of 2.63 pb^{-1} data collected at $\sqrt{s} = 2.2324$ GeV, which is 1.0 MeV above the $\Lambda\bar{\Lambda}$ mass threshold, the measurement of $e^+e^- \to \Lambda\bar{\Lambda}$ is present.[18] The Born cross section at this energy point is determined to be $305 \pm 45^{+66}_{-36}$ pb, where the first uncertainty is statistical and the second systematic. The result is larger than the traditional theory expectation for neutral baryon pairs. The theory predicts a vanishing cross section at threshold according to Eq. (1). The observed threshold enhancement implies a more complicated underlying physics scenario. The Born cross sections of process $e^+e^- \to \Lambda\bar{\Lambda}$ are also measured at $\sqrt{s} = 2.400$, 2.800 and 3.080 GeV, and are in good agreement with BABAR and DM2's results, but with improved precision. The comparison of the Born cross sections with previous experimental results as well as the pQCD prediction is shown in Fig. 2(a).

The process of $e^+e^- \to \Lambda_c\bar{\Lambda}_c$ is studies with the data collected at $\sqrt{s} = 4.5745$, 4.5800, 4.5900 and 4.5995 GeV.[19] At each energy point, ten Cabibbo-favored hadronic decays are used to reconstruct Λ_c^+. The total cross section is obtained from a weighted average over the 20 individual measurements. The most precise cross section measurement is achieved so far at $\sqrt{s} = 4.5745$ GeV, which is only 1.6 MeV above the threshold. The measured value is $236 \pm 11 \pm 46$ pb, which highlights the enhanced cross section near threshold and indicates the complexity of production behavior of the Λ_c. The resulting cross sections at the four energy points are shown in Fig. 2(b) together with the Belle results[20] for comparison. In addition, the higher statistics data samples at $\sqrt{s} = 4.5745$ and 4.5995 GeV enable the study of the polar angle distribution of Λ_c in the c.m. system. From these distributions, the $|G_E/G_M|$ ratios are extracted for the first time, to be $1.14 \pm 0.14 \pm 0.07$ and $1.23 \pm 0.05 \pm 0.03$.

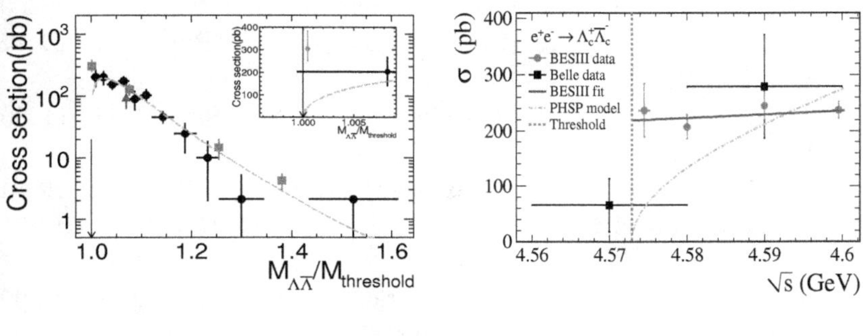

Fig. 2. Comparison of (a) the Born cross section for $e^+e^- \to \Lambda\bar{\Lambda}$, (b) the Born cross section of $e^+e^- \to \Lambda_c\bar{\Lambda}_c$.

5. Summary and Prospect

Proton FFs are measured at BESIII where both scan and ISR techniques are applied for proton. The remaining questions on proton FFs will be sovled, such as the discrepancy on $|G_E/G_M|$ ratio between previous results as well as the stage observed in the cross section of $e^+e^- \to p\bar{p}$. A precise measurement on neutron FF is onging. With the large data set, more precise results on hyperon FFs are expected to be obtained at BESIII, where we can achieve more precise cross section line-shape and test threshold effects. Besides, with larger data sample at each energy point, it is possible to completely determinine the G_E and G_M for hyperons, which will shed light on the understanding of polarization in nucleon.

References

1. N. Cabibbo and R. Gatto, Phys. Rev. **124**, 1577 (1961).
2. J. Schwinger, *Partice, Sources, andFields*) (Perseus Books Publishing, Massachuserrs, 1998), Vol.3; A. B. Arbuzov and T. V. Kopylova, J. High Energy Phys. 04, (2012) 009.
3. A. Z. Dubnickova *et al.*, Nuovo Cim. A **109** 241 (1996).
4. B. Delcourt *et al.* (DM1 Collaboration), Phys. Lett. B **86**, 395 (1979).
5. D. Bisello *et al.* (DM2 Collaboration), Nucl. Phys. B **224**, 379 (1983); Z. Phys. C **48**, 23 (1990).
6. A. Antonelli *et al.* (FENICE Collaboration), Nucl. Phys. B **517**, 3 (1998).
7. T. K. Pedlar *et al.* (CLEO Collaboration) Phys. Rev. Lett. **95**, 261803 (2005).
8. J. P. Lees *et al.* (BaBar Collaboration), Phys. Rev. D **87**, 092005 (2013); J. P. Lees *et al.* (BaBar Collaboration), Phys. Rev. D **88**, 072009 (2013).
9. M. Ablikim *et al.* (BESIII Collaboration), Phys. Rev. D **91**, 112004 (2015).
10. G. Bardin *et al.* (PS170 Collaboration), Nucl. Phys. B **411**, 3 (1994).
11. T. A. Armstrong *et al.* (E760 Collaboration), Phys. Rev. Lett. **70**, 1212 (1993).
12. M. Ambrogiani *et al.* (E835 Collaboration), Phys. Rev. D **60**, 032002 (1990); M. Andreotti *et al.*, Phys. Lett. B **559**, 20 (2003).
13. M. N. Achasov *et al.*, Phys. Rev. D **90**, 112007 (2014).
14. B. Aubert *et al.*, (BABAR Collaboration), Phys. Rev. D **76**, 092006 (2007).

15. M. Ablikim *et al.*, (BESIII Collaboration), arXiv:1905.09001 [hep-ex].
16. M. Ablikim *et al.*, (BESIII Collaboration), Phys. Rev. D **99**, 092002 (2019).
17. M. Ablikim *et al.*, (BESIII Collaboration), Phys. Rev. Lett. **123**, 122003 (2019).
18. M. Ablikim *et al.*, (BESIII Collaboration), Phys. Rev. D **97**, 032013 (2018).
19. M. Ablikim *et al.*, (BESIII Collaboration), Phys. Rev. Lett. **120**, 132001 (2018).
20. G. Pakhlova *et al.*, (Belle Collaboration), Phys. Rev. Lett. **101**, 172001 (2008).

Measuring space-time structure of Δ(1232) resonance with the HBT correlations in double pion photoproduction around 1 GeV *

Q. He[1,a,†], H. Fujimura[1,b], H. Fukasawa[1], R. Hashimoto[2,c], Y. Honda[1], T. Ishikawa[1], T. Iwata[2], S. Kaida[1], H. Kanda[4,h], J. Kasagi[1], A. Kawano[3], S. Kuwasaki[1], K. Maeda[4], S. Masumoto[5], M. Miyabe[1], F. Miyahara[1,d], K. mochizuki[1], N. Muramatsu[1], A. Nakamura[1], K. Nawa[1], S. Ogushi[1], Y. Okada[1], Y. Onodera[1], K. Ozawa[6], Y. Sakamoto[3], M. Sato[1], H. Shimizu[1], H. Sugai[1,e], K. Suzuki[1,f], Y. Tajima[2], S. Takahashi[1], Y. Taniguchi[1], Y. Tsuchikawa[1], H. Yamazaki[1,g], R. Yamazaki[1], and H. Yoshida[1] (the FOREST collaboration)

[1] *Research Center for Electron Photon Science, Tohoku University, Sendai 982-0826, Japan*

[2] *Department of Physics, University of Tokyo, Tokyo 113-0033, Japan*

[3] *Department of Physics, Yamagata University, Yamagata 990-8560, Japan*

[4] *Department of Physics, Tohoku University, Sendai 980-8578, Japan*

[5] *Department of Physics, University of Tokyo, Tokyo 113-0033, Japan*

[6] *Institute of Particle and Nuclear Studies (IPNS), High Energy Accelerator Research Organization (KEK), Tsukuba 305-0801, Japan*

[a] *Department of Nuclear Science and Engineering, Nanjing University of Aeronautics and Astronautics (NUAA), Nanjing 210016, China*

[b] *Department of Physics, Wakayama Medical University, Wakayama 641-8509, Japan*

[c] *Institute of Materials Structure Science, KEK, Tsukuba 305-0801, Japan*

[d] *Accelerator Laboratory, High Energy Accelerator Research Organization (KEK), Tsukuba 305-0801, Japan*

[e] *Department of Quantum Science and Energy Engineering, Tohoku University, Sendai 980-8579, Japan*

[f] *The Wakasa Wan Energy Research Center, Tsuruga 914-0192, Japan*

[g] *Radiation Science Center, High Energy Accelerator Research Organization (KEK), Tsukuba 305-0801, Japan*

[h] *Research Center for Nuclear Physics, Osaka University, Osaka 567-0047, Japan*

[†] *E-mail: heqh@nuaa.edu.cn*

Spatial-temporal properties of Δ(1232) resonance produced in double neutral pion photoproduction are investigated through HBT correlations of two identical pions. For this end, we develop a model in which the HBT correlations are measured in the reference frame of Δ at rest. A very preliminary result of the two pion correlation is obtained, showing the Gaussian radius of the intermediate resonance state (Δ (1232) dominant) is about 0.71±0.02 fm.

Keywords: Bose-Einstein correlations; HBT; space-time properties; photoproduction.

1. Introduction

Investigating space-time properties of baryon resonance may shed light on a dynamic process of hadron reactions and provide useful information for hadron structure. Since baryon resonances are very short-lived, it is impossible to directly measure their size. Thanks to the Hanbury Brown and Twiss (HBT) correlation method [1], [2], also known as the Bose-Einstein correlation, a viable indirect measurement is possible. The HBT effect was employed in measuring angular radii of astronomical objects in the 1950s [1], [2] and later was widely used as a useful tool to study the space-time structure of particle emitting volumes produced in high energy elementary-particle collisions and heavy-ion collisions [3]–[9].

In the first and second resonance region the $2\pi^0$ photoproduction is dominated by Δ(1232) sequential decay $\gamma p \to \Delta\pi^0 \to p\pi^0\pi^0$ [10]–[12]. We are trying to extract information on spatial extension of Δ(1232) via measuring $\pi^0\pi^0$ HBT correlations in the double π^0 photoproduction on the proton.

2. Methodology

To measure the Δ(1232) radius via $\pi^0\pi^0$ HBT correlations in $\gamma p \to p\pi^0\pi^0$, we need to establish the relation between the $\pi^0\pi^0$ correlations and the space-time structure of the Δ resonance, because in the reaction the first π^0 and Δ are produced at the same time at first and then the Δ decays into $\pi^0 p$ and thus two π^0s are not produced at the same time. For this end we develop a model in which the spatial extension of a hadron is equivalent to the three constituent valence quark distribution inside it [13]. The relation between the Δ size and two pion correlations is established based on the assumption that the emitting pion arises from an energetic

* This work is supported by the National Natural Science Foundation of China (No. 11805099), the Fundamental Research Funds for the Central Universities (Nos. 1006-XAA18059 and 1006-YAH17063), and the Ministry of Education and Science of Japan (No. 19002003) and JSPS KAKENHI (Nos. 24244022, 26400287).

quark somewhere inside the Δ domain (Fig. 1) and the wavefunctions of the first pion and Δ overlap each other in the original $\pi^0\Delta$ system. The originating place of the pion emitted from an energetic quark offers information on the pion emitting source distribution. The intrinsic spin of Δ induced relative angular momentum between the Δ decayed pion and nucleon is also taken into account.

In this model, the HBT effect is measured in the reference frame of Δ at rest. The space-time coordinates are defined so that the center of Δ is the origin of the coordinates O, when the first π^0 and Δ appear simultaneously at τ=0, where τ denotes the proper time at the rest frame of Δ. The first and second π^0s take place at $A(0,x)$ and $B(\tau,y)$, respectively and the decay time τ does not give any effect in the measurement as long as strong decay is concerned. If a cylindrical coordinates (r,θ,z) is employed where the z axis is parallel to the Δ decayed π^0's momentum vector p_2, the correlation function in the rest Δ reference frame may be given by:

$$C(q,p_2) = 1 + \lambda\, exp\left(-\frac{\alpha^2 q^2}{2}\right) exp\left(-\frac{\alpha^2 q_z^2}{2}\right) J_0(\beta q_r)$$
$$= 1 + \lambda\, exp(-\alpha^2 q_z^2)\, exp\left(-\frac{\alpha^2 q_r^2}{2}\right) J_0(\beta q_r) \qquad (1)$$

where q is the 3-d relative momentum of two pions, $\beta = 1/2p_2$, q_z the q projection on z axis, q_r the q projection on the r axis which is perpendicular to p_2 ($q^2 = q_z^2 + q_r^2$), λ the correlation strength (ranging from 0 to 1) and, α the Δ Gaussian radius. In Eq. (1), a Gaussian density profile of the Δ is assumed.

Figure 2 shows the correlation functions plotted in the (q_z, q_r) plane at fixed p_2=0.2 GeV. Due to low data statistics, we only measure $C(q,p_2)$ at fixed p_2:

$$C(q,p_2)|_{p_2} = C(q_z,q_r) = N_{ori}(q_z,q_r)/N_{ref}(q_z,q_r), \qquad (2)$$

where $N_{ori}(q_z,q_r) = \Delta N/\Delta q$ is the distribution function of the identical particles with HBT correlations from the same original events, and $N_{ref}(q_z,q_r)$ is the distribution function of the identical particles without HBT effects from so-called 'reference sample' generally produced via event mixing method [14], [15], in which two particles are selected from different events. Reliable event mixing in exclusive reactions with low multiplicity requires appropriate mixing cuts to preserve non-HBT correlations. Here we use the missing mass consistency (MMC) cut [16] and the energy hierarchy correspondence (EHC) cut that requires a boson with relatively higher/lower energy can only be swapped with a boson with relatively higher/lower energy [16]. To correct other mixing induced distortion, double ratio (D.R.) method is used to correct the correlation function:

$$C_{D.R.}(q_z, q_r) = \left(\frac{N_{ori}^{exp}(q_z, q_r)}{N_{ref}^{exp}(q_z, q_r)}\right) \Big/ \left(\frac{N_{ori}^{MC}(q_z, q_r)}{N_{ref}^{MC}(q_z, q_r)}\right), \quad (3)$$

where $N_{ori}(q_z, q_r)$ is the two-particle density of the original data, while $N_{ref}(q_z, q_r)$ the two-particle density from event mixing. The upper index "exp" and "MC" indicate experimental data and Monte Carlo simulation, respectively.

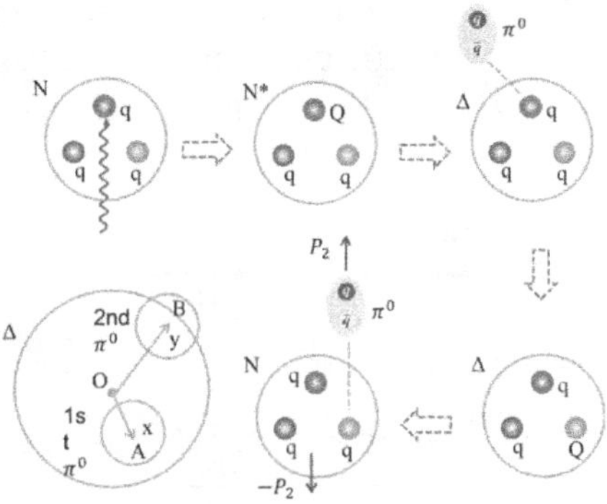

Fig. 1. A schematic illustration of the space-time coordinate system.

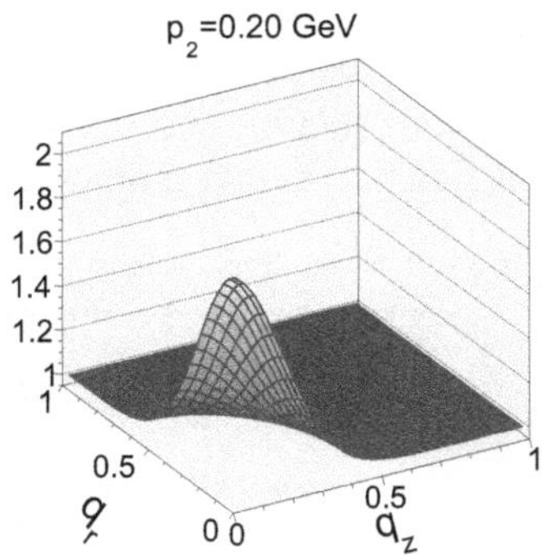

Fig. 2. 2-d plots of the correlation functions in the (q_z, q_r) plane at p_2=0.2 GeV.

3. Results

With this double ratio method, a very preliminary result of the correlation function at p_2=0.25-0.3 GeV is given in Fig. 3. Eq. (1) is used for fitting and the fit results are listed in Table 1. The datasets of the photoproduction $\gamma p \rightarrow p\pi^0\pi^0$ are from FOREST experiment at Research Center for Electron Photon Science

(ELPH) in Tohoku University, with a bremsstrahlung photon beam of E_γ=0.5-1.2 GeV in energy and a 4π EM calorimeter FOREST.

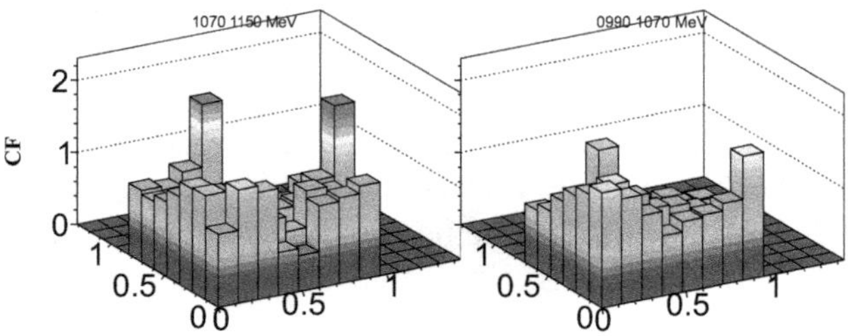

Fig. 3. Two-pion correlation functions plotted in the (q_z, q_r) plane at p_2=0.25-0.30 GeV from the FOREST experimental data $\gamma p \rightarrow p\pi^0\pi^0$ in the incident photon energy ranging from 0.9 to 1.15 GeV.

Simulation samples in the double ratio method are constructed by mixing $\gamma p \rightarrow \Delta\pi^0 \rightarrow p\pi^0\pi^0$ and $\gamma p \rightarrow p\pi^0\pi^0$ events with an optimal mixing ratio, determined by fitting the $p\pi^0$ invariant mass distributions of the mixed samples with all possible mixing ratio to the experimental data [17]. Because the misidentification of Δ from $\pi^0\pi^0 p$ in calculating two pion momenta at rest Δ frame may introduce systematic uncertainties, the same events construction and event mixing procedure are done in the Genat4 simulation for double ratio correction.

Table 1. Fit results of two pion HBT correlation functions.

E_γ (GeV)	N	λ	α (fm)	χ^2/ndf
1.07-1.15	0.85±0.01	1.05±0.07	0.67±0.02	173.1/43
0.99-1.07	0.83±0.01	1.24±0.07	0.76±0.02	89.5/36
Ave.	0.84±0.01	1.14±0.05	0.71±0.02	

4. Summary and discussion

The space-time structure of the excited baryon Δ (1232) is measured using the HBT correlations of two-pion from photoproduction with the incident photon energy of 0.5-1.2 GeV. A model is established in the reference frame of Δ at rest in order to relate the Δ size and two pion correlation. A very preliminary result of the two pion correlation is obtained, yielding the Gaussian radius of intermediate resonance state (Δ (1232) dominant) to be 0.71±0.02 fm.

References

[1] R. H. Brown and R. Q. Twiss, LXXIV. A new type of interferometer for use in radio astronomy, *London, Edinburgh, Dublin Philos. Mag. J. Sci.* **45**, 663 (1954).

[2] R. Hanbury Brown, A test of new type of new steller interferometer on SIRIUS, *Nature* **178**, 1046 (1956).

[3] C.-Y. Wong, *Introduction to High-Energy Heavy-Ion Collisions.* (Singapore, World Scientific, 1994).

[4] U. A. Wiedemann and U. Heinz, Particle interferometry for relativistic heavy-ion collisions, *Phys. Rep.* **319**, 145 (1999).

[5] R. M. Weiner, "Boson interferometry in high-energy physics," *Phys. Rep.* **327**, 249 (2000).

[6] M. Lisa, S. Pratt, R. Soltz, and U. Wiedemann, "Femtoscopy in Relativistic Heavy Ion Collisions: Two Decades of Progress," *Annu. Rev. Nucl. Part. Sci.* **55**, 357 (2005).

[7] D. Boal, C.-K. Gelbke, and B. Jennings, "Intensity interferometry in subatomic physics," *Rev. Mod. Phys.* **62**,553 (1990).

[8] W. Zajc *et al.*, Two-pion correlations in heavy ion collisions, *Phys. Rev. C* **29**, 2173 (1984).

[9] S. Pratt, Resolving the Hanbury Brown – Twiss Puzzle in Relativistic Heavy Ion Collisions, *Phys. Rev. C* **102**, 232301, (2009).

[10] M. Wolf *et al.*, "Photoproduction of neutral pion pairs from the proton," *Eur. Phys. J. A* **9**, 5 (2000).

[11] J. A. Gómez Tejedor and E. Oset, "Double pion photoproduction on the nucleon: study of the isospin channels," *Nucl. Phys. A* **600**, 413 (1996).

[12] U. Thoma *et al.*,N* and Δ* and decays into $N\pi^0\pi^0$, *Phys. Lett. B* **659**, 87 (2008).

[13] H. Shimizu, How to measure the Δ(1232) radius, ELPH Annual Report, 9 (2017).

[14] G. I. Kopylov and M. I. Podgoretskii, "Correlations of identical particles emitted by highly excited nuclei," *Yad. Fiz.* **15**, 392 (1972).

[15] G. I. Kopylov, "Like particle correlations as a tool to study the multiple production mechanism," *Phys. Lett. B* **50**, 472 (1974).

[16] Q. He *et al.*,An event mixing technique for Bose-Einstein correlations of two bosons in exclusive reactions, *Chinese Phys. C* **40**, 114002 (2016).

[17] Q. He, Bose-Einstein correlations between two neutral pions from photoproduction below 1.2 GeV, Tohoku University, Japan, 2014.

Two photon transition form factors of neutral pseudoscalar mesons

Minghui Ding* and Daniele Binosi
European Centre for Theoretical Studies in Nuclear Physics and Related Areas (ECT) and Fondazione Bruno Kessler*
Villa Tambosi, Strada delle Tabarelle 286, I-38123 Villazzano (TN) Italy
**E-mail: mding@ectstar.eu*

Khépani Raya and Lei Chang
School of Physics, Nankai University, Tianjin 300071, China

Craig D. Roberts
School of Physics, Nanjing University, Nanjing, Jiangsu 210093, China
Institute for Nonperturbative Physics, Nanjing University, Nanjing, Jiangsu 210093, China

Starting from the axial vector Ward-Green-Takahashi identity, and the necessity of preserving this identity when attempting to calculate the properties describing pseudoscalar mesons, we describe a Bethe-Salpeter kernel which takes into account the non-Abelian anomaly and therefore is suitable to describe the flavor mixing states η, η'. We then report results for the two photon transition form factors of η, η', and compare them with existing experimental data.

Keywords: Pseudoscalar meson; transition form factor; non-Abelian anomaly.

1. Ward-Green-Takahashi identity

Ward's identity[1] and its generalization by Green[2] and Takahashi,[3] ought to be preserved by any approach that is willing to capture QCD's fundamental features. In its axial vector formulation this identity reads

$$\begin{aligned} P^\mu \Gamma^a_{5\mu}(k;P) &= \mathcal{S}^{-1}(k_+) i\gamma_5 \mathcal{F}^a + i\gamma_5 \mathcal{F}^a \mathcal{S}^{-1}(k_-) \\ &\quad - 2i\mathcal{M}^{ab}\Gamma^b_5(k;P) - \mathcal{A}^a(k;P), \end{aligned} \tag{1}$$

where: P (k) is the total (relative) momentum between the amputated quark legs; $\{\mathcal{F}^a | a = 0, \ldots, N_f^2 - 1\}$ are the generators of $U(N_f)$ in the fundamental representation, orthonormalized according to $\text{tr}\mathcal{F}^a\mathcal{F}^b = \frac{1}{2}\delta^{ab}$; $\mathcal{M}^{ab} = \text{tr}_F\left[\{\mathcal{F}^a, \mathcal{M}^{\text{bm}}\}\mathcal{F}^b\right]$, with $\mathcal{M}^{\text{bm}} = \text{diag}[m_u^{\text{bm}}, m_d^{\text{bm}}, m_s^{\text{bm}}, \ldots]$ the current-quark bare masses matrix; $\mathcal{S} = \text{diag}[S_u, S_d, S_s, \ldots]$ the dressed-quark propagator matrix, with nonzero entries determined by the corresponding Dyson-Schwinger equation; $\Gamma^a_{5\mu}$ and Γ^a_5 the axial-vector and pseudoscalar vertex respectively, both satisfying the corresponding inhomogeneous Bethe-Salpeter equations; and, finally, $\mathcal{A}^a$ in an expression of the non-Abelian axial anomaly and can be defined through the topological charge density operator. Notice that only $\mathcal{A}^{a=0}$ is nonzero: therefore, flavor-nonsinglet neutral pseudoscalar mesons satisfy the standard identity, while flavor-singlet state satisfies the anomalous one.[4]

One of the far-reaching consequences of the Ward-Green-Takahashi (WGT) identity (1) is a mass formula for pseudoscalar mesons which is valid irrespectively from the magnitude of the current-quark masses. It reads

$$m_M^2 f_M^a = 2\,\mathcal{M}^{ab}\rho_M^b + \delta^{a0} n_M, \tag{2}$$

with: m_M the meson mass; f_M^a the decay constant; ρ_M^b the pseudoscalar projection of the meson's Bethe-Salpeter wave function onto the origin in configuration space; and n_M a quantity derived from topological charge density operator. This one single equation has a wide range of effects on Goldstone modes, excited states as well as system(s) described by the non-Abelian axial anomaly[4–6]; in particular, it shows that:

- The Nambu-Goldstone pion is a consequence of Dynamical Chiral Symmetry Breaking (DCSB).
- In the chiral limit, the decay constants of the pion's radial excitations are zero.
- η' is split from the octet pseudoscalars by an amount that depends on QCD's topological susceptibility.

2. Kernels for the bound-state equations

Yet another consequence of Eq. (1) is that it leads to a symmetry-preserving truncation scheme, the simplest of which is the Rainbow-Ladder (RL) approximation. RL truncation is known to be accurate for ground-state light-quark vector- and isospin-nonzero-pseudoscalar-mesons,[7–10] and has been used in a recent study on pion parton distribution functions.[11] Implemented in the Bethe-Salpeter and gap equations, RL truncation is expressed via the following kernels, respectively:

$$[\mathcal{K}_N(k,q;P)]_{\iota_1\iota_2}^{\iota_1'\iota_2'} = -\tfrac{4}{3}\mathcal{G}_{\mu\nu}(t)[i\gamma_\mu]_{\iota_1\iota_1'}[i\gamma_\nu]_{\iota_2'\iota_2}, \tag{3a}$$

$$\mathcal{G}_{\mu\nu}(t=k-q) = \tilde{\mathcal{G}}(t^2)T_{\mu\nu}(t), \tag{3b}$$

where: $t^2 T_{\mu\nu} = t^2\delta_{\mu\nu} - t_\mu t_\nu$; and $\tilde{\mathcal{G}}(t^2)$ is the interaction consistent with that determined in studies of QCD's gauge sector.[12]

However, RL truncation fails for the η- and η'-mesons as they do not produce vertices that satisfy the anomalous axial-vector WGT identity. Indeed, whilst the RL kernel is constrained by a large body of successful phenomenology, the form of the anomalous kernel is unknown. Considering the structure of the non-Abelian anomaly, it readily becomes apparent that no related contribution to the Bethe-Salpeter kernel can contain external quark or antiquark lines which are simply connected to the internal lines: purely gluonic configurations must mediate. Moreover, no finite sum of diagrams can be sufficient.[4] On general grounds, its contribution to the Bethe-Salpeter equation for pseudoscalar mesons must take the following form[13]:

$$[\mathcal{K}_A(k,q;P)]_{\iota_1\iota_2}^{\iota_1'\iota_2'} = -\sum_{i=1}^{4}\sum_{\mathsf{f}=l,s} a_i^{\mathsf{f}}(k,q;P) \times [\mathcal{F}^{\mathsf{f}}\mathcal{D}_i(q;P)]_{\iota_2'\iota_1'}[\mathcal{F}^{\mathsf{f}}\mathcal{D}_i(k;P)]_{\iota_1\iota_2}, \tag{4}$$

where: $\{\mathcal{D}_i(k;P)|i=1,\ldots,4\}$ constitute a Dirac basis of the pseudoscalar meson Bethe-Salpeter wave function; and $\{a_i^{\mathsf{f}}(k,q;P)|i=1,\ldots,4;\mathsf{f}=l,s\}$ are scalar functions. An educated Ansatz for $\mathcal{K}_A$ had been put forward in Ref. 13, which being essentially nonperturbative in nature, provided material support at infrared momenta. In this way the corresponding Bethe-Salpeter kernel becomes the sum of two terms: $\mathcal{K} = \mathcal{K}_N + \mathcal{K}_A$, and with the bound-state kernel thus defined, results for the $\gamma^*\gamma \to \eta,\eta'$ transition form factors have been obtained.

3. Two-photon transition form factors

The two-photon transition form factor provides a window for exploring the internal structure of neutral pseudoscalar meson.[14–16] The parameters in the gap and Bethe-Salpeter equations can be fixed by requiring that the solutions of the coupled-channels bound-state problems reproduce the empirical η,η' masses and the four phenomenologically-determined values of the light- and strange-quark η, η' decay constants.[13] We find $m_\eta = 0.56$ GeV and $m_{\eta'} = 0.96$ GeV, to be compared with the experimental values[17] 0.55 GeV and 0.96 GeV, respectively; other observables like decay constants, mixing angle and two-photon decay widths are reported in Ref. 13.

The transition form factors $F_{\eta,\eta'}(Q^2)$ obtained within the framework sketched above, are shown in Fig. 1 and compared with the asymptotic value of the π^0 form factor, $2f_\pi = 0.186\,\text{GeV}$, drawn as the dotted (red) curve in both panels. There are marked similarities between $F_\eta(Q^2)$ and the π^0 tran-

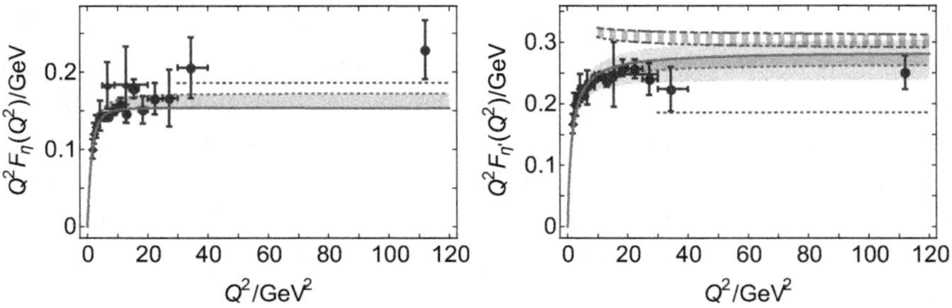

Fig. 1. $\gamma^*\gamma \to \eta, \eta'$ transition form factors: left panel, η; right panel, η'. Curves: solid (green), our prediction with complete evolution, with the shaded (green) band indicating an uncertainty in our prediction owing to variations in the value of f_η^s; dashed (blue), asymptotic limit, with the banded-shading (blue) region indicating the impact of uncertainty in f_η^s on the asymptotic behaviour. The broader, shaded (grey) band combines this with the uncertainty owing to omission of φ_M^0-φ_M^g mixing (φ_M^0 is the flavor singlet distribution amplitude and φ_M^g is the two-gluon distribution amplitude). In both panels, the dotted (red) curve is the π^0 asymptotic limit, $2f_\pi$. Experimental data are[18]: diamonds (blue) "CLEO"; circles (red) "BaBar".

sition form factor (see, for example, Ref. 14, Fig. 2); furthermore, our full $F_{\eta,\eta'}(Q^2)$ result (solid, green curve) agree well with existing data. Finally, at asymptotically large momentum transfers, *i.e.* on $\tau^2 := \Lambda_{\rm QCD}^2/Q^2 \simeq 0$, our full results meet the asymptotic trajectory (dashed blue curve).

The scope of the analysis sketched herein could also be extended to include the doubly off-shell transition form factors, for which the first data $\gamma^*(k_1)\gamma^*(k_2) \to \eta'$ now exist.[19]

Acknowledgment

Work partially supported by Jiangsu Province *Hundred Talents Plan for Professionals.*

References

1. J. C. Ward, An Identity in Quantum Electrodynamics, *Phys. Rev.* **78**, 182 (1950).
2. H. S. Green, A Pre-renormalized quantum electrodynamics, *Proc. Phys. Soc.* **A66**, 873-880 (1953).
3. Y. Takahashi, On the generalized Ward identity, *Nuovo Cim.* **6**, 371 (1957).

4. M. S. Bhagwat, L. Chang, Y. X. Liu, C. D. Roberts, and P. C. Tandy, Flavour symmetry breaking and meson masses, *Phys. Rev.* **C76**, 045203 (2007).
5. P. Maris, C. D. Roberts, and P. C. Tandy, Pion mass and decay constant, *Phys. Lett.* **B420**, 267-273 (1998).
6. A. Holl, A. Krassnigg, C. D. Roberts, and S. V. Wright, On the complexion of pseudoscalar mesons, *Int. J. Mod. Phys.* **A20**, 1778-1784 (2005).
7. T. Horn and C. D. Roberts, The pion: an enigma within the Standard Model, *J. Phys.* **G43**, no.7, 073001 (2016).
8. G. Eichmann, H. Sanchis-Alepuz, R. Williams, R. Alkofer, and C. S. Fischer, Baryons as relativistic three-quark bound states, *Prog. Part. Nucl. Phys.* **91**, 1-100 (2016).
9. C. S. Fischer, QCD at finite temperature and chemical potential from DysonSchwinger equations, *Prog. Part. Nucl. Phys.* **105**, 1-60 (2019).
10. V. D. Burkert, and C. D. Roberts, Colloquium : Roper resonance: Toward a solution to the fifty year puzzle, *Rev. Mod. Phys.* **91**, 011003 (2019).
11. M. Ding, K. Raya, D. Binosi, L. Chang, C. D. Roberts and S. M. Schmidt, Symmetry, symmetry breaking, and pion parton distributions, (arXiv:1905.05208 [nucl-th]).
12. D. Binosi, L. Chang, J. Papavassiliou and C. D. Roberts, Bridging a gap between continuum-QCD and ab initio predictions of hadron observables, *Phys. Lett.* **B742**, 183-188 (2015).
13. M. Ding, K. Raya, A. Bashir, D. Binosi, L. Chang, M. Chen and C. D. Roberts, $\gamma^*\gamma \to \eta, \eta'$ transition form factors, *Phys. Rev.* **D99**, 014014 (2019).
14. K. Raya *et al.*, Structure of the neutral pion and its electromagnetic transition form factor, *Phys. Rev.* **D93**, 074017 (2016).
15. K. Raya, M. Ding, A. Bashir, L. Chang, and C. D. Roberts, Partonic structure of neutral pseudoscalars via two photon transition form factors, *Phys. Rev.* **D95**, 074014 (2017).
16. J. Chen, M. Ding, L. Chang and Y. X. Liu, Two Photon Transition Form Factor of $\bar{c}c$ Quarkonia, *Phys. Rev.* **D95**, 016010 (2017).
17. M. Tanabashi *et al.*, Review of Particle Physics, *Phys. Rev.* **D98**, 030001 (2018).
18. P. del Amo Sanchez *et al.*, Measurement of the $\gamma\gamma^* \to \eta$ and $\gamma\gamma^* \to \eta'$ transition form factors, *Phys. Rev.* **D84**, 052001 (2011).
19. J. P. Lees *et al.*, Measurement of the $\gamma^*\gamma^* \to \eta'$ transition form factor, *Phys. Rev.* **D98**, 112002 (2018).

Positronium on the light front

Kaiyu Fu,* Hengfei Zhao,† and Xingbo Zhao$

Institute of Modern Physics, Chinese Academy of Sciences, Lanzhou 730000, China and School of Nuclear Science and Technology, University of Chinese Academy of Sciences, Beijing 100049, China
**E-mail: kaiyufu@impcas.ac.cn*
† E-mail: zhaohengfei@impcas.ac.cn
$ E-mail: xbzhao@impcas.ac.cn

James P. Vary‡

Department of Physics and Astronomy, Iowa State University, Ames, IA 50011, USA
‡ E-mail: jvary@iastate.edu

Basis Light-front Quantization (BLFQ) is a newly developed nonperturbative approach, aiming at solving relativistic bound systems based on the Hamiltonian formalism of light-front dynamics. In this work, we introduce its application to the positronium system at strong coupling, $\alpha = 0.3$, with a dynamical photon mediating the interaction between the positron and the electron. Nonperturbative mass renormalization is needed to cancel the fermion self-energy divergence. Here, we present the resulting mass spectrum, light-front wave functions (LFWFs), and the photon distribution inside positronium.

Keywords: BLFQ; positronium; light-front wave functions; photon distribution.

1. Introduction

Nonperturbative calculations are needed for understanding hadron strunctures in terms of fundamental quark and gluon degrees of freedom. BLFQ[1] is developed as a nonperturbative approach to bound states in a quantum field theory such as QCD. It combines the advantages of light-front dynamics[2] and modern developments of *ab initio* nuclear structure calculations.[3,4] BLFQ solves the Hamiltonian eigenvalue equation of the bound state in light-front coordinates with light-front gauge $A^+ = 0$. The resulting mass spectra and light-front wave functions (LFWFs) encode all the informations of the system. BLFQ has been successfully applied to the anomalous magnetic moment of the electron,[5,6] positronium,[7] quarkonium,[8,9] meson,[10–12] and proton[13] systems. In this work, we apply BLFQ to the positronium system in QED while retaining a photon as a dynamical degree of freedom. Positronium is an ideal test case for the inclusion of the higher Fock sector since QED is simpler than QCD. Fock sector dependent renormalization[14–16] is needed to cancel the fermion self-energy divergence. Here we present the resulting LFWF, mass spectrum, and photon distribution function.

2. Basis Light-Front Quantization

In BLFQ, we solve the eigenvalue equation $(P^- P^+ - P^{\perp 2})|\psi\rangle = M^2|\psi\rangle$, where $(P^-, P^+, P^\perp)$ is the energy momentum four-vector operator and M is the invariant mass, ψ is the mass eigenstate. In light-front gauge, $P^2 = P^- P^+ - P^{\perp 2}$ is the light-front Hamiltonian with the eigenvalue of mass square. P^+ and $P^\perp$ are the total longitudinal and transverse momenta of the system.

$$\begin{aligned} P^- = \int \mathrm{d}^2 x^\perp \mathrm{d}x^- \frac{1}{2}\bar{\Psi}\gamma^+ \frac{m_e^2 + \left(i\partial^\perp\right)^2}{i\partial^+}\Psi + \frac{1}{2}A^j \left(i\partial^\perp\right)^2 A^j \\ + ej^\mu A_\mu + \frac{e^2}{2} j^+ \frac{1}{\left(i\partial^+\right)^2} j^+, \end{aligned} \tag{1}$$

which is derived from the QED Lagrangian in light-front coordinates via the Legendre transformation. In BLFQ, the Hamiltonian eigenvalue equation is expressed in a truncated discretized basis and solved numerically by diagonalizing the resulting Hamiltonian matrix. We first expand the positronium state in a truncated Fock space, $|e^+e^-\rangle_{phys} = a|e^+e^-\rangle + b|e^+e^-\gamma\rangle$, where $|e^+e^-\rangle$ and $|e^+e^-\gamma\rangle$ are Fock sectors that we elect to retain. We then expand these Fock sectors into the three dimensional discretized basis. The transverse direction of the basis is a 2-dimensional harmonic oscillator, $\phi^b_{n,m}(\vec{p}_\perp)$, where n is the radial quantum number, m is the angular quantum number and b is a scale parameter. The longitudinal direction is the plane wave which is discretized by putting it in a "box" of Length L and applying periodic boundary conditions for bosons and antiperiodic boundary conditions for fermions. So, the longitudinal momentum $p^+ = \frac{2\pi}{L} j$, where j is an integer for bosons or a half-integer for fermions. The basis is made finite by restricting the quantum number in each basis state according to $\sum_i (2n_i + |m_i| + 1) \leq N_{max}$ and $\sum_i j_i = K$, where the sum is over Fock particles in each Fock basis. So, the longitudinal momentum fraction is $x = \frac{p^+}{P^+} = \frac{j}{K}$. N_{max} and K are basis truncation parameters. Large N_{max} and K mean high ultraviolet cutoff and low infrared cutoff.

3. Numerical results

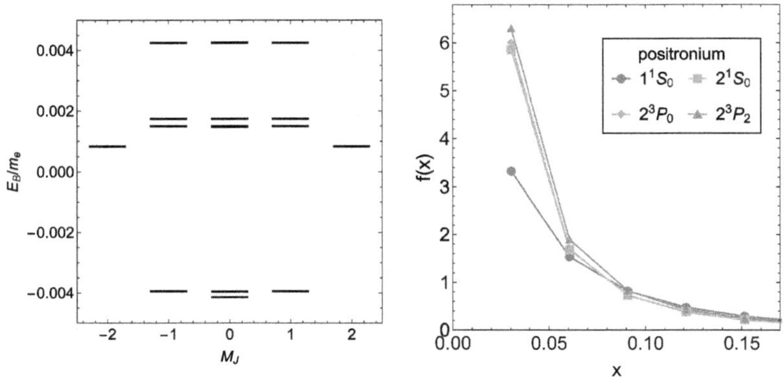

Fig. 1. Photon distribution function and interaction energy spectrum of positronium at $N_{max} = K - 1 = 32$, $\alpha = 0.3$. M_J is the total angular momentum projection of the system and m_e is the physical electron mass.

In Fig. 1, E_B is the interaction energy defined as the difference between the invariant mass of a bound state and the sum of masses of its component particles. The left panel in Fig. 1 is the energy spectrum with the identified lowest eight states. Using the approximate degeneracy and relations between parity and quantum numbers,[8] we can identify the lowest eight states. In the $M_J = 0$ column, the ordering of these states, from down to up, is 1^1S_0, 1^3S_1, 2^1S_0, 2^3S_1, 2^3P_0, 2^3P_1, 2^1P_1, 2^3P_2. Fig. 2 displays corresponding $|e^+e^-\rangle$ LFWFs of $M_J = 0$. The spectrum and LFWFs of $|e^+e^-\rangle$ reasonably agree with those from the effective one-photon-exchange calculation.[7] In addition, we obtain the LFWF's $|e^+e^-\gamma\rangle$ components which is not included in Ref. 7 and gives direct access to the photon content in positronium. $f(x)$ is the photon distribution function of positronium,

$$f(x) = \int |\psi(\beta, x)|^2 d\beta, \tag{2}$$

where x is the longitudinal momentum fraction of the photon and β represents the remaining variables in the LFWF of $|e^+e^-\gamma\rangle$. The right panel in Fig.1 shows that excited states have more photon content than the ground state in the small x region.

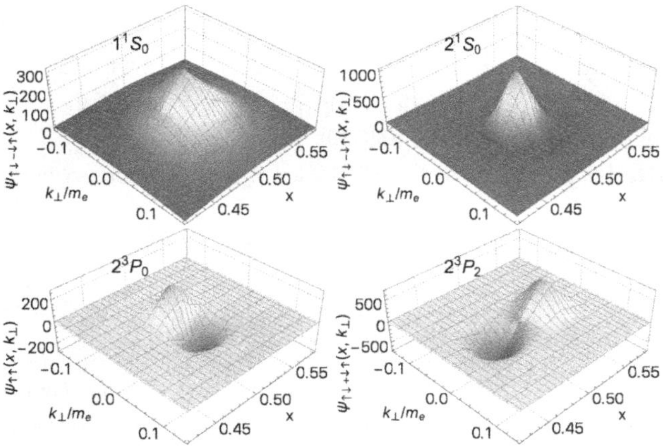

Fig. 2. LFWFs of $|e^+e^-\rangle$ at $N_{max} = K - 1 = 32$, $\alpha = 0.3$. Here x is the relative longitudinal momentum fraction and $k^\perp$ is the relative transverse momentum.

4. Conclusions

We apply BLFQ to the positronium system using the light-front QED Hamiltonian while retaining a photon as a dynamical degree of freedom. Sector-dependent renormalization appears to successful for canceling divergences. We obtain the reasonable spectrum and LFWFs along with the information of the photon content in positronium at strong coupling. By adding an effective confining potential between fermions, this positronium calculation can be straightforwardly extended to heavy quarkonium and light meson systems.

Acknowledgments

We thank J. Lan, S. Xu, Y. Li, and C. Mondal for many insightful discussions. This work of X. Z. is supported by new faculty startup funding by the Institute of Modern Physics, Chinese Academy of Sciences under the Grant No. Y632030YRC. J. P. V. is supported by the Department of Energy under Grants No. DE-FG02-87ER40371 and No. DE-SC0018223 (SciDAC4/NUCLEI). A portion of the computational resources were provided by the National Energy Research Scientific Computing Center (NERSC), which is supported by the Office of Science of the U.S. Department of Energy under Contract No. DE- AC02-05CH11231.

References

1. J. P. Vary, H. Honkanen, J. Li, P. Maris, S. J. Brodsky, A. Harindranath, G. F. de Teramond, P. Sternberg, E. G. Ng and C. Yang, *Phys. Rev. C* **81**, 035205 (Mar 2010).
2. S. J. Brodsky, H.-C. Pauli and S. S. Pinsky, *Physics Reports* **301**, 299 (1998).
3. P. Navrátil, J. P. Vary and B. R. Barrett, *Phys. Rev. Lett.* **84**, 5728 (Jun 2000).
4. P. Navrátil, J. P. Vary and B. R. Barrett, *Phys. Rev. C* **62**, 054311 (Oct 2000).
5. H. Honkanen, P. Maris, J. P. Vary and S. J. Brodsky, Phys. Rev. Lett. **106** (2011) 061603 doi:10.1103/PhysRevLett.106.061603 [arXiv:1008.0068 [hep-ph]].
6. X. Zhao, H. Honkanen, P. Maris, J. P. Vary and S. J. Brodsky, Phys. Lett. B **737**, 65 (2014) doi:10.1016/j.physletb.2014.08.020 [arXiv:1402.4195 [nucl-th]].
7. P. Wiecki, Y. Li, X. Zhao, P. Maris and J. P. Vary, *Phys. Rev. D* **91**, 105009 (May 2015).
8. Y. Li, P. Maris and J. P. Vary, *Phys. Rev. D* **96**, p. 016022 (Jul 2017).
9. J. Lan, C. Mondal, M. Li, Y. Li, S. Tang, X. Zhao and J. P. Vary, arXiv:1911.11676 [nucl-th].
10. S. Jia and J. P. Vary, *Phys. Rev. C* **99**, 035206 (Mar 2019).
11. J. Lan, C. Mondal, S. Jia, X. Zhao and J. P. Vary, arXiv:1907.01509 [nucl-th].
12. J. Lan, C. Mondal, S. Jia, X. Zhao and J. P. Vary, Phys. Rev. Lett. **122** (2019) no.17, 172001 doi:10.1103/PhysRevLett.122.172001 [arXiv:1901.11430 [nucl-th]].
13. C. Mondal, S. Xu, J. Lan, X. Zhao, Y. Li, D. Chakrabarti and J. P. Vary, arXiv:1911.10913 [hep-ph].
14. V. A. Karmanov, J.-F. Mathiot and A. V. Smirnov, *Phys. Rev. D* **77**, 085028 (Apr 2008).
15. V. A. Karmanov, J.-F. Mathiot and A. V. Smirnov, *Phys. Rev. D* **86**, 085006 (Oct 2012).
16. X. Zhao, H. Honkanen, P. Maris, J. P. Vary and S. J. Brodsky, *Physics Letters B* **737**, 65 (2014).

Configuration mixing of baryons in the large N_c limit of QCD

C. T. Willemyns
Service de Physique Nucléaire et Subnuclaire, Université de Mons,
Place du Parc 20, 7000 Mons, Belgium
E-mail: cintia.willemyns@umons.ac.be

N. N. Scoccola
Physics Department, Comisión Nacional de Energía Atómica,
Av. Libertador 8250, 1429 Buenos Aires, Argentina
CONICET,
Godoy Cruz 2290, 1425 Buenos Aires, Argentina
E-mail: scoccola@tandar.cnea.gov.ar

We study the configuration mixing effects on baryons in the large N_c limit. We present the results of a complete analysis of the spectrum of all the states in the $N = 2$ quark model band with $N_f = 3$.

Keywords: Baryon spectrum, $1/N_c$ expansion of QCD; configuration mixing; positive parity baryons; quark model.

1. Introduction

Baryons and their resonances can be analyzed as three quark systems. In the quark model (QM), these states belong to irreducible representations of $SU(6) \otimes O(3)$ where $SU(6)$ contains the spin group $SU(2)$ and the flavor group $SU(3)$ when $N_f = 3$. These QM symmetries are not QCD symmetries, therefore these states can mix in Nature.

Mixings between different representations of $SU(6) \otimes O(3)$, known as configuration mixings, can be studied in a large N_c QCD approach. In this approach, the $SU(3)$ gauge theory of QCD with three colors ($N_c = 3$) is generalized to an $SU(N_c)$ gauge theory.[1] It is then possible to build a $1/N_c$ expansion using effective quark operators to describe baryons.[2] It turns out that configuration mixing effects are not $1/N_c$ suppressed.[3–5]

Here we focus on mixings between states belonging to the band $N = 2$ of the QM as it is the first band to include more than one irreducible representation, it contains five multiplets: $[\mathbf{56}, 0^+]$, $[\mathbf{70}, 0^+]$, $[\mathbf{56}, 2^+]$, $[\mathbf{70}, 2^+]$, and $[\mathbf{20}, 1^+]$. Therefore, this band contains the lowest energy resonances with configuration mixing effects of order $\mathcal{O}(1)$.

The large N_c QCD theory does not have the QM symmetries, however in this limit a contracted spin-flavor symmetry $SU(2N_f)_c$ arises in the baryon sector. As a consequence of this symmetry baryons in the large N_c limit organize into towers labeled by the associated quantum number K.

2. Baryons in the large N_c limit

Only three spin-flavor representations arise when $N_c = 3$: completely symmetric (S), mixed symmetric (MS) and completely antisymmetric (A). Each symmetry corresponds to the $SU(6)$ multiplets **56**, **70** and **20** respectively.

The analysis of spin-flavor multiplets in large N_c QCD is a straightforward extension of methods used for $N_c = 3$. In this generalization of QCD with $N_c > 3$ the additional $N_c - 3$ quarks are considered in a completely symmetric spin-flavor configuration.

In the large N_c approach, multiplets have an infinite dimension, the physical baryons can be identified with states at the top of the flavor representations while the other states are spurious baryons that vanish when $N_c = 3$. Additionally, entire spin-flavor representations arise, that also decouple in the physical limit $N_c = 3$. Spurious states should be considered in the large N_c limit because states associated with physical baryons can result in a combination that contains spurious states.

All large N_c spin-flavor multiplets having states with the same quantum numbers as those associated with the physical states have to be considered, namely states with a (total) spin J, a hypercharge Y and an isospin number I that correspond to a baryon observed in Nature.

To do this analysis, we must decompose the $SU(6)$ spin-flavor representations into separate $SU(2)$ spin and $SU(3)$ flavor representations for N_c quark baryons. These $SU(6)$ decompositions can be found using a general method[6] and are explicitly shown in Eqs. (2) and (17) of Ref. 7. After studying the different multiplets of large N_c we find that for the band $N = 2$ we have 8 spin-flavor S multiplets, 32 spin-flavor MS multiplets and 10 spin-flavor A multiplets to consider.

Excited baryons with N_c quarks can be described as a symmetric or mixed-symmetric core of $N_c - 1$ quarks coupled to an orbitally excited quark.[7,8] The S and MS wave functions can be expressed as a linear combination of symmetric cores coupled to a quark:

$$|S, \mathbf{R}\rangle_{S,MS} = \sum_{\eta=\pm 1/2} c_{S,MS}(S, \mathbf{R}, \eta) \left|([S_c, \mathbf{R}_c]_S \times \ \mathbf{q})^{[S,\mathbf{R}]}\right\rangle, \qquad (1)$$

where $\mathbf{q}\equiv[1/2, \mathbf{3}]$. The notation adopted is $[S, \mathbf{R}]_{sym}$, with $\mathbf{R}$ being the flavor representation, S the spin and sym indicates the spin-flavor symmetry. Projections of S and $\mathbf{R}$ (S_z and Y, I, I_z respectively) also need to be taken into account and are omitted in expressions of the wave functions to lighten notation.

To build the antisymmetric wave functions, cores need to be mixed-symmetric. The A states can be expressed as:

$$|S, \mathbf{R}\rangle_A = \sum_i c_i \left|([S_{c_i}, \mathbf{R}_{c_i}]_{MS} \times \mathbf{q})^{[S,\mathbf{R}]}\right\rangle . \tag{2}$$

The mixed-symmetric cores, in turn, can be build as a combination of a symmetric state of $N_c - 2$ quarks coupled to a quark.

In both Eqs. (1) and (2) coefficients c_i have to be set to give the correct linear combinations to obtain the symmetries of each state.[7]

3. Operator expansion

The building blocks to find the expression of the operator expansion are the $SU(6)$ generators. These operators can be decomposed as one-body operators that act on the core $(S_c)_i$, $(T_c)_a$, $(G_c)_{ia}$ and operators denoted with lower case s_i, t_a, g_{ia} that act on the excited quark. These operators together with a generic spatial operator of rank k that we denote $\xi^{(k)}$ can be combined to obtain the operators of the expansion in $1/N_c$ to the lowest order.[8,9] The $\xi^{(k)}$ operator acts only on the excited quark as the core does not carry orbital angular momentum.

The Hamiltonian can be written as a linear combination of effective operators up to order $O(1)$ as:

$$H = \sum_{i=1}^{5} c_i^{\mathbf{T}_L, \mathbf{T}'_{L'}} O_i + \mathcal{O}(1/N_c), \tag{3}$$

where $\mathbf{T}_L$ stands for the $SU(6)$ irreducible representations with angular momentum L. The operators of this expansion are given by:

$$\begin{aligned} O_1 &= (N_c 1)^{[0,\mathbf{1}]} , \\ O_2 &= \left(\xi^{(1)} s\right)^{[0,\mathbf{1}]} , \\ O_3 &= \left(\xi^{(2)} \left(g G_c\right)^{[2,\mathbf{1}]}\right)^{[0,\mathbf{1}]} , \\ O_4 &= \tfrac{1}{N_c} \left(\xi^{(1)} \left(t G_c\right)^{[1,\mathbf{1}]}\right)^{[0,\mathbf{1}]} , \\ O_5 &= \tfrac{1}{N_c} \left(t T_c\right)^{[0,\mathbf{1}]} . \end{aligned} \tag{4}$$

4. Configuration mixing effects in the band $N = 2$ baryons

Using the expansion with the operators given in Eq. (4) states in the band $N = 2$ with same spin and isospin can mix. We obtained the mass matrices that we diagonalized. We find that all the states belonging to the S and MS multiplets of this band fall into one of the following nine towers:

$$\begin{aligned} m_0 &= \bar{c}_1^{S_0} N_c, \\ m_{1\pm} &= \bar{m}_1 \pm \delta_1, \\ m_{2\pm} &= \bar{m}_2 \pm \delta_2, \\ m_3 &= \bar{c}_1^{MS_2} N_c + c_2^{MS_2} - \tfrac{2}{7} c_3^{MS_2}, \\ m_{\frac{1}{2}} &= \bar{c}_1^{MS_0} N_c - 3 c_5^{MS_0}, \\ m_{\frac{3}{2}} &= \bar{c}_1^{MS_2} N_c - \tfrac{3}{2} \bar{c}_2^{MS_2} + 3 c_4^{MS_2} - 3 c_5^{MS_2}, \\ m_{\frac{5}{2}} &= \bar{c}_1^{MS_2} N_c + \bar{c}_2^{MS_2} - 2 c_4^{MS_2} - 3 c_5^{MS_2}, \end{aligned} \tag{5}$$

where

$$\begin{aligned} \delta_1 &= \sqrt{\left(\tfrac{1}{2}\left(\bar{c}_1^{MS_0} - \bar{c}_1^{MS_2}\right) N_c + \tfrac{3}{4} \bar{c}_2^{MS_2} + \tfrac{1}{2} c_3^{MS_2}\right)^2 + 2\left(c_3^{MS_0, MS_2}\right)^2}, \\ \delta_2 &= \sqrt{\left(\tfrac{1}{2}\left(\bar{c}_1^{S_2} - \bar{c}_1^{MS_2}\right) N_c + \tfrac{1}{4} \bar{c}_2^{MS_2} - \tfrac{1}{2} c_3^{MS_2}\right)^2 + 2\left(\bar{c}_2^{S_2, MS_2}\right)^2}, \end{aligned} \tag{6}$$

and where the coefficients $\bar{c}$ are redefinitions given by

$$\begin{aligned} \bar{c}_1^{\mathbf{T},\mathbf{T}'} &= c_1^{\mathbf{T},\mathbf{T}'} + \tfrac{1}{N_c} c_5^{\mathbf{T},\mathbf{T}'}, \\ \bar{c}_2^{S_2, MS_2} &= c_2^{S_2, MS_2} - c_4^{S_2, MS_2}, \\ \bar{c}_2^{MS_2} &= c_2^{MS_2} + c_4^{MS_2}. \end{aligned} \tag{7}$$

The $\delta_{1,2}$ in Eq. (5) parametrize the configuration mixing effects in the band $N = 2$. Energy differences between baryons in the same band are at most of order $\mathcal{O}(1)$, which means that $\left(c_1^{\mathbf{T}} - c_1^{\mathbf{T}'}\right) N_c \sim \mathcal{O}(1)$. Then, from Eq. (6) we conclude that configuration mixing effects are of order $\delta_{1,2} \sim \mathcal{O}(1)$.

We find that the operators $\mathcal{O}_2$ and $\mathcal{O}_3$ mix the $[\mathbf{56}, 2^+]$ and $[\mathbf{70}, 2^+]$ and the $[\mathbf{70}, 0^+]$ and $[\mathbf{70}, 2^+]$ multiplets respectively.

No configuration mixing is observed within the present scheme between the A multiplet and the S or MS multiplets. The operators in Eq. (4) cannot mix states with cores that have a different spin-flavor symmetry because these operators are built with generators of $SU(6)$ acting on the excited quark and on the core.

5. Conclusions

The large N_c method allows a study of configuration mixing effects by building excited baryons as cores of $N_c - 1$ quarks coupled to an excited quark.

With this approach we showed explicitly that the configuration mixing between the $[\mathbf{56}, 2^+]$ and $[\mathbf{70}, 2^+]$ and the $[\mathbf{70}, 0^+]$ and $[\mathbf{70}, 2^+]$ multiplets are of order $\mathcal{O}(1)$. Mixing between the A multiplet and the S and MS are subleading in $1/N_c$.

Acknowledgments

CTW thanks the F.N.R.S. for financial support to attend the conference. NNS acknowledges support by CONICET and ANPCyT (Argentina), under grants PIP17-700 and PICT17- 03-0571.

References

1. G. 't Hooft, A Planar Diagram Theory for Strong Interactions, *Nucl. Phys. B* **72**, 461 (1974).
2. E. Witten, Baryons in the 1/N expansion, *Nucl. Phys. B* **160**, 57 (1979).
3. J. L. Goity, $1/N_c$ countings in baryons, *Phys. Atom. Nucl.* **68**, 624 (2005).
4. J. L. Goity, *Large N(c) QCD 2004. Proceedings, International Workshop* (Trento, 2005).
5. T. D. Cohen, D. C. Dakin, A. Nellore and R. F. Lebed, Excited baryon decay widths in large N_c QCD, *Phys. Rev. D* **69**, 056001 (2004).
6. C. R. Hagen and A. J. Macfarlane, Reduction of Representations of SU_{mn} with respect to the Subgroup $SU_m \otimes SU_n$, *J. Math. Phys.* **6**, 1355 (1965).
7. C. T. Willemyns and N. N. Scoccola, Towers of baryons of the $\mathcal{N} = 2$ quark model band in the large N_c limit, *Phys. Rev. D* **98**, 034019 (2018)
8. C. E. Carlson, C. D. Carone, J. L. Goity and R. F. Lebed, Operator analysis of $l = 1$ baryon masses in large N(c) QCD, *Phys. Rev. D* **59**, 114008 (1999).
9. C. Willemyns and C. Schat, Towers of positive parity excited baryons and their mixing in the large N_c limit, *Phys. Rev. D* **95**, 094007 (2017).

Minkowski-space solutions of the Schwinger-Dyson equation for the fermion propagator with the rainbow-ladder truncation

Shaoyang Jia* and Pieter Maris
Department of Physics and Astronomy, Iowa State University, Ames, Iowa 50011, USA
**E-mail: sjia@iastate.edu*

Dyana C. Duarte, Tobias Frederico, Wayne de Paula
Instituto Tecnológico da Aeronáutica, DCTA, 12.228-900 São José dos Campos, Brazil

Emanuel Ydrefors
Institut de Physique Nucléaire, Université Paris-Sud, IN2P3-CNRS, 91406 Orsay Cedex, France
Instituto Tecnológico da Aeronáutica, DCTA, 12.228-900 São José dos Campos, Brazil

We solve the Minkowski-space Schwinger-Dyson equation (SDE) for the fermion propagator in quantum electrodynamics (QED) with massive photons. Specifically, we work in the quenched approximation within the rainbow-ladder truncation. Loop-divergences are regularized by the Pauli-Villars regularization. With moderately strong fermion-photon coupling, we find that the analytic structure of the fermion propagator consists of an on-shell pole and branch-cuts located in the timelike region. Such structures are consistent with the direct solution of the fermion propagator as functions of the complex momentum. Our method paves the way towards the calculation of the Minkowski-space Bethe-Salpeter amplitude using dressed fermion propagator.

Keywords: Fermion propagator; Schwinger-Dyson equation; Minkowski space.

1. Introduction to the Schwinger-Dyson equation for the fermion propagator in QED with massive photons

The Schwinger-Dyson equations (SDEs) are coupled integral equations of the Green's functions in a quantum field theory. They serve as recurrences relations of the n-point expansions of the generating functional, therefore are inherently nonperturbative. Albeit traditionally solved in the Euclidean space, we expect these integral equations to be well-defined for the Green's functions in the Minkowski momentum space[1,2] extending to the whole complex momentum plane.[3] In this proceeding, we solve the fermion propagator directly in the Minkowski space from its SDE.

Let us start with the Lagrangian of quantum electrodynamics (QED) with massive photons in the R_ξ gauge[4]

$$\mathcal{L} = \overline{\psi}(i\not{D} - m_{\rm B})\psi + \frac{1}{2}A_\mu \left[g^{\mu\nu}\left(\partial^2 + m_{\rm A}^2\right) - \left(1 - \frac{1}{\xi}\right)\partial^\mu\partial^\nu\right]A_\nu. \quad (1)$$

Here ψ is the fermion field with bare mass $m_{\rm B}$. A^μ is the massive gauge boson field with bare mass $m_{\rm A}$. The covariant derivative is defined as $D^\mu = \partial^\mu - ieA^\mu$ with e being the elementary charge. From Eq. (1), the bare propagator of the gauge boson is given by

$$D^0_{\mu\nu}(q) = \left[g_{\mu\nu} + (\xi - 1)q_\mu q_\nu/(q^2 - \xi m_{\rm A}^2)\right]/(q^2 - m_{\rm A}^2 + i\varepsilon). \quad (2)$$

In the quenched approximation, the gauge-boson propagator remains bare. The metric tensor is defined as $g^{\mu\nu} = \mathrm{diag}\{1, -1, -1, -1\}$ such that $q^2 > 0$ corresponds to q^μ being timelike.

The fermion propagator $S_{\rm F}(p)$ is to be solved from its SDE given by $S_{\rm F}^{-1}(p) = \not{p} - m_{\rm B} - \Sigma(p)$, with the fermion self-energy $\Sigma(p)$ defined as

$$\Sigma(p) \equiv \not{p}\Sigma_{\rm v}(p^2) + \Sigma_{\rm s}(p^2) = -ie^2 \int d\underline{k}\, \gamma^\nu\, S_{\rm F}(k)\Gamma^\mu(k,p)D_{\mu\nu}(q). \quad (3)$$

Here we define the integral measure as $\int d\underline{k} = \int d^4k/(2\pi)^4$. In Eq. (3), we have decomposed $\Sigma(p)$ into its Dirac vector part $\Sigma_{\rm v}(p^2)$ and Dirac scalar part $\Sigma_{\rm s}(p^2)$. Meanwhile, $\Gamma^\mu(k,p)$ is the one-particle-irreducible fermion-photon vertex, which in the rainbow-ladder truncation becomes γ^μ.

2. The spectral representation of the fermion propagator

Following the procedures in Ref. 5, the Källén-Lehmann spectral representation of the scalar propagator is generalized to the fermion propagator:

$$S_{\rm F}(p) \equiv S_{\rm v}(p^2) + S_{\rm s}(p^2) = \int_{m^2}^{+\infty} ds \frac{\rho_{\rm v}(s)}{p^2 - s + i\varepsilon} + \int_{m^2}^{+\infty} ds \frac{\rho_{\rm s}(s)}{p^2 - s + i\varepsilon}. \quad (4)$$

The parameter m is the fermion on-shell mass, which can be solved from $m\left[1 - \Sigma_{\rm v}(m^2)\right] = m_{\rm B} + \Sigma_{\rm s}(m^2)$. Equation (4) is equivalent to $S_{\rm F}(p) = \int dW\, \rho(W)/[\not{p} - W + i\,\mathrm{sign}(W)\,\varepsilon]$, with the support of the integral being $(-\infty, -m] \cup [m, +\infty)$ and $\rho(W)$ defined by $\rho(W) = \mathrm{sign}(W)[W\,\rho_{\rm v}(W^2) + \rho_{\rm s}(W^2)]$.

Equation (4) indicates that both $S_{\rm v}(p^2)$ and $S_{\rm s}(p^2)$ are holomorphic functions of the complex momentum except for $p^2 \geq m^2$. We further assume that singularities of $S_{\rm F}(p)$ specified by $\rho(W)$ are given by a mass-shell pole and branch-cuts along the positive-real axis. Such an assumption allows us to write $\rho_j(s) = Z\,\delta(s-m^2) + r_j(s)$ for $j \in \{\mathrm{v},\mathrm{s}\}$. The first term contributes to a simple pole structure of the propagator with residue $Z = \left[1-\Sigma_{\mathrm{v}}(m^2)\right]^{-1}$. After defining $A(p^2) = 1 - \Sigma\mathrm{v}(p^2)$ and $B(p^2) = -m_{\mathrm{B}} - \Sigma_{\mathrm{s}}(p^2)$, the second term $r_j(s)$ as the branch-cut part of the spectral functions can be calculated from

$$\begin{cases} r_{\mathrm{v}}(s) = -\dfrac{1}{\pi}\mathrm{Im}\left\{\dfrac{A(s)}{A^2(s)p^2 - B^2(s)}\right\}\Bigg|_{s\geq p_{\mathrm{th}}^2} \\ r_{\mathrm{s}}(s) = -\dfrac{1}{\pi}\mathrm{Im}\left\{\dfrac{-B(s)}{A^2(s)p^2 - B^2(s)}\right\}\Bigg|_{s\geq p_{\mathrm{th}}^2} \end{cases}. \tag{5}$$

Here p_{th}^2 is the threshold above which the imaginary parts of the fermion self-energy become nonzero. Specifically, we work with massive photons in the Feynman gauge such that there is a finite distance between the pole and the branch points.

3. Solution of the SDE in the quenched approximation

Within the quenched approximation and the Feynman gauge ($\xi = 1$), we calculate the fermion self-energy from its definition in Eq. (3). The momentum dependence of $\Sigma(p)$ is given by

$$\frac{\delta\,\Sigma(p)}{\delta\,\rho(W)} = -ie^2\int d\underline{l}\int dF^2\,\frac{-2y\not{p} + 4W}{[l^2 - (xW^2 + ym_{\mathrm{A}}^2 - xyp^2) + i\varepsilon]^2}, \tag{6}$$

with $\int dF^2 = \int_0^1 dx \int_0^1 dy\,\delta(1-x-y)$. Based on Eq. (6), the real part of the fermion self-energy is logarithmic divergent, which we remove using the Pauli-Villars regularization. Specifically, the regularized fermion self-energy $\Sigma_{\mathrm{PV}}(p)$ is defined as

$$\Sigma_{\mathrm{PV}}(p) = \Sigma(p) - (m_{\mathrm{A}} \to \Lambda), \tag{7}$$

with $\Lambda > m_{\mathrm{A}}$ being the mass of the regulator.

Because of Eq. (7), $\Sigma_{\mathrm{PV}}(p)$ vanishes asymptotically. This results in an additional benefit of the Pauli-Villars regularization that the direct implementation of the spectral representation for the self-energy is allowed. Consequently, the imaginary parts of the fermion self-energy uniquely determine $\Sigma_{\mathrm{PV}}(p)$ in the complex momentum plane. Furthermore, we can calculate the imaginary parts of the fermion self-energy in terms of the spectral functions of the fermion propagator using

$$-\frac{1}{\pi}\mathrm{Im}\{\Sigma_{\mathrm{PV}j}(p^2)\} = \frac{e^2}{(4\pi)^2}\left\{\int_{m^2}^{(\sqrt{p^2}-\sqrt{m_{\mathrm{A}}^2})^2} ds\,\mathcal{K}_j(p^2,s,m_{\mathrm{A}}) \right.$$
$$\left. \times\sqrt{(p^2-m_{\mathrm{A}}^2+s)^2-4p^2s} - (m_{\mathrm{A}}\to\Lambda)\right\}\rho_j(s), \qquad (8)$$

with $\mathcal{K}_{\mathrm{v}}(p^2,s,m_{\mathrm{A}}) = (p^2-m_{\mathrm{A}}^2+s)/p^4$ and $\mathcal{K}_{\mathrm{s}}(p^2,s,m_{\mathrm{A}}) = -4/p^2$.

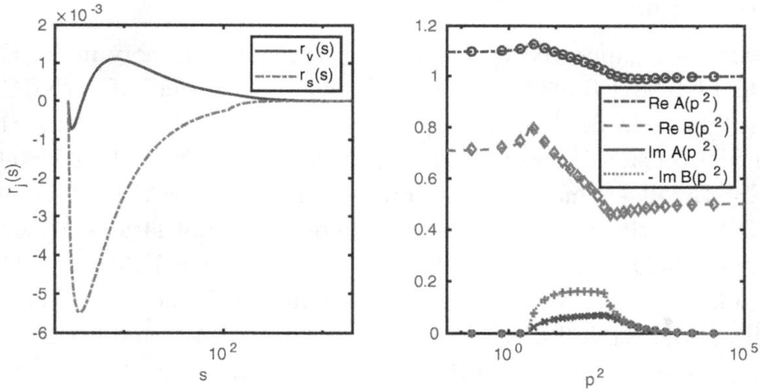

Fig. 1. Solutions of the fermion propagator SDE in massive QED. On the left panel, the blue solid line and the red dash-dot line are the Dirac-vector and the Dirac-scalar parts of the spectral function respectively. On the right panel, lines specified by the legends are the real and the imaginary parts of the inverse propagator. The discretized points correspond to solutions of the same SDE using the technique of un-Wick rotation.[3]

After regularization, the SDE for the fermion propagator becomes a numerically well-defined non-linear integral equation of $\rho_j(s)$. We obtain the numerical solution of the fermion spectral functions through iteration. As an example, we choose the bare mass $m_{\mathrm{B}} = 0.5$, the coupling constant $e^2/(4\pi) = 0.3$, the vector boson mass $m_{\mathrm{A}} = 1$, and the Pauli-Villars mass $\Lambda = 10$ to obtain the fermion on-shell mass of $m = 0.6502$ with residue $Z = 0.9096$. The corresponding branch-cut parts of the spectral functions are illustrated on the left panel of Fig. 1. With the same input parameters, we also compare the fermion self-energy in the timelike region with results obtained using the technique of un-Wick rotation,[3] as illustrated on the right panel of Fig. 1. Our results are in agreement within numerical uncertainties in the Euclidean space.

4. Conclusion

We have solved the SDE for the fermion propagator in the massive QED in the Landau gauge within the quenched approximation using the Pauli-Villars regularization. With the rainbow-ladder truncation, we find that the structure of the fermion propagator is consistent with a simple pole and bunch-cuts in the timelike region. Our solutions could serve as input conditions for the Minkowski-space Bethe-Salpeter equation of two-fermion bound states.

Acknowledgments

This work was supported by the US Department of Energy under Grants No. DE-FG02-87ER40371 and No. DE-SC0018223 (SciDAC-4/NUCLEI), and by Fundação de Amparo à Pesquisa do Estado de São Paulo, Brazil (FAPESP) Thematic grants No. 13/26258-4 and No. 17/05660-0, by CAPES, Brazil - Finance Code 001. DCD thanks FAPESP grant No. 2017/26111-4, the Simons Foundation under the Multifarious Minds Program grant 557037 and U.S. DOE under Grant No. DE-FG02- 00ER41132. TF thanks Conselho Nacional de Desenvolvimento Científico e Tecnológico (Brazil), Project INCT-FNA Proc. No. 464898/2014-5, and the Fullbright Visiting Professor Award. WP thanks CNPq under grants 438562/2018-6 and 313236/2018-6 and CAPES under the grant 88881.309870/2018-01. EY thanks FAPESP grants No. 2016/25143-7 and No. 2018/21758-2.

References

1. V. Sauli, Implications of analyticity to solution of Schwinger-Dyson equations in Minkowski space, *Few Body Syst.* **39**, 45 (2006).
2. S. Jia and M. R. Pennington, Exact Solutions to the Fermion Propagator Schwinger-Dyson Equation in Minkowski space with on-shell Renormalization for Quenched QED, *Phys. Rev.* **D96**, 036021 (2017).
3. T. Frederico, D. C. Duarte, W. de Paula, E. Ydrefors, S. Jia and P. Maris, Towards Minkowski space solutions of Dyson-Schwinger Equations through un-Wick rotation, in *Proc. Nuclear Theory in the Supercomputing Era - 2018 (NTSE-2018)*, (Daejeon, Korea, 2018). arXiv:1905.00703.
4. M. Peskin, *An Introduction To Quantum Field Theory* (CRC Press, Boca Raton, Florida, 2018).
5. R. Delbourgo and P. C. West, A Gauge Covariant Approximation to Quantum Electrodynamics, *J. Phys.* **A10**, 1049 (1977).

Unveiling the structure of pseudoscalar mesons

Khépani Raya* and Lei Chang
School of Physics, Nankai University, Tianjin 300071, China
**E-mail: khepani@nankai.edu.cn*

Minghui Ding and Daniele Binosi
European Centre for Theoretical Studies in Nuclear Physics and Related Areas (ECT) and Fondazione Bruno Kessler*
Villa Tambosi, Strada delle Tabarelle 286, I-38123 Villazzano (TN) Italy

Craig D. Roberts
School of Physics, Nanjing University, Nanjing, Jiangsu 210093, China
Institute for Nonperturbative Physics, Nanjing University, Nanjing, Jiangsu 210093, China

A valuable approach to the analysis of hadron physics observables is provided by QCD's equations-of-motion; namely, the Dyson-Schwinger equations. Drawing from a diverse collection of predictions, we revisit: $\gamma\gamma^* \to$ neutral pseudoscalar transition form factors, their corresponding valence-quark distribution amplitudes and a recent result on the pion distribution functions.

Keywords: Dyson-Schwinger equations; distribution amplitudes; distribution functions; transition form factors.

1. Introduction

In an exciting era of ongoing and forthcoming experiments, the Dyson-Schwinger equation (DSE) approach to QCD, which describes hadrons in terms of quarks and gluons, is placed as a promising tool for hadron physics.[1–3] The present manuscript sums up a collection of DSE results on the pseudoscalar mesons: valence-quark distribution amplitudes (PDAs),[4–6] $\gamma^*\gamma^* \to M$ transition form factors (TFFs)[6–9] and the pion distribution functions (PDF).[10]

2. Valence-quark distribution amplitudes

Given a pseudoscalar meson M with mass m_M, the PDA, $\phi_M(x;\zeta)$, is defined as the projection onto the light-front of its Bethe-Salpeter wave-

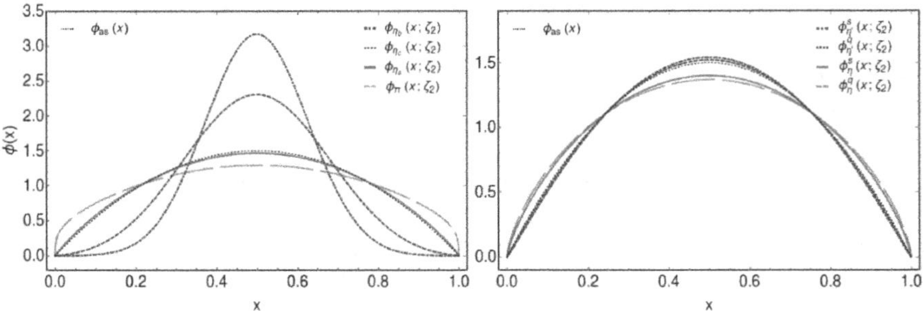

Fig. 1. PDAs at ζ_2 and the asymptotic profile,[11] $\phi_{\rm as}(x) := 6x(1-x)$. The η_s corresponds to a pseudoscalar with s-massive valence-quarks. Due to the dominance of dynamical chiral symmetry breaking (DCSB), $\phi_\pi(x;\zeta_2)$ is broader than $\phi_{\rm as}(x)$, while those corresponding to $\eta_{c,b}$, more influenced by the Higgs mechanism, are narrower. Lying close to $\phi_{\rm as}(x)$, we find the η_s, η and η' PDAs. There is an interplay between strong and weak mass generation being dominant and the s-quark is the boundary.[5,6]

function, $\chi_M(q;P) = S(q^+)\Gamma_M(q;P)S(q^-)$, such that:

$$\phi_M(x;\zeta) = \frac{Z_2}{f_M}\mathrm{tr}_{\rm CD}\int_q \delta_n^x(q^+)\gamma_5\gamma\cdot n\chi_M(q;P)\,, \quad \int_0^1 dx\,\phi_M(x;\zeta) = 1, \quad (1)$$
$$\delta_n^x(q^+) = \delta(n\cdot q^+ - x\, n\cdot P)\,,\; q^+ = q+\eta P\,,\; q^- = q-(1-\eta P)\,,$$

where $\int_q := \int \frac{d^4q}{(2\pi)^4}$, $\eta \in [0,1]$ defines the relative momentum between the quark/antiquark, f_M is the corresponding decay constant, Z_2 is the quark wavefunction renormalization constant and n is a light-like four vector ($n^2=0$, $n\cdot P = -m_M$). Computation of several Mellin moments allow to reconstruct the PDA.[4–6] A selection of PDAs is shown in Fig. 1.

3. $\gamma^*\gamma^* \to M$ transition form factors

In the impulse approximation (IA), the transition $\gamma^*\gamma^* \to M$ reads as[9]:

$$T_{\mu\nu}(Q_1,Q_2) = \frac{e^2}{4\pi^2}\epsilon_{\mu\nu\alpha\beta}Q_{1\alpha}Q_{2\beta}G_M(Q_1^2, Q_1\cdot Q_2, Q_2^2)$$
$$= \mathbf{e}_f^2\;\mathrm{tr}_{\rm CD}\int_q i\chi_\mu^f(q,q_1)\Gamma_M(q_1,q_2)S_f(q_2)i\Gamma_\nu^f(q_2,q)\,, \quad (2)$$

where Q_1, Q_2 are the momenta of the two photons. The kinematic arrangement is $q_1 = q+Q_1$, $q_2 = q-Q_2$; $\mathbf{e}_f^2$ is a charge factor associated with the meson's valence quarks. Γ_M is the pseudoscalar Bethe-Salpeter amplitude (BSA). The amputated (Γ_ν) and non amputated (χ_μ) quark-photon vertices can be properly expressed via the quark propagator dressing functions.[7] BSAs and quark propagators are obtained (at $\zeta=\zeta_2$) in the

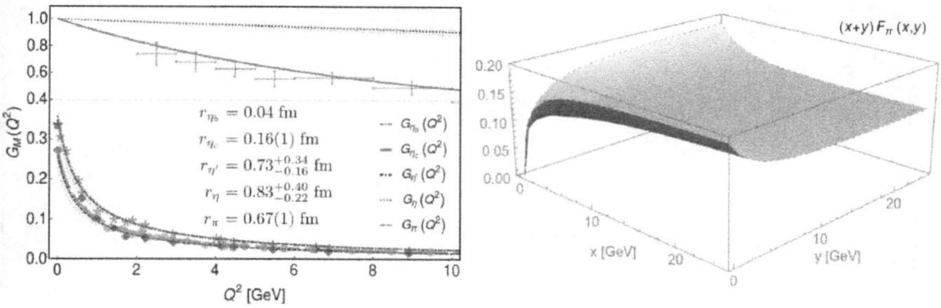

Fig. 2. **Left panel**. DSE predictions of $\gamma\gamma^* \to M$ TFFs and a variety of experimental data (CELLO, CLEO, Babar, Belle, L3 and PDG) for η_c (error bars), η' (asterisks), η (diamons) and π^0 (circles). See Ref. 9 for a complete list of data sources. The $\eta_{c,b}$ results are normalized to $G_M(0) = 1$ and the band surrounding the η_b corresponds to the non-relativistic QCD result of Ref. 13. Those form factors can be found, on a larger domain, in Refs. 6–8. **Right panel**. $\gamma^*\gamma^* \to \pi^0$ TFF.[9] The ultraviolet limit constraints[11] are preserved while simultaneously reproducing the empirical value at $Q^2 \to 0$.

rainbow-ladder approximation, self-consistent with the IA.[a] Via perturbation theory integral representations of those objects, the complete results basically require a series of perturbation-theory-like integrals. A similar approach was followed for the pion electromagnetic form factor.[12] The computed TFFs are displayed in Fig. 2.

4. Pion distribution functions

Detailed considerations have yielded the following symmetry-preserving expression[10,14] for the pion valence-quark PDF, $q_V(x;\zeta)$:

$$q_V(x;\zeta) = \mathrm{tr}_{\mathrm{CD}} \int_q \delta_n^x(q^+) n \cdot \partial_{q^+} [\Gamma_\pi(q^+; -P) S(q^+)] \Gamma_\pi(q^-; P) S(q^-) \,. \quad (3)$$

The renormalization scale ($\zeta = \zeta_H \sim 0.3$ GeV) is defined such that the fully dressed (valence-quarks) quasiparticles are the correct degrees of freedom,[15] $\int_0^1 dx\; q_V(x;\zeta_H) = 1$. We reconstruct $q_V(x;\zeta_H)$ from its Mellin moments and evolve through DGLAP equations to incorporate gluon and sea content in the manner that QCD prescribes.[10,15] This process is detailed in Ref. 10 and the obtained PDFs are shown in Fig. 3.

[a]The non-Abelian anomaly is introduced in the Bethe-Salpeter kernel of the $\eta-\eta'$ BSE[6] and a mild amendment of Eq. (2) is required.

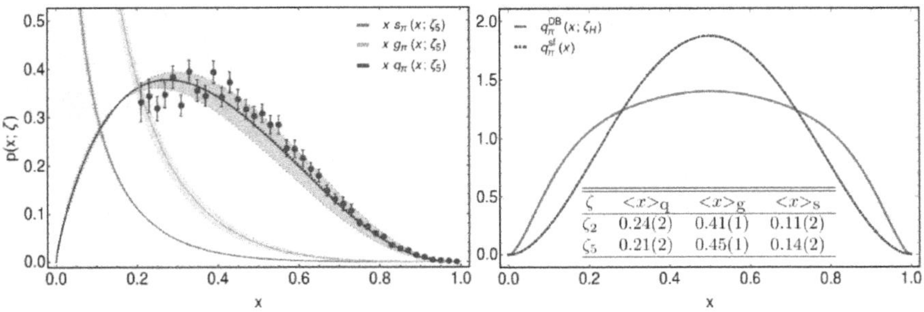

Fig. 3. **Left panel:** valence quark (q_π), gluon (g_π) and sea (s_π) pion PDFs at the experimental scale,[16,17] $\zeta = 5.2$ GeV $:= \zeta_5$. **Right panel:** valence-quark PDF at ζ_H. Here, $q_\pi^{\rm DB}(x;\zeta_2)$ is the DSE prediction from Ref. 10 and $q_\pi^{\rm sf}(x) = 30x^2(1-x)^2$ is a scale-free parton-like model,[14] the marked broadening of $q_\pi^{\rm DB}(x;\zeta_2)$ relative to $q_\pi^{\rm sf}(x)$ is a consequence of DCSB.

5. Conclusions and scope

We have recapitulated predictions for an array of pseudoscalar mesons properties obtained within a single unifying DSE framework. The examples chosen are illustrative, not exhaustive. Analyses of kaon PDFs and meson GPDs and TMDs are currently underway.

Acknowledgements

KR wants to acknowledge the local organizers for their hospitality. Work supported by: Jiangsu Province *Hundred Talents Plan for Professionals.*

References

1. Tanja Horn and Craig D. Roberts. The pion: an enigma within the Standard Model. *J. Phys.*, G43(7):073001, 2016.
2. Gernot Eichmann, Helios Sanchis-Alepuz, Richard Williams, Reinhard Alkofer, and Christian S. Fischer. Baryons as relativistic three-quark bound states. *Prog. Part. Nucl. Phys.*, 91:1–100, 2016.
3. Volker D. Burkert and Craig D. Roberts. Colloquium : Roper resonance: Toward a solution to the fifty year puzzle. *Rev. Mod. Phys.*, 91(1):011003, 2019.
4. Lei Chang, I. C. Cloet, J. J. Cobos-Martinez, C. D. Roberts, S. M. Schmidt, and P. C. Tandy. Imaging dynamical chiral symmetry breaking: pion wave function on the light front. *Phys. Rev. Lett.*, 110(13):132001, 2013.

5. Minghui Ding, Fei Gao, Lei Chang, Yu-Xin Liu, and Craig D. Roberts. Leading-twist parton distribution amplitudes of S-wave heavy-quarkonia. *Phys. Lett.*, B753:330–335, 2016.
6. Minghui Ding, Khépani Raya, Adnan Bashir, Daniele Binosi, Lei Chang, Muyang Chen, and Craig D. Roberts. $\gamma^*\gamma \to \eta, \eta'$ transition form factors. *Phys. Rev.*, D99(1):014014, 2019.
7. K. Raya, L. Chang, A. Bashir, J. J. Cobos-Martinez, L. X. Gutiérrez-Guerrero, C. D. Roberts, and P. C. Tandy. Structure of the neutral pion and its electromagnetic transition form factor. *Phys. Rev.*, D93(7):074017, 2016.
8. Khépani Raya, Minghui Ding, Adnan Bashir, Lei Chang, and Craig D. Roberts. Partonic structure of neutral pseudoscalars via two photon transition form factors. *Phys. Rev.*, D95(7):074014, 2017.
9. Khépani Raya, Adnan Bashir, and Pablo Roig. Contribution of neutral pseudoscalar mesons to a_μ^{HLbL} within a Schwinger-Dyson equations approach to QCD, arXiv:1910.05960 [hep-ph].
10. Minghui Ding, Khépani Raya, Daniele Binosi, Lei Chang, Craig D Roberts, and Sebastian M. Schmidt. Symmetry, symmetry breaking, and pion parton distributions, arXiv:1905.05208 [nucl-th]. 2019.
11. G. Peter Lepage and Stanley J. Brodsky. Exclusive Processes in Perturbative Quantum Chromodynamics. *Phys. Rev.*, D22:2157, 1980.
12. L. Chang, I. C. Cloet, C. D. Roberts, S. M. Schmidt, and P. C. Tandy. Pion electromagnetic form factor at spacelike momenta. *Phys. Rev. Lett.*, 111(14):141802, 2013.
13. Feng Feng, Yu Jia, and Wen-Long Sang. Can Nonrelativistic QCD Explain the $\gamma\gamma^* \to \eta_c$ Transition Form Factor Data? *Phys. Rev. Lett.*, 115(22):222001, 2015.
14. Lei Chang, Cédric Mezrag, Hervé Moutarde, Craig D. Roberts, Jose Rodríguez-Quintero, and Peter C. Tandy. Basic features of the pion valence-quark distribution function. *Phys. Lett.*, B737:23–29, 2014.
15. José Rodríguez-Quintero, Lei Chang, Khépani Raya, and Craig D. Roberts. Process-independent effective coupling and the pion structure function, arXiv:1909.13802 [hep-ph].
16. J. S. Conway et al. Experimental Study of Muon Pairs Produced by 252-GeV Pions on Tungsten. *Phys. Rev.*, D39:92–122, 1989.
17. Matthias Aicher, Andreas Schafer, and Werner Vogelsang. Soft-gluon resummation and the valence parton distribution function of the pion. *Phys. Rev. Lett.*, 105:252003, 2010.

Measurement of longitudinal spin asymmetries for weak boson production at STAR

Jinlong Zhang, for the STAR Collaboration

Department of Physics and Astronomy, Stonry Brook University, Stony Brook, New York 11794, USA
E-mail: jinlong.zhang@stonybrook.edu

The production of $W^{\pm}$ bosons in longitudinally polarized proton-proton collisions at RHIC provides a direct probe for the spin-flavor structure of the proton through the parity-violating single-spin asymmetry, A_L. At STAR, the leptonic decay channel $W \to e\nu$ can be effectively measured with the electromagnetic calorimeters and time projection chamber. STAR has measured the $A_L(W)$ as a function of the decay-electron's pseudorapidity from datasets taken in 2011 and 2012, which has provided significant constraints on the helicity-dependent PDFs of $\bar{u}$ and $\bar{d}$ quarks. In 2013 the STAR experiment collected an integrated luminosity of $\sim$250 pb^{-1} at $\sqrt{s} = 510$ GeV with an average beam polarization of $\sim$56%, which is more than three times larger than the total integrated luminosity of previous years. The final results from 2013 dataset for W-boson A_L will be reported. Also the impacts of STAR data on our knowledge of the sea-quark spin-flavor structure of the proton will be discussed.

Keywords: Proton spin structure; sea quarks; weak boson.

1. Introduction

The proton's spin structure has attracted both theoretical and experimental interest in the past few decades. Polarized inclusive deep-inelastic scattering (DIS) experiments have provided data showing that the quark and antiquark spins only contribute $\sim$30% of the proton spin.[1] In semi-inclusive DIS measurements, the flavor decomposition of quark spin contribution to proton spin can be accessed by identifying one or more hadrons in the final state. Fragmentation functions are required to relate the final state hadrons to the scattered quarks and antiquarks. Uncertainties of the flavor separated quark and antiquark spin contributions are still relatively large.[2,3]

In unpolarized Drell-Yan experiments, a flavor asymmetry between $\bar{u}$ and $\bar{d}$ has been observed.[4,5] It is natural to ask if such a flavor asymmetry also exists in the polarized sea. Different models developed to explain the unpolarized flavor asymmetry, however, gave very different predictions.[6] Pioneering measurements were made by COMPASS[7] but limited by precision. So, experimental input from the RHIC spin program becomes critical.

As one of the featured measurements of the RHIC spin program, $W^{\pm}$ boson production in polarized proton-proton collisions at RHIC were proposed as a unique tool to study the spin-flavor structure of the proton at a high scale, $Q \sim M_W$,[8,9] where Q describes the exchanged energy. Due to the parity violation of the weak interaction, $W^{\pm}$ bosons only couple to left-handed quarks and right-handed antiquarks. They naturally determine the helicity of the incident quarks. The charge of W boson selects a specific combination of the flavor of the incoming quarks, $u + \bar{d} \to W^{+}$ and $d + \bar{u} \to W^{-}$. Subsequently, their leptonic decays provide a fragmentation-function-free probe of the helicity-dependent Parton Distribution Functions (PDFs).

The longitudinal single-spin asymmetry is defined as $A_L = (\sigma_{+} - \sigma_{-})/(\sigma_{+} + \sigma_{-})$, where $\sigma_{+(-)}$ is the cross section when the polarized beam has positive (negative) helicity. At leading order, the A_L of $W^{\pm}$ are directly sensitive to $\Delta\bar{d}$ and $\Delta\bar{u}$,

$$A_L^{W^+} \propto \frac{\Delta\bar{d}(x_1)u(x_2) - \Delta u(x_1)\bar{d}(x_2)}{\bar{d}(x_1)u(x_2) + u(x_1)\bar{d}(x_2)}, \tag{1}$$

$$A_L^{W^-} \propto \frac{\Delta\bar{u}(x_1)d(x_2) - \Delta d(x_1)\bar{u}(x_2)}{\bar{u}(x_1)d(x_2) + d(x_1)\bar{u}(x_2)}, \tag{2}$$

where x_1 and x_2 are the momentum fractions carried by the scattering partons. The $A_L^{W^+}$ ($A_L^{W^-}$) approaches $\Delta u/u$ ($\Delta d/d$) in the very forward region of W rapidity, $y_W \gg 0$, and $-\Delta\bar{d}/\bar{d}$ ($-\Delta\bar{u}/\bar{u}$) in the very backward region of W rapidity $y_W \ll 0$.

First measurements of the W single-spin asymmetry at RHIC were reported by STAR[10] and PHENIX[11] collaborations from data collected during a successful commission run at $\sqrt{s} = 500$ GeV in 2009. In the following proton-proton running years, both STAR[12] and PHENIX[13] performed further measurements of W A_L with increased statistics and improved beam polarization. In 2013, STAR collected an integrated luminosity of ~250 pb^{-1} at $\sqrt{s} = 510$ GeV with an average beam polarization of ~56%. This is more than three times larger than the total integrated luminosity of previous years. In this contribution, we report the final results on W A_L from STAR data obtained in 2013 and the impact on flavor-separated light quark and antiquark polarization.[14]

2. Analysis

STAR measures the decay electrons (positrons) in $W \to e\nu$. The Time Projection Chamber (TPC) covering the full azimuth and a pseudorapidity range of $-1.3 < \eta < 1.3$, is the main tracking subsystem. It provides momenta and charge sign information for charged particles. Outside the TPC, the Barrel and Endcap Electromagnetic Calorimeters (BEMC and EEMC) covering full azimuth and pseudorapidity ranges of $-1 < \eta < 1$ and $1.1 < \eta < 2.0$ respectively, measure the energy of electrons and photons. A $W^{\pm} \to e^{\pm}\nu$ candidate event is characterized by a well isolated electron track carrying transverse energy, E_T^e, which exhibits the two-body decay "Jacobian Peak" near half of the $W^{\pm}$ mass, ~40 GeV. The undetected decay neutrinos lead to a large missing energy in the opposite azimuth of the $e^{\pm}$ candidates, so there will be a significant p_T imbalance when summing over all reconstructed final-state objects. In contrast, the p_T vector is well balanced for background events such as $Z/\gamma^* \to e^+e^-$ and QCD di-jet or multi-jet events. The W selection is achieved based on these isolation and p_T imbalance features.

STAR is not a 4π coverage detector. A di-jet event or $Z/\gamma^* \to e^+e^-$ event could have one of its jets or electrons outside the STAR acceptance. Such an event could be accepted if the detected jet or electron passes all the W selection criteria. In addition, a W boson can decay to $\tau + \nu$ and τ can further decay to electrons. We can not distinguish these feed down electrons from signal electrons. Contributions from Z/γ^* and τ which are well understood, are estimated from Monte Carlo (MC) simulation including all detector and luminosity effects. The QCD background is estimated using two procedures. The existing EEMC is used to assess the background from the corresponding uninstrumented acceptance region on the opposite side of the collision point. The remainder of the QCD background is estimated by normalizing the E_T spectrum of an pure QCD sample to the observed E_T spectrum in a QCD dominated interval.

3. Results

From the spin sorted yields of $W^{\pm}$ bosons, the longitudinal single-spin asymmetries were extracted in four pseudorapidity intervals, using

$$A_L = \frac{1}{\beta}\frac{1}{P}\frac{N_+/l_+ - N_-/l_-}{N_+/l_+ + N_-/l_-}, \tag{3}$$

where β quantifies the dilution due to background, P is the beam polarization, $N_+(N_-)$ is the W yield when the helicity of the polarized beam is

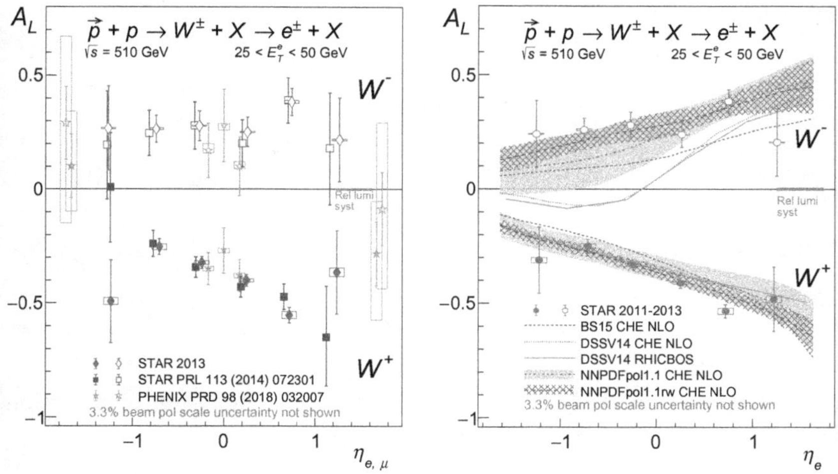

(a) Final 2013 results (b) Combined 2011-2013 results

Fig. 1: Longitudinal single-spin asymmetry, A_L, for $W^{\pm}$ production as a function of the lepton pseudorapidity, η_e, in comparison to theory predictions.

positive (negative), and $l_{\pm}$ are the relative luminosity correction factors.

From STAR 2011+2012 $W^{\pm}$ A_L results, it was noted that a larger $\Delta\bar{u}$ is preferred. With ~40% smaller uncertainties, the 2013 results confirmed the larger $\Delta\bar{u}$ preference, as shown in Fig 1a by the red data points. The combined 2011+2012 and 2013 results are shown in Fig 1b in comparison to the theory predictions. The 2011+2012 results have been included into the global QCD analysis by NNPDF group.[18] The constraints provided by these STAR data lead to a shift in the central value of $\Delta\bar{u}$ from negative to positive for $0.05 < x < 0.25$, which RHIC is sensitive to. The data favor $\Delta\bar{u} > \Delta\bar{d}$ which is opposite to the unpolarized distributions.

To quantitatively assess the impacts of STAR 2013 $W^{\pm}$ A_L results, a reweighting method[19] was implemented with the 100 publicly available replicas of NNPDFpol1.1 PDFs. The reweighted $W^{\pm}$ A_L predictions are shown in Fig 1b as the blue hatched band. Correspondingly, the impacts on the $\bar{u}$ and $\bar{d}$ polarization are shown in Fig 2a and 2b. The green and blue hatched bands are the distributions before and after reweighting with 2013 results respectively. Now, it can be found that $\Delta\bar{u}$ is positive and $\Delta\bar{d}$ is negative at medium x. And, the asymmetry between them, $\Delta\bar{u} - \Delta\bar{d}$ has similar size, but opposite sign compared to the flavor asymmetry of the unpolarized sea.

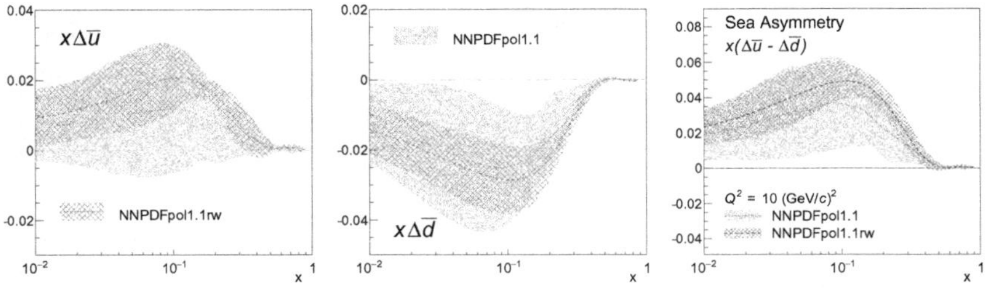

(a) $\Delta\bar{u}$ (b) $\Delta\bar{d}$ (c) $\Delta\bar{u} - \Delta\bar{d}$

Fig. 2: Impacts of STAR 2013 $W^{\pm}$ A_L results on light sea polarizations and the flavor asymmetry between $\Delta\bar{u}$ and $\Delta\bar{d}$.

References

1. B. W. Filipone and X. D. Ji, *Adv. Nucl. Phys.* **26** 1 (2001).
2. D. de Florian *et al*, *Phys. Rev. Lett.* **101**, 072001 (2008).
3. E. Leader *et al*, *Phys. Rev. D* **82**, 114018 (2010).
4. NA51 Collab. (A. Baldit *et al.*), *Phys. Lett. B* **332**, (1994) 244.
5. NuSea Collab. (R. S. Towell *et al.*), *Phys. Rev. D* **64**, 052002 (2001).
6. W. C. Chang and J. C. Peng, *Prog. Part. Nucl. Phys.* **79**, 95 (2014).
7. COMPASS Collab. (M. Alekseev *et al.*), *Phys. Lett.* **B693**, 227 (2010).
8. G. Bunce *et al.*, *Ann. Rev. Nucl. Par. Sci.* **50**, 525 (2000).
9. F. Tian *et al.*, *Nucl. Phys. A* **961**, 154 (2017);
   M. Y. Liu and B.-Q. Ma, *Phys. Rev. D* 98, 036024 (2018).
10. STAR Collab. (M. Aggarwal *et al.*), *Phys. Rev. Lett.* **106**, 062002 (2011).
11. PHENIX Collab. (A. Adare *et al.*), *Phys. Rev. Lett.* **106**, 062001 (2011).
12. STAR Collab. (L. Adamczyk *et al.*), *Phys. Rev. Lett.* **113**, 072301 (2014).
13. PHENIX Collab. (A. Adare *et al.*), *Phys. Rev. D* **93**, 051103 (2016).
14. STAR Collab. (J. Adam *et al.*), *Phys. Rev. D* **99**, 51102 (2019).
15. M. Cacciari, G. P. Salam, and G. Soyes, *JHEP* **4**, 063 (2008).
16. STAR Collab. (L. Adamczyk *et al.*), *Phys. Rev. D* **85**, 092010 (2012).
17. E. Aschenauer *et al*, arXiv: 1304.0079.
18. E. R. Nocera *et al. Nucl. Phys. B* **887**, 276 (2014).
19. R. D. Ball *et al. Nucl. Phys. B* **855**, 608 (2012).

Probing triple-parton interaction through three quarkonia associated production at the LHC

K. Y. Liu[1], J. H. Fu[1,2], Y. J. Li[1], G. Z. Xu[1], Y. J. Zhang[2], and C. Xiong[2*]

1 Department of Physics, Liaoning University, Shenyang 110036 , China
2 School of Physics, Beihang University, Beijing 100191, China
** Corresponding author: 09434@buaa.edu.cn*

Triple-parton interactions are very sensitive to the multidimensional tomography of protons. However this sensitivity is too rare to be observed until now. In this study, $\Upsilon + J/\psi + \phi$ production at LHCb at 13 TeV is proposed. The number of collected events is approximately 650. It is a very clean hard probe of multiple parton interaction, especially the unknown triple-parton interaction.

Keywords: Triple partons interaction; multiple parton interaction; quarkonia.

Introduction Multi-parton interaction (MPI) is becoming more important for testing standard model.[1–5] The leading MPI is double-parton scattering (DPS), which had been studied in many processes, such as associated quarkonia hadroproductions.[6–8] The next-to-leading MPI is triple-parton scattering (TPS), which is too complicated to measure for fewer production rates and more confused final states.[2,9] Under the assumption of no correlation between multiple partons,[2,7] the generic N-parton scattering (NPS) cross section can be written as[1]

$$\sigma^{\text{NPS}}_{f_1 \cdots f_N} = \frac{m}{N!} \frac{\prod_{i=1}^{N} \sigma^{\text{SPS}}_{f_i}}{\left(\sigma_{\text{eff, N}}\right)^{N-1}}. \tag{1}$$

The cross section $\sigma^{\text{SPS}}_{f_i}$ corresponds to the single-parton scattering (SPS) process of the inclusive production of the final state f_i. The effective cross section of NPS $\sigma_{\text{eff, N}}$ encodes the unknown information of the partons' transverse profiles, which can be determined through experimental measurements. In the cases of $N = 2$ and $N = 3$, $\sigma_{\text{eff, N}}$ are written as $\sigma_{\text{eff, DPS}}$ and $\sigma_{\text{eff, TPS}}$, respectively. Under several assumptions, Ref. 9 obtains $\sigma_{\text{eff, TPS}} = (0.82 \pm 0.11) \times \sigma_{\text{eff, DPS}}$.

Quarkonia production at B factories[10–19] and LHC[20–27] is a hot topic. And it had been studied to probe double-parton correlations.[7,28,29] The values of $\sigma_{\text{eff, DPS}}$ extracted from electro-weak boson processes is approximately 15 mb, and it is smaller than 10 mb from the quarkonia processes.[2,7] Theoretical studies of TPS are limited to open charm productions[9] and experimental studies are still ongoing. Triple-J/ψ hadroproduction was studied in Ref. 2. Thus, we consider $\Upsilon + J/\psi + \phi$ associated production at LHCb as a TPS hard probe in this work.

Frame of calculation In $\Upsilon + J/\psi + \phi$ hadroproduction, the typical Feynman diagrams of SPS, DPS, and TPS processes are shown in Fig. 1.

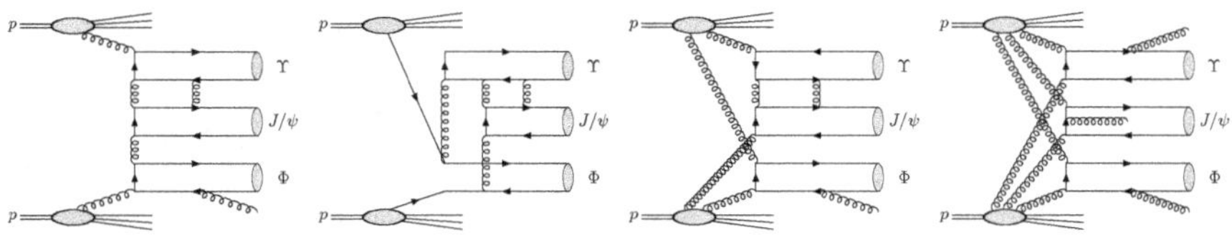

Fig. 1. Typical Feynman diagrams via SPS gg, SPS $s\bar{s}$, DPS, and TPS.

For the J^{PC} of Υ, J/ψ, and ϕ states are 1^{--}, so that at least three gluons connect with each state. Then, the SPS, DPS, and TPS processes via gluon fusion are $\mathcal{O}(\alpha_s^9)$. The $s - \bar{s}$ fusion processes are $\mathcal{O}(\alpha_s^8)$, but the parton distribution function of s and $\bar{s}$ is suppressed by a factor $1/100$ from that of the gluons. We the obtain following formula using Eq. (1):

$$\begin{aligned}
&\sigma^{\rm TPS}(pp \to \Upsilon + J/\psi + \phi + X) = \\
&\frac{\sigma^{\rm SPS}(pp \to \Upsilon + X) \times \sigma^{\rm SPS}(pp \to J/\psi + X) \times \sigma^{\rm SPS}(pp \to \phi + X)}{(\sigma_{\rm eff,\ TPS})^2}, \\
&\sigma^{\rm DPS}(pp \to \Upsilon + J/\psi + \phi + X) = \\
&\frac{\sigma^{\rm SPS}(pp \to \Upsilon + J/\psi + X)\sigma^{\rm SPS}(pp \to \phi + X)}{\sigma_{\rm eff,\ DPS}} + (\Upsilon \leftrightarrow J/\psi \leftrightarrow \phi). \quad (2)
\end{aligned}$$

J/ψ and Υ production can be described by the non-relativistic QCD frame (NRQCD).[30] However, ϕ production cannot be described by NRQCD due to the small s quark mass. Therefore, we estimate the ϕ associated production process as:

$$\sigma^{\rm SPS}(\Upsilon(J/\psi,\text{ or }\Upsilon J/\psi) + \phi + X) \approx \frac{1}{10}\sigma^{\rm SPS}(\Upsilon(J/\psi,\text{ or }\Upsilon J/\psi) + X).$$

Four SPS processes, namely, Υ, J/ψ, ϕ, and $\Upsilon + J/\psi$ occur.

The numerical result of $\Upsilon + J/\psi$[7] is used here. Moreover, Υ, J/ψ, and ϕ hadroproductions are directly estimated using experimental data.[31–33] The inclusive ϕ cross-section at $\sqrt{s} = 13$ TeV is approximately double of that at $\sqrt{s} = 7$ TeV. The SPS cross sections at LHCb at $\sqrt{s} = 13$ TeV are

$$\begin{aligned}
\sigma^{\rm SPS}(pp \to \Upsilon + J/\psi + X) &\approx 10 \text{ pb}, \\
\sigma^{\rm SPS}(pp \to \Upsilon + X) &\approx 0.2\ \mu\text{b}, \\
\sigma^{\rm SPS}(pp \to J/\psi + X) &\approx 15\ \mu\text{b}, \\
\sigma^{\rm SPS}(pp \to \phi + X) &\approx 3 \text{ mb}.
\end{aligned} \tag{3}$$

With a fixed $\sigma_{\text{eff, TPS}} = \sigma_{\text{eff, DPS}} = 10$ mb,

$$\begin{aligned}
\sigma^{\rm SPS}(pp \to \Upsilon + J/\psi + \phi + X) &\approx 1 \text{ pb}, \\
\sigma^{\rm DPS}(pp \to \Upsilon + J/\psi + \phi + X) &\approx 60 \text{ pb}, \\
\sigma^{\rm TPS}(pp \to \Upsilon + J/\psi + \phi + X) &\approx 90 \text{ pb}, \\
\sigma^{\rm total}(pp \to \Upsilon + J/\psi + \phi + X) &\approx 150 \text{ pb}
\end{aligned} \tag{4}$$

The branch rations are $Br(\Upsilon \to \mu^+\mu^-) = 0.024$, $Br(J/\psi \to \mu^+\mu^-) = 0.060$, and $Br(\phi \to K^+K^-) = 0.5$. The integrated luminosity of LHCb is approximately 6 fb^{-1}. We can estimate that the number of events is approximately 650, about 400 from TPS and 250 from DPS.

Summary We propose the use of the $\Upsilon + J/\psi + \phi + X$ production at LHCb at $\sqrt{s} = 13$ TeV to probe MPI. The TPS and DPS contributions are approximately 60% and 40%, respectively; that of SPS is less than 1%. Our result shows that it is a MPI dominant process. Although it is a very rare process, the data collected by LHCb is enough to probe TPS.

Acknowledgments

This work is supported by the National Natural Science Foundation of China (Grants No. 11722539, 11705078, and 11447018).

References

1. D. d'Enterria and A. Snigirev, Double, triple, and n-parton scatterings in high-energy proton and nuclear collisions, *Adv. Ser. Direct. High Energy Phys.* **29**, 159 (2018).
2. H.-S. Shao and Y.-J. Zhang, Triple prompt J/ψ hadroproduction as a hard probe of multiple-parton scatterings, *Phys. Rev. Lett.* **122**, 192002 (2019).
3. J. A. Grifols, J. F. Gunion and A. Mendez, Detection of a Charged Higgs via the Decays $H^\pm \to \Upsilon W^\pm$ and $H^\pm \to \theta W^\pm$, *Phys. Lett.* **B197**, 266 (1987).

4. Q.-H. Cao, Y. Liu, K.-P. Xie and B. Yan, Double parton scattering of weak gauge boson productions at the 13 TeV and 100 TeV proton-proton colliders, *Phys. Rev.* **D97**, 035013 (2018).
5. J.-P. Lansberg, New Observables in Inclusive Production of Quarkonia (2019).
6. V. M. Abazov *et al.*, Evidence for simultaneous production of J/ψ and Υ mesons, *Phys. Rev. Lett.* **116**, 082002 (2016).
7. H.-S. Shao and Y.-J. Zhang, Complete study of hadroproduction of a Υ meson associated with a prompt J/ψ, *Phys. Rev. Lett.* **117**, 062001 (2016).
8. J.-P. Lansberg, H.-S. Shao, N. Yamanaka and Y.-J. Zhang, Prompt J/ψ-pair production at the LHC: impact of loop-induced contributions and of the colour-octet mechanism (2019).
9. D. d'Enterria and A. M. Snigirev, Triple parton scatterings in high-energy proton-proton collisions, *Phys. Rev. Lett.* **118**, 122001 (2017).
10. K.-Y. Liu, Z.-G. He and K.-T. Chao, Problems of double charm production in e^+e^- annihilation at $\sqrt{s} = 10.6$-GeV, *Phys. Lett.* **B557**, 45 (2003).
11. K.-Y. Liu, Z.-G. He and K.-T. Chao, Inclusive charmonium production via double $c\bar{c}$ in e^+e^- annihilation, *Phys. Rev.* **D69**, 094027 (2004).
12. Y.-J. Zhang, Y.-j. Gao and K.-T. Chao, Next-to-leading order QCD correction to $e^+e^- \to J/\psi + \eta_c$ at $\sqrt{s} = 10.6$-GeV, *Phys. Rev. Lett.* **96**, 092001 (2006).
13. Y.-J. Zhang and K.-T. Chao, Double charm production $e^+e^- \to J/\psi + c\bar{c}$ at B factories with next-to-leading order QCD correction, *Phys. Rev. Lett.* **98**, 092003 (2007).
14. Y.-Q. Ma, Y.-J. Zhang and K.-T. Chao, QCD correction to $e^+e^- \to J/\psi + gg$ at B Factories, *Phys. Rev. Lett.* **102**, 162002 (2009).
15. Y.-J. Zhang, Y.-Q. Ma and K.-T. Chao, Factorization and NLO QCD correction in $e^+e^- \to J/\psi(\psi(2S)) + \chi_{c0}$ at B Factories, *Phys. Rev.* **D78**, 054006 (2008).
16. Y.-J. Zhang, Y.-Q. Ma, K. Wang and K.-T. Chao, QCD radiative correction to color-octet J/ψ inclusive production at B Factories, *Phys. Rev.* **D81**, 034015 (2010).
17. Y.-J. Li, G.-Z. Xu, K.-Y. Liu and Y.-J. Zhang, Search for $C = +$ charmonium and XYZ states in $e^+e^- \to \gamma + H$ at BESIII, *JHEP* **01**, 022 (2014).

18. G.-Z. Xu, Y.-J. Li, K.-Y. Liu and Y.-J. Zhang, $\alpha_s v^2$ corrections to η_c and χ_{cJ} production recoiled with a photon at e^+e^- colliders, *JHEP* **10**, 071 (2014).
19. Y.-J. Li, G.-Z. Xu, P.-P. Zhang, Y.-J. Zhang and K.-Y. Liu, Study of Color Octet Matrix Elements Through J/ψ Production in e^+e^- Annihilation, *Eur. Phys. J.* **C77**, 597 (2017).
20. R. Li, Y.-J. Zhang and K.-T. Chao, Pair Production of Heavy Quarkonium and B(c)(*) Mesons at Hadron Colliders, *Phys. Rev.* **D80**, 014020 (2009).
21. Y.-Q. Ma, K. Wang and K.-T. Chao, $J/\psi(\psi')$ production at the Tevatron and LHC at $\mathcal{O}(\alpha_s^4 v^4)$ in nonrelativistic QCD, *Phys. Rev. Lett.* **106**, 042002 (2011).
22. K.-T. Chao, Y.-Q. Ma, H.-S. Shao, K. Wang and Y.-J. Zhang, J/ψ Polarization at Hadron Colliders in Nonrelativistic QCD, *Phys. Rev. Lett.* **108**, 242004 (2012).
23. G.-Z. Xu, Y.-J. Li, K.-Y. Liu and Y.-J. Zhang, Relativistic Correction to Color Octet J/ψ Production at Hadron Colliders, *Phys. Rev.* **D86**, 094017 (2012).
24. Y.-J. Li, G.-Z. Xu, K.-Y. Liu and Y.-J. Zhang, Relativistic Correction to J/ψ and Υ Pair Production, *JHEP* **07**, 051 (2013).
25. H. Han, Y.-Q. Ma, C. Meng, H.-S. Shao, Y.-J. Zhang and K.-T. Chao, $\Upsilon(nS)$ and $\chi_b(nP)$ production at hadron colliders in nonrelativistic QCD, *Phys. Rev.* **D94**, 014028 (2016).
26. H.-S. Shao, Y.-Q. Ma, K. Wang and K.-T. Chao, Polarizations of χ_{c1} and χ_{c2} in prompt production at the LHC, *Phys. Rev. Lett.* **112**, 182003 (2014).
27. H. Han, Y.-Q. Ma, C. Meng, H.-S. Shao and K.-T. Chao, η_c production at LHC and indications on the understanding of J/ψ production, *Phys. Rev. Lett.* **114**, 092005 (2015).
28. R. Aaij *et al.*, Observation of J/ψ pair production in *pp* collisions at $\sqrt{s} = 7TeV$, *Phys. Lett.* **B707**, 52 (2012).
29. J.-P. Lansberg and H.-S. Shao, Production of $J/\psi + \eta_c$ versus $J/\psi + J/\psi$ at the LHC: Importance of Real α_s^5 Corrections, *Phys. Rev. Lett.* **111**, 122001 (2013).
30. G. T. Bodwin, E. Braaten and G. P. Lepage, Rigorous QCD analysis of inclusive annihilation and production of heavy quarkonium, *Phys. Rev.* **D51**, 1125 (1995), [Erratum: Phys. Rev.D55,5853(1997)].

31. R. Aaij *et al.*, Measurement of Υ production in pp collisions at $\sqrt{s}=$ 13 TeV, *JHEP* **07**, 134 (2018), [Erratum: JHEP05,076(2019)].
32. R. Aaij *et al.*, Measurement of forward J/ψ production cross-sections in pp collisions at $\sqrt{s} = 13$ TeV, *JHEP* **10**, 172 (2015), [Erratum: JHEP05,063(2017)].
33. R. Aaij *et al.*, Measurement of the inclusive ϕ cross-section in pp collisions at $\sqrt{s} = 7$ TeV, *Phys. Lett.* **B703**, 267 (2011).

Light meson parton distribution functions from basis light-front quantization and QCD evolution

Jiangshan Lan[a,b,c,*], Chandan Mondal[a,b], Shaoyang Jia[d],
Xingbo Zhao[a,b] and James P. Vary[d]
(BLFQ Collaboration)
[a]*Institute of Modern Physics, Chinese Academy of Sciences, Lanzhou 730000, China*
[b]*School of Nuclear Science and Technology, University of Chinese Academy of Sciences, Beijing 100049, China*
[c]*Lanzhou University, Lanzhou 730000, China*
[d]*Department of Physics and Astronomy, Iowa State University, Ames, IA 50011, USA*
[*]*jiangshanlan@impcas.ac.cn*

We investigate the parton distribution functions (PDFs) of the pion and kaon from the eigenstates of a light-front effective Hamiltonian in the constituent quark-antiquark representation suitable for low-momentum scale applications. By taking these scales as the only free parameters, the valence quark distribution functions of the pion, after QCD evolving, are consistent with the E615 experiment at Fermilab. In addition, the ratio of the up quark distribution in the kaon to that in the pion also agrees with the NA3 experimental result at CERN.

Keywords: BLFQ; LFWF; meson; PDF.

1. Introduction

Basis light-front quantization (BLFQ) is a nonperturbative approach with advantage of both light-front dynamics and ab initio nuclear structure methods, which has been developed for solving many-body bound state problems in quantum field theories.[1–4] It has been used to investigate a variety of systems and reactions.[5–11] Recently, the BLFQ approach using a Hamiltonian that includes the color singlet Nambu–Jona-Lasinio (NJL) interaction to account for the chiral dynamics has been applied to light mesons.[12–14] In this work, we present results for the parton distribution functions (PDFs) of the light mesons.

The probability of finding a valence quark with the longitudinal momentum fraction x inside the meson ($|q\bar{q}\rangle$) is given by[5]

$$f(x) = \frac{1}{4\pi\, x(1-x)} \sum_{rs} \int \frac{d\vec{\kappa}^{\perp}}{(2\pi)^2}\, \psi^*_{rs}(x, \vec{\kappa}^{\perp})\, \psi_{rs}(x, \vec{\kappa}^{\perp}), \tag{1}$$

where $\psi_{rs}(x, \vec{\kappa}^{\perp})$ is the valence light-front wave function (LFWF) of the meson. Correspondingly, the valence antiquark PDF is given by $f(1-x)$. The momentum sum rule is

$$\int_0^1 x\, f(x)\, dx + \int_0^1 x\, f(1-x)\, dx = 1, \tag{2}$$

which indicates that, at our model scale, the valence quark and antiquark carry the entire momentum of the meson. The normalization of our LFWFs ensures that valence quark PDFs are normalized.

2. Basis light-front quantization

LFWFs encode the structure of bound states, in the BLFQ approach, one obtains the LFWF by solving the time-independent light-front Schrödinger equation

$$H_{\text{eff}}|\Psi\rangle = M^2|\Psi\rangle, \tag{3}$$

where H_{eff} is the effective Hamiltonian of the system with the eigenvalue of the mass squared M^2. Generally, $|\Psi\rangle$ is expanded into all Fock sectors in the Hilbert space. In the valence Fock sector, the effective Hamiltonian for the light mesons with non-singlet flavor wave functions is given by[12]

$$H_{\text{eff}} = \frac{\vec{k}_\perp^2 + m_q^2}{x} + \frac{\vec{k}_\perp^2 + m_{\bar{q}}^2}{1-x} + \kappa^4 \vec{\zeta}_\perp^2 - \frac{\kappa^4\, \partial_x\big(x(1-x)\partial_x\big)}{(m_q + m_{\bar{q}})^2} + H^{\text{eff}}_{\text{NJL}}, \tag{4}$$

where m_q ($m_{\bar{q}}$) is the mass of the quark (antiquark), and κ is the strength of the confinement. $\vec{\zeta}_\perp \equiv \sqrt{x(1-x)}\vec{r}_\perp$ is the holographic variable,[15] with $\vec{k}_\perp$ being the conjugate variable of $\vec{r}_\perp$. The x-derivative is defined as $\partial_x f(x, \vec{\zeta}_\perp) = \partial f(x, \vec{\zeta}_\perp)/\partial x|_{\vec{\zeta}}$. The first two terms in Eq. (4) are the light-front kinetic energy for the quark and the antiquark. The third and the fourth terms are the confining potentials in the transverse direction based on the light-front holographic QCD (LFHQCD)[15] and a longitudinal confining potential[3] that reproduce 3D confinement in the nonrelativistic limit. Additionally, the $H^{\text{eff}}_{\text{NJL}}$ is the color-singlet NJL interaction to account for the chiral dynamics.[16]

3. Results

We evolve our initial PDFs[13] to the relevant experimental scales $\mu^2 = 16$ GeV^2 and $\mu^2 = 20$ GeV^2 for the pion and the kaon by the DGLAP equations,[17–19] respectively. Here, we adopt the higher order perturbative parton evolution toolkit to solve the DGLAP equations numerically.[20] We find that the initial scales increase when we progress from the leading order (LO) to the next-to-next-to-leading order (NNLO). Meanwhile, the evolved PDFs fit better to the experimental result demonstrated by smaller values of χ^2 per degree of freedom (d.o.f.) at higher orders, as shown in Table 1 as well as Figure 1.

Table 1. Initial scales and the $\chi^2/(\text{d.o.f.})$ at the first three orders of the DGLAP equation.

Order	$\mu^2_{0,\pi}$ [GeV^2]	$\mu^2_{0,K}$ [GeV^2]	E615[21] $\frac{\chi^2}{(d.o.f.)}$	NA3[22] $\frac{\chi^2}{(d.o.f.)}$
LO	0.120 ± 0.012	0.133 ± 0.013	6.71	0.88
NLO	0.205 ± 0.020	0.210 ± 0.021	4.67	0.56
NNLO	0.240 ± 0.024	0.246 ± 0.024	3.64	0.50

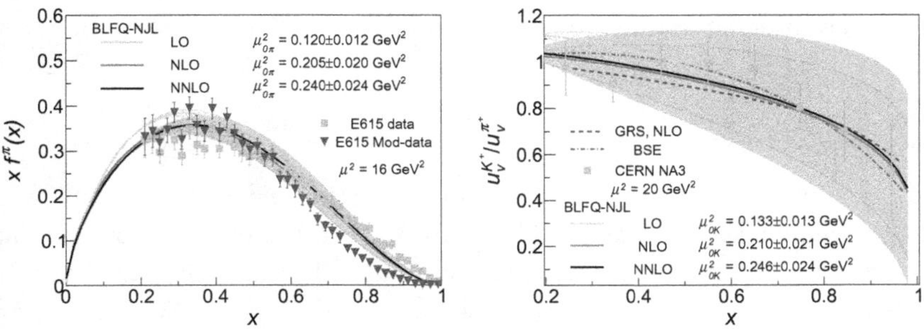

Fig. 1. (left) $xf^\pi(x)$ as a function of x for the pion. (right) The ratio of the u quark PDFs in the kaon to that in the pion.

In Table 1, the χ^2 is defined as the sum of the squared-difference of our results from the results of the E615 experiment at FNAL[21] and from the NA3 experiment at CERN,[22] both at the respective experimental scales. In Figure 1 (left panel), we show our results in each order from LO to NNLO compared with the E615 experimental result[21] and the modified result of the E615 experiment[23] for the pion. In Figure 1 (right panel) we compare our results for the ratio of the kaon to the pion up quark distributions with the NA3 experimental result[22] and the NLO Glück-Reya-Stratmann (GRS) model[24] as well as the prediction from the Bethe-Salpeter equation (BSE)[25] for the Kaon. The bands are due to a 10% uncertainty for initial scales. At large x, the behavior of our pion valence PDF is $(1-x)^{1.44}$. Furthermore, the various valence, gluon, and sea contributions to the first moment of the pion at 4 GeV^2 are $\{0.489, 0.398, 0.113\}$, respectively, which agree with the results from BSE model[26] $\{0.48(3), 0.41(2), 0.11(2)\}$. We further find our results agree well with other results.[21–26]

4. Conclusions

The PDFs of the light mesons have been presented using the BLFQ approach and QCD evolution. In our model, the initial scales of PDFs up to NNLO are $\mu_{0\pi}^2 = 0.240 \pm 0.024\ \text{GeV}^2$ for the pion and $\mu_{0K}^2 = 0.246 \pm 0.024\ \text{GeV}^2$ for the kaon. The result of the pion valence PDF has produced good agreement with the E615 experimental result[21] and the modified result of the E615 experiment.[23] In addition, the contributions to pion first moment at 4 GeV^2 is consistent with the results from the BSE model.[26] We also note that the ratio of the up quark PDFs in the kaon to that in the pion is in good agreement with the CERN-NA3 experimental result,[22] the result from GRS model[24] as well as the prediction from the BSE.[25]

Acknowledgments

CM is supported by the National Natural Science Foundation of China (NSFC) under the Grant No. 11850410436. XZ is supported by new faculty startup funding by the Institute of Modern Physics, Chinese Academy of Sciences and by Key Research Program of Frontier Sciences, CAS, Grant No ZDBS-LY-7020. SJ and JPV are supported in part by the Department of Energy under Grants No. DE-FG02-87ER40371 and No. DESC00018223 (SciDAC-4/NUCLEI). A portion of the computational resources were provided by the National Energy Research Scientific Computing Center (NERSC), which is supported by the Office of Science of the U.S. Department of Energy under Contract No. DE-AC02-05CH11231.

References

1. J. P. Vary *et al.*, Phys. Rev. C **81**, 035205 (2010).

2. P. Wiecki *et al.*, Phys. Rev. D **91**, no. 10, 105009 (2015).
3. Y. Li, P. Maris, X. Zhao and J. P. Vary, Phys. Lett. B (2016).
4. X. Zhao *et al.*, Phys. Lett. B **737**, 65 (2014).
5. Y. Li *et al.*, Phys. Rev. D **96**, no. 1, 016022 (2017).
6. S. Tang *et al.*, Phys. Rev. D **98**, no. 11, 114038 (2018).
7. G. Chen *et al.*, Phys. Rev. C **100**, no. 2, 025208 (2019)
8. M. Li *et al.*, Phys. Rev. D **100**, no. 3, 036006 (2019)
9. J. Lan *et al.*, arXiv:1911.11676 [nucl-th].
10. C. Mondal *et al.*, arXiv:1911.10913 [hep-ph].
11. W. Du *et al.*, arXiv:1911.10762 [nucl-th].
12. S. Jia and J. P. Vary, Phys. Rev. C **99**, no. 3, 035206 (2019).
13. J. Lan *et al.*, Phys. Rev. Lett. **122**, no. 17, 172001 (2019).
14. J. Lan *et al.*, arXiv:1907.01509 [nucl-th].
15. S. J. Brodsky *et al.*, Phys. Rept. **584**, 1 (2015).
16. S. Klimt *et al.*, Nucl. Phys. A **516**, 429 (1990).
17. V. N. Gribov and L. N. Lipatov, Sov. J. Nucl. Phys. **15**, 438 (1972).
18. G. Altarelli and G. Parisi, Nucl. Phys. B **126**, 298 (1977).
19. Y. L. Dokshitzer, Sov. Phys. JETP **46**, 641 (1977) [Zh. Eksp. Teor. Fiz. **73**, 1216 (1977)].
20. G. P. Salam and J. Rojo, Comput. Phys. Commun. **180**, 120 (2009)
21. J. S. Conway *et al.*, Phys. Rev. D **39**, 92 (1989).
22. J. Badier *et al.* [NA3 Collaboration], Z. Phys. C **18**, 281 (1983).
23. C. Chen *et al.*, Phys. Rev. D **93**, no. 7, 074021 (2016)
24. M. Glück, E. Reya and M. Stratmann, Eur. Phys. J. C **2**, 159 (1998)
25. T. Nguyen *et al.*, Phys. Rev. C **83**, 062201 (2011)
26. M. Ding *et al.*, arXiv:1905.05208 [nucl-th].

GPD measurements at COMPASS

P.-J. Lin
(on behalf of the COMPASS Collaboration)
Institut de Physique Nucléaire Orsay, CNRS-IN2P3,
Orsay, 91406, France
E-mail: po-ju.lin@cern.ch

Utilizing the 160 *GeV/c* polarized muon beams provided by the CERN SPS, COMPASS studies the Generalized Parton Distributions (GPDs) via Deeply Virtual Compton Scattering (DVCS) and Hard Exclusive Meson Production (HEMP). In this article, the GPD measurement program at COMPASS is introduced and some of the recent results are given.

Keywords: Generalized Parton Distributions; Deeply Virtual Compton Scattering; Hard Exclusive Meson Production; COMPASS

1. Introduction

Encapsulating the transverse position of partons as functions of their longitudinal momentum, Generalized Parton Distributions (GPDs) go beyond the 1-dimensional description of the partonic structure of the nucleon in terms of the parton distribution functions (PDFs), and thus provides a 3-dimentional picture of the nucleon. The GPDs can be accessed with hard exclusive reactions and there have been various experimental efforts put on the GPD measurements, including COMPASS. Measuring Deeply Virtual Compton Scattering (DVCS) and Hard Exclusive Meson Production (HEMP), COMPASS can study GPDs in the kinematic domain that is not covered by the previous experiments, and thus provide valuable inputs for the better understanding of GPDs.

2. The COMPASS Experiment

As a fixed target experiment located at CERN, COMPASS utilizes the M2 beamline of SPS that can provide various hadron and muon beams. In the GPD-measurement setting, the 160 GeV polarized μ^+ and μ^- beams are employed. The μ^+ and μ^- beams have opposite polarization and the absolute value of the polarization is about 80%. The muon beam is centered on a 2.5m long liquid hydrogen (LH_2) target, which is followed by the two-stage COMPASS spectrometer of large acceptance in angle and momentum.[1,2] Surrounding the LH_2 target is the recoiling proton detector, CAMERA. With the time-of-flight measurement, CAMERA determines the kinematic properties of a recoiling proton and plays an important role in ensuring the exclusivity of an event.

Following the pilot run in 2012 with a 4-week data taking, COMPASS had dedicated runs in 2016–17 with few detector upgrades and accumulated about 10 times more statistics than in the 2012 pilot run. With the Q^2 coverage from 1 to 5 $(\text{GeV}/c)^2$ and x_{Bj} from 0.005 to 0.1, COMPASS probes GPDs within the phase space where the sea quarks contribute significantly.

3. DVCS

DVCS is regarded as the golden channel of GPD study, not only because of its simplicity, but also due to its interference with the well-known Bethe–Heitler (BH) process with which an unique access to both the real and imaginary part of the DVCS amplitude is gained.

DVCS interaction contributes to the exclusive lepto-production of a single photon, $l + p \to l' + p' + \gamma$. The differential cross sections measured with the μ^+ and μ^- beams can be decomposed into terms of the BH process, the DVCS process, and the interference between them. Summing the cross sections of both the muon beams (beam charge and spin sum) gives:

$$2d\sigma \equiv d\sigma^{\overset{+}{\leftarrow}} + d\sigma^{\overset{-}{\rightarrow}} = 2(d\sigma^{BH} + d\sigma^{DVCS} - |P_\mu| d\sigma^{I}). \quad (1)$$

Here, the polarization of the incident muon is denoted as P_μ, and the charge and the helicity of the muon beam are labeled as $\pm$ and $\leftrightarrows$, respectively.

In a kinematic domain where the BH contribution is not dominant, it can be removed from the single proton production using Monte-Carlo simulation and Eq. (1). Moreover, integrating out the dependence on the azimuthal angle between the muon-scattering and the photon-production planes, we are left with the DVCS cross section which only depends on a term denoted as c_0^{DVCS}.[3] In the 2012-data analysis,[4] the extracted DVCS cross section as a function of t can be well described with an exponential function $e^{-B|t|}$. In the small x_{Bj} domain covered by COMPASS, c_0^{DVCS} is dominated by the imaginary part of the Compton Form Factor (CFF) $\mathcal{H}$, $\mathfrak{Im}\mathcal{H}$, which is the convolution of GPD H with the partonic level Compton interaction. The t-dependence slope B can be translated[4] to the average squared transverse extension of partons in the proton, as $\langle r_\perp^2 \rangle \approx 2\langle B(x_{Bj}) \rangle$. At $\langle x_{Bj} \rangle = 0.056$, the measured t-slope gives $B = (4.3 \pm 0.6_{stat} {}^{+0.1}_{-0.3}|_{sys})$ (GeV/c)^2, which corresponds to a transverse extension of the partons in the proton of $\sqrt{\langle r_\perp^2 \rangle} = (0.58 \pm 0.04_{stat} {}^{+0.01}_{-0.02}|_{sys} \pm 0.04_{model})$ fm.

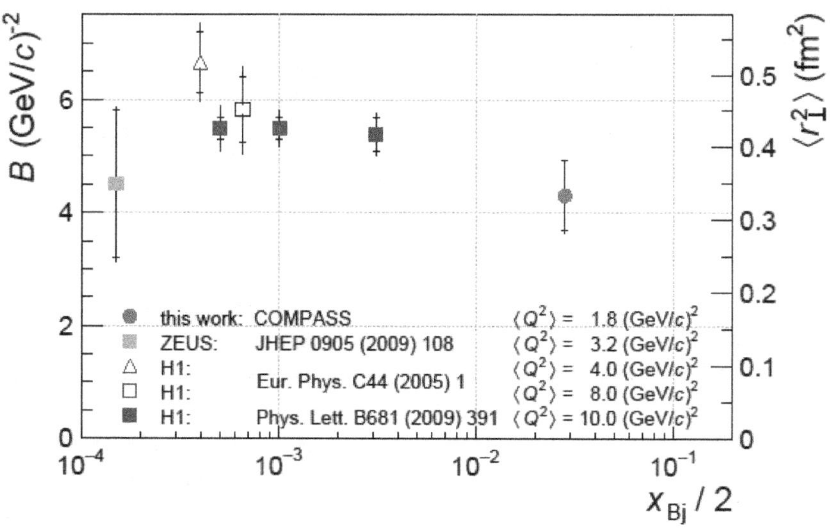

Fig. 1. The COMPASS extracted t-slope parameter,[4] or equivalently the average squared transverse extension of partons in the proton, together with the results from the measurements by H1 and ZEUS. The inner error bars represent the statistical uncertainty and the outer ones are the quadratic sum of the statistical and systematic uncertainties.

The t-slope acquired by COMPASS is shown together with other measurements in Fig. 1. With higher statistics accumulated, the same measurement but at three different x_{Bj} values around the COMPASS 2012 result is expected from the 2016–17 data.

4. HEMP

Due to their universality, the GPDs can also be explored with HEMP. With the difference in their quark content and quantum number, the exclusive production of different mesons acts naturally as flavor filter for GPDs.

HEMP is sensitive to not only the chiral-even GPDs but also the chiral-odd (transversity) ones, which cannot be accessed via DVCS.

4.1. *Exclusive π^0 Production*

If the beam polarization magnitude is the same for measurements with μ^+ and μ^- beams, the spin-independent reduced cross section for the π^0 production is

$$\begin{aligned}\frac{d^2\sigma_{\gamma^* p}}{dt d\phi_\pi} = \frac{1}{2\pi}\bigg[&\frac{d\sigma_T}{dt} + \epsilon\frac{d\sigma_L}{dt} + \epsilon\,\cos(2\phi_\pi)\frac{d\sigma_{TT}}{dt} \\ &+ \sqrt{2\epsilon(1+\epsilon)}\,\cos(\phi_\pi)\frac{d\sigma_{LT}}{dt}\bigg],\end{aligned} \tag{2}$$

where ϕ_π is the azimuthal angle between the muon-scattering and the π^0 production plane, and ϵ is the virtual-photon polarization parameter. The subscript of each cross-section term indicates the contribution from either the transversely polarized (σ_T) or the longitudinally polarized (σ_L) virtual photon, or the interference between differently polarized ones (σ_{TT}, σ_{LT}).

The terms involving transversely polarized virtual photons, σ_T, σ_{TT}, and σ_{LT}, arise from the coupling between the chiral-odd GPDs to the twist-3 pion wave function, and are attributed to GPDs $\tilde{H}$, $\tilde{E}$, H_T, and $\overline{E}_T = 2\tilde{H}_T + E_T$. Among them, the σ_{TT} term is of special interest since it is related only to $\overline{E}_T$, which is not yet well known and is expected to give a non-negligible contribution in the π^0 production.

In the average of kinematics of $\langle Q^2\rangle = 2.0$ (GeV$/c)^2$, $\langle\nu\rangle = 12.8$ GeV, and $\langle x_{Bj}\rangle = 0.093$, the dependence on $|t|$ of the cross section measured from 2012 data[5] is illustrated in Fig. 2. As can be seen in the plots, the results of COMPASS serve as important constraints for GPD modeling, in particular the chiral-odd ones. The values of the cross section extracted from the fit using Eq. (2) show a large contribution from the transversely polarized photons, and provides a clear evidence of the significance of the GPD $\overline{E}_T$.

4.2. *Exclusive ω Production*

From the angular distributions of particles produced in the exclusive ω production, a total number of 23 spin density matrix elements (SDMEs) have been extracted. From the 2012 data, the SDMEs can be sorted into

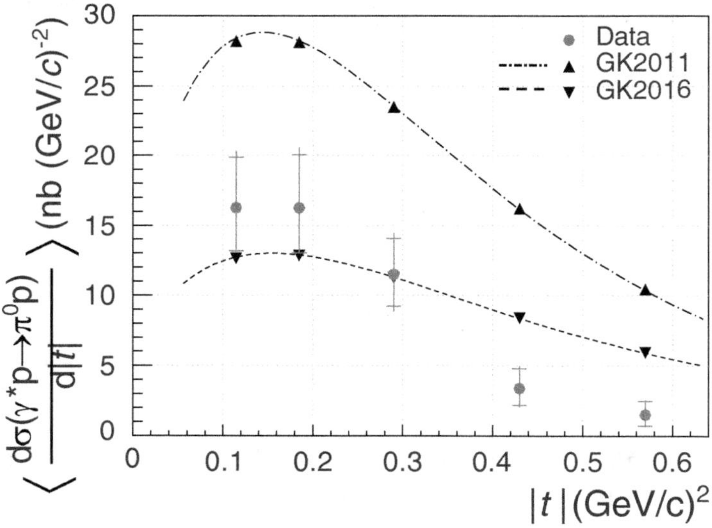

Fig. 2. The measured π^0 cross section as a function of $|t|$.[5] The inner error bars represent the statistical uncertainty and the outer ones are the quadratic sum of the statistical and systematic uncertainties. The data, given as the solid red circles, is compared with two predictions of the GK model,[6,7] denoted as the solid triangles.

5 different classes, depending on the helicity transitions from the virtual photon γ^* to the meson ω.

The hypothesis of s-channel helicity conservation (SCHC) implies the preservation of helicity in the $\gamma^* \to \omega$ transitions and some relations between SDMEs of a same class. For example, $r^1_{1-1} + \mathfrak{Im}\, r^2_{1-1}$, $\mathfrak{Re}\, r^5_{10} + \mathfrak{Im}\, r^6_{10}$, and $\mathfrak{Im}\, r^7_{10} - \mathfrak{Re}\, r^8_{10}$ should be equal to zero, which is observed within uncertainties in the COMPASS preliminary results. SCHC also implies that all SDMEs from the classes of transistions $\gamma^*_T \to \omega_L$, $\gamma^*_L \to \omega_T$, and $\gamma^*_T \to \omega_{-T}$ should be equal to zero. However, the values of certain SDMEs from the transition $\gamma^*_T \to \omega_L$ show a clear violation of SCHC. It has been suggested that the non-zero r^5_{00} is correlated to the GPDs as[8]

$$r^5_{00} \propto \mathfrak{Re}\big[\langle \overline{E}_T \rangle^*_{LT} \langle H \rangle_{LL} + \frac{1}{2} \langle H_T \rangle^*_{LT} \langle E \rangle_{LL}\big], \tag{3}$$

where the GPD H_T term is expected to bring considerable contribution, owing to the previous studies of the exclusive ρ^0 and ω production on the transversely polarized proton target at COMPASS.

5. Summary

Some of the results of GPD studies at COMPASS using the 2012 data are presented. COMPASS measures the transverse extension of partons in the proton by analyzing the DVCS cross-section sum of polarized μ^+ and μ^- beams. With both DVCS and HEMP, COMPASS provides constraints for GPD modeling in the previously unexplored kinematic domain. In addition to the already discussed π^0 and ω production, there are ongoing analysis on the exclusive production of ρ, ϕ, and J/ψ as well. With the future results of different mesons combined, the flavor decomposition of GPDs can hopefully be achieved.

Acknowledgment

The author would like to thank the laboratoire d'excellence Physique des deux infinis et des Origines (P2IO) for the funding support and doctor Nicole d'Hose for helpful discussions.

References

1. COMPASS Collaboration, The COMPASS experiment at CERN, *Nucl. Instrum. Methods A* **577**, 455 (2007).
2. COMPASS Collaboration, The COMPASS setup for physics with hadron beams, *Nucl. Instrum. Methods A* **779**, 69 (2015).
3. A.V. Belitsky, D. Müller, A. Kirchner, Theory of deeply virtual Compton scattering on the nucleon, *Nucl. Phys. B* **629**, 323 (2002).
4. COMPASS Collaboration, Transverse extension of partons in the proton probed in the sea-quark range by measuring the DVCS cross section, *Phys. Lett. B* **793**, 188 (2019).
5. COMPASS Collaboration, Measurement of the cross section for hard exclusive π^0 leptoproduction, *arXiv preprint arXiv:1903.12030* (2019).
6. S. Goloskokov and P. Kroll, Transversity in hard exclusive electroproduction of pseudoscalar mesons, *Eur. Phys. J. A* **47**, 112 (2011).
7. S. Goloskokov and P. Kroll, *private communications* (2016).
8. S. Goloskokov and P. Kroll, Transversity in exclusive vector-meson leptoproduction, *Eur. Phys. J. C* **74**, 2725 (2014).

Access to decoupled information on Generalized Parton Distributions (GPDs) via Double Deeply Virtual Compton Scattering (DDVCS)

S. Zhao

Institut de Physique Nucléaire d'Orsay
CNRS/IN2P3, Université Paris Sud & Paris Saclay
15 rue Georges Clémenceau, 91406 Orsay, France
E-mail: zhao@ipno.in2p3.fr

The Generalized Parton Distributions (GPDs) constitute an appropriate framework for a universal description of the partonic structure of the nucleon. Double Deeply Virtual Compton Scattering (DDVCS) process provides the only experimental way to measure independently the dependences of the GPDs on the average and transferred momentum. This proceeding discusses the feasibility of a DDVCS experiment in the context of JLab 12 GeV, model-predicted pseudo-data, and extraction of the relevant GPDs information based on a fitter algorithm.

Keywords: GPDs; DDVCS; experiment; feasibility.

1. Introduction

The electroproduction of a lepton pair $eN \to eNl\bar{l}$, which is sensitive to the DDVCS amplitude, provides the only experimental framework for a decoupled measurement of GPDs(x, ξ, t) as a function of both the average momentum fraction x and the transferred one ξ.[1–3] For instance, the Compton form factor (CFF) $\mathcal{H}$ associated with the GPD H and accessible in cross section or beam spin asymmetry experiments can be written

$$\mathcal{H}(\xi', \xi, t) = \mathcal{P} \int_{-1}^{1} H(x, \xi, t) \left[\frac{1}{x - \xi'} + \frac{1}{x + \xi'} \right] dx - i\pi H_{+}(\xi', \xi, t) \quad (1)$$

where $\mathcal{P}$ indicates the Cauchy principal value of the integral, and H_+ is the singlet GPD defined as $H_{+}(\xi', \xi, t) = H(\xi', \xi, t) - H(-\xi', \xi, t)$. The real part of the CFF is a complex quantity involving the convolution of parton propagators and the GPD values over x, but the imaginary part accesses the GPD values at $x = \pm\xi'$. The famous DVCS process has its limitation of $\xi' = \xi$. Because of the virtuality of final state photons, DDVCS provides a way to circumvent this limitation, allowing access to decoupled information at $\xi' \neq \xi$.

The DDVCS process is very challenging from the experimental point of view due to the small magnitude of the cross section, and requires high luminosity and full exclusivity of the final state. Taking advantage of the energy upgrade of the CEBAF accelerator, it is proposed to investigate the electroproduction of $\mu^+\mu^-$ di-muon pairs (avoiding complex antisymmetrization issues) and measure the beam spin asymmetry of the exclusive $ep \to e'p'\gamma^* \to e'p'\mu^+\mu^-$ reaction in the hard scattering regime.[4–7]

At sufficiently high virtuality of the initial space-like photon and small enough four-momentum transfer to the nucleon with respect to the photon virtuality ($-t \ll Q^2$), DDVCS can be seen as the absorption of a space-like photon by a parton of the nucleon, followed by the quasi-instantaneous emission of a time-like photon by the same parton, which finally decays into a di-muon pair (left panel of Fig. 1). Q^2 and Q'^2 represent the virtuality of the incoming space-like and outgoing time-like photons, respectively. The scaling variable ξ' and ξ write

$$\xi' = \frac{Q^2 - Q'^2 + t/2}{2Q^2/x_B - Q^2 - Q'^2 + t} \text{ and } \xi = \frac{Q^2 + Q'^2}{2Q^2/x_B - Q^2 - Q'^2 + t}. \quad (2)$$

2. Experiment projections

The experimental observables have been evaluated by a DDVCS event generator developed with the VGG model[8,9] at leading twist. Kinematic cuts to ensure the applicability of the GPD formalism have been applied.[7] We

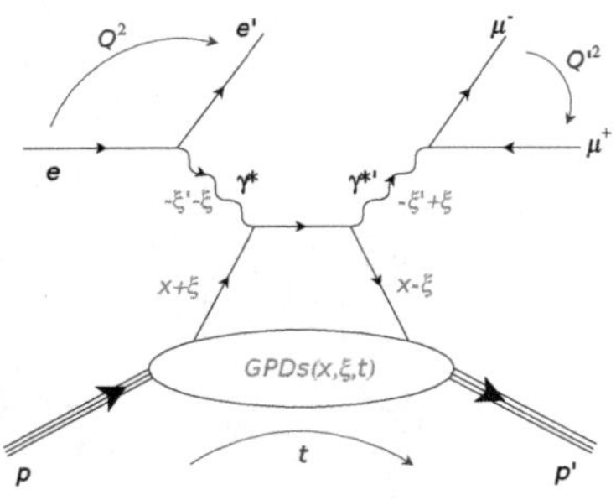

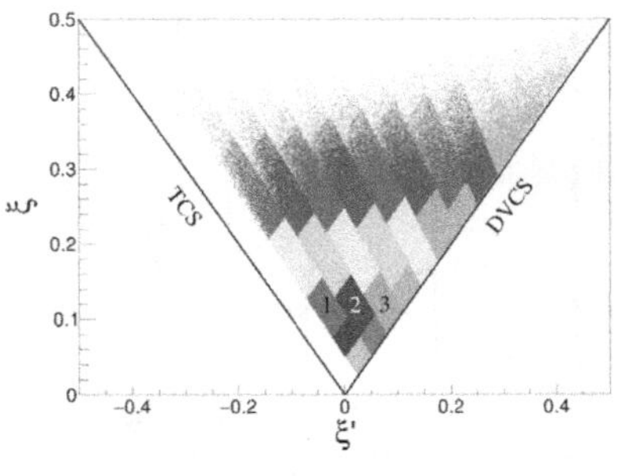

Fig. 1. Left panel: the handbag diagram symbolizing the DDVCS direct term with di-muon final states (there is also a crossed diagram where the final time-like photon is emitted from the initial quark). Right panel: The illustration of the grid of bins in (ξ', ξ) plane at forward limit ($t = 0$), where the solid lines indicate the DVCS ($\xi' = \xi$) and the Time-like Compton Scattering (TCS) limitation ($\xi' = -\xi$).

have adopted a 5-dimensional grid of bins in ξ', ξ, Q^2, t and ϕ covering the whole kinematic phase space of interest. The right panel of Fig. 1 depicts the grid of bins in (ξ', ξ), where three bins indicated by 1, 2 and 3 have approximately the same average $\xi = 0.11$ and different average ξ'. In the following, the experiment projections and the extraction of CFFs are discussed with respect to the three ξ' bins.

The projections have been performed in the ideal situation that all the particles of the final state can be detected with 100% efficiency. The count number has been calculated for a luminosity $\mathscr{L} = 10^{36}\text{cm}^{-2}\cdot\text{s}^{-1}$ considering 50 days running time. In this work, only unpolarized cross section σ_{UU} and beam spin cross section difference $\Delta\sigma_{\text{LU}}$ are discussed. Figure 2 shows them with statistic errors as a function of ϕ for the three ξ' bins, two bins in $-t$ $[0, 0.1, 0.3\ (\text{GeV}/c^2)^2]$ and three bins in Q^2 $[1, 1.5, 2.5, 3.5\ (\text{GeV}/c^2)^2]$. It indicates that it is possible to obtain DDVCS experimental observables with good precision.

3. CFF extraction

The fitting program is inspired from the DVCS one's,[10] corresponding to a quasi-model-independent way to extract CFFs. It consists in taking the eight CFFs as free parameters, and knowing the well-established BH and

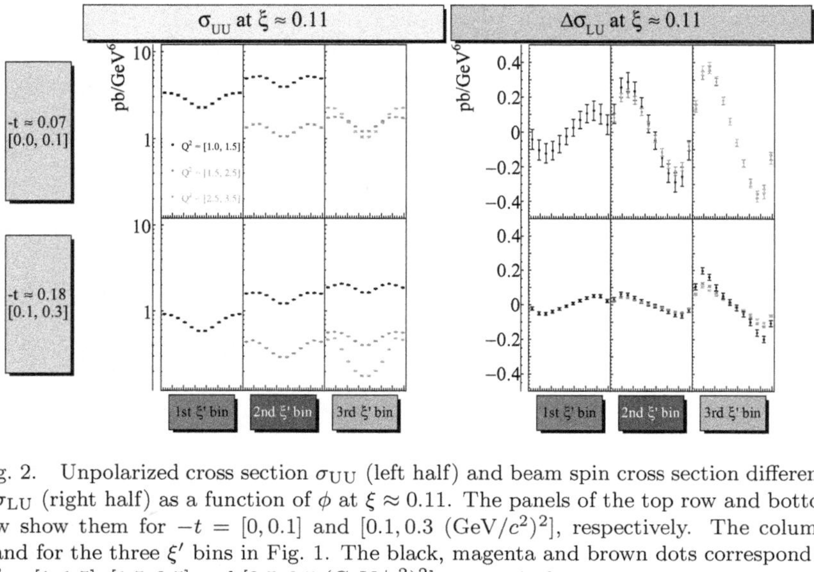

Fig. 2. Unpolarized cross section σ_{UU} (left half) and beam spin cross section difference $\Delta\sigma_{\text{LU}}$ (right half) as a function of ϕ at $\xi \approx 0.11$. The panels of the top row and bottom row show them for $-t = [0, 0.1]$ and $[0.1, 0.3\ (\text{GeV}/c^2)^2]$, respectively. The columns stand for the three ξ' bins in Fig. 1. The black, magenta and brown dots correspond to $Q^2 = [1, 1.5]$, $[1.5, 2.5]$ and $[2.5, 3.5\ (\text{GeV}/c^2)^2]$, respectively.

DDVCS leading-twist amplitudes, to fit, at a fixed kinematics, simultaneously the ϕ-distributions of several experimental observables. If only $\sigma_{\rm UU}$ and $\Delta\sigma_{\rm LU}$ are available, only CFF $\mathcal{H}$ can be well extracted, with $\sigma_{\rm UU}$ being particularly sensitive to the real part and $\Delta\sigma_{\rm LU}$ dominated by the imaginary part. Figure 3 shows the extracted CFF coefficient, defined as the fitted CFF value normalized by the one of the event generator. The dashed line indicates the coefficient being 1, which leads to the fact that the CFF is perfectly recovered. The imaginary part looks obviously well recovered, which means that we are able to extract the imaginary part of $\mathcal{H}$. With respect to the real part, there is a relatively large discrepancy between the fitted and the generated CFF, due to the complex component of $\sigma_{\rm UU}$ from different contributions. In order to perform a better extraction of the real part, the beam charge cross section is required since it involves only the interference contribution, which is linear in the CFFs.[3,7]

Eventually, the proton singlet GPD H_+ can be extrapolated from the imaginary part of CFF $\mathcal{H}$. Figure 4 shows $H_+(\xi', \xi, t)$ as a function of ξ' at fixed ξ for two bins in t and three bins in Q^2. The vertical solid line stands for the location accessed by DVCS and the colored points extracted from the DDVCS pseudo-data give the decoupled singlet GPD value.

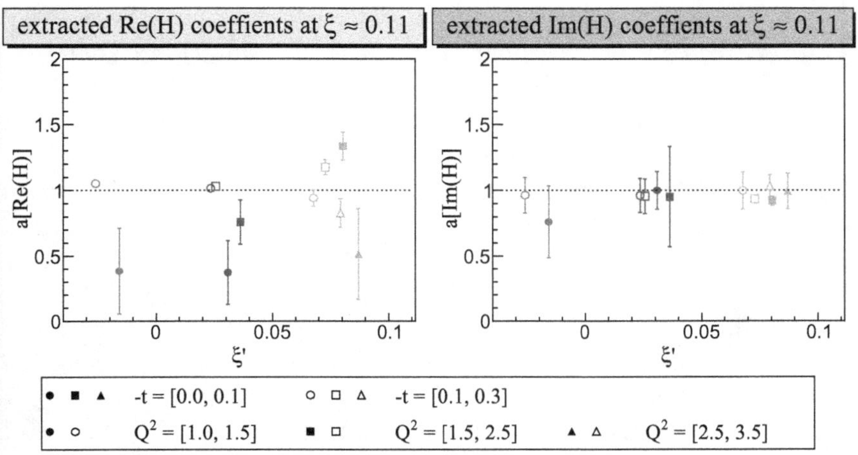

Fig. 3. The extracted CFF coefficients with respect to the observables in Fig. 2. The left panel is the real part of CFF $\mathcal{H}$ while the right one is the imaginary part.

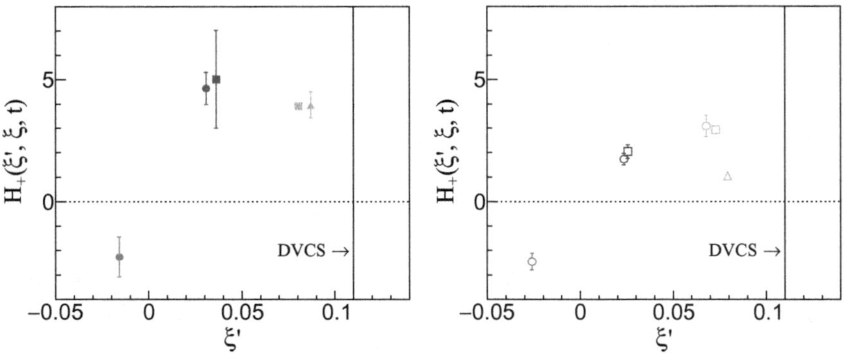

Fig. 4. The proton singlet GPD H_+. The left panel stands for $-t = [0, 0.1\ (\text{GeV}/c^2)^2]$ and the right one for $[0.1, 0.3\ (\text{GeV}/c^2)^2]$. The convention for points is the same as in Fig. 3.

4. Conclusion

Our study shows a high degree of feasibility of a DDVCS experiment at a relative challenging luminosity with exclusive final states completely detected. We also have the capability to access decoupled singlet GPD. The fitting program is still confronted with a severely underconstrained problem and running time difficulty, and further work is still ongoing.

References

1. M. Guidal and M. Vanderhaeghen, *Phys. Rev. Lett.* **90**, 012001 (2003).
2. A. V. Belitsky and D. Müller, *Phys. Rev. Lett.* **90**, 022001 (2003).
3. A. V. Belitsky and D. Müller, *Phys. Rev. D* **68**, 116005 (2003).
4. M. Boer, A. Camsonne, K. Gnanvo, E. Voutier, Z. Zhao, *et al.*, *Jefferson Lab Experiment* **LOI12–15–005** (2015).
5. (CLAS Collaboration) S. Stepanyan, *et al.*, *Jefferson Lab Experiment* **LOI12–16– 004** (2016).
6. I. V. Anikin, *et al.*, *Acta Phys. Pol. B* **49**, 741 (2018).
7. S. Zhao, *23rd Int. Spin Physics Symp. (Ferrara)* PoS(SPIN2018)068 (2019).
8. M. Vanderhaeghen, P. A. M. Guichon and M. Guidal, *Phys. Rev. Lett.* **80**, 5064 (1998).
9. M. Vanderhaeghen, P. A. M. Guichon and M. Guidal, *Phys. Rev. D* **60**, 094017 (1999)
10. M. Guidal, *Eur. Phys. J. A* **37**, 319 (2008).

GPD and TDA measurements based on hard exclusive pion electroproduction with CLAS at JLAB

S. Diehl[1,2,*], K. Joo[2], for the CLAS Collaboration

[1]*2nd Physics Institute, Justus Liebig University Giessen, 35410 Giessen, Germany*
[2]*Department of Physics, University of Connecticut, Storrs, CT 06269, USA*
**E-mail: stefan.diehl@exp2.physik.uni-giessen.de*

Besides a GPD based description, the collinear factorisation theorem of QCD also allows a TDA based description of the hard exclusive pion electroproduction. Both descriptions provide access to different kinematic regions and therefore give us insights into different aspects of the 3 dimensional nucleon structure. To understand the complete reaction mechanism and the validity of the different models, experimental studies over the full kinematic range have to be performed. In this work measurements of the beam-spin asymmetry have been performed over a wide range of kinematics in the deep inelastic regime for the hard exclusive e p → e n π^+ reaction using the CEBAF Large Acceptance Spectrometer (CLAS) and a 5.5 GeV polarized electron beam at Jefferson Lab (JLAB). The ϕ dependence of the BSA as well as the Q^2, x_B and t dependence of the extracted moments will be presented. A clear sign change of the BSA can be observed between pions emitted in forward and backward direction with a smooth transition around 90° in CM.

Keywords: Pion electroproduction; beam spin asymmetry; GPD; TDA; CLAS.

1. Introduction

Studying exclusive pion electroproduction can provide valuable insights into the 3 dimensional structure of the nucleon, since the size of the probe as well as the photon virtuality Q^2 and the momentum transfer to the nucleon t can be varied simultaneously. On the theoretical side, the colinear factorization theorem allows different descriptions of the hard exclusive pion production for different kinematic regimes.

A first description of the process is given by generalized parton distributions (GPD). GPDs are universal structure functions, which describe the hadronic structural information in terms of quark-gluon degrees of freedom [1-2]. They are a good tool to investigate the nature and the origin of the nucleon spin. In the impact parameter space, they can be interpreted as spatial femto-photographs of the nucleon structure in the transverse plane. To ensure factorization, the GPD based description requires large Q^2 and s values and a small t channel contribution. Experimentally this corresponds to a pion going in forward direction.

Besides the GPD based description the collinear factorization theorem also allows a description of the hard exclusive processes by Transition Distribution Amplitudes (TDAs) [3-6]. TDAs are universal structure functions which encode a physical picture, which is very similar to that of GPDs. They probe the partonic correlation between states of different baryonic charge and therefore give us access to non-minimal Fock components of the baryon light cone wave function. The TDA formalism is valid for large Q^2 and s and for a small u channel transfer, which correspond to large -t values and is therefore complementary to the GPD kinematic regime. Graphically this means that the proton takes most of the momentum of the virtual photon and is emitted in the forward direction, while the a pion from the mesonic cloud of the nucleon is emitted with low momenta in the backward direction. Therefore the process can be used to probe the structure of the mesonic cloud inside the nucleon.

A comparison between the GPD and the TDA mechanism for the exclusive pion production is given in Fig. 1.

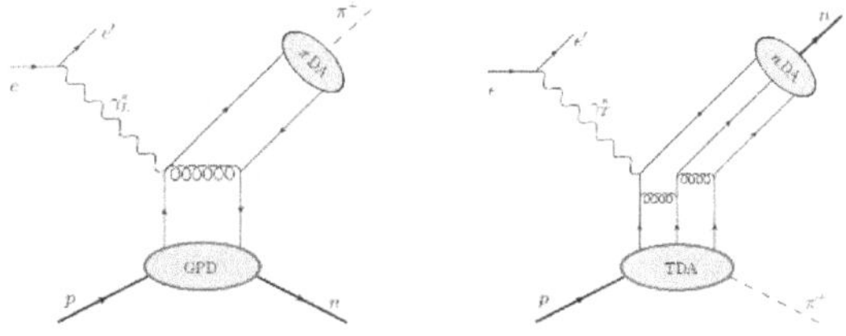

Fig. 1. Schematic diagram for the exclusive e p → e n π^+ reaction for a GPD based description (left) and for a TDA based description (right).

The cross section of the hard exclusive pion production in terms of transverse, longitudinal, transverse-transverse and longitudinal-transverse interference structure functions (T, L, TT, TL, and TL′) is given by:

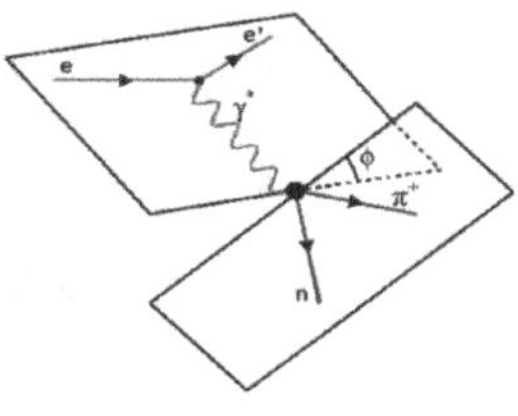

$$\frac{d\sigma}{d\Omega_{e'} d\epsilon_{e'} \Omega_\pi} = \Gamma \frac{d\sigma_\nu}{d\Omega_\pi} \qquad \Gamma = \frac{\alpha_{em}}{2\pi^2} \frac{E_{e'}}{E_0} \frac{W^2}{Q^2} \frac{1}{1-\epsilon}$$

$$\frac{d\sigma_\nu}{d\Omega_\pi} = \frac{d\sigma_T}{d\Omega_\pi} + \epsilon_L \frac{d\sigma_L}{d\Omega_\pi} + \sqrt{2\epsilon_L(1+\epsilon)} \frac{d\sigma_{TL}}{d\Omega_\pi} \cos(\phi) + \epsilon \frac{d\sigma_{TT}}{d\Omega_\pi} \cos(2\phi) + h\sqrt{2\epsilon_L(1-\epsilon)} \frac{d\sigma_{TL'}}{d\Omega_\pi} \sin(\phi)$$

It depends on the virtuality Q^2 and the transverse and longitudinal polarizations of the virtual photon (ε, ε_L) as well as on Bjorken x_B, the squared momentum transfer ***t*** to the proton, the angle ϕ between the leptonic and the hadronic planes and the beam polarization h. Measurements of the cross section in backward kinematics have been performed within the CLAS collaboration in Ref. [7] and show a good agreement with the $1/Q^8$ scaling behavior expected from the TDA mechanism. By substituting the terms in front of the ϕ dependences with the moments $A_{UU}^{\cos(\phi)}$, $A_{UU}^{\cos(2\phi)}$ and $A_{LU}^{\sin(\phi)}$ and introducing the unpolarized cross section $d\sigma_0$, the cross section can be simplified to:

$$d\sigma = d\sigma_0 \cdot (1 + h \cdot A_{LU}^{\sin(\phi)} \sin(\phi) + A_{UU}^{\cos(\phi)} \cos(\phi) + A_{UU}^{\cos(2\phi)} \cos(2\phi) \quad (1)$$

Based on the three moments, the beam-spin asymmetry can be defined as:

$$BSA = \frac{\sigma^+ - \sigma^-}{\sigma^+ + \sigma^-} = \frac{A_{LU}^{\sin(\phi)} \sin(\phi)}{1 + A_{UU}^{\cos(\phi)} \cos(\phi) + A_{UU}^{\cos(2\phi)} \cos(2\phi)} \quad (2)$$

where $\sigma\pm$ is the cross section for the two helicity states with the spin parallel (+) and anti-parallel (-) to the beam direction.

2. Experimental setup and particle identification

The analyzed data was recorded with the CEBAF Large Acceptance Spectrometer (CLAS) in hall B at Jefferson LAB in 2003, using a 5.5 GeV longitudinally polarized electron beam, interacting with an unpolarized hydrogen target. The average beam polarization was 74.9 ± 2.4(stat) %. The CLAS detector is build around a torus magnet, which divides the setup in six sectors. Each sector contains 3 regions of drift-chambers to determine the momentum of charged particles within the torus field, a Cherenkov counter and a time of flight system for the identification of charged particles and an electromagnetic calorimeter to identify electrons and to detect photons.

3. Analysis procedure

As a first step, the deep inelastic scattering region was selected with cuts on $Q^2 > 1$ GeV^2 and $W > 2$ GeV. To separate the forward and backward kinematic region, cuts on the Mandelstam variables t and u, and on the polar angle θ_{CM} of the π^+ in the centre of mass frame were applied. The forward direction is defined as $-t < 1.5$ GeV^2 and $\cos(\theta_{CM}) > 0$ and for the backward direction, cuts on $-u < 2.0$ GeV^2 and $\cos(\theta_{CM}) < 0$ are used.

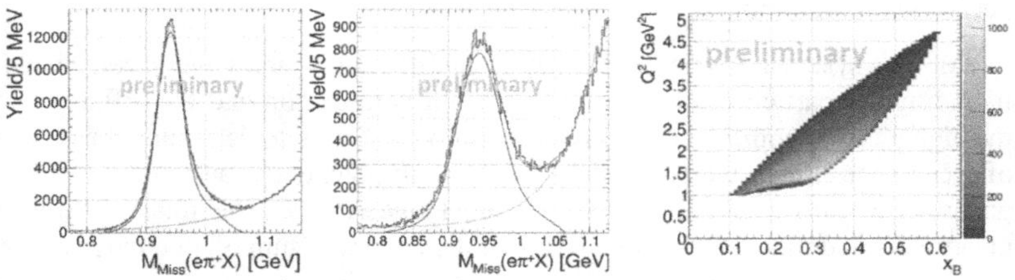

Fig. 3. Missing mass of $e\pi^+X$ in the forward region (left) and in the backward region (central). The black line shows the signal and the green line the fitted background contribution. The right plot shows the kinematic coverage of the exclusive $ep \to en\pi^+$ events in Q^2 and x_B.

Since for the reaction $ep \to en\pi^+$ only electrons and pions can be detected by CLAS, the exclusive events are selected by a cut on the missing neutron mass.

The fitted missing neutron peak and its background are shown in Fig. 3 (left and central) for the forward and backward region. The fit has been performed for each kinematic bin to determine the signal to background ratio. Fig. 3 (right) shows the kinematic coverage of the exclusive events in Q^2 and x_B.

4. Results

Experimentally, the beam spin asymmetry can be calculated as $(1/P_e)\cdot(N_i^+ - N_i^-)/(N_i^+ + N_i^-)$ with the average beam polarization P_e and the count rates $N^{\pm}$ in a specific bin with positive and negative helicity. The beam-spin asymmetry as a function of ϕ for the forward and backward kinematic region, integrated over the other kinematic variables (Q^2 and x_B) is shown in Fig. 4.

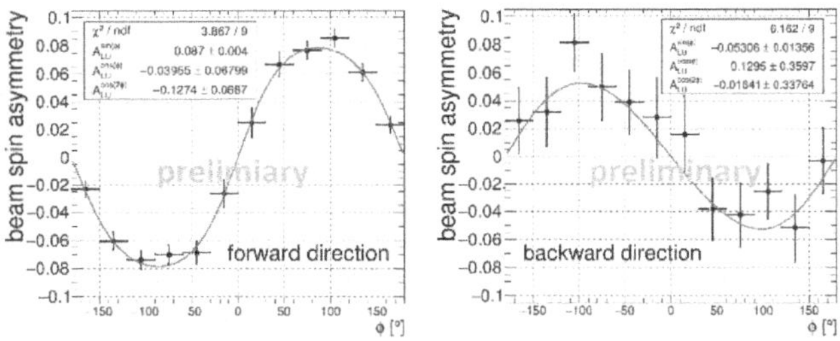

Fig. 4. Beam spin asymmetry in forward (left) and backward direction (right). Both plots are integrated over all Q^2 and x_B bins. The shown result is preliminary.

The ϕ dependence is fitted with Eq. 2. A clear sign change of $A_{LU}^{\sin(\phi)}$ can be observed between the forward and backward kinematic region. To investigate the transition between the forward and backward kinematics more closely, $A_{LU}^{\sin(\phi)}$ has been extracted for different bins in -t, Q^2 and x_B. Based on the signal to background ratio from the fit of the missing mass and based on the asymmetry of the background determined from a missing mass interval on the right side of the missing neutron peak in Fig. 3, a background subtraction has been performed in each bin.

For the systematic uncertainty, thirteen different sources were considered, including the particle identification, the beam polarization and the influence of the higher order moments. A fast MC simulation was used to determine the impact of acceptance effects, which was found to be negligible. Each source of systematic uncertainty stays well below the statistical uncertainty. The total systematic uncertainty is defined as the square root of the quadratic sum of all contributions. Fig. 5 shows the -t dependence of $A_{LU}^{\sin(\phi)}$ including the statis-tical and systematic uncertainty.

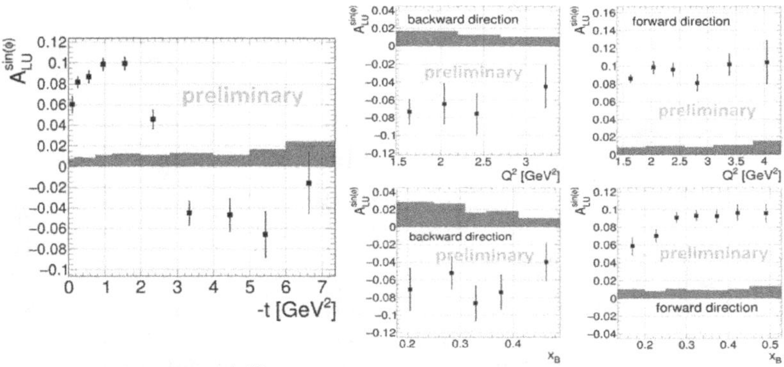

Fig. 5. Dependence of $A_{LU}^{\sin(\phi)}$ on -t (left), Q^2 (upper right) and x_B (lower right) [preliminary]. The data points show the statistical uncertainty, while the shaded area represents the systematic.

A clear transition from positive values up to a magnitude of 0.10 in the forward kinematic region (small -t) to negative values up to a magnitude of -0.06 in the backward kinematic region (large -t) can be observed. As expected the magnitude of $A_{LU}^{\sin(\phi)}$ decreases for very small and large -t values. The smooth transition happens around -t = 3 GeV², which corresponds to a θ_{CM} of 90°. Also the Q^2 and x_B dependence of $A_{LU}^{\sin(\phi)}$ in Fig. 5 clearly show, that $A_{LU}^{\sin(\phi)}$ is negative for all Q^2 and x_B bins in backward direction, while it is positive for all bins in forward direction. Since the forward kinematic region can be described by GPDs, while the backward kinematic region is described by TDAs, this sign change may indicate the transition between the GPD and TDA based formalisms.

Acknowledgments

The authors acknowledge very helpful discussions with B. Pire, K. Semenov-Tian-Shansky and P. Kroll as well as the outstanding efforts of the staff of the Accelerator and the Physics Divisions at JLAB in making this experiment possible. The work is partly supported by DOE grant no: DE-FG02-04ER41309.

References

[1] D. Mueller et al., Fortsch. Phys. 42, 101 (1994).
[2] X. Ji, Phys. Rev. Lett. 78, 610 (1997).
[3] L.L. Frankfurt et al., Phys. Rev. D 60, 014010 (1999).
[4] B. Pire and L. Szymanowski, Phys. Lett. B 622, 83 (2005).
[5] B. Pire and L. Szymanowski, Phys. Rev. D 71, 111501 (2005).
[6] J. P. Lansberg, B. Pire, L. Szymanowski, Phys. Rev. D 75, 074004 (2007).
[7] K. Park et al., CLAS Collaboration, Phys. Lett. B 780 340345, (2018).

Sea quark Sivers distribution functions

Fangcheng He[1,2*] and Ping Wang[1,3†]

[1] *Institute of High Energy Physics, Chinese Academy of Science, Beijing 100049, China*

[2] *School of Physics, University of Chinese Academy of Sciences, Beijing, China*

[3] *Theoretical Physics Center for Science Facilities, Chinese Academy of Sciences, Beijing 100049, China*

**hefc@ihep.ac.cn,*

†pwang4@ihep.ac.cn,

We investigate the Sivers distribution functions of sea quark in proton with chiral Lagrangian. The reasonable numerical results for the first momentum $x\Delta^N f_{\bar{q}}^{(1)}(x)$ are obtained without any fine tuning of the parameters. Our results are also comparable with the recent phenomenological extractions.

Keywords: Sivers function; chiral lagrangian; sea quark.

1. Introduction

The time-reversal odd distribution, Sivers distribution, is one of the eight transverse momentum dependent (TMD) parton distributions at leading twist. It describes the asymmetric distribution of unpolarized quarks in a transversely polarized parton parent hadron. It is crucial to explain the single-spin asymmetries (SSA) in semi-inclusive deep inelastic scattering (SIDIS), which have been observed decades ago.[1–3] For the Sivers distribution functions, most investigations are in the phenomenological models and focus only on the valence quark Sivers distribution functions. The Sivers distributions of sea quark are still lacking in most calculations. Sea quark Sivers functions are important to explaining azimuthal asymmetries for $\pi^{\pm}$ and $K^{\pm}$ production off a proton target in SIDIS and the asymmetrical cross sections for a vector boson in the polarized Drell-Yan process.[4,5] It is also helpful to testing the sign change of Sivers functions between SIDIS and Drell-Yan process. In our recent work,[6] we proposed a mechanism for the Sivers distribution function in a proton in the framework of chiral Lagrangian. The numerical results are obtained without any fine tuning of the parameters.

2. Sivers distribution function

According to the Trento convention, the unpolarized and Sivers distributions $f_1(x,\vec{k}_\perp)$ and $f_{1T}(x,\vec{k}_\perp)$ are defined as[7]

$$f_1^q(x,\vec{k}_\perp) + \frac{\epsilon^{ji}k_\perp^i S_\perp^j}{m_p} f_{1T}^q(x,\vec{k}_\perp) = \frac{1}{2}\int \frac{d\xi^- d^2\vec{\xi}_\perp}{(2\pi)^3} \times e^{-ixP^+\xi^- + i\vec{k}_\perp\cdot\vec{\xi}_\perp}\langle P,\vec{S}_\perp \mid \mathcal{O}^q \mid P,\vec{S}_\perp\rangle, \quad (1)$$

Where $\vec{S}_\perp$ is the transverse spin of a proton. The gauge invariant bilocal operator $\mathcal{O}_q$ is defined as

$$\mathcal{O}^q = \bar{q}(\xi^-,\vec{\xi}_\perp)\mathcal{L}^\dagger_{\xi_\perp}(\infty,\xi^-)\gamma^+\mathcal{L}_0(\infty,0)q(0,0), \quad (2)$$

Where $\mathcal{L}$ is the path-ordered light-cone color gauge link. In our approach, we use $SU(2)_V$ invariant bilocal pion operator to calculate the Sivers distribution of pion in a proton.[6]

$$\frac{\epsilon^{ji}k^i_{\perp\pi}S^j_\perp}{m_p} f_{1T}^{\pi/p}(z,\vec{k}_{\perp\pi}) = \frac{1}{2}\int \frac{dy^- d^2\vec{y}_\perp}{(2\pi)^3} \times e^{-i(zP^+y^- - \vec{y}_\perp\cdot\vec{k}_{\perp\pi})}\langle P,\vec{S}_\perp \mid \mathcal{O}^\pi_{Sivers} \mid P,\vec{S}_\perp\rangle. (3)$$

For π^+, the operator $\mathcal{O}^\pi_{Sivers}$ is written as

$$\mathcal{O}^{\pi^+}_{Sivers} = \left[\pi^-(y^-,\vec{y}_\perp)\partial^+\pi^+(0) - \partial^+\pi^-(y^-,\vec{y}_\perp)\pi^+(0)\right] \times \sqrt{2}g\{\int_0^\infty dz^-\rho^{0+}(z^-,0) + \int_\infty^{y^-} dz^-\rho^{0+}(z^-,\vec{y}_\perp)\}. \quad (4)$$

The matrix element in Eq. (3) can be calculated with chiral effective Lagrangian. The Sivers distribution function of sea quark $\bar{q}$ in proton can be obtained by the convolution form

$$k^i_\perp f_{1T}^{\bar{q}/p}(x,\vec{k}_\perp) = \int d^2\vec{k}_{\perp\pi}k^i_{\perp\pi} \int_x^1 \frac{dz}{z} f_{1v}^{\bar{q}/\pi}(\frac{x}{z},\vec{k}_\perp - \frac{x}{z}\vec{k}_{\perp\pi}) f_{1T}^{\pi/p}(z,\vec{k}_{\perp\pi}), \quad (5)$$

where $f_{1v}^{\bar{q}/\pi}(\frac{x}{z},\vec{k}_\perp - \frac{x}{z}\vec{k}_{\perp\pi})$ is the unpolarized valance quark TMD distribution in pion. The first moment of the Sivers distribution function is defined as[8]

$$\Delta^N f_{\bar{q}}^{(1)}(x) = \int d^2\vec{k}_\perp \frac{-k_\perp^2}{2m_p^2} f_{1T}^{\bar{q}/p}(x,\vec{k}_\perp) = \frac{1}{2m_p^2}\int_1^x d(\frac{x}{z}) f_{1v}^{\bar{q}/\pi}(\frac{x}{z}) \int d^2\vec{k}_{\perp\pi}\vec{k}^2_{\perp\pi} f_{1T}^{\pi/p}(z,\vec{k}_{\perp\pi}), \quad (6)$$

where $f_{1v}^{\bar{q}/\pi}(x)$ is valance quark $\bar{q}$ distribution in π and it can be obtained from the recent fit at $Q = 0.63$ GeV.[9]

3. Numerical results

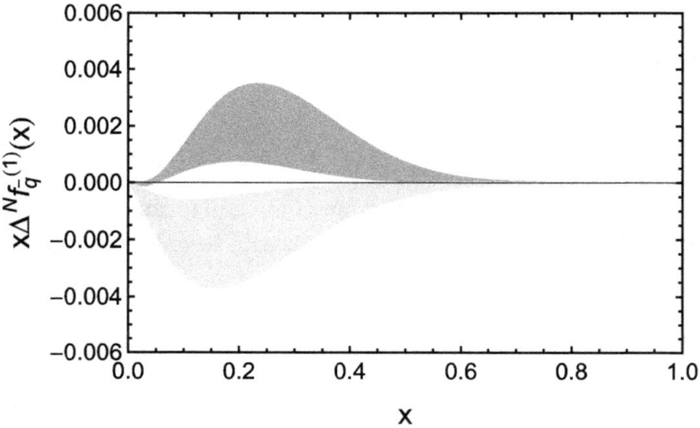

Fig. 1. The first momentum of sea quark Sivers distribution functions versus x at $Q = 0.63$ GeV. The green and yellow bands are the results of $x\Delta^N f_{\bar{d}}^{(1)}(x)$ and $x\Delta^N f_{\bar{u}}^{(1)}(x)$ with $0.8\ \text{GeV} \leq \Lambda_\pi \leq 1.2$ GeV and $1.6\ \text{GeV} \leq \Lambda_\rho \leq 2.0$ GeV.

The dipole regulator $\tilde{F}_j(k)$ $(j = \pi, \rho)$ is applied to deal with the ultraviolet divergence in our numerical calculation

$$\tilde{F}_j(k) = \left(\frac{M_j^2 - \Lambda_j^2}{k^2 - \Lambda_j^2}\right)^2 . \tag{7}$$

In Fig. 1, we plot the first moment of the Sivers distribution functions of $\bar{d}$ and $\bar{u}$. The green and yellow bands represent $x\Delta^N f_{\bar{d}}^{(1)}(x)$ and $x\Delta^N f_{\bar{u}}^{(1)}(x)$, respectively. Their absolute values are pretty small. The first moment of $\bar{d}$ in proton is positive. The maximum value of $x\Delta^N f_{\bar{d}}^{(1)}(x)$ is $0.0008 - 0.0035$ at x around 0.2. It then decreases with increasing x, and when x is larger than 0.6, $x\Delta^N f_{\bar{d}}^{(1)}(x)$ will tend to be zero. For $\bar{u}$ in proton, $x\Delta^N f_{\bar{u}}^{(1)}(x)$ is always negative. The maximum absolute value is about 0.0007–0.0037 at x around 0.15. Similar as for $x\Delta^N f_{\bar{d}}^{(1)}(x)$, when x is larger than 0.6, $x\Delta^N f_{\bar{u}}^{(1)}(x)$ will also approach to zero. Our result is consistent with the prediction in the large N_C limit where the absolute values of the

Sivers distribution functions of $\bar{d}$ and $\bar{u}$ are the same while their signs are opposite.[10] The phenomenological extractions at $Q = 1$ GeV also show that the central value of $x\Delta^N f_{\bar{d}}^{(1)}(x)$ is negative, while $x\Delta^N f_{\bar{u}}^{(1)}(x)$ is positive.[11]

4. Summary

We proposed a mechanism for the study of Sivers distribution function with chiral Lagrangian. The vector meson is introduced for the $SU(2)_V$ hidden symmetry. We construct the bilocal π operators $\mathcal{O}^{\pi}_{Sivers}$ which has the same symmetry as the Lagrangian. The hadronic matrix element of $\mathcal{O}^{\pi}_{Sivers}$ can be calculated to get the Sivers distribution of a pion in proton. By convoluting it with the valence quark distribution in pion, the Sivers distribution functions of $\bar{u}$ and $\bar{d}$ in proton are obtained. Numerical results show that the absolute values of $x\Delta^N f_{\bar{u}}^{(1)}(x)$ and $x\Delta^N f_{\bar{d}}^{(1)}(x)$ are close to each other, while their signs are opposite. Without fine tuning the parameters, our results are consistent with the prediction in the large N_C limit. It is also comparable with the recent experimental fit.

Acknowledgments

This work is supported by the National Natural Sciences Foundations of China under the grant No. 11475186, the Sino-German CRC 110 "Symmetries and the Emergence of Structure in QCD" project by NSFC under the grant No.11621131001, and the Key Research Program of Frontier Sciences, CAS under grant No. Y7292610K1.

References

1. D. W. Sivers, Single Spin Production Asymmetries from the Hard Scattering of Point-Like Constituents, *Phys. Rev.* **D41**, 83 (1990).
2. A. Bravar, Hadron azimuthal distributions and transverse spin asymmetries in DIS of leptons off transversely polarized targets from SMC, *Nucl. Phys. Proc. Suppl.* **79**, 520 (1999).
3. A. Airapetian *et al.*, Single spin azimuthal asymmetries in electroproduction of neutral pions in semiinclusive deep inelastic scattering, *Phys. Rev.* **D64**, 097101 (2001).
4. L. Adamczyk *et al.*, Measurement of the transverse single-spin asymmetry in $p^{\uparrow} + p \to W^{\pm}/Z^0$ at RHIC, *Phys. Rev. Lett.* **116**, 132301 (2016).

5. J. Huang, Z.-B. Kang, I. Vitev and H. Xing, Spin asymmetries for vector boson production in polarized p+p collisions, *Phys. Rev.* **D93**, 014036 (2016).
6. F. He and P. Wang, Sivers distribution functions of sea quark in proton with chiral Lagrangian, *Phys. Rev.* **D100**, 074032 (2019).
7. A. Bacchetta, U. D'Alesio, M. Diehl and C. A. Miller, Single-spin asymmetries: The Trento conventions, *Phys. Rev.* **D70**, 117504 (2004).
8. M. Anselmino, M. Boglione, U. D'Alesio, F. Murgia and A. Prokudin, Study of the sign change of the Sivers function from STAR Collaboration W/Z production data, *JHEP* **04**, 046 (2017).
9. M. Aicher, A. Schafer and W. Vogelsang, Soft-gluon resummation and the valence parton distribution function of the pion, *Phys. Rev. Lett.* **105**, 252003 (2010).
10. P. V. Pobylitsa, Transverse momentum dependent parton distributions in large N(c) QCD (2003).
11. A. Bacchetta and M. Radici, Constraining quark angular momentum through semi-inclusive measurements, *Phys. Rev. Lett.* **107**, 212001 (2011).

Three-dimensional imaging of proton in basis light-front quantization

Siqi Xu[1,2*], Chandan Mondal[1,2], Jiangshan Lan[1,2], Xingbo Zhao[1,2], James P. Vary[3]
(BLFQ Collaboration)
[1] *Institute of Modern Physics, Chinese Academy of Science, Lanzhou 730000, China*
[2] *School of Nuclear Science and Technology, University of Chinese Academy of Sciences, Beijing 100049, China*
[3] *Department of Physics and Astronomy, Iowa State University, Ames, Iowa 50011, USA*
**E-mail: xsq234@impcas.ac.cn*

We employ an effective Hamiltonian that includes the transverse and longitudinal confinement and the one-gluon exchange interaction with fixed coupling constant. By solving the eigenvalue equation in basis light-front quantization (BLFQ), we generate the light-front wavefunctions (LFWFs) for the nucleon in the valence quark Fock space. Fitting the model parameters, we obtain high quality descriptions of electromagnetic form factors and radius for proton while the results deviate somewhat from experimental data for neutron.

Keywords: BLFQ; LFWF; FFs; PDFs; GPDs.

1. Introduction

Basis light-front quantization (BLFQ) is a nonperturbative approach which is developed for solving bound state problems in quantum field theories.[1–13] This approach has been successfully applied to QED[3,4] and QCD[5–13] systems. In our work, we apply the BLFQ approach to the nucleon and study the electromagnetic form factors. As a Hamiltonian formalism, we adopt a light-front effective Hamiltonian, which includes the holographic QCD confinement potential supplemented by longitudinal confinement[5,6,14] along with the one-gluon exchange interaction with a fixed coupling constant. The light-front wave functions (LFWFs) are obtained by diagonalizing the effective Hamiltonian and used to calculate the electromagnetic form factors.

Electromagnetic form factors are crucial for probing the structure of the nucleon. In the light-front formalism, the Dirac and Pauli form factors, $F_1(Q^2)$ and $F_2(Q^2)$, are defined with the longitudinal vector current

(J^+)[15,16]

$$\langle P+q,\uparrow|\frac{J^+(0)}{2P^+}|P,\uparrow\rangle = F_1(Q^2), \tag{1}$$

$$\langle P+q,\uparrow|\frac{J^+(0)}{2P^+}|P,\downarrow\rangle = -(q^1-iq^2)\frac{F_2(Q^2)}{2M}, \tag{2}$$

where $Q^2 = -q^2 = q_\perp^2$ is the square of the momentum transfer, and M is the nucleon mass. The ket $|P,S_z\rangle$ represents the physical state that can be expanded in terms of the wave functions,[14]

$$|P,S_z\rangle = \int\prod_{i=1}^{3}\frac{dx_i d^2k_{i\perp}}{\sqrt{x_i}16\pi^3}16\pi^3\delta\left(1-\sum_{i=1}^{3}x_i\right)\delta^2\left(\sum_{i=1}^{3}k_{i\perp}\right)$$
$$\times\Psi^\Lambda(x_i,k_{i\perp},\lambda_i)\,|x_iP_i^+,k_{i\perp}+x_iP_\perp,\lambda_i\rangle\,. \tag{3}$$

Here, the S_z and λ_i are helicities of the nucleon and quarks respectively. The $x_i = \frac{k_i^+}{P^+}$ is the longitudinal momentum fraction of quarks. Thus, the flavor form factors can be written as the overlap of light-front wave functions. The nucleon Sachs form factors are written in the terms of Dirac and Pauli form factors,

$$G_E^i(Q^2) = F_1^i(Q^2) - \frac{Q^2}{4*M_i^2}F_2^i(Q^2), \quad G_M^i(Q^2) = F_1^i(Q^2) + F_2^i(Q^2). \tag{4}$$

The i = P or N represents the proton or neutron, and $F_{1/2}^i = \sum_f e_f F_{1/2}^{f/i}$ is the Dirac (Pauli) form factors of the nucleon.[17] And the electromagnetic radii of the nucleon can be obtained from

$$\langle r_E^2\rangle^i = -6\frac{dG_E^i(Q^2)}{dQ^2}\bigg|_{Q^2=0}, \qquad \langle r_M^2\rangle^i = -\frac{6}{G_M^i(0)}\frac{dG_M^i(Q^2)}{dQ^2}\bigg|_{Q^2=0}. \tag{5}$$

2. Hamiltonian Formalism

BLFQ solves the eigenvalue equation of the light-front Hamiltonian $P^-|\beta\rangle = P_\beta^-|\beta\rangle$, which leads to the eigenvalue P_β^- and the associated eigenvectors of the bound state. In our work, we only consider the lowest Fock-sector for the expansion of the nucleon, and employ an effective Hamiltonian P_{eff}^- which is given by

$$P_{\text{eff}}^- = \sum_i\frac{\mathrm{m_i^2}+\mathrm{p_{i\perp}^2}}{x_i} + \frac{1}{2}\sum_{i,j}\left(\kappa_T^4 x_i x_j r_{ij\perp}^2 + \frac{\kappa_L^4}{(\mathrm{m_i}+\mathrm{m_j})^2}\partial_{x_i}(x_i x_j\partial_{x_j})\right)$$
$$+\frac{1}{2}\sum_{i,j}\frac{C_F 4\pi\alpha_s}{Q^2}\bar{u}_{s_i'}(k_i')\gamma^\mu u_{s_i}(k_i)\bar{u}_{s_j'}(k_j')\gamma_\mu u_{s_j}(k_j), \tag{6}$$

where the $m_{i/j}$ is the constituent mass of quarks and the $i, j = 1, 2, 3$ label the Fock particles. For Each single-particle basis state, we employ the discrete plane-wave basis (k) in the longitudinal direction and 2D harmonic oscillator (2DHO) basis (n and m) in the transverse direction. Besides, a single quantum number (λ) presents the helicity degree of freedom.

For the nucleon, proton (or neutron) is the lowest eigenstate, denoted by $|P^{\Lambda}\rangle$, where the Λ indicates helicity of the nucleon. In momentum space, the LFWFs are written as

$$\Psi^{\Lambda}(x_i, k_{i\perp}, \lambda_i) = \sum_{\substack{n_1,m_1,n_2\\ m_2,n_3,m_3}} \Big(\psi^{\Lambda}(k_i, n_i, m_i, \lambda_i)$$

$$\times \prod_i \frac{\sqrt{2}}{b(2\pi)^{\frac{3}{2}}} \sqrt{\frac{n!}{(n+|m|)!}} e^{-p_\perp^2/(2b^2)} \left(\frac{|p_\perp|}{b}\right)^{|m|} L_n^{|m|}\left(\frac{p_\perp^2}{b^2}\right) e^{im\theta}\Big). \quad (7)$$

Here, b is an HO basis parameter with the dimension of mass, and $L_n^{|m|}(\frac{p_\perp^2}{b^2})$ is the generalized Laguerre polynomial.

3. Numerical Results

In this paper, we set the model parameters $m_{q/\mathrm{OGE}} = 0.2$ GeV, $m_{q/k} = 0.3$ GeV, $\kappa_T = 0.284$ GeV, $\kappa_L = 0.373$ GeV and $\alpha_s = 1.0 \sim 1.2$. In Fig 1(a), the Sachs form factors of the proton show an agreement with the experimental data, Except for G_M^{P} in the low Q^2 region. At $Q^2 = 0$, $G_M(0)$ gives the anomalous magnetic moments. Our calculations show that the anomalous magnetic moments of the proton ($G_M^{\mathrm{P}}(0) = 2.443 \pm 0.027$) is somewhat different with the experimental measurements ($G_M^{\mathrm{P}}(0) = 2.79$). In Fig 1(b), we show the Sachs form factors of the neutron and compare them with the experimental data revealing a significant difference. Especially, at $Q^2 = 0$, the $G_M^{\mathrm{N}}(0) = -1.405 \pm 0.026$ is disagrees with the experimental data ($G_M^{\mathrm{N}}(0) = -1.91$).

Table 1. Electromagnetic radii of the nucleon. Our results are compared with the experimental data.[17]

	$\langle r_E^P \rangle$ /(fm)	$\langle r_M^P \rangle$ /(fm)	$\langle r_E^N \rangle^2$ /(fm^2)	$\langle r_M^N \rangle$ /(fm)
BLFQ	0.85 ± 0.05	0.88 ± 0.03	-0.09 ± 0.17	0.90 ± 0.03
Exp. Data	0.833 ± 0.010	0.777 ± 0.016	-0.1161 ± 0.0022	$0.862^{+0.009}_{-0.008}$

We also calculate the electromagnetic radii of the nucleons, which we

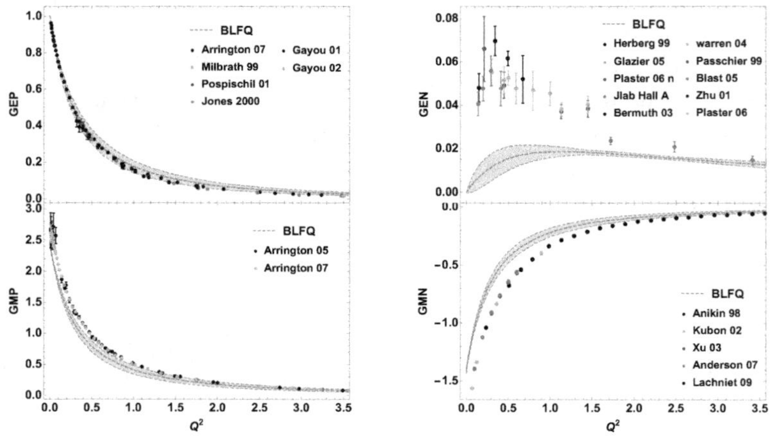

(a) (b)

Fig. 1. The Sachs form factors for the proton (a) and neutron (b). Nucleon Sach's FFs $G_E^{P/N}(Q^2)$ (upper panel) and $G_M^{P/N}(Q^2)$ (lower panel) are functions of Q^2. The bands are BLFQ results reflecting our α_s uncertainty of 10%. The experimental data are taken from Ref. 18.

show in Table 1. The BLFQ results are in a good agreement with the experimental data.[17]

4. Conclusion

In our work, we produce the light-front wave functions by solving the eigenvalue equation of light-front Hamiltonian, and evaluate the electromagnetic form factors of the nucleon. We observe the proton form factors are in a reasonable agreement with the experimental data. The neutron form factors show a significant issue in the low Q region. We also compare the electromagnetic radii of the nucleon with the experimental data.

Acknowledgment

We thank Henry Lamm, Wei Zhu for many useful discussions. CM is supported by the National Natural Science Foundation of China (NSFC) under the Grant No. 11850410436. XZ is supported by new faculty startup funding by the Institute of Modern Physics, Chinese Academy of Sciences and by Key Research Program of Frontier Sciences, CAS, Grant No ZDBS-LY-7020. JPV is supported by the Department of Energy under Grants

No. DE-FG02-87ER40371, and No. DE-SC0018223 (SciDAC4/NUCLEI). A portion of the computational resources were provided by the National Energy Research Scientific Computing Center (NERSC), which is supported by the Office of Science of the U.S. Department of Energy under Contract No.DE-AC02-05CH11231.

References

1. J. P. Vary *et al.*, Phys. Rev. C **81**, 035205 (2010).
2. H. Honkanen *et al.*, Phys. Rev. Lett. **106**, 061603 (2011).
3. P. Wiecki, Y. Li, X. Zhao, P. Maris and J. P. Vary, Phys. Rev. D **91**, no. 10, 105009 (2015).
4. D. Chakrabarti, X. Zhao, H. Honkanen, R. Manohar, P. Maris and J. P. Vary, Phys. Rev. D **89**, no. 11, 116004 (2014).
5. Y. Li, P. Maris and J. P. Vary, Phys. Rev. D **96**, no. 1, 016022 (2017).
6. Y. Li, P. Maris, X. Zhao and J. P. Vary, Phys. Lett. B **758**, 118 (2016).
7. J. Lan, C. Mondal, S. Jia, X. Zhao and J. P. Vary, Phys. Rev. Lett. **122**, no. 17, 172001 (2019).
8. S. Tang, Y. Li, P. Maris and J. P. Vary, Phys. Rev. D **98**, no. 11, 114038 (2018).
9. C. Mondal, S. Xu, J. Lan, X. Zhao, Y. Li, D. Chakrabarti and J. P. Vary, arXiv:1911.10913 [hep-ph].
10. W. Du, Y. Li, X. Zhao, G. A. Miller and J. P. Vary, arXiv:1911.10762 [nucl-th].
11. L. Adhikari, Y. Li, X. Zhao, P. Maris, J. P. Vary and A. Abd El-Hady, Phys. Rev. C **93**, no. 5, 055202 (2016).
12. L. Adhikari, Y. Li, M. Li and J. P. Vary, Phys. Rev. C **99**, no. 3, 035208 (2019).
13. M. Li, Y. Li, P. Maris and J. P. Vary, Phys. Rev. D **100**, no. 3, 036006 (2019).
14. S. J. Brodsky, G. F. de Teramond, H. G. Dosch and J. Erlich, Phys. Rept. **584**, 1 (2015).
15. S. J. Brodsky and S. D. Drell, Phys. Rev. D **22**, 2236 (1980).
16. G. D. Cates, C. W. de Jager, S. Riordan and B. Wojtsekhowski, Phys. Rev. Lett. **106**, 252003 (2011).
17. J. Beringer *et al.* [Particle Data Group], Phys. Rev. D **86**, 010001 (2012).
18. D. Chakrabarti and C. Mondal, Eur. Phys. J. C **73**, 2671 (2013).

Central exclusive meson production in proton–proton collisions in ALICE at the LHC

R. Schicker
(for the ALICE Collaboration)
Physikalisches Institut
University Heidelberg
69120 Heidelberg, Germany
E-mail: schicker@physi.uni-heidelberg.de

Central exclusive production at hadron colliders is characterised by the hadronic state produced at or close to midrapidity, and by the two forward scattered protons, or remnants thereof. No particles are produced between the midrapidity system and the forward going beam particles, and such events can hence be identified experimentally by a double-gap topology. At LHC energies, central exclusive production in proton–proton collisions is dominated by pomeron–pomeron fusion. The models to describe such reactions are reviewed, and the ongoing efforts in the ALICE Collaboration to analyse double-gap events taken in Run 2 at the LHC are presented.

Keywords: Central production; double-gap topology; pomeron; diffraction.

1. Diffractive event topologies

In the Regge approach, the hadronic interaction between strongly interacting particles is due to the exchange of trajectories. Such trajectories $\alpha(t)$ parameterise the almost linear dependence between the spin and the mass squared ($t = m^2$) of a particle and its higher spin excitations. A trajectory is characterised by two parameters, the intercept $1+\varepsilon$ and the slope α', and is expressed as $\alpha(t) = 1+\varepsilon+\alpha' t$. In Regge phenomenology, the contribution of a trajectory to the energy behaviour of the total cross section scales as $\sigma_{tot} \propto s^{\varepsilon}$. For meson trajectories, such as the ρ, ω, a, f, the value of $\varepsilon \sim -0.5$ and the corresponding contributions decrease as function of energy.[1] Such a behaviour is seen in the elastic and total hadron-hadron cross section which decrease from threshold up to a centre-of-mass energy $\sqrt{s} \sim 20$ GeV. At higher energies, however, elastic and total hadron-hadron cross sections increase.[2] Within the Regge approach, this increase is attributed to the existence of an additional exchange mechanism in both elastic and inelastic channels, which is refered to as the pomeron trajectory. The characterisation of the bound states underlying the pomeron trajectory, and their possible experimental verification, is a considerable challenge from the theoretical as well as the experimental perspective.

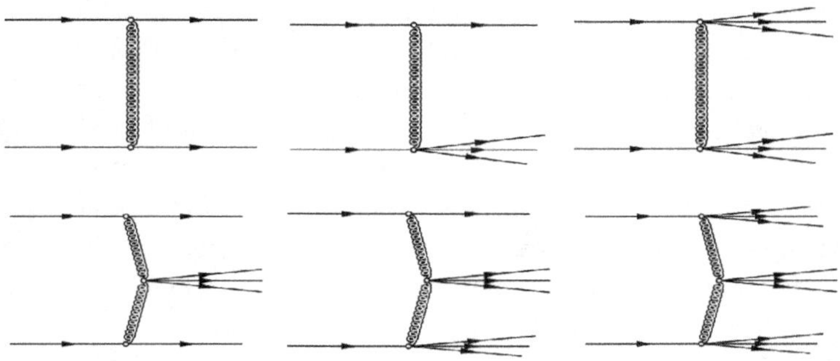

Fig. 1. Top row: Single pomeron elastic scattering (left), and single pomeron inelastic scattering (middle and right). Bottom row: Central exclusive production (left), central production with proton excitation (middle and right).

In the top row of Fig. 1, the diagrams of strong elastic and inelastic scattering due to single pomeron exchange are shown. The corresponding cross sections have been measured at LHC energies by the TOTEM and the ALICE Collaboration, respectively.[3,4] In the bottom row of Fig. 1, the double-pomeron exchange channel of exclusive production is shown on the left, and central production with proton excitation in the middle and on the right. The universality of pomeron exchange can be tested in these central production channels at high energies, where Regge meson contributions are negligible as compared to the pomeron exchange.

2. The spin structure of the soft pomeron

It is generally agreed that the pomeron exchange carries the vacuum internal quantum numbers of charge $Q = 0$, colour charge $Q_C = 0$, isospin $I = 0$ and charge conjugation $C = 1$. Under intense debate is, however, the spin structure of the pomeron exchange in the nonperturbative regime of QCD, the soft pomeron exchange.[5] In contrast to the vacuum expectation value of spin zero, a vector pomeron exchange has been successfully used to describe elastic proton−proton and proton−antiproton scattering data, as well as production of high-p_T jets and Drell-Yan pairs.[6,7] In this vector approach, the single pomeron exchange amplitudes of proton−proton and proton−antiproton elastic scattering carry, however, a relative minus sign. By use of the optical theorem, this minus sign propagates into an opposite sign between the proton−proton and proton−antiproton cross section.

Alternatively, a description of the soft pomeron as an effective rank-2 symmetric tensor has been proposed.[8] Here, the meson exchanges of positive and negative charge conjugation are treated as effective tensor and vector exchanges, respectively, with many effective couplings to hadrons derived from experimental data. The single spin asymmetry data of the STAR Collaboration have been analysed with the tensor pomeron model.[9,10] The results of this analysis disfavour the scalar pomeron, and are in very good agreement with the tensor pomeron. In the analysis of central production of scalar and pseudo-scalar mesons within the vector and tensor pomeron approach, the azimuthal opening angle between the two final-state protons was found to be a sensitive probe for the two approaches.[8]

3. DAMA model of central production

The double-differential cross section $d\sigma/dMdp_T$ for central exclusive meson production shown in the bottom left of Fig. 1 can be derived in a model based on a Dual Amplitude with Mandelstam Analyticity (DAMA).[11] For central production, the direct-channel pole decomposition of the DAMA amplitude is relevant.[12] In this approach, the pomeron flux in the proton is convoluted with the pomeron-pomeron-meson cross section derived by the optical theorem from the DAMA amplitude.[13] The DAMA amplitude is based on nonlinear, complex trajectories. The imaginary part of the trajectory is related to the decay width of the bound states of the trajectory, and is connected to the almost linear real part by a dispersion relation.

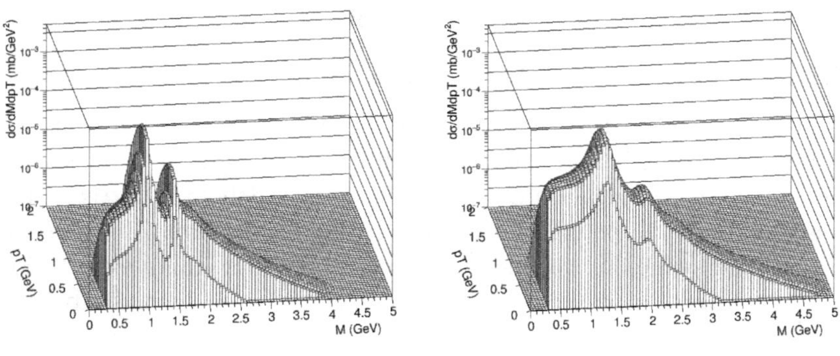

Fig. 2. Central exclusive cross section for resonances on the $f_0(980)$ trajectory (left), and on the $f_2(1270)$ trajectory (right) (Figs. from Ref. 13).

In Fig. 2 the cross section $d\sigma/dMdp_T$ for the resonances on the trajectory of the $f_0(980)$ is shown on the left, and the $f_2(1270)$ on the right.

4. Central exclusive production in ALICE

ALICE consists of a central barrel in the pseudorapidity range $|\eta| < 0.9$, and a forward muon spectrometer.[14] An inner tracking system (ITS), composed of silicon pixel, drift and strip detectors, measures the particle tracks in the innermost 6 layers of the central barrel. A cylindrical time projection chamber with a sensitive volume from r_{min}=88 cm to r_{max}=250 cm records track ionisation clusters. The specific Bethe-Bloch ionisation loss is used for particle identification, and the cluster space information is used together with the ITS space points to reconstruct the track momentum. Additional detectors outside of the central barrel are used for trigger purposes, and for classifying the event according to multiplicity. Each of the V0 and AD detectors consists of a pair of plastic scintillators which cover the range $2.8<\eta<5.1$, $4.9<\eta<6.1$, respectively, and $-3.7<\eta<-1.7$,$-7.0<\eta<-4.8$.

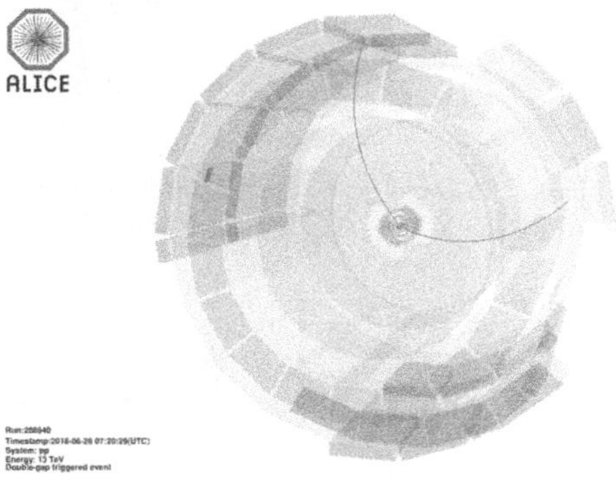

Fig. 3. Double-gap event in the ALICE central barrel.

The double-pomeron events of Fig. 1 consist of particles produced at midrapidity, and of the beam protons, or remnants thereof, at or close to beam rapidity. No particles are produced in between, which yields a rapidity region without particles, known as the gap. This rapidity gap is a signature of central production which can be used for filtering such events from minimum-bias events in the offline analysis, or for tagging at the trigger level. A double-gap event in the ALICE central barrel is shown in Fig. 3.

5. ALICE data on exclusive pion pair production

Exclusive production of $\pi^+\pi^-$ pairs has been studied within the tensor pomeron approach.[15] The dipion continuum and pion pairs from decays of scalar $f_0(500)$, $f_0(980)$ and tensor $f_2(1270)$ resonance were considered. The resulting mass distribution is shown in Fig. 4 on the left. Photoproduction of $\rho(770)$ is visible at LHC energies due to the large photon fluxes, shown in the middle. The uncorrected pion-pair mass distribution measured by ALICE in proton–proton collisions at $\sqrt{s} = 13$ TeV is shown on the right. An analogous analysis has been done for K^+K^--pairs.[16] The ALICE data sample shown in Fig. 4 will be increased in the LHC Run 3 by a factor of about 40. The study of the measured resonances by a Partial Wave Analysis in both the light quark and strangeness sector will become feasible.

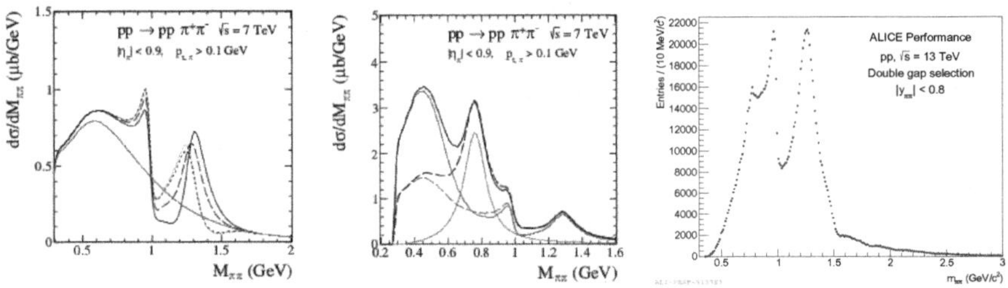

Fig. 4. Model predictions of Ref. 15 for central pion pair mass distribution (left and middle), and pion pair mass distribution measured by ALICE (right).

Acknowledgment

This work is partially supported by the German Federal Ministry of Education and Research under promotional reference 05P19VHCA1.

References

1. A. Donnachie, H.G. Dosch, P.V. Landshoff, and O. Nachtmann. *Pomeron physics and QCD*, Camb. Mono. Part. Nucl. Phys. **19** (2002).
2. M. Tanabashi *et al.*, (Part. Data Group), Phys.Rev.D**98**,(2018) 03001.
3. TOTEM Coll., (G. Antchev *et al.*), Eur.Phys.J. **C79**, (2019)no.10,861.
4. ALICE Coll., (B. Abelev *et al.*), Eur.Phys.J. **C73**, (2013) no.6, 2456.
5. C. Ewerz, M. Maniatis, O. Nachtmann, Annals Phys. 342 (2014) 31.
6. A. Donnachie, P.V. Landshoff, Phys.Lett. B727 (2013) 500.
7. A. Donnachie, P.V. Landshoff, Nucl. Phys. B **303** (1988) 634.
8. P. Lebiedowicz, O. Nachtmann, A. Szczurek, Ann. Phys. 344(2014)301.
9. L. Adamczyk *et al.*, STAR Collaboration, Phys. Lett. B **719** (2013) 62.

10. C. Ewerz, P. Lebiedowicz, O. Nachtmann, A. Szczurek, Phys. Lett. B763 (2016) 382.
11. A.J. Bugrij *et al.*, Fortschr. Phys. **21**, (1973) 427.
12. R. Fiore, L. Jenkovszky, R. Schicker, Eur.Phys.J.C76 (2016), no.1, 38.
13. R. Fiore, L. Jenkovszky, R. Schicker, Eur.Phys.J.C78 (2018) no.6, 468.
14. ALICE Coll., (B. Abelev *et al.*), Int.J.Mod.Phys. **A29**, (2014)1430044.
15. P. Lebiedowicz, O. Nachtmann, A. Szczurek, Phys.Rev. D93 (2016) no.5, 054015.
16. P. Lebiedowicz, O. Nachtmann, A. Szczurek, Phys.Rev. D98 (2018) 014001.

Simulation of exclusive π^0 electroproduction at EicC

Rong WANG

Institute of Modern Physics, Chinese Academy of Sciences, Lanzhou 730000, China
E-mail: rwang@impcas.ac.cn

The low-energy Electron-Ion Collider in China would provide enormous statistics of exclusive π^0 events. The kinematical coverage, statistical uncertainties of hard exclusive π^0 production at EicC are studied with MC simulations. The detections of the final particles are briefly discussed. These data at EicC would help constrain the chiral-odd GPDs around the small skewness ξ or the high Q^2, in addition to the current experiments at JLab.

Keywords: Electron-Ion Collider; π^0; deep virtual meson production; generalized parton distributions.

1. Electron-Ion Collider in China

Electron-Ion Collider (EIC), regarded as a "super electron microscope", is an ideal tool to measure the GPDs of the nucleon. Among the EIC plans worldwide, China's proposal is to build a polarized EIC by upgrading the HIAF (the heavy ion High Intensity Accelerator Facility) which is now under construction.[1] The parameters of Electron-Ion Collider in China (EicC) are as following: electron beam of maximum energy 3.5 GeV, proton beam of maximum energy 20 GeV, both beams polarized, luminosity of $2 \sim 4 \times 10^{33}$ cm^{-2}s^{-1} (see Fig. 1). The cross-section of exclusive meson production falls down rapidly with $1/Q^4$ behavior,[2] hence a low energy EIC has the great advantage in collecting the exclusive meson production data. Seizing this opportunity to unveil the nucleon structure, we discuss a simulation of hard exclusive π^0 production at EicC in the following sections.

2. Introduction of DVMP and GPDs

Similar to Deep Virtual Compton Scattering (DVCS) process, the hard exclusive meson production (or Deep Virtual Meson Production, DVMP) is

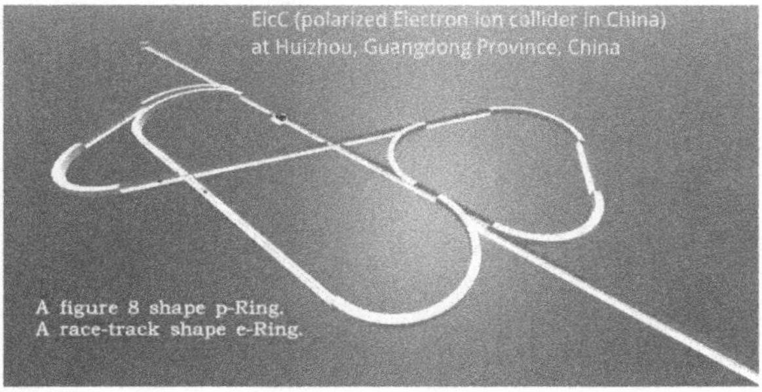

Fig. 1. The preliminary layout and design of the accelerators and the interaction point. (Merit form Yutie Liang.)

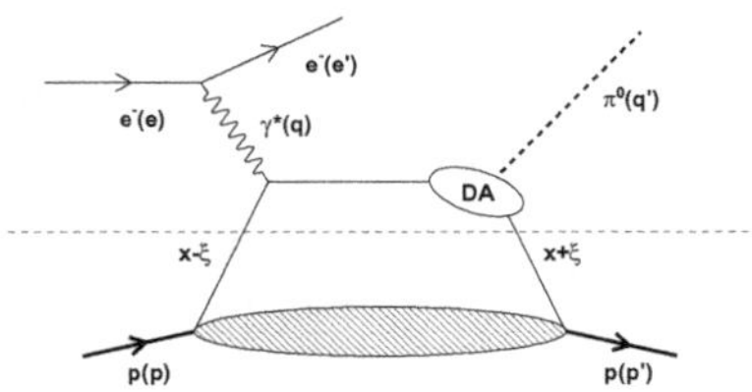

Fig. 2. The handbag Feynman diagram of π^0-DVMP process.

another channel which is sensitive to the Generalized Parton Distributions (GPDs) of the nucleon.[3–5] If the handbag framework is valid, the DVMP measurement could be used to check the GPD universality through comparisons with the DVCS data. Furthermore by studying channels of different mesons, we can easily do the flavor-separation of GPDs.[6–8]

Fig. 2 shows the schematic Feynman diagram of π^0-DVMP process. There are four kinematic variables in the π^0-DVMP channel, which are Q^2, x_B, t, and the azimuthal angle ϕ defined as the angle between the lepton plane and the π^0 production plane. The definitions of the invariant kinematical variables are shown in Eq. (1). In the old literatures decade ago,[6–8] the pseudoscalar meson DVMP processes are viewed to have the advantage in constraining the polarized GPDs. However, recent JLab experiments show that the transverse polarized photon plays a dominated role in the exclusive pseudoscalar meson production.[9–15] Coupling to the twist-3 distribution amplitude of π^0, the chiral-odd GPDs are involved in the measured transverse photon amplitude. It is time and an opportunity to extensively study the chiral-odd GPDs which is far-less known, for they are overwhelming in the pseudoscalar meson productions under a relatively low energy collision.

$$Q^2 \equiv -q^2, x_B \equiv \frac{Q^2}{2pq}, t \equiv (p-p')^2, \xi \equiv \frac{x_B}{2-x_B}\left(1+\frac{m_\pi^2}{Q^2}\right) \tag{1}$$

3. Simulation of π^0-DVMP at EicC

The MC simulation of π^0-DVMP is based on the following formulism on the differential cross-section, written as,

$$\frac{d^4\sigma}{dQ^2 dx_B dt d\phi_\pi} = \Gamma(Q^2, x_B, s)\frac{1}{2\pi}\Big[\sigma_T + \epsilon\sigma_L + \sqrt{2\epsilon(1+\epsilon)}cos(\phi_\pi)\sigma_{LT} + \epsilon cos(2\phi_\pi)\sigma_{TT}\Big], \tag{2}$$

$$\Gamma(Q^2, x_B, s) = \frac{\alpha y^2(1-x_B)}{2\pi x_B(1-\epsilon)Q^2}, \tag{3}$$

$$\epsilon^{-1} = 1 + 2\left(1+\frac{\nu^2}{Q^2}\right)\left(4\frac{\nu^2}{Q^2}\frac{1-y}{y^2} - 1\right), \tag{4}$$

in which y depends on Q^2 and x_B with the equation $Q^2 = xys$, and s is the invariant mass square of the collison. The unpolarized cross-section is decomposed into the un-seperated cross-section $\sigma_U = \sigma_T + \epsilon\sigma_L$, the interference cross-sections σ_{LT} and σ_{TT}. The parametrization of the reduced cross-sections are from Ref. 16.

Fig. 3 shows the invariant variable distributions of the simulated π^0-DVMP events at EicC machine. The kinematic coverage is much wider than that of JLab 12 GeV upgrade. Fig. 4 shows the energy and pseudorapidity distributions of the final particles. The scattered proton goes very backward in the DVMP channel, of which $|\eta|$ of some events reach 5. To have qualified exclusivity of the measurement, we will install a zero-degree detector on the beamline, such as a Roman pot. We also studied the two-photon separation problem in the barrel calorimeter. If we want 95% efficiency of two-photon separation, the position resolution should be better than 5 cm for a calorimeter located one meter away from the interaction point.

Using MC integral method, the total cross-section of π^0-DVMP at EicC is estimated to be 1.7 nb, in the region of $x_B < 0.85$, $1 < Q^2 < 100$ GeV^2,

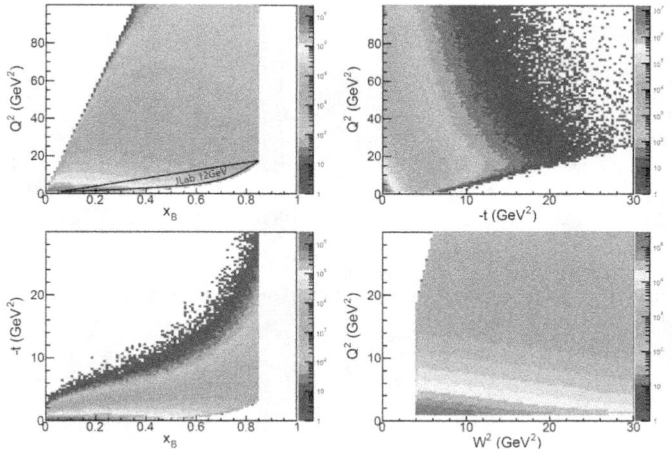

Fig. 3. The π^0-DVMP events in the planes of two invariant variables.

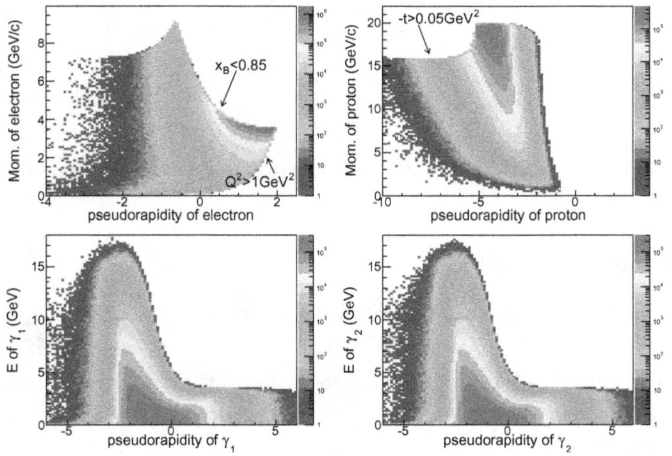

Fig. 4. The kinematics of the final state particles.

$0.05 < -t < 30$ GeV^2, and $W^2 > 4$ GeV^2. For a suggested experiment of 50 fb^{-1} integrated luminosity, we get around 84.5 million events in hard region. The relative statistic uncertainties of the reduced differential cross-section σ_U are shown in Fig. 5. The statistical uncertainties of double spin asymmetry A_{LL} are demonstrated in Fig. 6, assuming $A_{LL}^{const.} = 0.4$ and $A_{LL}^{cos\phi} = 0.2$.[12]

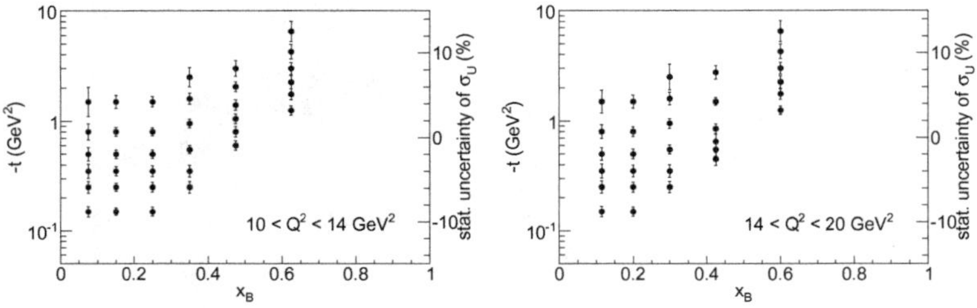

Fig. 5. The statistical uncertainties of the unseparated cross-section of a proposed EicC experiment of 50 fb^{-1} at high Q^2.

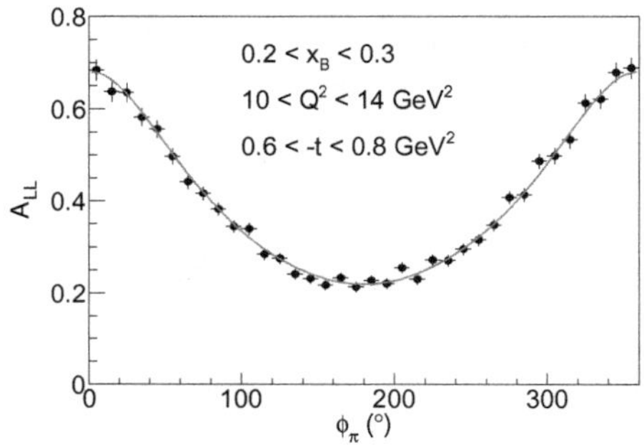

Fig. 6. The A_{LL} projection with statistical uncertainties of a proposed EicC experiment of 50 fb^{-1} at high Q^2.

4. Summary and outlook

Through the MC simulation, we show that we can get enormous amount of π^0-DVMP events at EicC machine, with years of running (50 fb^{-1}). These events are distributed in a much wider kinematical range (higher Q^2 and smaller x_B compared to the current JLab data), which are valuable for the chiral-odd GPD study. The statistical uncertainties of both the differential cross-section and asymmetry A_{LL} are small. The other types of asymmetries for the polarized experiments at EicC will be studied. Moreover, the detector capability and efficiency should also be evaluated in the future.

Acknowledgments

I thank Dr. Valery Kubarovsky for the parametrization of the differential cross-sections, and EicC working group for the suggestion and the fruitful discussions. This work is supported by the Key Research Program of the Chinese Academy of Sciences, Grant NO. XDPB09.

References

1. X. Chen, A Plan for Electron Ion Collider in China, *PoS* **DIS2018**, 170 (2018).
2. P. Kroll, Status of DVMP, DVCS and GPDs, *EPJ Web Conf.* **85**, 01005 (2015).
3. X.-D. Ji, Gauge-Invariant Decomposition of Nucleon Spin, *Phys. Rev. Lett.* **78**, 610 (1997).
4. X.-D. Ji, Deeply virtual Compton scattering, *Phys. Rev.* **D55**, 7114 (1997).
5. A. V. Radyushkin, Scaling limit of deeply virtual Compton scattering, *Phys. Lett.* **B380**, 417 (1996).
6. M. Vanderhaeghen, P. A. M. Guichon and M. Guidal, Deeply virtual electroproduction of photons and mesons on the nucleon: Leading order amplitudes and power corrections, *Phys. Rev.* **D60**, 094017 (1999).
7. K. Goeke, M. V. Polyakov and M. Vanderhaeghen, Hard exclusive reactions and the structure of hadrons, *Prog. Part. Nucl. Phys.* **47**, 401 (2001).
8. A. V. Belitsky and A. V. Radyushkin, Unraveling hadron structure with generalized parton distributions, *Phys. Rept.* **418**, 1 (2005).
9. E. Fuchey *et al.*, Exclusive Neutral Pion Electroproduction in the Deeply Virtual Regime, *Phys. Rev.* **C83**, 025201 (2011).
10. I. Bedlinskiy *et al.*, Measurement of Exclusive π^0 Electroproduction Structure Functions and their Relationship to Transversity GPDs, *Phys. Rev. Lett.* **109**, 112001 (2012).
11. I. Bedlinskiy *et al.*, Exclusive π^0 electroproduction at $W > 2$ GeV with CLAS, *Phys. Rev.* **C90**, 025205 (2014), [addendum: Phys. Rev. C90, no. 3, 039901 (2014)].
12. A. Kim *et al.*, Target and double spin asymmetries of deeply virtual π^0 production with a longitudinally polarized proton target and CLAS, *Phys. Lett.* **B768**, 168 (2017).
13. M. Defurne *et al.*, Rosenbluth separation of the π^0 electroproduction cross section, *Phys. Rev. Lett.* **117**, 262001 (2016).
14. M. Mazouz *et al.*, Rosenbluth separation of the π^0 Electroproduction Cross Section off the Neutron, *Phys. Rev. Lett.* **118**, 222002 (2017).
15. I. Bedlinskiy *et al.*, Exclusive η electroproduction at $W > 2$ GeV with CLAS and transversity generalized parton distributions, *Phys. Rev.* **C95**, 035202 (2017).
16. Model-for-the-pi0-eta-exclusive-cross-section `https://github.com/vkubarovsky/Model-for-the-pi0-eta-exclusive-cross-section`, Accessed: 2019-11-14.

Recent progress in basis light-front quantization

Xingbo Zhao,* Kaiyu Fu,$ Hengfei Zhao,† Jiangshan Lan,** Chandan Mondal,$$ Siqi Xu‡

Institute of Modern Physics, Chinese Academy of Sciences, Lanzhou 730000, China
School of Nuclear Science and Technology, University of Chinese Academy of Sciences, Beijing 100049, China
**E-mail: xbzhao@impcas.ac.cn,$ E-mail: kaiyufu@impcas.ac.cn*
*†E-mail: zhaohengfei@impcas.ac.cn,** E-mail: jiangshanlan@impcas.ac.cn*
$$ E-mail: mondal@impcas.ac.cn,‡ E-mail: xsq234@impcas.ac.cn

James P. Vary

Department of Physics and Astronomy, Iowa State University,
Ames, IA 50011, USA
E-mail: jvary@iastate.edu
(BLFQ Collaboration)

Basis Light-front Quantization (BLFQ) is a nonperturbative approach to quantum field theory. In this paper, we report our recent progress in applying BLFQ to the positronium system in QED and to the meson and the baryon system in QCD. We present preliminary results on the mass spectrum, light-front wave functions and other observables of these systems, where one dynamical gauge boson is retained for the positronium and meson systems.

Keywords: Light-front quantization; positronium; meson; baryon.

1. Introduction

Basis Light-front Quantization (BLFQ) has been developed as a nonperturbative approach to relativistic bound states.[1] It is based on the Hamiltonian formalism and the light-front quantum field theory. In BLFQ, the bound state problem is cast into an eigenvalue problem of the Hamiltonian:

$$P^- |\beta\rangle = P^-_\beta |\beta\rangle, \tag{1}$$

where the eigenvalues P^-_β correspond to the mass spectrum and the eigenvectors $|\beta\rangle$ encode their structural information. In this paper, we report our recent progress in applying BLFQ to the positronium system in QED and the meson and baryon system in QCD.

2. Positronium

The positronium ("Ps") is arguably the simplest bound state system in QED. In this work, we solve the positronium system from first principles - the QED Lagrangian.[2] In order to make the numerical calculation feasible, we perform basis truncation by retaining the two leading Fock sectors, that is, $|\text{Ps}\rangle = a|e^+e^-\rangle + b|e^+e^-\gamma\rangle$. In addition, we truncate the basis in the transverse (longitudinal) direction with the truncation parameter $N_{\max}$ (K).[1] Larger $N_{\max}$ (K) translates to more complete bases in the transverse (longitudinal) direction. We obtain our light-front QED Hamiltonian from the QED Lagrangian via the Legendre transformation. In our truncated basis the light-front QED Hamiltonian, using light-front gauge, takes the following form,

$$P^-_{\text{QED}} = \int \mathrm{d}^2x^\perp \mathrm{d}x^- \frac{1}{2}\bar{\Psi}\gamma^+ \frac{m_{e0}^2 + \left(i\partial^\perp\right)^2}{i\partial^+}\Psi + \frac{1}{2}A^j \left(i\partial^\perp\right)^2 A^j$$
$$+ej^\mu A_\mu + \frac{e^2}{2}j^+ \frac{1}{(i\partial^+)^2}j^+, \quad (2)$$

where ψ and A_μ are the fermion and gauge boson field operators, respectively, and $j^\mu = \bar{\Psi}\gamma^\mu\Psi$. The first two terms are their corresponding kinetic terms and the remaining terms describe their interaction. m_{e0} is the bare fermion mass. For numerical convenience, we take an artificially increased electromagnetic coupling constant $\alpha = 0.075$.

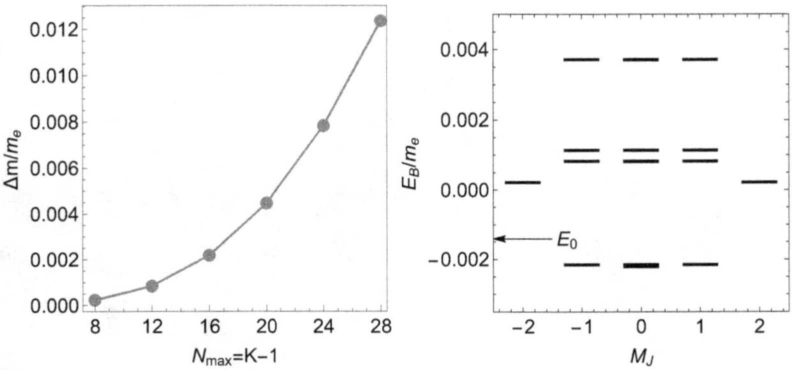

Fig. 1. Left panel: representative value of the mass counterterm (in units of the physical electron mass m_e) in the positronium problem as a function of basis truncation parameters $N_{\max} = K - 1$; right panel: the binding energy (E_B) spectrum of the positronium system at $N_{\max} = K - 1 = 28$ and $\alpha = 0.075$. E_0 is the ground state (1^1S_0) binding energy from nonrelativistic quantum mechanics with perturbative corrections.

In this calculation, we adopt the Fock-sector dependent renormalization,[3–5] according to which only the fermion mass in the $|e^+e^-\rangle$ sector needs to be renormalized, namely, the bare mass m_{e0} is different from the physical mass m_e. Different basis states take distinct values for the mass counterterm depending on their respective quanta available for self-energy fluctuation. The mass counterterms are determined from solving a series of single electron systems in the $|e\rangle + |e\gamma\rangle$ Fock sectors.[6]

The value of the mass counterterm $\Delta m = m_{e0} - m_e$ for a representative basis state in the positronium problem as a function of the truncation parameters is shown in the left panel of Fig. 1 and in the right panel, we present the binding energy spectrum of the positronium system, $E_B \equiv M_{\rm Ps} - 2\,m_e$, for different spin projections M_J. The arrow indicates the value of the ground state binding energy from nonrelativistic quantum mechanics,[7] which is close to our result. We note that the mass counterterm is typically on a larger scale than that of the binding energy. We also note that the scale of the binding energy and structures of multiplets in the spectrum are in reasonable agreement with the previous calculation based on an effective one-photon-exchange interaction between the e^+ and e^-.[8] The approximate degeneracy among different M_J substates and the information from the mirror parity and charge parity[10] allow us to identify the low-lying eigenstates. For the $M_J = 0$ states, from bottom to up, the lowest six states are 1^1S_0, 1^3S_1, 2^1S_0, 2^3S_1, 2^3P_0, and 2^3P_1. Fig. 2 illustrates the light-front wave function (LFWF) in the $|e^+e^-\rangle$ sector for three low-lying states with $M_J = 0$. Their shape and nodal structures are qualitatively similar to those based on the one-photon-exchange effective interaction.[8]

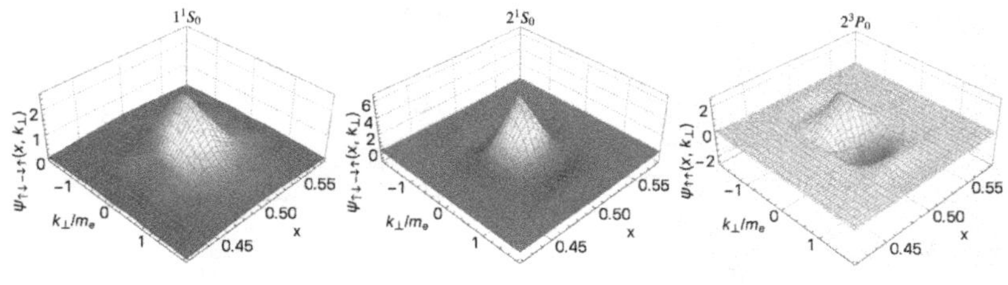

Fig. 2. The (normalized) LFWFs for the dominant spin component in the $|e^+e^-\rangle$ sector of the positronium system at $N_{\rm max} = K - 1 = 28$ and $\alpha = 0.075$. x is the longitudinal momentum fraction and $k_\perp$ represents the relative transverse momentum between e^+ and e^-.

3. Heavy meson

A similar calculation can be carried over to the heavy meson system in QCD. Like the positronium system, we retain the lowest two Fock sectors, $|q\bar{q}\rangle$ and $|q\bar{q}g\rangle$, in our basis. Our Hamiltonian contains two parts. From the QCD Lagrangian, we obtain the first part of the Hamiltonian,[9] P^-_{QCD}, which in our truncated Fock space takes a similar form to the QED Hamiltonian, P^-_{QED}, in Eq. (2), with the fermion field Ψ identified as the quark field, the gauge boson field A_μ identified as the gluon field $A^a_\mu T^a$ and the electric charge e replaced by the color charge g. In order to achieve a more accurate reproduction of the meson mass spectrum, we allow the quark mass appearing in the quark-gluon vertex interaction, $gj^\mu A^a_\mu T^a$, to be an independent phenomenological parameter, m'_q, from m_q in the kinetic energy term. We also apply a nonzero gluon mass m_g to ensure the low-lying states are dominated by the $|q\bar{q}\rangle$ sector.

In addition we include a phenomenological confining potential[10] in both the longitudinal and transverse directions in the $|q\bar{q}\rangle$ sector, which takes the following form,

$$P^-_{\text{C}} P^+ = \kappa^4 \vec{\zeta}_\perp - \frac{\kappa^4}{(m_q + m_{\bar{q}})^2} \partial_x (x(1-x)\partial_x), \tag{3}$$

where x is the longitudinal momentum fraction of the quark.[8] $\vec{\zeta}_\perp \equiv \sqrt{x(1-x)}\vec{r}_\perp$ is the holographic variable introduced by Brodsky and de Téramond,[11] and $\partial_x f(x, \vec{\zeta}_\perp) = \partial f(x, \vec{\zeta}_\perp)/\partial x|_{\vec{\zeta}}$. κ is the strength of the confinement and $m_q(m_{\bar{q}})$ is the mass of the quark (anti-quark). Thus, our total Hamiltonian is $P^- = P^-_{\text{QCD}} + P^-_{\text{C}}$.

With the charm quark mass $m_c = 1.561\,\text{GeV}$ (kinetic), $m'_c = 4.343\,\text{GeV}$ (interaction), the gluon mass $m_g = 0.4\,\text{GeV}$, the confining strength $\kappa = 1.14\,\text{GeV}$ and the strong coupling constant $g = 2.3$ for the charmonium, and similarly the bottom quark mass $m_b = 4.767\,\text{GeV}$ (kinetic), $m_f = 15\,\text{GeV}$ (interaction), $m_g = 0.4\,\text{GeV}$, $\kappa = 1.948\,\text{GeV}$ and $g = 1.6$ for the bottomonium, our resulting mass spectra for the low-lying $c\bar{c}$ and $b\bar{b}$ states agree with the experimental values reasonably well, as shown in the left and middle panel of Fig. 3, respectively. With the same parameter set used in the $c\bar{c}$ and $b\bar{b}$ systems we obtain the mass spectrum for the B_c system, as in the right panel of Fig. 3. In Fig. 4 we present the ground state LFWFs of the $c\bar{c}$, $b\bar{b}$ and B_c system in the $|q\bar{q}\rangle$ sector.

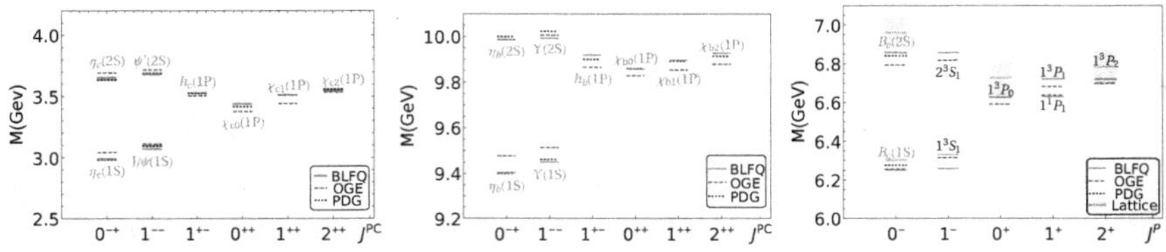

Fig. 3. Comparison of our BLFQ spectra at $N_{\max} = K - 1 = 8$ for charmonium (left), bottomonium (middle), and B_c meson (right) with the effective one-gluon-exchange (OGE) approach[10,12] and the experimental values (PDG).[13] Lattice results are from Refs. 14–16. The horizontal and vertical axes are the J^{PC} and invariant mass, respectively.

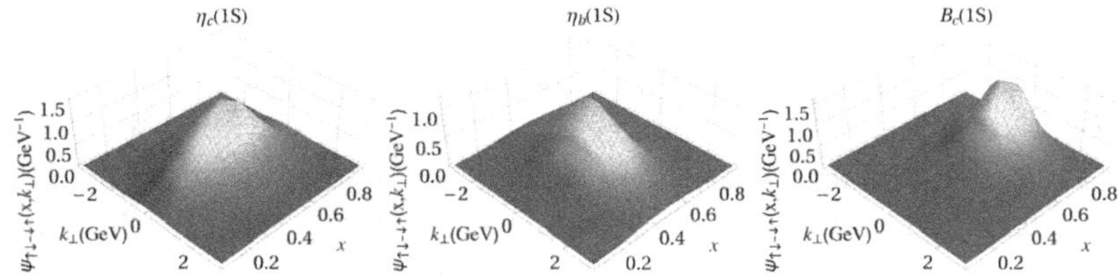

Fig. 4. The (normalized) LFWFs for the dominant spin component in the $|q\bar{q}\rangle$ sector of $\eta_c(1S)$ (left), $\eta_b(1S)$ (middle) and $B_c(1S)$ (right) at $N_{\max} = K - 1 = 8$. x is the longitudinal momentum fraction of the quark and $k_\perp$ represents the relative transverse momentum between the quark and the anti-quark.

4. Light meson

We perform a similar calculation in the light meson sector. With $m_u = m_d = 0.33\,\mathrm{GeV}$, $m'_u = m'_d = 3.38\,\mathrm{GeV}$, $\kappa = 0.77\,\mathrm{GeV}$ and g=1.74, we obtain the mass spectrum of the light mesons, as shown in the left panel of Fig. 5.

Although the masses for $a_0(980)$ and $b_1(1235)$ somewhat deviate from the experimental value,[13] the results for π, ρ, $a_1(1260)$, $\pi(1300)$, $a_2(1320)$, and $\pi_1(1400)$ are in good agreement with the experimental data.[13]

Next, we calculate the pion's parton distribution function (PDF), which represents the probability density of finding a parton with the longitudinal momentum fraction x inside the pion. We first calculate the initial scale PDF based on the LFWF from BLFQ in both the $|q\bar{q}\rangle$ and $|q\bar{q}g\rangle$ sectors. We then evolve the initial scale PDF from $\mu_0^2 = 0.31$ GeV^2 to the relevant experimental scales $\mu^2 = 16$ GeV^2 by the DGLAP equations[17–19] using the higher order perturbative parton evolution toolkit.[20]

In Fig. 5 (right panel), we compare our results with the E615 experimental result[21] and the modified result of the E615 experiment.[22] We find

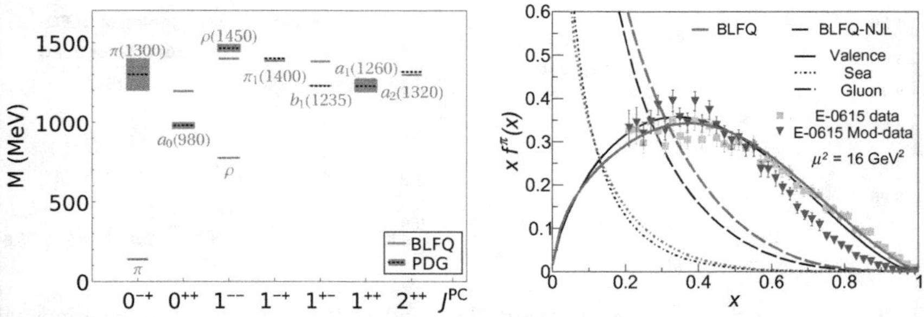

Fig. 5. Left panel: the mass spectrum of the light mesons; right panel: longitudinal momentum distribution $xf^{\pi}(x)$ as a function of longitudinal momentum fraction x of a parton in the pion.

that, for the valence PDF, our result shows slightly better agreement with the E615 experimental result[21] in the high-x region compared to the earlier calculation based an effective NambuJona-Lasinio interaction (BLFQ-NJL). For the gluon PDF, our result is larger than that obtained from the BLFQ-NJL model, and for the sea PDF, our result is close to that obtained from the BLFQ-NJL model.[23]

5. Baryon

We perform an initial calculation of the baryon system based on an effective interaction in the $|qqq\rangle$ sector.[25,26] This effective interaction consists of the (pairwise) one-gluon-exchange interaction and the (pairwise) longitudinal and transverse confining interactions. The parameters in the effective interaction are fixed by the mass and Dirac form factor of u, d flavor. For the baryon system, the ground state spin-1/2 (3/2) particle is the nucleon (Δ (1232)). Our preliminary results of their masses are compared with the experimental data in the left panel of Fig. 6. Our mass of Δ (1232) is about 170 MeV smaller than the experimental value. It remains to be seen that whether this mass discrepancy will decrease as the basis size increases. We also calculate the $I_{\frac{1}{2},\frac{1}{2}}(Q^2)$ elastic form factor[27] of the proton and Δ^+, which is compared in Fig 6. The larger slope of $I_{\frac{1}{2},\frac{1}{2}}$ for Δ^+ at $Q^2 \to 0$ suggests that the charge radius of Δ^+ is larger than that of the proton, which is as expected since Δ^+ is an excitation of the proton.

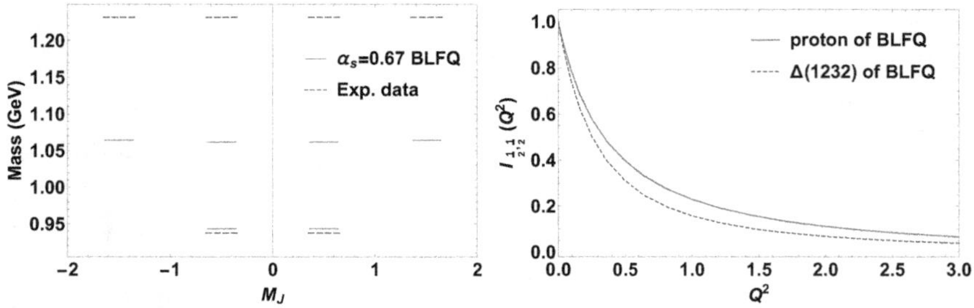

Fig. 6. Left panel: the masses of the nucleon and $\Delta(1232)$ compared to experimental data[13]; right panel: comparison of the $I_{\frac{1}{2},\frac{1}{2}}(Q^2)$ elastic form factor between the proton and Δ^+.

6. Conclusion

Through the applications to various bound state systems in QED and QCD we demonstrate that BLFQ is a versatile and powerful nonperturbative approach to quantum field theory for strongly interacting systems. We anticipate that in the future BLFQ will be a useful tool for understanding hadron mass spectrum and structure beyond the valence sector.

Acknowledgment

XZ is supported by Key Research Program of Frontier Sciences, CAS, Grant No ZDBS-LY-7020. CM is supported by the National Natural Science Foundation of China (NSFC) under the Grant No. 11850410436 and No. 11950410753. JPV is supported by the Department of Energy under Grants No. DE-FG02-87ER40371, and No. DE-SC0018223 (SciDAC4/NUCLEI). A portion of the computational resources were provided by the National Energy Research Scientific Computing Center (NERSC), which is supported by the Office of Science of the U.S. Department of Energy under Contract No.DE-AC02-05CH11231.

References

1. J. P. Vary *et al.*, Phys. Rev. C **81**, 035205 (2010).
2. K. Fu, H. Zhao, X. Zhao, J. P. Vary, to be published in the proceedings of Hadron 2019 conference (2020).
3. V. A. Karmanov, J.-F. Mathiot and A. V. Smirnov, Phys. Rev. D **77**, 085028 (2008).
4. V. A. Karmanov, J. F. Mathiot and A. V. Smirnov, Phys. Rev. D **86**, 085006 (2012).

5. X. Zhao, Few Body Syst. **56**, no. 6-9, 257 (2015).
6. X. Zhao, H. Honkanen, P. Maris, J. P. Vary and S. J. Brodsky, Phys. Lett. B **737**, 65 (2014).
7. H. A. Bethe and E. E. Salpeter, *Quantum Mechanics of One- and Two-Electron Atoms*, Springer, Heidelberg, 1957.
8. P. Wiecki, Y. Li, X. Zhao, P. Maris and J. P. Vary, Phys. Rev. D **91**, no. 10, 105009 (2015).
9. S. J. Brodsky, H. C. Pauli and S. S. Pinsky, Phys. Rept. **301**, 299 (1998).
10. Y. Li, P. Maris and J. P. Vary, Phys. Rev. D **96**, no. 1, 016022 (2017).
11. S. J. Brodsky, G. F. de Teramond, H. G. Dosch and J. Erlich, Phys. Rept. **584**, 1 (2015).
12. S. Tang, Y. Li, P. Maris and J. P. Vary, Phys. Rev. D **98**, no. 11, 114038 (2018).
13. M. Tanabashi *et al.* [Particle Data Group], Phys. Rev. D **98**, no. 3, 030001 (2018).
14. I. F. Allison *et al.* [HPQCD and Fermilab Lattice and UKQCD Collaborations], Phys. Rev. Lett. **94**, 172001 (2005).
15. E. B. Gregory *et al.*, Phys. Rev. Lett. **104**, 022001 (2010).
16. C. T. H. Davies, K. Hornbostel, G. P. Lepage, A. J. Lidsey, J. Shigemitsu and J. H. Sloan, Phys. Lett. B **382**, 131 (1996).
17. V. N. Gribov and L. N. Lipatov, Sov. J. Nucl. Phys. **15**, 438 (1972).
18. G. Altarelli and G. Parisi, Nucl. Phys. B **126**, 298 (1977).
19. Y. L. Dokshitzer, Sov. Phys. JETP **46**, 641 (1977).
20. G. P. Salam and J. Rojo, Comput. Phys. Commun. **180**, 120 (2009).
21. J. S. Conway *et al.*, Phys. Rev. D **39**, 92 (1989).
22. C. Chen, L. Chang, C. D. Roberts, S. Wan and H. S. Zong, Phys. Rev. D **93**, no. 7, 074021 (2016).
23. J. Lan, C. Mondal, S. Jia, X. Zhao and J. P. Vary, Phys. Rev. Lett. **122**, no. 17, 172001 (2019).
24. C. Alexandrou, T. Korzec, T. Leontiou, J. W. Negele and A. Tsapalis, PoS LATTICE **2007**, 149 (2007).
25. C. Mondal, S. Xu, J. Lan, X. Zhao, Y. Li, D. Chakrabarti and J. P. Vary, arXiv:1911.10913 [hep-ph].
26. S. Xu, C. Mondal, J. Lan, X. Zhao and J. P. Vary, to be published in the proceedings of Hadron 2019 conference (2020).
27. L. Adhikari, Y. Li, X. Zhao, P. Maris, J. P. Vary and A. Abd El-Hady, Phys. Rev. C **93**, no. 5, 055202 (2016).

Effective Light Front QCD Hamiltonian and spectral equation for quark-antiquark states

E.V. Prokhvatilov, V.A. Franke, M.V. Ioffe, I.A. Lebedev, A.M. Puchkov, P.A. Belov
High Energy and Elementary Particles Physics Department, St. Petersburg University, Russia

M. Yu. Malyshev
High Energy Physics Division, Petersburg Nuclear Physics Institute named by B. P. Konstantinov of National Research Center "Kurchatov Institute", Russia
E-mail: malyshev_my@pnpi.nrcki.ru

Yubing Dong
Institute of High Energy Physics, Chinese Academy of Sciences, China

The model for effective QCD Hamiltonian, describing bound states of "constituent" quark and antiquark, is proposed, using the Light Front (LF) formulation of QCD in the LF gauge. Zero mode of gluon field becomes independent canonical variable in this model unlike the known canonical formulations on the LF. This mode forms the "string" between quark and antiquark and bounds them. As quark and antiquark are separated by the "string", the interaction term, related to the interaction of fermion currents in the Hamiltonian, gives zero due to locality of this interaction. We introduce the nonlocal modification of this term and investigate the resulting spectral equation for mass squared. This equation is solved analytically in the large quark mass limit. However the rotation invariance turns out to be violated with our choice of modified interaction term. We show the way to restore the invariance and discuss the possibility to obtain the spectrum describing the experimental one.

Keywords: Quantization on the Light Front; gauge field theory; quark-antiquark model.

Quantum chromodynamics (QCD) is a model for the description of hadrons as bound states of quark and gluon fields. However this description requires nonperturbative methods due to strong coupling. Well known example of such methods is the formulation and calculations on the space-time lattice. We use alternative approach related to proposed by Dirac[1] formulation of field theory on the light front (LF). This formulation allows to simplify the description of quantum vacuum state, identifying it with the state having zero LF longitudinal momentum $p_- = (p_0 - p_3)/\sqrt{2} \geq 0$ in LF coordinates $x^\pm = (x^0 \pm x^3)/\sqrt{2}$, $x^\perp = (x^1, x^2)$, where x^0, x^1, x^2, x^3 are Lorentz coordinates and x^+ plays the role of time. The operator $P_+ = (P_0 + P_3)/\sqrt{2}$ plays the role of Hamiltonian.[2]

Zero ($p_- = 0$) mode of field is not well defined in the LF canonical formulation[3] due to the singularity at $p_- \to 0$. However this mode is responsible for vacuum effects (e.g. condensate description). It is not possible to ignore this mode in nonperturbative domain of QCD. The regularization $|x^-| \leq L$ with periodic bound conditions (p.b.c.) for fields separates zero mode, $p_- = 0$, from nonzero modes $p_- = \pi n/L$, $n \neq 0$. Zero mode ($n = 0$) has canonically conjugated momentum equal to zero and is not independent dynamical variable. So zero mode should be expressed through other, nonzero modes via canonical constraints. These constraints are complicated and not uniquely defined as quantized.[3]

So we investigate the limit transition to the LF formulation from usual formulation of field theory on space-like hyperplanes close to the LF. With this aim we introduce the LF approximating coordinates, $y^0 = x^+ + (\eta^2/2)x^-$, $y^3 = x^-$, $y^\perp = x^\perp$, with small parameter $\eta > 0$, and y^0 playing the role of time. The components q_μ of the momentum in these coordinates are: $q_0 = p_+$, $q_3 = p_- - (\eta^2/2)p_+$, $q_\perp = p_\perp$.

We use the QCD Hamiltonian on the plane $y^0 = 0$ with $|y^3| \leq L$ and p. b. c. for gluon fields (antiperiodic for quark fields),[4] and extract the term $H_{(0)}$ with only zero ($q_3 = 0$) mode of gluon field.

If we denote by $\pi_k^a(x^\perp)$ the canonical momentum conjugated to zero mode of gluon field $A_{k(0)}^a(x^\perp)$ ($a = 1, \ldots, N^2 - 1$ for SU(N) color group symmetry, $k = 1, 2$ are transverse Lorentz indices) we can obtain: $H_{(0)} = \frac{1}{4L\eta^2} \int d^2x^\perp \pi_k^a \pi_k^a$. This expression explicitly shows why in $\eta \to 0$ limit, i.e. on the LF, we need to have $\pi_k^a = 0$ as a condition for finiteness of LF "energy". However, considering the QED(1+1) model we find that nonperturbative vacuum effects may be taken into account only if we change the form of limit transition to the LF for zero mode terms in the Hamiltonian, namely, one should take $\eta \to 0$ and $L \to \infty$ simultaneously at L_η fixed.[4] So we investigate the possibility to use such limit in our case. We require the finite contribution of $H_{(0)}$ to mass squared $m^2 = 2q_0 q_3 + \eta^2 q_0^2 - q_\perp^2$ in $\eta \to 0$ limit. The contribution of only zero mode term is equal to $\eta^2 H_{(0)}^2 = \left(\frac{1}{4L\eta} \int d^2x^\perp \pi_k^a \pi_k^a \right)^2$. One can keep this term finite in the limit $\eta \to 0$, $L \to \infty$, $L_0 = 4\eta L$ fixed. For terms with nonzero modes we use the limit, $\eta \to 0$ at fixed L, which agrees with canonical formulation on the

LF. The $H_{(0)}$ contribution to the $2q_3q_0$ term turns out to be not boost invariant as the consequence of nonsymmetry in limit transition for zero and nonzero modes. We assume the possibility to drop this contribution. Further we investigate the simplest model which includes the above mentioned zero mode contribution in the definition of the mass squared operator on the LF: $m^2_{eff} = \left(\frac{1}{L_0}\int d^2x^\perp \pi^a_k\pi^a_k\right)^2 + 2P_+P_- - P_\perp^2$, where P_+, P_-, $P_\perp$ are canonical expressions for momentum operators in terms of fields on the LF including zero mode of gluon field as independent canonical variable. We can generate this expression for m^2_{eff} using the effective LF Hamiltonian:

$$P_+^{eff} = \frac{1}{2P_-}\left(\frac{1}{L_0}\int d^2x^\perp \pi^a_k\pi^a_k\right)^2 + P_+. \tag{1}$$

The vacuum state $|0\rangle$ is defined as follows: $P_-|0\rangle = \pi^a_k(x^\perp)|0\rangle = 0$.

Let us apply the proposed effective LF Hamiltonian to the description of quark-antiquark bound states using "constituent" quarks and antiquarks and corresponding fields. We connect quark and antiquark with the "string" containing only zero mode of gluon field $A^a_{k(0)}$. By "string" we mean the following "path ordered" exponent: $U_{x^\perp,x'^\perp} = Pexp\left(-ig\int_{x^\perp}^{x'^\perp}\sum_{k=1,2} dz^k A_{k(0)}(z^\perp)\right)$. We take the stright line path for the string in transverse coordinates. These states are invariant w.r. to the gauge symmetry transformations remaining after fixing the LF gauge ($A^a_- = 0$).

Then we act by the effective Hamiltonian P_+^{eff} on quark-antiquark state,

$$|f\rangle = \sum_{m,m'>0}\int d^2x^\perp d^2x'^\perp f_{m,m'}(x'^\perp - x^\perp) b^+_m(x^\perp) U_{x^\perp,x'^\perp} d^+_{m'}(x'^\perp)|0\rangle,$$

$P_-|f\rangle = p_n|f\rangle$, $P_\perp|f\rangle = 0$, where b^+_n and d^+_n correspond to quark and antiquark creation operators respectively. Also we project our effective Hamiltonian onto the space of such states.

The used projection of the Hamiltonian contains only fermion field and gluon zero mode. It contains the interaction between fermion currents, but this interaction, being local in $x^\perp$, does not give nonzero result on our basis states, except for the case when quark and antiquark are not separated.[5] To restore the interaction between separated quark and antiquark we introduce nonlocal modification of this term keeping the existing gauge invariance.[5] Here we introduce new parameter R which characterizes the size of the nonlocality.[5]

The action of the 1st term in Eq. (1) on the $U_{x^\perp, x'^\perp}$ in the state is singular and requires the discretization of the path from $x^\perp$ to $x'^\perp$ with the intervals of small size Δ. Then we get the contribution containing parameter $L_0\Delta$ which is taken finite in $\Delta \to 0$, $L_0 \to \infty$ limit.

In result we get the following spectral equation in $L_0 \to \infty$ limit, introducing new variable $\xi = \frac{p_m}{p_n}$, $1-\xi = \frac{p_{m'}}{p_n}$ and $f(\xi, x^\perp)$ instead of $f_{m,m'}(x^\perp)$:

$$\left(\frac{M^2 - \nabla_\perp^2}{\xi(1-\xi)} + \frac{g^4\left(N-\frac{1}{N}\right)^2 \rho^2}{4(L_0\Delta)^2} - \frac{g^2\left(N-\frac{1}{N}\right)}{4\pi^2 R^2} \left(\frac{1}{\xi(1-\xi)} \right) - C \right) f(\xi, x^\perp)$$

$$- \frac{g^2\left(N-\frac{1}{N}\right)}{2\pi^2 R^2} P \int_0^1 \frac{d\xi' \; f(\xi', x^\perp)}{(\xi'-\xi)^2} = m^2 f(\xi, x^\perp), \qquad (2)$$

where M is quark mass, $\nabla_\perp^2 = \partial_1^2 + \partial_2^2$, $\rho^2 = (x^1)^2 + (x^2)^2$. We avoid the singularity at $\xi' = \xi$ in the integral by introducing into equation the principal value symbol P and the inclusion of renormalization constant C.

To estimate the spectrum related to Eq. (2) we investigate the large quark mass limit. We rewrite the spectral equation in terms of dimensionless variables: $\overline{x} = x^\perp/R_0$, $\overline{m} = mR_0$, $\overline{M} = MR_0$, $\beta = \frac{g^2\left(N-\frac{1}{N}\right)}{2L_0\Delta} R_0^2$, $\gamma = \frac{g^2\left(N-\frac{1}{N}\right)}{2\pi^2}$, $\overline{\nabla}_\perp^2 = \nabla_\perp^2 R_0^2$ where $R_0 = (R^2/M)^{1/3}$, and change the variable ξ to $\omega = 2\xi - 1$, $-1 \le \omega \le 1$. We get

$$\left(\frac{\overline{M}^2 - \overline{\nabla}_\perp^2 - \frac{\gamma}{2}}{1-\omega^2} + \frac{\beta^2}{4}\overline{x}^2 - \frac{C}{4} \right) \overline{f}(\omega, \overline{x})$$

$$- \frac{\gamma}{2\overline{M}} P \int_{-1}^1 \frac{d\omega' \; \overline{f}(\omega', \overline{x})}{(\omega'-\omega)^2} = \frac{\overline{m}^2}{4} \overline{f}(\omega, \overline{x}).$$

Large quark mass limit corresponds to $\omega^2 \ll 1$. It is convinient to introduce new variable $s = \overline{M}\omega$, $-\overline{M} \le s \le \overline{M}$, and expand the equation in $\overline{M}^{-1}$. Assume also that R_0 is independent of M. We get the following result (introducing the new parameter $C' = C - 4\overline{M}^2$, C' finite in $\overline{M} \to \infty$ limit):

$$\left(-\overline{\nabla}_\perp^2 + s^2 + \frac{\beta^2}{4}\overline{x}^2 - \frac{C'}{4} \right) \widetilde{f}(s, \overline{x}) - \frac{\gamma}{2} P \int_{-\infty}^{\infty} \frac{ds' \; \widetilde{f}(s', \overline{x})}{(s'-s)^2} = \frac{\overline{m}^2}{4} \widetilde{f}(s', \overline{x}).$$

After Fourier transformation in s, $\varphi(z, \overline{x}) = \int ds \exp{(isz)} \widetilde{f}(s, \overline{x})$, we get the following equation:

$$\left(-\overline{\nabla}_\perp^2 - \partial_z^2 + \frac{\beta^2}{4}\overline{x}^2 + \frac{\pi}{2}\gamma|z| - \frac{C'}{4} \right) \varphi(z, \overline{x}) = \frac{\overline{m}^2}{4} \varphi(z, \overline{x}). \qquad (3)$$

We can write $\varphi(z,\overline{x}) = \varphi_1(\overline{x})\varphi_2(z)$. Then the spectrum of this equation is the sum of harmonic oscillator spectrum for $\varphi_1(\overline{x})$: $\overline{m}_1^2(n_1,n_2) = 4\beta\sum_{k=1,2}\left(n_k+\frac{1}{2}\right) - C'$, and the Airy equation spectrum for $\varphi_2(z)$: $\overline{m}_2^2(n_3) = 4\left(\frac{\pi\gamma}{2}\right)^{\frac{2}{3}}|\zeta_{n_3}|$, where $n_i \in \mathbb{Z}$, $n_i \geq 0$, $i = 1,2,3$, ζ_n are zeros of corresponding Airy eigenfunctions ($|\zeta_n| \approx \left(\frac{3\pi}{4}\left(n+\frac{1}{2}\right)\right)^{\frac{2}{3}}$).

The obtained equation has no explicit rotational symmetry and we cannot classify the eigenstates using total orbital momentum l. The simplest possibility to restore rotation invariance of our equation is to change the term with $|z|$ in such a way that we could get the equation for 3-dimensional harmonic oscillator. This could be done by the corresponding modification of the interaction term in effective LF Hamiltonian. Then we would have 3-dim harmonic oscillator specrum $m^2 \sim (2n + l + const)$. However the known fit to experiment is different,[6] $m^2 \sim (n + l + const)$. A possible explanation of this difference can be found in semiclassical quantization of the string with quark and antiquark at ends.[6] There it is shown that 3-dim harmonic oscillator-like spectrum corresponds to oscillating quark and antiquark, while the spectrum used to fit the experiment corresponds to the oscillation of massive gluon between quark and antiquark. This implies the inclusion of nonzero mode gluon into quark-antiquark state that can be directly done in our effective model. So the present paper can be considered as the motivation to such investigations.

Let us remark that the equation similar to Eq. (2) can be met in the paper[7] and series of other papers cited in this article. There the holographic idea, related with the AdS/QCD is applied to obtain LF Schrödinger-like equation for quark-antiquark states. However their equation differs from our spectral equation, e.g. in the choice of variables describing transverse coordinates and in the form of the potential.

Acknowledgments

We thank the organizers of the Hadron2019 conference, also we thank S. S. Afonin for useful discussions. The work of E. V. Prokhvatilov, M. Yu. Malyshev, I. A. Lebedev and Yubing Dong is partially supported by grant Y8299220K5 of Chinese Academy of Sciences.

Bibliography

1. P. A. M. Dirac, Forms of relativistic dynamics, *Rev. Mod. Phys.* **21**, 392 (1949).

2. B. L. G. Bakker and at al., Light-front quantum chromodynamics: A framework for the analysis of hadron physics, *Nucl. Phys. B – Proc. Suppl.* **251–252**, 165 (2014), arXiv:1309.6333 [hep-ph].
3. V. A. Franke, Yu. V. Novozhilov, S. A. Paston and E. V. Prokhvatilov, *Focus on quantum field theory* (Nova science publishers, New York, 2005), New York, ch. Quantization of Field Theory on the Light Front, pp. 23–81, arXiv:hep-th/0404031.
4. E. V. Prokhvatilov and V. A. Franke, Limit transition to light-front coordinates in field theory and QCD Hamiltonian, *Sov. J. Nucl. Phys.* **49**, 688 (1989).
5. M. Y. Malyshev, E. V. Prokhvatilov and V. A. Franke, Effective Hamiltonian for QCD on the Light Front, *Phys. Part. Nucl. Lett.* **16**, 533 (2019).
6. S. S. Afonin and I. V. Pusenkov, Universal description of radially excited heavy and light vector mesons, *Phys. Rev. D* **90**, 094020 (2014), arXiv:1411.2390 [hep-ph].
7. S. J. Brodsky, G. F. de Teramond, G. H. Dosch and J. Erlich, Light-front holographic QCD and emerging confinement, *Phys. Rep.* **584**, 1 (2015), arXiv:1407.8131v2 [hep-ph].

Holograhic Wigner distributions for the pion

Mohammad Ahmady[a,*], Chandan Mondal[b,†], Ruben Sandapen[c,†],
James P. Vary[d,§] and Xingbo Zhao[a,e,‖]

[a] *Mount Allison University, Sackville, New Brunswick, Canada, E4L 1E6*
[b] *Institute of Modern Physics, Chinese Academy of Sciences, Lanzhou-730000, China*
[c] *Acadia University, Wolfville, Nova-Scotia, Canada, B4P 2R6*
[d] *Iowa State University, Ames, IA 50011, USA*
[e] *School of Nuclear Science and Technology, University of Chinese Academy of Sciences, Beijing 100049, China*
[*] *mahmady@mta.ca*
[†] *mondal@impcas.ac.cn*
[†] *ruben.sandapen@acadiau.ca*
[§] *jvary@iastate.edu*
[‖] *xbzhao@impcas.ac.cn*

We study the Wigner distributions of the pion using a holographic light-front pion wavefunction with dynamical spin effects to reveal its multidimensional structure.

Keywords: Wigner distributions; light-front holography; light mesons.

1. Introduction

Wigner distributions in QCD, commonly known as phase-space distributions, were first introduced by Ji.[1] After appropriate phase-space reductions, these distributions reduce to generalized parton distributions (GPDs) and transverse momentum dependent parton distributions (TMDs) which are measurable in high energy experiments (For a review on these distributions and the experiments to measure them, see Refs. 2–5). GPDs allow us to have a three dimensional picture of the hadron in position space.[6] On the other side, TMDs contains three dimensional information regarding the spin-spin and spin-orbit correlations in momentum space.[7] Wigner distributions for spin-$\frac{1}{2}$ systems have been investigated in different models e.g., light-cone chiral quark soliton model,[8] light-front dressed quark model,[9] light-cone spectator model,[10] AdS/QCD inspired quark-diquark model[11,12] as well as light front QED model.[13]

Here, we study the Wigner distributions for different quark polarizations in the pion using a holographic light-front pion wavefunction which includes dynamical spin effects. It has been observed that such effects allow for an excellent simultaneous description of a wide range of data: the decay constant, charge radius, spacelike EM and transition form factors, as well as, after QCD evolution, both the parton distribution function and the parton distribution amplitude data with a single universal AdS/QCD scale.[14] Recently, this spin-improved holographic wavefunction has been used to predict the leading twist TMDs of the pion.[15]

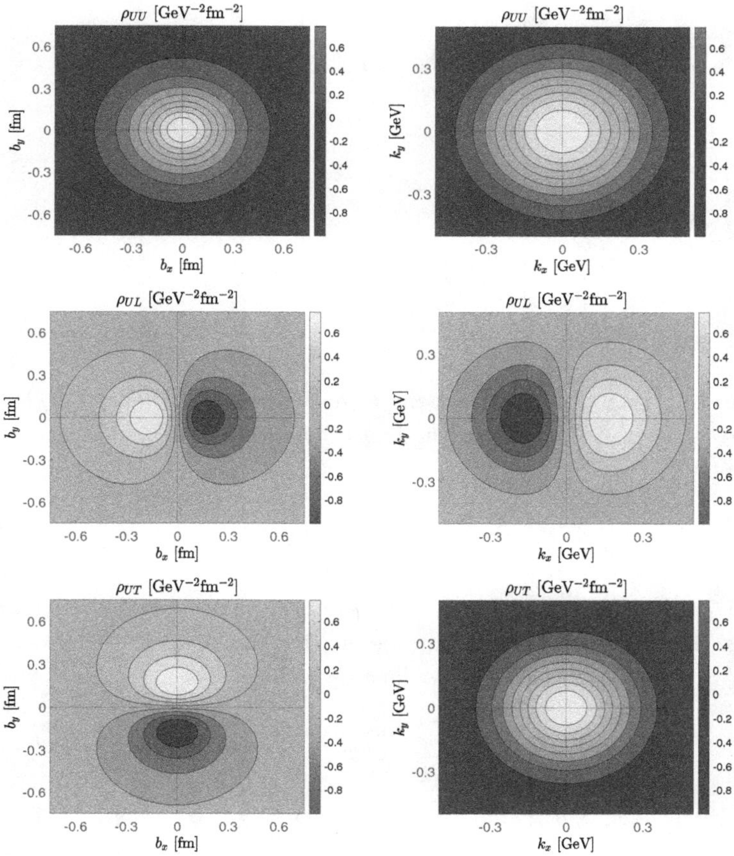

Fig. 1. Wigner distribution $\rho(\boldsymbol{b}_\perp, \boldsymbol{k}_\perp)$ of the unpolarized quark (*upper panels*), the longitudinal polarized quark (*middle panels*), and the transversely polarized quark (*lower panels*) inside the pion. (*Left panels*) the distributions are in the impact-parameter space with fixed transverse momentum $\boldsymbol{k}_\perp = k_\perp \hat{\boldsymbol{e}}_y$ and $k_\perp = 0.3$ GeV. (*Right panels*) the distributions are in the transverse-momentum space with fixed impact parameter $\boldsymbol{b}_\perp = b_\perp \hat{\boldsymbol{e}}_y$ and $b_\perp = 0.3$ fm.

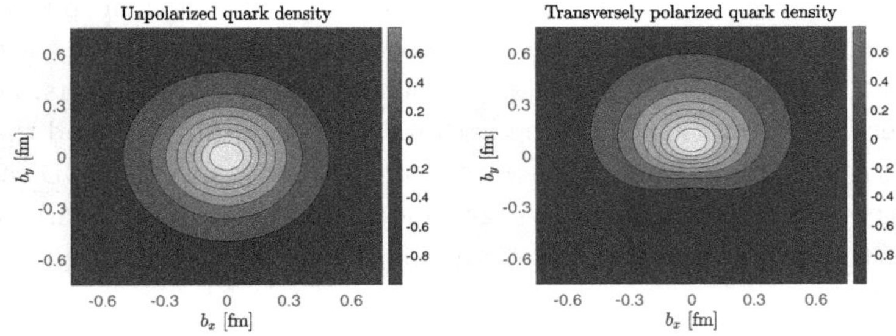

Fig. 2. Transverse spin densities of pion for unpolarized quark (left panel) and transversely polarized (along $\hat{x}$) quark (right panel).

2. Wigner distributions

Wigner distributions are defined as[8]

$$\rho^{[\Gamma]}(x, \boldsymbol{b}_\perp, \boldsymbol{k}_\perp) = \int \frac{\mathrm{d}^2\Delta_\perp}{(2\pi)^2}\, e^{-i\boldsymbol{\Delta}_\perp\cdot\boldsymbol{b}_\perp}\, W^{[\Gamma]}(x, \boldsymbol{\Delta}_\perp, \boldsymbol{k}_\perp), \tag{1}$$

where $W^{[\Gamma]}(x, \boldsymbol{\Delta}_\perp, \boldsymbol{k}_\perp)$ is the generalized correlator at $\xi = \Delta^+/P^+ = 0$, and $\boldsymbol{b}_\perp$ is the impact parameter in the position space conjugate to $\boldsymbol{\Delta}_\perp$. Explicitly,

$$W^{[\Gamma]} = \int \frac{dz^-\, d^2\boldsymbol{z}_\perp}{2(2\pi)^3}\, e^{ik\cdot z}\, \langle p' \,|\, \bar{\psi}(-\tfrac{1}{2}z)\,\Gamma\,\mathcal{W}\,\psi(\tfrac{1}{2}z)\,|\, p\rangle \Big|_{z^+=0}, \tag{2}$$

with $\Gamma \equiv \{\gamma^+, \gamma^+\gamma^5, i\sigma^{j+}\gamma_5\}$. We define the initial and final momenta of the pion in a symmetric frame as $p' = (p^+, p'^-, \frac{\Delta_\perp}{2})$ and $p'' = (p^+, p''^-, -\frac{\Delta_\perp}{2})$, respectively. For instance,

$$\begin{aligned}
W^{[\gamma^+]}(x, \boldsymbol{\Delta}_\perp, \boldsymbol{k}_\perp) &= \sum_{h',h,\bar{h}} \Psi^*_{h'\bar{h}}(x, \boldsymbol{k}''_\perp)\, \chi^\dagger_{h'}\, \chi_h\, \Psi_{h\bar{h}}(x, \boldsymbol{k}'_\perp), \\
W^{[\gamma^+\gamma_5]}(x, \boldsymbol{\Delta}_\perp, \boldsymbol{k}_\perp) &= \sum_{h',h,\bar{h}} \Psi^*_{h'\bar{h}}(x, \boldsymbol{k}''_\perp)\, \chi^\dagger_{h'}\, \sigma_3\, \chi_h\, \Psi_{h\bar{h}}(x, \boldsymbol{k}'_\perp), \\
W^{[i\sigma^{+j}\gamma_5]}(x, \boldsymbol{\Delta}_\perp, \boldsymbol{k}_\perp) &= \sum_{h',h,\bar{h}} \Psi^*_{h'\bar{h}}(x, \boldsymbol{k}''_\perp)\, \chi^\dagger_{h'}\, \sigma_j\, \chi_h \Psi_{h\bar{h}}(x, \boldsymbol{k}'_\perp),
\end{aligned}$$

where σ_i are the Pauli spin matrices and χ_h is the helicity spinor. The arguments $\boldsymbol{k}'_\perp$ and $\boldsymbol{k}''_\perp$ of the light-front wavefunctions are given by $\boldsymbol{k}'_\perp = \boldsymbol{k}_\perp - (1-x)\frac{\boldsymbol{\Delta}_\perp}{2}$, and $\boldsymbol{k}''_\perp = \boldsymbol{k}_\perp + (1-x)\frac{\boldsymbol{\Delta}_\perp}{2}$. One then can classify the unpolarized, longitudinally polarized and transversely polarized Wigner distributions in pion as: $\rho_{UU}(x, \boldsymbol{b}_\perp, \boldsymbol{k}_\perp) = \rho^{[\gamma^+]}(x, \boldsymbol{b}_\perp, \boldsymbol{k}_\perp)$, $\rho_{UL}(x, \boldsymbol{b}_\perp, \boldsymbol{k}_\perp) = \rho^{[\gamma^+\gamma_5]}(x, \boldsymbol{b}_\perp, \boldsymbol{k}_\perp)$, and $\rho_{UT}(x, \boldsymbol{b}_\perp, \boldsymbol{k}_\perp) = \rho^{[i\sigma^{j+}\gamma_5]}(x, \boldsymbol{b}_\perp, \boldsymbol{k}_\perp)$, respectively.

We compute the pion Wigner distributions using the spin-improved holographic light-front wavefunctions given by[14]

$$\Psi_{h,\bar{h}}(x, \boldsymbol{k}) = \left[(M_\pi x\bar{x} + Bm_f)h\delta_{h,-\bar{h}} - Bk_\perp e^{-ih\theta_{k_\perp}}\delta_{h,\bar{h}}\right] \frac{\Psi(x, k_\perp^2)}{x\bar{x}}. \quad (3)$$

We refer to B as the dynamical spin parameter. $B \to 0$ means no spin-orbit correlations as in the original holographic wavefunction,[16] while $B \geq 1$ corresponds to a maximal spin-orbit correlations. With $B \geq 1$, $m_{u/d} =$ 330 MeV and a universal AdS/QCD scale, $\kappa = 523$ MeV, we successfully predict simultaneously the pion decay constant, charge radius, spacelike electromagnetic and transition form factors, the pion parton distribution functions after taking into account perturbative QCD evolution.

3. Results

In Fig. 1, we show the first Mellin moments of Wigner distributions ρ_{UU}, ρ_{UL} and ρ_{UT} for the pion in the upper, central and lower panels. The left panels plot the distributions in the impact-parameter space with fixed transverse momentum $\boldsymbol{k}_\perp = k_\perp \hat{\boldsymbol{e}}_y$ and $k_\perp = 0.3$ GeV, while the right panels plot the distributions in the transverse-momentum space with fixed impact parameter $\boldsymbol{b}_\perp = b_\perp \hat{\boldsymbol{e}}_y$ and $b_\perp = 0.3$ fm. We find no distortions in unpolarized quark distributions in both the transverse momentum space and the impact parameter space. They both are circularly symmetric. However, we observe the dipolar distortion patterns for the longitudinally polarized quark in both spaces and for the transversely polarized quark in impact parameter space only. For the longitudinally polarized quark, the polarity of the impact space distribution is opposite to that in momentum space. When the quark is transversely polarized along x-direction, the deformation in $b_\perp$ space appears in y-direction. These deformation patterns are similar to those of the valence quark distributions of the proton considered as a quark-diquark system.[11]

Lattice QCD calculations give access to x moments of quark spin densities. To compare with lattice results, we compute the spin density for the transversely polarized quark:

$$\rho_T(\boldsymbol{b}_\perp) = \frac{1}{2}\int dx\, d^2\boldsymbol{k}_\perp\, x\, \left[\rho_{UU}(x, \boldsymbol{b}_\perp, \boldsymbol{k}_\perp) + s_T\, \rho_{UT}(x, \boldsymbol{b}_\perp, \boldsymbol{k}_\perp)\right]. \quad (4)$$

Fig. 2 shows that the unpolarized density is axially symmetric with the peak at the center of pion ($b_\perp = 0$), while due to the dipolar distortion from ρ_{UT}, the resulting ditribution for a transversely polarized (along +ve $\hat{x}$) quark gets shifted toward positive $\hat{y}$. Our predictions for spin distributions are in qualitative agreement with lattice results.[17]

Acknowledgements

MA and RS are supported by NSERC (Canada) Grants: SAPIN-2017-00033 and SAPIN-2017-00031, respectively. CM is supported by NSFC (China) under the Grant No. 11850410436. JPV is supported by the DoE under Grants No. DE-FG02-87ER40371, and No. DE-SC0018223 (SciDAC4/NUCLEI). XZ is supported by new faculty startup funding by IMP-CAS by Key Research Program of Frontier Sciences, CAS Grant No. ZDBS-LY-7020.

References

1. X. Ji, Phys. Rev. Lett. **91** 062001 (2003).
2. X. Ji, J. Phys. G **24** 1181 (1998).
3. K. Goeke *et al.*, Prog. Part. Nucl. Phys. **47** 401 (2001).
4. P. Mulders and R. Tangerman, Nucl. Phys. B **461**, 197 (1996).
5. D. Boer and P. J. Mulders, Phys. Rev. D **57**, 5780 (1998).
6. M. Burkardt, Int. J. Mod. Phys. A **18**, 173 (2003).
7. C. Lorcé and B. Pasquini, Phys. Rev. D **84**, 034039 (2011).
8. C. Lorcé and B. Pasquini, Phys. Rev. D **84**, 014015 (2011).
9. A. Mukherjee *et al.*, Phys. Rev. D **90**, 014024 (2014).
10. T. Liu and B.-Q. Ma, Phys. Rev. D **91**, 034019 (2015).
11. D. Chakrabarti, T. Maji, C. Mondal and A. Mukherjee, Eur. Phys. J. C **76**, no. 7, 409 (2016).
12. D. Chakrabarti, T. Maji, C. Mondal and A. Mukherjee, Phys. Rev. D **95**, no. 7, 074028 (2017).
13. N. Kumar and C. Mondal, Nucl. Phys. B **931**, 226 (2018).
14. M. Ahmady, C. Mondal and R. Sandapen, Phys. Rev. D **98**, 034010 (2018).
15. M. Ahmady, C. Mondal and R. Sandapen, Phys. Rev. D **100**, no. 5, 054005 (2019).
16. S. J. Brodsky *et al.*,Phys. Rept. **584**, 1 (2015).
17. D. Brommel *et al.*, Phys. Rev. Lett. **101**, 122001 (2008).

Valence structures of light and strange mesons from the basis light-front quantization framework

Shaoyang Jia* and James P. Vary

Department of Physics and Astronomy, Iowa State University, Ames, Iowa 50011, USA

**E-mail: sjia@iastate.edu*

We apply the basis light-front quantization (BLFQ) framework to solve for the structures of mesons with light and strange valence quarks. Our approach treats mesons as relativistic bound states with quarks confined in both the transverse direction and the light-front longitudinal direction. The spin-orbit interactions of these confined quarks are further specified by the Nambu–Jona-Lasinio model. We address the $\mathrm{U}(1)_{\mathrm{A}}$ axial anomaly by including the Kobayashi-Maskawa-'t Hooft interaction regularized by our basis. We present the structures of the pion, the kaon, the eta meson, and the eta-prime meson in terms of their valence light-front wave functions obtained from the eigenvalue problem of our light-front Hamiltonian.

Keywords: Hadron struecture; meson; light-front quantization.

1. Introduction to basis light-front quantizaton

The light-front quantization framework[1] for quantum field theories specifies the commutation relation for field operators at equal light-front time $x^+ = x^0 + x^3$. Orthogonal to x^+, the light-front longitudinal coordinate is defined as $x^- = x^0 - x^3$. The momenta conjugate to the light-front time and longitudinal coordinate are given by $p^\mp = p^0 \mp p^3$, leaving the transverse momentum specified by $\vec{p}^{\perp} = (p^1, p^2)$. Because of the light-front quantization conditions, the Hamiltonian of the field theory is a well-defined hermitian operator in the Hilbert space covering all relevant Fock sectors. The resulting Hamiltonian dynamics allows non-relativistic many-body methods to be applied to problems in quantum field theories.[2]

Bound states structures are obtained from the wave functions as solutions of the light-front Schrödinger equation

$$\mathcal{H}_{\mathrm{eff}}|\Psi\rangle = M^2|\Psi\rangle. \tag{1}$$

Here the effective Hamiltonian is defined as $\mathcal{H}_{\text{eff}} = P^+\mathcal{P}^- - \vec{P}^{\perp 2}$, where the light-front energy operator $\mathcal{P}^-$ is obtained from the Lagrangian using the Legendre transform.[1] While $|\Psi\rangle$ is the light-front state vector for the bound state with mass M and conserved light-front 3-momentum $(P^+, \vec{P}^\perp)$. We truncate the meson state vector to the valence Fock sector such that[3]

$$|\Psi(P^+, \vec{P}^\perp)\rangle = \sum_{\text{p},\text{q},r,s} \int_0^1 \frac{dx}{4\pi x(1-x)} \int \frac{d\vec{\kappa}^\perp}{(2\pi)^2} \psi_{\text{pq}rs}(x, \vec{\kappa}^\perp) \times b^\dagger_{\text{p}r}(xP^+, \vec{\kappa}^\perp + x\vec{P}^\perp)\, d^\dagger_{\text{q}s}((1-x)P^+, -\vec{\kappa}^\perp + (1-x)\vec{P}^\perp)|0\rangle, \quad (2)$$

with $b^\dagger_{\text{p}r}$ and $d^\dagger_{\text{q}s}$ being the creation operators for the quark and the anti-quark respectively. The valence light-front wave function (LFWF) $\psi_{\text{pq}rs}(x, \vec{\kappa}^\perp)$ depends on the quark longitudinal momentum fraction x and the relative transverse momentum $\vec{\kappa}^\perp$. The non-italic letters in the subscripts are flavor labels, leaving the italic subscripts specifying spins.

Let us adopt the kinetic energy, the transverse confinement potential, and longitudinal confinement potential collectively in H_0 defined in Eq. (10) of Ref. 3 to construct the basis functions in the leading Fock sector of mesons.[4] We then decompose the valence LFWF in basis functions:

$$\psi_{\text{pq}rs}(x, \vec{\kappa}^\perp) = \sum_{n,m,l} \psi_{\text{pq}}(n, m, l, r, s)\, \phi_{nm}\left(\vec{\kappa}^\perp / \sqrt{x(1-x)}\right) \chi_l(x). \quad (3)$$

Each term in Eq. (3) is an eigenfunction of H_0. The transverse basis function ϕ_{nm}, the longitudinal basis function χ_l, and the corresponding eigenvalue are identical to those in Refs. 3,4.

With the basis-diagonal Hamiltonian H_0 supplemented by an effective one-gluon exchange interaction, one can solve for the valence structures of heavy quakonia and B_c mesons.[4,5] However for light and strange mesons, we supplement H_0 by chiral interactions instead of gauge interactions.

2. The three-flavor Nambu–Jona-Lasinio interactions

When the explicit contributions to the structures of QCD bound states from gluons are omitted, the dynamics of quarks are specified by the global chiral symmetry, resulting in quark interactions given by the Nambu–Jona-Lasinio (NJL) model. Explicitly, the Lagrangian in the scalar channel of the color-singlet NJL model is given by[6]

$$\mathcal{L}^{(4)}_{\text{NJL,SU(3)}} = \overline{\psi}(i\partial\!\!\!/ - m)\psi + G_\pi \sum_{i=0}^{8} \left[\left(\overline{\psi}\lambda^i\psi\right)^2 + \left(\overline{\psi}i\gamma_5\lambda^i\psi\right)^2 \right]. \quad (4)$$

Here the flavor decomposition of the fermion field is given by $\psi = (\mathrm{u}, \mathrm{d}, \mathrm{s})^{\mathrm{T}}$. Meanwhile, λ^i are the flavor-space Gell-Mann matrices.

In order to account for the breaking of the $\mathrm{U}(1)_{\mathrm{A}}$ symmetry in the NJL model, the Kobayashi-Maskawa-'t Hooft determinant terms in the form of

$$\mathcal{L}_{\mathrm{det}} = G_{\mathrm{D}} \left[\det \overline{\psi}(1+\gamma_5)\psi + \det \overline{\psi}(1-\gamma_5)\psi\right] \tag{5}$$

are supplemented to Eq. (4). Specifically, determinants in Eq. (5) are taken in the flavor space, resulting in 6-fermion interactions. Working in the meson valence Fock sector we need to reduce interactions in Eq. (5) into those with up to 4 fermions, where the following divergent integral is encountered:

$$\begin{aligned}\mathcal{R}(\alpha, \beta, \kappa, L_{\mathrm{max}}, N_{\mathrm{max}}) \equiv & \int_0^{+\infty} \frac{dp_1^+}{4\pi p_1^+} \int \frac{d\vec{p}_1^{\perp}}{(2\pi)^2} \int_0^{+\infty} \frac{dp_2^+}{4\pi p_2^+} \int \frac{d\vec{p}_2^{\perp}}{(2\pi)^2} \\ & \times (4\pi p_1^+)\delta(p_1^+ - p_2^2)\,(2\pi)^2\delta\left(\vec{p}_1^{\perp} - \vec{p}_2^{\perp}\right)\Big|_{(\text{basis-regularized})}.\end{aligned} \tag{6}$$

To regularized Eq. (6), recall that in the practice of solving for the meson LFWF numerically, the following basis cut-offs are implemented[3]:

$$0 \le l \le L_{\mathrm{max}}, \quad 0 \le n \le N_{\mathrm{max}}, \quad \text{and} \quad -M_{\mathrm{max}} \le m \le M_{\mathrm{max}}. \tag{7}$$

After expanding the delta functions in terms of basis functions within the basis cutoffs defined by Eq. (7), Eq. (6) becomes finite. Further, because the integrals over longitudinal momentum fractions carry no dimension while the divergence does, the limit of $L_{\mathrm{max}} \to +\infty$ can be safely reached. Consequently, we adopt $\mathcal{R}(\alpha, \beta, \kappa, +\infty, N_{\mathrm{max}}) = \kappa^2(N_{\mathrm{max}}+1)/(8\pi^2)$ as the basis-regularized version of Eq. (6).

3. Results for the π^0, ρ^0, η, and η'

After implementing the basis regularization, we obtain the following reduction of the 6-fermion interaction to the 4-fermion interaction:

$$\det \overline{\psi}(1 \pm \gamma^5)\psi \to -2\mathbf{m}_{\mathrm{a}}\mathcal{R}_{\mathrm{a}}\, \epsilon_{\mathrm{abc}}\epsilon_{\mathrm{agh}}\, \overline{\psi}_{\mathrm{b}}(1 \pm \gamma^5)\psi_{\mathrm{g}}\, \overline{\psi}_{\mathrm{c}}(1 \pm \gamma^5)\psi_{\mathrm{h}}, \tag{8}$$

where $\mathcal{R}_{\mathrm{a}}$ is the $\mathcal{R}(\alpha, \beta, \kappa, L_{\mathrm{max}}, N_{\mathrm{max}})$ for the quark of flavor "a". We work in the SU(2) isospin symmetric limit, which prompts the decomposition of the valence LFWF into isospin symmetric and anti-symmetric components. The Hamiltonian then becomes block-diagonal for the isospin scalar states and the isospin vector states both in the flavor-nonet channel.

Similar to Ref. 3, we ignore instantaneous interactions and self-energy corrections. The chiral interaction term of the Hamiltonian in the BLFQ-

Table 1. Input parameters for the BLFQ-NJL model of the η and η' mesons.

m_{l}	κ_{ll}	m_{s}	κ_{ss}	G_π	$N_{\max}$	$M_{\max}$	$L_{\max}$
300 MeV	400 MeV	450 MeV	400 MeV	1.7 GeV^{-2}	8	2	8

Table 2. Masses, mixing angles, and decay constants of the η and η' mesons.

	m_η (MeV)	$m_{\eta'}$ (MeV)	θ_8	θ_0	f_8 (MeV)	f_0 (MeV)
BLFQ	554	939	4.23°	16.4°	211	89.4
Ref. 7	547.86	957.8	-	-	-	-
Ref. 8	-	-	-21.2°	-9.2°	164.1	152.3

NJL model then becomes

$$H_{\text{NJL}}^{\text{eff}} = \int dx^2 \int d\vec{x}^{\perp} \Big\{ - G_\pi P^+ \sum_{i=0}^{8} \left[\left(\overline{\psi}\gamma^i\psi\right)^2 - \left(\overline{\psi}\gamma^5\lambda^i\psi\right)^2 \right] \\ - G_{\text{D}} P^+ \left[\det \overline{\psi}\left(1+\gamma^5\right)\psi + \det \overline{\psi}\left(1-\gamma^5\right)\psi\right] \Big\}. \tag{9}$$

Equation (9) supplements the basis-diagonal Hamiltonian H_0 to account for the chiral dynamics that is relevant for the light and the strange mesons. We then solve for the eigenstates from Eq. (1) using the combined Hamiltonian in our basis representation. For the isospin-vector mesons, we choose the strength of the determinant interaction such that $m_{\text{s}}\mathcal{R}_{\text{s}}G_{\text{D}} = -G_\pi$. Together with the same m_{l} and κ_{ll} as those in Ref. 3, the LFWF of the neutral pion is identical to that of the charged pion in momentum space.

To solve for the flavor wave function of the η and η', we apply the two-angle mixing scheme to calculate the decay constants and mixing angles.[9–11] Also with the condition that $m_{\text{s}}\mathcal{R}_{\text{s}}G_{\text{D}} = -G_\pi$, we obtain preliminary results for the masses, mixing angles, and decay constants of the η and η' mesons as listed Table 2 using input parameters in Table 1.

4. Conclusion

We have expanded the basis light-front quantization framework with confinement and the Nambu–Jona-Lasinio interactions previously applied to the charged light and strange mesons[3] to describe the valence structures of flavor-nonet mesons. Specifically, we have developed the regularization of divergences in the reduction of multi-fermion vertices consistent with our basis truncation. With a rough estimate of the model parameters, we obtained preliminary results for the masses, decay constants, and mixing angles of the η and η' mesons. With future searches in the model parameter space, we expect to find a consistent description of the valence structures for η, η', ω, and ϕ mesons within the BLFQ-NJL model.

Acknowledgment

This work was supported by the Department of Energy under Grants No. DE-FG02-87ER40371 and No. DE-SC0018223 (SciDAC4/NUCLEI).

References

1. S. J. Brodsky, H.-C. Pauli and S. S. Pinsky, Quantum chromodynamics and other field theories on the light cone, *Phys. Rept.* **301**, 299 (1998).
2. J. P. Vary, H. Honkanen, J. Li, P. Maris, S. J. Brodsky, A. Harindranath, G. F. de Teramond, P. Sternberg, E. G. Ng and C. Yang, Hamiltonian light-front field theory in a basis function approach, *Phys. Rev.* **C81**, 035205 (2010).
3. S. Jia and J. P. Vary, Basis light front quantization for the charged light mesons with color singlet Nambu–Jona-Lasinio interactions, *Phys. Rev.* **C99**, 035206 (2019).
4. Y. Li, P. Maris and J. P. Vary, Quarkonium as a relativistic bound state on the light front, *Phys. Rev.* **D96**, 016022 (2017).
5. S. Tang, Y. Li, P. Maris and J. P. Vary, B_c mesons and their properties on the light front, *Phys. Rev.* **D98**, 114038 (2018).
6. S. Klimt, M. F. M. Lutz, U. Vogl and W. Weise, Generalized SU(3) Nambu–Jona-Lasinio Model. part. 1. mesonic modes, *Nucl. Phys.* **A516**, 429 (1990).
7. M. Tanabashi *et al.*, Review of Particle Physics, *Phys. Rev.* **D98**, 030001 (2018).
8. T. Feldmann, P. Kroll and B. Stech, Mixing and decay constants of pseudoscalar mesons, *Phys. Rev.* **D58**, 114006 (1998).
9. J. Schechter, A. Subbaraman and H. Weigel, Effective hadron dynamics: From meson masses to the proton spin puzzle, *Phys. Rev.* **D48**, 339 (1993).
10. T. Feldmann and P. Kroll, Mixing of pseudoscalar mesons, *Phys. Scripta* **T99**, 13 (2002).
11. R. Escribano and J.-M. Frere, Study of the eta - eta-prime system in the two mixing angle scheme, *JHEP* **06**, 029 (2005).

Session 7

Hadrons in Hot and Huclear Environment Including Hypernuclei

Light flavor baryon production from small to large collision systems at ALICE

D. Colella for the ALICE Collaboration

Istituto Nazionale di Fisica Nucleare,
via E. Orabona 4
Bari, 70125, Italy
domenico.colella@cern.ch

Studies of light hadron and nuclei production are fundamental to characterize the hot and dense fireball created in ultra-relativistic heavy ion collisions and to investigate hadronisation mechanisms at the LHC. Observables investigated as a function of the charged particle multiplicity in proton-proton and proton-lead collisions have shown features not expected and qualitatively similar to what has been observed in larger size colliding systems. The ALICE experiment, exploiting its excellent tracking and PID capabilities, has performed an extensive and systematic study of strange and non-strange hadrons, short-lived hadron resonances and light (anti-)(hyper-)nuclei. A critical overview of these results will be presented through comparison with the statistical hadronisation model.

Keywords: Hot matter; QGP; SHM; light flavour.

1. Introduction

To the best of our knowledge, hot and dense matter is produced in heavy-ion collisions at the Large Hadron Collider (LHC) energies. The system cools down and undergoes a transition to hadron gas. While the particle yields are fixed at the moment when the rate of inelastic collisions becomes negligible (chemical freeze-out), the transverse momentum distributions continue to change until elastic interactions cease (kinetic freeze-out). Studying hadrons and light nuclei containing light flavor and strange quarks is fundamental to better characterize the scenario described above. The yield of light flavor hadrons containing strange quarks is measured to study the dependency of the strangeness enhancement by event activity, colliding system, colliding energy and strangeness content of the particles and to verify if it is present also for particles with hidden strangeness. Production yield of short-lived resonances with lifetimes comparable to that of the fireball is measured to study rescattering and regeneration processes in the dense hadronic medium. The yield of nuclei and anti-nuclei in heavy-ion collisions provide information about the chemical freeze-out temperature and it is very intriguing to understand how such loosely bound systems can survive the hot fireball.

The ALICE Collaboration performed a systematic study of non-strange hadrons (π^+, K^+, p), single and multi-strange weak decaying hadrons (K^0_S, Λ, Ξ^-, Ω^-), short lived resonances ($\rho(770)^0$, $K^*(892)^0$, $\phi(1020)$, $\Sigma^*(1385)^+$, $\Lambda(1520)$, $\Xi(1530)^0$), light nuclei (d, t, ^{3}He and $^3_\Lambda$H) and the corresponding antiparticles, in many colliding systems at various energies provided by the LHC.[1–11] The resonances will be referred to only by their commonly used symbol.

2. Results and observations

Particle yields in 0-10% central Pb–Pb collisions at $\sqrt{s_{NN}} = 5.02$ TeV are compared to predictions from three Statistical Hadronization Models (SHMs), each based on a grand-canonical ensemble, and are shown in the left panel of Figure 1. The models assume hadron production at chemical equilibrium and reproduce most of the measured yields within uncertainties. The estimated chemical freeze-out temperature is about 153 MeV, same as the value obtained in describing the data in Pb–Pb collisions at $\sqrt{s_{NN}} = 2.76$ TeV.[11,12] A remarkable exception is the K^{*0}, for which the deficit with respect to the predicted yield can be explained by loss of K^{*0} signal in the hadronic phase. There is also a tension for protons and multi-strange baryons whose explanation requires additional effects (baryon annihilation, interactions in the hadron gas or feed-down from excited hadronic states, etc.).[3,6] It is notable that the models also reproduce production of nuclei and hyper-nuclei although their binding energies are much smaller than the extracted values of the chemical freeze-out temperature. Overall, SHMs are quite successful and describe yields of production of hadrons which vary over seven orders of magnitude.

One of the most remarkable results established at the LHC is the smooth evolution of particle chemistry from small to large systems as a function of charged particle multiplicity.[1] The particle yields normalized to charged-pion yield evolve monotonically and continuously as a function of charged particle multiplicity independent of collision energy and system size, as shown in Figure 1 (right). These data also demonstrate, for the first time, the presence of strangeness enhancement in high multiplicity collisions of small systems, the effect being stronger for particles with higher strangeness

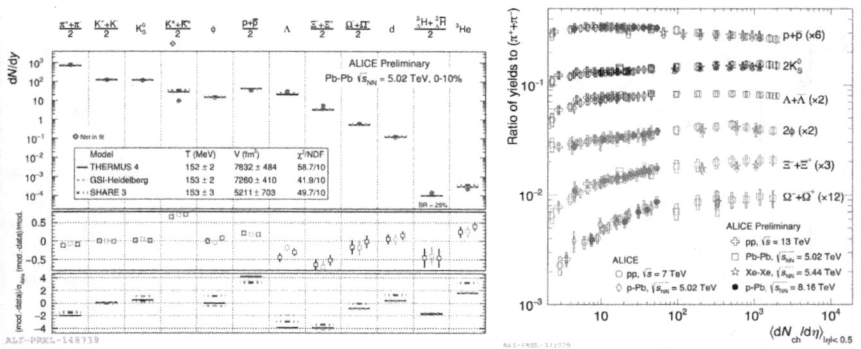

Fig. 1. (left) Particle yields compared to three grand-canonical SHM predictions in 0-10% central Pb–Pb collisions at $\sqrt{s_{\rm NN}} = 5.02$ TeV. (right) Ratios of particle yields measured as a function of multiplicity across different collision systems.

content. The original interpretation of this phenomenon, considered as a signature of QGP formation in heavy-ion collisions, is not anymore straightforward. In the strangeness canonical approach, the multiplicity dependence can be explained by requiring local strangeness conservation while the bulk of the particles can still be described in the grand-canonical ensemble. Deviation from this description is present for the K^{*0}, while a better description is obtained for Ξ and Ω applying core-corona corrections. An interesting deviation, is also present for the ϕ meson: being a strangeness-neutral particle, a flat multiplicity dependence is predicted, but a trend is observed.

Effect of particle rescattering and regeneration in the hadronic phase can be probed by measuring the yield of hadronic states having a lifetime comparable to the one of the fireball. In this picture the resonance yields in the final state are determined by the resonance yields at chemical freeze-out, their lifetimes, hadronic phase lifetime and by scattering cross sections. Figure 2 (left) shows ratios of the resonance yields to yields of stable hadrons having similar constituent quarks, for a set of particles covering a wide range of lifetimes from 1.3 fm/c for the ρ^0 meson up to 46.4 fm/c for the ϕ meson. The ratios are suppressed in central heavy-ion collisions for resonances with lifetimes $\tau < 20$ fm/c compared to pp and peripheral heavy-ion collisions. For longer-lived resonances, as the ϕ, ratios are not suppressed. These results support the existence of a hadronic phase that lives long enough to cause a significant reduction of the reconstructed yields of short lived resonances. Resonance measurements allow for estimates of the hadronic

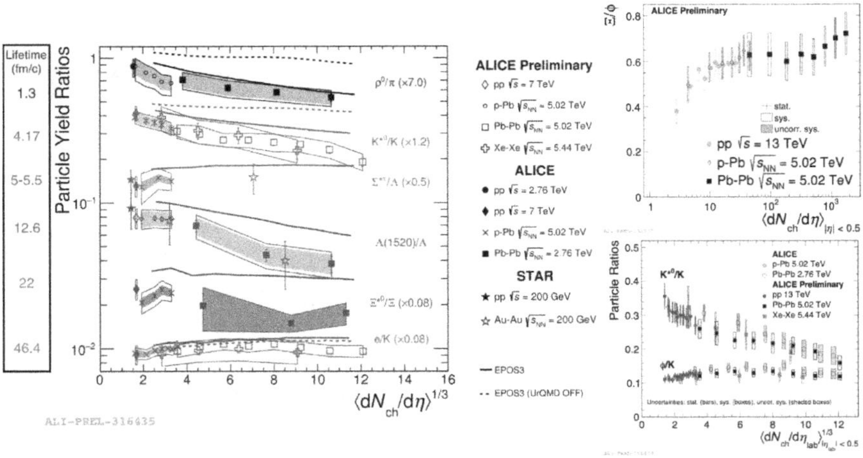

Fig. 2. (left) ρ^0/π, K^{*0}/K, $\Sigma^{\pm}/\Lambda$, $\Lambda(1520)/\Lambda$, Ξ^0/Ξ and ϕ/K yield ratios as a function of multiplicity. (right) Ξ/ϕ and ϕ/K yield ratios as a function of multiplicity.

phase lifetime: 10 fm/c for the most central Pb–Pb collisions and 1-2 fm/c for the most peripheral Pb–Pb collisions. These values are estimated from the EPOS+UrQMD model.[13]

In order to investigate the behavior of hidden strangeness, the following two ratios have been measured as a function of charged particle multiplicity and are shown in Figure 2 (right): $\Xi(|S|{=}2)/\phi(|S|{=}0)$ and $\phi(|S|{=}0)/K(|S|{=}1)$. The multiplicity evolution of these two ratios suggests that the ϕ meson behaves as if it had between 1 and 2 units of strangeness, so that the Ξ is more enhanced than ϕ, which is more enhanced than K.

Light nuclei are characterised by a low binding energy (EB $\sim$ 1 MeV) compared to the temperature of the chemical freeze-out, ($T_{ch} \sim$ 160 MeV). Therefore, in principle one would not expect to observe any nucleus. The experimental results on their production are usually interpreted within two approaches. According to the SHM, (anti-)nuclei are produced at the chemical freeze-out in statistical equilibrium, along with all the other hadrons. As already shown this model nicely predict the yields for d, ${}^3_\Lambda$H and ^{3}He in 0-10% central Pb–Pb collisions. In the coalescence picture, nucleons that are close to each other in phase space after chemical freeze-out can merge and form a nucleus via coalescence. In Figure 3 the coalescence parameter for the deuterons (B_2) is shown as a function of charged particle multiplicity in many colliding systems.[2,10] This is the main observable of the model, related to the probability to form a nucleus via coalescence. B_2

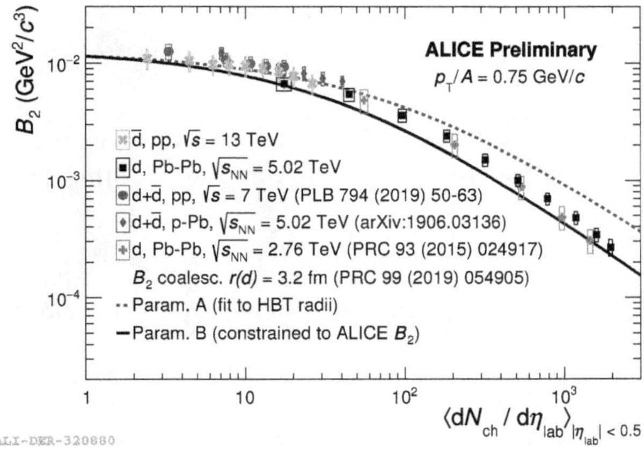

Fig. 3. Anti-deuteron coalescence parameter B_2 at fixed $p_T/A = 0.75$ GeV/c as a function of the average charged-particle multiplicity.

is independent of the collision system or the energy, and depends only on the charged particle density. This suggest a common production mechanism that depends only on the system size. In the current state more data and more precise model calculations are needed to discern the production mechanisms of light nuclei.

References

1. ALICE Collaboration, S. Acharya *et al.* Phys. Rev. C 99 (2019) 024906.
2. ALICE Collaboration, S. Acharya *et al.* Phys. Rev. C 97 (2018) 024615.
3. ALICE Collaboration, B. Abelev *et al.* Phys. Lett. B 728 (2014) 2538.
4. ALICE Collaboration, J. Adam *et al.* Euro. Phys. J. C (2016) 76-245.
5. ALICE Collaboration, J. Adam *et al.* Phys. Lett. B 758 (2016) 389401.
6. ALICE Collaboration, B. Abelev *et al.* Phys. Rev. C 88 (2013) 044910.
7. ALICE Collaboration, B. Abelev *et al.* Phys. Rev. Lett. 111 (2013) 222301.
8. ALICE Collaboration, B. Abelev *et al.* Phys. Lett. B 728 (2014) 216227.
9. ALICE Collaboration, B. Abelev *et al.* Phys. Rev. C 91 (2015) 024609.
10. ALICE Collaboration, J. Adam *et al.* Phys. Rev. C 93 (2016) 024917.
11. ALICE Collaboration, S. Acharya *et al.* Nucl. Phys. A 971 (2018) 120.
12. M. Floris Nucl. Phys. A 931 (2014) 103-112
13. A.G. Knospe *et al.* Phys. Rev. C 93 (2016) 014911.

Study of $\bar{K}N$ interaction from the hadron-hadron correlation in high-energy nuclear collisions

Y. Kamiya

CAS Key Laboratory of Theoretical Physics, Institute of Theoretical Physics, Chinese Academy of Sciences, Zhong Guan Cun East Street 55, Beijing 100190, China
E-mail: yukikamiya@mail.itp.ac.cn

T. Hyodo

Yukawa Institute for Theoretical Physics, Kyoto University, Kyoto 606-8502, Japan
Department of Physics, Tokyo Metropolitan University, Hachioji 192-0397, Japan
E-mail: @laboratory.com

K. Morita

RIKEN Nishina Center, Wako 351-0198, Japan
National Institutes for Quantum and Radiological Science and Technology, Rokkasho Fusion Institute, Rokkasho, Aomori, 039-3212, Japan
E-mail: morita.kenji@qst.go.jp

A. Ohnishi

Yukawa Institute for Theoretical Physics, Kyoto University, Kyoto 606-8502, Japan
E-mail: ohnishi@yukawa.kyoto-u.ac.jp

We study the hadron-hadron momentum correlation function of a K^-p pair from high-energy nuclear collisions. Based on the correlation formula for coupled-channel systems, a scheme to calculate the correlation consistently including the coupled-channel effects and the Coulomb interaction in a consistent manner is constructed. With the chiral SU(3) based realistic $\bar{K}N$-$\pi\Sigma$-$\pi\Lambda$ coupled-channel potentials, we calculate the K^-p correlation function and discuss the coupled-channel effect on the correlation function.

Keywords: $\bar{K}N$ interaction; hadron correlation; chiral SU(3) dynamics.

1. Introduction

Originally, the measurement of the hadron-hadron correlation from the high-energy nuclear collisions is utilized to study the property of the emission source of hadrons. In these studies, the information of the hadron interaction determined from the scattering experiments are utilized as input of the calculation of the correlation. Given the property of emission source, the correlation function can be used to extract the information of the interaction. The most advantageous feature of this method is that we can investigate the systems that are difficult to approach with the direct scattering experiments. Among these systems, one of the most interesting system is $\bar{K}N$. Recently, ALICE group has reported their analysis of the K^-p correlation function from high-multiplicity events of pp collisions.[1] The data show great precision from the very low-energy region so that the cusp structure at the $\bar{K}^0n$ threshold can be confirmed.

The theoretical studies on the K^-p correlation function are found in Refs. 2–4. With these studies, it has been pointed out that the coupled-channel effect, the Coulomb interaction, and threshold energy differences among the isospin multiplets must be treated carefully for the detailed study of the $\bar{K}N$ interaction from the correlation function data. However, there is no work which consistently includes these effects.

In this study, we investigate the K^-p correlation function with treating the coupled-channels effects, the Coulomb interaction, and the threshold energy differences in a consistent manner. We start our discussion from the coupled-channels version of the correlation function formula.[4,5] The careful treatment on the boundary condition of the coupled-channel wave function is discussed. Employing the realistic $\bar{K}N$-$\pi\Sigma$-$\pi\Lambda$ coupled-channels potential,[6] the K^-p correlation is calculated in the charged basis. From the results of the correlation and its breakdown to coupled-channel components, we discuss the importance of inclusion of the coupled-channel effect and the Coulomb interaction. All the contents in this paper are based on Ref. 7.

2. Formalism

Let us consider the correlation function of two asymptotically observed particles, a and b with relative momentum $\boldsymbol{q}$ in the pair rest frame. We consider the case where this particle pair couples to other hadron channels. Then the correlation function including coupled-channel effect is given as[4,5]:

$$C(\boldsymbol{q}) = \int d^3r \sum_j \omega_j S_j(\boldsymbol{r}) |\Psi_j^{(-)}(\boldsymbol{q};\boldsymbol{r})|^2 \,, \tag{1}$$

where the wave function $\Psi_j^{(-)}$ in the j-th channel is written as a function of the relative coordinate $\boldsymbol{r}$ in that channel, with outgoing boundary condition for the measured channel. $S_j(\boldsymbol{r})$ and ω_j are the (normalized) relative source function and its weight in the j-th channel. In general, these $S_j(\boldsymbol{r})$ and ω_j

can be different among coupled-channels, but for simplicity we use the common source of the spherical Gaussian with radius R given as $S_j(r) = S_R(r) \equiv \exp(-r^2/4R^2)/(4\pi R^2)^{3/2}$, and the common weight ($\omega_j = 1$).

Focusing on the small $q = |\boldsymbol{q}|$ region, we can assume that the dominant modification by the strong interaction of the wave function $\Psi_j^{(-)}(\boldsymbol{q};\boldsymbol{r})$ in Eq. (1) appears only in its s-wave part. With this assumption, the j-th channel component of the wave function with outgoing boundary condition for the charged particle pair is given as

$$\Psi_j^{(-)}(\boldsymbol{q};\boldsymbol{r}) = [\phi^C(\boldsymbol{q};\boldsymbol{r}) - \phi_0^C(qr)]\delta_{1j} + \psi_j^{(-)}(q;r), \tag{2}$$

where $r = |\boldsymbol{r}|$ and $\phi^C(\boldsymbol{q};\boldsymbol{r})$ is the free Coulomb wave function, $\phi_0^C(qr)$ is its s-wave component, and $\psi_j^{(-)}(q;r)$ represents the s-wave function that includes both strong and the Coulomb potential effects in the j-th channel.

To obtain the wave function $\psi_j^{(-)}(q;r)$, we need to solve the coupled-channels Schrödinger equation $H\Psi = E\Psi$ with Hamiltonian $H = K + V_{\text{Strong}} + V_{\text{Coulomb}}$, where $K = \text{diag}(-\nabla^2/(2\mu_j) + \Delta_j, \cdots)$ is the kinetic term, and V_{Strong} (V_{Coulomb}) is the strong (Coulomb) interaction potential with the reduced mass μ_j and the threshold energy difference relative to channel 1 Δ_j. V_{Coulomb} is diagonal and finite for the Coulomb attractive or repulsive channels while the other components are zero.

The out-going boundary condition at $r \to \infty$ for $\psi_j^{(-)}$ is given as

$$\psi_j^{(-)}(q;r) \to \frac{1}{2iq_j}\left[\delta_{1j}\,\frac{u_j^{(+)}(q_j r)}{r} + A_j(q)\,\frac{u_j^{(-)}(q_j r)}{r}\right], \tag{3}$$

for open channels ($E > \Delta_i$), where $u_j^{(+)}$ and $u_j^{(-)}$ are the outgoing and incoming waves with a coefficient $A_j(q)$. In absence of the Coulomb interaction, the $u_j^{(\pm)}/r$ are the spherical waves, $u_j^{(\pm)}(qr)/r = e^{\pm iqr}/r$. Including the Coulomb interaction we have $u_j^{(\pm)}(qr) = \pm e^{\mp i\sigma_j}[iF(qr) \pm G(qr)]$ with $\sigma_j = \arg\Gamma(1 + i\eta_j)$, $\eta_j = -\mu_j\alpha/q_j$, and the regular (irregular) Coulomb functions $F(qr)$ [$G(qr)$]. For closed channels ($E < \Delta_i$), the asymptotic form is given by substituting $q_j = -i\kappa_j = -i\sqrt{2\mu_j(\Delta_j - E)}$ as $\psi_j^{(-)}(r) \to A_j(q)\,u_j^{(-)}(-i\kappa_j r)/(2\kappa_j r) \propto e^{-\kappa_j r}/\kappa_j r$.

3. Result

We now employ the effective $\bar{K}N$-$\pi\Sigma$-$\pi\Lambda$ coupled-channel potential[6] for the calculation of the wave function $\psi_j^{(-)}$. This potential has been constructed to reproduce the coupled-channel amplitude of the chiral SU(3) dynamics at next-to-leading order,[8] which satisfies the constraint on the K^-p scattering

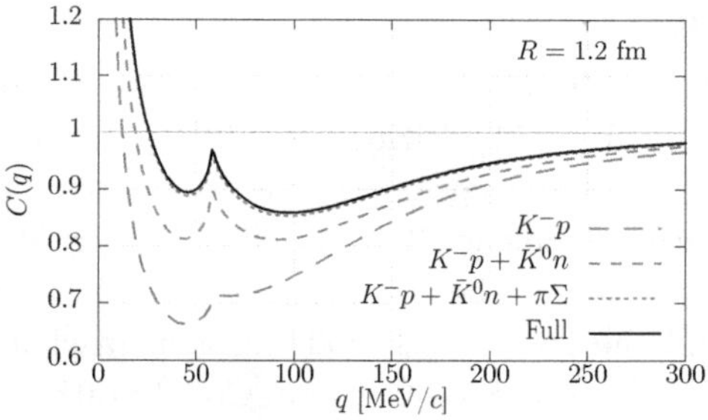

Fig. 1. K^-p correlation function with $R = 1.2$ fm. The long-dashed line denotes the result with K^-p component only. The short-dashed, dotted, and solid lines show the results in which the contributions from $\bar{K}^0n$, $\bar{K}^0n$ and $\pi\Sigma$, and from all coupled-channel components are added, respectively.

length by the SIDDHARTA kaonic hydrogen data.[9] We work within the particle basis and the six-channel (K^-p, $\bar{K}^0n$, $\pi^-\Sigma^+$, $\pi^0\Sigma^0$, $\pi^+\Sigma^-$ and $\pi^0\Lambda$) coupling is treated explicitly while the contribution from the other channels are implicitly included in the construction of the potential.

Now we show our results of the K^-p correlation function with the source size of $R = 1.2$ fm in Fig. 1. The strong enhancement due to the Coulomb interaction appears at the low-momenta. We can see the clear cusp structure at the $\bar{K}^0n$ threshold at $q \simeq 58$ MeV/c. From the breakdowns into the coupled-channel contribution, we can understand that $\bar{K}^0n$ and $\pi\Sigma$ sources give the crucial role to enhance the K^-p correlation while the contribution from the $\pi^0\Lambda$ source is almost negligible. We also find that the cusp structure becomes more prominent by the inclusion of $\bar{K}^0n$ component.

4. Conclusion

We have studied the K^-p correlation function from the high-energy nuclear collisions. Based on the coupled-channels correlation function formula,[4,5] a scheme to calculate the correlation consistently including all the effects of coupled channels, the Coulomb potential and threshold differences has been developed. The realistic coupled-channels potential,[6] which is constructed based on the next-to-leading order chiral SU(3) dynamics, has been employed. In accordance with the ALICE data,[1] the calculated K^-p correlation shows $\bar{K}^0n$ cusp and the strong enhancement at the low-momentum region due to the Coulomb interaction. The decomposition of the contributions from the coupled-channel sources reveals that the coupled-channel effect gives the crucial role to enhance the correlation.

Acknowledgments

This work is supported in part by the Grants-in-Aid for Scientific Research from JSPS (Nos. 19H05151, 19H05150, 19H01898, and 16K17694), by the Yukawa International Program for Quark-hadron Sciences (YIPQS), by the Polish National Science Center NCN under Maestro grant EC-2013/10/A/ST2/00106, by the National Natural Science Foundation of China (NSFC) and the Deutsche Forschungsgemeinschaft (DFG) through the funds provided to the Sino-German Collaborative Research Center (NSFC Grant No. 11621131001, DFG Grant No. TRR110), by the NSFC under Grant No. 11847612 and No. 11835015, by the Chinese Academy of Sciences (CAS) under Grant No. QYZDB-SSW-SYS013 and No. XDPB09, and the CAS President's International Fellowship Initiative (PIFI) under Grant No. 2020PM0020.

References

1. S. Acharya *et al.*, Scattering studies with low-energy kaon-proton femtoscopy in proton-proton collisions at the LHC (2019).
2. A. Ohnishi, K. Morita, K. Miyahara and T. Hyodo, Hadronhadron correlation and interaction from heavyion collisions, *Nucl. Phys.* **A954**, 294 (2016).
3. S. Cho *et al.*, Exotic Hadrons from Heavy Ion Collisions, *Prog. Part. Nucl. Phys.* **95**, 279 (2017).
4. J. Haidenbauer, Coupled-channel effects in hadron-hadron correlation functions, *Nucl. Phys.* **A981**, 1 (2019).
5. R. Lednicky, V. V. Lyuboshits and V. L. Lyuboshits, *Phys. Atomic Nuclei* **61**, p. 2950 (1998).
6. K. Miyahara, T. Hyodo and W. Weise, Construction of a local $\bar{K}N - \pi\Sigma - \pi\Lambda$ potential and composition of the $\Lambda(1405)$, *Phys. Rev.* **C98**, p. 025201 (2018).
7. Y. Kamiya, T. Hyodo, K. Morita, A. Ohnishi and W. Weise, K^-p correlation function from high-energy nuclear collisions and chiral SU(3) dynamics, arXiv:1911.01041 [hep-ph].
8. Y. Ikeda, T. Hyodo and W. Weise, Chiral SU(3) theory of antikaon-nucleon interactions with improved threshold constraints, *Nucl. Phys.* **A881**, 98 (2012).
9. M. Bazzi *et al.*, Kaonic hydrogen X-ray measurement in SIDDHARTA, *Nucl. Phys.* **A881**, 88 (2012).

Dense nuclear matter based on a chiral model with parity doublet structure

Masayasu Harada

Department of Physics; Nagoya University, Nagoya 464-8602, Japan
E-mail: harada.masayasu@nagoya-u.jp

In this talk, I summarize our recent works in Refs. 1 and 2, in which we constructed an extended parity doublet model to study four nucleons, $N(939)$, $N(1440)$, $N(1535)$ and $N(1650)$. We obtained several constraints to two chiral invariant masses from the properties of nucleons in the vacuum and the nuclear matter. Our results show that the smaller chiral invariant masses cause the larger tidal deformability, and that the constraint from the tidal deformability obtained from the neutron star merger observed through the gravitational wave GW170817 is satisfied when the chiral invariant mass of nucleon is larger than about 600 MeV.

Keywords: Parity doublet model; neutron star; high density nuclear matter.

1. Introduction

Investigation of the origin of hadron masses is one of the interesting problem in QCD. One key word will be the spontaneous chiral symmetry breaking caused by the chiral condensate made from a quark and an anti-quark. The chiral condensate is expected to generate a part of hadron masses and causes the mass difference between chiral partners. Another candidate for the origin is called the "chiral invariant mass", which may be caused by the gluon condensate. It is interesting to ask how much mass of nucleon is from the spontaneous chiral symmetry breaking, or what amount of chiral invariant mass is included in the nucleon mass.

A clue was obtained from the analysis of high density nuclear matter from a Skyrme-crystal model in Refs. 3 and 4. There, it was shown that the nucleon mass decreases with increasing density below the twice of the normal nuclear density, while it becomes stable in the high density region above twice of normal nuclear density. It was discussed that this behavior indicates existence of a chiral invariant mass.

For a quantitative analysis of the chiral invariant mass, it is convenient to use chiral models for nucleons including the chiral invariant mass. In the parity doublet model,[5,6] the chiral invariant mass is formulated, where the excited nucleon $N(1535)$ is regarded as the chiral partner to the ordinary nucleon, and both of them have the chiral invariant mass.

Recently in Refs. 7, 8, 9 and 2, we studied nuclear matter using models with parity doublet structure. It was shown that the chiral invariant mass of nucleon is constrained by the saturation properties at normal nuclear density, and neutron star properties such as the tidal deformability observed through the gravitational wave GW170817.[10–13]

2. An Extended Parity Doublet Model for Nucleons

In the model constructed in Ref. 1, we included two representations, $\psi \in [(1,2) \oplus (2,1)]$ and $\eta \in [(2,3) \oplus (3,2)]$ under the $SU(2)_L \times SU(2)_R$ symmetry to study four nucleons, $N(939)$, $N(1440)$, $N(1535)$ and $N(1650)$. The model includes two chiral invariant masses: $m_0^{(1)}$ for ψ and $m_0^{(2)}$ for η; six independent Yukawa interactions and four terms with one derivative.

	Mass	Width $[\Gamma_{N^* \to N\pi}]$	g_A
$N(939)$	939	-	1.272
$N(1440)$	1430	228	-
$N(1535)$	1535	68	-0.25-0.25
$N(1650)$	1655	84 $[N(939)\pi]$ 22 $[N(1440)\pi]$	0.55

Table 1. Physical inputs used to determine the couplings of nucleons to scalar and pseudoscalar mesons. Units of masses and widths are in MeV.

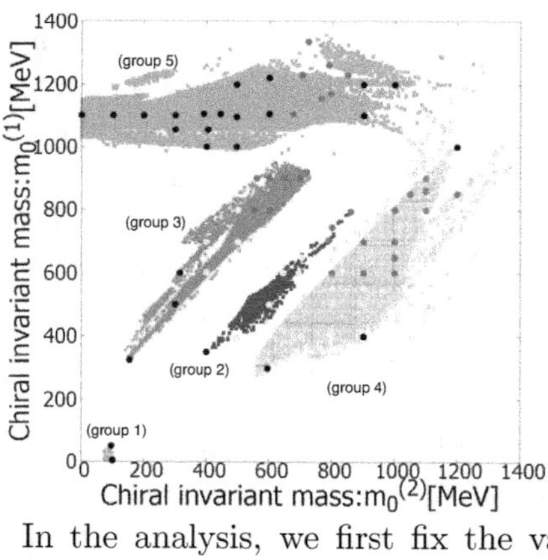

Fig. 1. Allowed region for two chiral invariant masses. Painted regions indicate the solutions which reproduce the physical inputs at vacuum. For combinations of two chiral invariant masses indicated by black points, the saturation properties of normal nuclear matter are not satisfied. For the yellow points the obtained tidal deformability does not satisfy the constraint by GW170817, while the predictions for red points satisfy the constraints (see Sec. 3).

In the analysis, we first fix the values of two chiral invariant masses to some constants, and then we use eleven physical inputs shown in Table 1 to restrict the remaining ten parameters. Note that we adopt the restriction of $-0.25 \le g_A(1535) \le 0.25$ from the lattice analysis[14] showing $g_A(1535) \sim \mathcal{O}(0.1)$. Using these physical inputs, we find that solutions are categorized into five groups as shown in Fig. 1.

3. Constraints from Nuclear Matter and Neutron Stars

In the analysis in Ref. 2, following Ref. 7, we adopt the potential terms for the σ with the σ^6 term and include the ω and ρ mesons into our model using the hidden local symmetry (For reviews, see e.g., Refs. 15 and 16. Extension to extreme conditions can be seen in e.g., Refs. 17, 18 and 19.). We calculate the thermodynamic potential in the nuclear medium using the mean field approximation. Then, we adjust model parameters to reproduce the saturation properties of nuclear matter: the nuclear saturation density $\rho_0 = 0.16\,\mathrm{fm}^{-3}$, the binding energy $E_{\mathrm{bind}} = -16\,\mathrm{MeV}$, the incompressibility $K = 240\,\mathrm{MeV}$ and the symmetry energy at normal nuclear matter $E_{\mathrm{sym}} = 31\,\mathrm{MeV}$ for given values of two chiral invariant masses $m_0^{(1)}$ and $m_0^{(2)}$. It was shown that the saturation properties cannot be reproduced for the combinations indicated by the black dots in Fig. 1, and they are reproduced for the combinations indicated by yellow and red points.

We next obtained constraint to the chiral invariant masses from the tidal deformability of neutron stars observed through the gravitation wave GW170817.[10–13] In our analysis, we require that the binary dimensionless tidal deformability satisfies $\tilde{\Lambda} < 800$ for the Chirp mass of $M_{\mathrm{Chirp}} = 1.118 M_\odot$. Figure 2 shows typical examples of the tidal deformability for several choices of chiral invariant masses. In the figure, the vertical axis shows the value of the tidal deformability and the horizontal axis shows the ratio of masses of two neutron stars attending the neutron star merger observed through the gravitational wave.

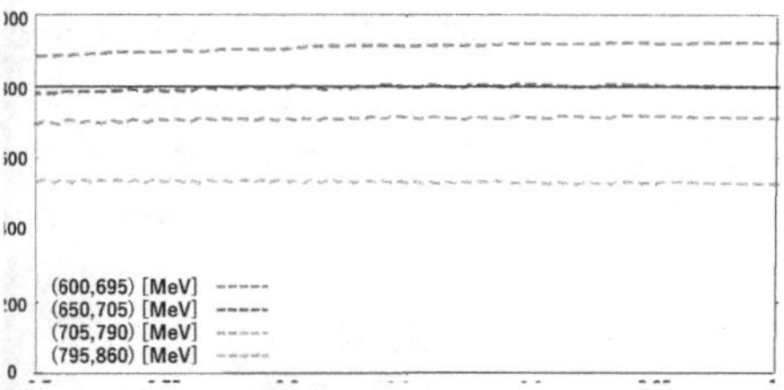

Fig. 2. Binary dimensionless tidal deformability ($\Lambda_{\mathrm{bi}} \equiv \tilde{\Lambda}$) for several choices of two chiral invariant masses. The horizontal axis shows the ratio of two masses of neutron stars.

Here we use the constraint of the mass ratio, $0.7 < M_2/M_1 < 1$.[10,13] The dashed curves, from up to down, are for $(m_0^{(1)}, m_0^{(2)}) = (600, 695)$, $(650, 705)$, $(705, 790)$, $(795, 860)\,\mathrm{MeV}$, respectively. The black solid line at $\tilde{\Lambda} = 800$ is the upper bound of the constraint from the observation of GW170817. The figure shows that the choice of $(m_0^{(1)}, m_0^{(2)}) = (600, 695)$ is excluded, while the other combinations are allowed. We checked many combinations of chiral invariant masses, and the summary of our results is shown in Fig. 1, where the combinations of two chiral invariant masses indicated by yellow dots are excluded, while the combinations by red dots are allowed. One can see that the constraint from the tidal deformability is satisfied when both the chiral invariant masses are larger than about 600 MeV.

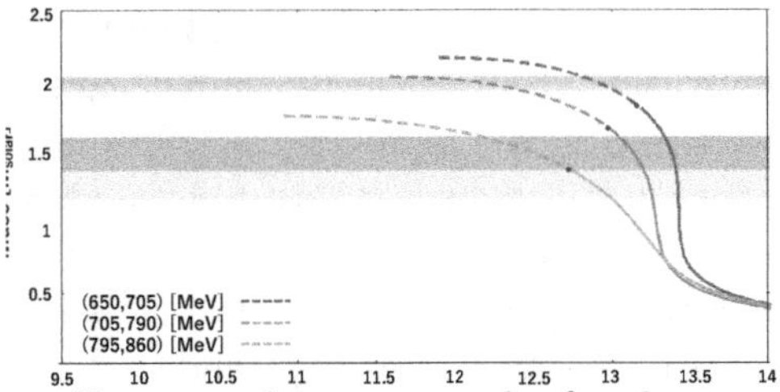

Fig. 3. Relation between the mass and radius of neutron stars for several combinations of two chiral invariant masses: $(m_0^{(1)}, m_0^{(2)}) = (650, 705)$, $(705, 790)$, $(795, 860)$ MeV.

Next, we show our results for the mass-radius relation for neutron stars. Figure 3 shows the mass-radius relation, where the solid curves imply that the central density of the neutron star is smaller than three times of normal nuclear density, while the dashed curves show that the central density is larger than three times of normal nuclear density. Comparing Fig. 3 with Fig. 2, one can easily see that the larger tidal deformability corresponds to the larger radius and the smaller chiral invariant masses. The reason why smaller chiral invariant masses are excluded is understood as follows: The masses of nucleons are roughly given as sums of the chiral invariant mass and the mass generated by the spontaneous chiral symmetry breaking caused by the VEV of σ. This implies that the coupling of σ to nucleon is larger for smaller chiral invariant mass. Then, the attractive force mediated by σ contribution is larger for smaller chiral invariant masses. On the other hand, the repulsive force mediated by ω contribution is larger for larger σ contribution to satisfy the saturation properties at normal nuclear density. The attractive force by σ contribution becomes smaller for larger density associated with the partial chiral symmetry restoration, while the repulsive force by ω contribution becomes larger. The larger repulsive force makes the radius and tidal deformability larger. As a result, the smaller chiral invariant masses cause the larger tidal deformability. That is why smaller chiral invariant masses are excluded.

Next, we show our results for the symmetry energy and the slope parameter in Fig. 4. In our analysis, we included ρ meson into the model. The formulas for calculating the symmetry energy and the slope parameter are given in Eqs. (35) and (38) in Ref. 2. In Fig. 4, we show the density dependences of $E_{\rm sym}$ and L. In this figure the upper three curves show the slope parameter, and the lower three show the symmetry energy. This figure shows that the slope parameter has very little dependence on the chiral invariant masses, and the symmetry energy does not depend on the choice of chiral invariant masses. Furthermore, both increase linearly with the density. We have these properties since both of the symmetry energy and the slope parameter are dominated by ρ meson contribution proportional to $\frac{g_{\rho NN}^2}{m_\rho^2}$.

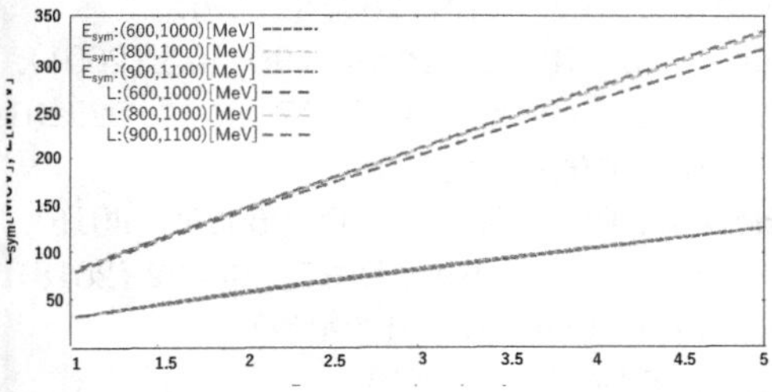

Fig. 4. Predicted symmetry energy (lower three curves) and slope parameter (upper three curves) for $(m_0^{(1)}, m_0^{(2)}) = (900, 1100)$, $(800, 1000)$, $(600, 1000)$MeV, from up to down.

4. Summary

In Refs. 1 and 2, we constructed an extended parity doublet model to study four nucleons, $N(939)$, $N(1440)$, $N(1535)$ and $N(1650)$. We obtained several constraints to two chiral invariant masses from the properties of nucleons at vacuum and the nuclear matter. Our results show that combinations indicated by the red points in Fig. 1 are allowed. This shows that the constraint from the tidal deformability is satisfied when both the chiral invariant masses are larger than about 600 MeV.

In the present model, only the iso-singlet scalar meson σ is included in the scalar meson sector. It is interesting to extend the sector to include iso-triplet scalar mesons and the ones including strange quarks based on chiral models such as a model proposed in Ref. 20. The present model may not be applicable in high density region realized in super-heavy neutron stars with $2M_\odot$. This may be changed by Including effects of quark degrees (See e.g., Refs. 21 and 22.). It is also interesting to study in-medium modifications of hadrons including a heavy quark as in Refs. 23, 24, 25, 26, 27.

Acknowledgment

The work of M.H. is supported in part by JPSP KAKENHI #16K05345.

References

1. T. Yamazaki and M. Harada, Phys. Rev. D **99**, 034012 (2019).
2. T. Yamazaki and M. Harada, Phys. Rev. C **100**, 025205 (2019).
3. Y. L. Ma, M. Harada, H. K. Lee, Y. Oh, B. Y. Park and M. Rho, Phys. Rev. D **88**, 014016 (2013) Erratum: [Phys. Rev. D **88**, 079904 (2013)].
4. Y. L. Ma, M. Harada, H. K. Lee, Y. Oh, B. Y. Park and M. Rho, Phys. Rev. D **90**, 034015 (2014).
5. C. E. Detar and T. Kunihiro, Phys. Rev. D **39**, 2805 (1989).
6. D. Jido, M. Oka and A. Hosaka, Prog. Theor. Phys. **106**, 873 (2001).
7. Y. Motohiro, Y. Kim and M. Harada, Phys. Rev. C **92**, no. 2, 025201 (2015) Erratum: [Phys. Rev. C **95**, 059903 (2017)].
8. Y. Takeda, Y. Kim and M. Harada, Phys. Rev. C **97**, 065202 (2018).
9. Y. Takeda, H. Abuki and M. Harada, Phys. Rev. D **97**, 094032 (2018).
10. B. P. Abbott *et al.*, Phys. Rev. Lett. **119**, 161101 (2017).
11. B. P. Abbott *et al.*, Astrophys. J. **848**, L12 (2017).
12. B. P. Abbott *et al.*, Phys. Rev. Lett. **121**, 161101 (2018).
13. B. P. Abbott *et al.* [LIGO Scientific and Virgo Collaborations], Phys. Rev. X **9**, no. 1, 011001 (2019).
14. T. T. Takahashi and T. Kunihiro, Phys. Rev. D **78**, 011503 (2008).
15. M. Bando, T. Kugo and K. Yamawaki, Phys. Rept. **164**, 217 (1988).
16. M. Harada and K. Yamawaki, Phys. Rept. **381**, 1 (2003).
17. M. Harada and K. Yamawaki, Phys. Rev. Lett. **86**, 757 (2001).
18. M. Harada and C. Sasaki, Phys. Lett. B **537**, 280 (2002).
19. M. Harada, Y. Kim and M. Rho, Phys. Rev. D **66**, 016003 (2002).
20. Y. Kuroda, M. Harada, S. Matsuzaki and D. Jido, arXiv:1910.09146 [hep-ph].
21. G. Baym, T. Hatsuda, T. Kojo, P. D. Powell, Y. Song and T. Takatsuka, Rept. Prog. Phys. **81**, no. 5, 056902 (2018).
22. M. L. Marczenko, D. Blaschke, K. Redlich and C. Sasaki, Phys. Rev. D **98**, no. 10, 103021 (2018).
23. D. Suenaga, B. R. He, Y. L. Ma and M. Harada, Phys. Rev. C **89**, 068201 (2014).
24. D. Suenaga, B. R. He, Y. L. Ma and M. Harada, Phys. Rev. D **91**, no. 3, 036001 (2015).
25. D. Suenaga and M. Harada, Phys. Rev. D **93**, no. 7, 076005 (2016).
26. M. Harada, Y. L. Ma, D. Suenaga and Y. Takeda, Progress of Theoretical and Experimental Physics, 2017, 11, 113D01 (2017).
27. D. Suenaga, S. Yasui and M. Harada, Phys. Rev. C **96**, 015204 (2017).

Mass modifications of light mesons in nuclear matter in the three-flavor extended Linear Sigma Model

Daiki Suenaga

Key Laboratory of Quark and Lepton Physics (MOE) and Institute of Particle Physics, Central China Normal University, Wuhan 430079, China
E-mail: suenaga@mail.ccnu.edu.cn

Phillip Lakaschus

Institute for theoretical physics, Max-von-Laue Str. 1, D-60438 Frankfurt am Main, Germany
E-mail: lakaschus@th.physik.uni-frankfurt.de

We investigate mass modifications of scalar, pseudo-scalar, vector, and axial-vector mesons in nuclear matter in a unified way within the three-flavor extended Linear Sigma Model (eLSM). Nuclear matter is constructed by the two-flavor Parity Doublet Model (PDM) at one loop of the nucleon together with meson mean fields. As a result, we find masses of all mesons except pion, kaon, the lightest scalar-isoscalar meson, ρ meson, and ω meson are reduced in nuclear matter. Also, our results suggest that the value of chiral invariant mass (M_0) should be $M_0 \approx 0.8$ GeV.

Keywords: Chiral symmetry; nuclear matter; meson mass.

1. Introduction

Chiral symmetry plays an important role in a comprehensive understanding of light mesons. Although it is well known that chiral symmetry is spontaneously broken for providing nucleon mass in the vacuum, this symmetry is believed to get restored in nuclear matter, and correspondingly, light mesons properties can be significantly affected. Therefore, we study mass modifications of light mesons in nuclear matter from the view point of chiral symmetry.

In this write-up, we review the main points of our recent paper.[1] In Sec. 2, we introduce our model employed to study meson masses in nuclear matter. And in Sec. 3, we show the results and conclusions of our work.

2. Model

In the present study, the light mesons are introduced by the three-flavor extended Linear Sigma Model (eLSM) comprehensively,[2] and nuclear matter is described by one-loop approximation of the nucleon together with the meson mean fields by means of two-flavor Parity Doublet Model (PDM).[3]

2.1. *Three-flavor extended Linear Sigma Model*

The reduced Lagrangian of three-flavor eLSM[1] is of the form

$$\begin{aligned}
\mathcal{L}_{\rm eLSM}^{\rm red} &= {\rm Tr}[(D_\mu\Phi)^\dagger(D^\mu\Phi)] - m_0^2{\rm Tr}[\Phi^\dagger\Phi] - \lambda_1\left({\rm Tr}[\Phi^\dagger\Phi]\right)^2 \\
&- \lambda_2{\rm Tr}[(\Phi^\dagger\Phi)^2] + {\rm Tr}[H(\Phi^\dagger+\Phi)] - \frac{1}{4}{\rm Tr}[L_{\mu\nu}L^{\mu\nu} + R_{\mu\nu}R^{\mu\nu}] \\
&+ {\rm Tr}\left[\left(\frac{m_1^2}{2}+\Delta\right)(L_\mu^2+R_\mu^2)\right] + c_1\left({\rm det}\Phi - {\rm det}\Phi^\dagger\right)^2 \\
&+ h_2{\rm Tr}[L_\mu^2\Phi\Phi^\dagger + R_\mu^2\Phi^\dagger\Phi] + 2h_3{\rm Tr}[L_\mu\Phi R^\mu\Phi^\dagger] \\
&+ g_{4p}\Big({\rm Tr}[L_\mu L_\nu L^\mu L^\nu] + {\rm Tr}[R_\mu R_\nu R^\mu R^\nu]\Big) \\
&+ g_{4p}\Big({\rm Tr}[L_\mu L^\mu L_\nu L^\nu] + {\rm Tr}[R_\mu R^\mu R_\nu R^\nu]\Big)\,. \qquad (1)
\end{aligned}$$

In this Lagrangian, Φ includes the scalar and pseudo-scalar nonets while L_μ and R_μ include vector and axial-vector nonets, respectively, as

$$\Phi = \frac{1}{\sqrt{2}}\begin{pmatrix} \frac{\sigma_N + a_0^0 + i(\eta_N+\pi^0)}{\sqrt{2}} & a_0^+ + i\pi^+ & K_0^{*+} + iK^+ \\ a_0^- + i\pi^- & \frac{\sigma_N - a_0^0 + i(\eta_N-\pi^0)}{\sqrt{2}} & K_0^{*0} + iK^0 \\ K_0^{*-} + iK^- & \bar{K}_0^{*0} + i\bar{K}^0 & \sigma_S + i\eta_S \end{pmatrix},$$

$$L_\mu(R_\mu) = \frac{1}{\sqrt{2}}\begin{pmatrix} \frac{\omega_N+\rho^0}{\sqrt{2}} \pm \frac{f_{1N}+a_1^0}{\sqrt{2}} & \rho^+ \pm a_1^+ & K^{*+} \pm K_1^+ \\ \rho^- \pm a_1^- & \frac{\omega_N-\rho^0}{\sqrt{2}} \pm \frac{f_{1N}-a_1^0}{\sqrt{2}} & K^{*0} \pm K_1^0 \\ K^{*-} \pm K_1^- & \bar{K}^{*0} \pm \bar{K}_1^0 & \omega_S \pm f_{1S} \end{pmatrix}_\mu . \qquad (2)$$

The covariant derivative $D_\mu\Phi$ is defined by $D_\mu\Phi = \partial_\mu\Phi - ig_1(L_\mu\Phi - \Phi R_\mu)$, and $L_{\mu\nu} = \partial_\mu L_\nu - \partial_\nu L_\mu$ ($R_{\mu\nu} = \partial_\mu R_\nu - \partial_\nu R_\mu$) is the field strength. Δ and H are 3×3 matrices responsible for the current quark mass effects as $\Delta = (0, 0, \delta_S)$ and $H = (h_{0N}, h_{0N}, h_{0S})$. The chiral transformation laws for Φ, L_μ and R_μ are $\Phi \to g_L\Phi g_R^\dagger$, $L_\mu \to g_L L_\mu g_L^\dagger$ and $R_\mu \to g_R R_\mu g_R^\dagger$, respectively ($g_{L(R)} \in U(3)_{L(R)}$). The c_1 term in Eq. (1) is present due to the $U(1)_A$ anomaly. Under the spontaneous breakdown of chiral symmetry in medium, σ_N, σ_S and $\omega_N^{\mu=0}$ possess their mean fields: $\phi_N \equiv \langle\sigma_N\rangle$, $\phi_S \equiv \langle\sigma_S\rangle$, $\bar{\omega}_N \equiv \langle\omega_N^{\mu=0}\rangle$, and correspondingly, meson properties are significantly affected even at tree level.

Most of the parameters in Eq. (1) can be fixed by fitting masses and decay widths of the light mesons in the vacuum. The parameters determined in the previous work is indicated in Table 1.

Table 1. Parameters of the three-flavor eLSM.

Parameters in eLSM	Values
C_1 [GeV^2]	-0.9183
C_2 [GeV^2]	0.4135
c_1 [GeV^{-2}]	450.5
δ_S [GeV^2]	0.1511
g_1	5.8433
$\phi_N\|_{\rm vac}$ [GeV]	0.1646
$\hat{\phi}_S\|_{\rm vac}$ [GeV]	0.1262
h_{0N} [GeV^3]	0.001135
h_{0S} [GeV^3]	0.02138
h_2	9.880
h_3	4.867
λ_2	68.30

Table 2. Input parameters of the present work.

Inputs	Values
m_+ [GeV]	0.939
m_- [GeV]	1.535
$g_A^{N_+}$	1.267
$g_A^{N_-}$	0.2±0.3
$\Gamma_{N^*\to N\eta}$ [GeV]	0.065
$P\|_{\rho_0}$	0
$\mu_B\|_{\rho_0}$ [GeV]	0.923
K [GeV]	0.24

2.2. *Two-flavor Parity Doublet Model*

In order for constructing nuclear matter, we employ the following two-flavor PDM

$$\begin{aligned}
\mathcal{L}_N &= \bar{\psi}_{1r}(i\partial\!\!\!/ + \mu_B\gamma_0 + g_V\tilde{R}\!\!\!/)\psi_{1r} + \bar{\psi}_{1l}(i\partial\!\!\!/ + \mu_B\gamma_0 + g_V\tilde{L}\!\!\!/)\psi_{1l} \\
&+ \bar{\psi}_{2r}(i\partial\!\!\!/ + \mu_B\gamma_0 + h_V\tilde{L}\!\!\!/)\psi_{2r} + \bar{\psi}_{2l}(i\partial\!\!\!/ + \mu_B\gamma_0 + h_V\tilde{R}\!\!\!/)\psi_{2l} \\
&+ \tilde{g}\mathrm{Tr}[\tilde{R}_\mu + \tilde{L}_\mu]\left(\bar{\psi}_{1r}\gamma^\mu\psi_{1r} + \bar{\psi}_{1l}\gamma^\mu\psi_{1l} + \bar{\psi}_{2r}\gamma^\mu\psi_{2r} + \bar{\psi}_{2l}\gamma^\mu\psi_{2l}\right) \\
&- M_0\left[\bar{\psi}_{1l}\psi_{2r} - \bar{\psi}_{1r}\psi_{2l} - \bar{\psi}_{2l}\psi_{1r} + \bar{\psi}_{2r}\psi_{1l}\right] \\
&- k_1(\det\tilde{\Phi} + \det\tilde{\Phi}^\dagger)\left[\bar{\psi}_{1l}\psi_{2r} - \bar{\psi}_{1r}\psi_{2l} - \bar{\psi}_{2l}\psi_{1r} + \bar{\psi}_{2r}\psi_{1l}\right] \\
&- k_2\left(\det\tilde{\Phi} - \det\tilde{\Phi}^\dagger\right)\left[\bar{\psi}_{1l}\psi_{2r} + \bar{\psi}_{1r}\psi_{2l} + \bar{\psi}_{2l}\psi_{1r} + \bar{\psi}_{2r}\psi_{1l}\right] \\
&- G_1\left[\bar{\psi}_{1r}\tilde{\Phi}^\dagger\psi_{1l} + \bar{\psi}_{1l}\tilde{\Phi}\psi_{1r}\right] - G_2\left[\bar{\psi}_{2r}\tilde{\Phi}^\dagger\psi_{2l} + \bar{\psi}_{2l}\tilde{\Phi}\psi_{2r}\right] ,
\end{aligned} \tag{3}$$

in which $\psi_{1l(1r)}$ and $\psi_{2l(2r)}$ are *naive-assigned* and *mirror-assigned* nucleons, whose chiral transformation laws are $\psi_{1l(1r)} \to \tilde{g}_{L(R)}\psi_{1l(1r)}$, $\psi_{2l(2r)} \to \tilde{g}_{R(L)}\psi_{2l(2r)}$, respectively ($\tilde{g}_{L(R)} \in U(2)_{L(R)}$). We assume that the interaction manners among the nucleons and the mesons are governed by two-flavor chiral symmetry, so that the meson fields $\tilde{\Phi}$, $\tilde{L}_\mu$ and $\tilde{R}_\mu$ in Eq. (3) include only two-flavor quarkonia states as $\tilde{\Phi} = \sigma_N + i\pi^a\tau^a$, $\tilde{V}_\mu \equiv \tilde{L}_\mu + \tilde{R}_\mu = \frac{1}{2}(\omega_N + \rho^a\tau^a)_\mu$, and $\tilde{A}_\mu \equiv \tilde{L}_\mu - \tilde{R}_\mu = \frac{1}{2}(f_{1N} + a_1^a\tau^a)_\mu$. μ_B is a baryon number chemical potential put to access to finite baryon density. One of the most interesting characteristics of the two-flavor PDM is the presence of the so-called *chiral invariant mass* (M_0 in the forth line in Eq. (3)). Namely, the nucleon can possess its mass without violating chiral symmetry. In fact, by diagonalizing the mass matrix, we can get two mass eigenvalues $m_\pm = \frac{1}{2}\left(\sqrt{(G_1+G_2)^2\phi_N^2 + 4\left(M_0 + \frac{k_1}{2}\phi_N^2\right)^2} \mp (G_2 - G_1)\phi_N\right)$ under the spontaneous breakdown of chiral symmetry which clearly shows that, at the chiral restoration point $\phi_N = 0$, nucleon mass reads $m_\pm = M_0$. The corresponding eigenstates $N_\pm$ carry positive and negative parities, such that we assign N_+ to the nucleon while N_- to $N^*(1535)$. In terms of Eq. (3), we should notice that the unfamiliar k_1 and k_2 terms are put for taking into account the direct effects to the nucleons induced by the $U(1)_A$ axial anomaly.

Our goal of the current work is to investigate mass changes of the light mesons in nuclear matter, then, we need to construct the matter. We compute a grand potential (Ω), [or equivalently, an effective action ($\Gamma_{\rm eff}$)] describing nuclear matter by one loops of the nucleon together with the meson mean fields ϕ_N, ϕ_S and $\bar{\omega}$ by combining the eLSM (1) and the PDM (3). Hence, the meson two-point functions can be obtained by taking second functional derivatives of $\Gamma_{\rm eff}$ with respect to the corresponding meson fields.

To fix the remaining parameters, we adopt the nucleon mass, $N^*(1535)$ mass, axial charges, $N^*(1535) \to N\eta$ decay width as inputs. Besides, the saturation property of the nuclear matter at $\rho_0 = 0.16$ fm^{-3}, the incompressibility are employed as inputs. These inputs are summarized in Table 2. However, even if we employ these inputs, there still remain two undetermined parameters. In the following argument, we will fix $k_1 = 0$ and $M_0 = 0.8$ GeV, which yields $G_1 = 3.922$, $G_2 = 7.542$, $k_2 = -1.499$ GeV^{-1}, $g_V = 3.847$, $h_V = -9.186$, $\tilde{g} = -2.623$, $\lambda_1 = -22.98$, and $g_{4p} = 42.01$.

3. Results and conclusions

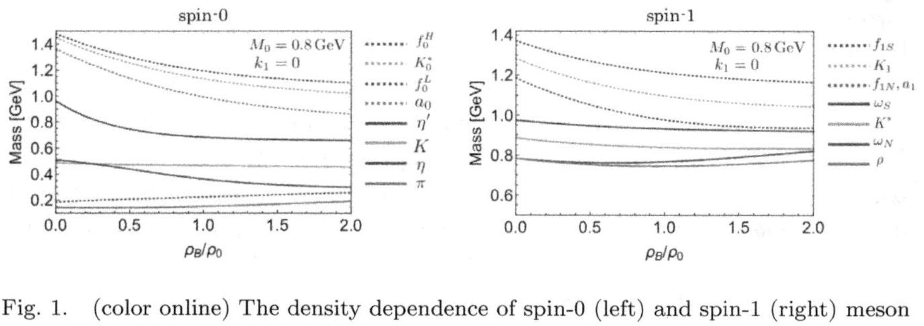

Fig. 1. (color online) The density dependence of spin-0 (left) and spin-1 (right) meson masses with $M_0 = 0.8$ GeV and $k_1 = 0$.

In Fig. 1, we show the resultant mass modifications of the light mesons in nuclear matter with $M_0 = 0.8$ MeV and $k_1 = 0$. For spin-0 mesons, all meson masses except pion, kaon, and f_0^L (the lightest scalar-isoscalar meson) ones decrease as the density increases. Especially, the mass reduction of η' meson at the normal nuclear density (ρ_0) is about 200 MeV, which is larger than the previous works.[4] In terms of f_0^L meson, we find $m_{f_0^L} \approx 0.27$ GeV, while experimentally the lightest scalar-isoscalar meson mass ($f_0(500)$) lies in the range of $m_{f_0(500)} = 0.4$ - 0.5 GeV with a large width of $\Gamma_{f_0(500)} = 0.4$ - 0.7 GeV. To resolve this discrepancy, we may need to incorporate an additional scalar-isoscalar degree of freedom such as a tetraquark state or a meson molecule. It has been shown that such a state affects the properties of the medium.[5] For spin-1 mesons, we find all meson masses except ρ and ω_N mesons ones decrease at finite density. Particularly, ω_N meson mass does not change at ρ_0 from the vacuum one, which is consistent with the experimental result.[4] This favourable result can be obtained only when we take $M_0 \approx 0.8$ GeV for any choices of k_1, which suggests the value of the chiral invariant mass should be $M_0 \approx 0.8$ GeV.

The mass modifications obtained in the present work provide useful information of partial restoration of chiral symmetry to existing/forthcoming experiments. Furthermore, the fitted model parameters such as M_0 will lead to a better understanding of hadrons from the view point of chiral symmetry.

Acknowledgment

D.S. is supported by NSFC Grant No. 20201191997.

References

1. D. Suenaga and P. Lakaschus, arXiv:1908.10509 [nucl-th].
2. D. Parganlija, P. Kovacs, G. Wolf, F. Giacosa and D. H. Rischke, Phys. Rev. D **87**, no. 1, 014011 (2013) doi:10.1103/PhysRevD.87.014011.
3. C. E. DeTar and T. Kunihiro, Phys. Rev. D **39**, 2805 (1989). doi:10.1103/PhysRevD.39.2805
4. V. Metag, M. Nanova and E. Y. Paryev, Prog. Part. Nucl. Phys. **97**, 199 (2017) doi:10.1016/j.ppnp.2017.08.002.
5. S. Gallas, F. Giacosa and G. Pagliara, Nucl. Phys. A **872** (2011) 13 doi:10.1016/j.nuclphysa.2011.09.008 [arXiv:1105.5003 [hep-ph]].

The chiral phase transition temperature in (2+1)-flavor QCD

Heng-Tong Ding[1], Prasad Hegde[2], Olaf Kaczmarek[1,3], Frithjof Karsch[3,4], Anirban Lahiri[3], Sheng-Tai Li[1,5,*], Swagato Mukherjee[4], Hiroshi Ohno[6], Peter Petreczky[4], Christian Schmidt[3] and Patrick Steinbrecher[4]
(for the HotQCD collaboration)

[1] *Key Laboratory of Quark & Lepton Physics (MOE) and Institute of Particle Physics, Central China Normal University, Wuhan 430079, China.*
[2] *Center for High Energy Physics, Indian Institute of Science, Bangalore 560012, India*
[3] *Fakultät für Physik, Universität Bielefeld, D-33615 Bielefeld, Germany*
[4] *Physics Department, Brookhaven National Laboratory, Upton, NY 11973, USA*
[5] *Institute of Modern Physics,Chinese Academy of Sciences, Lanzhou 730000, China*
[6] *Center for Computational Sciences, University of Tsukuba, Tsukuba, Ibaraki 305-8577, Japan*
**E-mail: stli@impcas.ac.cn*

The chiral phase transition temperature T_c^0 is a fundamental quantity of QCD. To determine this quantity we have performed simulations of (2 + 1)-flavor QCD using the Highly Improved Staggered Quarks (HISQ/tree) action on N_τ=6, 8 and 12 lattices with aspect ratios N_σ/N_τ ranging from 4 to 8. In our simulations we fix the strange quark mass to its physical value $m_s^{\rm phy}$, and vary the values of two degenerate light quark masses m_l from $m_s^{\rm phy}/20$ to $m_s^{\rm phy}/160$ which correspond to a Goldstone pion mass m_π ranging from 160 MeV to 55 MeV in the continuum limit. We employ two estimators T_{60} and T_δ to extract the chiral phase transition temperature T_c^0, after taking the chiral limit, thermodynamic limit and continuum limit, we present our current estimate for $T_c^0 = 132^{+3}_{-6}$ MeV.

Keywords: Chiral phase transition; chiral limit; continuum limit; thermodynamic limit; QCD phase diagram; critical point.

1. Introduction

One of ultimate goals of lattice QCD calculations is to map out the QCD phase diagram.[1] It is well established for the (2 + 1)-flavor QCD at the physical point ($m_{u,d}^{\rm phy}$, $m_s^{\rm phy}$) the transition is a rapid crossover[2,3] and the pseudo-critical temperature $T_{\rm pc}$ is found to be 156.5(1.5) MeV.[4] In the limit of vanishing u and d quark masses we have a true chiral phase transition[5] and there are indications from lattice QCD calculations that this transition is 2nd order belonging to the $O(4)$ universality class[6,7] which is shown in the Columbia plot, the left plot of Fig. 1. The $U_A(1)$ symmetry remains broken at the chiral phase transition temperature, T_c^0. We present our current estimate for the chiral phase transition temperature $T_c^0 = 132^{+3}_{-6}$ MeV,[6,8] which is consistent with QCD-inspired model calculations and is around 25 MeV smaller than $T_{\rm pc}$.[9,10]

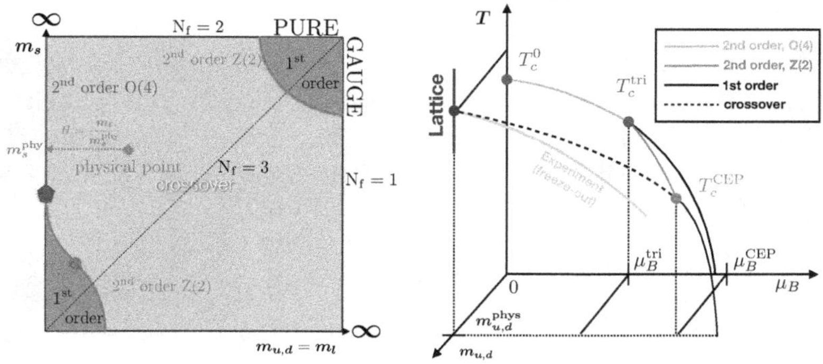

Fig. 1. Left: The Columbia plot in the case $m_s^{\rm phy} > m_s^{\rm tri}$. Right: The QCD phase diagram in the $(T, \mu_B, m_{u,d})$ plane.

We expect that the 2nd order phase transition for vanishing u, d quark masses will eventually change to a 1st order phase transition as we increase the baryon chemical potential, μ_B. This means that there is a tri-critical point $(\mu_B^{\rm tri}, T_c^{\rm tri})$ as shown in the right plot of Fig. 1. At finite μ_B it is obtained that $T_c^0(\mu_B)$ decreases with increasing μ_B up to leading order in μ_B.[11,12] At the physical point with finite μ_B, if one expects a first order phase transition at large μ_B, then there exists a critical end point $(\mu_B^{\rm CEP}, T_c^{\rm CEP})$. Based on the Random Matrix Model and NJL model, the tri-critical temperature $T_c^{\rm tri}$ is larger than the critical end point temperature $T_c^{\rm CEP}$, i.e. $T_c^{\rm tri} - T_c^{\rm CEP}\,(m_q) \propto m_q^{2/5}$.[13–15] So the chiral phase transition temperature T_c^0 can be considered as an upper bound of the critical end point $T_c^{\rm CEP}$ in the QCD phase diagram at finite μ_B, e.g. $T_c^0 > T_c^{\rm tri} > T_c^{\rm CEP}$.

2. Lattice setup

We employ the Highly Improved Staggered Quarks and tree-level improved gauge action (HISQ/tree) in our simulations of (2 + 1)-flavor QCD. We fix the strange quark mass to its physical quark mass value $m_s^{\rm phy}$, and vary the light quark masses values from $m_s^{\rm phy}/160$ to $m_s^{\rm phy}/20$ which correspond to 55 MeV $\le m_\pi \le$ 160 MeV. To perform the continuum and thermodynamic extrapolations, simulations have been performed on lattices with temporal extent $N_\tau = 6$, 8 and 12 and with spatial volumes in the range $4N_\tau \le N_\sigma$

$\leq 8N_\tau$. We have performed simulations for at least 10000 time units (TUs) at each temperature for each data sets with each configuration separated by every 5 TUs. To compute the chiral condensate and its susceptibility, we used 50 random noise vectors on each gauge field configuration and constructed unbiased estimators for various traces.

3. Results

We study the subtracted chiral condensate M and its susceptibility χ_M which by construction are free from UV divergences. We normalize the chiral observables with proper power of the kaon decay constant $f_K = (156.1/\sqrt{2})$ MeV,

$$M = m_s\left(\langle\bar{\psi}\psi\rangle_l - \frac{2m_l}{m_s}\langle\bar{\psi}\psi\rangle_s\right)/f_K^4,\ \chi_M \equiv \frac{\partial M}{\partial H} \equiv m_s^2\chi_{l,\,\mathrm{subtot}}\,/f_K^4. \quad (1)$$

To extract the chiral phase transition temperature, we proposed two novel estimators $T_{60}(V,H)$ and $T_\delta(V,H)$, see Refs. 6, 7 and 8 for more details.

$$\begin{aligned} T_X(H,L) = T_c^0\left(1+\left(\frac{z_X\,(z_L)}{z_0}\right)H^{1/\beta\delta}\right) \\ + c_X H^{1-1/\delta+1/\beta\delta}, \quad X = \delta, 60. \end{aligned} \quad (2)$$

These two novel estimators converge to T_c^0 in the chiral limit and thermodynamic limit based on $O(4)$ universality class as shown in Eq. 2. Where $z_X(z_L)$ can be obtained from the ratio of finite size scaling functions determined by 3-D $O(4)$ universality class in Ref. 16, e.g. $f_\chi\,(z_{60}, z_L)\,/f_\chi\,(z_p, z_L) = 0.6$, see Ref. 8 for more details.

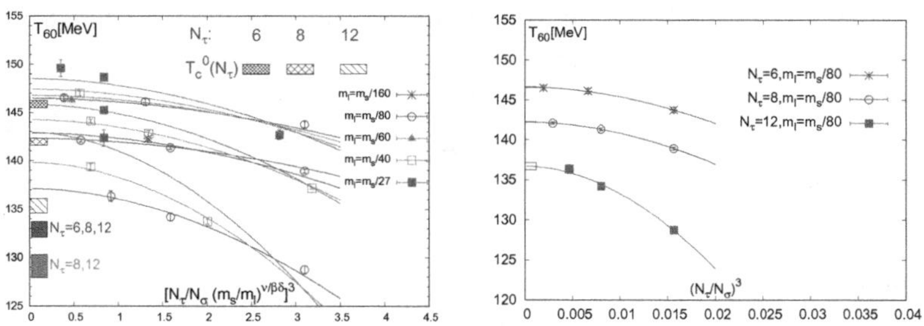

Fig. 2. Left: Fit T_{60} using fit ansatz Eq. 2 for each fixed N_τ, for $H \leq 1/27$ with all available volumes used. Grid black blocks show chiral limit results at fixed N_τ and colored blocks show the continuum extrapolated results. Right: Fit T_{60} at fixed quark mass for each N_τ.

As shown in the left plot of Fig. 2, for each fixed N_τ we have performed a finite-size scaling fit using the fit ansatz Eq. 2, and we extract the T_c^0 by performing the continuum extrapolation using these fit results. In the right plot of Fig. 2, we perform the $O(4)$ thermodynamic extrapolation for fixed quark mass $m_l = m_s^{\rm phy}/80$ and then perform the continuum extrapolation, and after performing these procedures for all the quark masses, we obtain T_c^0 by performing the chiral extrapolation.

In Fig. 3, we summarize all the results of T_c^0 obtained after taking care of the chiral, thermodynamic and continuum limits. Our current estimate for chiral phase transition temperature is $T_c^0 = 132^{+3}_{-6}$ MeV, where a 2-3 MeV shift comes from the continuum extrapolation including or excluding our coarsest lattices $N_\tau = 6$.

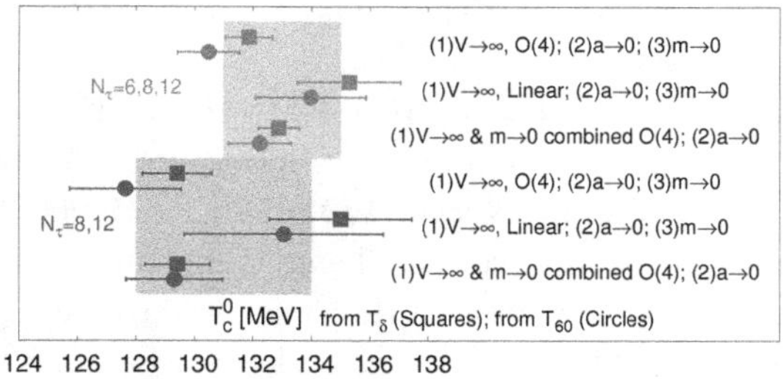

Fig. 3. The T_c^0 obtained by averaging all the fit results.[8]

Acknowledgments

This work was supported in part by the Deutsche Forschungsgemeinschaft (DFG) through the grant 315477589-TRR 211, the grant 05P2018 (ErUM-FSP T01) of the German Bundesministerium für Bildung und Forschung, grant 283286 of the European Union, the National Natural Science Foundation of China under grant numbers 11947237, 11775096 and 11535012, Furthermore, this work was supported through Contract No. DE-SC0012704 with the U.S. Department of Energy, through the Scientific Discovery through Advanced Computing (SciDAC) program funded by the U.S. Department of Energy, Office of Science, Advanced Scientific Computing Research and Nuclear Physics and the DOE Office of Nuclear Physics funded BEST topical collaboration, and a Early Career Research Award of the Science and Engineering Research Board of the Government of India. Numerical calculations have been made possible through PRACE grants at CSCS, Switzerland, and at CINECA, Italy as well as grants at the Gauss Centre for Supercomputing and NIC-Jülich, Germany. These grants provided access to resources on Piz Daint at CSCS, Marconi at CINECA as well as on JUQUEEN and JUWELS at NIC. Additional calculations have been performed on GPU clusters of USQCD, at Bielefeld University, the PC^2 Paderborn University and the Nuclear Science Computing Center at Central China Normal University (NSC^3), Wuhan, China. Some data sets have also partly been produced at the TianHe II Supercomputing Center in Guangzhou.

References

1. H.-T. Ding, F. Karsch and S. Mukherjee, Thermodynamics of strong-interaction matter from Lattice QCD, *Int. J. Mod. Phys.* **E24**, 1530007 (2015).
2. Y. Aoki, G. Endrődi, Z. Fodor, S. D. Katz and K. K. Szabó, The order of the quantum chromodynamics transition predicted by the standard model of particle physics, *Nature* **443**, 675 EP (10 2006).
3. A. Bazavov, T. Bhattacharya, M. Cheng, C. DeTar, H.-T. Ding, S. Gottlieb, R. Gupta, P. Hegde, U. M. Heller, F. Karsch, E. Laermann, L. Levkova, S. Mukherjee, P. Petreczky, C. Schmidt, R. A. Soltz, W. Soeldner, R. Sugar, D. Toussaint, W. Unger and P. Vranas, Chiral and deconfinement aspects of the qcd transition, *Phys. Rev. D* **85**, 054503 (Mar 2012).
4. A. Bazavov *et al.*, Chiral crossover in QCD at zero and non-zero chemical potentials, *Phys. Lett.* **B795**, 15 (2019).
5. R. D. Pisarski and F. Wilczek, Remarks on the chiral phase transition in chromodynamics, *Phys. Rev. D* **29**, 338 (Jan 1984).
6. H.-T. Ding, P. Hegde, O. Kaczmarek, F. Karsch, A. Lahiri, S.-T. Li, S. Mukherjee and P. Petreczky, Chiral phase transition in $(2+1)$-flavor QCD, *PoS* **LATTICE2018**, 171 (2019).
7. H. T. Ding, P. Hegde, F. Karsch, A. Lahiri, S. T. Li, S. Mukherjee and P. Petreczky, Chiral phase transition of (2+1)-flavor QCD, *Nucl. Phys.* **A982**, 211 (2019).
8. H. T. Ding *et al.*, Chiral Phase Transition Temperature in (2+1)-Flavor QCD, *Phys. Rev. Lett.* **123**, 062002 (2019).
9. J. Berges, D.-U. Jungnickel and C. Wetterich, Two flavor chiral phase transition from nonperturbative flow equations, *Phys. Rev. D* **59**, 034010 (Jan 1999).
10. J. Braun, B. Klein, H. J. Pirner and A. H. Rezaeian, Volume and quark mass dependence of the chiral phase transition, *Phys. Rev.* **D73**, 074010 (2006).

11. O. Kaczmarek, F. Karsch, E. Laermann, C. Miao, S. Mukherjee, P. Petreczky, C. Schmidt, W. Soeldner and W. Unger, Phase boundary for the chiral transition in $(2+1)$-flavor QCD at small values of the chemical potential, *Phys. Rev.* **D83**, 014504 (2011).
12. P. Hegde and H.-T. Ding, The curvature of the chiral phase transition line for small values of μ_B, *PoS* **LATTICE2015**, 141 (2016).
13. Y. Hatta and T. Ikeda, Universality, the QCD critical / tricritical point and the quark number susceptibility, *Phys. Rev.* **D67**, 014028 (2003).
14. M. A. Halasz, A. D. Jackson, R. E. Shrock, M. A. Stephanov and J. J. M. Verbaarschot, Phase diagram of qcd, *Phys. Rev. D* **58**, 096007 (Sep 1998).
15. M. Buballa and S. Carignano, Inhomogeneous chiral phases away from the chiral limit, *Phys. Lett.* **B791**, 361 (2019).
16. J. Engels and F. Karsch, Finite size dependence of scaling functions of the three-dimensional $o(4)$ model in an external field, *Phys. Rev. D* **90**, 014501 (Jul 2014).

Λ^* matter and its stability

J. Hrtánková[a,*], M. Schäfer[a,b], and J. Mareš[a]

[a] *Nuclear Physics Institute CAS, 250 68 Rez, Czech Republic*
[*] *E-mail: hrtankova@ujf.cas.cz*
[b] *Faculty of Nuclear Sciences and Physical Engineering, Czech Technical University in Prague, 115 19 Prague 1, Czech Republic*

A. Gal, E. Friedman, and N. Barnea
Racah Institute of Physics, The Hebrew University, 91904 Jerusalem, Israel

We performed calculations of nuclear systems composed solely of Λ^* hyperons, aiming at exploring the possibility of existence of absolutely stable Λ^* matter. We considered Λ^* interaction strengths compatible with the $\Lambda^*\Lambda^*$ binding energy $B_{\Lambda^*\Lambda^*}$ given by the $\bar{K}N$ interaction model by Yamazaki and Akaishi.[1] We found that the binding energy per Λ^* saturates at values well below 100 MeV for mass number $A \geq 120$. The Λ^* matter is thus highly unstable against strong interaction decay.

Keywords: Strange matter; Λ^* resonance; SVM; RMF.

1. Introduction

This contribution concerns our recent study of Λ^* nuclei,[2] which was stirred up by a conjecture about absolutely stable charge-neutral baryonic matter composed solely of $\Lambda(1405)$ (Λ^*) hyperons.[3]
We calculated Λ^* few-body systems within the Stochastic Variational Method (SVM),[4] as well as Λ^* many-body systems within the Relativistic Mean Field (RMF) approach.[5] The meson-exchange Λ^* potentials applied in our work were fitted to reproduce the $\Lambda^*\Lambda^*$ binding energy $B_{\Lambda^*\Lambda^*} = 40$ MeV, given by the phenomenological $\bar{K}N$ interaction model.[1] We recall that the $\bar{K}N$ potentials used by Akaishi and Yamazaki,[1,3] fitted for $I = 0$ to the mass and width of the $\Lambda(1405)$ resonance, fail to reproduce K^- single-nucleon absorption fractions deduced from K^- capture bubble chamber experiments.[6] Nevertheless, we employed these very strong potentials in order to demonstrate that while solving the A-body Schrödinger equation for purely attractive $\Lambda^*\Lambda^*$ interactions will inevitable lead to collapse, with the binding energy per particle diverging as A increases, this

scenario promoted in Ref. 3 is unlikely in standard many-body approaches. In the following sections, we discuss only briefly our main results; more details can be found in Ref. 2.

2. Λ^* Few-Body Systems

We started our study of Λ^* nuclei by calculations of few-body systems within the Stochastic Variational Method[4] for the meson-exchange potentials of the Dover-Gal form[7]:

$$V_{\Lambda^*\Lambda^*}(r) = g^2_{\omega\Lambda^*}\,(1 + \tfrac{1}{8}\tfrac{m^2_\omega}{M^2_{\Lambda^*}})\,Y_\omega(r) - g^2_{\sigma\Lambda^*}\,(1 - \tfrac{1}{8}\tfrac{m^2_\sigma}{M^2_{\Lambda^*}})\,Y_\sigma(r) + g^2_{\omega\Lambda^*}\tfrac{1}{6}\tfrac{m^2_\omega}{M^2_{\Lambda^*}}\,Y_\omega(r)(\vec{\sigma}_1\cdot\vec{\sigma}_2)\;, \tag{1}$$

or the Machleidt form[8]:

$$V_{\Lambda^*\Lambda^*}(r) = g^2_{\omega\Lambda^*}(1 + \tfrac{1}{2}\tfrac{m^2_\omega}{M^2_{\Lambda^*}})\,Y_\omega(r) - g^2_{\sigma\Lambda^*}\,(1 - \tfrac{1}{4}\tfrac{m^2_\sigma}{M^2_{\Lambda^*}})\,Y_\sigma(r) + g^2_{\omega\Lambda^*}\tfrac{1}{6}\tfrac{m^2_\omega}{M^2_{\Lambda^*}}\,Y_\omega(r)(\vec{\sigma}_1\cdot\vec{\sigma}_2)\;, \tag{2}$$

where $M_{\Lambda^*} = 1405$ MeV, m_i are the meson masses, $g_{i\Lambda^*} = \alpha_i g_{iN}$ are the corresponding coupling constants with g_{iN} taken from the HS model,[9] and $Y_{i=\sigma,\omega}(r) = \exp(-m_i r)/(4\pi r)$. In the above expressions, the mass correction factors ($\sim m_i/M_{\Lambda^*}$ and $m_i^2/M^2_{\Lambda^*}$) as well as the spin-spin interaction terms ($\sim (\vec{\sigma}_1\cdot\vec{\sigma}_2)$) are included.

In the calculations we fit either the value of α_σ and kept α_ω fixed to 1 or vice versa in order to get the binding energy of the $\Lambda^*\Lambda^*$ system $B_{\Lambda^*\Lambda^*} = 40$ MeV. We present here only selected results for $\alpha_\sigma \neq 1$.

In Fig. 1, left panel, we show the binding energy per Λ^*, B/A, as a function of mass number in few-body Λ^* nuclei, calculated within the SVM approach for the Machleidt potential (1). When the spin-spin interaction is omitted, the binding energy per particle is rapidly increasing with A, reaching $B/A \approx 130$ MeV for A=6. The mass corrections have almost no effect on the calculated values of B/A. On the other hand, when the spin-spin interaction is taken into account, the increase of B/A is considerably less steep. The corresponding rms radius of the considered Λ^* nuclei is presented in the right panel. The rms radius is extremely small, hardly exceeding the value 0.8 fm even if the spin-spin interaction is included.

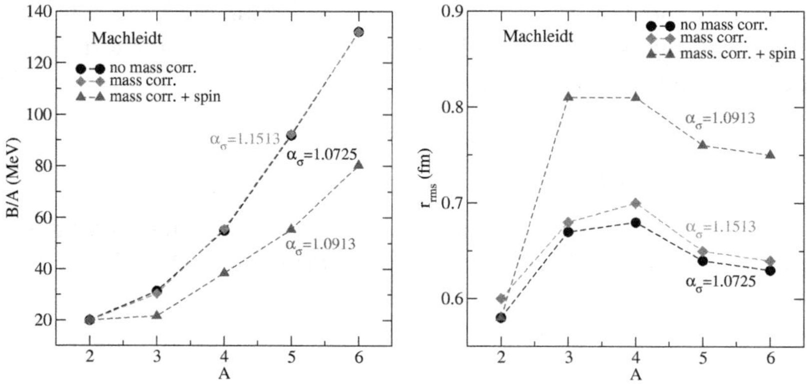

Fig. 1. Binding energy of Λ^* nuclei per particle, B/A (left panel) and rms radius (right panel) of few-body Λ^* systems as a function of mass number A, calculated using the Machleidt potential with and without mass corrections, as well as including spin-spin interaction.

3. Λ^* Many-Body Systems

As the next step, we explored many-body systems composed solely of Λ^* hyperons within the RMF framework,[5] where the interaction among Λ^*'s is mediated by the exchange of the scalar σ and vector ω meson fields. The underlying Lagrangian density is of the form

$$\mathcal{L} = \bar{\Lambda}^* \left[\mathrm{i}\gamma^\mu D_\mu - (M_{\Lambda^*} - g_{\sigma\Lambda^*}\sigma) \right] \Lambda^* + (\sigma, \omega_\mu \text{ free-field terms}) \ , \quad (3)$$

where $D_\mu = \partial_\mu + \mathrm{i}\, g_{\omega\Lambda^*}\,\omega_\mu$. It is to be noted that the isovector-vector $\vec{\rho}$ and Coulomb fields were not taken into account since the Λ^* is a neutral $I = 0$ baryon. First calculations were perfomed using the linear HS model[9] with the coupling constants scaled by $\alpha_{\scriptscriptstyle 1}$, $g_{i\Lambda^*} = \alpha_i g_{iN}$, determined by fitting $B_{\Lambda^*\Lambda^*}$ (see previous section). For comparison, we performed also calculations using the nonlinear NL-SH model.[10] The corresponding scaling parameter α_σ was fitted to yield the binding energy of the $8\Lambda^*$ system calculated within the HS model. We explored Λ^* nuclei with closed shells and solved self-consistently the coupled system of the Klein-Gordon equations for meson fields and the Dirac equation for Λ^*.

The results of our RMF calculations are summarized in Fig. 2. In the left panel, the binding energy per particle, B/A, is plotted as a function of mass number A, calculated within the RMF HS model with the properly rescaled σ meson coupling constant corresponding to the Λ^* potentials (1) and (2). For comparison, B/A calculated within the RMF NL-SH model in

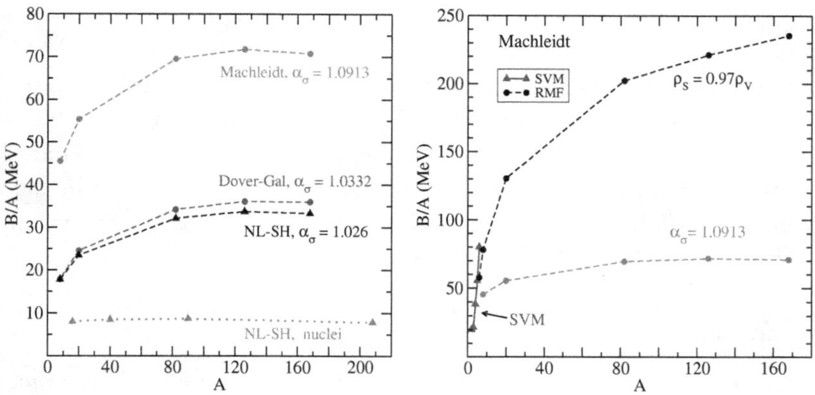

Fig. 2. Left panel: Binding energy of Λ^* nuclei per particle, B/A as a function of mass number A, calculated within the HS and NL-SH models; B/A in atomic nuclei ('nuclei') is shown for comparison. Right panel: Comparison of B/A calculated in Λ^* nuclei within the HS model for the Machleidt potential (red line) with a similar calculation using $\rho_{\rm s} = 0.97\rho_{\rm v}$ (black line); B/A in few body systems, calculated within the SVM is shown for comparison. See text for details.

Λ^* nuclei as well as in ordinary nuclei is shown as well. The binding energy per Λ^* saturates with the number of constituents for $A \geq 120$ in all versions considered and reaches tens of MeV depending on the potential used. Calculations with the rescaled ω coupling constant yield similar saturation curves for B/A in Λ^* nuclei.

The observed saturation originates from the Lorentz covariance which introduces two types of baryon densities — the scalar density $\rho_{\rm s}$ associated with the attractive σ field and the vector (baryon) density $\rho_{\rm v}$ associated with the repulsive ω field. In dense matter, the scalar density decreases with respect to the vector density since $\rho_{\rm s} \sim M^*/E^* \rho_{\rm v}$ where $\frac{M^*}{E^*} < 1$, and $M^* = M - g_{\sigma B}\langle\sigma\rangle$ is baryon effective mass. As a consequence, the attraction from the scalar field is reduced considerably at higher densities. This is illustrated in Fig. 2 (right panel), where we present the RMF calculation of B/A in Λ^* nuclei, in which we replaced the scalar density $\rho_{\rm s}$ by a density equal to $0.97\rho_{\rm v}$ (this corresponds to $\rho_{\rm s}/\rho_{\rm v}$ in ^{16}O). The binding energy per Λ^* (denoted '$\rho_{\rm s} = 0.97\rho_{\rm v}$') is rapidly increasing in this case, similar to the SVM calculations (also shown for comparison), and does not seem to saturate within the explored mass range, unlike B/A evaluated using the 'dynamical' scalar density $\rho_{\rm s}$ (denoted '$\alpha_\sigma = 1.0913$'). It is to be noted that the central density of calculated Λ^* nuclei saturates as a function of A as well, reaching about twice nuclear matter density.

Finally, we introduced the Λ^* absorption and explored how the Λ^* decay width changes in the medium. We considered the two-body decay $\Lambda^*\Lambda^* \rightarrow \Lambda\Lambda$ in the $1s$ state, described by the imaginary part of an optical potential in a '$t\rho$' form with the amplitude fitted to assumed width $\Gamma_{\Lambda^*\Lambda^*} = 100$ MeV at threshold, taking into account phase space suppression. We found that the conversion widths, despite being suppressed to some extent in the Λ^* nuclei (by 28% in A=8 systems and by less than 1% in A=168 systems), remain considerable and the $\Lambda^*\Lambda^*$ pairs will thus inevitably decay.

4. Summary

We performed calculations of Λ^* nuclei with various Λ^* interaction strengths compatible with the value $B_{\Lambda^*\Lambda^*} = 40$ MeV of the YA model[1] in order to demonstrate that the Λ^* stable-matter scenario[3] is not supported by standard many-body approaches. We found that the binding energy per Λ^* in many-body systems saturates in all cases for $A \geq 120$ at values far below ≈ 290 MeV, which is the energy required to reduce the $\Lambda(1405)$ mass in the medium below the mass of the lightest hyperon $\Lambda(1116)$. The Λ^* matter is thus highly unstable against strong interaction decay.

Acknowledgment

This work was partly supported by the Czech Science Foundation GACR grant 19-19640S.

References

1. T. Yamazaki and Y. Akaishi, Phys. Rev. **C 76**, 045201 (2007).
2. J. Hrtánková, N. Barnea, E. Friedman, A. Gal, J. Mareš and M. Schäfer, Phys. Lett. **B 785**, 90 (2018).
3. Y. Akaishi and T. Yamazaki, Phys. Lett. **B 774**, 552 (2017).
4. K. Varga and Y. Suzuki, *Stochastic Variational Approach to Quantum-Mechanical Few-Body Problems* (Springer 1998).
5. B. D. Serot and J. D. Walecka, Adv. Nucl. Phys. **16**, 1 (1986).
6. E. Friedman and A. Gal, Nucl. Phys. **A 959**, 66 (2017).
7. C. B. Dover and A. Gal, Prog. Part. Nucl. Phys. **12**, 171 (1984).
8. R. Machleidt, Adv. Nucl. Phys. **19**, 189 (1989).
9. C. J. Horowitz and B. D. Serot, Nucl. Phys. **A 368**, 503 (1981).
10. M. M. Sharma, M. A. Nagarajan, P. Ring, Phys. Lett. **B 312** (1993) 377.

Capturing some medium effects in the dilaton to study hadrons in AdS/QCD models

Alfredo Vega* and M. A. Martin Contreras
Instituto de Física y Astronomía y Centro de Astroffisica de Valparaíso, Universidad de Valparaíso, Avenida Gran Bretaña 1111, Valparaíso, Chile
**E-mail: alfredo.vega@uv.cl*

In AdS/QCD models medium effects are usually introduced in the metric, but recently some authors have started to consider the possibility to catch media properties in the dilaton field. In this talk we present a couple of examples about this topic.

Keywords: AdS/QCD; hadron melting temperature; form factors.

1. Introduction

The understanding of how temperature or a very dense medium can affect hadronic properties is a topic that has attracted the interest of several physicists, who in their attempt to clarify the hadron phenomenology under these extreme conditions, have built research facilities to do complex experiments (e.g., see Ref. 1). They have also developed theoretical tools such as lattice QCD, sum rules or potential models to approach hadron phenomenology in these extreme conditions. Together with these approaches, in the last 20 years, a set of techniques has been added based on gauge/gravity dualities (e.g., see Refs. 2–5), where temperature and medium effects are introduced by considering proper metrics in the model formulation.

From the gravitational point of view, the metric and the dilaton field define the background, and both are related dynamically as solutions of the equations of motion obtained from the Hilbert-Einstein action variation,[6–8] but in the phenomenological AdS/QCD models, it is common to use static dilatons (e.g., see Refs. 9–13) to softly break the conformal symmetry, and in general the dilaton is not considered as a part of the background able to catch medium properties that affect the hadrons studied in these models.

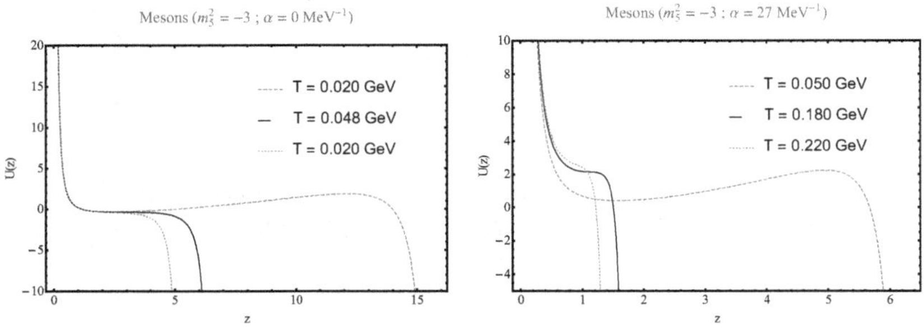

Fig. 1. These plots show the holographic potential calculated by using (1). On the right side, the traditional case ($\alpha = 0$) is considered, where the continuous line is calculated with the prediction of the model for melting temperature for this case, which is 48 MeV. On the left side, we plot potentials considering a thermal quadratic dilaton with $\alpha = 27$ MeV^{-1} (that was fitted to obtain 180 MeV for mesonic melting temperature).

Recently some authors have started to consider the possibility to catch media properties in the dilaton field,[14–17] and here we discuss two examples about the of this idea. In Section 2 we consider a temperature dependent dilaton and his effect on hadron melting temperature.[14] In Section 3 we introduce a dilaton which depend on density and use it together a well known model to calculate nucleon form factors inside nucleus.[17] Finally Section 4 present some perspectives of this kind a dilatons.

2. Hadron melting temperature in models whit thermal a dilatons

In this example we consider an scalar field in a background which considers a black hole and a dilaton field. In this case it is possible to obtain an equation like $-u''(z) + U(z)\,u(z) = 0$ (for details e.g., Ref. 14), where $U(z)$ is the thermal holographic potential given by

$$U(z) = \frac{e^{2\,A(z)}}{f(z)}\,m_5^2 - \frac{B'(z)\,f'(z)}{2\,f(z)} + \frac{f''(z)}{2\,f(z)} - \frac{f'(z)^2}{4\,f(z)^2} - \frac{\omega^2}{f(z)^2} + \frac{B'(z)^2}{4} - \frac{B''(z)}{2}. \tag{1}$$

From this holographic potential we will obtain the melting temperature of some scalar hadrons by studying when well disappears. In this case, we use the hadron mass calculated at $T = 0$.

Here we consider only the thermal version of the usual quadratic dilaton

$$\phi_1(z,T) = \kappa^2(1 + \alpha\,T)z^2. \tag{2}$$

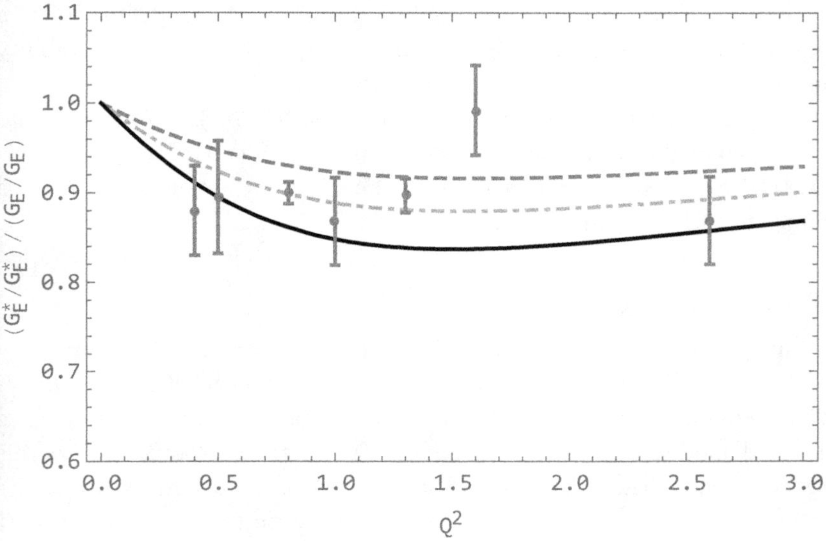

Fig. 2. Rouble ratio $(G_E^*/G_M^*)/(G_E/G_M)$ for protons. Densities considered are $\rho_B/\rho_0 = 1$ (continuos line), $\rho_B/\rho_0 = 0.8$ (dot-dashed line) and $\rho_B/\rho_0 = 0.6$ (dashed line). Experimental data for protons are from Refs. 21–23.

Notice that in general the thermal dilatons are extensions of those used at zero temperature. As the metric changes, the spectrum is not directly obtained by changing κ only. In order to know how the spectrum and the Regge trajectories are changed in this case, it is necessary to study the related spectral functions, but melting temperatures can be estimated in a easy way analyzing the potential as you can see in Fig. 1.

3. Soft wall model for the nuclear medium

In Ref. 18, the authors use AdS/QCD soft wall-like models with an AdS metric and a quadratic dilaton to calculate proton electromagnetic form factors. This model can be extended in a simply way to study properties of protons inside nuclei.

By space reasons here we can not give details, but as you can see in Ref. 17 the quadratic dilaton parameter is re-scaled as

$$\kappa^* = \sqrt{\left(1 - 0.21\frac{\rho_B}{\rho_0}\right)}\,\kappa. \tag{3}$$

This allows to reproduce the scaling behaviour for the proton mass inside the nuclei, as it was suggested in Ref. 19. Its use in AdS/QCD predicts a Regge behavior for hadrons inside atomic nuclei and offer us the possibility to use the expressions calculated in Ref. 18.

In Fig. 2 we show our results for double ratio $(G_E^*/G_M^*)/(G_E/G_M)$ and compare against measurements made at JLab,[21,22] and Mainz Microtron (MAMI)[23] for polarization transfer double ratio for a bound proton, which is connected with this quantity.

4. Final comments

In most of AdS/QCD models the medium properties are encoded exclusively in the metric. Although dilatons used in most of AdS/QCD models are static, it should be remembered that they are background fields (which in dynamical models are related or coupled to metric). Motivated by this fact, we consider the dilaton could catch part of environment properties therefore they could be a complement to the metric tensor capturing part of the medium properties. In this form they can help us to describe hadron properties in a dense medium and/or at finite temperature.

Acknowledgements

The authors acknowledge the financial support of FONDECYT (Chile) under Grants No. 1180753 (A. V) and No. 3180592 (M. A. M).

References

1. E. V. Shuryak, Nucl. Phys. A **750**, 64 (2005) doi:10.1016/j.nuclphysa.2004.10.022 [hep-ph/0405066].
2. J. M. Maldacena, Int. J. Theor. Phys. **38**, 1113 (1999) [Adv. Theor. Math. Phys. **2**, 231 (1998)] doi:10.1023/A:1026654312961, 10.4310/ATMP.1998.v2.n2.a1 [hep-th/9711200].
3. E. Witten, Adv. Theor. Math. Phys. **2**, 253 (1998) doi:10.4310/ATMP.1998.v2.n2.a2 [hep-th/9802150].
4. E. Witten, Adv. Theor. Math. Phys. **2**, 505 (1998) doi:10.4310/ATMP.1998.v2.n3.a3 [hep-th/9803131].
5. S. S. Gubser, I. R. Klebanov and A. M. Polyakov, Phys. Lett. B **428**, 105 (1998) doi:10.1016/S0370-2693(98)00377-3 [hep-th/9802109].
6. D. Li, S. He, M. Huang and Q. S. Yan, JHEP **1109**, 041 (2011) doi:10.1007/JHEP09(2011)041 [arXiv:1103.5389 [hep-th]].

7. S. He, S. Y. Wu, Y. Yang and P. H. Yuan, JHEP **1304**, 093 (2013) doi:10.1007/JHEP04(2013)093 [arXiv:1301.0385 [hep-th]].
8. R. Zöllner and B. Kämpfer, arXiv:1807.04260 [hep-th].
9. A. Karch, E. Katz, D. T. Son and M. A. Stephanov, Phys. Rev. D **74**, 015005 (2006) doi:10.1103/PhysRevD.74.015005 [hep-ph/0602229].
10. H. J. Kwee and R. F. Lebed, JHEP **0801**, 027 (2008) doi:10.1088/1126-6708/2008/01/027 [arXiv:0708.4054 [hep-ph]].
11. S. J. Brodsky and G. F. de Teramond, Phys. Rev. D **77**, 056007 (2008) doi:10.1103/PhysRevD.77.056007 [arXiv:0707.3859 [hep-ph]].
12. A. Vega, I. Schmidt, T. Gutsche and V. E. Lyubovitskij, Phys. Rev. D **83**, 036001 (2011) doi:10.1103/PhysRevD.83.036001 [arXiv:1010.2815 [hep-ph]].
13. T. Branz, T. Gutsche, V. E. Lyubovitskij, I. Schmidt and A. Vega, Phys. Rev. D **82**, 074022 (2010) doi:10.1103/PhysRevD.82.074022 [arXiv:1008.0268 [hep-ph]].
14. A. Vega and M. A. Martin Contreras, Nucl. Phys. B **942**, 410 (2019) doi:10.1016/j.nuclphysb.2019.03.014 [arXiv:1808.09096 [hep-ph]].
15. T. Gutsche, V. E. Lyubovitskij, I. Schmidt and A. Y. Trifonov, Phys. Rev. D **99**, no. 5, 054030 (2019) doi:10.1103/PhysRevD.99.054030 [arXiv:1902.01312 [hep-ph]].
16. T. Gutsche, V. E. Lyubovitskij, I. Schmidt and A. Y. Trifonov, Phys. Rev. D **99**, no. 11, 114023 (2019) doi:10.1103/PhysRevD.99.114023 [arXiv:1905.02577 [hep-ph]].
17. A. Vega and M. A. Martin Contreras, arXiv:1812.00642 [hep-ph].
18. T. Gutsche, V. E. Lyubovitskij, I. Schmidt and A. Vega, Phys. Rev. D **86**, 036007 (2012) doi:10.1103/PhysRevD.86.036007 [arXiv:1204.6612 [hep-ph]].
19. G. E. Brown, V. Koch and M. Rho, Nucl. Phys. A **535**, 701 (1991). doi:10.1016/0375-9474(91)90483-M
20. K. Saito, K. Tsushima and A. W. Thomas, Phys. Rev. C **55**, 2637 (1997) doi:10.1103/PhysRevC.55.2637 [nucl-th/9612001].
21. S. Strauch *et al.* [Jefferson Lab E93-049 Collaboration], Phys. Rev. Lett. **91**, 052301 (2003) doi:10.1103/PhysRevLett.91.052301 [nucl-ex/0211022].
22. M. Paolone *et al.*, Phys. Rev. Lett. **105**, 072001 (2010) doi:10.1103/PhysRevLett.105.072001 [arXiv:1002.2188 [nucl-ex]].
23. S. Dietrich *et al.*, Phys. Lett. B **500**, (2001) 47 - 52 doi:10.1016/S0370-2693(01)00052-1

Fate of heavy quark bound states inside quark-gluon plasma

Xiaojun Yao*, Weiyao Ke, Yingru Xu, Steffen Bass, Thomas Mehen and Berndt Müller

Department of Physics, Duke University
Durham, NC 27708, USA
**Center for Theoretical Physics, Massachusetts Institute of Technology*
Cambridge, MA 02139, USA
**E-mail: xiaojun.yao@duke.edu*

Transport equations have been applied successfully to describe the quarkonium evolution inside the quark-gluon plasma, which include both plasma screening effects and recombination. We demonstrate how the quarkonium transport equation is derived from QCD by using the open quantum system framework and effective field theory. Weak coupling and Markovian approximations used in the derivation are justified from a separation of scales. By solving the equations numerically, we study the Upsilon production in heavy ion collisions.

Keywords: Heavy quarkonium, quark-gluon plasma, heavy ion collisions, plasma screening, open quantum system, Boltzmann transport equation.

1. Introduction

Heavy quarkonium is a bound state of a heavy quark-antiquark pair ($Q\bar{Q}$). Inside the quark-gluon plasma (QGP), which is a deconfined phase of nuclear matter, the attractive potential between the $Q\bar{Q}$ is screened. The plasma screening becomes stronger as the temperature increases. As a result, at sufficiently high temperature, the potential is too weak to support the bound state formation.[1,2] In other words, quarkonium "melts". This effect is called static plasma screening effect. It leads to suppression of quarkonium production in heavy ion collisions, if the QGP is formed. Thus, quarkonium suppression is used as a signature of the QGP formation in heavy ion collisions.

Experimental measurements confirmed the quarkonium suppression in heavy ion collisions at both RHIC and LHC. To explain the data quantitatively, one has to take into account another plasma screening effect, called dynamical screening effect. It describes the quarkonium dissociation in dynamical scattering processes. It happens when enough energy is transferred from the medium to the quarkonium state to excite it to the continuum.

It generates a thermal width of quarkonium, which increases with the temperature. These two screening effects are connected in the sense that both of them are generated from the thermal loop correction to the quarkonium propagator.

In addition to the plasma screening effects, another crucial process in understanding the quarkonium in-medium evolution is the recombination of unbound $Q\bar{Q}$'s.[3,4] When the medium temperature is below the melting temperature of quarkonium, an unbound $Q\bar{Q}$ may radiate out a certain amount of energy and form a bound state. The Q and $\bar{Q}$ in the recombination may come from the same initial hard scattering vertex or different initial hard vertices. The latter case is negligible when only a few heavy quarks are produced in the collision, but will be enhanced as the heavy quark number increases. The recombination is crucial to explain the collision energy dependence of J/ψ suppression. J/ψ is less suppressed as the collision energy increases from RHIC to LHC because of the enhanced recombination (see Ref. 5 for a review).

To understand the quarkonium in-medium dynamics, one has to consider both plasma screening effects and recombination. Transport equations have been applied successfully.[6–12] However, the connection between the transport equation and the underlying theory QCD is not clear. Furthermore, it is not clear when the transport equation is a valid description. In this proceeding, we will explain how to derive the Boltzmann transport equation for quarkonium in-medium dynamics from QCD using the open quantum system framework. The transport equation will be briefly reviewed in Sec. 2 and the derivation will be explained in Sec. 3. We will discuss the relation between the validity condition of transport equations and a hierarchy of scales. Finally, phenomenological results on Upsilon suppression will be shown in Sec. 4 and conclusions will be drawn in Sec. 5.

2. Boltzmann Transport Equation

The Boltzmann transport equation for quarkonium can be written as:

$$(\partial_t + \dot{\boldsymbol{x}} \cdot \nabla_{\boldsymbol{x}}) f(\boldsymbol{x}, \boldsymbol{k}, t) = \mathcal{C}^+(\boldsymbol{x}, \boldsymbol{k}, t) - \mathcal{C}^-(\boldsymbol{x}, \boldsymbol{k}, t)\,, \tag{1}$$

in which $f(\boldsymbol{x}, \boldsymbol{k}, t)$ is the phase space distribution function of a quarkonium state (each state has its own f). The left hand side describes the free streaming of quarkonium and $\dot{\boldsymbol{x}} \equiv \frac{\mathrm{d}\boldsymbol{x}}{\mathrm{d}t}$. The right hand side contains two collision terms: recombination $\mathcal{C}^+(\boldsymbol{x}, \boldsymbol{k}, t)$ and dissociation $\mathcal{C}^-(\boldsymbol{x}, \boldsymbol{k}, t)$. The static screening effect has been accounted for in the quarkonium wavefunction, which is used in the calculation of $\mathcal{C}^{\pm}$.

3. Derivation of Transport Equation

We will use the framework of open quantum system. In this framework, the whole system under study consists of a sub-system and an environment. The whole system evolves unitarily. The evolution is also time-reversible if the underlying theory respects time-reversal symmetry such as QCD. If we focus on just the sub-system, we will integrate out the environment degrees of freedom. Then the evolution equation of the sub-system is non-unitary and time-irreversible. In our case, the sub-system consists of bound and unbound $Q\bar{Q}$'s while the environment is the QGP.

If we assume the sub-system and the environment interact weakly, we can expand the evolution equation to second order in perturbation and obtain the Lindblad equation:

$$\rho_S(t) = \rho_S(0) - i\Big[tH_S + \sum_{a,b} \sigma_{ab}(t) L_{ab}, \rho_S(0)\Big] + \sum_{a,b,c,d} \gamma_{ab,cd}(t)\Big(L_{ab}\rho_S(0)L^\dagger_{cd} - \frac{1}{2}\{L^\dagger_{cd}L_{ab}, \rho_S(0)\}\Big), \quad (2)$$

where definitions of each term can be found in Ref. 13. It can be shown that the Lindblad equation leads to the Boltzmann transport equation under the Markovian approximation if a Wigner transform is applied to the sub-system density matrix. The Wigner transform is defined by

$$f(\boldsymbol{x}, \boldsymbol{k}, t) \equiv \int \frac{\mathrm{d}^3k'}{(2\pi)^3} e^{i\boldsymbol{k}'\cdot\boldsymbol{x}} \Big\langle \boldsymbol{k} + \frac{\boldsymbol{k}'}{2} \Big| \rho_S(t) \Big| \boldsymbol{k} - \frac{\boldsymbol{k}'}{2} \Big\rangle, \quad (3)$$

which connects the density matrix and the phase space distribution. A one-to-one correspondence between terms in the Lindblad equation and those in the transport equation can be established. More specifically, the σ_{ab} term gives the static screening; The $L_{ab}\rho_S(0)L^\dagger_{cd}$ term corresponds to recombination while the $\{L^\dagger_{cd}L_{ab}, \rho_S(0)\}$ term leads to dissociation; Remaining terms give the free streaming. Detailed expressions of $\mathcal{C}^\pm$ can be found in Refs. 13 and 14.

The weak coupling and Markovian approximations can be justified from a separation of scales: $M \gg Mv \gg Mv^2 \gtrsim T \gtrsim m_D$. Here M denotes the heavy quark mass; v is the relative velocity between the heavy quark pair inside quarkonium; T is the plasma temperature and m_D denotes the Debye mass. Under this hierarchy of scales, the effective field theory potential nonrelativistic QCD (pNRQCD) can be constructed from QCD by a sequence of renormalization group flow, matching calculations and non-relativistic expansions.[15,16] In pNRQCD, a $Q\bar{Q}$ color singlet S turns to an octet O (and vice versa) by interacting with gluons in a dipole interaction: $O^\dagger \boldsymbol{r}\cdot\boldsymbol{E}S+h.c.$, where $\boldsymbol{E}$ is the chromo-electric field. Since the typical energy of medium gluons is T, the interaction vertex scales as $rT \sim \frac{T}{Mv} \sim v$ where r is the quarkonium size. Since v is assumed small, the interaction is weak. Furthermore, due to the weak interaction, the sub-system relaxation time can be shown to be much bigger than the environment correlation time, which justifies the Markovian approximation. The Markovian approximation is coarse graining, which means the finer detail of the environment dynamics is not resolved by the sub-system evolution.

In a nutshell, whenever the separation of scales $M \gg Mv \gg Mv^2 \gtrsim T \gtrsim m_D$ is true, the Boltzmann transport equation for quarkonium is valid. In practice, we have $Mv^2 \sim 500$ MeV for both charmonium and bottomonium and $T \lesssim 500$ MeV in current heavy ion experiments. The derivation here explains why the transport equation phenomenology works well.

4. Phenomenological Results

In this section, we will show phenomenological results based on the quarkonium transport equation derived in the last section. After quarkonium dissociation, a valid description of the in-medium heavy quark dynamics is the transport equation of open heavy quarks.[17] Thus, we have to couple the transport equations of quarkonium with those of open heavy quarks.[18] We will focus on the bottomonium system since the separation of scales works better for the bottom quark.

We will solve the coupled transport equations by Monte Carlo simulations. Details can be found in Ref. 11.

In heavy ion experiments, two observables of quarkonium are of interest: the nuclear modification factor R_{AA} and the azimuthal anisotropy coefficients v_n. The former is defined by the ratio of the quarkonium production cross section in heavy ion collisions and that in proton-proton collisions, scaled by the effective number of binary nucleon-nucleon collisions in heavy ion collisions such that $R_{AA} = 1$ without any medium effect. The latter is defined by Fourier decomposition of the quarkonium azimuthal distribution into $\sum_n 2v_n \cos(2n\phi)$, where ϕ is the azimuthal angle of quarkonium with respect to the reaction plane that is defined event-by-event. Our calculation results for 5.02 TeV Pb-Pb collisions are shown in Fig. 1, which are in good agreement with the experimental measurements.

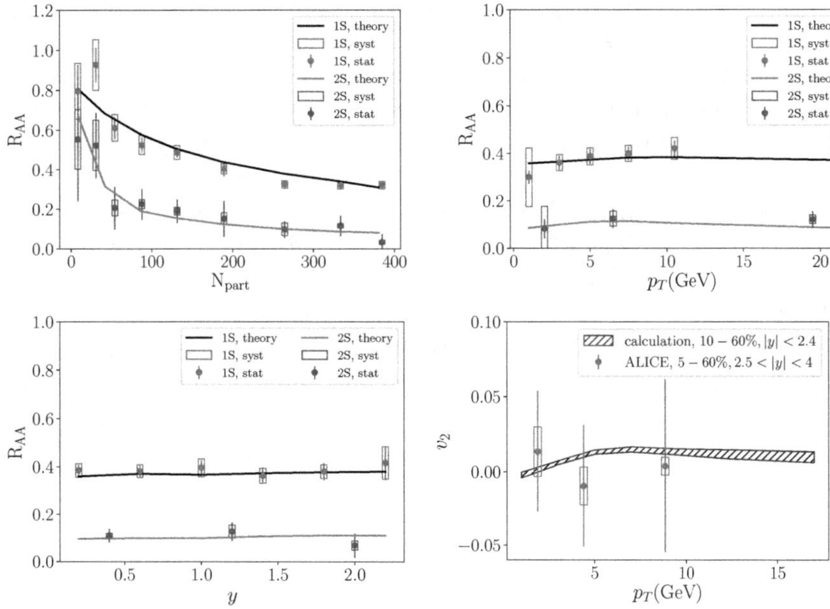

Fig. 1. Results of R_{AA} and v_2 in 5.02 TeV Pb-Pb collisions compared with CMS measurements of R_{AA}[19] and ALICE measurements of v_2.[20] $N_{\rm part}$, p_T and y denote the collision centrality (related to the impact parameter), quarkonium transverse momentum and rapidity.

5. Conclusions

We showed how to derive the quarkonium transport equation inside the QGP in a theoretically controlled way with a separation of scales. This provides a theoretical justification of applying transport equation in phenomenology of quarkonium production in heavy ion collisions. Phenomenological results on Upsilon production were also presented which agree well with data. Extension to the study of doubly heavy baryon can also be made.[21] Quarkonium transport equations in a different hierarchy of scales ($T \gg Mv^2$) have been discussed in Ref. 22,23.

Acknowledgments

This work is supported by U.S. Department of Energy research grants DE-FG02-05ER41367 and DE-FG02-05ER41368. XY also acknowledges support from DE-SC0011090, Brookhaven National Laboratory and Department of Physics, Massachusetts Institute of Technology.

References

1. T. Matsui and H. Satz, Phys. Lett. B **178**, 416 (1986).
2. F. Karsch, M. T. Mehr and H. Satz, Z. Phys. C **37**, 617 (1988).
3. R. L. Thews, M. Schroedter and J. Rafelski, Phys. Rev. C **63**, 054905 (2001).
4. A. Andronic, P. Braun-Munzinger, K. Redlich and J. Stachel, Phys. Lett. B **652**, 259 (2007).
5. A. Andronic *et al.*, Eur. Phys. J. C **76**, no. 3, 107 (2016).
6. L. Grandchamp, R. Rapp and G. E. Brown, Phys. Rev. Lett. **92**, 212301 (2004).
7. L. Yan, P. Zhuang and N. Xu, Phys. Rev. Lett. **97**, 232301 (2006).
8. B. Krouppa, R. Ryblewski and M. Strickland, Phys. Rev. C **92**, no. 6, 061901 (2015).
9. X. Du, R. Rapp and M. He, Phys. Rev. C **96**, no. 5, 054901 (2017).
10. J. Zhao and B. Chen, Phys. Lett. B **776**, 17 (2018).
11. X. Yao, W. Ke, Y. Xu, S. Bass and B. Müller, Nucl. Phys. A **982**, 755 (2019).
12. X. Yao, W. Ke, Y. Xu, S. Bass and B. Müller, PoS HardProbes **2018**, 157 (2018).
13. X. Yao and T. Mehen, Phys. Rev. D **99**, no. 9, 096028 (2019).
14. X. Yao and B. Müller, Phys. Rev. D **100**, no. 1, 014008 (2019).
15. N. Brambilla, A. Pineda, J. Soto and A. Vairo, Nucl. Phys. B **566**, 275 (2000).
16. S. Fleming and T. Mehen, Phys. Rev. D **73**, 034502 (2006).
17. W. Ke, Y. Xu and S. A. Bass, Phys. Rev. C **98**, no. 6, 064901 (2018).
18. X. Yao and B. Müller, Phys. Rev. C **97**, no. 1, 014908 (2018).
19. A. M. Sirunyan *et al.* [CMS Collaboration], Phys. Lett. B **790**, 270 (2019).
20. S. Acharya *et al.* [ALICE Collaboration], Phys. Rev. Lett. **123**, no. 19, 192301 (2019).
21. X. Yao and B. Müller, Phys. Rev. D **97**, no. 7, 074003 (2018).
22. N. Brambilla, M. A. Escobedo, J. Soto and A. Vairo, Phys. Rev. D **97**, no. 7, 074009 (2018).
23. T. Miura, Y. Akamatsu, M. Asakawa and A. Rothkopf, arXiv: 1908.06293 [nucl-th].

Studying the ϕ meson in nuclear matter from simulated pA reactions

Philipp Gubler

Advanced Science Research Center, Japan Atomic Energy Agency, Tokai, Ibaraki 319-1195, Japan
E-mail: gubler@post.j-parc.jp

This article summarizes recent progress in studying the behavior of the ϕ meson in nuclear matter. Both a purely theoretical work, invstigating the ϕ meson in a dense medium of infinite size and a more phenomenologially oriented study, simulating pA reactions, which are used to experimentally explore the modification of the ϕ meson in nuclear matter, are discussed.

Keywords: ϕ meson; nuclear matter; pA reactions; numerical simulations.

1. Introduction

The ϕ meson, a vector meson mainly composed of a strange quark and an anti-strange quark, could change its properties when inserted into a nuclear medium. Theoretically, such a modifcation (in form of a negative mass shift) was predicted in Ref. 1 in a QCD sum rule appraoch. Later, works based on hadronic models on the other hand found a large broadening of the ϕ meson in nuclear matter with only a small mass shift2] More recent works about this topic can be found in Refs. 3, 4 (QCD sum rules) and Refs. 5–8 (models). Here, it is worth mentioning especially Ref. 3, where the extent of a mass shift of the ϕ meson in nuclear matter was shown to be correlated with the strange sigma term $\sigma_{sN} = m_s\langle N|\bar{s}s|N\rangle$. σ_{sN} determines the rate of the reduction of the strange quark chiral condensate in nuclear matter within the linear density approximation,

$$\langle \bar{s}s \rangle_\rho \simeq \langle 0|\bar{s}s|0\rangle + \rho\langle N|\bar{s}s|N\rangle = \langle 0|\bar{s}s|0\rangle + \rho\frac{\sigma_{sN}}{m_s}. \tag{1}$$

Because σ_{sN} is still not yet determined with good precision even from present-day lattice QCD calculations (see the discussion in Ref. 9 and the references cited therein), no precise predictions about the ϕ meson mass shift can be made. An experimental measurement of this mass shift could, conversely, constrain the value of σ_{sN}3]

From an experimental point of view, several collaborations investigated the ϕ meson properties in nuclei. Among the various experiments, the E325 experiment at KEK 10] deserves special attention, as it is so far the only one which could measure both the mass and width of the ϕ in nuclear matter at the same time. By measuring the dilepton spectrum of 12 GeV p-Cu reactions in the ϕ meson energy region, a negative mass shift and increased width at normal nuclear matter density ρ_0 was reported,10]

$$\frac{m_\phi(\rho)}{m_\phi(0)} = 1 + k_1 \frac{\rho}{\rho_0}, \qquad \frac{\Gamma_\phi(\rho)}{\Gamma_\phi(0)} = 1 + k_2 \frac{\rho}{\rho_0}, \tag{2}$$

with

$$k_1 = 0.034^{+0.006}_{-0.007}, \qquad k_2 = 2.6^{+1.8}_{-1.2}. \tag{3}$$

In these proceedings, some new developments in the study of the ϕ meson in nuclear matter will be briefly reviewed. In the first part, the work of Ref. 4, which was completed recently, and which deals with the momentum dependence of the ϕ meson energy (e.g. its dispersion relation) in nuclear matter, will be discussed. In the second part, some first results of an ongoing work simulating the p-Cu reaction of the E325 experiment, using a transport approach, will be presented. The last section consists of a short summary and conclusions.

2. The dispersion relation of the ϕ meson in nuclear matter

In this section, we will review the work presented in Ref. 4, in which the momentum dependence of the ϕ meson in nuclear matter is investigated, making use of the method of QCD sum rules. The same problem was studied already more than 20 years ago in Ref. 11, but is revisited and updated in Ref. 4 because of recent theoretical and experimental developments. One important theoretical step was Ref. 12, where the operator product expansion (OPE) of the vector channel, which is the basic input in the QCD sum rule calculation for the ϕ meson, was computed for the first time for the complete set of scalar and non-scalar operators up to mass dimension 6. This allows for a systematic study of the corresponding sum rule, including a reliable error analysis. On the experimental side, it is worth mentioning the E16 experiment at J-PARC, which will start running in 2020.13] One goal of this experiment is to measure the in-medium dispersion relation of the ϕ meson at normal nuclear matter density. It is hence timely to provide an updated theoretical theoretical prediction for this experiment.

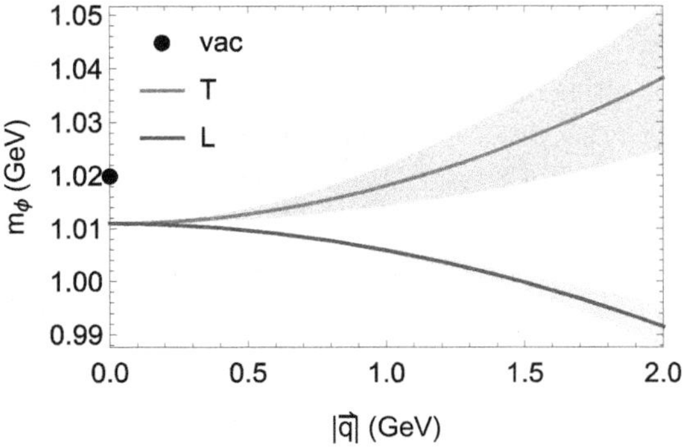

Fig. 1. The ϕ meson mass, shown as a function of momentum $|\boldsymbol{q}|$ for longitudinal (blue curve) and transverse (red curve) modes, at normal nuclear matter density ρ_0. Adapted from Ref. 4.

Because of limitations of space, we will here refraim from discussing any of the technical aspects of the QCD sum rule method, and simply discuss the obtained final results. One important aspect of the in-medium dispersion relations of vector particles is the fact that, they can generally be split depending on the (longitudinal or transverse) polarization of the particle. Indeed, we find in our QCD sum rule study that the behavior of the longitudinal and transverse modes behave differently. This is illustrated in Fig. 1, where the mass of the two modes are shown as a function of momentum $|\boldsymbol{q}|$. It is seen in the figure that longlitudinal and transverse modes move away from each other with increasing momentum. We refer the interested reader to Ref. 4, where the consequences of this result are discussed in detail.

3. Numerical simulations of pA reactions

To be able to understand the experimentally measured dilepton spectra from pA reactions and compare them to theoretical results, a detailed understanding the dynamics governing the reaction is necessary. For this purpose, we use the HSD transport approach14] to carry out simulations of pA reactions with an incoming proton energy of 12 GeV, corresponding to the E325 experiment at KEK10] HSD is a covariant, microscopic off-shell transport model, in which one can specify vector meson spectral functions and their density dependence as an input. In our simulation, we have for the ϕ meson spectral function so far used four different scenarios: "no density dependence", "broadening", "mass shift" and "mass shift +

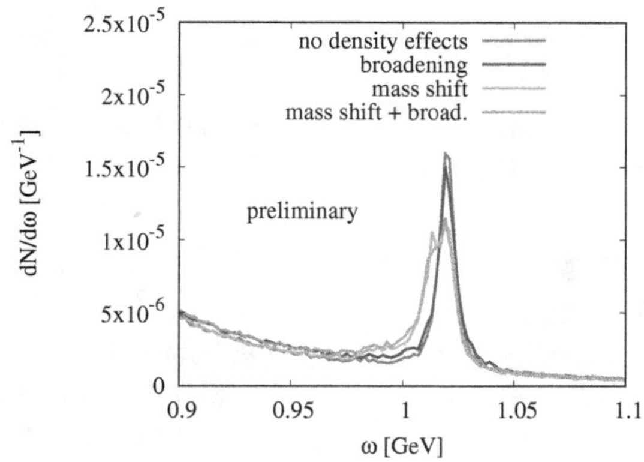

Fig. 2. Dilepton spectrum in the ϕ meson region for the four scenarios mentioned in the text, for p-Cu reactions with a 12 GeV projectile proton.

broadening". The amount of the mass shift and broadening is chosen such that it corresponds to the findings reported by the E325 experiment, given in Eqs.(2-3).

We will show here some first, still preliminary results of the simulation. Focusing on the di-lepton spectrum in the ϕ meson peak region, the four scenarios are compared in Fig.2. A visible enhancement of the spectrum in the energy region below the ϕ meson peak appears only for the "mass shift" and "mass shift + broadening" cases.

4. Summary and conclusions

Theoretical studies investigating the behavior of the ϕ meson in nuclear matter were reviewed in these proceedings. We have discussed how the effects of non-zero momentum can modify the ϕ meson spectrum, and have pointed out that the peak positions of the longitudinal and transverse components behave differently as a function of momentum. First attempts in simulating pA reactions probed in actual experiments for studying the behavior of the ϕ meson in nuclear matter were also presented. Our preliminary results show that a mass shift of the ϕ in nuclear matter has a visible effect on the dilepton spectrum.

Acknowledgments

The author thanks HyungJoo Kim and Elena Bratkovskaya for the fruitful collaborations that are mentioned in these proceedings. The author is furthermore supported by KAKENHI under Contract No. JP18K13542 and the Leading Initiative for Excellent Young Researchers (LEADER) of the Japan Society for the Promotion of Science (JSPS).

References

[1] T. Hatsuda and S. H. Lee, QCD sum rules for vector mesons in the nuclear medium, *Phys. Rev.* **C46**, R34 (1992).

[2] F. Klingl, T. Waas and W. Weise, Modification of the phi meson spectrum in nuclear matter, *Phys. Lett.* **B431**, 254 (1998).

[3] P. Gubler and K. Ohtani, Constraining the strangeness content of the nucleon by measuring the ϕ meson mass shift in nuclear matter, *Phys. Rev.* **D90**, 094002 (2014).

[4] H. Kim and P. Gubler, The ϕ meson with finite momentum in a dense medium, arXiv:1911.08737 [hep-ph].

[5] D. Cabrera, A. N. Hiller Blin and M. J. Vicente Vacas, ϕ meson self-energy in nuclear matter from ϕN resonant interactions, *Phys. Rev.* **C95**, 015201 (2017).

[6] P. Gubler and W. Weise, Moments of ϕ meson spectral functions in vacuum and nuclear matter, *Phys. Lett.* **B751**, 396 (2015).

[7] J. J. Cobos-Martínez, K. Tsushima, G. Krein and A. W. Thomas, ϕ meson mass and decay width in nuclear matter and nuclei, *Phys. Lett.* **B771**, 113 (2017).

[8] D. Suenaga and P. Lakaschus, Comprehensive study of mass modifications of light mesons in nuclear matter in the three-flavor extended Linear Sigma Model, arXiv:1908.10509 [nucl-th].

[9] P. Gubler and D. Satow, Recent Progress in QCD Condensate Evaluations and Sum Rules, *Prog. Part. Nucl. Phys.* **106**, 1 (2019).

[10] R. Muto *et al.*, Evidence for in-medium modification of the phi meson at normal nuclear density, *Phys. Rev. Lett.* **98**, 042501 (2007).

[11] S. H. Lee, Vector mesons in-medium with finite three momentum, *Phys. Rev.* **C57**, 927 (1998), [Erratum: Phys. Rev.C58,3771(1998)].

[12] H. Kim, P. Gubler and S. H. Lee, Light vector correlator in medium: Wilson coefficients up to dimension 6 operators, *Phys. Lett.* **B772**, 194 (2017), [Erratum: Phys. Lett.B779,498(2018)].

[13] S. Ashikaga *et al.*, Measurement of Vector Meson Mass in Nuclear Matter at J-PARC, *JPS Conf. Proc.* **26**, 024005 (2019).

[14] W. Cassing and E. L. Bratkovskaya, Parton-Hadron-String Dynamics: an off-shell transport approach for relativistic energies, *Nucl. Phys.* **A831**, 215 (2009).

Poster Session

Study of $e^+e^- \to \pi^+\pi^-\pi^0$ process using initial state radiation at KLOE

Bo Cao

Department of Physics and Astronomy, Uppsala University,
Uppsala, 75236, Sweden.
E-mail: bo.cao@physics.uu.se

Xiao Lin Kang

LNF-INFN, Via E. Fermi 40, Frascati (RM), 00044, Italy
E-mail: xiaolin.kang@lnf.infn.it

(On behalf of the KLOE-2 Collaboration)

We study the initial state radiation (ISR) process $e^+e^- \to \pi^+\pi^-\pi^0(\gamma)$ using a data sample of 1.62 fb^{-1} e^+e^- collected with KLOE detector at ϕ-meson peak. By reconstructing the 3π invariant mass spectrum, this analysis aims to determine the line shape of the process $e^+e^- \to \pi^+\pi^-\pi^0$ in the energy range [720, 820] MeV. Further goals of this analysis is to extract the cross section for the vector resonances processes $e^+e^- \to \omega/\phi \to 3\pi$, and the continuum process $e^+e^- \to \gamma^* \to 3\pi$. Products of branching fractions $\mathcal{B}(\phi/\omega \to e^+e^-)\mathcal{B}(\phi/\omega \to 3\pi)$ can also be determined.

Keywords: KLOE/KLOE-2; $e^+e^- \to \pi^+\pi^-\pi^0\gamma(\gamma)$.

1. Introduction

The anomaly of the magnetic moment of the muon, a_μ, is one of the most precisely known quantities in particle physics. The theoretical prediction of the Standard Model (SM),[1] $a_\mu^{\rm SM}$, is ~ 4 standard deviations lower than the experimental measurement, $a_\mu^{\rm exp}$, performed at BNL,[2] providing one of the strongest indication of physics beyond the SM. Currently, the uncertainties of $a_\mu^{\rm exp}$ and $a_\mu^{\rm SM}$ are of comparable magnitude. With the new ongoing g-2 experiment at Fermilab,[3] the experimental error is expected to be improved by a factor of four. A large part of the uncertainty on the theoretical estimates comes from the leading order hadronic vacuum contribution, $a_\mu^{\rm had,LO}$, which is not calculable by perturbative QCD but has to be evaluated with a dispersion integral using measured hadronic cross sections as input. The process $e^+e^- \to \pi^+\pi^-$ contributes approximately for 70%

of the total estimate of $a_\mu^{\text{had,LO}}$ and is also the largest contributor to its uncertainty. The KLOE experiment[4] extracted the most precise $\pi\pi$ cross section determination in the energy range of $(0.10, 0.95)$ GeV^2. The second largest contributor to $a_\mu^{\text{had,LO}}$ uncertainty is the process $e^+e^- \to \pi^+\pi^-\pi^0$ in the low energy region, which was studied by SND,[5] CMD-2,[6] and BaBar experiments.[7] The high luminosity e^+e^- collisions data collected by KLOE at the ϕ meson peak provide a unique opportunity to study this process using the initial state radiation (ISR) method, allowing to largely reduce the experimental errors of the line shape measurements.

2. Experiment

The KLOE detector, operated at the Frascati e^+e^- collider DAΦNE at the ϕ meson mass peak, is composed of a cylindrical drift chamber (DC) and an electromagnetic calorimeter (EMC) surrounded by a superconducting coil providing a 0.52 Tesla axial magnetic field. The drift chamber provides a spatial resolution $\sim$ 150 μm in the transverse plane r-ϕ and $\sim$ 2 mm along the z-direction. The momentum resolution for large angle tracks is $\sigma_{p_\perp}/p_\perp \sim 0.4\%$. The calorimeter consists of a cylindrical barrel and two end-caps providing a solid angle coverage of 98%. The energy and time resolutions for photons are $\sigma_E/E = 5.7\%/\sqrt{E\,(\text{GeV})}$ and $\sigma_t = 54\ \text{ps}/\sqrt{\text{E}\,(\text{GeV})} \oplus 100$ ps, respectively.

3. Analysis

In this analysis, a $\sim$ 1.62 fb^{-1} data sample collected in 2004 and 2005 was used. Approximately 10% of the data sample is used to tune the analysis itself. Signal events $e^+e^- \to \pi^+\pi^-\pi^0\gamma$ are selected by requiring two charged tracks with opposite curvatures and three clusters. Tracks are assumed to be charged pions and a vertex is required to be reconstructed in a cylindrical volume with dimension $\rho_{\text{vtx}} = \sqrt{x^2+y^2} < 4$ cm and $|z_{\text{vtx}}| < 10$ cm. Prompt clusters are defined by requiring their Time-of-Flight (ToF) T_{clu} to satisfy $|T_{\text{clu}} - R_{\text{clu}}/c| < \min(2,\ 5\sigma_t)$ ns, where R_{clu} is the distance from the position of the cluster to the interaction point, c is the speed of light and σ_t is the EMC time resolution at the cluster energy. Energy deposit of cluster in the calorimeters is required to be $E_{\text{clu}} > 15$ MeV and angular acceptance is $|\cos\theta| < 0.92$.

Most of the signal events pass the $K_S K_L$ selection criteria of the KLOE event classification algorithm. A kinematic fit is applied to such sample, imposing the energy-momentum conservation together with the ToF con-

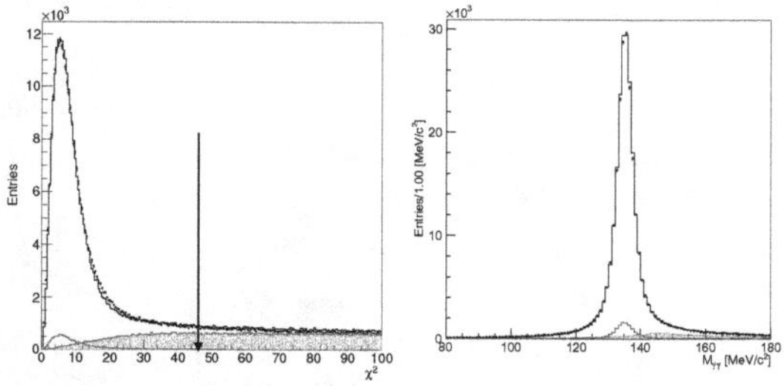

Fig. 1. χ^2 distribution of the kinematic fit (left) and $\gamma\gamma$ invariant mass distributions of best π^0 photon pair (right) for data (dots), signal (blue) and $\phi \to K_L K_S$ (shaded) events. The sum of all MC simulated backgrounds and the signal is also shown as black histograms.

straint on the three prompt neutral clusters. The χ^2 distribution is shown in the left plot in Figure 1. $K_L K_S$ events can be significantly rejected by requiring $\chi^2 < 46$. The π^0 candidate is reconstructed by building up a pseudo $\chi^2_{\gamma\gamma}$ for all the $\gamma\gamma$ pairs, using the energies and the positions of the neutral clusters. The combination with the smallest $\chi^2_{\gamma\gamma}$ is taken to be the π^0 photon pair, and the unpaired photon is asigned to be the ISR one.

To further suppress the background processes, a variable M is constructed which satisfys $\sqrt{s} - \sqrt{M^2 + |\vec{p}_{\pi^+}|^2} - \sqrt{M^2 + |\vec{p}_{\pi^-}|^2} - |\vec{p}_\phi - \vec{p}_{\pi^+} - \vec{p}_{\pi^-}| = 0$. Here $\vec{p}_\phi$, $\vec{p}_{\pi^+}$, and $\vec{p}_{\pi^-}$ are the three-momentum of ϕ, π^+, and π^-, respectively; $\sqrt{s}$ is the center of mass energy; all of them are in ϕ rest frame. Most of the $\phi \to \rho\pi$ events are rejected by the requirement $M > 300\ \text{MeV/c}^2$. The invariant mass spectrum of $\pi^+\pi^-\pi^0$, $M_{3\pi}$, for the surviving events is shown as the black dots in Figure 2. A Maximum Likelihood (ML) fit to data is performed by using the Monte Carlo (MC) simulated background and signal shapes with free fractions for the different contributions. The fit result is shown as the black histograms in Figure 2.

4. Conclusion and outlook

After applying the above selection criteria, invariant mass spectrum of 3π is shown in Figure 2. The remain backgrounds are mainly from $\phi \to \eta\gamma$, $K_L K_S$, and K^+K^-. It is remarkable that the contributions from continuum processes ($e^+e^- \to e^+e^-$, $\mu^+\mu^-$ and $\pi^+\pi^-$) are negligible. In the ω mass region, a good agreement between data and MC simulations is achieved after the ML fit with minimum degrees of freedom. The ongoing analysis aims to evaluate the line shape of $e^+e^- \to 3\pi$ in the ω mass

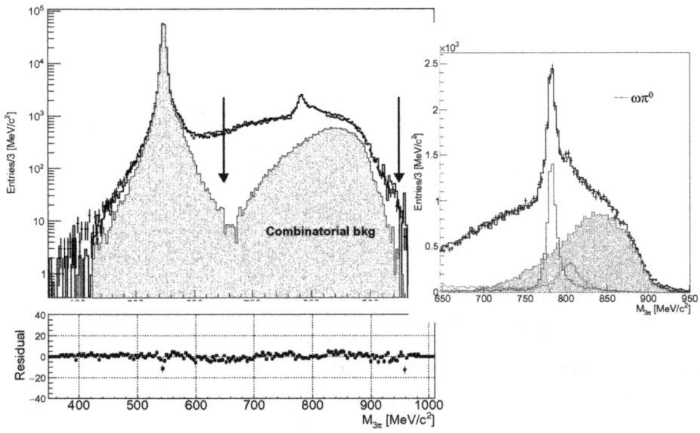

Fig. 2. $\pi^+\pi^-\pi^0$ invariant mass distribution for data (dots) with a superimposed ML fit using MC shapes (black histogram). The shaded histogram includes $\phi \to \eta\gamma$ background, peaked at the η mass, and the combinatorial background, concentrated at higher masses. The right plot is a zoom in the ω mass region where the signal distribution is in blue and the $\phi \to \omega\pi^0$ is in red. The lower plot shows the residuals between data and the MC-based ML fit.

region, extract the cross sections of $e^+e^- \to \phi/\omega \to 3\pi$ and the non-resonant process $e^+e^- \to \gamma^* \to 3\pi$. The product of branching fractions $\mathcal{B}(\phi/\omega \to e^+e^-)\mathcal{B}(\phi/\omega \to 3\pi)$ will also be determined.

References

1. F. Jegerlehner, *EPJ Web of Conferences* **166**, 00022 (2018).
2. G. W. Bennett *et al.*, (Muon g-2 Collaboration), *Phys. Rev. Lett.* **89**, 101804 (2002); *Phys. Rev. Lett.* **92**, 161802 (2004); *Phys. Rev. D* **73**, 072003 (2006).
3. J. Grange *et al.*, arXiv:1501.06858 (2015).
4. A. Anastasi *et al.*, (KLOE-2 Collaboration), *J. High Energy Phys.* **03**, 173 (2018).
5. M. N. Achasov *et al.*, (SND Collaboration), *Phys. Rev. D* **68**, 052006 (2003).
6. R. R. Akhmetshin *et al.*, (CMD-2 Collaboration), *Phys. Lett. B* **578**, 285 (2004).
7. B. Aubert *et al.*, (BaBar Collaboration), *Phys. Rev. D* **70**, 072004 (2004).

Decays of P_c into $J/\psi N$ and $\eta_c N$ with heavy quark spin symmetry

Shuntaro Sakai

CAS Key Laboratory of Theoretical Physics, Institute of Theoretical Physics, Chinese Academy of Sciences, Zhong Guan Cun East Street 55, Beijing 100190, China

Hao-Jie Jing and Feng-Kun Guo

CAS Key Laboratory of Theoretical Physics, Institute of Theoretical Physics, Chinese Academy of Sciences, Zhong Guan Cun East Street 55, Beijing 100190, China

School of Physical Sciences, University of Chinese Academy of Sciences, Beijing 100049, China

The decay properties of hidden-charm pentaquarks, P_c, are studied with a formulation respecting heavy quark spin symmetry. As a result of the symmetry, seven $\bar{D}^{(*)}\Sigma_c^{(*)}$ molecules emerge as found in some previous studies, and the partial decay widths of these states to $\eta_c N$ can be comparable or larger compared with those to $J/\psi N$.

Keywords: Hidden-charm pentaquark; decay properties; heavy quark spin symmetry.

1. Introduction

After the recent report of three P_c peaks around 4.4 GeV in the $J/\psi p$ invariant mass distribution in $\Lambda_b \to K^- J/\psi p$,[1] numerous theoretical studies on these peaks have been carried out. Because the observed peaks are close to the $\bar{D}\Sigma_c$ and $\bar{D}^*\Sigma_c$ thresholds, the explanation of these peaks as $\bar{D}^{(*)}\Sigma_c$ molecules sounds a plausible approach. However, with naive approach respecting heavy quark spin symmetry (HQSS), a symmetry of the QCD Lagrangian emerging in the limit of the infinitely large quark mass, seven states composed of $\bar{D}^{(*)}\Sigma_c^{(*)}$ appears[2,3] while in Ref. 1 only three peaks are claimed. Here, we investigate another conclusion of the $\bar{D}^{(*)}\Sigma_c^{(*)}$ molecular description of P_c with HQSS: the molecular P_c that can decay into the $\eta_c N$ channel has a comparable branching fraction to that of the $J/\psi N$ channel. The decay properties of P_c are studied, e.g., in Refs. 4, 5 after the report

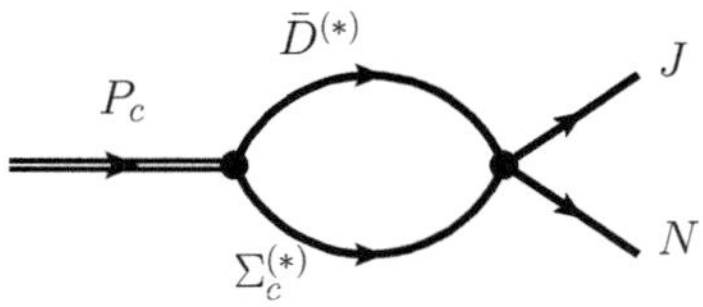

Fig. 1. The diagram for $P_c \to JN$ with $J = \eta_c$ or J/ψ.

of Ref. 1. The details of this study and references of related works can be found in Ref. 6.

2. Setup

For the decay of P_c into $J/\psi N$ and $\eta_c N$, we consider the diagram in Fig. 1. The amplitude from this diagram consists of the long-distance properties of P_c, which is now dominantly determined by the binding energy of the P_c molecules. The coupling strength of the P_c to $\bar{D}^{(*)}\Sigma_c^{(*)}$ is obtained from the residue of the P_c pole in the $\bar{D}^{(*)}\Sigma_c^{(*)}$ amplitude. The $\bar{D}^{(*)}\Sigma_c^{(*)}$ scattering t matrix is obtained with the scattering equation, $t = (1 - vG)^{-1}v$, where v and G are the $\bar{D}^{(*)}\Sigma_c^{(*)}$ interaction kernel and $\bar{D}^{(*)}\Sigma_c^{(*)}$ two-body loop function, respectively. The observed two P_c peaks around 4.45 GeV near the $\bar{D}^*\Sigma_c$ threshold are assumed to arise from the $\bar{D}^*\Sigma_c$ molecules. The couplings of P_c to $\bar{D}^{(*)}\Sigma_c^{(*)}$, $g^i_{\rm eff}$, are given by $g^i_{\rm eff} g^j_{\rm eff} = \lim_{w \to w_{P_c}} (w - w_{P_c}) t_{ij}(w)$, where w_{P_c} is the P_c pole position, and the index i specifies the channel, $\bar{D}^{(*)}\Sigma_c^{(*)}$. The interaction kernel v respecting HQSS can be constructed as in Refs. 2, 3. Two pairs of $(\bar{D}, \bar{D}^*)$ and (Σ_c, Σ_c^*) are settled into a doublet of HQSS, respectively, and the interactions of $\bar{D}^{(*)}\Sigma_c^{(*)}$ are related with each other from HQSS. At the leading order, the constructed s-wave $\bar{D}^{(*)}\Sigma_c^{(*)}$ Lagrangian[2] has two free parameters which are fixed to reproduce the positions of two peaks near the $\bar{D}^*\Sigma_c$ threshold in the $J/\psi p$ distribution.[1] As in Refs. 2,3, seven P_c states related by HQSS are obtained, and these seven states appear close to the thresholds, $(\bar{D}\Sigma_c)_{1/2}$, $(\bar{D}\Sigma_c^*)_{3/2}$, $(\bar{D}^*\Sigma_c)_{1/2,3/2}$, and $(\bar{D}^*\Sigma_c^*)_{1/2,3/2,5/2}$ (where the subindices denote the total spin of $\bar{D}^{(*)}\Sigma_c^{(*)}$), with binding energies of typically several to tens of MeV. The obtained pole of the $\bar{D}\Sigma_c$ molecule located around 4.31 GeV may be associated with the $P_c(4312)$ seen by LHCb.[1] The s-wave coupling between $\bar{D}^{(*)}\Sigma_c^{(*)}$ and JN ($J = \eta_c$ or J/ψ), $g_{\bar{D}^{(*)}\Sigma_c^{(*)},JN}$, also obtained in a fashion respecting HQSS associating the $(\eta_c, J/\psi)$ pair as a doublet of HQSS, which gives results consistent with those in Ref. 7.

With these amplitudes, the $P_c \to JN$ amplitude, $t_{P_c,JN}$, is given by

$$t_{P_c,JN} = g_{P_c,\bar{D}^{(*)}\Sigma_c^{(*)}} \tilde{G} g_{\bar{D}^{(*)}\Sigma_c^{(*)},JN}, \tag{1}$$

with $\tilde{G}$ the $\bar{D}^{(*)}\Sigma_c^{(*)}$ two-body loop function. Including the appropriate phase-space factor, one can evaluate the $P_c \to JN$ branching fractions, $\Gamma_{P_c,JN}$ for both $J = \eta_c$ and J/ψ. In this study, we focus on the production ratio relative to the $\bar{D}\Sigma_c$-molecule one without the need to fix a common constant appearing in the $\bar{D}^{(*)}\Sigma_c^{(*)} \to JN$ amplitudes.

3. Results and discussions on the $P_c \to \eta_c N, J/\psi N$ decays

With the setup given above, we have partial widths of six P_c states, P_{ci}, to the s-wave $\eta_c N$ or $J/\psi N$. The pole positions of P_{ci}, the values of the coupling of P_{ci} to $\bar{D}^{(*)}\Sigma_c^{(*)}$, and the transition amplitudes of $\bar{D}^{(*)}\Sigma_c^{(*)} \to JN$ can be found in Ref. 6. Defining the ratios $r_i = \Gamma_{P_{ci}\to J/\psi N}/\Gamma_{P_{c1}\to J/\psi N}$ and $\tilde{r}_i = \Gamma_{P_{ci}\to \eta_c N}/\Gamma_{P_{c1}\to J/\psi N}$, we can see qualitative consequences of the molecular description of the P_c with HQSS on the decay properties. The P_{ci} states with $i = 1$ to 7 correspond to the $(\bar{D}\Sigma_c)_{1/2}$, $(\bar{D}\Sigma_c^*)_{3/2}$, $(\bar{D}^*\Sigma_c)_{1/2}$, $(\bar{D}^*\Sigma_c)_{3/2}$, $(\bar{D}^*\Sigma_c^*)_{1/2}$, $(\bar{D}^*\Sigma_c^*)_{3/2}$, and $(\bar{D}^*\Sigma_c^*)_{5/2}$ molecules, respectively. The ratios r_i and $\tilde{r}_i$ in case 1 and 2 ($P_{c3} = P_c(4440)$ and $P_{c4} = P_c(4457)$ in case 1 and the opposite assignment in case 2) are given by

$$\begin{aligned}
\text{case 1}: \ & r_2 = (4.4, 4.4), \ r_3 = (9.3, 9.6), \ r_4 = (1.2, 1.2), \\
& r_5 = (2.7, 2.8), \ r_6 = (5.1, 5.4), \\
& \tilde{r}_1 = (2.9, 2.9), \ \tilde{r}_3 = (0.4, 0.4), \ \tilde{r}_5 = (9.8, 10.3),
\end{aligned} \tag{2}$$

$$\begin{aligned}
\text{case 2}: \ & r_2 = (4.6, 4.7), \ r_3 = (10.1, 10.5), \ r_4 = (1.9, 1.9), \\
& r_5 = (3.3, 3.5), \ r_6 = (13.9, 14.4), \\
& \tilde{r}_1 = (4.0, 4.0), \ \tilde{r}_3 = (2.4, 2.5), \ \tilde{r}_5 = (10.2, 10.7),
\end{aligned} \tag{3}$$

where the first and second numbers in parentheses are obtained using a cut off $\Lambda = 0.7$ and 1 GeV in the $\bar{D}^{(*)}\Sigma_c^{(*)}$ loop, respectively. These ratios r_i and $\tilde{r}_i$ are found to be stable against varying the cutoff around several hundred MeV. These r_i's are close to or larger than 1, that is, six P_c peaks are expected naively because a prominent peak of $P_c(4312)$ in the $J/\psi p$ spectrum was clearly seen in experiment. It is notable that the $\tilde{r}_i$'s for the decay to $\eta_c N$ is also larger than 1 for some P_{ci}. Then, assuming a similar background and production mechanism to those in $J/\psi N$, the P_c peaks in the $\eta_c N$ channel are also expected (see also Ref. 8 for a discussion on $P_c \to \eta_c N$). While there is room for further efforts (the production mechanism, details of the $\bar{D}^{(*)}\Sigma_c^{(*)}$ interaction, or kinematic effects, and they have been addressed intensively in recent works), the discussion on the properties of the molecular P_c expected from HQSS in this work should be useful to examine the nature of the peaks in the $J/\psi p$ distribution.

Acknowledgments

This work is supported in part by the National Natural Science Foundation of China (NSFC) under Grant No. 11621131001, No. 11847612 and No. 11835015, by the Chinese Academy of Sciences (CAS) under Grant No. QYZDB-SSW-SYS013 and No. XDPB09, and by the CAS Center for Excellence in Particle Physics (CCEPP), by the 2019 International Postdoctoral Exchange Program, and by the CAS President's International Fellowship Initiative (PIFI) under Grant No. 2019PM0108.

References

1. R. Aaij *et al.*, Observation of a narrow pentaquark state, $P_c(4312)^+$, and of two-peak structure of the $P_c(4450)^+$, *Phys. Rev. Lett.* **122**, 222001 (2019).
2. M.-Z. Liu, Y.-W. Pan, F.-Z. Peng, M. Sánchez Sánchez, L.-S. Geng, A. Hosaka and M. Pavon Valderrama, Emergence of a complete heavy-quark spin symmetry multiplet: seven molecular pentaquarks in light of the latest LHCb analysis, *Phys. Rev. Lett.* **122**, 242001 (2019).
3. C. W. Xiao, J. Nieves and E. Oset, Heavy quark spin symmetric molecular states from $\bar{D}^{(*)}\Sigma_c^{(*)}$ and other coupled channels in the light of the recent LHCb pentaquarks, *Phys. Rev.* **D100**, 014021 (2019).
4. C.-J. Xiao, Y. Huang, Y.-B. Dong, L.-S. Geng and D.-Y. Chen, Exploring the molecular scenario of Pc(4312) , Pc(4440) , and Pc(4457), *Phys. Rev.* **D100**, 014022 (2019).
5. Y.-H. Lin and B.-S. Zou, Strong decays of the latest LHCb pentaquark candidates in hadronic molecule pictures, *Phys. Rev.* **D100**, 056005 (2019).
6. S. Sakai, H.-J. Jing and F.-K. Guo, Decays of P_c into $J/\psi N$ and $\eta_c N$ with heavy quark spin symmetry, *Phys. Rev.* **D100**, 074007 (2019).
7. M. B. Voloshin, Some decay properties of hidden-charm pentaquarks as baryon-meson molecules, *Phys. Rev.* **D100**, 034020 (2019).
8. J.-J. Wu, R. Molina, E. Oset and B. S. Zou, Prediction of narrow N^* and Λ^* resonances with hidden charm above 4 GeV, *Phys. Rev. Lett.* **105**, 232001 (2010).

Strangeness nuclear physics at $\overline{\rm P}$ANDA

Patrick Achenbach*, Sebastian Bleser, Michael Bölting, Josef Pochodzalla, Falk Schupp and Marcell Steinen

on behalf of the $\overline{\rm P}$ANDA Collaboration

Institut für Kernphysik & Helmholtz Institute Mainz, Johannes Gutenberg-Universität, 55099 Mainz, Germany
**E-mail: achenbach@uni-mainz.de*
www.kernphysik.uni-mainz.de & www.hi-mainz.de

$\overline{\rm P}$ANDA at FAIR will address the physics of strangeness in nuclei by several novel measurements. These studies are only made possible by the one-of-a-kind combination of the stored antiproton beam at FAIR and the modular $\overline{\rm P}$ANDA detector which will be complemented by a carbon filament target for hyperon production and a secondary target, in which low-momentum hyperons can be stopped. An additional germanium detector array for high-resolution γ-spectroscopy offers the unique possibility to search for X-rays from very heavy hyperatoms. It will also allow to extend the studies on $\Lambda\Lambda$ hypernuclei by determining excited states of these nuclei for the first time. Furthermore, the exclusive production of hyperon–antihyperon pairs close to their threshold in antiproton–nucleus collisions offers a so-far unexplored opportunity to elucidate the behaviour of antihyperons in cold nuclei.

Keywords: Single and double Λ hypernuclei; hyperatoms; hyperons and antihyperons; nucleon–hyperon interactions; γ-spectroscopy.

1. The $\overline{\rm P}$ANDA Experiment at FAIR

$\overline{\rm P}$ANDA at FAIR will address the physics of strangeness in nuclei by several novel measurements. These studies will only be made possible by the one-of-a-kind combination of the stored antiproton beam at FAIR and the modular $\overline{\rm P}$ANDA detector. The large cross sections for the production of associated hyperon–antihyperon pairs[1] combined with the high luminosities that can be achieved even with very thin primary targets will lead to the formation of utilizable numbers of hyperon–antihyperon pairs, Ξ^- hyperatoms, and single and double Λ hypernuclei.[2,3] In this contribution we focus on the setup and the physics potential of the Ξ^- hyperatom study.

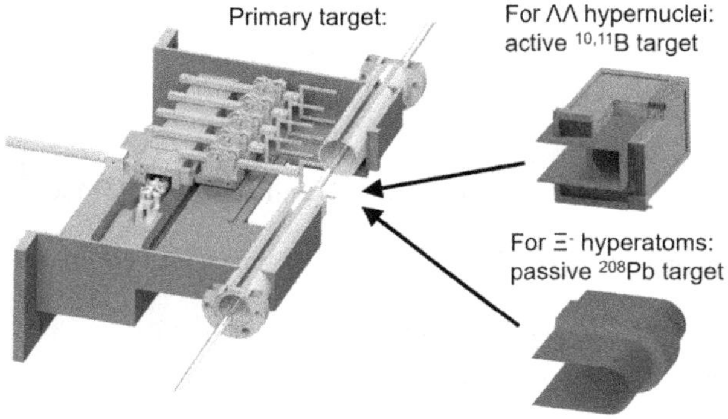

Fig. 1. Left: Details of the primary target station which will house several μm-sized carbon filaments in holding frames (beige) that can be precisely positioned in the halo of the antiproton beam and remotely replaced by spare targets to secure long-term reliability. The vacuum chamber (blue) has an opening in which the secondary targets can be inserted. Right: Secondary targets made from silicon detectors (orange) and boron layers (pink) for $\Lambda\Lambda$ hypernuclei studies (top) or alternatively from a passive sheet of lead (pink) for Ξ^- hyperatom studies (bottom). The secondary targets will be an integral part of the vacuum system.[3]

2. Dedicated Setup for Hyperatoms and Hypernuclei

In addition to the general purpose $\overline{\text{P}}$ANDA setup, the hyperatom and hypernuclear studies require i) dedicated primary targets to produce Ξ^- hyperons, ii) secondary targets for stopping a low-momentum fraction of these Ξ^- hyperons, and iii) a high purity germanium (HPGe) array for high-resolution γ-spectroscopy. The secondary targets will be mounted close to the beam line and will cover a large solid angle to reach maximum stopping probability for the short-lived hyperons. Figure 1 shows details of the primary target and two different secondary targets optimized for either hyperatom or hypernuclear studies. The so-called PANGEA hodoscope of HPGe triple detectors[4] will be placed at backward angles with respect to the antiproton beam.

3. X-rays from Heavy Ξ^- Hyperatoms

Strongly interacting, negatively charged particles such as Ξ^- hyperons could occupy atomic orbits in which their wavefunctions have significant overlap with the nuclear wavefunction, so that they are influenced by the strong interaction and can be captured by the nucleus. This leads to energy levels that are shifted and broadened with respect to the pure elec-

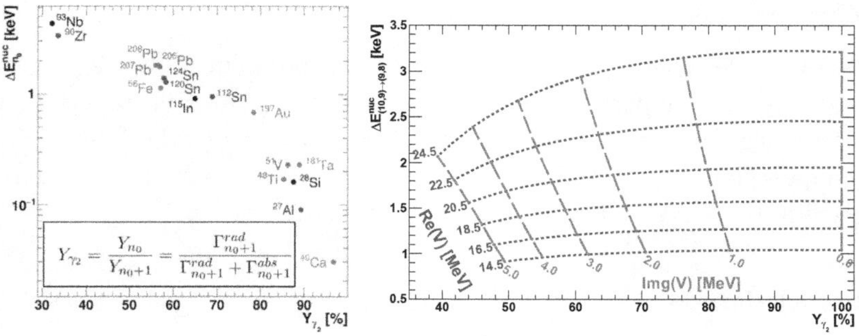

Fig. 2. Left: Overview of X-ray energy shifts and relative yields for possible target materials in a hyperatom experiment. Elements with small energy shifts or low relative yields are difficult to study experimentally. Right: Mapping of the real and imaginary parts of the $\Xi^- - {}^{208}$Pb optical potential on the energy shift of the $(10, 9) \rightarrow (9, 8)$ transition and the relative yield $Y_{\gamma 2}$ of this transition with respect to the $(11, 10) \rightarrow (10, 9)$ transition. The calculations were performed by M. Steinen[2] with a program provided by E. Friedman.[5,6]

tromagnetic situation. In Ξ^- hyperatoms the electric dipole transition ($\Delta n = \Delta l = 1$) leading to the lowest atomic state (n_0, l_0) not fully absorbed by the strong interaction with the nucleus is the last transition which could be observed with a reasonable X-ray branching ratio. This level is shifted by the energy $\Delta E^{\rm nuc}_{n_0}$ and broadened by the nuclear absorptive width $\Gamma^{\rm abs}_{n_0}$ due to the strong interaction. The shifts, widths, and relative yields of the transition lines may be expressed in terms of real and imaginary parts of the nuclear optical potential. To probe the nuclear periphery of Ξ^- hyperatoms firstly the two X-ray energies E_{γ_1} and E_{γ_2} of the unshifted $(n_0 + 2, l_0 + 2) \rightarrow (n_0 + 1, l_0 + 1)$ and the shifted $(n_0 + 1, l_0 + 1) \rightarrow (n_0, l_0)$ transition could be measured. Secondly, the intensity of the last X-ray transitions will be reduced by the strong interaction and the relative yield Y_{γ_2} of the last X-ray to the second to last X-ray transition could be measured, which is dependent on the radiative and nuclear absorptive widths of the $(n_0 + 1, l_0 + 1)$ state. Figure 2 shows these strong interaction effects in different possible target elements and a quantitive study of such effects in dependence of the optical potential for $\Xi^- - {}^{208}$Pb hyperatoms. It can be demonstrated that the Ξ^- hyperatom X-ray spectroscopy at $\overline{\rm P}$ANDA will, e.g., allow to constrain the neutron skin of lead nuclei.[7]

4. Conclusions

$\overline{\text{P}}$ANDA is a versatile experiment with a broad and unique physics program. Hypernuclei and hyperatoms are unique femto-laboratories for strong interaction studies. Furthermore, strangeness nuclear physics is embedded in the quest for the determination of the equation-of-state of nuclei with a direct link to dense stellar systems such as neutron stars.

Acknowledgments

Work supported by Deutscher Akademischer Austauschdienst (DAAD) PPP Japan 2017 57345296 and as JSPS Joint Research Project. This project has received funding from the European Union's Horizon 2020 research and innovation programme under grant agreement No. 824093.

References

1. M. F. M. Lutz *et al.* ($\overline{\text{P}}$ANDA Collaboration), Physics performance report for $\overline{\text{P}}$ANDA: Strong interaction studies with antiprotons (2009), arxiv:0903.3905.
2. M. Steinen, Study of heavy Ξ^- hyperatoms at $\overline{\text{P}}$ANDA, PhD thesis, Johannes Gutenberg-Universität, Mainz (unpublished).
3. S. Bleser, High resolution γ-spectroscopy of $\Lambda\Lambda$-hypernuclei at $\overline{\text{P}}$ANDA, PhD thesis, Johannes Gutenberg-Universität, Mainz (unpublished).
4. M. Doncel, B. Cederwall, A. Gadea, J. Gerl, I. Kojouharov, S. Martin, R. Palit and B. Quintana, Performance and imaging capabilities of the DEGAS high-resolution γ-ray detector array for the DESPEC experiment at FAIR, *Nucl. Instrum. Methods Phys. Res., Sect. A* **873**, 36 (2017).
5. C. J. Batty, E. Friedman and A. Gal, Experiments with Ξ^- atoms, *Phys. Rev. C* **59**, 295 (1999).
6. E. Friedman, private communication.
7. J. Pochodzalla, P. Achenbach, S. Bleser, M. Bölting, F. Schupp and M. Steinen, Strangeness nuclear physics at $\overline{\text{P}}$ANDA in a nutshell, in *Write-ups for workshop on the project for the extended hadron experimental facility of J-PARC*, 7–22 (2019), arXiv:1906.02357.

Status of hypertriton binding energy measurements at the Mainz Microtron

Patrick Achenbach*, Sergio Alves Garre, Philipp Eckert, Saül Garcia Orrit, Philipp Herrmann, Pascal Klag, Marten Mildeberger and Josef Pochodzalla

on behalf of the A1 Collaboration

Institut für Kernphysik, Johannes Gutenberg–Universität, 55099 Mainz, Germany
**E-mail: achenbach@uni-mainz.de*
www.kernphysik.uni-mainz.de

Masashi Kaneta, Yoshihiro Konishi, Sho Nagao, Satoshi N. Nakamura and Yuichi Toyama

Department of Physics, Graduate School of Science, Tohoku University, Sendai, 980-8578, Japan

Toshiyuki Gogami

Department of Physics, Graduate School of Science, Kyoto University, Kyoto, 606-8502, Japan

In recent years the method of decay-pion spectroscopy was pioneered at the Mainz Microtron (MAMI). This method has the potential to achieve ground state mass measurements of light hypernuclei with unprecedented precision. It is aimed at statistical and systematic uncertainties in the Λ binding energy of about 20 keV. Especially the hypertriton provides several important benchmarks for the strong interaction theory dealing with strange baryons, comparable to the role of deuterium for conventional baryon interactions. Preparations for the high-precision decay-pion spectroscopy of hypertriton are ongoing at MAMI, in particular the use of a lithium target is foreseen in the future. This target will be actively cooled to withstand the beam heating. The experiment will be complemented by extensive calibrations to reach the precision goal.

Keywords: Hypernuclei; Λ binding energy; nucleon–hyperon interactions; charged–particle spectroscopy; mesonic weak decay.

1. Hypernuclear Decay-pion Spectroscopy at MAMI

In recent years the method of hypernuclear decay-pion spectroscopy was pioneered at the Mainz Microtron (MAMI).[1,2] In experiments on determining the binding energy of the ${}^{4}_{\Lambda}$H hypernucleus, binding energy resolution and precision could be improved significantly compared to older emulsion measurements. This was possible by using the high-resolution spectrometer facility at MAMI in combination with a dedicated kaon tagger.

Presently, the largest systematic error in the decay-pion spectroscopy at MAMI originates from the uncertainty in the absolute beam energy affecting the absolute momentum calibration of the spectrometers by $\delta p \approx 100\,\mathrm{keV}/c$. For improved beam energy measurements a novel optical interferometry method was established at MAMI: Two undulators act as sources of synchrotron radiation and a monochromator serves as Fourier spectrum analyzer. The phase difference of the light is changed by varying the distance between the sources. By measuring the period of the oscillation the relativistic Lorentz factor of the beam can then be calculated with unprecedented precision.[7] Extensive calibrations for the decay-pion spectroscopy aim at systematic uncertainties in the Λ binding energy of about 20 keV.

2. The Lightest Hypernucleus: The Hypertriton

Especially the hypertriton provides several important benchmarks for the strong interaction theory dealing with strange baryons, comparable to the role of deuterium for conventional baryon interactions. However, emulsion data of hypertriton binding energies are consistent with zero and span over a wide range characterized by a FWHM of 2.1 MeV.[3] Recently, the hypertriton binding energy has been studied in heavy ion collisions by the STAR Collaboration, and a tension between their mass value and the emulsion value exists.[4] It could be argued that there are underestimated sources of systematic uncertainties in emulsion data: i) past changes in the assumed masses might have affected the emulsion data significantly, and ii) it was shown with precision measurements that in p-shell hypernuclei a discrepancy exists of the order of 0.5 MeV between the emulsion data and the SKS-KEK data obtained with the (π^+, K^+) reaction[5] and also between the SKS-KEK data and the FINUDA data.[6] As the knowledge of hypertriton properties remains limited, preparations for the high-precision decay-pion spectroscopy of hypertriton are ongoing at MAMI.

3. Lithium Target for Increased Hypertriton Formation

An increased yield for hypertriton formation is expected from statistical model predictions for solid-state Li targets.[3] When an electron beam of $10\,\mu\mathrm{A}$ current is traversing a 50 mm long target the heat load from ionization energy-loss will be of the order of 50 W. Consequently, a novel 1.5 mm

Fig. 1. Photos of the Li target for hypertriton production, attached to a copper frame in which a cooling liquid is circulating. The left photo shows a view from backward spectrometer direction. The blue arrow indicates the MAMI electron beam direction. The right photo shows a view from this beam direction.

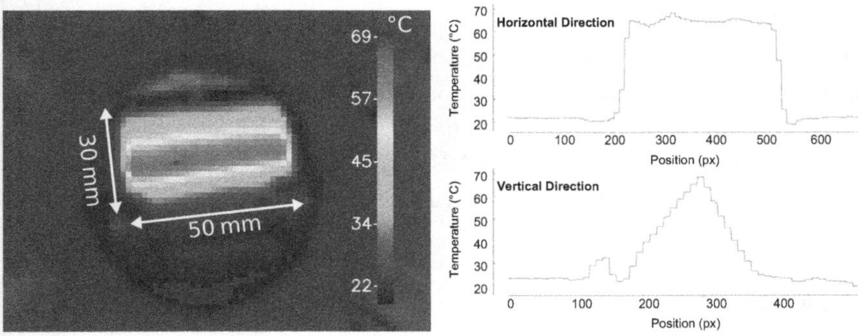

Fig. 2. Left: Thermal image of the Li target during heating with a MAMI electron beam of 10 μA current and 855 MeV energy. The target was installed in a vacuum chamber and the beam was coming from the left. The image is distorted because it had to be taken through a GaAs lens and additional optical elements. Right: Measured heat distributions in the horizontal and vertical directions of the target. The position scale is shown in camera pixels and the temperature scale was calibrated with a PT100 sensor.

wide Li target inside a copper frame for an efficient heat transport with an active cooling system was developed and tested.[8,9] Fig. 1 shows two photos. It was operated with a MAMI electron beam of 855 MeV energy and currents between 2 μA and 10 μA. The precise alignment of beam direction and longitudinal target axis was achieved with a beam wobbler and a remotely operatated rotating stage for the target. The beam diameter was measured to be $\varnothing \leq 0.5$ mm, so that the full intensity could be used with the target. During the test thermal images were taken with a long-wave infrared imager.[10] An image for heating with $10\,\mu$A beam current is shown in Fig. 2. The temperature did not exceed 70° C, safely below the melting point at 180° C. The horizontal distribution of the heat was uniform inside the target length and the temperature in both vertical directions was constantly and rapidly decreasing from the point-like beam spot. No thermal damages have been observed during these tests.

4. Conclusion

The high-resolution decay-pion spectroscopy was pioneered at the Mainz Microtron MAMI and a new measurement with a lithium target is foreseen to determine the hypertriton binding energy with a precision of 20 keV.

Acknowledgments

Work supported by Deutsche Forschungsgemeinschaft (DFG) Research Grant PO 256/7-1, by Deutscher Akademischer Austauschdienst (DAAD) PPP Japan 2017 57345295, as JSPS Joint Research Project, by Grant-in-Aid for Scientific Research on Innovative Areas "Toward new frontiers: Encounter and synergy of state-of-the-art astronomical detectors and exotic quantum beams", by JSPS KAKENHI Nos. JP17H01121 and JP18H05459, and by the Graduate Program on Physics for the Universe, Tohoku University (GP-PU).

References

1. A. Esser *et al.* (A1 Collaboration), Observation of $^{4}_{\Lambda}H$ hyperhydrogen by decay-pion spectroscopy in electron scattering, *Phys. Rev. Lett.* **114**, 232501 (2015).
2. F. Schulz *et al.* (A1 Collaboration), Ground-state binding energy of $^{4}_{\Lambda}\mathrm{H}$ from high-resolution decay-pion spectroscopy, *Nucl. Phys. A* **954**, 149 (2016).
3. P. Achenbach, S. Bleser, J. Pochodzalla and M. Steinen, High-precision measurement of the hypertriton mass, *Proceedings of Science PoS(Hadron2017)* **207**, 1 (2018).
4. J. Adam *et al.* (STAR Collaboration), Precise measurement of the mass difference and the binding energy of hypertriton and antihypertriton (2019), arxiv:1904.10520.
5. T. Gogami *et al.* (HKS (JLab E05-115) Collaboration), High resolution spectroscopic study of $^{10}_{\Lambda}\mathrm{Be}$, *Phys. Rev. C* **93**, 034314 (2016).

6. E. Botta, T. Bressani and A. Feliciello, On the binding energy and the charge symmetry breaking in $A \leq 16$ Λ-hypernuclei, *Nucl. Phys. A* **960**, 165 (2017).
7. P. Klag *et al.*, Novel optical interferometry of synchrotron radiation for absolute electron beam energy measurements, *Nucl. Instrum. Methods Phys. Res., Sect. A* **910**, 147 (2018).
8. M. Mildeberger, Entwicklung eines Li-Targets für die Pion-Spektroskopie von Hyperkernen, bachelor's thesis, Johannes Gutenberg–Universität, Mainz (2019).
9. P. Eckert, Untersuchung der ^{12}C(e,e'$\pi^+\pi^+$)^{12}Be*-Reaktion am MAMI-C, master's thesis, Johannes Gutenberg–Universität, Mainz (2019).
10. SparkFun Electronics, FLIR Radiometric Lepton Dev Kit (2019), www.sparkfun.com.

On holographic relation between radial meson trajectories and deconfinement temperature

S. S. Afonin

Saint Petersburg State University, 7/9 Universitetskaya nab.,
St. Petersburg, 199034, Russia
E-mail: s.afonin@spbu.ru

A. D. Katanaeva

University of Barcelona, Martí i Franquès 1,
Barcelona, 08028, Catalonia, Spain
E-mail: katanaeva@fqa.ub.edu

The interrelation between the deconfinement temperature of hadron medium and parameters of radial Regge trajectories within the bottom-up holographic models for QCD is scrutinized. We show that the lattice data on the deconfinement temperature can yield a powerful restriction on the spectrum of excited mesons and glueballs within the framework of holographic approach. The best phenomenological agreement and theoretical self-consistency are achieved if the scalar meson $f_0(1500)$ is considered as the lightest glueball.

Keywords: Holographic QCD; deconfinement temperature.

One of primary questions in the QCD phase diagram is to calculate the critical temperature T_c at which hadronic matter undergos a transition to a deconfined phase. Within the bottom-up holographic approach this type of studies was initiated by Herzog in Ref. 1 and continued by many authors (see, *e.g.*, a brief review in Ref. 2). In this approach, the gravity part of a $5D$ model is assumed to come from a dual description of gluodynamics and can be used to study thermodynamic properties of original $4D$ gauge theory. The deconfinement is related to the Hawking–Page phase transition between a low temperature thermal Anti-de Sitter (AdS) space and a high temperature black hole in the AdS/QCD models.

The estimation of T_c, in general, is model dependent. Traditionally one fixes the hadron parameters from the vector meson sector due to a relatively rich experimental data on light vector mesons. However, there seems to be no theoretical reason why the vector meson spectra should be preferred.

In the present study, we will argue that the scalar glueball (and its radial excitations) is much better option for fixation of the model parameters. Our main arguments can be shortly formulated as follows. (i) Phase diagram can be studied in pure gluodynamics. Since the holographic approach is defined in the large-N_c limit of gauge theories where the glueballs dominate over the usual mesons and baryons the gluodynamics must dictate the overall mass scale and thereby the major contribution to the deconfinement temperature T_c. (ii) Using isospectrality concept[3] we show that the predicted values of T_c are more stable for scalar glueballs than for vector mesons. (iii) Phenomenological lattice reasons: Numerical values of T_c determined from the scalar glueballs on the lattices fit much better the lattice results for T_c.

These argument are scrutinized in our paper[2] (see also Refs. 4 and 5).

Let us introduce the $5D$ holographic action with a universal gravitational part and some matter part,

$$S = \int d^4x dz \sqrt{-g} f^2(z) \left(\mathcal{L}_{gravity} + \mathcal{L}_{matter}\right), \tag{1}$$

$$\mathcal{L}_{gravity} = -\frac{1}{2k_g}\left(\mathcal{R} - 2\Lambda\right). \tag{2}$$

Here g_{MN} ($g = \det g_{MN}$) represents an AdS related metric, k_g is a factor proportional to the $5D$ Newton constant, $\mathcal{R}$ is the Ricci scalar and Λ is the cosmological constant. The choice of the dilaton background $f(z)$ dictates a particular holographic model. They differ as well by the interval the z coordinate spans. We assume $z \in [0, z_{max}]$, though $z_{max} = \infty$ is possible and will be of the main interest in the present work (the soft-wall (SW) background).

The holographic calculation of critical temperature is related to the leading contribution in the large-N_c counting, that is the $\mathcal{L}_{gravity}$ part scaling as $\frac{1}{2k_g} \sim N_c^2$ while $\mathcal{L}_{matter}$ scales as N_c. According to Ref. 1 the deconfinement in AdS/QCD occurs as the Hawking–Page phase transition that is a transition between different gravitational backgrounds. We call the order parameter of this transition ΔV.

V's are the free action densities evaluated on different backgrounds corresponding to two phases. The confined phase is given by the thermal AdS of radius R and defined by the general AdS line element

$$ds^2_{Th} = \frac{R^2}{z^2}\left(dt^2 - d\vec{x}^2 - dz^2\right), \tag{3}$$

with the time direction is restricted to the interval $[0, \beta]$. The metric of the Schwarzschild black hole in AdS describes the deconfined phase and is given by

$$ds^2_{BH} = \frac{R^2}{z^2}\left(h(z)dt^2 - d\vec{x}^2 - \frac{dz^2}{h(z)}\right), \tag{4}$$

where $h(z) = 1 - (z/z_h)^4$ and z_h denotes the horizon of the black hole. The corresponding Hawking temperature is related to the horizon as $T = 1/(\pi z_h)$.

The cosmological constant in $5D$ AdS is $\Lambda = -6/R^2$ and both these metrics are the solutions of the Einstein equations. They provide the same value of the Ricci scalar $\mathcal{R} = -20/R^2$. Hence, the free action densities differ only in the integration limits,

$$V_{\text{Th}}(\epsilon) = \frac{4R^3}{k_g}\int_0^{\beta} dt \int_{\epsilon}^{z_{max}} dz f^2(z) z^{-5}, \tag{5}$$

$$V_{\text{BH}}(\epsilon) = \frac{4R^3}{k_g}\int_0^{\pi z_h} dt \int_{\epsilon}^{\min(z_{max}, z_h)} dz f^2(z) z^{-5}. \tag{6}$$

The two geometries are compared at $z = \epsilon$ where the periodicity in the time direction is locally the same, *i.e.* $\beta = \pi z_h \sqrt{h(\epsilon)}$. Then, we may construct the order parameter for the phase transition,

$$\Delta V = \lim_{\epsilon \to 0} \left(V_{\text{BH}}(\epsilon) - V_{\text{Th}}(\epsilon)\right). \tag{7}$$

The thermal AdS is stable when $\Delta V > 0$, otherwise the black hole is stable. The condition $\Delta V = 0$ defines the critical temperature T_c at which the transition between the two phases happens. Eqn. (7) yields z_h as a function of the model dependent parameters – z_{max} and/or those possibly introduced in $f(z)$. We must invoke the matter sector $\mathcal{L}_{matter}$ to give physical meaning to these parameters and to connect T_c to a particular type of a holographic model.

As was recently noticed in Ref. 3, the soft-wall background is not fixed by the form of linear spectrum as one can find an infinite number of one-dimensional potentials leading to identical spectrum of normalized modes. The corresponding family of potentials is referred to as isospectral potentials.

In brief, the problem for mass spectrum in the bottom-up holographic models can be reduced to a one-dimensional Schrödinger equation

$$-\psi_n''(z) + \widehat{\mathcal{V}}(z)\psi_n(z) = M^2(n)\psi_n(z), \tag{8}$$

where $\widehat{\mathcal{V}}(z)$ is the Schrödinger potential which depends on the 5D dilaton background $f^2(z)$, metric, and spin. A particular form of the Schrödinger potential defines the eigenvalues of Eqn. (8) and hence the mass spectrum $M(n)$. In the case of SW models it is a potential similar to the one that appears when considering the radial part of the wavefunction of a $2D$ harmonic oscillator system.

According to Ref. 3 and references therein, there exists the following isospectral transformation between $\mathcal{V}_J(z)$ and $\widehat{\mathcal{V}}_J(z)$,

$$\widehat{\mathcal{V}}_J(z) = \mathcal{V}_J(z) - 2\frac{d^2}{dz^2}\ln[I_J(z) + \lambda]. \tag{9}$$

This technique allows one to generate a family of dilaton functions $f(z)$ appearing in $\widehat{\mathcal{V}}_J(z)$, each member assigned to the value of the parameter λ. The case of $\lambda = \infty$ corresponds to the original $\mathcal{V}_J(z)$. The function $I_J(z)$ is defined through the ground eigenstate of $\mathcal{V}_J$, ψ_0, and is given by

$$I_J(z) \equiv \int_0^z \psi_0^2(z')dz'. \tag{10}$$

Different λ provide slightly different form of the potential, but the eigenvalues of Eqn. (8) and, hence, the spectrum remain the same.

The main problem we studied can be formulated as follows: Does isospectrality entail isothermality (i.e. identical predictions for T_c)? Generically the answer is negative. But we found one important exception: If model parameters are fixed from the scalar glueball channel within the generalized SW holographic model of Ref. 6 (which is able to reproduce accurately the glueball radial spectrum) then an isospectral family of models leads to almost identical predictions for the deconfinement temperature. The typical predictions lie in the range $T_c \simeq 175 \pm 15$ MeV which agrees very well with modern unquenched lattice estimations. The further details are contained in Refs. 7, 8.

References

1. C. P. Herzog, *Phys. Rev. Lett.* **98**, 091601 (2007).
2. S. S. Afonin and A. D. Katanaeva, *Eur. Phys. J. C* **74**, 3124 (2014).
3. V. Vega and P. Cabrera, *Phys. Rev. D* **93**, 1140265 (2016).
4. A. Katanaeva and S. Afonin, *EPJ Web Conf.* **137**, 13006 (2017).
5. A. Katanaeva and S. Afonin, *AIP Conf. Proc.* **1701**, 100010 (2016).
6. S. S. Afonin, *Phys. Lett. B* **719**, 399 (2013).
7. S. S. Afonin and A. D. Katanaeva, *Phys. Rev. D* **98**, 114027 (2018).
8. A. D. Katanaeva and S. S. Afonin, *Theor. Math. Phys.* **200**, 1383 (2019).

A novel approach on τ lepton decay and its application to polarization amplitudes

L. R. Dai*, S. Q. Liu, Z. Y. Liu

Department of Physics, Liaoning Normal University, Dalian, 116029, China
**E-mail: dailr@lnnu.edu.cn*
www.lnnu.edu.cn

E. Oset, R. Pavao, S. Sakai

Departamento de Física Teórica and IFIC, Centro Mixto
Universidad de Valencia-CSIC, Institutos de Investigacíon de Paterna,
Aptdo. 22085, 46071 Valencia, Spain

A novel approach on τ decay is given and the analytical amplitudes are shown. An interesting application to the different polarization amplitudes for different models beyond the Standard Model is presented, finding one magnitude sensitive and useful to test different models.

Keywords: Tau decay; polarization amplitude; angular momentum algebra.

1. Introduction

τ lepton decay is of important and interesting to learn about weak interaction as well as strong interaction.[1] τ decays into ν_τ and a pair of mesons make up for a sizeable fraction of the τ decay width in PDG[2] in which several modes are well measured for $\tau^- \to \nu_\tau$PP and $\tau^- \to \nu_\tau$PV, here P denotes the pseudoscalar meson and V the vector meson. However, surprisingly, there are no $\tau^- \to \nu_\tau$VV reported. Actually, the pseudoscalar and vector mesons differ only by the spin arrangement of the quarks, hence it should be possible to relate the rates of decay for two pseudoscalar mesons and the related pseudoscalar-vector or vector modes, for instance, $\tau^- \to \nu_\tau K^0 K^-, \nu_\tau K^0 K^{*-}, \nu_\tau K^{*0} K^-, \nu_\tau K^{*0} K^{*-}$.

Now we present a novel approach with no any free parameter, the analytical amplitudes on τ decay is obtained in the standard model (SM)[3] and beyond the standard model (BSM) with $(\gamma^\mu - \alpha\gamma^\mu\gamma_5)$ structure[4] ($\alpha = 1$ for the SM). We evaluate different ratios for the τ reaction and find some observables which are very sensitive to the value of α, which should stimulate experimental work to investigate possible physics BSM.

2. Formalism

The first step is to look at the $\tau^- \to \nu_\tau q\bar{q}$ decay for the Cabibbo-favored $d\bar{u}$ production. We can obtain the Cabibbo-suppressed mode substituting the d quark by an s quark. However, we are interested in the production of two mesons, not just one, as it would come from the mechanism when $q\bar{q}$ merge into a meson. The procedure to produce two mesons is hadronization by creating a new $q\bar{q}$ pair with the quantum numbers of the vacuum. This is depicted in Fig. 1. where $JM(J'M')$ are the final angular momenta and

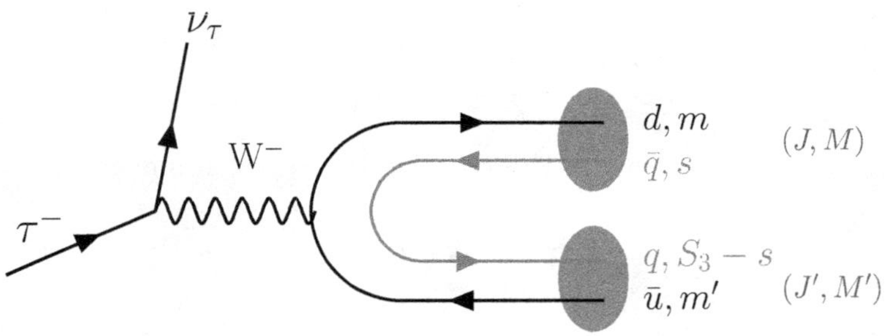

Fig. 1. Hadronization of the primary $d\bar{u}$ pair to produce two mesons, s is the third component of the spin of $\bar{q}$ propagating as a particle, while $S_3 - s$ is the third component of the spin of q, where S_3 is the third component of the total spin S of $\bar{q}q$.

third component of the two mesons produced. The sum over the M' and M components was done in [3] and keep track of the individual M' and M contributions in [4].

The Hamiltonian of the weak interaction is given by

$$H = \mathcal{C} L^\alpha Q_\alpha \,, \tag{1}$$

where the $\mathcal{C}$ contains the couplings of the weak interaction, playing no role since we are only concerned about ratios of rates. The leptonic current is given by

$$L^\mu = \langle \bar{u}_\nu | \gamma^\mu - \gamma^\mu \gamma_5 | u_\tau \rangle, \tag{2}$$

and the quark current

$$Q^\mu = \langle \bar{u}_d | \gamma^\mu - \gamma^\mu \gamma_5 | v_{\bar{u}} \rangle. \tag{3}$$

The final transition amplitudes for the $J=1, J'=0$ (vector- pseudoscalar) are obtained by
1) $M=0$

$$\overline{\sum}\sum |t|^2 = \frac{1}{m_\tau m_\nu}\frac{1}{6}\left(\frac{1}{4\pi}\right)^2 \left\{\left(E_\tau E_\nu + p^2\right) + 2\alpha^2\left(E_\tau E_\nu - p^2\right)\right\}, \quad (4)$$

2) for $M=1$

$$\overline{\sum}\sum |t|^2 = \frac{1}{m_\tau m_\nu}\frac{1}{6}\left(\frac{1}{4\pi}\right)^2 \{(E_\tau E_\nu + p^2) + 2\alpha(E_\nu + E_\tau)p + \left[2E_\tau E_\nu + (E_\nu - E_\tau)p\right]\alpha^2\}, \quad (5)$$

3) for $M=-1$

$$\overline{\sum}\sum |t|^2 = \frac{1}{m_\tau m_\nu}\frac{1}{6}\left(\frac{1}{4\pi}\right)^2 \{(E_\tau E_\nu + p^2) - 2\alpha(E_\nu + E_\tau)p + \left[2E_\tau E_\nu - (E_\nu - E_\tau)p\right]\alpha^2\}, \quad (6)$$

where p is the momentum of the τ in the M_1M_2 rest frame given by

$$p = \frac{\lambda^{1/2}(m_\tau^2, m_\nu^2, M_{\text{inv}}^{2(M_1M_2)})}{2M_{\text{inv}}^{(M_1M_2)}}, \quad E_\tau = \sqrt{m_\tau^2 + p^2}, \quad E_\nu = p. \quad (7)$$

3. Results of polarization amplitudes

The differential width is given for $\tau^- \to \nu_\tau K^{*0} K^-$ decay

$$\frac{d\Gamma}{dM_{\text{inv}}^{(K^{*0}K^-)}} = \frac{2\,m_\tau 2\,m_\nu}{(2\pi)^3}\frac{1}{4m_\tau^2}\, p_\nu \widetilde{p}_1 \overline{\sum}\sum |t|^2, \quad (8)$$

where p_ν is the neutrino momentum in the τ rest frame and $\widetilde{p}_1$ the momentum of K^{*0} in the $K^{*0}K^-$ rest frame. The total differential width is given by

$$R = \frac{d\Gamma}{dM_{\text{inv}}^{(K^{*0}K^-)}}|_{M=+1} + \frac{d\Gamma}{dM_{\text{inv}}^{(K^{*0}K^-)}}|_{M=0} + \frac{d\Gamma}{dM_{\text{inv}}^{(K^{*0}K^-)}}|_{M=-1}. \quad (9)$$

Our results show that the polarization amplitudes vary very much with different values of α. In particular we find a magnitude, $\frac{1}{R}\frac{d\Gamma}{dM_{\text{inv}}^{(K^{*0}K^-)}}|_{M=+1} - \frac{1}{R}\frac{d\Gamma}{dM_{\text{inv}}^{(K^{*0}K^-)}}|_{M=-1}$, which is extremely sensitive to α. In Fig. 2 we show this magnitude, finding that this magnitude changes even sign for some value of α which turns out to be most suited to investigate possible physics BSM.

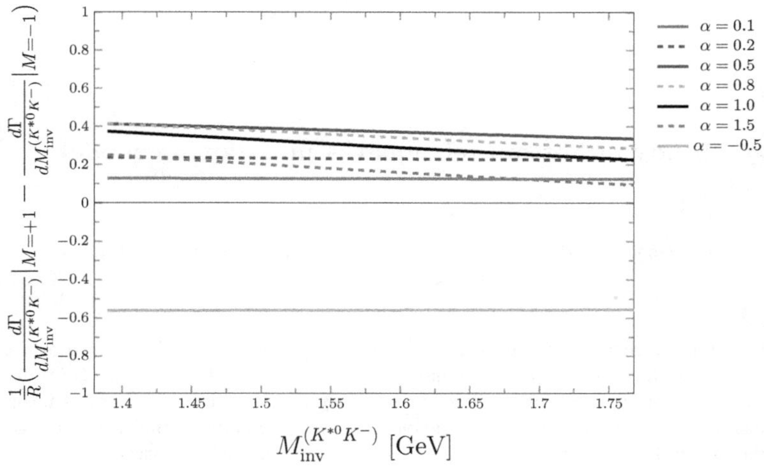

Fig. 2. The difference of $\frac{d\Gamma}{dM_{\rm inv}^{(K^{*0}K^-)}}|_{M=+1} - \frac{d\Gamma}{dM_{\rm inv}^{(K^{*0}K^-)}}|_{M=-1}$ as a function of α, divided by the total differential width.

Acknowledgments

LRD acknowledge supports from the National Natural Science Foundation of China (Grant Nos. 11975009, 11575076) and Spanish Ministerio de Economia y Competitividad and European FEDER funds under Contracts No. FIS2017-84038-C2-1-P B and No. FIS2017-84038-C2-2-P B, and the Generalitat Valenciana in the program Prometeo II-2014/068, and the project Severo Ochoa of IFIC, SEV-2014-0398 (EO).

References

1. M. Antonelli, D. M. Asner, D. Bauer et al, *Phys. Rept.* **494**, 197 (2010).
2. M. Tanabashi *et al.* [Particle Data Group], *Phys. Rev. D* **98**, 030001 (2018).
3. L. R. Dai, R. Pavao, S. Sakai and E. Oset, *Eur. Phys. J. A* **55**, 20 (2019).
4. L. R. Dai and E. Oset, *Eur. Phys. J. A* **54**, 219 (2018).

Study of bottom tetraquarks in a diquark model

Zahra Ghalenovi

Department of Physics, Kosar University of Bojnord, Iran
E-mail: z_ghalenovi@kub.ac.ir

The mass spectrum of the ground-state tetraquarks with hidden bottom are studied within a diquark-antidiquark configuration. The considered Hamiltonian is a combination of the linear confining, Coulombic and spin-spin interaction terms. Using the perturbation theory, we calculate the heavy meson spectra and then estimate the masses of the hidden bottom tetraquaks in the non-relativistic limit. We find that the exotic state $Z_b(10650)$ can be considered as diquarkantidiquark bound state.

Keywords: Diquark picture; potential model; non-relativistic limit.

1. Introduction

Recently, significant experimental progress has been achieved regarding the discoveries of the new bottomonium- and charmonium-like exotic states called X, Y, Z particles. In 2011, the Belle collaboration observed the $Z_b(10650)$ in the $\pi^{\pm}h_b(1, 2P)$ mass spectrum for the first time with favored quantum number $I^G(J^P) = 1^+(1^+)$[1] and later, updated the value of the mass and width $M_{Z_b}(10650) = (10652.2 \pm 1.5)$MeV and $\Gamma_{Z_b(10650)} = (11.5 \pm 2.2)$MeV.[2] The possible assignments of the $Z_b(10650)$ state are the tetraquark state,[3,4] molecular state,[5,6] re-scattering effect,[7,8] etc. In the compact tetraquark picture,the constituent objects of the system are a diquark and an antidiquark. In the present study, we interpret the exotic states as tetraquark systems. We solve the tetraquark problem in a two-step procedure: first, we use a quark-quark interaction Hamiltonian in order to obtain the mass of a constituent 'good diquark'. Second, we regard the diquarks as point-like objects and use a diquark-antidiquark interaction Hamiltonian in order to obtain the tetraquark masses. This work is organized as follows. In Sec. 2 we introduce our potential model and solve the Schrödinger equation. Our predictions for tetraquark masses are presented in Sec. 3. The conclusion is presented in Sec. 4.

2. Potential model

In this paper we use a phenomenological potential model to estimate the tetraquark masses. The interaction Hamiltonian for the quark-quark (diquark-antidiquark) interaction leading to the formation of diquarks (tetraquarks) is given by

$$H^{qq}(r) = V^{qq}(r) + H^{qq}_{hyp}(r) \ , \tag{1}$$

where the unperturbed potential consists of three parts:

$$V^{qq}(r) = \lambda r - \frac{k_s \alpha_s}{r} - C, \tag{2}$$

the first term is a confining potential and the second term $-\tau/r$ is a Coulomb-like potential due to one-gluon exchange processes. λ and C are constant. The hyperfine interaction is given by:

$$H_{hyp} = \frac{8\pi k_s \alpha_s}{3m_1 m_2} \frac{e^{\frac{-r}{\sigma_s}}}{4\pi\sigma_s^2 r}(\vec{s}_1 \cdot \vec{s}_2) \ , \tag{3}$$

where s_i is the spin operator of the i-th quark and σ_s is a constant. Since the introduced potential depends on r only, the Schrödinger equation for two-body problem with the unperturbed potential $V^{qq}(r)$ for the ground-state, $l = 0$, is given by

$$\left[\frac{d^2}{dr^2} + \frac{2}{r}\frac{d}{dr}\right]\psi(r) = -2m[E - V(r)]\psi(r) \ , \tag{4}$$

where $\psi(r)$ is the radial wave function and m is the reduced mass. The transformation $\varphi(r) = r\psi(r)$ reduces Eq. (4) to the form

$$\frac{d^2}{dr^2}\varphi(r) + \left[2mE - 2m\lambda r + \frac{2mk_s\alpha_s}{r}\right]\varphi(r) = 0 \ . \tag{5}$$

The Schrödinger equation (5) has no analytical solution. Upon perturbation theory, for the states with $l = 0$, the linear term can be considered as the parent (for details see Ref. 9). Therefore, Eq. (5) has the new form

$$\frac{d^2}{dr^2}\varphi(r) + [2mE - 2m\lambda r]\,\varphi(r) = 0 \ . \tag{6}$$

Due to the familiar boundary condition $\varphi \longrightarrow 0$ as $r \longrightarrow \infty$, one can get the radial wavefunction of two-body system as following

$$\varphi(r) = NAi\left[(2m\lambda)^{1/3}(r - \frac{E}{b})\right] \ , \tag{7}$$

where N is the normalization constant. There is another boundary condition $u(0) = 0$, which gives us the unperturbed energy of the system

$E = 2.3194 \left(\frac{\lambda^2}{2m}\right)^{1/3}$.[9] After having solved the Schrödinger equation, one can calculate the Coulombic and spin-spin interactions perturbatively. We add the energy coming from C constant to the perturbed energy and therefore, the perturbed energy of the considered system will be given by

$$E_{perturbed} = \langle -\frac{2mk_s\alpha_s}{r} \rangle + \langle V_{ss} \rangle + C \ . \tag{8}$$

3. Diquark model

A diquark (an antidiquark) is a cluster of two quarks (antiquarks) which can form a bound state. The diquark (antidiquark) masses (m_D) is obtained as a combination of the perurbed and unperturbed energies and also the constituent quark masses, $M_D = m_{q1}+m_{q2}+E_{qq}+E_{perturbed}+C$. We can evaluate the mass of the tetraquarks containing a diquark and an antidiquark by following the same step: $M_{tetraquark} = m_D + m_{\bar{D}} + E_{D\bar{D}} + E_{perturbed} + 2C$. Note that once we consider the tetraquarks states the values of λ, k_s and C parameters used in the diquark systems should be multiplied by two.[12] The values of potential parameters and quark masses are chosen such that we get the best meson spectra. To obtain the mass of a particular meson we use $M_{q\bar{q}} = m_q + m_{\bar{q}} + E_{q\bar{q}} + E_{perturbed} + 2C$. In Table 1 our results for the charm and bottom meson masses are compared to the existing experimental data.[13]

1. The heavy meson masses (in MeV).

Meson	Theory	Exp	Meson	Theory	Exp
D	1869	1869	B	5268	5279
D^*	2010	2010	B^*	5321	5324
D_s	1989	1968	B_s	5355	5366
D_s^*	2107	2112	B_s^*	5400	5414
η_c	2985	2983	η_b	9467	9399
B_c	6298	6288	Υ_b	9489	9460

In Table 2 our theoretical predictions for the ground-state tetraquarks are listed and compared with the predictions obtained by others[12,14,15] and in which S and A denote the scalar and axial vector diquarks (antidiquarks). Our predictions are close to the ones obtained by Ref. 14. Our results show the bottomonium-like structure $Z_b(10650)$ with $M = 10652.2$ MeV and $J = 1$ reported by Belle collaboration,[2] can be considered as AA or SA $bs\bar{b}\bar{s}$ system. The authors of Ref. 15 suggested the $Z_b(10650)$ state as the first radial excitation of AA $bq\bar{b}\bar{q}$ tetraquark state.

2. The predicted hidden bottom tetraquark masses (in MeV).

	J	Content	Theory	14	15	12
	0	SS	10409	10471	10309	10521
	1	SA	10424	10492	10316	
$bq\bar{b}\bar{q}$	0	AA	10399	10473	10143	
	1	AA	10419	10494	10233	
	2	AA	10458	10534	10413	
	0	SS	10658	10662		10973
	1	SA	10670	10682		
$bs\bar{b}\bar{s}$	0	AA	10643	10671		
	1	AA	10667	10686		
	2	AA	10701	10716		

4. Conclusion

We have calculated the masses of the ground-state heavy and light bottom tetraquarks in the framework of a non-relativistic model with the lager potential as well as spin-spin interaction. The perturbation theory have used to solve two-body equation of the considered systems. Then we have obtained the masses of tetraquark states. We found that the $Z_b(10650)$ state can be considered as the ground-state AA or SA $bs\bar{b}\bar{s}$ tetraquark.

References

1. Belle Collaboration, arXiv:1105.4583
2. Belle Collaboration, Phys. Rev. Lett. **108**, 122001 (2012).
3. ZG. Wang, Eur. Phys. J. C **79**, 489 (2019).
4. M. Tanabashi et al., Phys. Rev. D **98**, 030001 (2018).
5. Y. Yang, J. Ping, C. Deng, H.S. Zong, J. Phys. G **39**, 105001 (2012).
6. X.W. Kang, Z.H. Guo, J.A. Oller, Phys. Rev. D **94**, 014012 (2016).
7. D.Y. Chen, X. Liu, S.L. Zhu, Phys. Rev. D **84**, 074016 (2011).
8. G. Li, F.L. Shao, C.W. Zhao, Q. Zhao, Phys. Rev. D **87**, 034020 (2013).
9. I.J.R. Aitchison and J.J. Dudek. Eur. J. Phys. **23**, 605 (2002).
10. Z. Ghalenovi et. al., Int. J. Mod. Phys. E **21** (2012) 1250057.
11. Z. Ghalenovi and M. Moazzen Sorkhi, Eur. Phys. J. Plus **133**, 301 (2018).
12. Z. Ghalenovi, F. Giacosa and D. H. Rischke, Acta. Phys. Polon. B **47**, 5 (2016). arXiv:1507.03345
13. Particle Data Group, Phys. Rev. D **98**, 030001 (2018).
14. D. Ebert et. al., Mod. Phys. Lett. A **24**, 567 (2009).
15. S. Patel and P.C. Vinodkumar, Eur. Phys. J. C **76**, 356 (2016).

Initial-final state strong interaction corrections to the $\bar{B} \to D\bar{\nu}\ell$ $(\bar{\nu}_\tau \tau)$ decays

N. Ikeno

Department of Agricultural, Life and Environmental Sciences, Tottori University, Tottori 680-8551, Japan,
Departamento de Física Teórica and IFIC, Centro Mixto Universidad de Valencia-CSIC Institutos de Investigación de Paterna, Aptdo.22085, 46071 Valencia, Spain
E-mail: ikeno@tottori-u.ac.jp

L. R. Dai

Department of Physics, Liaoning Normal University, Dalian 116029, China
E-mail: dailr@lnnu.edu.cn

E. Oset

Departamento de Física Teórica and IFIC, Centro Mixto Universidad de Valencia-CSIC Institutos de Investigación de Paterna, Aptdo.22085, 46071 Valencia, Spain
E-mail: oset@ific.uv.es

We report the result for the calculation of the strong interaction effects between the B and D mesons in the $\bar{B} \to D\bar{\nu}\ell$ reaction, as a crossing process of reactions with BD in the final state, where the strong interaction between the mesons leads to a bound BD state. We find corrections to the tree level amplitude, and see the effect of the corrections studied in the R_D ratio for the rates of $\bar{B} \to D\bar{\nu}_\tau\tau$ and $\bar{B} \to D\bar{\nu}_e\ell$ decays.

Keywords: Strong interaction, semileptonic decay.

1. Introduction

Different experiments have reported values for the semileptonic B decay ratios,[1]

$$R_D = \frac{BR(B^- \to D\bar{\nu}_\tau\tau)}{BR(B^- \to D\bar{\nu}_\ell\ell)} \qquad (\text{with } \ell = e \text{ or } \mu)\,, \tag{1}$$

which exceed the values provided by the Standard Model (SM) ($R_D = 0.312(19)$). Recently, the Belle data[2] have reduced the value of R_D

($R_D^{\rm Belle} = 0.307 \pm 0.037 \pm 0.016$) and the discrepancies with the SM are significantly reduced. Some theoretical approaches are also performed in other frameworks writing the weak transition amplitudes in terms of form factors, or using lattice QCD calculation, or heavy quark effective theory.[3]

On the other hand, it is unclear which effects from strong interaction are accounted for in parameterized form factors. In Ref. 4, the final interaction of BD is relatively strong and leads to a bound state with binding of 15 – 38 MeV. Thus, we study the strong interaction effect between BD in the semileptonic decays and see the effects of this interaction in the R_D ratio.[5]

2. Formalism

We use the formalism of the semileptonic decay $\bar{B} \to \bar{\nu}\ell D$ in Ref. 5 which was extended in Ref. 6. The $\bar{B} \to D\bar{\nu}\ell$ transition would correspond to the crossing process in Fig. 1, where instead of having $W \to BD$ we have $\bar{B} \to DW$. The BD state will interact by exchanging mesons, and in the intermediate states one can have other meson pairs that couple to BD, essentially, $B_s D_s$. The exchange of the vector mesons is the essential ingredient in Ref. 4.

The crossing process to Fig. 1 is given in Fig. 2. In Fig. 2 (b), we take into account the diagrams for the strong interaction of $\bar{B}D$ with intermediate B, D pseudoscalars and with intermediate B^*, D^* vector mesons. We define the correction factors R_P and R_V for the strong interaction with intermediate pseudoscalars and vector mesons, respectively.[5]

In order to see how the changes obtained influence the $\bar{\nu}\ell$ invariant mass distribution $d\Gamma/dM_{\rm inv}^{(\nu\ell)}$, we evaluate the differential invariant mass distribution as,[5]

$$\frac{d\Gamma}{dM_{\rm inv}^{(\nu\ell)}} = \frac{1}{(2\pi)^3}\frac{1}{M_B^2} p_D \tilde{p}_\nu \overline{\sum}\sum |t|^2 \tag{2}$$

where M_B is the mass of the B meson, p_D is the D momentum in the B rest frame and $\tilde{p}_\nu$ the ν momentum in the $\nu\ell$ rest frame. The amplitude $|t|^2$ includes the correction factors R_P and R_V. In the calculation, we evaluate separately the spatial components of R_P^0, R_V^0 and R_P^3, R_V^3. After integrating Eq. (2), we obtain the width Γ and different ratios of rates R_D. Detailed calculations are shown in Ref. 5.

3. Numerical results

We see the results for the correction factors $R_{\rm p}^0$, $R_{\rm p}^3$, $R_{\rm v}^0$, $R_{\rm v}^3$ as a function of $M_{\rm inv}(\nu\ell)$ for $\bar{\nu}\ell$ production. We found that the factors $R_{\rm p}^0$ and $R_{\rm p}^3$ reach sizes

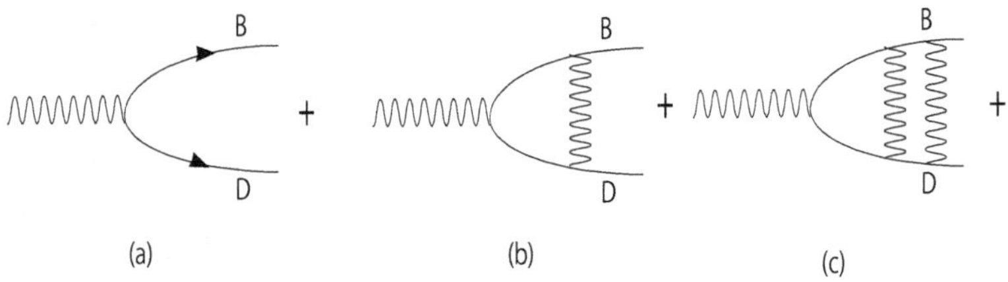

Fig. 1. (a) $W \to BD$ bare process. (b)(c) interaction of the BD state by meson exchange.

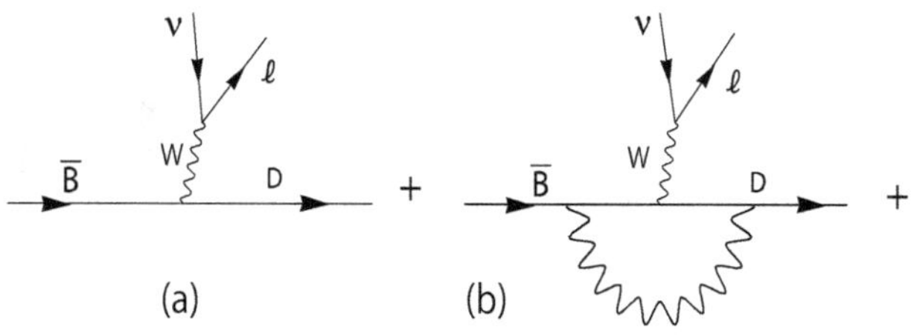

Fig. 2. Crossing process to Fig. 1(a) tree level, (b) meson exchange between the initial and final meson.

of as much as 25% around $M_{\rm inv}(\nu\ell) \simeq 0$. On the other hand, the factors $R^0_{\rm v}$ and $R^3_{\rm v}$ are comparatively very small and can be neglected. This means that the intermediate B, D pseudoscalar mesons are the relevant elements in the corrections.

In order to see which is the effect of the corrections done in the ratio R_D of Eq. (1), we show the branching ratios R_D for different values of $q_{\rm max}$ in Table 1. As we see the results from the tree level in Table 1, we obtain $R_D \simeq 0.23$. This is a bit short of the SM value $R_D \simeq 0.31$ quoted in Section 1, but a fair result considering that it is a pure theoretical result with no free parameters and no fit to data.[6] Taking $q_{\rm max} \simeq 700$ MeV, close to values used in Ref. 4, we have $R_D \simeq 0.204$. From the results of Table 1, we found that the strong interaction corrections are responsible for a 10% change of this ratio.

4. Summary

We have performed a theoretical calculation of the strong interaction corrections between the initial and final meson in the $\bar{B} \to D\bar{\nu}\ell$ decay. This is the analogy of the final state interaction in processes where a BD pair is produced at the end. The existence of calculations in which the strong interaction between B and D leads to a bound state indicates that the same

Table 1. Branching ratios R_D changing with $q_{\max}$. The case a, b, c, d correposnd to the calculations with and without the corrections done here; The case a represents tree level; b with a factor of $(1+R_P^0+R_V^0)^2$; c with a factor of $(1+R_P^3+R_V^3)^2$; d with both factors $(1+R_P^0+R_V^0)^2$ and $(1+R_P^3+R_V^3)^2$.

	a	b	c	d
$q_{\max} = 0.7$ GeV	0.228	0.240	0.194	0.204
$q_{\max} = 0.8$ GeV	0.228	0.243	0.185	0.196
$q_{\max} = 1.0$ GeV	0.228	0.250	0.166	0.181

interaction in the crossed channel $\bar{B} \to D\bar{\nu}\ell$ should be also relevant. We have performed this evaluation using the same ingredients as those used to bind the BD states and we obtain corrections to the tree level $\bar{B} \to D\bar{\nu}\ell$ amplitudes of the order of $15 - 25\%$, which are relevant in a theoretical calculation. W

We used the results obtained here to see the effects of these strong corrections in the R_D ratio for $\bar{\nu}_\tau\tau$ and $\bar{\nu}\ell$ production and we found effects of the order of 10%. This means that if one wishes to do a theoretical calculation of this ratio with the precision of 1.5% claimed in fits to data within the Standard Model, the effects studied here must be necessarily considered.

Acknowledgments

N. I. acknowledges the supports from JSPS Overseas Research Fellowships and JSPS KAKENHI Grant Number JP19K14709. LRD acknowledges the supports from the National Natural Science Foundation of China (Grant Nos. 11975009, 11575076).

References

1. R. X. Shi, L. S. Geng, B. Grinstein, S. Jäger and J. Martin Camalich, arXiv:1905.08498 [hep-ph].
2. A. Abdesselam et al. (Belle), (2019), arXiv:1904.08794 [hep-ex].
3. D. L. Yao, P. Fernandez-Soler, F. K. Guo and J. Nieves, arXiv:1906.00727 [hep-ph].
4. S. Sakai, L. Roca and E. Oset, Phys. Rev. D **96** (2017) 054023.
5. N. Ikeno, L. R. Dai, E. Oset, arXiv:1907.11128(2019) [hep-ph].
6. L. R. Dai, X. Zhang and E. Oset, Phys. Rev. D **98** (2018) 036004.

Testing the nature of $h_1(1380)$ resonance in $\chi_{cJ} \to \phi K^* \bar{K}, \phi h_1(1380)$

Xiang-Kai Luo, Sheng-Juan Jiang and Wei-Hong Liang*
Department of Physics, Guangxi Normal University, Guilin 541004, China
**E-mail: liangwh@gxnu.edu.cn*

S. Sakai
Institute of Theoretical Physics, CAS, Zhong Guan Cun East Street 55 100190 Beijing, China

E. Oset
Departamento de Física Teórica and IFIC, Centro Mixto Universidad de Valencia - CSIC, Institutos de Investigación de Paterna, Aptdo. 22085, 46071 Valencia, Spain

We have done a theoretical analysis of the date on the $\chi_{cJ} \to \phi K^* \bar{K} \to \phi K \pi \bar{K}$ reaction, and by considering the $K^* \bar{K}$ final state interaction we produce the $h_1(1380)$ which is clearly seen in BESIII experiment.

Keywords: $h_1(1380)$ state; final state interaction; molecular picture.

1. Formalism

We report on a study of the $\chi_{cJ} \to \phi K^* \bar{K} \to \phi K \pi \bar{K}$ done in Ref. 1. The first step in our analysis is to provide a picture for $\chi_{cJ} \to VVP$, with V and P denoting vector meson and pseudoscalar meson, respectively. Since we are interested only in the shape of the final $K^* \bar{K}$ mass distribution, we can ignore the strength of this vertex, but we must relate the different possible trios with two vectors and one pseudoscalar. We propose now the structure $H = \mathcal{C}\langle VVP \rangle$,

$$H = \mathcal{C}\phi \left[K^{*+}K^- + K^{*0}\bar{K}^0 + K^{*-}K^+ + \bar{K}^{*0}K^0 + \phi \left(\frac{-\eta}{\sqrt{3}} + \sqrt{\frac{2}{3}}\eta' \right) \right]. \quad (1)$$

The spin-parity structure of the vertices for the different χ_{cJ} states are:

1) $\chi_{c0} \to \phi K^* \bar{K}$ $[\chi_{c2} : I^G\left(J^{PC}\right) = 0^+\left(0^{++}\right)]$

The structure must be of the type

$$(\vec{\epsilon}_\phi \times \vec{\epsilon}_{K^*}) \cdot \vec{p}_\phi. \quad (2)$$

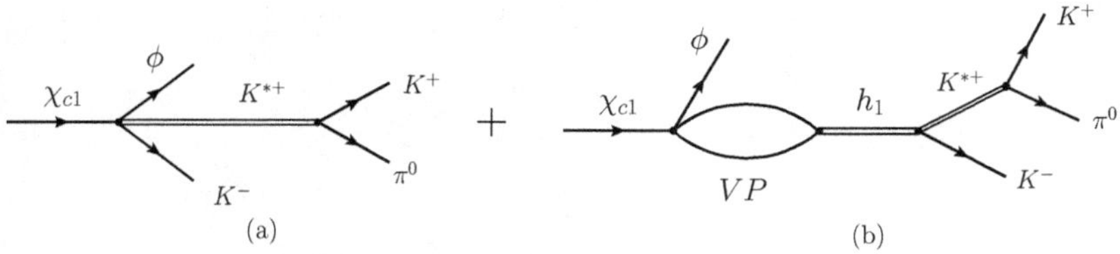

Fig. 1. Mechanisms for $\chi_{c1} \to \phi K^{*+}K^- \to \phi K^+\pi^0 K^-$ reaction, (a) tree level; (b) rescattering of $VP(K^*\bar{K}, \bar{K}^*K, \phi\eta)$ producing the $h_1(1380)$ resonance that decays into $K^{*+}K^-$ and later in $K^+\pi^0K^-$.

If we wish to produce the $h_1(1380)$ resonance, we will have to consider the diagram of Fig. 1(b) and sum coherently in the loop over the states of Eq. (1), and have the combination $2\,(\vec{\epsilon}_{K^*} \times \vec{\epsilon}_\phi) \cdot \vec{p}_K + 2\,(\vec{\epsilon}_\phi \times \vec{\epsilon}_{K^*}) \cdot \vec{p}_K - \frac{1}{\sqrt{3}}\,(\vec{\epsilon}_{\phi_1} \times \vec{\epsilon}_{\phi_2}) \cdot \vec{p}_\eta - \frac{1}{\sqrt{3}}\,(\vec{\epsilon}_{\phi_2} \times \vec{\epsilon}_{\phi_1}) \cdot \vec{p}_\eta = 0$. And we see that in the coherent sum the terms cancel and there is no $h_1(1380)$ production.

2) $\chi_{c1} \to \phi K^*\bar{K}$ $\quad \left[\chi_{c1} : I^G\left(J^{PC}\right) = 0^+\left(1^{++}\right)\right]$

Considering the symmetry of the two vectors we will have the combinations for the tree amplitude of Fig. 1(a), $\mathcal{C}'[(\vec{\epsilon}_{\chi_{c1}} \cdot \vec{p}_{K^*}) + (\vec{\epsilon}_{\chi_{c1}} \cdot \vec{p}_\phi)](\vec{\epsilon}_\phi \cdot \vec{\epsilon}_{K^*})$. In the coherent sum of Fig. 1(b) for the loop of VP we will get

$$(\vec{\epsilon}_\phi \cdot \vec{\epsilon}_{K^*})\,(\vec{\epsilon}_{\chi_{c1}} \cdot \vec{p}_\phi)\,. \tag{3}$$

This term will then interfere with the s-wave term of the tree level which has the same structure.

3) $\chi_{c2} \to \phi K^*\bar{K}$ $\quad \left[\chi_{c2} : I^G\left(J^{PC}\right) = 0^+\left(2^{++}\right)\right]$

Given the symmetry between the two vectors and what was found in points 1) and 2), it is clear that we should now combine the ϕ and K^* spin to $J = 2$ to avoid the cancellations found in point 1) where $J = 1$. The spin 2 of the χ_{c2} can be combined with the $\vec{p}_\phi$ to give another tensor of rank two to be contracted with the $J = 2$ object constructed with the two vectors. We would have remaining terms in the tree level and the loops involving $\vec{p}_\phi$ that would produce the $h_1(1380)$ resonance and some interference between them.

Next we want to have $K^+K^-\pi^0$. We must consider that in the experiment the K^* is seen as a $K\pi$ state, we can have $\chi_{c1} \to \Phi K^{*+}, K^{*+} \to K^+\pi^0$. Since we are concerned only about the shape of the distributions, we consider only the first mechanism that we depict in Fig. 1. The amplitude corresponding to the mechanisms of Fig. 1, summing over the K^* and h_1 polarizations, we find

$$t = A\,(\vec{\epsilon}_{\chi_{c1}} \cdot \vec{p}_\phi)\,\vec{\epsilon}_\phi \cdot (\vec{p}_K - \vec{p}_\pi)\,D_{K^*}\left(M_{inv}\left(K^+\pi^0\right)\right)T, \tag{4}$$

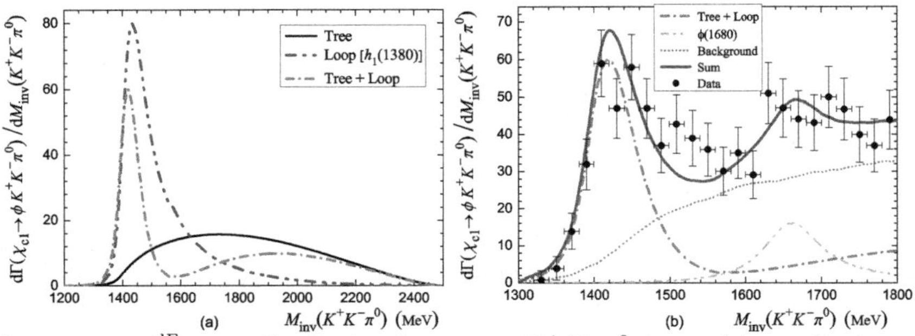

Fig. 2. $\frac{\mathrm{d}\Gamma}{\mathrm{d}M_{\text{inv}}(K^{*+}K^-)}$ distribution for $\chi_{c1} \to \phi K^+ K^- \pi^0$ decay: (a) contributions of the terms in Eq. (5); (b) comparison of our results with the data of Ref. 2.

$$T = h_{K^{*+}K^-} + D_{h_1}(M_{\text{inv}}(K^{*+}K^-))\,[\,h_{K^*\bar{K}}\, g^2_{h_1,K^*\bar{K}}\; G_{K^*\bar{K}}(M_{\text{inv}}(K^{*+}K^-)) + h_{\phi\eta}\, g_{h_1,\phi\eta}\, g_{h_1,K^*\bar{K}}\; G_{\phi\eta}(M_{\text{inv}}(K^{*+}K^-))\,]\,. \tag{5}$$

When we sum and average over polarizations in $|t|^2$, we find at the end

$$\overline{\sum}\sum |t|^2 = B\,\vec{p}_\phi^{\,2}\,\tilde{p}_{K^+}^{2}\,|D_{K^*}|^2\,|T|^2, \tag{6}$$

where in B we concentrate the different constant factors. The double differential mass distribution is given by

$$\frac{\mathrm{d}^2\Gamma}{\mathrm{d}M_{\text{inv}}(K^{*+}K^-)\mathrm{d}M_{\text{inv}}(K^+\pi^0)} = \frac{1}{(2\pi)^5} p_\phi p_{K^-} \tilde{p}_{K^+} \frac{1}{4M^2_{\chi_{c1}}} \overline{\sum}\sum |t|^2. \tag{7}$$

2. Results and conclusion

The information on the couplings and G function we take from Ref. 3, where the $h_1(1380)$ resonance dynamically generated. We integrate the differential width of Eq. (7) over $M_{\text{inv}}(K^+\pi^0)$ and compare the $\frac{\mathrm{d}\Gamma}{\mathrm{d}M_{\text{inv}}(K^{*+}K^-)}$ distribution with experiment. The results are shown in Fig 2.

In Fig. 2(a) we can see the contribution to $\frac{\mathrm{d}\Gamma}{\mathrm{d}M_{\text{inv}}(K^{*+}K^-)}$ from the first term of T in Eq. (5) (tree level), the second term that contains the $h_1(1380)$ propagator ("Loop" in the figure) and the coherent sum. The interesting finding is that the tree level, which by itself could be considered a background and contributes basically according to phase space, interferes destructively with the $h_1(1380)$ signal and the resulting shape is quite different from the one of the $h_1(1380)$ itself, which has a much broader shape. It is interesting to see that our contribution of the $h_1(1380)$ alone has basically the same shape as the one of Ref. 2 in Fig. 7 of that work. The interference between the tree level mechanism and the $h_1(1380)$ contribution is then missed in that analysis. It is clear that our approach will show a dip in the region of 1550-1600 MeV of the $K^+K^-\pi^0$ invariant mass. Something that in the fit to the date in the experiment of Ref. 2 was difficult to obtain.

What we see in Fig. 2(b) is that our approach produces naturally a clear dip in the region of 1550-1600 MeV, while the fit of Ref. 2 gives a distribution that is above the data in that region. In our approach, the relative strength and sign are given once we assume that the $h_1(1380)$ is generated from the interaction between pseudoscalar and vector. This is why the dip which we predict for this distribution is tied to the nature of the $h_1(1380)$ as a dynamically generated resonance, and the fact that this feature is present in the experiment provides a great support for the molecular picture of the $h_1(1380)$.

In summary, based on the picture that the $h_1(1380)$ is generated from the s-wave VP interaction, we have carried out a study of the $\chi_{cJ} \to \phi K^* \bar{K} \to \phi K^+\pi^0 K^-$ reaction and obtained a fair reproduction of the shape of the experimental data. Another remarkable feature of the study was that we could determine the relative strength between the tree level contribution to the process and the one that contains the $h_1(1380)$ production, and we found a destructive interference between the two processes that significantly distorts the $h_1(1380)$ signal with respect to the Breit-Wigner shape and produces a dip in the $K^+\pi^0 K^-$ mass distribution around the region of 1550-1600 MeV. This dip is present in the experiment and not reproduced in a picture that sums incoherently the h_1 Breit-Wigner distribution with a smooth background, providing a strong support to the molecular picture of the $h_1(1380)$ resonance.

Acknowledgment

This work is partly supported by the National Natural Science Foundation of China under Grants No. 11975083, No. 11847317 and No. 11565007.

References

1. S. J. Jiang, S. Sakai, W. H. Liang and E. Oset, Phys. Lett. B **797**, 134831 (2019).
2. M. Ablikim *et al.* [BESIII Collaboration], Phys. Rev. D **91**, no. 11, 112008 (2015).
3. L. Roca, E. Oset and J. Singh, Phys. Rev. D **72**, 014002 (2005).

Production of four charged pions with the CMD-3 detector at the VEPP-2000 collider

A. A. Korobov[ab] R. R. Akhmetshin[ab] A. N. Amirkhanov[ab] A. V. Anisenkov[ab] V. M. Aulchenko[ab] V. Sh. Banzarov[a] N. S. Bashtovoy[a] D. E. Berkaev[ab] A. E. Bondar[ab] A. V. Bragin[a] S. I. Eidelman[abe] D. A. Epifanov[ab] L. B. Epshteyn[abc] A. L. Erofeev[ab] G. V. Fedotovich[ab] S. E. Gayazov[ab] A. A. Grebenuk[ab] S. S. Gribanov[ab] D. N. Grigoriev[abc] F. V. Ignatov[ab] V. L. Ivanov[ab] S. V. Karpov[a] A. S. Kasaev[a] V. F. Kazanin[ab] I. A. Koop[ab] A. N. Kozyrev[ac] E. A. Kozyrev[ab] P. P. Krokovny[ab] A. E. Kuzmenko[ab] A. S. Kuzmin[ab] I. B. Logashenko[ab] A. P. Lysenko[a] P. A. Lukin[ab] K. Yu. Mikhailov[a] V. S. Okhapkin[a] A. V. Otboev[a] E. A. Perevedentsev[ab] Yu. N. Pestov[a] A. S. Popov[ab] G. P. Razuvaev[ab] A. A. Ruban[a] N. M. Ryskulov[a] A. E. Ryzhenenkov[ab] A. V. Semenov[ab] A. I. Senchenko[a] Yu. M. Shatunov[a] V. E. Shebalin[ab] D. N. Shemyakin[ab] B. A. Shwartz[ab] D. B. Shwartz[ab] A. L. Sibidanov[ad] E. P. Solodov[ab] V. M. Titov[a] A. A. Talyshev[ab] A. I. Vorobiov[a] Yu. V. Yudin[ab]

Budker Institute of Nuclear Physics, SB RAS, Novosibirsk, 630090, Russia[a]
Novosibirsk State University, Novosibirsk, 630090, Russia[b]
Novosibirsk State Technical University, Novosibirsk, 630092, Russia[c]
University of Victoria, Victoria, BC, Canada V8W 3P6[d]
Lebedev Physical Institute RAS, Moscow, 119333, Russia[e]

A study of the process $e^+e^- \to \pi^+\pi^-\pi^+\pi^-$ is carried out with the detector CMD-3 at the VEPP-2000 collider. An integrated luminosity of 68.3 pb^{-1} was collected in 2011-2018 in the energy range 650-1060 MeV. The cross section of the reaction has been measured. This process gives a large contribution to uncertainty of the hadronic contribution to the muon anomalous magnetic moment. This measurement can be also used to test the relation between the cross section of $e^+e^- \to \pi^+\pi^-\pi^+\pi^-$ and the spectral function for the $\tau^- \to \pi^-\pi^0\pi^0\pi^0$ decay predicted by the conservation of vector current (CVC).

Keywords: Low energy QCD; electron-positron annihilation; hadrons, muon g-2.

1. Introduction

Since 2010 the CMD-3[1] and SND[2] experiments have been carried out in the energy range 320-2000 MeV at VEPP-2000 collider. The most precise previous studies of this process were performed at CMD-2,[3] SND[4] and

BaBar.[5] Initial State Radiation technique(ISR) has been used in the BaBar measurement to achieve 3% accuracy of the cross-section. A data sample of 68.3 pb^{-1} collected in 2013-2018 by CMD-3 in the energy range 650-1060 MeV was used in our work. CMD-3 installed in one of two VEPP-2000 interaction regions is a universal cryogenic magnetic detector. The cylindrical drift chamber (DC)[6] and double-layer multiwire proportional Z-chamber are parts of its tracking system, both also used for a trigger. The liquid xenon (LXe) barrel calorimeter with a radiation length of 5.4 X_0 provides 1-2 mm spatial resolution.[7] The barrel CsI crystal calorimeter with a radiation length of 8.1 X_0 is placed outside and the endcap BGO calorimeter with a thickness of 13.4 X_0 is installed inside the solenoid.[8]

The luminosity has been measured using events of Bhabha scattering at large angles[9] and $e^+e^- \to \gamma\gamma$ with about 1% accuracy. The c.m. energy has been measured by using the Back-Scattering-Laser-Light system[10] with about 0.06 MeV systematic uncertainty. To obtain a detection efficiency, Monte Carlo (MC) simulation of our detector based on the GEANT4[11] package has been developed.

2. Simulation

In the previous CMD-2 study[12] the $a_1\pi$ dominance has been shown in the energy range up to 1400 MeV. The VDM model with a mixture $a_1\pi$ and ρf_0 and $a_2\pi$ of intermediate states has been used to simulate $e^+e^- \to \pi^+\pi^-\pi^+\pi^-$ events. One hundred thousand $e^+e^- \to \pi^+\pi^-\pi^+\pi^-$ events have been simulated with radiative corrections at each energy point.

3. Selection of $e^+e^- \to \pi^+\pi^-\pi^+\pi^-$ events

$e^+e^- \to \pi^+\pi^-\pi^+\pi^-$ candidates should have four tracks in the DC with the following requirements: $d < 0.25$ cm and $dz < 12$ cm, where d and dz are a minimum distance from a track to the beam axis in the transverse plane and a minimum distance from a track to the center of the interaction region along the beam axis, respectively. Reconstructed momenta and angles of the tracks for four-track candidates are used for further selection. In this region several physical reactions could mimic the studied process in the detector. The background strongly depends on the c.m energy. In the energy range up to 1 GeV the $e^+e^- \to \pi^+\pi^-\pi^0$, $e^+e^- \to e^+e^-e^+e^-$ and $e^+e^- \to e^+e^-$ with extra tracks from decays or nuclear interaction of pions as well as from a conversion of the photons from the π^0 decay in the detector material are the main sources of background.

To reduce the background from the$e^+e^- \to \pi^+\pi^-\pi^0$ reaction, we require a pion hypothesis for each track in e/π separation. After this requirement the remaining contribution from three pions to the number of four-track candidates is less than 5%.To determine the number of signal events the following criteria are used:

- $30 > 2E_{beam} - \sum_{i=1}^{4} \sqrt{p_i^2 + m_\pi^2} > -20$ MeV
- $|\sum_{i=1}^{4} \vec{p_i}| < 100$ MeV

4. Electron-pion separation

Ionization losses in DC, and track energy deposition in calorimeters are used to suppress background from the final states containing electrons and positrons. The whole momentum range is divided into two parts: below and above 100 MeV/c. If the track momentum p_i is below 100 MeV/c, ionization losses in DC are used to separate electrons and pions. Pion hypothesis for such tracks is assumed, if p_i is above 100 MeV/c and $E_{\rm cal} - 12/11 * p_i + 70 < 0$, where $E_{\rm cal}$ is track energy deposition in the calorimeter. A clean sample of $e^+e^- \to \pi^+\pi^-\pi^0$ is used to test the e/π separation procedure. We use the data for the negative track energy deposition in the calorimeter because there is a difference between simulation and experiment for this parameter at the same momentum. We assume that this difference is related to simulation of nuclear interaction. The following criteria are used in experimental data:

- Track is referred to as a pion if $E_{cal} - 12/11 * p + 70 < 0$ and $p > 100$ MeV/c
- Track is referred to as a pion if $|f(x) - \frac{dE}{dx}| < 1500$ and $p < 100$ MeV/c, where $f(p) = \frac{5.58*10^9}{(p+40)^3} + 2.21 * 10^3 - 3.77 * 10^{-1} * p$

The following criteria are used in MC data:

- Track is referred to as a pion according to pion identification dependence on track momentum separately for π^+ and π^-.
- Track is referred to as a pion if $|f(x) - \frac{dE}{dx}| < 1500$ and $p < 100$ MeV, where $f(p) = \frac{5.58*10^9}{(p+40)^3} + 2.21 * 10^3 - 3.77 * 10^{-1} * p$

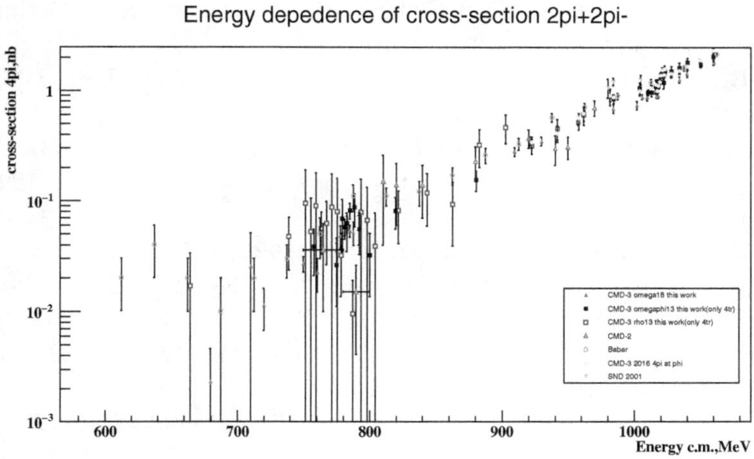

Fig. 1. Comparison of our preliminary results on $e^+e^- \to \pi^+\pi^-\pi^+\pi^-$ cross section with the previous experiments:open circle - BaBar,[5] open triangle - CMD-2,[3] full square-this work (1st scan), open square - this work (2nd scan),full triangle-this work (3rd scan) full circle - CMD-3 2016,[13] full triangle - SND.[4]

5. Preliminary results

We used the following formula to measure $e^+e^- \to \pi^+\pi^-\pi^+\pi^-$ cross-section: $\sigma = \frac{N_{sig}}{L*\epsilon*(1+\delta)}$, where L is integrated luminosity at this energy, ϵ is detection efficiency, $1+\delta$ is the radiative correction calculated according to Ref. 14. In the energy range from threshold up to 1 GeV the cross section rapidly rises. We have managed to suppress the $e^+e^- \to \pi^+\pi^-\pi^0$ background around ω mass. Energy dependence of the $e^+e^- \to \pi^+\pi^-\pi^+\pi^-$ cross section is shown in Fig. 1. Our results are consistent with the previous experiments. The statistical uncertainties only are shown. A study of systematic effects is in progress.

Acknowledgments

We thank the VEPP-2000 personnel for the excellent machine operation. This work is partially supported by the Russian Foundation for Basic Research grants RFBR 17-02-00327-a.

References

1. B. I. Khazin et al. (CMD-3 Collaboration), Nucl. Phys. B (Proc. Suppl.) **181-182**, 376 (2008).

2. M. N. Achasov et al. (SND Collaboration), Nucl. Instrum. Methods Phys. Res., Sect. **A449**, 125 (2000).
3. R. R. Akhmetshin et al. (CMD-2 Collaboration), Phys. Lett. **B595**, 101 (2004).
4. M. N. Achasov et al. (SND Collaboration), JETP **96**, 789 (2003).
5. B. Aubert et al. (BaBar Collaboration), Phys. Rev. **D85**, 112009 (2012).
6. F. Grancagnolo et al., Nucl. Instr. Meth. A**623**, 114 (2010).
7. A. V. Anisyonkov et al., Nucl. Instr. Meth. A**598**, 266 (2009).
8. D. Epifanov (CMD-3 Collaboration), J. Phys. Conf. Ser. **293**, 012009 (2011); V. M. Aulchenko et al., JINST **10**, P10006 (2015); V. E. Shebalin et al., JINST **9**, C10013 (2014).
9. R. R. Akhmetshin et al., Nucl. Phys. B (Proc. Suppl.) **225-227**, 69 (2012); G. V. Fedotovich et al., Phys. At. Nucl. **78**, 591 (2015).
10. V. Abakumova et al., Phys. Rev. Lett. **110**, 140402 (2013).
11. S. Agostinelli et al. (GEANT4 Collaboration), Nucl. Instr. Meth. **A506**, 250 (2003).
12. R. R. Akhmetshin et al. (CMD-2 Collaboration), Phys. Lett. **B466**, 392 (1999).
13. R. R. Akhmetshin et al. (CMD-2 Collaboration), Phys. Lett. **B768**, 345 (2017).
14. E. A. Kuraev and V. S. Fadin, Sov. J. Nucl. Phys. **41**, 466 (1985); S. Actis et al., Eur. Phys. J. **C66**, 585 (2010).

Holographic description of decay constants in AdS/QCD models: A summary and perspectives

Miguel Angel Martin Contreras* and Alfredo Vega†
Instituto de Física, Universidad de Valparaíso,
Valparaíso, ZIP 2340000, Chile
**E-mail: miguelangel.martin@uv.cl*
†*E-mail: alfredo.vega@uv.cl*

In this talk we discuss the holographic approximation to the calculation of mesonic decay constants in the context of the bottom-up AdS/QCD models such as the soft wall-like approaches. We show some of the technical difficulties and some possible solutions to them.

Keywords: AdS/QCD; holographic QCD; meson phenomenology.

1. Introduction

One of the most important hadronic properties is the decay constant, related to the transition to vacuum of a given mesonic state.

Holographically speaking,[1,2] the decay constants can be obtained from the spectral decomposition of the holographic 2-point function, which is constructed from the bulk-to-boundary propagator $V(z, q)$ and the on-Shell boundary action. We will focus on vector mesons following the prescription given by Ref. 3.

2. Geometric background

As our starting point, we will consider an AdS-like geometry defined by the 5-line element

$$dS^2 = e^{2\,A(z)}\left[dz^2 + \eta_{\mu\nu}\,dx^\mu\,dx^\nu\right], \tag{1}$$

where $\eta_{\mu\nu}$ is a 5-dimensional Minkowski metric. Since we want AdS-like geometries, we will impose that the warp factor, at the conformal boundary, behaves as $\log(R/z)$ when $z \to 0$.

3. Decay constant holographic construction

In general, vector mesons are dual to vector bulk fields $\psi_m(z,x)$, with an action given by[4]

$$I_{\text{Vector}} = -\frac{1}{4\,g_V^2}\int d^5x\,\sqrt{-g}\,e^{-\Phi(z)}\,F_{mn}\,F^{mn}, \tag{2}$$

where $\Phi(z)$ is a generic dilaton that could be quadratic-like, static or dynamical. If we want to reconstruct linear Regge trajectories, is enough to impose that the dilaton should be quadratic-like. From this action is always possible to write down a boundary action for the bulk vector fields in terms of its *Schwinger-like source expansion.*[2] The 2-point function is obtained by taking the second functional derivative with respect to these Schwinger-like sources. In the most general AdS/QCD model case we can obtain the following 2-point function

$$\begin{aligned}\Pi\left(-q^2\right) &= -\left.\frac{e^{-B(z)}}{g_V^2\left(-q^2\right)}\,\partial_z\,V\left(z,q\right)\right|_{z\to 0} = \sum_{n=1}^{\infty}\frac{e^{-B(z)}\,e^{-A(z)}\left.\left[\psi_n''(z,q)\right]^2\right|_{z\to 0}}{g_V^2\,(m_n^2)(-q^2+m_n^2+i\epsilon)}\\ &\equiv \sum_{n=1}^{\infty}\frac{F_n^2}{-q^2+m_n^2+i\epsilon}.\end{aligned}$$

The pole expansion of the 2-point function allows us to extract the decay constants as the residues of such expansion. Therefore, we have for a generic bottom-up AdS/QCD model

$$F_n^2 = \frac{1}{g_V^2}\lim_{z\to 0}\,e^{-\phi(z)}\,\left|\psi_n''(z,q)\right|^2. \tag{3}$$

The g_V constant, that gives the units in the action, is fixed by matching the holographic 2-point function with the corresponding one in large N QCD at large Q^2. The main ingredient is the Sturm-Liouville solution of the equations of motion, the holographic eigenfunction $\psi(z,q)$.

3.1. *Eigenfunction derivative calculation*

To calculate the decay constants F_n we need to estimate the holographic eigenfunction ψ. To do this, we recall the holographic behavior of these kind of solutions near to the conformal boundary, allowing us to expand the eigenfunction in a power series near to the boundary as

$$\psi_n(z,q) = \mathcal{C}_{\Delta-S}(q,n,S)\, z^{\Delta-S} + \sum_{m>\Delta+S}^{\infty} \mathcal{C}_m(q,n,S)\, z^m, \tag{4}$$

where $\mathcal{C}_m(q,n,S)$ is a constant that depends on the spin and the mesonic radial excitation level. In our case at hand, $S = 1$. The second derivative can be calculated as

$$\psi_n''(z,q) = \mathcal{C}_{\Delta-S}(q,n,S)\,(\Delta - S)(\Delta - S - 1)\, z^{\Delta-S-2} + \ldots \tag{5}$$

Since all of the terms with power higher than $\Delta - S - 2$ are zero when we evaluate the limit $z \to 0$, only the $\mathcal{C}_{\Delta-S}$ coeffcient survives. This behavior is prove to be general in Ref. 3. Finally, the decay constants aare given by

$$F_n^2 = \frac{e^{\Phi_0}}{g_V^2}\, \mathcal{C}_{\Delta-S}^2(q,n,S)\,(\Delta - S)^2(\Delta - S - 1)^2. \tag{6}$$

Thus, the decay constant depends on the first coefficient on the power series only. This is a powerful form of testing AdS/QCD model easily.

4. Testing the decay constant expression

In order to prove our decay constant formula (6), it is instructive to work with $f_n = F_n/M_n$ instead of F_m. Recall that M_n is mass spectrum calculated from e.o.m. for the bulk vector field. From experiments, it is expected that decay constants f_n should be decreasing with the excitation number. Therefore we can write that

$$R_{\text{Decay}} = \frac{f_{n+1}}{f_n} = \frac{M_n\, F_{n+1}}{M_{n+1}\, F_n} < 1. \tag{7}$$

This phenomenological evidence is translated into the following holographic recipe:

$$R_{\text{AdS/QCD model}} = \frac{M_n\, \mathcal{C}_{\Delta-S}(q,n+1,S)}{M_{n+1}\, \mathcal{C}_{\Delta-S}(q,n,S)} < 1. \tag{8}$$

Calculating this ratio $R_{\text{AdS/QCD model}}$ will allow us to test if the AdS/QCD model at hand has a good behavior with the decay constants.

We will check this Eq. (8) by evaluating some AdS/QCD models found in the literature. The following table summarizes the results for four different approaches. Notice that only the soft wall model with UV cutoff has the right phenomenological behavior.

Table 1. Comparison of the decay ratio (8) for four bottom-up models: the hard wall, soft wall, soft wall with UV cutoff and the Li-Huang models.

Ratio	**HWM**[8]	**SWM**[4]	**SWM with UV Cutoff**[7]	**LHM**[9]
f_1/f_2	> 1	1	< 1	> 1

5. Conclusion

The expression (6) provides a good estimation tool to test easily the phenomenological behavior of the mesonic decay constants. This procedure is independent of the bottom-up AdS/QCD model used to construct the dual state to the mesons.

Acknowledgement

The authors acknowledge the financial support by FONDECYT (Chile) under Grants No. 1180753 (A. V.) and No. 3180592 (M. A. M. C.).

References

1. J. Erdmenger, N. Evans, I. Kirsch and E. Threlfall, "Mesons in Gauge/Gravity Duals - A Review," Eur. Phys. J. A **35**, 81 (2008).
2. J. M. Maldacena, "The Large N limit of superconformal field theories and supergravity," Int. J. Theor. Phys. **38**, 1113 (1999) [Adv. Theor. Math. Phys. **2**, 231 (1998)].
3. M. A. Martin Contreras and A. Vega, "Decay constants in AdS/QCD: a different approach," arXiv:1910.10922 [hep-th].
4. A. Karch, E. Katz, D. T. Son and M. A. Stephanov, "Linear confinement and AdS/QCD," Phys. Rev. D **74**, 015005 (2006).
5. J. Erlich, E. Katz, D. T. Son and M. A. Stephanov, "QCD and a holographic model of hadrons," Phys. Rev. Lett. **95**, 261602 (2005)s.
6. S. S. Afonin, "Holographic like models as a five-dimensional rewriting of large-Nc QCD," Int. J. Mod. Phys. A **25**, 5683 (2010).
7. N. R. F. Braga, M. A. Martin Contreras and S. Diles, "Decay constants in soft wall AdS/QCD revisited," Phys. Lett. B **763**, 203 (2016).
8. H. Boschi-Filho and N. R. F. Braga, "Gauge/string duality and scalar glueball mass ratios," JHEP **0305**, 009 (2003).
9. D. Li and M. Huang, "Dynamical holographic QCD model for glueball and light meson spectra," JHEP **1311**, 088 (2013).

The bottomed strange molecules with isospin 0

Zhi-Feng Sun

School of Physical Science and Technology, Lanzhou University,
Lanzhou 730000, China
**E-mail: sunzf@lzu.edu.cn*

Ju-Jun Xie

Institute of Modern Physics, Chinese Academy of Sciences,
Lanzhou 730000, China
**E-mail: xiejujun@impcas.ac.cn*

E. Oset

Departamento de Fsica Terica and IFIC, Centro Mixto Universidad de
Valencia-CSIC, Institutos de Investigacíon de Paterna,
Aptdo. 22085, 46071 Valencia, Spain
**E-mail: oset@ific.uv.es*

Using the chiral unitary approach, we study the possibility of the existence of bottomed strange molecular states with isospin 0. We find three bound states with spin-parity 0^+, 1^+ and 2^+ generated by the $\bar{K}^*B^*$ and ωB_s^* interaction, among which the state with spin 2 can be identified as $B_{s2}^*(5840)$. In addition, we also study the $\bar{K}^*B$ and ωB_s interaction and find a bound state which can be associated to $B_{s1}(5830)$. Besides, the $\bar{K}B^*$ and ηB_s^* and $\bar{K}B$ and ηB_s systems are studied, and two bound states are predicted. We expect that further experiments can confirm our predictions.

Keywords: Bottomed meson; Strange meson; Molecular state.

1. Introduction

The chiral unitary approach incorporates two points, i.e., the local hidden and chiral symmetries[1–4] as well as the Bethe-Salpeter equation, which has been instrumental in explaining many properties of hadronic physics[5–10]. In Ref.[9], the chiral unitary approach was used to study the interactions of ρ, ω and D^*, and three states with spin $J = 0, 1, 2$ were predicted, among which the second and third states are identified with $D^*(2640)$ and $D_2^*(2460)$, respectively. The other one predicted, $D(2600)$, was found later by BaBar Collaboration[11] and has been reconfirmed by LHCb Collaboration[12,13]. In

Ref.[10], the $\rho(\omega)B^*(B)$ interaction was studied, and $B_1(5721)$ and $B_2^*(5747)$ are explained as $\rho(\omega)$ and ρB molecules. There are two bottomed strange states, $B_{s1}(5830)$ and $B_{s2}^*(5840)$, listed in Particle Data Group[14], of which the masses are

$$M[B_{s1}(5830)] = 5828.70 \pm 0.20 MeV, \tag{1}$$

$$M[B_{s2}^*(5840)] = 5839.85 \pm 0.12 MeV, \tag{2}$$

respectively.

In this work, we extrapolate the chiral unitary approach to the systems containing bottomed and strange quarks. The paper is organized as follows. After this short introduction, in sections II and III we will show the chiral unitary approach and how to obtain the potentials of the $B^*\bar{K}^*$ and $B\bar{K}^*$ systems. And then we will give the T matrix of the Bethe-Salpeter equation simplified by the on-shell factorization. Finally, we make a short summary and discussion.

2. The chiral unitary approach

In order to describe the interactions between bottomed and strange mesons, we use the Lagrangians constructed by considering the local hidden and the chiral symmetries. It should be noted that, in this work, to deal with the amplitudes involving heavy mesons, we use the local hidden gauge Lagrangians extrapolated to SU(4) case, but for the exchange of light vectors one is only making use of the relevant SU(3) subgroup.

We also need to discuss the justification of the couplings for the vertices containing bottomed mesons and light vector meson. If $\bar{s}$ and $\bar{b}$ quarks are seen as spectators, the topology of the vertices $K^{(*)}K^{(*)}\rho$ and $B^{(*)}B^{(*)}\rho$ are the same. If we compare these two vertices on both the quark and the hadron level, we find the following relations

$$\frac{t_{B^{(*)}}}{t_{K^{(*)}}} = \frac{m_{B^{(*)}}}{m_{K^{(*)}}}. \tag{3}$$

And then we found that the Lagrangians we use take automatically into account all the elements discussed above.

3. The $B^*\bar{K}^*$ and $B\bar{K}^*$ interactions

With the Lagrangians mentioned above, we can calculate the amplitudes corresponding to the processes $B^*\bar{K}^* \to B^*\bar{K}^*$, $B^*\bar{K}^* \to B_s^*\omega$, $B_s^*\omega \to B^*\bar{K}^*$, $B_s^*\omega \to B_s^*\omega$, $B\bar{K}^* \to B^*\bar{K}^*$, $B\bar{K}^* \to B_s\omega$ and $B_s\omega \to B\bar{K}^*$.

For the $B^*\bar{K}^*$ sector, we project the polarization vector products into different spin states, i.e., $\mathcal{P}(0) = \frac{1}{3}\epsilon^i\epsilon^i\epsilon^j\epsilon^j$, $\mathcal{P}(1) = \frac{1}{2}(\epsilon^i\epsilon^j\epsilon^i\epsilon^j - \epsilon^i\epsilon^j\epsilon^j\epsilon^i)$, $\mathcal{P}(1) = \frac{1}{2}(\epsilon^i\epsilon^j\epsilon^i\epsilon^j + \epsilon^i\epsilon^j\epsilon^j\epsilon^i) - \frac{1}{3}\epsilon^i\epsilon^i\epsilon^j\epsilon^j$. The spin of $B\bar{K}^*$ system can only be 1, with the projection of the polarization vector products $\mathcal{P} = \epsilon^i\epsilon^i$. And then we get all the amplitudes with isospin $I = 0$ projected to spin 0, 1, 2 for $B^*\bar{K}^*$ systems and to spin 1 for $B\bar{K}^*$ system.

4. T-matrix

Using the Bethe-Salpeter equation in its on-shell factorized form, we obtain the T-matrix

$$T = (I - VG)^{-1}V, \tag{4}$$

where V is the transition amplitudes projected to s-wave, and G is the two-meson loop function, i.e.,

$$G = \int_0^{q_{max}} \frac{q^2 dq}{(2\pi)^2} \frac{\omega_1 + \omega_2}{\omega_1\omega_2[(P^0)^2 - (\omega_1 + \omega_2)^2 + i\epsilon]}. \tag{5}$$

P is the total four-momentum of the two mesons in the loop, m_1 and m_2 are the masses, q_{max} stands for the cut off, $\omega_i = \sqrt{\vec{q}_i^{\,2} + m_i^2}$.

5. Summary and discussion

By solving the Bethe-Salpeter equation with the amplitudes that we obtain, we get the masses of the $B^*\bar{K}^*$ systems. We found three states with spin 0, 1, 2 which have masses 5908.5 – 5894.4 MeV, 5912.1 – 5898.2 MeV and 5847.8 – 5831.7 MeV, respectively. The state with spin 2 can be identified with $B^*_{s2}(5840)$. However, the $B_{s1}(5830)$ can not be explained as the state with spin 1, since its mass is smaller than that of $B^*_{s2}(5840)$. By studying the $B\bar{K}^*$ system, we get a bound state with a mass 5822.3 – 5806.9 MeV which tells us that $B_{s1}(5830)$ can be explained as a $B\bar{K}^*$ molecule. In addition, we also studied $\bar{K}B^*/\eta B^*_s$ and $\bar{K}B/\eta B_s$ interaction, and predicted two bound state with masses 5671.2 – 5663.6 MeV and 5475.4 – 5457.5 MeV, respectively. These results are listed in Table. 1. We expect further experiments to confirm our predictions.

Acknowledgments

This work is supported by the National Science Foundation for Young Scientists of China under Grants NO. 11705069.

Table 1. Summary of our results where the cut off is in the range of 1055 $\sim$ 1085 MeV and the masses in this table is in the unit of MeV.

State mass	$I(J^P)$	Main component	Exp.
5475.4 $\sim$ 5457.5	$0(0^+)$	$\bar{K}B$	-
5671.2 $\sim$ 5663.6	$0(1^+)$	$\bar{K}B^*$	-
5822.3 $\sim$ 5806.9	$0(1^+)$	$\bar{K}^*B$	$B_{s1}(5830)$
5908.5 $\sim$ 5894.4	$0(0^+)$	$\bar{K}^*B^*$	-
5912.1 $\sim$ 5898.2	$0(1^+)$	$\bar{K}^*B^*$	-
5847.8 $\sim$ 5831.7	$0(2^+)$	$\bar{K}^*B^*$	$B_{s2}^*(5840)$

References

1. M. Bando, T. Kugo, S. Uehara, K. Yamawaki and T. Yanagida, Phys. Rev. Lett. **54**, 1215 (1985).
2. M. Bando, T. Kugo and K. Yamawaki, Phys. Rept. **164**, 217 (1988).
3. U. G. Meissner, Phys. Rept. **161**, 213 (1988).
4. M. Harada and K. Yamawaki, Phys. Rept. **381**, 1 (2003)
5. R. Molina, D. Nicmorus and E. Oset, Phys. Rev. D **78**, 114018 (2008)
6. L. S. Geng and E. Oset, Phys. Rev. D **79**, 074009 (2009)
7. J. J. Xie, M. Albaladejo and E. Oset, Phys. Lett. B **728**, 319 (2014)
8. X. L. Ren, L. S. Geng, E. Oset and J. Meng, Eur. Phys. J. A **50**, 133 (2014)
9. R. Molina, H. Nagahiro, A. Hosaka and E. Oset, Phys. Rev. D **80**, 014025 (2009)
10. P. Fernandez-Soler, Z. F. Sun, J. Nieves and E. Oset, Eur. Phys. J. C **76**, no. 2, 82 (2016)
11. P. del Amo Sanchez *et al.* [BaBar Collaboration], Phys. Rev. D **82**, 111101 (2010)
12. R. Aaij *et al.* [LHCb Collaboration], JHEP **1309**, 145 (2013)
13. R. Aaij *et al.* [LHCb Collaboration], Phys. Rev. D **94**, no. 7, 072001 (2016)
14. M. Tanabashi et al. (Particle Data Group), Phys. Rev. D 98, 030001 (2018)

Study of charm baryons in a hypercentral quark model

Zahra Ghalenovi

Department of Physics, Kosar University of Bojnord, Iran
E-mail: z_ghalenovi@kub.ac.ir

Asadolah Tavakolinezhad

Department of Basic Sciences, Shahrood Branch, Islamic Azad University, Shahrood, Iran

We study the single charm baryons in a non-relativistic quark model. We use the hypercentral approach to simplify three-body equation of the baryonic system. In our study, the spin and isospin dependent interactions are calculated perturbatively. The mass spectra of the single charm baryons are calculated. We also predict the radiative decay width and the transition magnetic moments of the Σ_c baryons. A comparison of our results with the experimental data and the predictions of recent models is presented.

Keywords: Potential model; hypercentral approach; non-relativistic limit; charm baryons.

1. Introduction

The investigation of hadrons containing heavy quarks is of great interest in understanding the dynamics of QCD at the hadronic scale. There is renewed interest both experimentally and theoretically in the properties of heavy flavour baryons. In this study we consider the three-body problem of the baryons containing one charm quark. Our work is presented in a two-step procedure: first, we use the hypercentral model in order to solve the three-body Schrödinger equation by performing an ansatz method and obtain the wavefunction and eigenvalues of the barionic system. Second, we introduce the perturbative hyperfine interaction and then obtain the baryon masses, transition magnetic moments and radiative decay widths.

2. Theoretical formalism

To describe the baryon as a bound state of three constituent quarks we use the hypercentral model as a useful method in describing baryonic systems.

In the hypercentral model, if the energy of the system depends only on the quarks distance, the Schrödinger equation has the form of

$$[\frac{d^2}{dx^2}+\frac{5}{x}\frac{d}{dx}-\frac{\gamma(\gamma+4)}{x^2}]\psi_\gamma(x)=-2m[E_\gamma-V(x)]\psi_\gamma(x) \tag{1}$$

where γ is the hyperangular momentum and m is the reduced mass (for details see Refs. 1–3). The considered potential $V(x)$ is a combination of the confining harmonic oscillator and Coloumbic terms $V(x)=ax^2-\frac{c}{x}$. We consider the ground-state baryons and to solve Eq. (1) assume the wave function $\psi(x)$ as follows

$$\psi(x)=x^{-5/2}e^{-\frac{1}{2}\alpha x^2+\delta lnx}, \tag{2}$$

where α and δ are constant. By substitution of $\psi(x)$ into Eq. (1) and comparing the coefficient of x on both sides of the equation, the parameters $\alpha=-\frac{m\omega^2}{2}$ and $\delta=\frac{5}{2}$ are found, where $\omega=\sqrt{\frac{2a}{m}}$ is the frequency of oscillating. Using the obtained parameters we can get $E=3\omega-\frac{2mc^2}{25}$. The pertubative hyperfine interaction is a combination of spin-spin, spin-isospin and isospin-isospin interaction terms ($H_{hyp}=H_s+H_t+H_{st}$)[1–3]:

$$H_s=A_s(\frac{1}{\sqrt{\pi}\sigma_s})^3exp[\frac{-x^2}{\sigma_s^2}](\vec{s_1}.\vec{s_2}), \tag{3}$$

$$H_t=A_t(\frac{1}{\sqrt{\pi}\sigma_t})^3exp[\frac{-x^2}{\sigma_t^2}](\vec{t_1}.\vec{t_2}), \tag{4}$$

$$H_{st}=A_{st}(\frac{1}{\sqrt{\pi}\sigma_{st}})^3exp[\frac{-x^2}{\sigma_{st}^2}](\vec{s_1}.\vec{s_2})(\vec{t_1}.\vec{t_2}). \tag{5}$$

s_i and t_i are the spin and isospin operator of the i th quark and A_i and σ_i are constant.

3. Results and discussion

Now we are able to calculate the masses and decay widths of the baryons using the obtained eigenfunctions and energy eigenvalues. Baryon mass is obtained by sum of the quark masses and the perturbative hyperfine interaction. The light quark masses and potential parameters are taken from Ref. 4. The charm quark mass ($m_c=1573$) is obtained by fitting to the mass of Σ_c baryon. Our results for the baryon masses are listed in Table 1 and compared with experimental data.[5] The radiative decay width for transitions between $J=\frac{3}{2}^+$ and $J=\frac{1}{2}^+$ baryons is given by[6,7]:

$$\Gamma_R=\frac{q^3}{4\pi}\frac{2}{2J+1}\frac{e^2}{m_p^2}|\mu_{\frac{3}{2}^+\to\frac{1}{2}^+}|^2, \tag{6}$$

where m_p is the proton mass and q is the photon energy. The radiative transition magnetic moments $\mu_{\frac{3}{2}^+\to\frac{1}{2}^+}$ are computed as

$$\mu_{\frac{3}{2}^+\to\frac{1}{2}^+} = \Sigma_i < \phi_{sf}^{\frac{3}{2}^+} |\mu_i \sigma_i| \phi_{sf}^{\frac{1}{2}^+} > . \quad (7)$$

$< \phi_{sf}^J|$ represents the spin flavour wave function of the baryons with spin J and μ_i is the magnetic moment of ith quark is defined as

$$\mu_i = \frac{e_i}{2m_i^{eff}} \quad (8)$$

Here, e_i and σ_i represent the charge and the spin of the ith quark and the effective quark mass m_i^{eff} is defined as

$$m_i^{eff} = m_i(1 + \frac{E + < H_{hyp} >}{\sum_i m_i}). \quad (9)$$

In order to evaluate $B_{3/2} \to B_{1/2}\gamma$ transition magnetic moments, we take the geometric mean of effective mass of the constituent quarks inside the initial- and final-state baryons. The calculated transition magnetic moments and radiative decay widths of the Σ_c baryons are listed in Tables 2 and 3 respectively and compared with other theoretical results.

Table 1. Single charm baryon masses (in MeV).

Baryon	Present work	Exp
Σ_c	2452	2452
Σ_c^*	2526	2518
Λ_c	2274	2286
Ξ_c	2559	2470
Ξ_c^*	2637	2645
Ω_c	2668	2695
Ω_c^*	2748	2766

Table 2. Magnitude of the transition magnetic moments ($\mu_{\frac{3}{2}^+\to\frac{1}{2}^+}$) in μ_N.

Decay Mode	Present work	[6]	[8]
$\Sigma_c^{*+} \to \Lambda_c^+\gamma$	1.56	1.85	1.75
$\Sigma_c^+ \to \Lambda_c^+\gamma$	1.58	1.34	1.27
$\Sigma_c^{*++} \to \Sigma_c^{++}\gamma$	1.30	1.08	0.99
$\Sigma_c^{*+} \to \Sigma_c^+\gamma$	0.07	0.008	0.008
$\Sigma_c^{*0} \to \Sigma_c^0\gamma$	1.15	1.06	1.02

Table 3. Radiative decay widths Γ_R (in KeV).

Decay Mode	Present work	[6]	[8]
$\Sigma_c^{*+} \to \Lambda_c^+ \gamma$	163.9	154.4	143.9
$\Sigma_c^{+} \to \Lambda_c^+ \gamma$	58.7	60.55	58.1
$\Sigma_c^{*++} \to \Sigma_c^{++} \gamma$	2.98	1.15	0.85
$\Sigma_c^{*+} \to \Sigma_c^+ \gamma$	0.01	~ 0	~ 0
$\Sigma_c^{*0} \to \Sigma_c^0 \gamma$	2.32	1.12	1.20

4. Conclusion

We consider the ground-state heavy baryons containing single charm quark. A potential model was introduced and the Schrödinger equation was solved using the hypercentral approach. Using the obtained wave function and energy eigenvalues, we could obtain the charm baryon masses. Finally, we calculated the transition magnetic moments and radiative decay widths of the Σ_c baryons. Our results are in good agreement with the experimental date and also the predictions presented by others.

References

1. M. M. Giannini, E. Santopinto and A. Vassalo, Prog. Part. Nucl. Phys. **50**, 263 (2003).
2. Z. Ghalenovi, A. A. Rajabi, S. X. Qin and D. H. Rischke, Mod. Phys. Lett. A **29**, 1450106 (2014).
3. Z. Ghalenovi, F. Giacosa and D. H. Rischke, Acta Phys. Polon. B **47**, 5 (2016).
4. Z. Ghalenovi and A. Tavakolinezhad, J. Phys. G Conf. Ser. **347**, 012015 (2012).
5. Partignani et. al. (Particle Data Group), Phys. Rev. D **98**, 030001 (2018).
6. A. Majethiya, B. Patel, and P. C. Vinodkumar, Eur. Phys. J A **42**, 213 (2009).
7. Z. Ghalenovi and M. Moazzen Sorkhi, Eur. Phys. J. Plus, **133**, 301 (2018).
8. T. M. Aliev and A. Ozpineci, Nucl. Phys. B **806**, 137 (2009).

A generalization of ideas to get meson wave functions from holographic models

Alfredo Vega* and M. A. Martin Contreras
Instituto de Física y Astronomía y Centro de Astroffisica de Valparaíso,
Universidad de Valparaíso,
Avenida Gran Bretaña 1111, Valparaíso, Chile
**E-mail: alfredo.vega@uv.cl*

Based on holographic correspondence Brodsky and de Teramond suggested a procedure to relate AdS modes with meson wave functions. In this poster we consider an extension of this to obtain meson wave functions for models which consider different dilatons and/or asymptotically AdS spaces.

Keywords: Light front wave function; AdS/QCD.

1. Introduction

The hadronic wave function in terms of quark and gluon degrees of freedom plays an important role in predictions for QCD processes, but in a direct extraction several problems are encountered. There are several non-perturbative approaches to obtain properties of distribution amplitudes and/or hadronic wave functions from QCD. Some time ago, based on the holographic correspondence, Brodsky and de Teramond suggest a matchig procedure to obtain mesonic Light-Front Wave Functions (LFWF) in terms of field duals to hadrons and electromagnetic current in AdS side.[1,2]

The holographic LFWF for hard wall models and soft wall with quadratic dilations was obtained in the original works about this topic. Later authors had considered some improvements considering aspects in QCD side (e.g., see Refs. 3–7), and they used this kind of wave functions to study different hadron properties, but was not considered the possibility to obtain LFWF for other AdS/QCD models, which is the road that we start to explore in this work.

2. Meson wave function in holographical models

The comparison of form factors calculated both in the light-front formalism and in AdS/QCD models offers the possibility to relate AdS modes to light-front wave functions (LFWF). Below we briefly discuss this matching procedure.

In the light-front formalism the electromagnetic form factor of pion can be written as

$$F(Q^2) = 2\pi \int_0^1 dx \, \frac{1-x}{x} \int_0^\infty d\zeta \, \zeta \, J_0\left(\zeta Q \sqrt{\frac{1-x}{x}}\right) \widetilde{\rho}(x,\zeta), \tag{1}$$

where $\widetilde{\rho}(x,\zeta)$ is the effective transverse distribution of partons; Q^2 is the spacelike momentum transfer squared; J_0 is the Bessel function.

Here we introduced the variable

$$\zeta = \sqrt{\frac{x}{1-x}} \left| \sum_{j=1}^{n-1} x_j b_j \right|, \tag{2}$$

which represents the x-weighted transverse impact coordinate of the spectator system.

On the other side the corresponding expression for scalars in AdS with a dilaton $\varphi(z)$ is

$$F(Q^2) = \int_0^\infty dz \, \Phi_{nJ}(z) J_\kappa(Q^2, z) \Phi_{nJ}(z) \,, \tag{3}$$

where $\Phi(z)$ corresponds to modes that represent hadrons, $J(Q^2, z)$ is the dual mode to the electromagnetic current.

The key step is to write the electromagnetic current as[3]

$$J(Q^2, z) = \int_0^1 dx \, g(Q^2, x) \, J_0\left(\zeta Q \sqrt{\frac{1-x}{x}}\right), \tag{4}$$

This allows us compare both form factors, and redefining some quantities we can establish a matching to obtain a relationship between AdS modes and the LFWF

$$|\widetilde{\psi}_{q_1\bar{q}_2}(x,\zeta)|^2 = A^2 \, x(1-x) \, g(Q^2, x) \, \frac{|\Phi(\zeta)|^2}{2\pi\zeta} \,. \tag{5}$$

Here A is constrained by the probability condition $P_{q_1\bar{q}_2} = \int_0^1 dx \int d^2 b \, |\widetilde{\psi}_{q_1\bar{q}_2}(x,b)|^2 \leq 1$ with $P_{q_1\bar{q}_2}$ being the probability of finding the valence Fock state $|q_1\bar{q}_2\rangle$ in the meson M.

In general is not difficult obtain $\Phi(\zeta)$ numerically at least in AdS/QCD models, but obtain $g(Q^2, x)$ is a problem not addresed yet, therefore (5) has not been uses to other AdS/QCD models.

3. Extension to other AdS/QCD models

Basically we need to determine $g(Q^2, x)$ in (4) to use it in (5).

If we consider the change of variable

$$\alpha = \sqrt{\frac{1-x}{x}} \tag{6}$$

in (4), we get

$$J(Q^2, z) = \int_0^\infty d\alpha \, \frac{2\alpha}{(1+\alpha^2)^2} \, g(\alpha) J_0(\zeta Q \alpha), \tag{7}$$

which looks like Hankel transform of order zero.

By using,

$$\int_0^\infty J_\nu(kr)\, J_\nu(k'r) r \, dr = \frac{1}{k}\delta(k - k'), \tag{8}$$

it is possible to find

$$\frac{2\alpha}{(1+\alpha^2)^2} \frac{1}{Q^2\alpha} \, g(Q^2, \alpha) = \int_0^\infty J(Q^2, z)\, J_0(zQ\alpha)\, z \, dz. \tag{9}$$

The last expression let us get $g(Q^2, x)$ function associated with a $J(Q^2, z)$ for different AdS/QCD models.

It is easy to check that $g(Q^2, x) = 1$ in hard wall model, but in traditional soft wall modell with quadratic dilaton the situation is different, because we obtain a more complex expression that we don't include here.

3.1. *Perspectives*

Exist plenty AdS/QCD models considering different dilatons and/or different asymtotically AdS metrics which try to catch aspect of hadronic phenomenology which hard wall or soft wall with quadratic dilaton not consider. The approach considered here could be interesting to known the LFWF associated to these models.

Additionally, a similar matching procedure has been used to get GPDs,[8–10] and ideas exposes in this work can be uses in this case also.

Acknowledgment

The authors acknowledge the financial support of FONDECYT (Chile) under Grants No. 1180753 (A. V) and No. 3180592 (M. A. M).

References

1. S. J. Brodsky and G. F. de Teramond, Phys. Rev. Lett. **96**, 201601 (2006) doi:10.1103/PhysRevLett.96.201601 [hep-ph/0602252].
2. S. J. Brodsky and G. F. de Teramond, Phys. Rev. D **77**, 056007 (2008) doi:10.1103/PhysRevD.77.056007 [arXiv:0707.3859 [hep-ph]].
3. A. Vega, I. Schmidt, T. Branz, T. Gutsche and V. E. Lyubovitskij, Phys. Rev. D **80**, 055014 (2009) doi:10.1103/PhysRevD.80.055014 [arXiv:0906.1220 [hep-ph]].
4. J. R. Forshaw and R. Sandapen, Phys. Rev. Lett. **109**, 081601 (2012) doi:10.1103/PhysRevLett.109.081601 [arXiv:1203.6088 [hep-ph]].
5. S. S. Chabysheva and J. R. Hiller, Annals Phys. **337**, 143 (2013) doi:10.1016/j.aop.2013.06.016 [arXiv:1207.7128 [hep-ph]].
6. T. Gutsche, V. E. Lyubovitskij, I. Schmidt and A. Vega, Phys. Rev. D **87**, no. 5, 056001 (2013) doi:10.1103/PhysRevD.87.056001 [arXiv:1212.5196 [hep-ph]].
7. T. Gutsche, V. E. Lyubovitskij, I. Schmidt and A. Vega, Phys. Rev. D **89**, no. 5, 054033 (2014) Erratum: [Phys. Rev. D **92**, no. 1, 019902 (2015)] doi:10.1103/PhysRevD.92.019902, 10.1103/PhysRevD.89.054033 [arXiv:1306.0366 [hep-ph]].
8. A. Vega, I. Schmidt, T. Gutsche and V. E. Lyubovitskij, Phys. Rev. D **83**, 036001 (2011) doi:10.1103/PhysRevD.83.036001 [arXiv:1010.2815 [hep-ph]].
9. A. Vega, I. Schmidt, T. Gutsche and V. E. Lyubovitskij, Phys. Rev. D **85**, 096004 (2012) doi:10.1103/PhysRevD.85.096004 [arXiv:1202.4806 [hep-ph]].
10. D. Chakrabarti and C. Mondal, Phys. Rev. D **88**, no. 7, 073006 (2013) doi:10.1103/PhysRevD.88.073006 [arXiv:1307.5128 [hep-ph]].

Reggeized model for $K^{*+}\Lambda$ photoproduction on proton

Ai-Chao Wang and Fei Huang*

School of Nuclear Science and Technology, University of Chinese Academy of Sciences, Beijing 100049, China
**E-mail: huangfei@ucas.ac.cn*

The interaction mechanism of $K^*\Lambda$ photoproduction is re-analyzed within a Regge-inspired effective Lagrangian approach. The result shows that the differential cross-section data for $\gamma p \to K^{*+}\Lambda$ can be satisfactorily described by introducing the only $N(2060)5/2^-$ resonance in the s channel, which is quite different from our earlier work performed in an effective Lagrangian approach [A. C. Wang *et al.*, Phys. Rev. C 96, 035206 (2017)], where the amplitudes are computed by evaluating Feynman diagrams and it is found that apart from the $N(2060)5/2^-$ resonance, at least one additional resonance is indispensable for reproducing the data. Although the individual terms of the background contributions on the angular distributions are found to be highly model dependent, the extracted mass of $N(2060)5/2^-$ from different models tends to be coincident, independent on how the t-channel amplitudes are constructed, whereas the width does not.

Keywords: K^* photoproduction; effective Lagrangian approach; nucleon resonances.

1. Introduction

With the present situation that our current knowledge of most of the excited nucleon states (N^*'s) is mainly coming from πN scattering and pion photoproduction, the photoproduction reaction of $\gamma p \to K^{*+}\Lambda$ provides another alternative tool to investigate N^*'s. In fact, this reaction is more suitable than the pion production to study the N^*'s in the less-explored higher energy region due to its higher threshold energy. Another superiority of $K^{*+}\Lambda$ photoproduction to investigate N^*'s is that $K^{*+}\Lambda$ couples to N^*'s with isospin 1/2 only, thus this reaction acts as an "isospin filter" and is more selective to distinguish certain resonances than pion production reactions.

Theoretically, our group have successfully described the differential cross-section data for $\gamma p \to K^{*+}\Lambda$ in Ref. 1, where the contribution of t-channel meson exchanges is obtained by calculating the corresponding

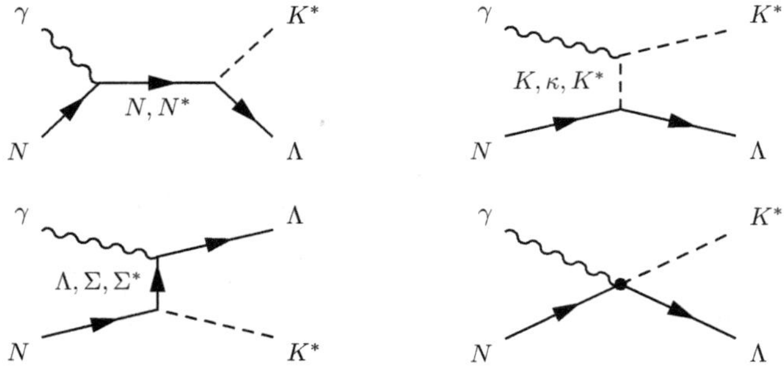

Fig. 1. Structure of the amplitude for $\gamma N \to K^*\Lambda$. Time proceeds from left to right.

Feynman amplitudes. In literature, the Regge model and the interpolated Regge model are also commonly used to analyze various reactions [see Ref. 2 for a detailed discussion]. Here, we re-analyze the differential cross-section data on $\gamma p \to K^{*+}\Lambda$ within an effective Lagrangian approach, where the t-channel amplitudes are examined one by one in Feynman type, Regge type, or interpolated Regge type. The purpose is to analyze how strongly the reaction mechanism, the resonance contents and the associated resonance parameters extracted from the $\gamma p \to K^{*+}\Lambda$ reaction depend on the particular way to construct the t-channel amplitudes.

2. Formalism

We consider the following contributions in constructing the s-, t-, and u-channel amplitudes for the $\gamma N \to K^*\Lambda$ reaction: (a) N and N^*'s exchanges in the s channel, (b) K, κ, and K^* meson exchanges in the t channel, and (c) Λ, Σ, and $\Sigma^*(1385)$ hyperon exchanges in the u channel. These contributions together with the interaction current are schematically shown in Fig. 1. The full reaction amplitude can then be expressed as

$$M^{\nu\mu} = M_s^{\nu\mu} + M_t^{\nu\mu} + M_u^{\nu\mu} + M_{\text{int}}^{\nu\mu}, \tag{1}$$

with $M_s^{\nu\mu}$, $M_t^{\nu\mu}$, and $M_u^{\nu\mu}$ standing for s-, t- and u-channel amplitudes, respectively, and $M_{\text{int}}^{\nu\mu}$ representing the generalized interaction current which is introduced to keep the full amplitude $M^{\nu\mu}$ gauge invariant. The explicit expression of $M_{\text{int}}^{\nu\mu}$ and the effective Lagrangians, propagators and form factors introduced to evaluate the Feynman diagrams of $M_s^{\nu\mu}$, $M_t^{\nu\mu}$, and $M_u^{\nu\mu}$ can be found in Ref. 1. Here we just present the formalism that is relevant to the Reggeized treatment of the t-channel K, K^* and κ exchanges.

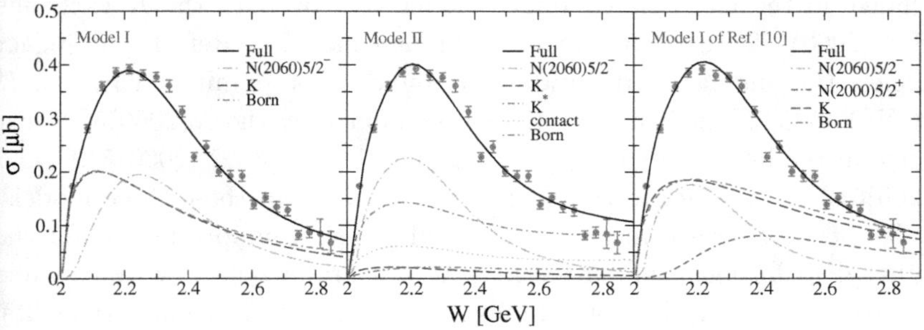

Fig. 2. Total cross sections with individual contributions for $\gamma p \to K^{*+}\Lambda$. The left graph is for model I (Regge model), the middle one corresponds to model II (interpolated Regge model), and the right one is for model I of Ref.[1] (Feynman model). The scattered symbols are data from the CLAS Collaboration.[3]

The t-channel amplitudes in Regge model are constructed by make the following substitution of the form factor in Feynman model:

$$f_t \quad \to \quad \mathcal{F}_t = \left(t - m^2\right) \mathcal{P}_{\text{Regge}}(t) f_t. \tag{2}$$

The explicit expression of $\mathcal{P}_{\text{Regge}}(t)$, the t-channel propagator in Regge model, can be found in Ref. 2. Note that to keep the full amplitude gauge invariant, one also needs to replace f_t with $\mathcal{F}_t$ in the interaction current $M^{\nu\mu}_{\text{int}}$. Similarly, the t-channel amplitudes in the interpolated Regge model are constructed by make the following substitution of the form factor:

$$f_t \quad \to \quad \mathcal{F}_{R,t} = \mathcal{F}_t R + f_t \left(1 - R\right), \tag{3}$$

where $\mathcal{F}_t$ is given in Eq. (2) and $R = R_s R_t$ with

$$R_s = \frac{1}{1 + e^{-(s - s_R)/s_0}}, \qquad R_t = \frac{1}{1 + e^{-(t + t_R)/t_0}}. \tag{4}$$

The s_R, s_0, t_R, and t_0 are parameters introduced to ensure a smooth transition between Feynman amplitude and Regge amplitude.[2]

3. Results and discussion

In all the Feynman model, the Regge mode, and the interpolated Regge model, the differential cross-section data for $\gamma p \to K^{*+}\Lambda$ can be well described by introducing the $N(2060)5/2^-$ resonance in the s channel. However, the reaction mechanisms are found to be highly model dependent. Figure 2 shows the calculated total cross sections with individual contributions. One sees that in the near threshold region, the angular distributions are dominated by the K exchange in the Regge model, by the $N(2060)5/2^-$

exchange in the interpolated Regge model, and by both the K exchange and $N(2060)5/2^-$ exchange in the Feynman model of Ref. 1. At higher energies, the angular distributions are found to be dominated by the K and $N(2060)5/2^-$ exchanges in the Regge model, by the $N(2060)5/2^-$ exchange in the interpolated Regge model, and by the K, $N(2060)5/2^-$ and $N(200)5/2^+$ exchanges in the Feynman model. In all these three models, the $N(2060)5/2^-$ resonance is needed and plays an important role for the angular distributions. The fitted masses of it are in a narrow range while the fitted widths vary in a broad range in different models and various fits. This finding might be a hint that the $N(2060)5/2^-$ is really needed and plays a significant role in the reaction $\gamma p \to K^{*+}\Lambda$, independent of how the t-channel amplitudes are constructed. The mass of this resonance can be basically determined by the present data, while the width cannot.

4. Summary

The differential cross-section data for $\gamma p \to K^{*+}\Lambda$ are re-analyzed within an effective Lagrangian approach with the t-channel amplitudes being constructed in Regge type and interpolated Regge type. The results show that the only $N(2060)5/2^-$ resonance is needed to describe the data. The individual terms of the background contributions on the angular distributions are highly model dependent. Nevertheless, the extracted mass of $N(2060)5/2^-$ from different models tends to be coincident, independent on how the t-channel amplitudes are constructed, whereas the width does not.

Acknowledgments

This work is partially supported by the National Natural Science Foundation of China under Grants No. 11475181 and No. 11635009, the Youth Innovation Promotion Association of Chinese Academy of Sciences under Grant No. 2015358, and the Key Research Program of Frontier Sciences of Chinese Academy of Sciences under Grant No. Y7292610K1.

References

1. A. C. Wang, W. L. Wang, F. Huang, H. Haberzettl, and K. Nakayama, Phys. Rev. C **96**, 035206 (2017).
2. A. C. Wang, W. L. Wang, F. Huang, and G. X. Peng, arXiv: 1902.05267.
3. W. Tang *et al.* [CLAS Collaboration], Phys. Rev. C **87**, 065204 (2013).

Measurements of $e^+e^- \to K_S^0 K^\pm \pi^\mp \pi^0/\eta$ at center-of-mass energies from 3.90 to 4.60 GeV

Chengwei Wang (for the BESIII Collaboration)

Nanjing University
Nanjing 210093, People's Republic of China
E-mail: wangcw@ihep.ac.cn

The process $e^+e^- \to K_S^0 K^\pm \pi^\mp \pi^0/\eta$ has been studied at center-of-energies between 3.90 to 4.60 GeV using 5.2 fb^{-1} luminosity data samples collected with the BESIII detector. In this proceeding, we present the results about the cross sections and the upper limits of charmonium resonance $Y(4260)$ and $Z_c^{0,\pm}$, that are searched through decay modes $K_S^0 K^\pm \pi^\mp \pi^0/\eta$.

Keywords: Charmonium-like states; $Y(4260)$; $Z_c^{0,\pm}$; BESIII.

1. Introduction

In recent years, many exotic charmonium-like states, also referred to as XYZ states, have been discovered at many experiments, which are incompatible with the naive quark model. $Y(4260)$, a very important role in the family of XYY states was first observed in initial state radiation (ISR) process $e^+e^- \to \gamma_{\rm ISR}\pi^+\pi^- J/\psi$ by BarBar[1] and then confirmed by CLEO[2] and Belle.[3] In the following years, $Y(4260)$ is also observed in final states, such as $\omega\chi_{c0}$,[4] $\pi^+\pi^- J/\psi$,[5] $\pi^+\pi^-\psi(3686)$,[6] $\pi^+\pi^- h_c$,[7] and $\pi^+ D^0 D^{*-}$.[8] Another member of XYZ states $Z_c(3900)^\pm$ has been discovered by BESIII[9] and Belle,[10] and shortly after confirmed by CLEO-c.[11] $Z_c(3900)^\pm$ is a manifestly exotic state and it should be a good candidate tetra-quark states. A neutral partner $Z_c(3900)^0$ was also observed by BESIII,[12] confirming earlier evidence report by CLEO-c.[11]

There are many interpretations on the structure of $Y(4260)$ and $Z_c(3900)$, but none of them have reached consensus on corresponding nature. Considering no light hadron decay modes have been observed via $Y(4260)$ and $Z_c(3900)$ before, thus, it is possible to utilize the decay modes with the final states $K_S^0 K^\pm \pi^\mp \pi^0/\eta$ for their potentially large branching fractions (BFs) to search those two states. In this poster, we mainly base on the published work[13] to present a measurement of the Born cross section (σ_B) for $e^+e^- \to K^0_S K^\pm \pi^\mp \pi^0/\eta$ and report upper limits of $e^+e^- \to Y(4260) \to K^0_S K^\pm \pi^\mp \pi^0/\eta$, $e^+e^- \to \pi^{\mp,0} Z_c(3900)^{\pm,0} \to K^0_S K^\pm \pi^\mp \pi^0$, and $e^+e^- \to \pi^\mp Z_c(3900)^\pm \to K^0_S K^\pm \pi^\mp \eta$.

2. Data analysis

After some common selections, there is no peaking background to exist. Counting on a two-dimension scatter is used to estimate signal yields. The Born cross section is calculated from

$$\sigma_{\rm B} = \frac{N_{\rm sig}}{\mathcal{L} \cdot \mathcal{B} \cdot \epsilon \cdot (1+\delta^{\rm ISR}) \cdot \frac{1}{|1-\Pi(s)|^2}}, \tag{1}$$

where $\mathcal{L}$ is the integrated luminosity, ϵ is the detection efficiency, $\mathcal{B}$ is the product of the BF of $K^0_S \to \pi^+\pi^-$ and that of $\pi^0/\eta \to \gamma\gamma$, $\frac{1}{|1-\Pi(s)|^2}$ is the vacuum polarization correction factor, and $(1+\delta^{\rm ISR})$ is the ISR correction factor. Here, to estimate the detection efficiency, we generate a mixing MC sample including intermediate resonances with couplings tuned to match those appear in the data sample and is weighted according to the momentum distributions observed in the data sample.

3. Results

Since there is no obvious structure in the line shapes of the Born cross section for $e^+e^- \to K^0_S K^\pm \pi^\mp \pi^0/\eta$ as shown in Fig. 1, the corresponding upper limits are determined by fitting to line shapes with $f(\sqrt{s}) = \frac{p_0}{(\sqrt{s})^{p_1}} + \frac{12\pi\Gamma_{e^+e^-}\mathcal{B}\Gamma_{tot}}{(s-M^2)^2+M^2\Gamma_{tot}^2}$, where the first item describes the continuum process, and the second is the resonance $Y(4260)$. Finally, we get the the upper limits of $\Gamma_{e^+e^-}\mathcal{B}$ at the 90% confidence level (C.L.) 0.050 and 0.19 eV for the π^0 mode and the η mode, respectively.

There is also no obvious $Z_c(3900)$ signal in the invariant mass distributions of $K^0_S K^\pm \pi^{\mp,0}$ and $K^0_S K^\pm \eta$. Thus, we set the upper limits of production cross section $\sigma_{\rm B}(e^+e^- \to \pi\ Z_c(3900))$, with $Z_c(3900) \to K^0_S\ K\ \pi/\eta$ by fitting to the invariant mass of $K^0_S K \pi/\eta$ at the 90% C.L., where, the $Z_c(3900)$ signal is described by MC simulated shape, and the background is described by a second polynomial function. The Fig. 2 shows the fit results.

No matter whether it is energy dependent or not, all of the systematic uncertainties have been considered in all fitting process when set upper limits.

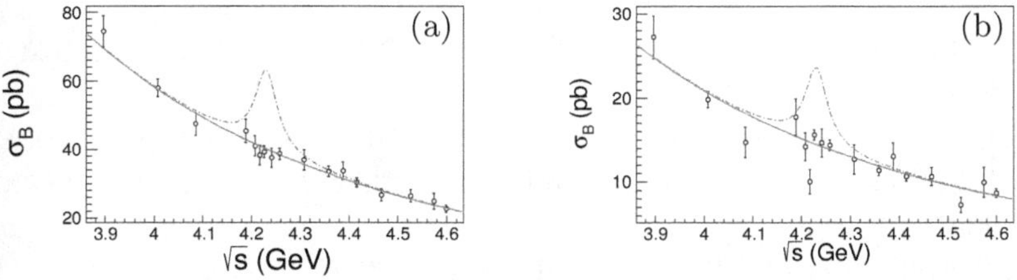

Fig. 1. Line shapes of Born cross sections for $e^+e^- \to K_S^0 K\pi\pi^0$ (a), and $e^+e^- \to K_S^0 K\pi\eta$ (b). The dots with error bars are the measured Born cross sections. The solid red lines are the fitted results and the pink dash-dotted lines are the MC shape of the $Y(4260)$ with an arbitrary scale factor.

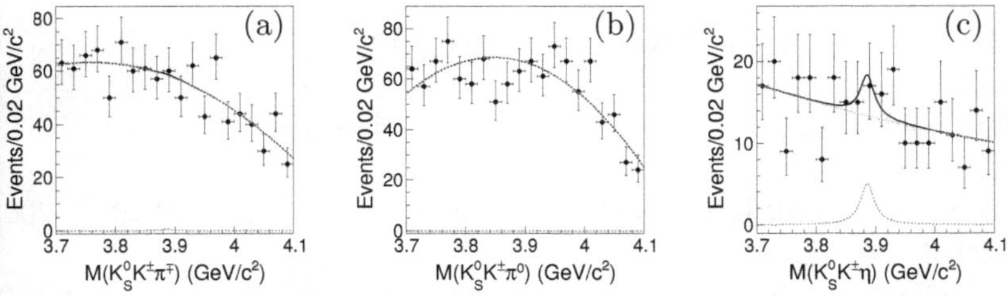

Fig. 2. The optimal fit for (a) $Z_c(3900)^0 \to K_S^0 K^\pm \pi^\mp$, (b) $Z_c(3900)^\pm \to K_S^0 K^\pm \pi^0$, and (c) $Z_c(3900)^\pm \to K_S^0 K^\pm \eta$ (at $\sqrt{s} = 4.258$ GeV as examples) . The black error bars are the data, the black solid lines are the fit curve, the green dashed lines denote the background shape, and the red dashed lines denote the signal shape.

4. Summary

The Born cross sections for $e^+e^- \to K_S^0 K^\pm \pi^\mp \pi^0/\eta$ have been measured by BESIII at center-of-mass energies from 3.90 to 4.60 GeV. Since no clear structure is observed, the upper limits of the product $\Gamma_{e^+e^-}\mathcal{B}$ ($Y(4260) \to K_S^0 K^\pm \pi^\mp \pi^0$) and $\Gamma_{e^+e^-}\mathcal{B}(Y(4260) \to K_S^0 K^\pm \pi^\mp \eta)$ are estimated to be less than 0.05 eV and 0.19 eV at 90% C.L., respectively. There also no obvious $Z_c(3900)$ signal in the process $e^+e^- \to \pi Z_c(3900)$, $Z_c(3900) \to K_S^0 K\pi/\eta$ is observed. The 90% C.L. upper limits on the cross sections are given at $\sqrt{s} = 4.226$, 4.258, 4.358, 4.416, and 4.600 GeV. It necessary to collect more data by BESIII in future to explore the deep mechanism of absence of a signal for light hadron decay modes with the annihilation of $c\bar{c}$ in the $Y(4260)$ and $Z_c(3900)$.

References

1. B. Aubert *et al.* [BaBar Collaboration], Phys. Rev. Lett. **95**, 142001 (2005).
2. Q. He *et al.* [CLEO Collaboration], Phys. Rev. D **74**, 091104 (2006).
3. C. Z. Yuan *et al.* [Belle Collaboration], Phys. Rev. Lett. **99**, 182004 (2007).
4. M. Ablikim *et al.* [BESIII Collaboration], Phys. Rev. Lett. **114**, no. 9, 092003 (2015).
5. M. Ablikim *et al.* [BESIII Collaboration], Phys. Rev. Lett. **118**, no. 9, 092001 (2017).
6. M. Ablikim *et al.* [BESIII Collaboration], Phys. Rev. D **96**, no. 3, 032004 (2017); Phys. Rev. D **99**, no. 1, 019903 (2019).
7. M. Ablikim *et al.* [BESIII Collaboration], Phys. Rev. Lett. **118**, no. 9, 092002 (2017).
8. M. Ablikim *et al.* [BESIII Collaboration], Phys. Rev. Lett. **122**, no. 10, 102002 (2019).
9. M. Ablikim *et al.* [BESIII Collaboration], Phys. Rev. Lett. **110**, 252001 (2013).
10. Z. Q. Liu *et al.* [Belle Collaboration], Phys. Rev. Lett. **110**, 252002 (2013).
11. T. Xiao, S. Dobbs, A. Tomaradze and K. K. Seth, Phys. Lett. B **727**, 366 (2013).
12. M. Ablikim *et al.* [BESIII Collaboration], Phys. Rev. Lett. **115**, no. 11, 112003 (2015).
13. M. Ablikim *et al.* [BESIII Collaboration], Phys. Rev. D **99**, no. 1, 012003 (2019).

The $\phi(2170)$ production in the process $\gamma p \to \eta\phi p$

Guan-Ying Wang, Chen-Guang Zhao, En Wang*, De-Min Li, and Guan-Nan Li
School of Physics and Microelectronics, Zhengzhou University, Zhengzhou, Henan 450001, China
**E-mail: wangen@zzu.edu.cn*

We have studied the $\gamma p \to \eta\phi p$ reaction within the effective Lagrangian approach, and our results show that there may be a peak, at least a bump structure around 2180 MeV associated to the resonance $\phi(2170)$ in the $\eta\phi$ mass distribution. We suggest to search for the resonance $\phi(2170)$ in this reaction, which would be helpful to shed light on its nature.

Keywords: $\phi(2170)$; photo-production; effective Lagrangian approach.

1. Introduction

The state $\phi(2170)$ was observed by the Babar Collaboration via the process $e^+e^- \to \gamma\phi f_0(980)$,[1] and later confirmed by Belle, BESII, and BESIII Collaborations.[2–5] However, the available information of the $\phi(2710)$, only obtained from the e^+e^- collision experiments, is not enough to distinguish different interpretations, such as $c\bar{c}$, tetraquark state, hybrid, $\Lambda\bar{\Lambda}$ or $\phi f_0(980)$ molecule. The information about the $\phi(2170)$ production in other processes will be helpful to shed light on its nature.

As we know, the associate production of hadrons by photon has been extensively studied since it provides an excellent tool to learn details of the hadron spectrum.[6–9] The intense photon beams can be used to study the strangeonium-like states because of the strong affinity of the photon for $s\bar{s}$. It should be pointed out that, in Fig. 25 of Ref. 10, the $K\bar{K}$ distribution of the reaction $\gamma N \to K^+K^-N$ shows an enhancement around 2150 MeV exists, which could be associated to a resonance with the same quantum numbers as $\phi(1680)$, i.e. $J^{PC} = 1^{--}$. Thus, it is natural to associate this structure to the $\phi(2170)$, which implies that the $\phi(2170)$ photoproduction should be accessible experimentally. In addition, $\Gamma(\phi(2170) \to \eta\phi)/\Gamma(\phi(2170) \to \phi f_0(980)) = (1.7\pm0.7\pm1.3)/(2.5\pm0.8\pm0.4)^{11}$ indicates that the coupling of the $\phi(2170)$ to the $\phi f_0(980)$ channel is of the same order of magnitude as its coupling to $\eta\phi$ channel, which suggests that the

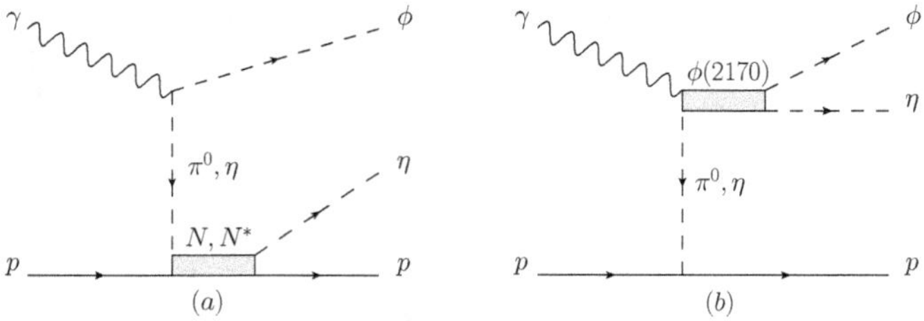

Fig. 1. Feynman diagrams for the $\gamma p \to \eta\phi p$ reaction. (a) the contribution of the t-channel π^0 and η exchanges with the intermediate states N and N^*. (b) the contribution of the intermediate states $\phi(2170)$ production.

$\phi(2170)$ has a sizeable coupling to the $\eta\phi$ channel. All the above factors encourage us to study the $\phi(2170)$ production in the reaction of $\gamma p \to \eta\phi p$ within the effective Lagrangian approach.

2. Formalisms

In this section, we will present the mechanisms for the reaction,

$$\gamma(p_1, s_1) + p(p_2, s_2) \to \phi(p_3, s_3) + \eta(p_4) + p(p_5, s_5), \tag{1}$$

by considering the tree level diagram as depicted in Fig. 1. We consider the background contribution of the t-channel π^0 and η exchanges with the final state ηp producing through the intermediate states N and N^*, as shown in Fig. 1(a). The $\phi(2170)$ can be directly produced by t-channel π^0 and η exchanges, and then decays to $\eta\phi$, which is shown in Fig. 1(b).

Then the differential cross section for the reaction $\gamma p \to \eta\phi p$ can be expressed as,

$$d\sigma(\gamma p \to \eta\phi p) = \frac{1}{8E_\gamma}\bar{\sum}|\mathcal{M}_{\text{total}}|^2 \times \frac{d^3p_3}{2E_3}\frac{d^3p_4}{2E_4}\frac{m_p d^3p_5}{E_5}\delta^4(p_1 + p_2 - p_3 - p_4 - p_5), \tag{2}$$

with

$$\mathcal{M}_{\text{total}} = \mathcal{M}_N^\pi + \mathcal{M}_{N^*}^\pi + \mathcal{M}_N^\eta + \mathcal{M}_{N^*}^\eta + \mathcal{M}_{\phi^*}, \tag{3}$$

where E_3, E_4, and E_5 are the energies of the ϕ, η, and outgoing proton, respectively, and E_γ is the photon energy in the laboratory frame. The details of the scattering amplitudes are given in Ref. 12.

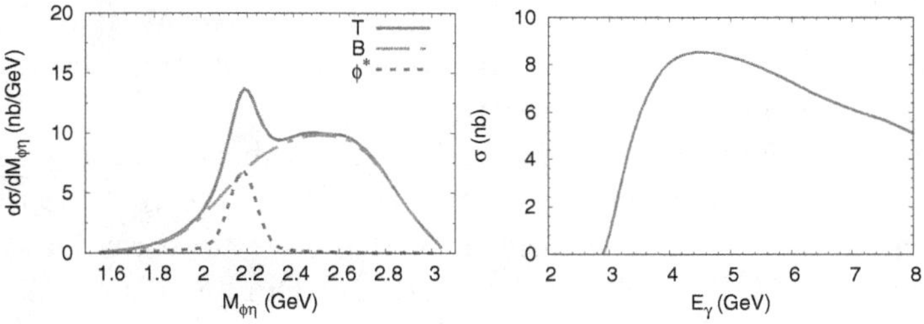

Fig. 2. Left: The $\eta\phi$ mass distribution of the $\gamma p \to \eta\phi p$ reaction with $E_\gamma = 8$ GeV in the presence of $\Gamma(\phi(2170) \to \eta\phi) \simeq 6.6$ MeV. The curves labeled as 'B' and 'ϕ^*'stand for the contributions of the background and the intermediate $\phi(2170)$ production, respectively. The curve labeled as 'T' corresponds to the total contributions. Right: Total cross section of the $\gamma p \to \phi\eta p$ reaction.

3. Result and Discussion

With the above formalisms, we calculate the total and differential cross sections for the $\gamma p \to \eta\phi p$ reaction by using a Monte Carlo multi-particle phase space integration program. The $\eta\phi$ mass distribution of the $\gamma p \to \eta\phi p$ reaction with $E_\gamma = 8$ MeV is shown in the left panel of Fig. 2. As we can see, there is a peak structure around 2180 MeV, which is associated to the resonance $\phi(2170)$. As shown in Ref. 12, there still exists a bump structure even with the low limit of the couplings of the $\phi(2170)$.

It should be pointed out that for the photo-production, there is a contribution from Pomeron exchange, whose effect is dominant at large center-of-mass energy and forward angle. However, in this paper only the $\eta\phi$ mass distribution is relevant to the signal of $\phi(2170)$, and the comprehensive mechanism involved the Pomeron exchange dose not change too much the shape of the $\eta\phi$ mass distribution.

In the right panel of Fig. 2, we show the total cross section of the $\gamma p \to \eta\phi p$ reaction. Very recently, the reaction of $\gamma p \to \eta\phi p$ is also suggested to study the nucleon resonances production in Ref. 13 where the total cross section at $E_\gamma = 3.8$ GeV is around $8 \sim 10$ nb, which is consistent with our prediction.

4. Conclusions

Motivated by the small enhancement around 2150 MeV in the K^+K^- mass distribution of the $\gamma p \to K^+K^-p$ reaction measured by Omega Photon Collaboration, and the clues that the branching ratio $\text{Br}(\phi(2170) \to \eta\phi)$ is of the same order as $\text{Br}(\phi(2170) \to \phi f_0(980))$, we propose to search for the resonance $\phi(2170)$ in the $\gamma p \to \eta\phi p$ reaction. Our calculations show that there will be a peak, at least a bump structure around 2180 MeV in the $\eta\phi$ mass distribution of $\gamma p \to \phi\eta p$ reaction.

Finally, it should be noted that the GlueX Collaboration has proposed to search for the $\phi(2170)$ in the photoproduction,[14] and the $\gamma p \to \eta\phi p$ reaction has been selected as a particularly suitable process to search for strangeonium states by the CLAS12 Collaboration.[15] Our predictions should be useful for the future experimental study.

Acknowledgements

This work is partly supported by the National Natural Science Foundation of China under Grant Nos. 11505158, 11605158. It is also supported by the Academic Improvement Project of Zhengzhou University.

References

1. B. Aubert *et al.* [BaBar Collaboration], Phys. Rev. D **74**, 091103 (2006).
2. C. P. Shen *et al.* [Belle Collaboration], Phys. Rev. D **80**, 031101 (2009).
3. M. Ablikim *et al.* [BES Collaboration], Phys. Rev. Lett. **100**, 102003 (2008).
4. M. Ablikim *et al.* [BESIII Collaboration], Phys. Rev. D **91**, 052017 (2015).
5. M. Ablikim *et al.* [BESIII Collaboration], Phys. Rev. D **99**, 012014 (2019).
6. J. J. Xie, E. Wang, and J. Nieves, Phys. Rev. C **89**, 015203 (2014).
7. E. Wang, J. J. Xie, and J. Nieves, Phys. Rev. C **90**, 065203 (2014).
8. E. Wang, J. J. Xie, W. H. Liang, F. K. Guo, and E. Oset, Phys. Rev. C **95**, 015205 (2017).
9. Y. Y. Wang, L. J. Liu, E. Wang, and D. M. Li, Phys. Rev. D **95**, 096015 (2017).
10. J. Busenitz *et al.*, Phys. Rev. D **40**, 1 (1989).
11. M. Tanabashi *et al.* [Particle Data Group], Phys. Rev. D **98**, 030001 (2018).
12. C. G. Zhao, G. Y. Wang, G. N. Li, E. Wang and D. M. Li, Phys. Rev. D **99**, no. 11, 114014 (2019).
13. J. Q. Fan, S. F. Chen, and B. C. Liu, Phys. Rev. C **99**, 025203 (2019).
14. A. AlekSejevs *et al.* [GlueX Collaboration], arXiv:1305.1523 [nucl-ex].
15. A. Filippi [CLAS Collaboration], EPJ Web Conf. **96**, 01013 (2015).

Canonical interpretation of the $X(4140)$ state within the $^3P_0$ model

Wei Hao, Guan-Ying Wang, En Wang*, De-Min Li and Guan-Nan Li

School of Physics and Microelectronics, Zhengzhou University, Zhengzhou, Henan 450001, China

**E-mail: wangen@zzu.edu.cn*

Recently, LHCb collaboration has confirmed the state $X(4140)$, with a much larger width $\Gamma = 83 \pm 21^{+21}_{-14}$ MeV than the previous experimental measurements, which has confused the understanding of its nature. We have investigated the possibility of the $\chi_{c1}(3P)$ interpretation for the $X(4140)$, considering the mass and the strong decay properties.

Keywords: $X(4140)$; $^3P_0$ model; quark model.

1. Introduction

Recently, the LHCb measurements has confirmed the $X(4140)$ state with high statistic data,[1,2] with a mass $4146.5 \pm 4.5^{+4.6}_{-2.8}$ MeV and a width $83 \pm 21^{+21}_{-14}$ MeV, much larger than the previous experimental measurements,[3] and the quantum numbers were determined to be $J^{PC} = 1^{++}$. There are many different suggestions about the structure of the $X(4140)$,[4,5] especially because of the large discrepancy of the width.

Indeed, it is natural and necessary to exhaust the possible $q\bar{q}$ description of the observed states before restoring to the more exotic assignments. In this work, taking the meson wave functions by solving the relativistic/non-relativistic Schrödinger equation, we investigate the decay properties of the $X(4140)$ as the assignment of charmonium state in the $^3P_0$ model, and provide more information about the decay modes for searching for the $X(4140)$, which could be useful to extract the more precision width.

2. The $^3P_0$ decay model

The $^3P_0$ model, also known as the quark-pair creation model, has been widely applied to study strong decays of hadrons with considerable success.[6–9] In the model, the transition operator T of the decay $A \to BC$ in the $^3P_0$ model can be written,

$$T = -3\gamma \sum_m \langle 1m1-m|00\rangle \int d^3\boldsymbol{p}_3 d^3\boldsymbol{p}_4 \delta^3(\boldsymbol{p}_3+\boldsymbol{p}_4) \times \mathcal{Y}_1^m\left(\frac{\boldsymbol{p}_3-\boldsymbol{p}_4}{2}\right) \chi^{34}_{1,-m}\phi^{34}_0\omega^{34}_0 b^\dagger_3(\boldsymbol{p}_3) d^\dagger_4(\boldsymbol{p}_4), \tag{1}$$

where γ is a dimensionless parameter corresponding the strength of quark-antiquark $q_3\bar{q}_4$ pair producing from the vacuum. The decay width $\Gamma(A \to BC)$ can be expressed in terms of the partial wave amplitude,

$$\Gamma(A \to BC) = \frac{\pi|\boldsymbol{P}|}{4M_A^2}\sum_{LS}|\mathcal{M}^{LS}(\boldsymbol{P})|^2, \tag{2}$$

where the explicit expressions for $\mathcal{M}^{LS}(\boldsymbol{P})$ can be found in Refs. 7–9.

In this work, we take into account two choices of the wave functions for mesons: Case A: the modified non-relativistic quark model (MNRQM), Case B: the modified Godfrey-Isgur model (MGI). The details of model and the wave functions can be found in Ref. 10.

3. Results and discussions

The mass spectra of the charmonium states predicted by two kinds of quark models are shown in Ref. 10. According to the PDG,[3] the mass of the $X(4140)$ $[I^G(J^{PC}) = 0^+(1^{++})]$ is 4146.8 ± 2.4 MeV, which is consistent with the predicted mass of $\chi_{c1}(3P)$ in both models within the uncertainties of the models. Next, we will calculate the strong decay widths of the $X(4140)$ state as the $\chi_{c1}(3P)$ assignment.

In the $^3P_0$ model, the strength of quark-antiquark pair creating from the vacuum, is taken to be $\gamma = 4.52$ in Case A, and $\gamma = 5.90$ for Case B, by fitting to the total widths of the well established charmonium states, $\psi(3770)$ (1^3D_1), $\psi(4040)$ (3^3S_1), $\psi(4160)$ (2^3D_1), and $\chi_{c2}(2P)$.

With the above parameters, we have calculated the partial decay widths and total decay width, as shown in Table 1 for both Case A and Case B. The total widths are 12.63 MeV for Case A, and 31.34 MeV for Case B, which are consistent with the average value $\Gamma = 22^{+8}_{-7}$ MeV of PDG, considering the uncertainties of the $^3P_0$ model. It should be pointed out that the decay modes $D\bar{D}^*$ and $D^*\bar{D}^*$ have large decay widths. We suggest to search for this state in those two channels, and to measure the width precisely, which can be shed light on its nature.

We also show the dependence of the $\chi_{c1}(3P)$ decay width on the masses in Fig. 1. Indeed, if the small width of the $X(4140)$ is confirmed in future

Table 1. Decay widths of the $\chi_{c0}(3P)$, $\chi_{c1}(3P)$ and $\chi_{c2}(3P)$ states (in MeV). The mass of the $\chi_{c1}(3P)$ is taken to be the one of the $X(4140)$, and the masses of the $\chi_{c0}(3P)$, and $\chi_{c2}(3P)$ are taken from Ref. 10, respectively for Case A and Case B.

State	Channel	Mode	Γ (Case A)	Γ (Case B)
$\chi_{c0}(3P)$	$0^+ \to 0^-0^-$	$D\bar{D}$	10.58	0.22
		$D_s^+D_s^-$	0.37	1.87
	$0^+ \to 1^-1^-$	$D^*\bar{D}^*$	16.28	35.95
Total Width			27.23	38.03
$\chi_{c1}(3P)$	$1^+ \to 0^-1^-$	$D\bar{D}^*$	4.54	14.48
		$D_s\bar{D}_s^*$	1.23	0.70
	$1^+ \to 1^-1^-$	$D^*\bar{D}^*$	6.86	16.17
Total Width			12.63	31.34
$\chi_{c2}(3P)$	$2^+ \to 0^-0^-$	$D\bar{D}$	7.71	8.79
		$D_s^+D_s^-$	0.63	0.10
	$2^+ \to 0^-1^-$	$D\bar{D}^*$	20.04	11.34
		$D_s\bar{D}_s^*$	0.17	0.13
	$2^+ \to 1^-1^-$	$D^*\bar{D}^*$	11.33	26.87
Total Width			39.89	47.23

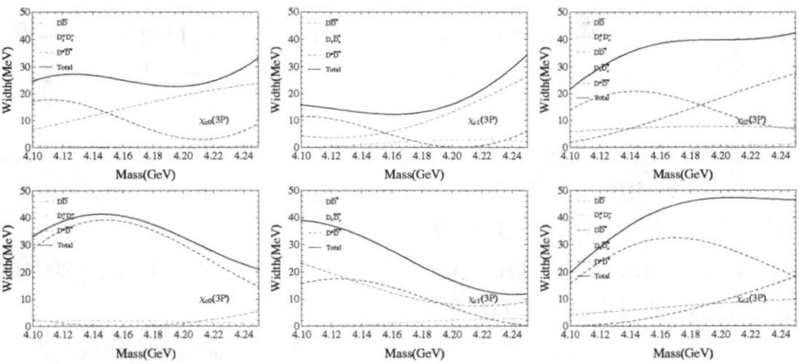

Fig. 1. The dependence of the width of $\chi_{c0}(3P)$, $\chi_{c1}(3P)$ and $\chi_{c2}(3P)$ on the initial state mass with the wave functions of Case A (up panels) and Case B (low panels).

high-precision measurements, the $X(4140)$ could be explained as the charmonium state $\chi_{c1}(3P)$. In addition, the processes of the $B^+ \to J/\psi\phi K^+$ and $e^+e^- \to \gamma J/\psi\phi K^+$ decay were investigated in Refs. 11, 12, where the existences of the $X(4140)$, with the small width, and the molecular state $X(4160)$ is supported by the low $J/\psi\phi$ invariant mass distributions.[1,2,13]

Finally, the strong decay width of the $\chi_{c0}(3P)$ and $\chi_{c2}(3P)$ states in Table 1, and the mass dependence of the total widths are also shown in Fig. 1.

4. Summary

We have investigated the strong decay properties of the $X(4140)$ with the assignment of the $\chi_{c1}(3P)$ states in the $^3P_0$ model, taking two cases for the wave functions. The total decay width of the $\chi_{c1}(3P)$ is predicted to be 12.63 MeV for Case A, and 31.34 MeV for Case B, both of which are in agreement with the PDG average width of $X(4140)$. Thus, we conclude that, the $X(4140)$, with a small width, could be explained as the charmonium state $\chi_{c1}(3P)$, and the high-precision measurement about the $X(4140)$ could shed light on the nature of the $X(4140)$.

Acknowledgements

This work is partly supported by the National Natural Science Foundation of China under Grant Nos. 11505158, 11605158, the Key Research Projects of Henan Higher Education Institutions (No. 20A140027), and the Academic Improvement Project of Zhengzhou University.

References

1. R. Aaij *et al.* [LHCb Collaboration], Phys. Rev. D **95**, 012002 (2017).
2. R. Aaij *et al.* [LHCb Collaboration], Phys. Rev. Lett. **118**, 022003 (2017).
3. M. Tanabashi *et al.* [Particle Data Group], Review of Particle Physics, Phys. Rev. D **98**, no. 3, 030001 (2018).
4. D. Y. Chen, Eur. Phys. J. C **76**, no. 12, 671 (2016).
5. Z. G. Wang and Z. Y. Di, Eur. Phys. J. C **79**, no. 1, 72 (2019).
6. L. Micu, Nucl. Phys. B **10**, 521 (1969).
7. F. E. Close and E. S. Swanson, Phys. Rev. D **72**, 094004 (2005).
8. D. M. Li and E. Wang, Eur. Phys. J. C **63**,297 (2009).
9. Q. F. Lü, T. T. Pan, Y. Y. Wang, E. Wang, and D. M. Li, Phys. Rev. D **94**,074012 (2016).
10. W. Hao, G. Y. Wang, E. Wang, G. N. Li and D. M. Li, arXiv:1909.13099 [hep-ph].
11. E. Wang, J. J. Xie, L. S. Geng and E. Oset, Phys. Rev. D **97**, no. 1, 014017 (2018).
12. E. Wang, J. J. Xie, L. S. Geng and E. Oset, Chin. Phys. C **43**, no. 11, 113101 (2019).
13. M. Ablikim *et al.* [BESIII Collaboration], Phys. Rev. D **97**, no. 3, 032008 (2018).

Discerning the two $K_1(1270)$ poles in $D^0 \to \pi^+ VP$ decay

G. Y. Wang

School of Physics and Microelectronics, Zhengzhou University, Zhengzhou, Henan, 450001, China

L. Roca

Departamento de Física. Universidad de Murcia. E-30100 Murcia. Spain.

E. Oset

Departamento de Física Teórica e IFIC, Centro Mixto Universidad de Valencia-CSIC, Institutos de Investigacion de Paterna, Apdo 22085, 46071 Valencia, Spain

Within the chiral unitary approach, the axial-vector resonance $K_1(1270)$ has been predicted to manifest a two-pole nature. The lowest pole has a mass of 1195 MeV and a width of 246 MeV and couples mostly to $K^*\pi$, and the highest pole has a mass of 1284 MeV and a width of 146 MeV and couples mostly to ρK. We analyze theoretically how this double-pole structure can show up in the $D^0 \to \pi^+ VP$ decays by looking at the vector-pseudoscalar (VP) invariant mass distribution for different VP channels, exploiting the fact that each pole couples differently to different VP pairs. We find that the final $\bar{K}^*\pi$ and $\rho\bar{K}$ channels are sensible to the different poles of the $K_1(1270)$ resonance and hence are suitable reactions to analyze experimentally the double pole nature of this resonance.

1. Introduction

One of the novel things of the work of Ref. 1 is that two states appeared related to the $K_1(1270)$, which coupled differently to the different channels. These states were further studied in Ref. 2 in connection with experimental information which provided support for two states, one with 1195 MeV, coupling mostly to $K^*\pi$, and another one with 1284 MeV, coupling mostly to ρK. The two peaks were well differentiated in reactions leading to $K^*\pi$ and ρK in the final state.

In the present work we pay attention to the $D^0 \to \pi^+ VP$ decay, with V a vector meson and P a pseudoscalar one. We shall see that we can relate the rates of production of $K^*\pi$ and ρK, and the differential mass distributions peak at different masses as a consequence of the different weight of each of the two $K_1(1270)$ resonances in each of the reactions.

2. Axial-vector mesons in the chiral unitary approach

The chiral unitary approach for the vector-pseudoscalar interaction in s-wave,[1,2] where most of the low lying axial-vector resonances are obtained dynamically without the need to include them as explicit degrees of freedom, see Refs. 1, 2 for further details and explanations.

3. Formalism for $D^0 \to \pi^+ VP$ decay

3.1. *Tree level production*

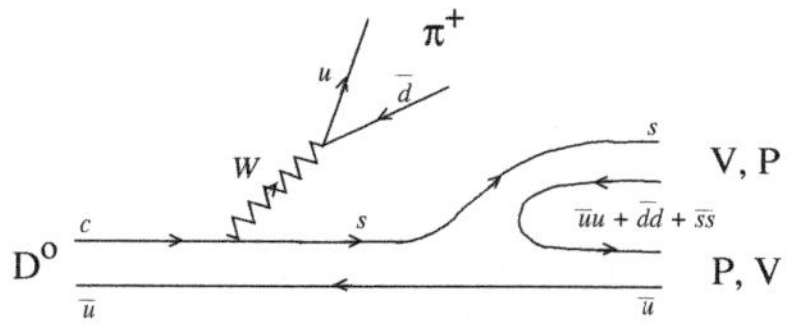

Fig. 1. Elementary $D^0 \to \pi^+ VP$ process at the quark level.

For the different VP pairs that can be produced, we can determine their relative strength by using the $SU(3)$ argument. For the final hadronic state we find

$$\sum_i h_i\,(V_i P_i)\Big|_{(I=\frac{1}{2},I_3=-\frac{1}{2})} \equiv \sqrt{\frac{3}{2}}\bar{K}^*\pi - \sqrt{\frac{3}{2}}\rho\bar{K} + \frac{1}{\sqrt{2}}\omega\bar{K} + \phi\bar{K}. \tag{1}$$

The tree level amplitude will have the structure

$$t_{\rm tree} = a h_i \vec{\epsilon}\cdot\vec{p}_{\pi^+} + b h_i \vec{\epsilon}\cdot\vec{p}_P. \tag{2}$$

3.2. *VP final state interaction*

From Refs. 1, 2, the two poles for the axial-vector $K_1(1270)$ resonance are dynamically generated from the VP interaction. Therefore we have to implement the final state interaction of the VP pair produced in the elementary $D^0 \to \pi^+ VP$ mechanism of Fig. 1. This is depicted in Fig. 2.

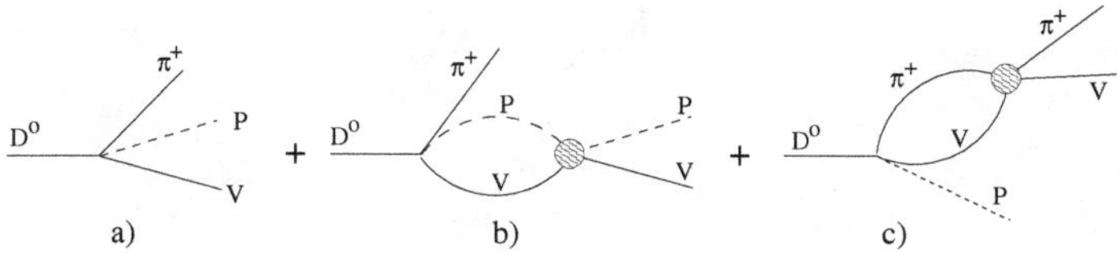

Fig. 2. *VP* final state interaction.

The amplitudes are of the form

$$\mathcal{M}_i = A\vec{\epsilon}\cdot\vec{p}_{\pi^+} + B\vec{\epsilon}\cdot\vec{p}_P. \tag{3}$$

with $\vec{p}_{\pi^+}$ and $\vec{p}_P$ the momentum of the π^+ and the pseudoscalar P_i in the D_0 rest frame.

4. Results

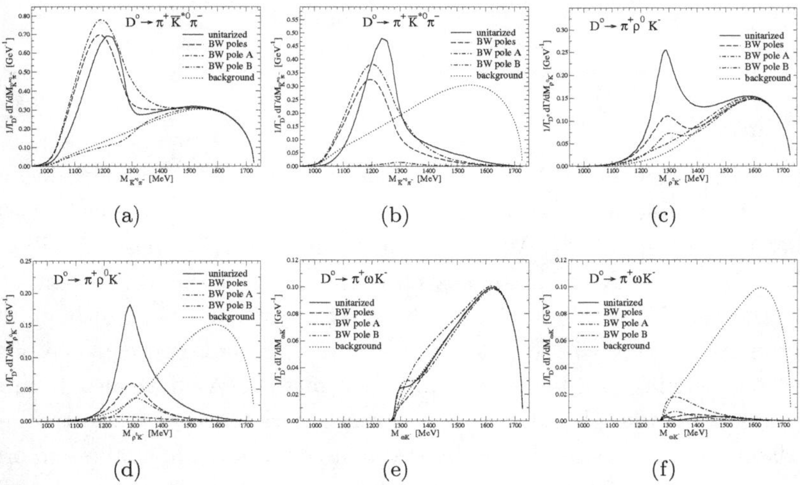

Fig. 3. VP invariant mass distribution in the $D^0 \to \pi^+ VP$ decay for $D^0 \to \pi^+\bar{K}^{*0}\pi^-$, $D^0 \to \pi^+\rho^0 K^-$ and $D^0 \to \pi^+\omega K^-$. (a),(c),(e) considering the interaction with the tree level elementary mechanism of Fig. 1. (b),(d),(f) are without the interference with the background contributions.

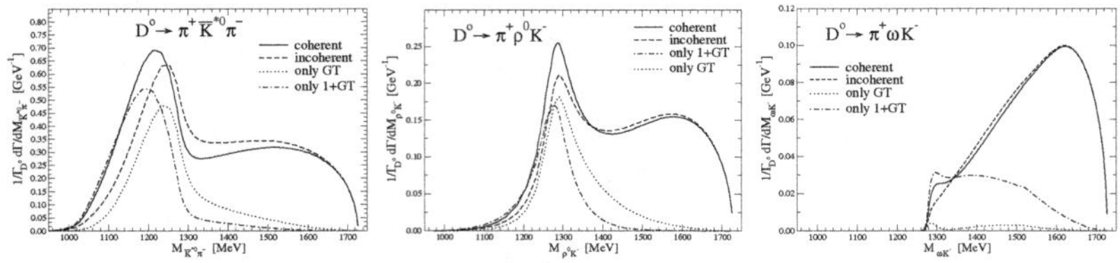

(a) (b) (c)

Fig. 4. VP invariant mass distribution in the $D^0 \to \pi^+ VP$ decay for $D^0 \to \pi^+ \bar{K}^{*0} \pi^-$, $D^0 \to \pi^+ \rho^0 K^-$ and $D^0 \to \pi^+ \omega K^-$. Solid line: full results using the unitarized $I = 1/2$ amplitude with the other contributions added coherently. Dashed line: same but adding the background incoherently.

Table 1. Different contributions to the branching ratios for $D^0 \to \pi^+ VP$ decay. The results in parenthesis are without the background mechanisms. Pole $A(B)$ means that the $B(A)$ pole are removed from the model. All the results are in %.

	unitarized	BW poles	pole A	pole B	background
$\bar{K}^{*0}\pi^-$	23 (8.6)	26 (6.6)	28 (8.6)	14 (0.24)	14
$\rho^0 K^-$	7.9 (2.7)	5.8 FIT (0.97)	5.3 (0.19)	5.1 (0.57)	4.8
ωK^-	2.8 (0.07)	2.8 FIT (0.10)	3.0 (0.28)	2.8 (0.088)	2.8

5. Summary

We have studied theoretically the suitability of the $D^0 \to \pi^+ VP$ decay to check the double pole nature of the $K_1(1270)$ resonance predicted by the chiral unitary approach. We have found that the $\bar{K}^*\pi$ mass distribution in the $D^0 \to \pi^+ \bar{K}^*\pi$ channel is clearly dominated by the lowest mass pole and the $D^0 \to \pi^+ \rho \bar{K}$ channel by the highest mass pole, although the interference between both poles is relevant. This is a consequence of the reactions peaking at different VP invariant masses. We have also discussed the important role played by the coherent interference with the tree level mechanism which should be considered in an experimental analysis in order to properly extract the resonance properties. Devoted experimental studies of this reaction could help to shed light on the double pole structure of this resonance.

References

1. L. Roca, E. Oset and J. Singh, Phys. Rev. D **72**, 014002 (2005).
2. L. S. Geng, E. Oset, L. Roca and J. A. Oller, Phys. Rev. D **75** (2007) 014017.

CPSIA information can be obtained
at www.ICGtesting.com
Printed in the USA
BVHW041413060820
585641BV00004B/11